Textbook of
Organic Chemistry

C N Pillai

Retd. Professor, Department of Chemistry
Indian Institute of Technology Madras
Chennai

CRC Press
Taylor & Francis Group
Boca Raton London New York

CRC Press is an imprint of the
Taylor & Francis Group, an **informa** business

Visit the Taylor & Francis Web site at
http://www.taylorandfrancis.com

and the CRC Press Web site at
http://www.crcpress.com

Distributed in India, China, Pakistan, Bangladesh, Sri Lanka, Nepal, Bhutan, Indonesia, Malaysia, Singapore and Hong Kong by
Orient Blackswan Private Limited

Typeset at
Krishtel eMaging Solutions Pvt. Ltd., Chennai 600 017

Preface

Many students, when first exposed to organic chemistry, find the subject intimidating when faced by the innumerable structural formulae. It is true that many things have to be memorised. The subject will be more interesting if the student is made to appreciate that much of organic chemistry can be logically deduced from fundamentals. Most of the facts and concepts of organic chemistry are amenable to reasonable explanations and deductions. I feel that if the subject is presented in the class room with emphasis on these aspects, organic chemistry will be more interesting to the student. In this book, an attempt has been made to follow this ideal.

Textbook of Organic Chemistry is directed towards students who have already had exposure to the basics of chemistry, including an introduction to organic chemistry. This book conforms to the syllabus of Indian universities at the undergraduate level, but can be quite useful to students at a more advanced level also. The choice of examples under special topics like natural products is somewhat arbitrary. The syllabi of different universities, in this respect, may differ.

At the end of each chapter exercises are provided, not only for testing the understanding of the student, but also to strengthen the concepts discussed in the text. 'Challenging Questions' are meant for those students who want to delve deeper into the subject. Topics of current interest that are related to the subject matter of the chapter are suggested for preparing project reports, which could be done individually or in groups. The data for such reports are readily available from websites and specialised books.

Most of the manuscript has been read carefully and critically by Dr S Ananthan, Reader in Chemistry, Presidency College, Chennai. His suggestions for improvement of the treatment of certain topics and corrections – not only trivial errors but also the more serious factual errors – have been invaluable. I record my profound thanks to him.

My editor at Universities Press, Dr Gita S Dattatri, has gone through the manuscript at every stage of its development. Her contribution to the book, not only as the editor, but also as a critical reader, has been invaluable and is gratefully acknowledged. I also wish to place on record my appreciation of the excellent cooperation extended by the typesetters, Krishtel eMaging Solutions Pvt. Ltd., Chennai.

The book is submitted to the students and teachers of organic chemistry in Indian Universities. Corrections, criticisms and suggestions for improvement are welcome.

Chennai   *C N Pillai*

Contents

1 Basic Concepts of Bonding in Organic Chemistry

OBJECTIVES In this chapter, you will learn about,

- organic chemistry
- hybridisation
- electronic effects—resonance, inductive, electromeric
- reactive intermediates—carbocations, carbanious, free radicals
- steric effects

1.1 WHAT IS ORGANIC CHEMISTRY?

Historically, the term '*organic chemicals*' was used to denote a class of chemicals isolated from plants or animals. By the eighteenth century, a large number of such materials were known and identified as chemical compounds. These include compounds like sugar, alcohol, acetic acid, urea and camphor. Their behaviour and characteristics were different from those of chemicals isolated from minerals that were classified as inorganic. The term organic was originally used to emphasise their origin from living things. It has been known for a long time now, that all of these chemicals can be prepared in the laboratory. However, the classification of chemicals as organic and inorganic has continued. This classification seems justified and has been accepted by the scientific community. As chemistry evolved, organic chemists seem to have developed their own language and way of thinking.

So, what is organic chemistry? Organic chemistry has been famously defined as *the chemistry of carbon compounds*, though compounds of carbon such as the carbonates and carbides are not included. Carbon compounds do form a large group; there are millions of them, more than all other known compounds put together. Such a large number of compounds are possible due to the unique chain-forming property of carbon, referred to as *catenation*. In certain polymers there are thousands of carbons linked one after the other through carbon–carbon bonds. Because of its *tetravalency* (valency = 4), one carbon can be bonded to four other atoms, which could either be carbons or other atoms. Carbon can form stable covalent bonds with itself and with almost all other nonmetals, metalloids and some metals. This property of carbon arises because of its position in the top middle of the periodic table—neither strongly electronegative nor strongly electropositive. It has little tendency to donate electrons or to accept electrons, but shares them very well to form covalent bonds. Almost all the bonds in organic molecules – from the simple ones like methane to the complex molecules like steroids, tetracycline, chlorophyll, proteins, nucleic acids, carbohydrates like starch and cellulose, oils and fats, rubbers (both natural and synthetic), synthetic polymers like PVC or phenol-formalydehyde resins, and all others classified as organic compounds – are covalent. In fact, organic chemistry is the chemistry of the covalent bond involving carbon and other atoms—notably hydrogen, oxygen, nitrogen, the halogens, sulphur, phosphorus and to a lesser extent, most other elements. The architecture of the molecule, involving information about how atoms are bonded and how they are oriented, is called the structure of the molecule.

The concerns of the organic chemist include,

- the preparation of molecules with specific structures in the laboratory(synthesis),
- determination of the structure of molecules; natural and synthetic(analysis and structure determination), and
- understanding the principles governing the behaviour, properties and reactions of molecules (physical organic chemistry).

1.2 STRUCTURE OF MOLECULES—HYBRIDISATION AND GEOMETRY

1.2.1 *sp³* HYBRIDISATION—METHANE AND ETHANE

At this stage, let us recapitulate the electronic configuration and the atomic orbitals of carbon. Carbon with an atomic number of 6 has the electronic configuration, $1s^2, 2s^2, 2p^2$.

This electronic configuration predicts divalency for carbon and the formation of compounds such as CH_2. Such a compound is indeed known but is highly reactive and has an extremely short life time. As we know, the stable compounds of carbon are those with carbon having a valency of 4 as in CH_4, methane, which has four equivalent C—H bonds. The tetravalency of carbon and the equivalence of the bonds can be readily accounted for by invoking the concepts of *promotion* and *hybridisation*. After *promotion* of one 2s election to $2p_z$, we have the configuration,

This is compatible with a valency of 4, but not compatible with the fact that the four C—H bonds in methane are equivalent. Namely, they have the same bond lengths and bond energies, and are disposed symmetrically around the carbon atom (same H—C—H bond angle). This equivalence is obtained by *hybridisation* of the four atomic orbitals. Hybridisation means mixing together of a certain number of atomic orbitals which are not of the same energy or directional property, to obtain an equal number of hybrid orbitals of identical energy and directional property. The orbitals obtained by mixing one s orbital and three p orbitals are called sp^3 *hybrid orbitals*. So, after hybridisation we have,

The shape of the sp^3 hybrid orbital is different from that of the s or the p orbital.

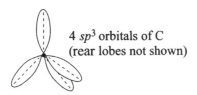

The sp^3 orbital has an axis and a direction. The four sp^3 orbitals of carbon are so oriented that they are as far separated as possible from each other (recall the VSEPR theory).

4 sp^3 orbitals of C
(rear lobes not shown)

When one of the sp^3 orbitals of carbon overlaps with the s orbital of a hydrogen, we get a C—H σ-bond, due to sp^3—s overlap. The overlap, in order to be maximum, is along the direction of the sp^3 axis. Such a bond where the electron density has cylindrical symmetry about the internuclear axis is called a σ-bond.

The four C—H bonds thus formed are identical and are symmetrically oriented around the carbon. If a three dimensional figure is imagined using the lines connecting the four hydrogen atoms of methane, it will be a *regular tetrahedron*.

The four sp^3 orbitals of carbons have their axes pointing towards the four corners of a regular tetrahedron. The four H—C—H *bond angles* are equal and have the value expected from this geometric figure, about 109.5°. The C—H *bond length* in methane is 101 pm. Each of the C—H bonds in methane is an electron pair bond. Representations such as in 1.1a and 1.1b are familiar to us. 1.1a is called the *valence bond structure* and 1.1b, the *Lewis structure*.

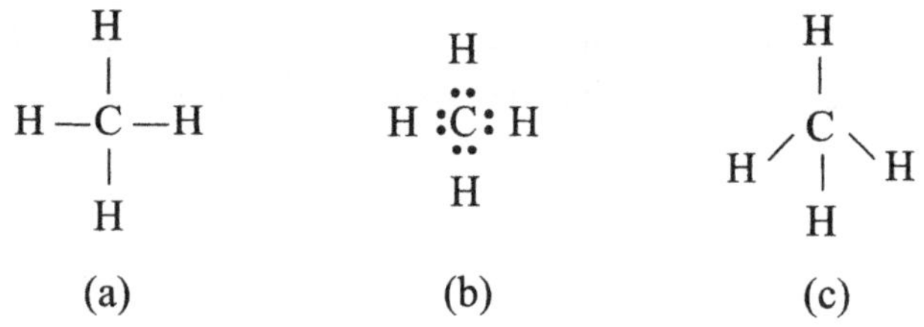

H
|
H —C —H
|
H

(a)

H
..
H :C: H
..
H

(b)

H
|
C
H ⟋ | ⟍ H
|
H

(c)

$$(1.1)$$

Structures 1.1a and 1.1b may give a false impression that all the atoms – hydrogens and carbon – are in one plane, which is of course not true. The molecule has a tetrahedral shape. A better representation is that shown in 1.1c. Since this is more difficult to draw, it is acceptable to draw methane as in 1.1a. 1.1b emphasises the facts that the σ-bond consists of two electrons and that the carbon in methane has a complete octet of electrons. Both points are useful for understanding and interpreting reactions.

We have seen how a σ-bond (the C—H bond) is formed by the overlap of an *s* orbital of hydrogen and an sp^3 hybrid orbital of carbon. The condition for a bond to be called a true σ-bond is that the overlap of the two atomic orbitals concerned should be along the same axis so that the resultant molecular orbital will have cylindrical symmetry around the axis of overlap. σ-Bonds can also be formed by the overlap of other types of orbitals. For example, overlap of one sp^3 orbital with another sp^3 orbital gives rise to a σ-bond. This is what happens in ethane (C_2H_6), the next higher member of the alkane family. Each carbon is sp^3 hybridised and forms three σ-bonds with H atoms as in methane and one σ-bond with the other carbon by sp^3–sp^3 orbital overlap (1.2).

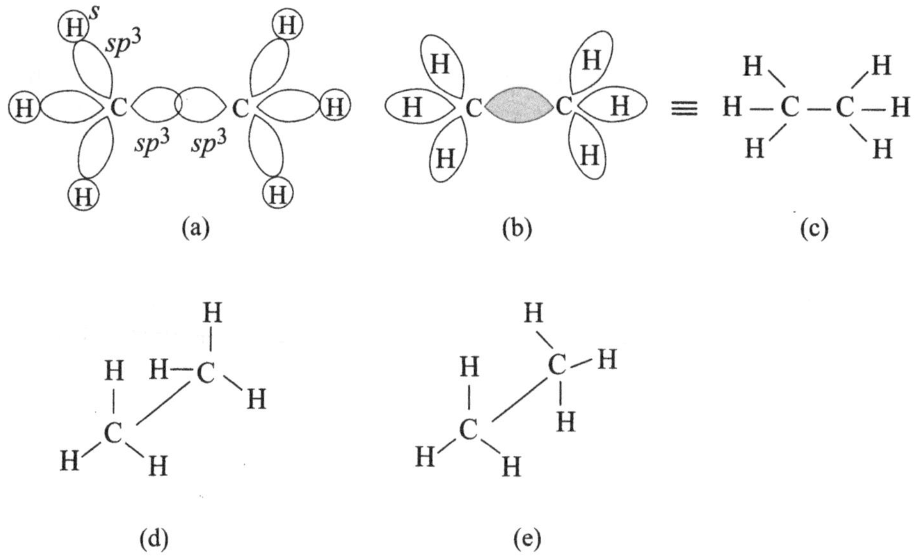

(a)   (b)   (c)

(d)   (e)

$$(1.2)$$

Note: The small rear lobes of the hybrid orbitals are not shown, to make the drawings clearer.

Free rotation is possible around the C—C σ-bond. The orientation of the C—H bonds on one carbon in space, relative to those on the other carbon can be any one of an infinite number of *rotational* forms. These rotational forms which are not isomers are called *conformations*. Due to the thermal energy present in the molecule, there is constant motion and the conformations readily get interconverted (1.2d and 1.2e). Larger molecules involving chains and rings are formed in the same way. The C—C bond length in such alkanes is about 154 pm.

1.2.2 sp^2 HYBRIDISATION—ETHENE (ETHYLENE)

Let us consider double and triple bonds as in ethene (ethylene) and ethyne (acetylene), where the carbons are not tetrahedral. In ethene, C_2H_4, each carbon atom is bonded to only three other atoms—two hydrogens and one carbon. The carbon and the three atoms attached to it are in one plane. This is not what we expect from sp^3 hybridisation. The shape of ethene is in conformity with the hybridisation of one s and two p orbitals, leaving behind an unhybridised p orbital. The hybrid orbitals are called sp^2 orbitals.

The three equal sp^2 hybrid orbitals will have maximum separation when they are planar and are at an angle of $120°$ (1.3a).

$$(1.3)$$

(a)   (b)

In ethene, two of the sp^2 hybrid orbitals overlap with the s orbitals of two hydrogen atoms to form sp^2–s sigma bonds which have the same shape but are not identical to the sp^3–s σ-bonds of methane. The remaining sp^2 orbital overlaps with one sp^2 orbital of another similar carbon (sp^2–sp^2 overlap) to form a C—C σ-bond of ethene (1.3b). Each carbon of ethene has – in addition to the sp^2 orbitals – one unhybridised p orbital whose axis is perpendicular to the sp^2 plane (1.4).

$$(1.4)$$

These p orbitals overlap laterally. This lateral overlap gives rise to two regions of electron density above and below the plane of the molecule, corresponding to the two lobes of the p orbitals (1.5).

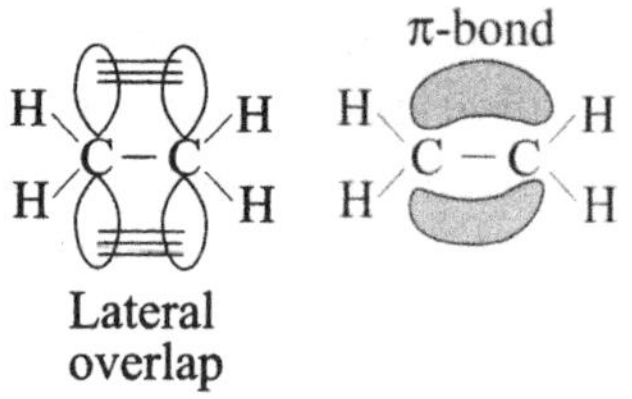

Lateral
overlap

(1.5)

Thus, the lateral overlap of *p* orbitals gives rise to a bond which does not have the cylindrical symmetry of the σ-bond. Such a bond is called a π-*bond*. The double bond of ethene is composed of one σ-bond and one π-bond. The six atoms of ethene, four Hs and two Cs, are in the same plane (coplanar). *Ethene is therefore a planar molecule.*

The usual way of representing ethene (1.6a) with a double bond is acceptable, but it should be kept in mind that the two bonds of the double bond are not equivalent.

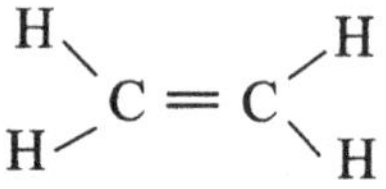

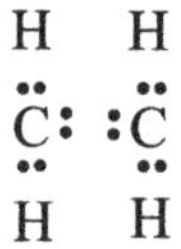

(1.6)

(a)   (b)

1.2.3 *sp* HYBRIDISATION—ETHYNE (ACETYLENE)

Yet another kind of hybridisation is seen in molecules like ethyne (acetylene), C_2H_2, containing a triple bond between the carbon atoms (1.7).

$$H-C \equiv C-H \text{ (ethyne)}$$ (1.7)

The triple bond is composed of two π-bonds and one σ-bond which implies that there are two unhybridised *p* orbitals on each of the carbon atoms, and hence only two hybrid orbitals. These arise by *sp hybridisation*.

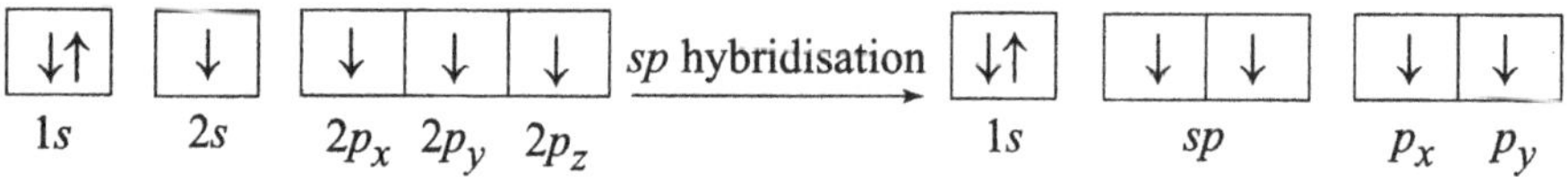

The *sp* orbitals are the result of the mixing of one 2*s* orbital and one 2*p* orbital, leaving two of the *p* orbitals unhybridised. (Labelling them as p_x and p_y is only for convenience. They could as well be labelled p_y and p_z, or p_x and p_z. The *x*, *y* and *z* orbitals are energy-degenerate orbitals.) The *sp* orbitals are at 180° to each other, which is the angle that gives them maximum separation. In more common terminology, the *sp* orbitals are collinear (along the same line) but pointing in opposite directions. The two unhybridised *p* orbitals are mutually perpendicular (along the *X* and *Y* axes) and the straight line of the *sp* hybrid orbitals is perpendicular to these two, along the *Z* axis as shown below.

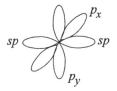

In forming ethyne, each of the carbons uses one of the sp orbitals to form the σ-bond with one hydrogen and the remaining two sp orbitals on the two carbons overlap with each other to form a C—C σ-bond (1.8).

$$ \text{(1.8)} $$

For maximum sp–sp overlap, the two sp orbitals are oriented along the same straight line with the result that the four atoms of ethyne are collinear. *Ethyne is a linear molecule.* The remaining two unhybridised p orbitals on each carbon are shown in 1.9 and their overlap forms two π-bonds.

$$ \text{H} - \text{C} - \text{C} - \text{H} \ \rightarrow \ \text{H} - \text{C} \equiv \text{C} - \text{H} $$

(a)   (b)   (1.9)

The four lobes of the two π-bonds in 1.9b overlap and the π-electron cloud is like a cylinder enveloping the C—C axis (1.10).

$$ \text{H} - \text{C} - \text{C} - \text{H} $$

(1.10)

1.2.4 CYCLOALKANES

Structure 1.11a represents 1-hexene C_6H_{12}, one of the many isomers of molecules having this molecular formula. The bonds in hexene and other alkenes are similar to those in ethene. 1.11b is *a diene* with two double bonds.

$$
\begin{array}{c}
\text{H} \quad \text{H} \quad \text{H} \quad \text{H} \quad \text{H} \quad \text{H} \\
| \quad\; | \quad\; | \quad\; | \quad\; | \quad\; | \\
\text{C} = \text{C} - \text{C} - \text{C} - \text{C} - \text{C} - \text{H} \\
| \qquad\quad | \quad\; | \quad\; | \quad\; | \\
\text{H} \qquad\quad \text{H} \quad \text{H} \quad \text{H} \quad \text{H}
\end{array}
\qquad \text{(a)}
$$

(1.11)

$$
\begin{array}{c}
\text{H} \quad \text{H} \quad \text{H} \quad \text{H} \quad \text{H} \quad \text{H} \\
| \quad\; | \quad\; | \quad\; | \quad\; | \quad\; | \\
\text{C} = \text{C} - \text{C} = \text{C} - \text{C} - \text{C} - \text{H} \\
| \qquad\qquad\qquad\quad | \quad\; | \\
\text{H} \qquad\qquad\qquad\quad \text{H} \quad \text{H}
\end{array}
\qquad \text{(b)}
$$

Structures 1.12a, b and c represent cyclohexane, C_6H_{12}. It is isomeric with hexene, 1.11a, but does not have a double bond and is not an alkene. Cyclohexane is an example of a *cycloalkane* formed by the linking of the two ends of a carbon chain. The bonding in cycloalkanes is the same as in alkanes. Structures 1.12a–1.12c are different ways of drawing the structure of cyclohexane. In 1.12c, only the framework is shown. Each corner of the hexagon represents a carbon atom. The hydrogens are not shown. It is assumed that each carbon atom carries the required number of hydrogens, two each in this case, to satisfy tetravalency.

$$(1.12)$$

1.12d is cyclohexene, a *cycloalkene*. Making the same assumptions as for the cyclohexane structures, cyclohexene may also be drawn as 1.12e.

1.2.5 ARENES—BENZENE

Now, let us study a class of cyclic compounds,which apparently contain double bonds, but are not alkenes. They are the *aromatic hydrocarbons*, a representative example of which is benzene C_6H_6. The history of the determination of the structure of benzene is one fascinating aspect of organic chemistry. The emperical formula of benzene was known from the quantitative analysis of its carbon and hydrogen composition. This coupled with the molecular weight of benzene gave the molecular formula of benzene C_6H_6.

1.2.5.1 The Kekulé Structure of Benzene

Organic chemistry had to wait till 1872 to get a reasonably satisfying structure for benzene, which could explain most of its properties. The structure that we use today (1.13a or 1.13b) was proposed by August Kekulé (1829–96), a German chemist in 1872.

$$(1.13)$$

He proposed that benzene exists as a mixture of two molecules which are in rapid equilibrium with each other. This structure could explain most of the reactions of benzene but not all. Importantly, this structure could not explain the fact that benzene is not an alkene—it does not behave like an alkene with three double bonds (a triene). The differences between alkenes and benzene in chemical reactions are discussed subsequently. A notable difference between an alkene and benzene is in the carbon–carbon bond length. Some bond lengths are listed in Table 1.1.

Table 1. 1 Some carbon–carbon bond lengths

Molecule	Bond length (pm)
$CH_3 — CH_3$	153
$CH_2 = CH_2$	134
$CH \equiv CH$	120
Benzene	139

The significant thing to be noted here is that the $C — C$ bond length decreases as we go from single bond to double bond to triple bond. In benzene, all the six $C — C$ bonds are of the same length—between the single bond length and double bond length. Structure 1.13 predicts three single bonds of about 153 pm length and three shorter double bonds. In reality, it is not so.

Simple valence bond theory is insufficient to understand the nature of bonding in benzene. An understanding of the structural difference between benzene and its analogues – called arenes or aromatic compounds on the one hand and alkenes on the other – could be obtained only with the advent of the molecular orbital theory of bonding and the theory of resonance. The problem encountered in benzene also introduces us to the concept of delocalised bonds. In the valence bond theory, we talk about localised bonds, namely bonds between atoms represented by pairs of electrons involved in the bonds and confined (localised) to the space between the atoms concerned. Thus, a $C — C$ single bond involves one pair of electrons and a $C — C$ double bond involves two pairs of electrons. In both cases, the electrons are localised between the carbon atoms as in 1.14.

$$-\overset{|}{\underset{|}{C}} - \overset{|}{\underset{|}{C}} - \qquad \overset{|}{C} = \overset{|}{C}$$

$$-\overset{|}{\underset{|}{C}} : \overset{|}{\underset{|}{C}} - \qquad \overset{|}{C} :: \overset{|}{C} \qquad\qquad (1.14)$$

(a)   (b)

Extending this to benzene, we have 1.13a and 1.13b called *Kekulé structures*, which we find inadequate. The six carbon atoms of benzene are sp^2 hybridised as expected from the Kekulé structures. The molecule has to necessarily be planar. The involvement of the six unhybridised p orbitals (1.15) is what needs examination.

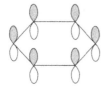

$$(1.15)$$

1.2.5.2 Molecular Orbitals of Benzene

The formation of the molecular orbitals from the six atomic (p) orbitals is represented in the MO energy diagram (1.16).

6 p atomic orbitals $\rightarrow$ 6π molecular orbitals

$$\text{(1.16)}$$

ψ_1, ψ_2 and ψ_3 are bonding orbitals. ψ_1 is a fully delocalised orbital and is of the lowest energy. ψ_2 and ψ_3 are of higher energy but degenerate. These three orbitals accommodate the six π- electrons. This cannot be represented satisfactorily using the valence bond structure. The top view of the bonding molecular orbitals is shown in 1.17 as seen from above the plane of the benzene ring. Each orbital has another lobe below the plane of the ring, which is not shown.

ψ_2 ⬡⬡ ⬡ ψ_3

$$\text{(1.17)}$$

ψ_1 ◯

The total view of the electron cloud is that of ring shaped regions above and below the benzene hexagon as in 1.18. It will be more accurate to represent benzene as 1.19 rather than as either of the Kekulé structures.

$$\text{(1. 18 and 1.19)}$$

We say that the π-bonds of benzene are delocalised. That is, not localised between specific pairs of carbons as in 1.13. In the chemical literature, benzene is represented using both notations in 1.19 and 1.13. The inaccurate Kekulé structure is often useful in understanding the reactions of benzene where it gets transformed into nonaromatic structures.

1.2.5.3 Valence Bond Structure of Benzene—Resonance

While the MO theory gives an accurate representation of the structure of benzene, valence bond structures are more familiar to organic chemists who are used to them. The theory of resonance introduced mainly by Linus Pauling around 1930, offers an attractive alternate way of looking at the structure of benzene and other such molecules containing delocalised bonds.

The unsatisfactory nature of the concepts of localised double bonds in benzene will be clear when we examine the overlap of the p orbitals (1.20a and 1.20b).

(1.20)

(a)   (b)

Why should the p orbitals on C1 overlap with the p-orbitals on C2 only as in (1.20a); why not with the p orbitals on C_6 which are oriented in the same way (1.20b)? The theory of resonance takes into account this dilemma and handles it satisfactorily. According to the theory of resonance, the two Kekulé structures of benzene are two extreme localised but unreal ways of representing benzene. The actual molecule of benzene has a structure which is neither of the two, but the sum total of both.

Linus Pauling (1901–94), an American chemist, who worked in the California Institute of Technology, made important contributions in many areas of chemistry and physics. He received the Nobel prize (chemistry) in 1954 and a second Nobel for Peace in 1962 for his campaign against nuclear weapons.

In resonance terminology, 1.13a and 1.13b are *contributing structures* or *canonical structures*. The actual structure is a *hybrid* of both. Kekulé's original idea was that benzene is a mixture of two molecules which are in rapid equilibrium with each other due to the shifting of bonds; these are represented by double arrows for the forward and the reverse reactions, representing equilibrium (1.13). This is now replaced by a different symbol (1.21).

(1.21)

(a)   (b)

The equilibrium arrows have been replaced by a , *resonance arrow*, which is a double-headed arrow ($\longleftrightarrow$). In organic chemistry, a clear distinction is made between double arrows (equilibrium) and a double-headed arrow (resonance). Resonance *does not* imply equilibrium or interconversion between two different molecules.

Structure 1.19 is an accepted way of representing the resonance hybrid. The following statement describes resonance. *Whenever a molecule cannot be accurately represented by a single valence bond structure, or more than one valence bond structure can be drawn for a molecule without violating the rules for electron pair bonds, the individual valence bond structures are unreal. The actual molecule is a resonance hybrid of all those structures (contributing or canonical structures).* The very fact that the electron distribution or bonding in the actual molecule (hybrid) is different from that in the canonical structures means that the hybrid represents a lower energy state than any of the canonical forms. The hybrid represents the most stable arrangement of all the particles (atoms, electrons) constituting the molecule.

1.2.5.4   Resonance Energy—Heats of Hydrogenation

Resonance is a stabilising phenomenon. The lower the energy content of a system (in this case the ensemble of atoms and electrons which form the molecule), the more stable it is. The actual molecule, benzene, is more stable than either of the canonical structures 1.13a or 1.13b. A quantifiable measure of this stability, called the resonance energy, can be obtained by elaborate calculations. In the case of benzene, a reasonably good value for resonance energy can be obtained experimentally. The experiment involves the measurement of heats of hydrogenation. Hydrogenation of alkenes, namely addition of hydrogen to an alkene to give an alkane is an exothermic reaction.

$$\text{\Large }>\!\!C\!=\!C\!<\ +\ H_2\ \rightarrow\ -\overset{|}{\underset{H}{C}}-\overset{|}{\underset{H}{C}}-\ +\ \text{Heat}$$

Heats of hydrogenation are expressed in $kJ\,mol^{-1}$ or $kcal\,mol^{-1}$ ($1\,kcal = 4.184\,kJ$). In this book, the SI unit, kJ, is used. Occasionally, the equivalent values in kcal are also given. The heat of hydrogenation of alkenes is fairly constant, about $-126\,kJ$ ($-30\,kcal\,mol^{-1}$), per double bond and is additive as seen from Fig. 1.1. The heat of hydrogenation of cyclohexene is $-120\,kJ\,mol^{-1}$; that of 1, 3-cyclohexadiene, with two double bonds is expected to be $-2 \times 120 = -240\,kJ\,mol^{-1}$. Experimentally, it is found to be $-233\,kJ\,mol^{-1}$. The difference between the expected (calculated) value and the experimental value, $7\,kJ\,mol^{-1}$, is not due to experimental error, but is real and significant. We shall come back to this difference presently (Section 1.3.1.2). A localised hypothetical Kekulé benzene, which can be called 1, 3, 5-cyclohexatriene, with three double bonds is expected to have a heat of hydrogenation of $-360\,kJ\,mol$. Experimentally, benzene is found to have a heat of hydrogenation of only $-208\,kJ\,mol^{-1}$. The difference between the calculated and the experimental value, $360 - 208 = 152\,kJ\,mol^{-1}$ (Fig. 1.1) is the energy difference between a

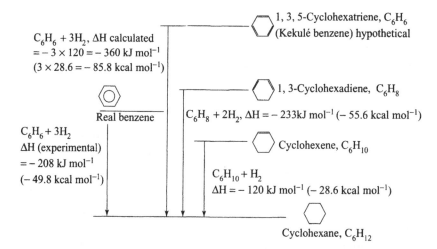

Fig. 1.1   Resonance energy of benzene from heats of hydrogenation

Kekulé structure of benzene (which we shall now call a canonical structure of benzene) and the actual benzene molecule, which is the resonance hybrid. This is the *resonance energy* of benzene. The actual benzene molecule is more stable than the hypothetical cyclohexatriene by $152 \, kJ \, mol^{-1}$

This large stability of the benzene molecule explains why it tends to retain its resonance stabilised structure during its reactions; a quality which was recognised early by chemists, causing benzene to be classified in a separate family of hydrocarbons called arenes. A large group of compounds are known based on benzene and similar highly resonance-stabilised systems and are called aromatics or aromatic compounds.

The structure of benzene involving a cyclic array of alternate single and double bonds (in its canonical structures) and an array of six unhybridised p orbitals arranged over a planar six-membered ring, gives rise to the delocalisation leading to its stability. Benzene is said to possess the property called *aromaticity*.

1.3 ELECTRONIC EFFECTS

1.3.1 RESONANCE

1.3.1.1 General Introduction

We have seen from the discussion above, that when there is conjugation, which occurs in molecules containing alternate double bonds, covalent bonds are not simple and delocalisation of electrons has to be taken into account. This delocalisation causes the electron density in molecules to be not necessarily what is expected at a particular site in a molecule and has a profound influence on the reactions of such molecules. Consider the case of acrolein (1.22). On examining the localised structure 1.22a, one cannot imagine that C3 has a lower electron density than C2. When the canonical structure 1.22b is superimposed on 1.22a, the resultant hybrid has a residual positive charge on C3.

$$\begin{array}{cc} \underset{H}{\overset{H}{\diagdown}}\overset{3}{C}=\overset{2}{C}\overset{H}{\diagup} & \underset{H}{\overset{H}{\diagdown}}\overset{+}{C}-\overset{H}{C}\overset{H}{\diagup} \\ \quad \overset{1}{C}=\ddot{O}: \longleftrightarrow \quad C-\ddot{O}: \\ H \diagup & H \diagup \end{array} \tag{1.22}$$

(a)   (b)

The molecule does undergo reactions which can be predicted from such structures. This is a simple illustration of the influence of the resonance effect on reactivity. The term *mesomerism* is often used synonymously with resonance. Resonance or mesomerism is an inherent aspect of the structure of the molecule. It exists even when the molecule is not subjected to any external influence and is a property of the ground state of the molecule. This and similar effects which affect the electron distribution in molecules – in ways which are not visible in simple valence bond structures – are called *electronic effects* in molecules. These electronic effects profoundly influence properties and reactivities. These effects may be the result of inherent ground state phenomena or may be caused by external influences.

1.3.1.2 Conjugated Molecules

Molecules such as 1,3-butadiene (1.23), 1,3-cyclohexadiene (1.24) and 1,3,5-hexatriene (1.25) are called *conjugated polyenes*. In these, single and double bonds alternate.

$$CH_2 = CH - CH = CH_2 \qquad\qquad (1.23)$$

$$(1.24)$$

$$CH_2 = CH - CH = CH - CH = CH_2 \qquad\qquad (1.25)$$

Molecules such as 1,4-cyclohexadiene (1.26) and 1,4-hexadiene (1.27) are not conjugated.

$$(1.26)$$

$$CH_2 = CH - CH_2 - CH = CH - CH_3 \qquad\qquad (1.27)$$

Conjugation can not only be seen between carbon–carbon multiple bonds, but can also involve other multiple bonds, such as those between $C=C$ and $C=O$ double bonds. Acrolein (1.22) is a conjugated molecule.

It must be obvious by now that conjugation exists when there is an unhybridised *p* orbital on an atom, which is flanked by two (or more) other atoms also bearing unhybridised *p* orbitals as in 1.28 and 1.29.

1,3 Butadiene   (1.28)

Acrolein   (1.29)

The question that we asked ourselves about the overlapping *p* orbitals of benzene (1.20) arises in these cases also. Can the two orbitals in the middle (1.30, 1.31) not overlap with each other?

$$(1.30 \text{ and } 1.31)$$

Indeed they can and they do. The π-bond of butadiene is delocalised as in benzene and the electron cloud encompasses all four carbons; it is not localised as in 1.28 but delocalised as shown below.

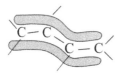

For this to happen, all the atoms of 1,4 butadiene – the four carbons and the six hydrogens – should be coplanar. Even though rotation around the C2—C3 sigma bond is possible, the most stable conformation of 1,4-butadiene is the one which is planar. It is this planar conformation alone that has the extra resonance stabilisation. Different orientations of the atoms in a molecule obtained by rotation around a single bond are called conformations (see the conformations of ethane (1.1c, 1.1d, 1.1e). 1,3-Butadiene is represented as a resonance hybrid of three canonical structures (1.32).

$$
\begin{array}{cccc}
\text{(a)} & \text{(b)} & \text{(c)} & \text{(d)}
\end{array}
$$
(1.32)

The charges shown on C1 and C4 are formal charges. Since the molecule is symmetric, the probability of having a + or − charge on either carbon is the same; this is not the case with acrolein (1.22).

(1.33)

Because of the greater electronegativity of oxygen compared to carbon, the negative charge prefers to reside as in 1.22b and not as in 1.33. Since 1.33 is an improbable structure, it is not counted as a significant canonical structure.

A consequence of conjugation or delocalisation – which we shall now refer to as resonance – in 1,3-butadiene (1.32) is that it has resonance stabilisation. The hybrid 1.32d is more stable (possesses less energy) than any of the other canonical structures (1.32a, 1.32b or 1.32c). Similarly 1,3-cyclohexadiene is also a conjugated molecule and has to be correctly represented as a delocalised hybrid and not as the localised structure. The discrepancy in the heat of hydrogenation of 6.7 kJ mol^{-1} (which we noted in our discussion of Fig. 1.2, Section 1.4.2.8), is actually a measure of the stabilisation energy of the conjugated molecule. Some other examples of conjugated molecules which exhibit resonance are shown below (1.34: carbonate ion, 1.35: formate ion, 1.36: formic acid).

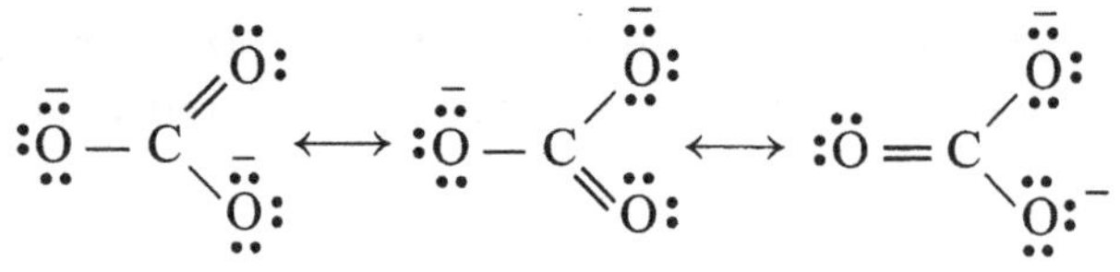

$$\ddot{\overset{-}{\text{O}}} - \text{C}\overset{\displaystyle\overset{-}{\ddot{\text{O}}:}}{\underset{\ddot{\text{O}}:}{\diagup}} \longleftrightarrow \overset{-}{:}\ddot{\text{O}} - \text{C}\overset{\displaystyle\ddot{\overset{-}{\text{O}}}:}{\underset{\text{O}:}{\diagdown}} \longleftrightarrow :\ddot{\text{O}} = \text{C}\overset{\displaystyle\ddot{\overset{-}{\text{O}}}:}{\underset{\ddot{\text{O}}:^{-}}{\diagdown}} \tag{1.34}$$

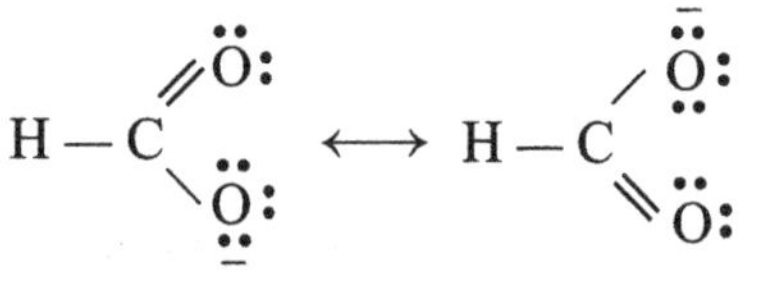

$$\text{H} - \text{C}\overset{\displaystyle\ddot{\text{O}}:}{\underset{\ddot{\text{O}}:}{\diagup}} \longleftrightarrow \text{H} - \text{C}\overset{\displaystyle\ddot{\overset{-}{\text{O}}}:}{\underset{\ddot{\text{O}}:}{\diagup}} \tag{1.35}$$

$$\text{H} - \text{C}\overset{\displaystyle\ddot{\text{O}}:}{\underset{\ddot{\text{O}} - \text{H}}{\diagup}} \longleftrightarrow \text{H} - \text{C}\overset{\displaystyle:\ddot{\overset{-}{\text{O}}}:}{\underset{\overset{+}{\text{O}} - \text{H}}{\diagup}} \tag{1.36}$$

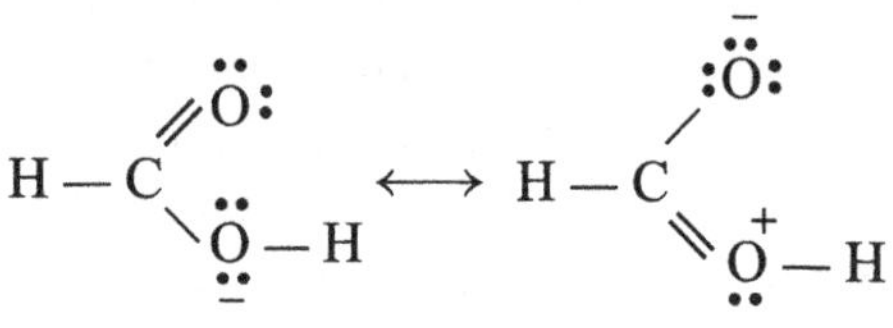

One of the important rules of resonance is that if the canonical structures are identical as in benzene (2 identical canonical structures), carbonate ion (1.34, 3 identical canonical structures) and formate ion (1.35, two identical canonical structures), the canonical structures contribute equally to the hybrid. When the canonical structures are identical or nearly so, the resonance energy is high.

1.3.1.3 Drawing Resonance Structures

Some more examples of molecules where resonance exists are given as structures 1.37 to 1.41. Let us examine these structures more closely. Formaldehyde (1.37), is representative of all molecules containing the carbonyl group $-\text{C}=\text{O}$. The charged canonical structure shows the inherent polarity of the carbonyl group. The lone pairs on the oxygen are shown. This is to help in keeping track of the number of electrons. These structures illustrate the use of curved arrows to indicate the formal shift of electron pairs. The curved arrow in formaldehyde starts from the double bond and ends at the oxygen. This means that a pair of electrons from the double bond (the π-electrons) has been shifted to the oxygen. This gives oxygen 3 lone pairs of electrons and a negative charge called a formal charge. The carbon is left with only 6 electrons (3 pairs) around it, giving it a positive charge. (A formula for calculating formal charges is given in Section 1.4.2.3.)

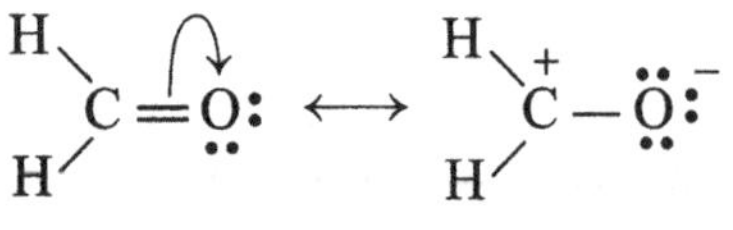

$$\overset{\text{H}}{\underset{\text{H}}{\diagdown\diagup}}\text{C} = \ddot{\text{O}}: \longleftrightarrow \overset{\text{H}}{\underset{\text{H}}{\diagdown\diagup}}\overset{+}{\text{C}} - \ddot{\overset{-}{\text{O}}}: \qquad \text{Formaldehyde} \tag{1.37}$$

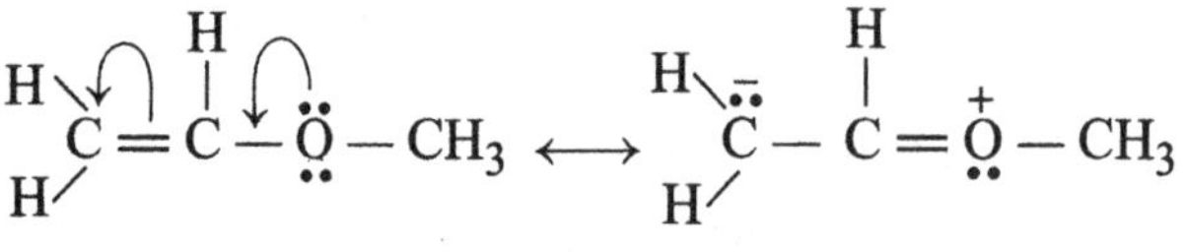

       (a)   (b)

$$\overset{\text{H}}{\underset{\text{H}}{\diagdown\diagup}}\text{C} = \overset{\text{H}}{\underset{}{\text{C}}} - \ddot{\text{O}} - \text{CH}_3 \longleftrightarrow \overset{\text{H}}{\underset{\text{H}}{\diagdown\diagup}}\overset{-}{\text{C}} - \overset{\text{H}}{\underset{}{\text{C}}} = \overset{+}{\text{O}} - \text{CH}_3 \qquad \text{Methyl vinyl ether} \tag{1.38}$$

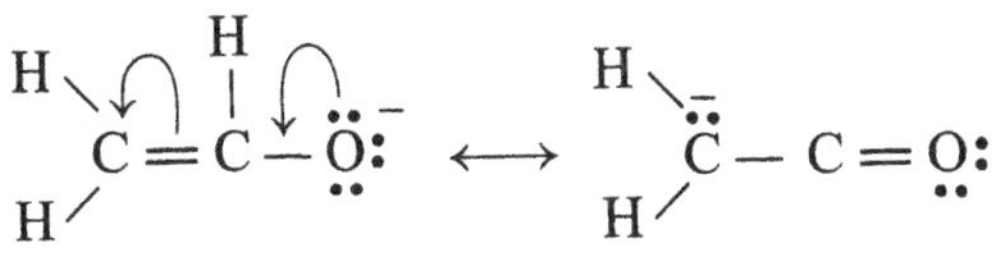

Anion of vinyl alcohol   (1.39)

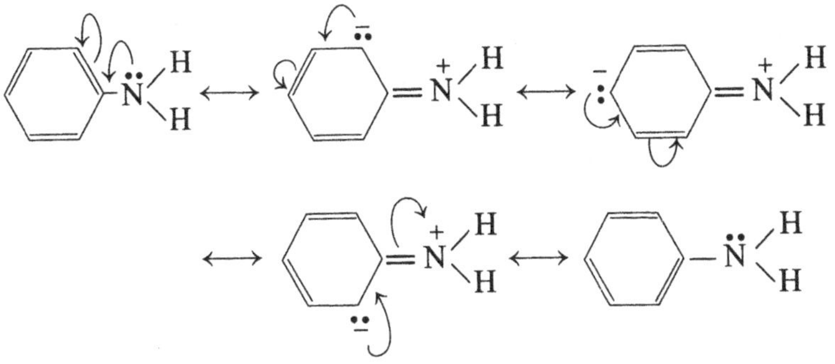

Aniline   (1.40)

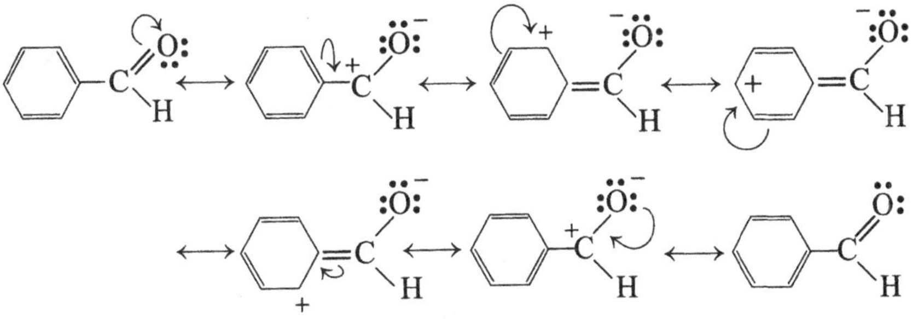

Benzaldehyde

(1.41)

In 1.38, the involvement of the lone pair of electrons of oxygen in resonance is shown. In these structures, all the lone pairs of electrons on the atoms like oxygen and nitrogen are specifically shown. In structures 1.38–1.40, the lone pairs of electrons on O and N are delocalised. In some of the canonical structures, they are in double bond formation. In such molecules, O and N are considered to be sp^2 hybridised. In molecules where they are not involved in double bond formation, like H_2O and NH_3, O and N are sp^3 hybridised. In 1.38, the delocalisation of one of the pairs on oxygen is shown, with the help of curved arrows. The relocation of bonds and lone pairs can be easily understood if the curved arrows are followed carefully. The formal charges in the canonical structure also follow and can be checked using the formula given in Section 1.4.2.3. In 1.39, the stabilisation of an anion by delocalisation is shown. In structures 1.40 and 1.41, the involvement of the conjugated double bonds of benzene is shown. The usefulness of the 'unreal' Kekulé structures of benzene in such situations is obvious. Notice that the first and last canonical structures of 1.40 and 1.41 are in fact the two Kekulé structures of the molecules. Groups such as NH_2 release electrons into the benzene ring (or any other system with which it is conjugated) thereby increasing the electron density in the ring. Groups such as the carbonyl (as in benzaldehyde) withdraw electrons from the benzene ring, thereby decreasing the electron density in the ring. The former type of groups – the electron-releasing groups – are said to have a $+R$ effect, and the latter – the electron-withdrawing groups – are said to have a $-R$ effect. A partial list of groups with their electron-releasing or electron-withdrawing effects is given in Table 1.3.

1.3.1.4 Rules for Resonance

(i)   Any compound for which more than one acceptable Lewis structure can be drawn cannot correctly be described by any single structure, but has a structure which is a hybrid of all the possible structures. The contributions of each canonical structure to the hybrid is the extent to which the properties of the hybrid correspond to those of that canonical structure as predicted by analogy with other similar real structures. That canonical structure which is closest to the actual molecule in energy, will contribute most to the hybrid.

(ii)   The stability of the resonance hybrid is greater than that of any of the canonical structures.

(iii) The greater the number of canonical structures, the greater the stability of the molecule.

(iv)   The number of unpaired electrons in any canonical structure should not be different from that in the actual molecule. The number of unpaired electrons can be experimentally determined. For example, the O_2 molecule in the ground state has two unpaired electrons, as determined experimentally (1.42a). An excited (high energy) form of O_2 called singlet oxygen is known, which has no unpaired electrons (1.42b).

$$\ddot{\underset{\cdot\cdot}{O}} : \ddot{\underset{\cdot\cdot}{O}} : \quad \text{or} \quad \dot{O} - \dot{O} \qquad\qquad : \ddot{\underset{\cdot\cdot}{O}} :: \ddot{\underset{\cdot\cdot}{O}} : \quad \text{or} \quad O = O \tag{1.42}$$

$$\qquad\qquad (a) \qquad\qquad\qquad\qquad\qquad\qquad (b)$$

1.42a and 1.42b are not canonical structures; they are two different molecules.

(v)   Canonical structures can differ from each other only in the position of electrons (bonding electrons and non-bonding electrons). They cannot differ too much in the relative position of the atoms or in the hybridisation of the atoms. There can be minor differences in bond lengths between the hybrid and its canonical structure as in the case of the $C-C$ bond lengths in the real molecule of benzene and its canonical structures (Kekulé structures, 1.43).

134 pm   139 pm
153 pm   139 pm   (1.43)

Expected for   Actual bond length
Kekulé structure   in benzene

However, 1.44a and 1.44b are two different molecules. They cannot be canonical structures of the same molecule because, (i) the relative position of the carbon atoms are different in the two structures and (ii) the hybridisation of the carbons are different.

In 1.44a, all the carbons are sp^2 hybridised. In 1.44b, two are sp^2 hybridised and the other two are sp^3 hybridised.

$$
\begin{array}{cc}
\text{(a)} & \text{(b)}
\end{array}
\tag{1.44}
$$

(vi) Resonance is possible only when the conjugated portion of the molecule is planar and only when the p orbitals involved in resonance have their axis parallel to each other in order to enable maximum lateral overlap. (See the discussion on the need for planarity in 1,3-butadiene for resonance stabilisation, Section 1.3.1.2. Also, see Section 1.5.2 for another example which illustrates this rule.)

1.3.1.5 Stability of Canonical Structures

(i) A structure where unlike charges are separated is less stable than one where there is no charge separation. In 1.37a is more stable than b as is the case of 1.38 to 1.41, the charged structures are less stable than the uncharged ones.

(ii) A structure where like-charges are close to each other is less stable than one where they are well separated.

(iii) In structures with charges, those with negative charges on the most electronegative atoms will be more stable than others. Consider the case of acrolein (1.33), Both 1.22b and 1.33 are valid Lewis structures. But the contributions by 1.33 to the hybrid is negligible.

(iv) Structures where the atoms of the second row of the periodic table (Li to Ne) have more than 8 electrons in the valence shell are unstable and can be neglected. Structures such as 1.45b are considered invalid (nitrogen has 10 electrons).

$$
\begin{array}{cc}
\text{(a)} & \text{(b)}
\end{array}
\tag{1.45}
$$

1.3.2 INDUCTIVE EFFECT

Another ground state electronic effect is caused by differences in the electronegativities of the atoms involved in a covalent bond. In a diatomic molecule containing the same element like H_2,

Cl_2 or O_2, the electron density in the orbitals involved in the bond or bonds is symmetrically placed between the two nuclei. In other words, the bonding electrons are shared equally by the two atoms. Such a molecule has no *dipole moment* and *polarity*. This is in contrast to a molecule like HCl where the two atoms differ in electronegativity. Chlorine is more electronegative than hydrogen. The bonding pairs of electrons are attracted towards the chlorine and pulled away from the hydrogen. The electron cloud of the σ-bond in HCl may have a shape similar to that shown below.

This causes the rod like molecules to have a negative end and a positive end. In organic chemistry terminology, there is a charge separation in the molecule and the chlorine bears a partial negative charge and the hydrogen, a partial positive charge; the symbol δ is used to denote partial charge.

HCl is a polar molecule; it has a dipole moment. The extreme state of this charge separation is ionisation. In the gas phase, HCl is not ionic; it is a covalent molecule. The theory of resonance can be used to gain another view of the charge separation (1.46).

$$H - Cl \longleftrightarrow H^+ \ Cl^- \tag{1.46}$$

The actual HCl molecule is a hybrid of the two structures and what we see is an electronic effect. This effect causes a charge separation and a polarity in the bond. This is called the *inductive effect*. The direction of the inductive effect is often represented by an arrow over the bond.

Now, let us look at a molecule like carbon tetrachloride, CCl_4. This has a tetrahedral structure similar to that of methane. Each carbon–chlorine bond is polar and has charge separation (1.47).

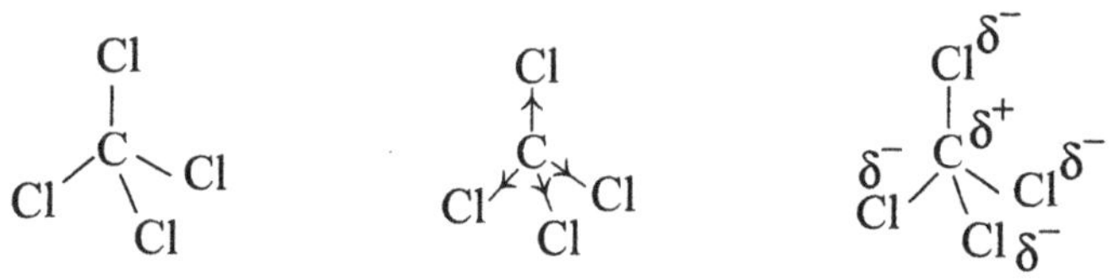

$$\tag{1.47}$$

However, since the molecule is symmetrical, there is no dipole moment. The force with which chlorine attracts the valence electrons away from hydrogen in HCl and from carbon in CCl_4, is called the inductive effect of chlorine. Inductive effect, caused by electronegativity is a *permanent electronic effect* in a molecule and can exist even without conjugation, unlike resonance. (Representing inductive effect using the symbols of resonance as in 1.46 is a liberty we have taken and is not the norm. It also goes to show that these are only concepts that help in understanding phenomena and should not be rigidly compartmentalised.)

Consider chloromethane, notice that we have reverted to the habit of drawing a 'flat' structure, knowing fully well that it is a tetrahedral molecule. The inductive effect of chlorine causes charge separation as in 1.48b.

$$
\begin{array}{ccc}
\underset{\text{(a)}}{\overset{\displaystyle H}{\underset{\displaystyle H}{H-C-Cl}}}
&
\underset{\text{(b)}}{\overset{\displaystyle H}{\underset{\displaystyle H}{H-\overset{\delta^+}{C}\overset{\delta^-}{-Cl}}}}
&
\underset{\text{(c)}}{\overset{\displaystyle H^{\delta\delta^+}}{\underset{\displaystyle H\,\delta\delta^+}{{}^{\delta\delta^+}H-\overset{\delta^+}{C}\overset{\delta^-}{-Cl}}}}
\end{array}
\qquad (1.48)
$$

The partial positive charge induced on the carbon by the chlorine causes this carbon to attract the bonding electrons of the $C-H$ bond more than in, say, CH_4. This causes the hydrogens also to acquire partial positive charges. The symbol $\delta\delta$ has been used to imply that the electron pull in the $C-H$ bond is more feeble than that in the $C-Cl$ bond. *The further away we move from the atom causing the inductive effect, the more feeble is the effect.* In chloroform (1.49) – trichloromethane in the IUPAC nomenclature – there are three electronegative chlorine atoms, each exerting its inductive effect.

$$
\overset{\displaystyle Cl^{\delta^-}}{\underset{\displaystyle Cl^{\delta^-}}{\overset{\delta^+}{H}-\overset{\delta^+}{C}\!-\!Cl^{\delta^-}}}
\qquad (1.49)
$$

The concerted pull by three chlorines makes the carbon in this molecule more positive than it is in CCl_4 and the hydrogen, in turn more positive. *More the number of atoms inducing the inductive effect, stronger is the effect.* Inductive effect can be transmitted through connecting covalent bonds, whether single or multiple, through long distances along a molecule, but the effect will become more and more feeble as the distance increases. Resonance effects on the other hand are transmitted only through conjugated double bonds and can be fully transmitted through any distance as long as conjugation is maintained. For example, the partial positive charge on C3 in 1.22b is quite pronounced and is responsible for many characteristic reactions of acrolein. Halogens and other electronegative atoms withdraw electrons by inductive effect. This is designated as the $-I$ *effect*. Alkyl groups exert an electron-releasing inductive effect called the $+I$ effect (Section 1.4.3.4).

1.3.3 ELECTROMERIC EFFECTS: INDUCED ELECTRONIC EFFECT

In addition to the ground state electronic effect discussed so far, changes in electron density distribution in molecules can occur under external influence. For example, a halogen molecule, say bromine, is symmetric and has no dipole moment. The bonding electrons of the $Br-Br$ σ-bond are equally shared by the two atoms. Three reactions illustrating the creation of polarity in the nonpolar bromine molecule (induced polarity) are given below (Eqs. 1.50–1.52).

$$
Br-Br+I^- \longrightarrow \left[\overset{\delta^-}{Br} \twoheadleftarrow \overset{\delta^+}{Br} \ldots I^- \right] \longrightarrow Br^-\ Br-I
\qquad (1.50)
$$

$$Br-Br + \left|\substack{\diagdown C \diagup \\ \| \\ \diagup C \diagdown}\right| \longrightarrow \left[Br \twoheadleftarrow Br \cdots \substack{\diagdown C \diagup \\ \| \\ \diagup C \diagdown}\right] \longrightarrow \overset{-}{Br} \; \overset{+}{Br} \substack{\diagdown C \diagup \\ | \\ \diagup C \diagdown} \qquad (1.51)$$

$$Br-Br + \underset{\underset{Br}{|}}{\overset{\overset{Br}{|}}{Al}}-Br \longrightarrow \left[Br \twoheadrightarrow Br \cdots \underset{\underset{Br}{|}}{\overset{\overset{Br}{|}}{Al}}-Br\right] \longrightarrow \overset{+}{Br} \; \overset{-}{Br}-\underset{\underset{Br}{|}}{\overset{\overset{Br}{|}}{Al}}-Br \qquad (1.52)$$

In Eq. 1.50, in the vicinity of an electron rich ion I^-, a polarity is induced in the bromine molecule (structure in the square brackets), finally leading to the product; the inter-halogen compound IBr. In Eq. 1.51, the π-electron cloud of the alkene is an electron rich centre. When the bromine molecule is in its vicinity, a polarity is induced in the bromine molecule which facilitates the formation of the product, the cyclic bromonium ion. Equation 1.52 shows how the opposite kind of polarity can be induced in bromine under the influence of an electron attracting or electron deficient reagent $AlBr_3$ which is a Lewis acid. In all these examples, the structures in the square brackets are not the final products but transitional structures, called *transition states*. At this stage, the reaction can go either in the forward or in the reverse direction. The induced polarity that we are considering here is not a permanent feature, but is a *time variable effect*. This is also called *inductomeric effect*.

Let us consider Eq. 1.51 in greater detail. Neither the halogen, Br_2, nor the alkene (ethene, for example) is polar in the ground state. Polarity has been induced in each of them by the other. Also noteworthy is that one cannot induce a polarity in the H_2 molecule by ethene, even though H_2 appears to be similar to Br_2—both are diatomic molecules connected by a single bond. Also, one cannot induce a polarity in ethane by bromine, even though both ethane and ethene are hydrocarbons. The point that is being made here is, that both molecules involved in this interaction should be *polarisable*. Ethene (alkene in general) is polarisable because the electrons of the π-bond are less tightly held by the nuceli than those of σ-bonds. The π-electron cloud can be more readily distorted (polarised) by external electron attracting reagents than σ-electrons. Similarily in halogens, each halogen atom has 3 pairs of electrons in its valence shell which are not involved in bonding. These are called lone pairs or non-bonded pairs of electrons. The orbitals bearing these electrons can be utilised for further bonding (Eq. 1.52). Also, because of the strong electronegativity of the halogens, the σ-electrons themselves can be polarised (pushed in one direction) by another reagent which comes with electrons (iodide ion or alkene). Thus, the concept of *polarisability* is important.

Polarisability is important even in conjugated systems. Look at the structure of 1,3 butadiene (1.32). Even though canonical structures with charge separation (1.32b and 1.32c) can be written, the hybrid has no polarity. However, under the influence of a reagent, polarity corresponding to (b) or (c) can manifest. The following situation (Eq. 1.53) arises when a proton approaches one end of the molecule.

$$
\begin{array}{c}
\text{C}=\text{C} \\
\quad\quad \text{C}=\text{C}
\end{array}
+ \text{H}^+ \longrightarrow
\left[
\begin{array}{c}
\overset{+}{\text{C}}-\text{C} \\
\quad\quad \text{C}=\text{C}\cdots\text{H}^+
\end{array}
\right]
\longrightarrow
\begin{array}{c}
\overset{+}{\text{C}}-\text{C} \\
\quad\quad \text{C}-\text{C}-\text{H}
\end{array}
$$

(1.53)

In the transition state, the polarised form of butadiene is involved.

Drawing resonance structures—analysis of the structure of diazomethane

Diazomethane is a highly reactive gaseous molecule. Its molecular formula has been established from elemental analysis and molecular weight determination to be CH_2N_2. Using this information, we can draw several possible skeletal structures of diazomethane (1–6).

$$
\begin{array}{cccccc}
\text{H} & \text{H} & \text{H} & \text{H H H} & & \text{H} \\
| & | & | & | | | & \text{H} \quad \text{N} & \text{N} \\
\text{H}-\text{N}-\text{C}-\text{N} & \text{H}-\text{C}-\text{N}-\text{N} & \text{N}-\text{C}-\text{N} & \text{N}-\text{C}-\text{N} & \text{C} \quad | & \text{H}-\text{C} \quad | \\
& & & & \text{H} \quad \text{N} & \text{N} \\
(1) & (2) & (3) & (4) & (5) & (6)
\end{array}
$$

Before we can proceed further, we have to know which atom is attached to which other atom. The complexity of even such a 'simple' molecule containing only 5 atoms, is obvious. The earlier chemists had to depend entirely on chemical reactions to choose between different possible structures. Today we have physical tools such as spectroscopy and diffraction studies which give direct information on the structure of molecules. Diazomethane is known to have a structure corresponding to 2. (Molecules corresponding to most of the other skeletons have also been identified.)

Knowing the skeletal structure, we can draw valence bond structures. In the following discussion, formal charges are calculated according to the formula given in Section 1.4.2.3. The total number of valence electrons in the molecules is 16. This is a neutral molecule, not an ion. It also does not have any unpaired (odd) electrons. All the 16 electrons should be present in all the canonical structures as 8 pairs.

$$
\begin{aligned}
2\,\text{H} &= 2 \text{ electrons} \\
1\,\text{C} &= 4 \text{ electrons} \\
2\,\text{N} &= 10 \text{ electrons} \\
\hline
&\quad 16 \text{ electrons}
\end{aligned}
$$

We start with the most obvious structure—A.

$$
\begin{array}{c}
\text{H} \\
\diagdown \\
\text{C} = \overset{+}{\text{N}}_1 = \overset{\cdot\cdot}{\underset{\cdot\cdot}{\text{N}}}{:}_2 \\
\diagup \\
\text{H} \\
(A)
\end{array}
$$

Tally the electrons:

6 bonds	− 12	electrons
2 lone pairs	− 4	electrons
	16	electrons

Check formal charges:

$$C,\, 4 - 0 - \tfrac{8}{2} = 0$$

$$N_1,\, 5 - 0 - \tfrac{8}{2} = +1$$

$$N_2,\, 5 - 4 - \tfrac{4}{2} = -1$$

We can use curved arrows to get more structures.

$$
\overset{\text{H}}{\underset{\text{H}}{\diagdown\diagup}}\text{C} = \overset{+}{\text{N}} = \overset{\cdot\cdot}{\text{N}}{:} \quad \longleftrightarrow \quad \overset{\text{H}}{\underset{\text{H}}{\diagdown\diagup}}\text{C} - \overset{+}{\text{N}} = \overset{\cdot\cdot}{\underset{\cdot\cdot}{\text{N}}}{:}
$$

(A)   (B)

$$
\overset{\text{H}}{\underset{\text{H}}{\diagdown\diagup}}\text{C} = \overset{+}{\text{N}} = \overset{-}{\text{N}}{:} \quad \longleftrightarrow \quad \overset{\text{H}}{\underset{\text{H}}{\diagdown\diagup}}\overset{-}{\text{C}} - \overset{+}{\text{N}} \equiv \text{N}{:} \qquad (C)
$$

$$
\overset{\text{H}}{\underset{\text{H}}{\diagdown\diagup}}\text{C} = \overset{+}{\text{N}} = \overset{\cdot\cdot}{\underset{\cdot\cdot}{\text{N}}}{:} \quad \longleftrightarrow \quad \overset{\text{H}}{\underset{\text{H}}{\diagdown\diagup}}\text{C} = \overset{\cdot\cdot}{\text{N}} - \overset{\cdot\cdot}{\text{N}}{:} \qquad (D)
$$

These – A, B, C and D – are the four 'good' structures. How about the following?

$$
\overset{\text{H}}{\underset{\text{H}}{\diagdown\diagup}}\text{C} = \overset{+}{\text{N}} = \overset{\cdot\cdot}{\underset{}{\text{N}}}{:}^{-} \quad \longleftrightarrow \quad \overset{\text{H}}{\underset{\text{H}}{\diagdown\diagup}}\overset{+}{\text{C}} - \overset{+}{\underset{\cdot\cdot}{\text{N}}} - \overset{\cdot\cdot}{\underset{\cdot\cdot}{\text{N}}}{:}^{=} \qquad (E)
$$

$$
\overset{\text{H}}{\underset{\text{H}}{\diagdown\diagup}}\text{C} = \overset{+}{\text{N}} = \overset{\cdot\cdot}{\text{N}}{:}^{-} \quad \longleftrightarrow \quad \overset{\text{H}}{\underset{\text{H}}{\diagdown\diagup}}\overset{\cdot\cdot}{\text{C}}{}^{-} - \overset{++}{\text{N}} = \overset{\cdot\cdot}{\text{N}}{:}^{-} \qquad (F)
$$

$$
\overset{\text{H}}{\underset{\text{H}}{\diagdown\diagup}}\text{C} = \overset{+}{\text{N}} = \overset{\cdot\cdot}{\text{N}}{:}^{-} \quad \longleftrightarrow \quad \overset{\text{H}}{\underset{\text{H}}{\diagdown\diagup}}\text{C} = \overset{++}{\text{N}} - \overset{\cdot\cdot}{\underset{\cdot\cdot}{\text{N}}}{:}^{=} \qquad (G)
$$

$$
\overset{\text{H}}{\underset{\text{H}}{\diagdown\diagup}}\text{C} = \overset{+}{\text{N}} = \overset{\cdot\cdot}{\underset{}{\text{N}}}{:} \quad \longleftrightarrow \quad \overset{\text{H}}{\underset{\text{H}}{\diagdown\diagup}}\text{C} = \text{N} \equiv \text{N}{:} \qquad (H)
$$

E, F and G are 'bad' structures, mainly because they have two like-charges on the same atom. Their contribution to the hybrid will be negligible. In structure H, although there are no charges, it is untenable because the middle nitrogen has 10 electrons around it.

Now let us consider the relative merits of structures A–D. We shall apply the following criteria.

(i)   More number of covalent bonds—greater stability.

(ii)   Greater separation of unlike charges—less stability.

(iii) $+$ Charge on the more electropositive carbon atom and $-$ charge on the more electronegative nitrogen is preferred, compared to the reverse situation.

(iv)   Structures possessing atoms with complete octet of electrons are preferred over those with incomplete octet of electrons.

Now, let us evaluate the good structures, A, B, C and D.

Structure A: 4 Covalent bonds (other than $C-H$ bonds); charges separated by 1 bond

Structure B: 3 Covalent bonds; charges separated by 2 bonds; C has only sextet of electrons.

Structure C: 4 Covalent bonds; charges separated by 1 bond; negative charge on carbon and positive charge on nitrogen.

Structure D: 3 Covalent bonds; the terminal N has only sextet; no charge separation.

Overall, the contribution is expected to be maximum from canonical structure A.

Taken in isolation, in A and C, the middle nitrogen is sp hybridised and the $C-N-N$ portion of the molecule should be linear. The middle N in B and D appears to be sp^2 hybridised; in which case, the $C-N-N$ bond angle should be about $120°$. The molecule is known to be linear. The middle nitrogen in all the structures is sp hybridised as a result of resonance.

1.4   REACTIVE INTERMEDIATES

1.4.1   ELECTROPHILES AND NUCLEOPHILES

This is an appropriate place to introduce two new terms, *nucleophile* and *electrophile*.

In Eq. 1.50, the reagent I^- has a negative charge. It has a liking (affinity) for positive centres such as positive ions (cations), electron deficient molecules like Lewis acids, and atoms in a neutral molecule where there is a partial positive charge like C3 of acrolein (1.22b). Species which have an affinity for positive centers are called nucelophiles. (To aid memory, the nucleus is a positive centre.) All anions (like I^-), neutral molecules with lone pairs of electrons like Br_2, H_2O or NH_3 (Eq. 1.54) or polarisable bonds like the π-bond of alkenes (Eq. 1.51) are nucleophiles.

$$H^+ + H_2\overset{\displaystyle ..}{\underset{\displaystyle ..}{O}}: \longrightarrow \overset{\displaystyle H \; \overset{+}{} \; H}{\underset{\displaystyle H}{\overset{|}{\underset{..}{O}}}} \qquad (a)$$

$$H^+ + :NH_3 \longrightarrow H-\overset{+}{\underset{\displaystyle \diagdown H}{N}}\overset{\diagup H}{}-H \qquad (b)$$

$$(1.54)$$

All species which have an affinity for electron rich centres (negative or negatively polarised centres) are electrophiles. In Eq. 1.50, Br_2 is an electrophile, though in Eq. 1.52, it is a nucelophile. In

Eq. 1.52, $AlBr_3$ is an electrophile. Electrophiles include cations, electron deficient molecules like Lewis acids and positive sites in neutral molecules like the C3 of acrolein (Eq. 1.22). Note that if a reaction involves a nucleophile, there will be a complementary electrophile also in that reaction.

1.4.2   DISSOCIATION OF BONDS

1.4.2.1   Homolysis and Heterolysis

The majority of organic reactions can be classified as those involving the making and breaking of covalent bonds. The electronic effects discussed above will help considerably in understanding these reactions. Some further terms related to the breaking of bonds will be introduced at this stage. A covalent bond involving an electron pair can be broken in two ways—*homolytic scission* and *heterolytic scission*. These are shown in Eq. 1.55 for a neutral molecule A–B.

$$A:B \longrightarrow A:^- + B^+ \quad (a)$$

$$A:B \longrightarrow A^+ + :B^- \quad (b) \qquad\qquad (1.55)$$

$$A:B \longrightarrow A\cdot + \cdot B \quad (c)$$

(a) and (b) where one of the atoms retains the bonding pair of electrons and the other is left with an incomplete octet, is called heterolytic scission or *heterolysis*. From the neutral molecule AB, the two fragments formed are ions. In Eq. 1.55c, each atom retains one of the electrons from the bonding pair. Each fragment is neutral and carries a single unpaired electron, sometimes called an odd electron. The fragments are called free radicals and the bond breaking is called *homolysis* or homolytic scission.

Heterolysis is characteristic of highly polar bonds of the type in structures 1.46 and 1.48. Where there is already a charge separation in the ground state, appropriate conditions like media of high dielectric constant and solvents which can solvate the ions formed, facilitate heterolysis. Homolysis is characteristic of weak bonds, that is, those with relatively low bond dissociation energy. Thus, the covalent bond in I_2 with a bond dissociation energy of $36\,kcal\,mol^{-1}$, can break homolytically more easily than the $C-C$ bond in ethane which has a bond dissociation energy of $88\,kcal\,mol^{-1}$. Heating helps homolysis and even the $C-C$ bond in ethane can break to give two $\cdot CH_3$ radicals if heated to a high enough temperature (above 600°C). The energy required for breaking the bond can also be supplied photochemically, that is, by shining light of the appropriate wave length on the molecule. Radiation in the ultraviolet or visible region is required for such photochemical reactions. Some bond dissociation energies are listed in Table 1.2.

1.4.2.2   Carbocations and Carbanions

Heterolytic dissociations (Eq. 1.56) give rise to ions.

$$H_3C - Cl \rightarrow H_3C^+ + :Cl^- \quad (a)$$

$$H_3C - Li \rightarrow H_3C:^- + Li^+ \quad (b) \qquad\qquad (1.56)$$

Table 1. 2 Bond dissociation energies (example, $H - CH_3 \rightarrow H\bullet + \bullet CH_3$; ΔH = Bond dissociation energy, $D_{H-CH_3} = 104\,kcal\,mol^{-1}$).

Bond broken	D (kJ mol⁻¹)	D (kcal mol⁻¹)	Bond broken	D (kJ mol⁻¹)	D (kcal mol⁻¹)
$H-H$	435	104	$Cl-CH_3$	349	84
$H-Cl$	431	103	$Br-CH_3$	293	70
$H-Br$	366	88	$I-CH_3$	234	56
$H-I$	297	71	$Br-Br$	192	46
$H-CH_3$	435	104	$Cl-Cl$	242	58
$H-C_2H_5$	410	98	$I-I$	150	36
$H-CH(CH_3)_2$	397	94	$HO-OH$	146	35
$H-C(CH_3)_3$	380	91	$CH_2=CH_2$	600	146
CH_3-CH_3	368	88	$CH\equiv CH$	826	200
$H-C_6H_5$	468	112			
$H-CH=CH_2$	435	104			

The inorganic anions and cations, Cl^- and Li^+ are familiar species. If the medium is suitable (high dielectric constant, solvent capable of solvating the ions), they can exist indefinitely. The organic ions CH_3^+ and CH_3^- are of a different category. Carbon can neither sustain a positive charge nor a negative charge comfortably. An ion with a positive charge on carbon is called a *carbocation*. These cations have also been called *carbonium ions*. The latter name is no longer used. The addition of the suffix, '-onium' to a cation is familiar in chemistry as in ammonium, oxonium, hydronium, sulphonium and bromonium. A distinguishing feature of all these ions is that they are formed when an electrophile forms a bond with the central atoms, utilising a lone pair of electrons on it, increasing its valency.

$$H^+ + :NH_3 \rightarrow H-NH_3^+ \quad \text{(ammonium)}$$

$$H^+ + :O \overset{H}{\underset{R}{<}} \rightarrow H-\overset{+}{O}\overset{H}{\underset{R}{<}} \quad \text{(oxonium)}$$

(1.57)

This is not the case in carbocations. Hence, cations of the type R_3C^+ are called carbocations. Another name, carbenium ion, has been proposed but is not widely used. Ions of the type $^+CH_5$ are now known to be formed by the protonation of hydrocarbons by super acids. These are now called carbonium ions.

Ions with a negative charge on carbon are called carbanions, illustrated by $:CH_3^-$, a methyl carbanion. Both carbocations and carbanions are high energy species and are highly reactive. Their formation during reactions has been well established but they survive in the reaction medium for very short periods of time and rapidly undergo further reactions. Hence, they are referred to as *reactive intermediates or transient intermediates*. Often, their formation or presence has to be inferred by indirect means.

1.4.2.3   Calculating the Formal Charge on an Atom

A simple way to obtain the formal charge on an atom is to use the formula,

$$Charge = N - U - 1/2\,S \quad where,$$

N = number of valence electrons in the neutral atom. For carbon it is 4.

U = number of unshared electrons. (No. of electrons present as lone
    pairs + no. of odd electrons)

S = number of shared electrons (Electrons in the bonds)

For the carbanion, $:CH_3$, charge on $C = 4 - 2 - 6/2 = -1$
For the carbocation, CH_3, charge on $C = 4 - 0 - 6/2 = +1$
For the free radical, $\cdot CH_3$, charge on $C = 4 - 1 - 6/2 = 0$
For the hydronium ion, $H_3O:$, charge on $O = 6 - 2 - 6/2 = +1$
For the ammonium ion, H_4N, charge on $N = 5 - 0 - 8/2 = +1$

This formula cannot be applied to CH_5 where the bonding is of a different kind.

1.4.2.4   Stability of Ions

Transient species like carbocations carbanions and carbon free radicals are involved as intermediates in many reactions. Their relative stabilities, in many cases, control the course of such reactions.

First let us consider the carbocations. Since carbon is not a strongly electropositive atom, it is not comfortable with a positive charge, unlike ions such as Na^+ or H^+. Even these ions of electropositive atoms are commonly encountered only in solution or in solid matrices as in ionic crystals. In these environments the charge is not localised on a particular atom. Na^+ in water has a large number of water molecules surrounding it by solvation. These solvent molecules share the charge. Using an expression that is by now familiar to us, the charge is delocalised. A proton in water solution is actually covalently bound to a water molecule – forming the hydronium ion – in which the positive charge is shared by three hydrogens. This is shown in 1.58 using resonance notation.

$$\tag{1.58}$$

The hydronium ion in turn, is further solvated by more water molecules, further delocalising the charge. *An ion becomes stabilised when the charge is effectively delocalised.*

In the case of Na^+, the charge is delocalised by solvation. That is, the charge is shared by molecules which are not part of the basic structure of the ion. The stabilisation is external. In the case of H_3O^+, the charge is delocalised within the molecule. When the stabilisation is internal, it becomes part of the structure of the molecule. In addition to this internal stabilisation, the ion is further stabilised by solvation, namely external stabilisation. In the case of carbon ions – cations and anions – solvation is indeed an important factor. But here, we are more concerned with internal or structural factors which stabilise them.

Delocalisation of charge in these ions is due to the electronic factors that we have discussed; resonance effects as a result of conjugation and inductive effects are important in considering stability.

1.4.2.5 Shape of Carbocations

A carbocation, for example CH_3^+, has a trivalent carbon. Only three bonding atomic orbitals of carbon are utilised, the fourth bonding orbital is vacant. Notionally, the unhybridised orbital which is vacant, has its axis perpendicular to the sp^2 plane. Thus, we represent the CH_3^+ cation as shown below. *Carbocations are planar.*

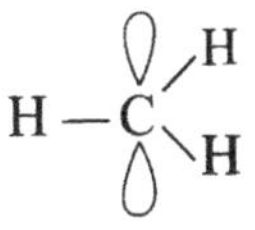

1.4.2.6 Stability of Carbocations—Resonance Effects

Consider the allyl cation, $CH_2 = CH - CH_2^+$. Its structure can be represented as follows (1.59).

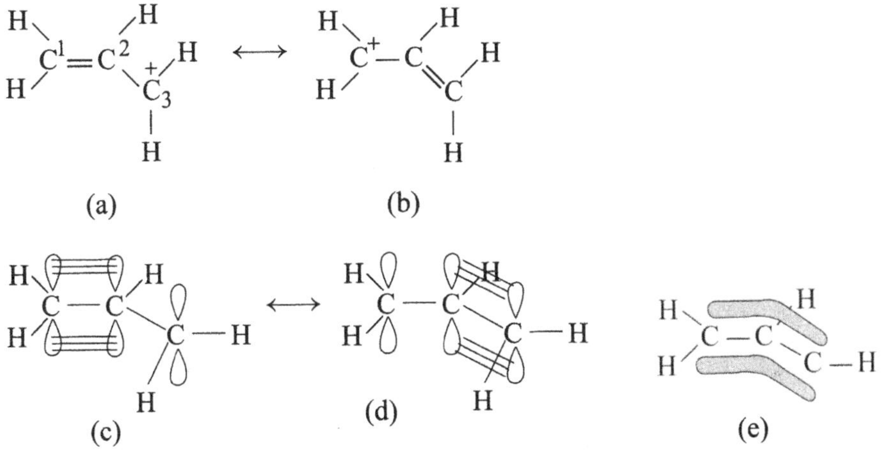

(1.59)

Examination of 1.59a shows that the atoms attached to C1 and C2 are all in the same plane [cf. planarity of alkenes (1.5)]. The two hydrogens of C3 need not be in the same plane, because the C2−C3 bond is a single bond. However, 1.59a is not a unique valence bond structure for this molecule. It can also exist as 1.59b where C2 and C3 are coplanar. The hybrid molecule is planar with all the atoms – carbons and hydrogens – being in the same plane. The charge on the allyl cation is not concentrated or localised on C3, but is delocalised. This delocalisation makes it more stable than a formally similar structure [propyl cation (1.60)] where this delocalisation is not possible.

$$\text{Propyl cation} \tag{1.60}$$

In the allyl cation, the vacant p orbital on the carbon which formally bears the positive charge is in conjugation with the p orbitals of the double bond as in 1.59c and 1.59d. When the carbocation is in conjugation with the electrons of a neighbouring p orbital and if the geometry permits, there can be conjugation and delocalisation. Some examples are given below (1.61–1.63).

$$CH_2 = CH - \overset{+}{C}H - CH_3 \longleftrightarrow \overset{+}{C}H_2 - CH = CH - CH_3 \tag{1.61}$$

$$\tag{1.62}$$

$$\tag{1.63}$$

The curved arrow in 1.61 implies that an electron pair from the C1—C2 double bond is shifted to the C2—C3 position, leaving C1 with the positive charge. In 1.62, conjugation of the vacant p orbitals with the benzene ring is shown. The canonical structures of 1.62, called the benzyl cation, can be considered to be formed by the electron shifts shown by the curved arrows. These electron shifts of course do not take place in a step-wise manner. The arrows are used only to help keep track of the electrons and bonds. The hybrid has a structure similar to 1.63 with the charge highly delocalised but with higher charge densities at the positions marked δ^+ in comparison with the other carbon atoms. Overlap of the vacant p orbitals can be with any other occupied p orbital, not necessarily only with that of a double bond. This is illustrated in 1.64 where a nearby oxygen atom stabilises the carbocation.

$$CH_3 - \overset{..}{\underset{..}{O}} - \overset{+}{C}H_2 \longleftrightarrow CH_3 - \overset{+}{\underset{..}{O}} = CH_2 \qquad \tag{1.64}$$

(a)   (b)

Resonance structures such as 1.65b are not considered even though the valence bond structure appears valid.

$$\tag{1.65}$$

(a)   (b)

This is because a positive charge on monovalent oxygen gives rise to a high energy situation and this structure will contribute very little to the actual structure. In fact, the cation 1.65a is not only not stabilised by resonance, it is extremely unstable because of the inherent polarity of the $C = O$ group, where C is positive and O is negative.

$$\ce{>C=O} \longleftrightarrow \ce{>\overset{+}{C}-O^-}$$

In 1.66, the presence of two like charges on adjacent atoms makes it very unstable.

$$\bar{O} - \overset{+}{C} \underset{\displaystyle \overset{+}{C}}{\overset{\displaystyle H}{}} \quad (1.66)$$

In 1.64, we saw that the proximity of the lone pair on the oxygen makes the neighbouring carbocation stable due to delocalisation of the charge. However, this molecule presents a situation where two opposing electronic effects exist. Oxygen is an electronegative atom and it has an electron withdrawing inductive effect as shown in 1.67.

$$CH_3 \rightarrow\!\!\!\!-O \leftarrow\!\!\!\!- \overset{+}{C}H_2 \quad (1.67)$$

The consequence of this is that the positive charge on the carbon, instead of getting delocalised (spread out) becomes more localised (increased). This is a destabilising effect. Opposing effects existing in the same molecule is not uncommon. In most of such situations, the resonance effect has greater weightage than the inductive effect.

The cation 1.64 is indeed more stable than a cation which does not have an oxygen, for example, $CH_3 - CH_2 - CH_2^+$.

1.4.2.7 Hyperconjugation

Alkyl groups represent a category which have electronic effects different from the ones dealt with so far. We have seen how the π-electrons of a double bond or the pair of electrons in the p orbitals of a heteroatom like oxygen can conjugate with the vacant p orbital of a carbocation and stabilise it. An alkyl group like CH_3 has neither π-bonds nor lone pairs of electrons. However, alkyl groups containing $C - H$ bonds stabilise carbocations by a special kind of conjugation called *hyperconjugation*. Let us try to visualise what is involved in hyperconjugation (1.68).

$$(1.68)$$

What this amounts to is that, each $C-H$ bond on the carbon attached to the C^+ contributes its σ-electron density to stabilise the charge. As seen from the canonical structures, the positive charge is delocalised into the CH_3 group attached to the C^+. More the number of such adjacent $C-H$ bonds, the greater is the hyperconjugative stabilisation. Thus, the order of stability of alkyl carbocations is,

$$\overset{+}{C}H_3 \;<\; \overset{+}{C}H_2-CH_3 \;<\; \overset{+}{C}H(CH_3)_2 \;<\; \overset{+}{C}(CH_3)_3$$
$$\text{(a)} \qquad\quad \text{(b)} \qquad\qquad \text{(c)} \qquad\qquad \text{(d)}$$

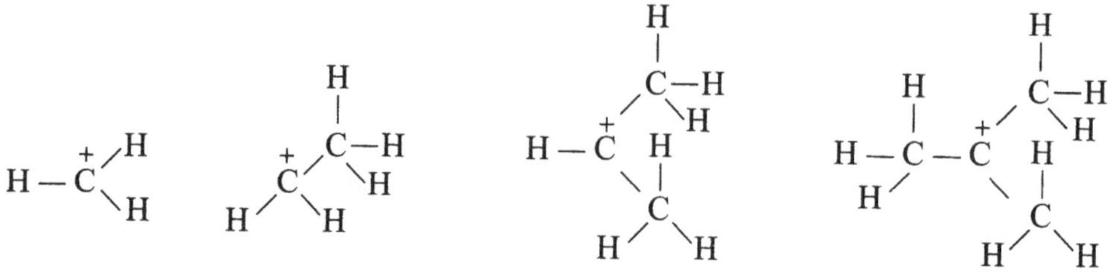

$$\text{(a)} \qquad\qquad \text{(b)} \qquad\qquad \text{(c)} \qquad\qquad\qquad \text{(d)}$$

$$(1.69)$$

The groups featured in 1.69 are methyl (a), ethyl (b), 2-propyl or isopropyl (c), and 2-methylpropyl or *tert*-butyl (d). (a) and (b) are primary alkyl groups, (c) is secondary and (d) is tertiary. A primary carbon is one which is attached to one other carbon (except methyl where it is attached to none), a secondary is one which is attached to two other carbons and a tertiary is one which is attached to three other carbons. If the carbon is attached to four other carbons, it is called quaternary as in 2,2-dimethylpropane or neopentane.

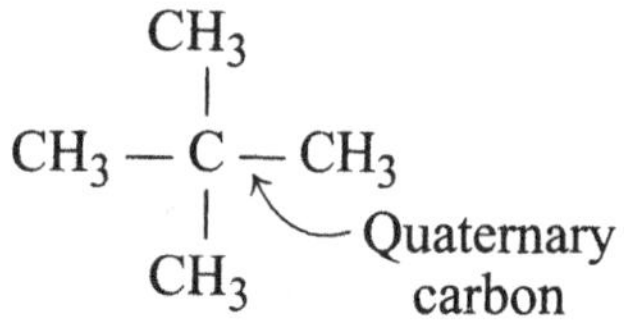

It is obvious that there can be no radical or ion which is quaternary.

In the series methyl, ethyl, isopropyl and *tert*-butyl carbocations, the number of $C-H$ bonds which can enter into hyperconjugation with the vacant p orbital on the positive carbon atom are 0, 3, 6 and 9 respectively; correspondingly, these carbocations are increasingly more stable in that order. An added feature of the alkyl groups is that they have an electron releasing inductive effect, which has the same consequence as hyperconjugation on the stability of cations. In the series under discussion, the inductive effect can be represented as shown below.

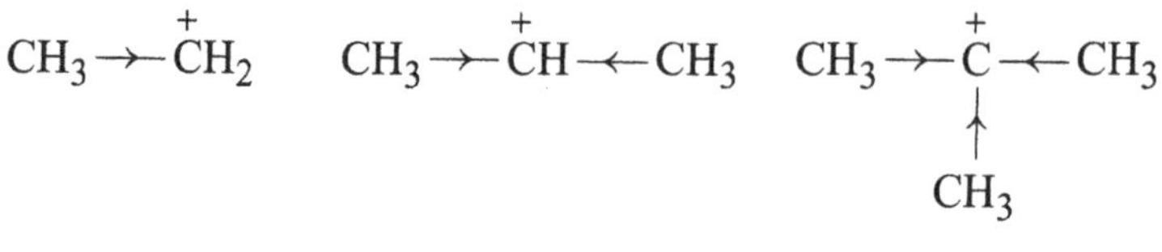

This effect tends to neutralise the positive charge. Since each ion has a unit positive charge and the charge cannot disappear, neutralisation of charges in this context means dissipation or delocalisation of charge. The net charge, $+1$, remains unaltered. Thus, the inductive effect of the alkyl groups also contributes to the increasing stabilisation of carbocations in the series $1° < 2° < 3°$. (A new method of representation has been introduced here with $1°, 2°$ and $3°$ denoting primary, secondary and tertiary respectively.)

This is an instance where the resonance effect and inductive effect reinforce each other in the stabilisation of carbocations. Recall that in the case of 1.64, the two electronic effects worked in opposition to each other. Alkyl groups are among the few groups which release or push away electrons along the connecting bond by inductive effect. This electron-releasing inductive effect is referred to as the $+I$ *effect*. Similarly, the electron-releasing resonance effect, as that of alkyl groups, is referred to as the $+R$ effect.

1.4.2.8 Hyperconjugation in Alkenes

Hyperconjugation of alkyl groups is manifested not only in conjugation with the vacant p orbital of carbocations but also with the occupied p orbitals of π-bonds of double bonds. This is illustrated with propene.

All three $C-H$ bonds of the methyl group of propene enter into hyperconjugation. This hyperconjugation stabilises the alkene. This statement can be understood better if we take two isomeric alkenes, 1-butene and 2-butene (1.70a and 1.70b), into consideration.

$$CH_2 = CH - CH_2 - CH_3 \qquad CH_3 - CH = CH - CH_3$$
$$\text{(a)} \qquad\qquad\qquad \text{(b)} \qquad\qquad\qquad\qquad (1.70)$$

These alkenes are stabilised by hyperconjugation as shown.

1-Butene

The total number of canonical structures for 1-butene is 3.

2-Butene

$$\left[H_b - \underset{\underset{H_c}{|}}{\overset{\overset{H_a}{|}}{C}} - C \equiv \underset{\underset{H}{|}}{C} - CH_3 \longleftrightarrow H_b - \underset{\underset{H_c}{|}}{\overset{\overset{H_a^+}{|}}{C}} = \underset{\underset{H}{|}}{C} - \overset{..}{\underset{\underset{H}{|}}{C}} - CH_3 \longleftrightarrow \begin{array}{l} \text{2 more} \\ \text{structures} \\ \text{corresponding} \\ \text{to } H_b \text{ and } H_c \end{array} \right]$$

$$\left[H_3C - \underset{\underset{H}{|}}{C} = \underset{\underset{H}{|}}{C} - \overset{\overset{H}{|}}{\underset{\underset{H}{|}}{C}} - H \longleftrightarrow H_3C - \overset{..}{\underset{\underset{H_c}{|}}{C}} - \underset{\underset{H}{|}}{C} = \underset{\underset{H}{|}}{\overset{\overset{H^+}{}}{C}} - H \longleftrightarrow \begin{array}{l} \text{2 more similar} \\ \text{structures} \end{array} \right]$$

The total number of canonical structures for 2-butene is 7 (6 involving hyperconjugation and one original structure). One can predict that 2-butene is more stable that 1-butene. This can be verified using heats of hydrogenation data (Fig. 1.3, cf. heats of hydrogenation of benzene Fig. 1.1). This aspect is discussed in greater detail in Chapter 4, Section 4.4.4.1.

The lower heat of hydrogenation of 2-butene is a measure of the greater resonance (hyperconjugation) stability of 2-butene over 1-butene (what is listed in Fig. 1.2 is *trans* 2-butene, one of the two geometrical isomers of 2-butene. Geometrical isomerism is discussed in Chapter 22). Generally, more the number of alkyl groups attached to the alkene carbons, the greater the stability of the alkene.

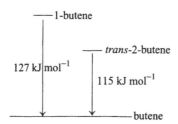

Fig. 1.2 Heats of hydrogenation of butenes

1.4.2.9 Shape of Carbanions

A carbanion has a lone pair of electrons on the carbon. Thus $:CH_3$ has a lone pair of electrons and a negative charge and is isoelectronic with $\overset{..}{N}H_3$, ammonia. The hybridisation of carbon in $\overset{..}{C}H_3$ as well as nitrogen in ammonia is sp^3. The shape of the molecule is pyramidal with the carbon at the apex and the hydrogens at the base. The axis of the lone pair is an sp^3 hybrid orbital on the apex carbon and the three C—H bonds point towards the four corners of a tetrahedron.

1.4.2.10 Stabilisation of Carbanions by Resonance

While sp^3 is the hybridisation in an isolated carbanion, when it is attached to an alkene, due to conjugation, the negative carbon gets rehybridised to sp^2 with the lone pair now being in an unhybridised p orbital. Thus, the allyl anion, $CH_2 = CH - \overset{\bullet\bullet}{CH_2}$ may be represented as follows.

When the orbital bearing the lone pair of the carbanion is in conjugation with the π-bond of an alkene, there is delocalisation and resonance stabilisation. The benzyl carbanion is stabilised in the same manner as the carbocation.

What about $CH_3 - \overset{\bullet\bullet}{\underset{\bullet\bullet}{O}} - \overset{-}{CH_2}$ in analogy with the corresponding cation $CH_3 - \overset{\bullet\bullet}{\underset{\bullet\bullet}{O}} - \overset{+}{CH_2}$? The latter, we found, is stabilised by conjugation with the p orbital of oxygen (1.64). However, there is no such conjugation in the carbanion. The p orbital on the carbon and on the oxygen are fully occupied and do not overlap. However, the electron-withdrawing inductive effect of the oxygen delocalises the negative charge to some extent (1.71) and confers some stability to the carbanion.

$$CH_3 - O \leftarrow CH_2^- \tag{1.71}$$

Alkyl groups on the carbanion have the opposite effect. Unlike in carbocations, alkyl groups destabilise carbanions by inductive effect. Hyperconjugation does not exist because the orbital of the carbanion is full and cannot accept electrons from the $C-H$ bond. However, inductive effect does exist. The alkyl group acts to increase the negative charge on the negative centre by pushing the electron density towards it. This is a destabilising effect. Not surprisingly, the order of stability of carbanions is,

$$CH_3 > 1° > 2° > 3°$$

This is just the opposite of the order of stability of carbocations.

1.4.2.11 Inductive Effect and Hybridisation

Inductive effect of the alkyl group (methyl, ethyl) as we have seen in connection with the stability of carbocations, is an electron-releasing effect, $(+I$ effect). Alkenyl and aryl groups behave differently.

Ethenyl (vinyl) (a)   Phenyl, (C_6H_5-) (b)

(1.72)

Electronegativity of carbon changes with hybridisation. The increasing order of electronegativity is $sp^3 < sp^2 < sp$. This is because, the s character of the hybrid orbital increases in that order and the electron density in the sp orbital is closer to the nucleus than in the sp^2 orbital which in turn is closer than in the sp^3 orbital. Phenyl and vinyl groups which are attached through the sp^2 hybridised carbon exert electron-withdrawing inductive $(-I)$ effect. This is an additional effect by which the allyl and benzyl carbanions are stabilised.

1.4.2.12 Stabilisation of the Carbanion by the Carbonyl Group

We have seen how the allyl carbanion (1.59) is stabilised by conjugation. When the conjugation is with a carbonyl group, this stabilisation is much stronger.

(1.73)

The contribution by the canonical structure 1.73b to this anion is more than the contribution by the canonical structure of the allyl carbanion because the negative charge in this case, is on the more electronegative oxygen.

The carbonyl is an electron-withdrawing group by resonance. Some other such groups are cyano (1.74) and nitro (1.75).

$$\left[:N \equiv C - \overset{H}{\underset{H}{C}}: \quad \longleftrightarrow \quad :\overset{..}{N} = C = \overset{H}{\underset{H}{C}} \right] \tag{1.74}$$

$$\left[\overset{:\overset{..}{O}}{\underset{:\overset{..}{O}}{}} \overset{+}{N} - \overset{H}{\underset{H}{C}}: \quad \longleftrightarrow \quad \overset{:\overset{..}{O}}{\underset{:\overset{..}{O}}{}} \overset{+}{N} = \overset{H}{\underset{H}{C}} \right] \tag{1.75}$$

1.4.2.13 Shape of Free Radicals

Free radicals with half-filled p orbitals are very similar to carbocations in shape and stability. Simple radicals like $\cdot CH_3$ are planar with the odd electron occupying an unhybridised p orbital of the sp^2 bybridised carbon. However, the situation is not as rigid as it was for carbocations, where deviation from planarity is not tolerated.

1.4.2.14 Stability of Free Radicals

A free radical has an incomplete octet of valence electrons and behaves as an electron deficient species (cf. carbocations). Electron-releasing electronic effects (resonace and inductive) stabilise free radicals. The following structures (Strs. 1.76) are illustrative of resonance stabilisation of free radicals. A curved half-arrow 'fish hook' is used to show the movement of a single electron. (cf. a full curved arrow is used for the movement of a pair electrons.)

$$\left[CH_2 = CH - \overset{\cdot}{C}H_2 \quad \longleftrightarrow \quad \overset{\cdot}{C}H_2 - CH = CH_2 \right] \text{allyl radical} \qquad (a)$$

$$\left[\underset{}{\bigcirc}\overset{\cdot}{C}H_2 \longleftrightarrow \underset{}{\bigcirc}= CH_2 \longleftrightarrow \cdot\underset{}{\bigcirc}= CH_2 \longleftrightarrow \underset{}{\bigcirc}= CH_2 \right.$$

$$\tag{1.76}$$

$$\left. \longleftrightarrow \underset{}{\bigcirc}- \overset{\cdot}{C}H_2 \right] \text{benzyl radical} \qquad (b)$$

$$\left[CH_3 - \overset{..}{O} - \overset{\cdot}{C}H_2 \quad \longleftrightarrow \quad CH_3 - \overset{\cdot}{O} = CH_2 \right] \qquad (c)$$

Hyperconjugation also operates in a manner similar to the case of carbocations as shown.

$$\left[\overset{\cdot}{C}H_2 - \overset{H}{\underset{H}{C}} - H \quad \longleftrightarrow \quad CH_2 = \overset{H\cdot}{\underset{H}{C}} - H \quad \longleftrightarrow \quad \text{2 more structures} \right]$$

The more the number of such $C-H$ bonds in conjugation with the orbital of the odd electron, the more stable the radical. So, as with carbocations, the order of stability for alkyl free radicals is:

$$\text{methyl} < 1° < 2° < 3°$$

Inductive effect (+I) of the alkyl groups also reinforces this order as can be seen.

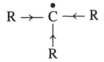

A list of some common functional groups and their electronic effects (resonance and inductive effects) are tabulated (Table 1.3).

Table 1. 3   Some common functional groups and their electronic effects

Group		Resonance effect		Inductive effect	
Name (example)	Structure	Electron-releasing (+R)	Electron-withdrawing (−R)	Electron-releasing (+I)	Electron-withdrawing (−I)
Alkyl (methyl)	$-CH_3$	✓	✗	✓	✗
Aryl[1] (phenyl)	⟨phenyl⟩	✓	✓	✗	✓
Alkenyl[1] (vinyl)	$-CH_2=CH_2$	✓	✓	✗	✓
Carbonyl[2]	$-\overset{\vert}{C}=O$	✗	✓	✗	✓
Nitro	$-N^+\!\!\begin{smallmatrix}O\\O^-\end{smallmatrix}$	✗	✓	✗	✓
Cyano	$-C\equiv N$	✗	✓	✗	✓
Hydroxy	$-OH$	✓	✗	✗	✓
Alkoxy	$-OR$				
Amino	$-NH_2$ $-NR_2$	✓	✗	✗	✓
Halide	$-F, -Cl,$ $-Br, -I$	✓	✗	✗	✓
Ammonium	$-\overset{+}{N}H_3$ $-\overset{+}{N}R_3$	✗	✗	✗	✓

[1]Can release or withdraw electrons, depending on the nature of the other group attached to it.
[2]Carbonyl group is present in aldehydes, ketones, carboxylic acids and esters.

1.5 STERIC EFFECTS

From the above discussion, we have seen how electronic effects of different kinds, which are not obvious from the valence bond structures of molecules, influence the stability and reactivity of molecules. Another effect which influences stability and reactivity is the *steric effect*. This effect arises from the fact that an atom occupies a certain space—an atom is a '*space-filling*' object. This is more relevant when we consider *groups* which are made up of more than one atom. The atoms or groups require a certain amount of space, not only to just remain in place but also to allow some movement. These are movements due to vibrations and rotations. In other words, atoms and groups require some 'elbow room' so to say! The space so required is referred to as the effective volume. Any reduction in this causes strain in the molecule. Some examples which illustrate the importance of steric effects are given below.

1.5.1 STERIC EFFECT IN THE IONISATION OF ALKYL HALIDES: RELIEF OF STERIC STRAIN

There are some reactions of alkyl halides (generally represented as RX) where the important step involves a heterolytic dissociation of the $R-X$ bond into R^+ and X^-. R^+, the carbocation, is a short lived intermediate in such reactions and undergoes further reactions to give stable products. An example is the elimination of HX from such a molecule to give an alkene as from *tert*-butyl chloride (1.77).

$$
CH_3-\underset{\underset{CH_3}{|}}{\overset{\overset{CH_3}{|}}{C}}-Cl \quad \xrightarrow{\text{slow}} \quad CH_3-\underset{\underset{CH_3}{|}}{\overset{\overset{CH_3}{|}}{C^+}} + :Cl^- \qquad \text{(a)}
$$

$$
CH_3-\underset{\underset{CH_3}{|}}{\overset{\overset{CH_3}{|}}{C^+}} \quad \xrightarrow[\text{(solvent)}]{\text{fast}} \quad CH_2{=}C{\overset{\diagup CH_3}{\diagdown CH_3}} + H^+ \text{(solvent)} \qquad \text{(b)}
$$

(1.77)

The concentration of the *tert*-butyl cation will not build up in the medium, because its decomposition (by loss of proton, deprotonation) will take place as soon as it is formed. Looking more closely at the ionisation step (1.77a), among different alkyl halides – such as CH_3-Cl, CH_3CH_2-Cl, $(CH_3)_2CH-Cl$ and $(CH_3)_3C-Cl$ – this step will take place at rates which increase as the stability of R^+ increases. Thus, we can predict that the rates of step (1.77a) will increase in the order,

$$CH_3Cl < CH_3CH_2Cl < (CH_3)_2CHCl < (CH_3)_3CCl$$

Experimentally, this has indeed been found to be true. However, attributing this trend entirely to the electronic effects (hyperconjugation and inductive effect) which stabilise the carbocations is not correct. The central carbon of the alkyl halide is sp^3 hybridised and tetrahedral with four groups (hydrogen or alkyl and halogen) around it. Upon ionisation, it becomes sp^2 hybridised (trigonal planar) with only three groups around it. The groups in the carbocation are further away from each other (at 120° angular separation) than the groups in the alkyl halides (at 109.5°).

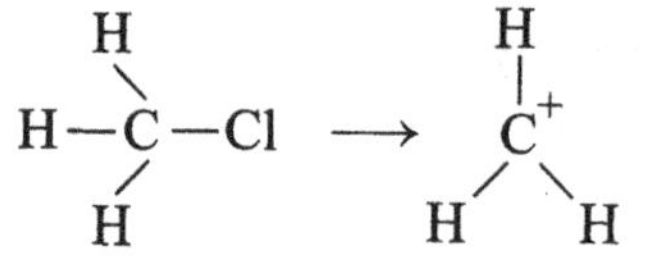

(a) 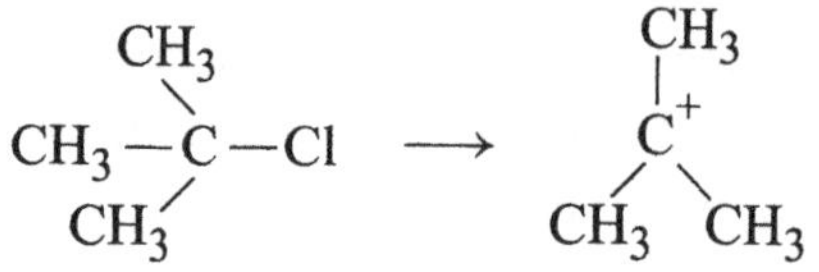   (b)

In a molecule like *tert*-butyl chloride where the relatively bulky methyl groups (relative to hydrogen in $CH_3 - Cl$) are close to each other, there is a certain steric strain which is relieved when the molecule ionises and the groups move further apart from each other. The greater rate of ionisation of *tert*-butyl chloride is partly due to this relief of steric strain. This is an example of steric effect affecting reactivity.

1.5.2 STERIC EFFECT AFFECTING CONJUGATION: STERIC INHIBITION OF RESONANCE

In aniline we have seen how the nitrogen lone pair is delocalised into the bezene ring (1.40). In the actual aniline molecule, positions 2, 4 and 6 of the benzene ring have more electron density than the other positions (1.78) a feature of the molecule which has profound effect on its reactivity.

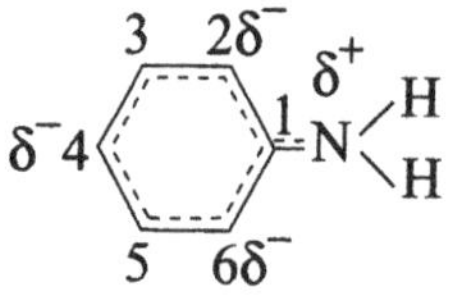

(1.78)

The situation is the same in N, N-dimethyl aniline (1.79).

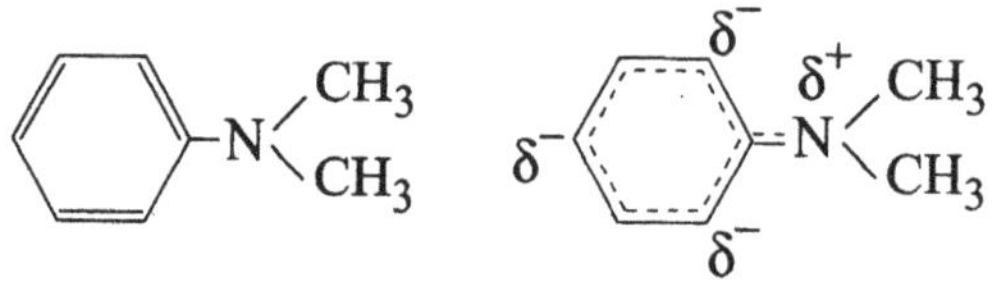

(1.79)

However, the situation dramatically changes in 2, 6-dimethyl-N,N-dimethyl aniline (1.80).

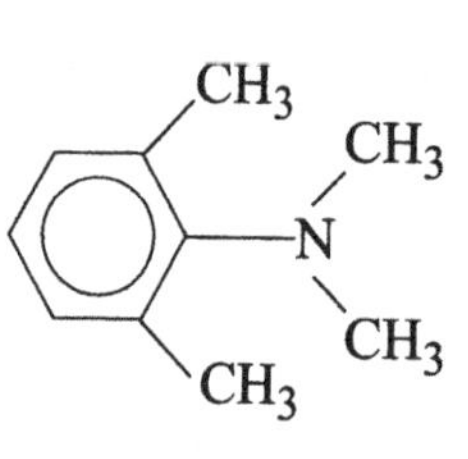

(1.80)

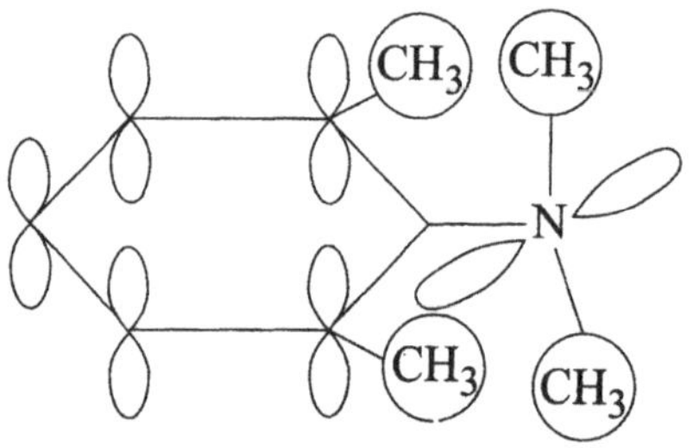

(a)   (b)

In this molecule, because of the proximity of the methyl groups on the nitrogen with the methyl groups at the 2 and 6 positions of the ring, the molecule assumes a conformation which has less steric strain. This is achieved by the $-N(CH_3)_2$ group rotating in such a way that its plane is almost perpendicular to that of the benzene ring. This causes the p orbital of the nitrogen bearing the lone pair of electrons to orient its axis orthogonal to the axis of the p orbitals of the ring (1.80b). Under these conditions, conjugation of the type shown in 1.79 is not possible. The 2,4 and 6 positions of the benzene ring do not acquire the extra negative charge as in structures 1.78 and 1.79. This is an illustration of the phenomenon of *steric inhibition of resonance*.

1.5.3 STERIC HINDRANCE OF ESTER HYDROLYSIS

Esters of the type 1.81a are hydrolysed when heated with aqueous alkali. The first step in this reaction involves the attack of the carbonyl carbon by the OH^- (1.81b).

$$R-\overset{\overset{\textstyle O}{\|}}{C}-O-R' + OH^- \longrightarrow R-\overset{\overset{\textstyle O}{\|}}{C}-O^- + ROH \qquad R-\overset{\overset{\textstyle O}{\|}}{\underset{\underset{\textstyle :\bar{O}H}{}}{C}}-OR' \longrightarrow R-\overset{\overset{\textstyle O^-}{|}}{\underset{\underset{\textstyle OH}{|}}{C}}-OR'$$

$$\text{(a)} \qquad\qquad\qquad\qquad\qquad\qquad \text{(b)} \qquad\qquad \text{(c)}$$

$$(1.81)$$

Such a reaction (*nucleophilic attack* by the OH^-) is possible because of the polarity of the $C{=}O$ group (1.82).

$$R-\overset{\overset{\textstyle O}{\|}}{C}-OR' \longleftrightarrow R-\overset{\overset{\textstyle O^-}{|}}{\underset{\underset{\textstyle +}{}}{C}}-OR' \qquad\qquad (1.82)$$

In this step, the sp^2 carbon of the $C{=}O$ changes over to become an sp^3 carbon (see 1.81c). Such a change (the reverse of $sp^3 \rightarrow sp^2$, which we saw in the ionisation of alkyl halides, Section 1.5.1), causes an increase in the crowding around the reaction centre.

If the R group is already bulky, this steric effect will be felt more strongly than if the R group is relatively small. In other words, in ester hydrolysis, the bulkier the R group the slower is the reaction. Experiments support this prediction. Methyl acetate (1.83a) is readily hydrolysed but methylpivalate (1.83b) is very difficult to hydrolyse by heating with aqueous alkali.

$$CH_3-\overset{\overset{\textstyle O}{\|}}{C}-O-CH_3 \qquad\qquad CH_3-\overset{\overset{\textstyle CH_3}{|}}{\underset{\underset{\textstyle CH_3}{|}}{C}}-\overset{\overset{\textstyle O}{\|}}{C}-O-CH_3$$

$$(1.83)$$

$$\text{(a)} \qquad\qquad\qquad\qquad\qquad\qquad \text{(b)}$$

Several examples of steric effects emphasising the importance of size and shape, and the arrangement of atoms of a molecule in space will be encountered as we go along. This area is broadly covered under the topic called *stereochemistry*.

KEY POINTS

Hybridisation–Shapes of Molecules

- Tetravalency of carbon
- Valence electrons of carbon – $2s^2, 2p^2$
- Promotion $2s^1, 2p^3$
- sp^3 hybridisation : Four sp^3 orbitals, tetrahedral, as in CH_4
- sp^2 hybridisation : Three sp^2 orbitals + one unhybridised p orbital, trigonal planar as in ethene; one $C-C$ σ-bond and one $C-C$ π-bond. Ethene is planar
- sp hybridisation : Two sp orbitals + two unhybridised p orbitals, linear as in ethyne; two π-bonds and one σ-bond, in the triple bond, ethyne is linear
- Kekulé structure of benzene is inadequate. Benzene has six sp^2 hybridised carbons in a regular hexagon, planar molecule, six unhybridised p orbitals
- Concepts of delocalisation and conjugation
- Introduction to the theory of resonance
- Resonance stabilisation
- Resonance energy
- Heats of hydrogenation

Electronic Effects

- **Resonance:** Whenever a molecule can be represented by more than one valid Lewis structure, none of them accurately describes the molecule. The actual structure of that molecule will be a weighted average of all the structures and is called the resonance hybrid. The individual unreal Lewis structures are called canonical structures. The constituents of a molecule which include the nuclei and the electrons, will assume that configuration which has least energy. This configuration may not be readily represented by a Lewis structure. The theory of resonance helps us to get a picture of the real structure of the molecule. Certain rules are to be followed in drawing the canonical structures.

- **Inductive effect:** This arises from the polarity of the covalent bond caused by the electronegativity differences between the atoms constituting a bond. It is transmitted through bonds, but diminishes as we go further away from the electronegative atom. The more the number of such strongly electronegative atoms, the stronger the effect.

- **Inductomeric effect:** Due to polarisability, bonds or molecules which are not polar in the ground state can be made polar (polarised) under the influence of an external regent.

Steric Effects

Steric effects play an important role in deciding the reactivity of molecule. Steric effects arise from the size of atoms and groups, and the shapes of molecules. Three specific examples are given as illustration.

- Relief of steric strain—illustrated by the role in the ionisation,

$$R - X \rightarrow R^+ + X^-$$

$(CH_3)_3C - Cl$ ionises faster than CH_3Cl, not only because the cation from the former being tertiary is more stable, but also because the carbocation is less sterically crowded than the parent alkyl halide.

- Steric inhibition of resonance—the amino group ($-NR_2$) when attached to a benzene ring, enters into resonance by electron release.

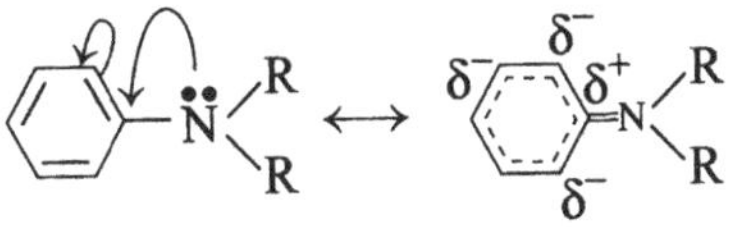

(I)

This resonance requires the benzene ring and the $-NR_2$ group to be coplanar. In the molecule below, this coplanarity is not possible due to steric hindrance.

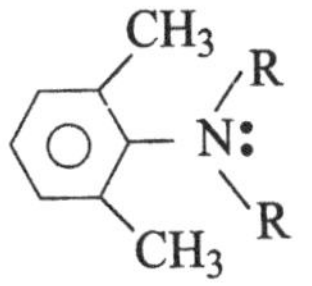

(II)

The reactivity of I is different from that of II due to the steric inhibition of resonance

- Steric hindrance in ester hydrolysis—in the alkaline hydrolysis of esters, the following step occurs.

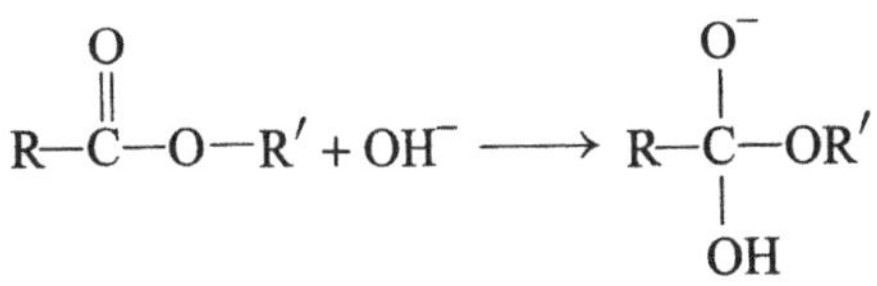

In going from I to II, the crowding around the central carbon increases. This is reflected in the fact that the ester where $R = CH_3$ is hydrolysed faster than that where $R = (CH_3)_3C-$.

Reactive Intermediates

- Electrophiles are reagents which are 'electron-loving', that is, those which have an affinity for negative centres. These are themselves electron-poor. Nucleophiles are 'nucleus-loving', that is, those which have affinity for positive centres. These are themselves electron-rich.

- During reactions, covalent bonds break or dissociate heterolytically or homolytically. In heterolysis the bonding electron pair is retained by one of the atoms, usually the more electronegative one, and as a result ions are formed.

$$A : B \rightarrow A^+ + : B^-$$

- In homolysis, each atom retains one electron. The species formed are neutral and are called free radicals.

$$A : B \rightarrow A^{\bullet} + {}^{\bullet}B$$

- Carbocations are species where a carbon carries a positive charge. Carbanions are those where carbon carries a negative charge and a lone pair of electrons. Free radicals (of carbon) have an odd electron (unpaired electron) on carbon. These are illustrated for the methyl cation, anion and radical.

     Planar   Pyramidal   Planar

- The formal charge on atom can be calculated using the formula:
  Charge $= N - U - 1/2S$
  Where $N =$ number of valence electrons, $U =$ number of unshared electrons (lone pair electrons and odd electrons) and $S =$ number of shared electrons.

- Carbocations, carbanions and carbon free radicals are short-lived species, also referred to as reactive intermediates or transient intermediates.

- Effects which help to dissipate the charge density stabilise the ions. These effects may be resonance or inductive effects. Electron-releasing effects stabilise carbocations and electron-withdrawing effects stabilise carbanions.

- Carbon free radicals are also stabilised by resonance and inductive effects in the same way as carbocations.

- For the alkyl groups, the order of stability is:

Carbocations: methyl $< 1° < 2° < 3°$

Free radicals: methyl $< 1° < 2° < 3°$

Carbanions: methyl $> 1° > 2° > 3°$

($1°, 2°, 3°$ refer to primary, secondary, and tertiary)

EXERCISES

SECTION I

1. Show the hybridisation of each carbon in the following molecules.

(a) $CH_2 = CH - CH = CH_2$

(f) $CH_2 = \underset{\underset{CH_3}{|}}{C} - C \overset{\displaystyle O}{\underset{\displaystyle H}{\diagup\diagdown}}$

(b) 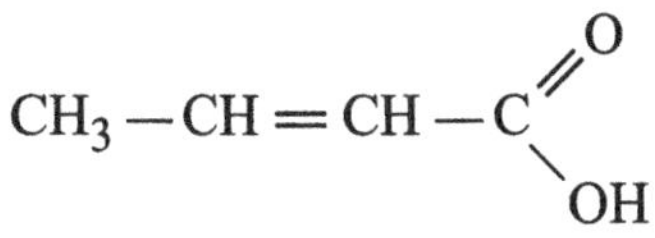 $CH_3 - CH = CH - C \overset{\displaystyle O}{\underset{\displaystyle OH}{\diagup\diagdown}}$

(g) $CH_2 = CH - C \equiv N$

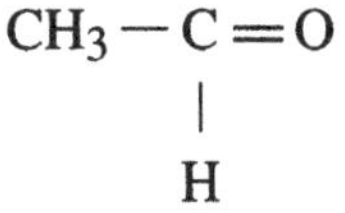

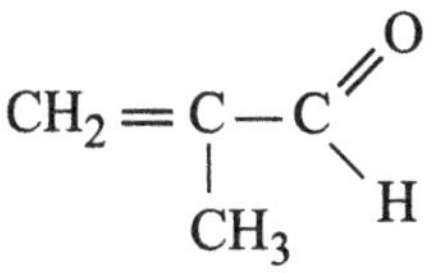

(c) $CH_3 - C \equiv C - CH = CH_2$

(h) $O = C = C = C = O$

(d) $CH_2 = C = CH_2$

(i) 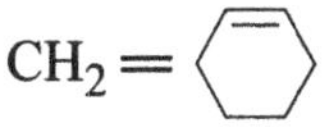 $CH_2 = \langle \overline{} \rangle$

(e) $CH_3 - \underset{\underset{H}{|}}{C} = O$

(j) 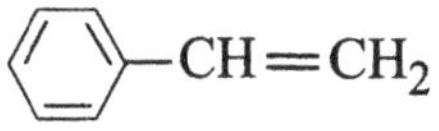 $\langle \overline{} \rangle - CH = CH_2$

2. Predict the specified bond angles.

(a) $\overset{1}{C}H_2 = \overset{2}{C}H - \overset{3}{C}H_3$ ($C_1 - C_2 - C_3$ angle)

(d) $H - C \equiv N$ ($H - C - N$ angle)

(b) $\overset{3}{C}H_3 - \overset{2}{C} \equiv \overset{1}{C} - H$ ($C_1 - C_2 - C_3$ angle)

(e) $CH_2 = C = O$ ($C - C - O$ angle)

(c) $CH_3 - \overset{\displaystyle O}{\overset{\displaystyle ||}{C}} - CH_3$ ($C - C - C$ angle)

(f) CO_2 ($O - C - O$ angle)

3. In the following equations, of the two reactants on the left hand side, one is a nucleophile and the other an electrophile. Identify each.

(a) $NH_3 + H^+ \rightarrow NH_4^+$

(b) $H_2O + \overset{+}{C}H_3 \rightarrow CH_3 - \overset{+}{O} \overset{\displaystyle H}{\underset{\displaystyle H}{\diagup\diagdown}}$

(c) 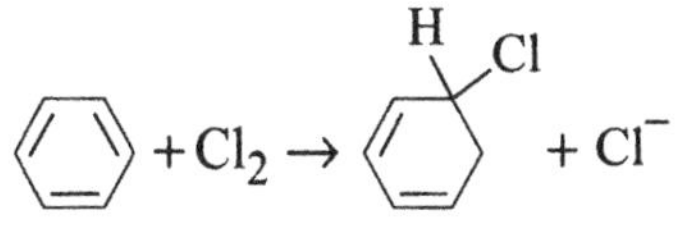 $\langle \overline{} \rangle + Cl_2 \rightarrow \langle \overline{} \rangle \overset{\displaystyle H}{\underset{}{\diagup}} \overset{\displaystyle Cl}{} + Cl^-$

(d) $BF_3 + CH_3-O-CH_3 \rightarrow (CH_3)_2 \overset{+}{O} - \overset{-}{B}F_3$

(e) $CH_3-Cl + \overset{-}{O}H \rightarrow [HO\cdots CH_3 \cdots Cl]^-$

(f) $CH_2=CH-CH=O + :R^- \rightarrow R-CH_2-CH=CH-\overset{-}{O}$

4.   Rearrange the following in the increasing order of stability.

(a) $CH_2=CH_2, CH_2= C\begin{smallmatrix} \nearrow CH_3 \\ \searrow CH_3 \end{smallmatrix}, CH_2=CH-CH_3$

(b) $CH_3-O-\overset{+}{C}H_2, \underset{\underset{O}{\parallel}}{CH_3-C}-\overset{+}{C}H_2, CH_3-O-\overset{+}{C}H-CH_3$

(c) $ClC\overset{\cdot\cdot}{\overline{\underline{H}}}_2, Cl_2C\overset{\cdot\cdot}{\overline{\underline{H}}}, Cl_3\overset{\cdot\cdot}{\overline{C}}{}^-, F_3\overset{\cdot\cdot}{\overline{C}}{}^-$

(d)   ⬡$-\overset{\cdot\cdot}{\overline{C}}H_2$, ⬡$-\overset{\cdot\cdot}{\overline{C}}H_2$, ⬡$-\overset{-}{C}H-CH_3$

(e) $CH_3-\overset{\cdot}{C}H_2$, $CH_3-\overset{\cdot}{C}H-$⬡, ⬡$-\overset{\cdot}{C}H-$⬡

5.   In each of the following pairs, predict which one will undergo heterolytic dissociation at the highlighted bond, faster. (Hint: Heterolysis is faster when the ions formed are more stable.)

(a) $CH_3-\underset{\underset{CH_3}{|}}{CH}-CH_2-Cl$ and $CH_3-\underset{\underset{CH_3}{|}}{\overset{\overset{Cl}{|}}{C}}-CH_3$

(b) $CH_3-CH_2-\underset{\underset{H}{|}}{\overset{+}{O}}-H$ and ⬡$-CH_2-\underset{\underset{H}{|}}{\overset{+}{O}}-H$

(c) $H-CH_2-\overset{\overset{O}{\parallel}}{C}-CH_3$ and $H-CH_2-CH_2-CH_3$

(d) $H-CCl_3$ and $H-CF_3$

(e) $H-CH_2-CF_3$ and $H-CF_3$

SECTION II

1. Draw resonance (canonical) structures (in some cases, curved arrows are shown which will be useful for guidance). Show all the lone pairs of electrons and charges.

(a) Nitromethane, $CH_3-\overset{+}{N}\overset{O}{\underset{O^-}{\diagup}}$

(h) Nitrobenzene, $\langle O \rangle -\overset{+}{N}\overset{O}{\underset{O^-}{\diagup}}$

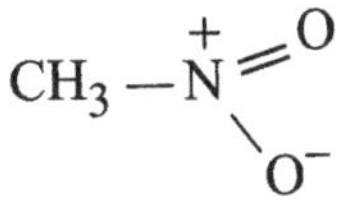

(b) Methyl nitrite, $CH_3-O-N=0$

(i) Anilinium ion, $\langle \rangle -\overset{+}{N}H_3$

(c) Phenol, $\langle O \rangle -OH$

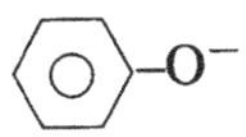

(j) 4-Nitrophenoxide ion, $\bar{O}-\langle O \rangle -\overset{+}{N}\overset{O}{\underset{O^-}{\diagup}}$

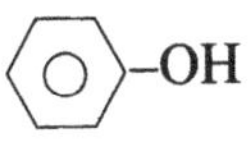

(d) Phenoxide ion, $\langle O \rangle -O^-$

(k) $\langle O \rangle -CH_3$

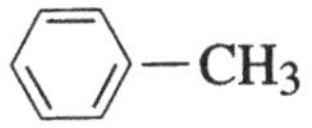

(hyperconjugation should be considered)

(e) Pyridine, $\langle O \rangle_{N:}$

(l) Ozone

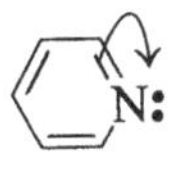

(f) Diazomethane,CH_2-N-N
(draw valid Lewis structures)

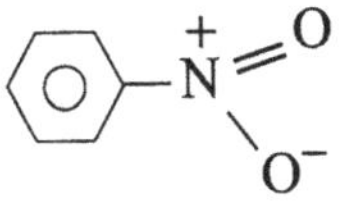

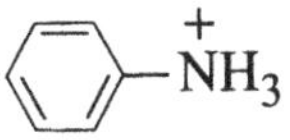

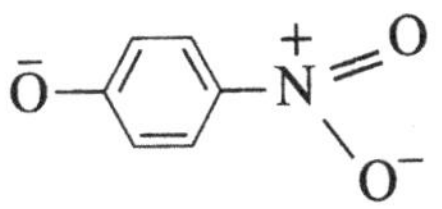

(m) Azide ion, N_3^-

(g) Hydrazoic acid,
HN$_3$, (H—N—N—N, linear)

2. When the following molecules are heated, which bond is likely to break easily in a homolytic manner? Give reasons for your answer.

(a) 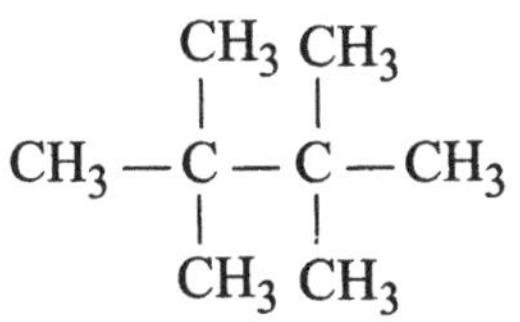 $CH_3-\overset{\overset{\displaystyle CH_3}{|}}{\underset{\underset{\displaystyle CH_3}{|}}{C}}-\overset{\overset{\displaystyle CH_3}{|}}{\underset{\underset{\displaystyle CH_3}{|}}{C}}-CH_3$

(c) $CH_3-\overset{\overset{\displaystyle CH_3}{|}}{\underset{\underset{\displaystyle CH_3}{|}}{C}}-O-O-H$

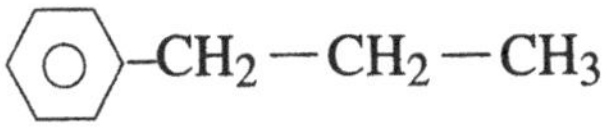

(b) $\langle O \rangle -CH_2-CH_2-CH_3$

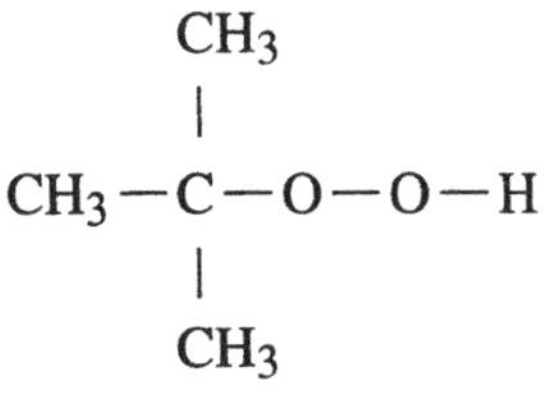

(d) $CH_2=CH-CH_2-CH_2-CH=CH_2$

3. If the structure of benzene is one of the Kekulé structures, how many isomeric dimethylbenzenes are possible? Draw their structures.

4. The molecule, CH_2 (methylene, a carbene) is known as a transient intermediate. Draw the likely structures of the molecule and predict its shape.

5. The compound (A) C_6H_6 is known as an unstable molecule. It correspondes to one of the early structures proposed for benzene. What are the hybridisations of the carbons? Predict the shape of the molecule. Can it be a resonance (canonical) structure of benzene? Discuss.

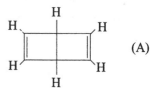

(A)

CHALLENGING QUESTIONS

1. Cyclopropane seems to contain only single bonds. It is believed that the $C-C$ bond involves $sp^3 - sp^3$ overlap, but not along their axis, thus allowing only partial overlap. Draw a structure of cyclopropane, showing the orbital overlap.

2. In question 1d, (Section I) we have seen the hybridisation of the carbons in 1,2-propadiene (allene) $CH_2 = C = CH_2$. Predict the shape of the total molecule (including the orientation of the hydrogens).

3. Consider the molecule, cyclobutyne.

$$H_2C - CH_2$$
$$|\qquad|$$
$$C \equiv C$$

Give reasons describing why this molecule cannot exist as a stable molecule.

PROJECT

Prepare a report on the history of the 'benzene problem'. The early chemists had difficulty in understanding the unusual behaviour and reactions of benzene. Many ingeneous structures were proposed to account for the then known reactions of benzene.

2 Nomenclature of Organic Compounds

OBJECTIVES   In this chapter, you will learn about,

- how organic compounds are classified
- what functional groups are
- the rules of IUPAC nomenclature
- naming of aliphatic compounds
- naming of aromatic compounds
- naming of alicyclic compounds

2.1 INTRODUCTION

The way we write the structures of organic molecules – which we now take for granted – is a triumph of logical thinking coupled with intuition on the part of early chemists such as August Kekulé, Archibald Scott Couper and Alexander Butlerov. Consider the simple structures such as those of methane (2.1a), ethane (2.1b), ethene (2.1c), ethyne (2.1d), ethanol (2.1e), acetic acid (ethanoic acid, 2.1f) and benzene (2.1g).

$$\tag{2.1}$$

Some basic assumptions are involved in writing these structures. Some of which are,

    (i)   carbon is tetravalent,

    (ii) carbon can form single, double or triple bonds with itself or with other atoms,

    (iii) other common atoms also have their valences ($H = 1, O = 2$, halogen $= 1, N = 3$) and

    (iv) structures can involve not only chains but also rings.

Such basic concepts and the practice of using short straight lines to indicate bonds continue to be the basis of visualising the *structures* of all molecules, simple ones like methane and complex ones like DNA. Looking at the structures of the vast number of known organic compounds, some patterns become evident. (i) Many combinations of atoms occur repeatedly. (ii) Some such combinations remain intact during transformations (reactions) or undergo transformations in a characteristic and predictable manner.

From these generalisations, the concepts of *families* of organic compounds and *functional groups* have evolved.

2.1.1 FAMILIES OF ORGANIC COMPOUNDS; CLASSIFICATION OF ORGANIC COMPOUNDS

Organic compounds can be broadly classified into the following types based on the nature of the carbon skeletons.

    (i)   Aliphatic, acyclic or open-chain

    (ii) Aliphatic, cyclic or alicyclic

    (iii) Aromatic

    (iv) Heterocyclic—(a) aromatic and (b) nonaromatic.

2.1.1.1 Aliphatic Compounds

This term is complementary to the term 'aromatic compound'. Aliphatic compounds involve carbon chains connected by single or multiple bonds. The carbons may be connected through an atom other than carbon, called a heteroatom. Some examples are listed below.

    (a) $H_3C-CH_2-CH_2-CH_3$ (Straight chain)

    (b) $H_3C-CH-CH_2-CH_3$ (Chain with branch)
               |
             CH_3

    (c) $H_3C-C=CH-C\equiv C-H$ (Branched chain with multiple bonds)
              |
           CH_3

    (d) $H_3C-CH_2-O-CH_2-CH_3$ (Chain with a hetero atom)

(e) $H_3C - CH_2 - CH - CH_3$ (Chain with a hetero atom)
$\qquad\qquad\qquad\quad |$
$\qquad\qquad\qquad NH - CH_3$

$\qquad\qquad\qquad\qquad\qquad O$
$\qquad\qquad\qquad\qquad\qquad ||$
(f) $HS - CH_2 - CH_2 - C - OH$ (Chain with hetero atoms)

2.1.1.2 Alicyclic Compounds

When two carbon atoms of an aliphatic chain join through a covalent bond to form a cyclic structure (cycle, ring), it is called an alicyclic compound. These may have the same structural features (multiple bonds, branching or heteroatoms) as in open-chain aliphatic compounds (2.2a and 2.2b). Chemically they behave in a similar manner as the aliphatic compounds.

$$ \tag{2.2} $$

(a) Cyclohexane   (b) Cyclohexanol

2.1.1.3 Aromatic Compounds

Aromatic compounds are also cyclic, but have delocalised bonds giving rise to the property called aromaticity (see Chapter 1, Section 1.2.5). Chemically they behave differently from alicyclic compounds. Thus, benzene (2.3a) and cyclohexane (2.2a) have clear differences. Phenol (2.3b) and cyclohexanol (2.2b) likewise, show differences in reactions.

$$ \tag{2.3} $$

(a) Benzene   (b) Phenol

2.1.1.4 Heterocyclic Compounds

Both alicyclic and aromatic compounds may have heteroatoms in the ring. Such compounds are called heterocylic compounds. Typical examples are piperidine (2.4a) and pyridine (2.4b).

$$ \tag{2.4} $$

(a) Piperidine   (b) Pyridine
(non-aromatic heterocycle)   (aromatic heterocycle)

2.1.1.5 Functional Groups

Functional group is the name given to a group of atoms, or even a single atom, which is a reactive centre in a molecule and whose reactions are very similar, regardless of which molecule it is present in. Some examples are (i) double bond in an alkene, (ii) the carbonyl group in an aldehyde or ketone, (iii) hydroxyl group in an alcohol (2.5).

$$
\underset{\text{Alkene}}{R-CH=CH-R} \qquad \underset{\text{Carbonyl}}{H-\overset{\overset{\displaystyle O}{\|}}{C}-R} \qquad \underset{\text{Hydroxyl}}{R-OH} \tag{2.5}
$$

The same functional group when present in different classes of organic compounds may show differences. For example, the OH group (hydroxyl) when present in non-aromatic compounds such as aliphatic and alicyclic, behaves differently from the hydroxyl group present in aromatic compounds. We classify organic compounds into families on the basis of the functional groups present. The most popular way of studying organic compounds systematically, is based on families and functional groups. The common functional groups and the families to which they belong are listed in Table 2.1.

Table 2. 1 Families and functional groups

Family	Functional group (structure)	Typical example	Name (of the example)	Suffix
Alkane	$C-H$ and $C-C$ bonds	CH_3-CH_3	Ethane	–ane
Alkyl	$R-$	CH_3-CH_2-	Ethyl	–yl
Alkene	$\ce{>C=C<}$	$CH_2=CH_2$	Ethene (ethylene)	–ene
Alkyne	$-C\equiv C-$	$HC\equiv CH$	Ethyne (acetylene)	–yne
Arene			Benzene	
Aryl	$Ar-$	C_6H_5-	Phenyl	–yl
Halide	$R-X, Ar-X$ $(X=F, Cl, Br, I)$	CH_3-Cl	Chloromethane	
Alcohol	$R-OH$	CH_3-OH	Methanol	–ol
Phenol	$Ar-OH$	C_6H_5-OH	Phenol	

Table 2. 1 (Continued)

Family	Functional group (structure)	Typical example	Name (of the example)	Suffix
Ether	$R-O-R$ $R-O-Ar, Ar-O-Ar$	CH_3-O-CH_3	Dimethyl ether	ether
Amine	$R-N\diagdown,\quad Ar-N\diagdown$	CH_3-NH_2	Methylamine	amine
Nitrile	$R-C\equiv N, Ar-C\equiv N$	$CH_3C\equiv N$	Acetonitrile	nitrile
Nitro compounds	$R-\overset{+}{N}\overset{\diagup O}{\diagdown O^-}$ $Ar-\overset{+}{N}\overset{\diagup O}{\diagdown O^-}$	$C_6H_5-\overset{+}{N}\overset{\diagup O}{\diagdown O^-}$	Nitrobenzene	—
Carbonyl compounds Aldehydes	$R-\overset{H}{\underset{}{C}}=0$ $Ar-\overset{H}{\underset{}{C}}=0$	$CH_3-\overset{O}{\overset{\|}{C}}-H$	Acetaldehyde (ethanal)	–al
Ketones	$R-\overset{O}{\overset{\|}{C}}-R'$ $R-\overset{O}{\overset{\|}{C}}-Ar'$ $Ar-\overset{O}{\overset{\|}{C}}-Ar'$	$CH_3-\overset{O}{\overset{\|}{C}}-CH_3$	Acetone (propanone)	–one
Carboxylic acids	$R-\overset{O}{\overset{\|}{C}}-OH$ $Ar-\overset{O}{\overset{\|}{C}}-OH$	$CH_3-\overset{O}{\overset{\|}{C}}-OH$	Acetic acid (ethanoic acid)	–oic acid

Table 2. 1   (Continued)

Family	Functional group (structure)	Typical example	Name (of the example)	Suffix
Ester	$\underset{\underset{\displaystyle Ar-C-O-R'}{\overset{\displaystyle O}{\|}}}{\overset{\displaystyle O}{\overset{\|}{R-C-O-R'}}}$	$CH_3-\overset{\overset{\displaystyle O}{\|}}{C}-O-CH_3$	Methyl acetate (methyl ethanoate)	–oate
Amide	$R-\overset{\overset{\displaystyle O}{\|}}{C}-N\big\langle$	$CH_3-\overset{\overset{\displaystyle O}{\|}}{C}-NH_2$	Acetamide (ethanamide)	amide
Carboxylic acid halide (acyl halide)	$\underset{\underset{\displaystyle Ar-C-X}{\overset{\displaystyle O}{\|}}}{\overset{\displaystyle O}{\overset{\|}{R-C-X}}}$	$CH_3-\overset{\overset{\displaystyle O}{\|}}{C}-Cl$	Acetylchloride (ethanoyl chloride)	–oyl halide
Carboxylic acid anhydride	$R-\overset{\overset{\displaystyle O}{\|}}{C}-O-\overset{\overset{\displaystyle O}{\|}}{C}-R'$	$CH_3-\overset{\overset{\displaystyle O}{\|}}{C}-O-\overset{\overset{\displaystyle O}{\|}}{C}-CH_3$	Acetic anhydride (ethanoic anhydride)	–oic anhydride

2.2   INTRODUCTION TO NOMENCLATURE

When new compounds were discovered, they were given names based on their source or properties in much the same way as elements were named. As the number of organic compounds grew, the need for systematic nomenclature arose. These attempts were based on functional groups and families. For many compounds, the earlier arbitrary names – called trivial or common names – are still in use. Thus, many organic compounds have a trivial and a systematic name [(2.6)].

$$CH_3-CH_2-OH \;—\; \text{trivial name} \qquad : \text{Ethyl alcohol}$$
$$\text{systematic name} : \text{Ethanol}$$

$$CH_3-CO-CH_3 \;—\; \text{trivial name} \qquad : \text{Acetone}$$
$$\text{systematic name} : \text{Propanone}$$

$$CH_3COOH \qquad\; —\; \text{trivial name} \qquad : \text{Acetic acid}$$
$$\text{systematic name} : \text{Ethanoic acid} \qquad\qquad (2.6)$$

Attempts at systematic nomenclature started in the 1890s and have gone through many refinements. The system of nomenclature currently accepted by the scientific community is the one proposed and monitored by the International Union of Pure and Applied Chemistry (IUPAC). Given the IUPAC name of a molecule, anyone who understands the rules will arrive at the same structure. A given compound can have one or a few acceptable IUPAC names. The lack of uniqueness of names arises because several names which are entrenched by popular usage, are approved by IUPAC as acceptable. (An illustration is the nomenclature of 2.18).

2.2.1 ALKANE FAMILY

All aliphatic compounds are considered to be derived from alkanes by branching, by introduction of functional groups like double bonds or heteroatoms in the chain, by replacement of the hydrogen atom by functional groups or by joining the ends of chains to form cyclic structures. All the unbranched alkanes, from the one carbon methane to all the other ones have been assigned names. The names of all other aliphatic and alicyclic molecules are based on these. The first few alkanes have trivial names. The names of the higher alkanes are generated by combining Greek or Latin prefixes to denote the number of carbons and the suffix '–ane', to denote the family. Such names for the lower ones and some representative higher ones are listed in Table 2.2. Removal of one hydrogen from an alkane chain gives an *alkyl group* (2.7).

$$CH_3-CH_3 \text{ (Ethane)} \qquad\qquad CH_3-CH_2- \text{ (Ethyl)}$$

$$CH_3-CH_2-CH_3 \text{ (Propane)} \qquad CH_3-CH_2-CH_2- \text{ (Propyl)}$$

$$CH_3-(CH_2)_8-CH_3 \text{ (Decane)} \qquad CH_3-(CH_2)_8-CH_2- \text{ (Decyl)} \qquad (2.7)$$

2.2.1.1 Branched Alkanes

Branching is possible in alkanes containing 4 or more carbons. The 4-carbon alkane, C_4H_{10}, has two isomers—2.8a and 2.8b.

$$CH_3-CH_2-CH_2-CH_3 \qquad\qquad CH_3-CH-CH_3$$
$$\qquad\qquad\qquad\qquad\qquad\qquad\qquad\qquad\qquad\qquad | $$
$$\text{(a)} \quad n\text{-Butane} \qquad\qquad\qquad\qquad\qquad CH_3 \qquad (2.8)$$

$$\text{(b)} \quad \text{Isobutane}$$

The trivial names for these are *n*-butane and isobutane. The prefix '*n*' has been used to denote a normal, unbranched chain and the prefix 'iso' to denote an isomer. For the 4-carbon system this is adequate since there are no other isomers. As the number of carbons increases, the number of isomers increases and trivial names become inadequate.

2.2.1.2 IUPAC Nomenclature of Alkanes

An IUPAC name has four components—locant(s), name(s) of substituent(s), parent name, name(s) of functional groups(s). This is illustrated by the systematic name of isobutyl alcohol (2.9). The locant numerals are separated from words by hyphens.

Table 2. 2   Names of alkanes

Carbon number	Structure	Name
1	CH_4	Methane
2	$CH_3 - CH_3$	Ethane
3	$CH_3 - CH_2 - CH_3$	Propane
4	$CH_3 - (CH_2)_2 - CH_3$	Butane
5	$CH_3 - (CH_2)_3 - CH_3$	Pentane
6	$CH_3 - (CH_2)_4 - CH_3$	Hexane
7	$CH_3 - (CH_2)_5 - CH_3$	Heptane
8	$CH_3 - (CH_2)_6 - CH_3$	Octane
9	$CH_3 - (CH_2)_7 - CH_3$	Nonane
10	$CH_3 - (CH_2)_8 - CH_3$	Decane
11	$CH_3 - (CH_2)_9 - CH_3$	Undecane
12	$CH_3 - (CH_2)_{10} - CH_3$	Dodecane
13	$CH_3 - (CH_2)_{11} - CH_3$	Tridecane
19	$CH_3 - (CH_2)_{17} - CH_3$	Nonadecane
20	$CH_3 - (CH_2)_{18} - CH_3$	Eicosane (Icosane)
30	$CH_3 - (CH_2)_{28} - CH_3$	Triacontane
100	$CH_3 - (CH_2)_{98} - CH_3$	Hectane

$$CH_3 - CH - CH_2 - OH$$
$$|$$
$$CH_3$$

(2.9)

$$\underset{a}{2} \text{-} \underset{b}{methyl} \text{-} \underset{c}{1} \text{-} \underset{d}{propan} \underset{e}{ol}$$

a : locant of substituent
b : substituent (prefix)
c : locant of functional group
d : parent name
e : functional group (family) (suffix)

For alkanes, the suffix family name, – *ane*, is built into the parent names. Substituents present on the parent alkane chain are specified by prefixes and their location, by (locant) numbers*.

The steps involved in the nomenclature of alkanes are given below and illustrated by the alkane C_7H_{16}, 3-methylhexane (2.10)

$$CH_3 - CH_2 - CH - CH_2 - CH_2 - CH_3$$
$$|$$
$$CH_3$$

(2.10)

*See key points, nomenclature, for a more detailed picture.

Step 1: Identify the longest carbon chain. Three carbon chains can be seen in this molecule, one with 6 carbons (hexane), another with 5 carbons (pentane) and a third one with 4 carbons (butane) (2.11a–2.11c).

$$CH_3-CH_2-CH-CH_2-CH_2-CH_3 \qquad (a)$$
$$| $$
$$CH_3$$

$$CH_3-CH_2-CH-CH_2-CH_2-CH_3 \qquad (b)$$
$$|$$
$$CH_3$$

$$CH_3-CH_2-CH-CH_2-CH_2-CH_3 \qquad (c) \qquad (2.11)$$
$$|$$
$$CH_3$$

We select the longest chain, hexane. The molecule is a methylhexane—a hexane with a methyl substituent on it. Hexane is the parent alkane.

Step 2: Number the carbon atoms of the parent chain, in such a way that the substituent is closer to the end from where the numbering starts. 2.12a is right, 2.12b is wrong.

$$\overset{1}{CH_3}-\overset{2}{CH_2}-\overset{3}{CH}-\overset{4}{CH_2}-\overset{5}{CH_2}-\overset{6}{CH_3} \qquad (a)$$
$$|$$
$$CH_3$$

$$\overset{6}{CH_3}-\overset{5}{CH_2}-\overset{4}{CH}-\overset{3}{CH_2}-\overset{2}{CH_2}-\overset{1}{CH_3} \qquad (b) \qquad (2.12)$$
$$|$$
$$CH_3$$

The molecule is named 3-methylhexane, not 4-methylhexane. Methyl is the substituent which is included in the name as a prefix and written as one word along with the parent name. The location of the substituent is included in the name as a number just before the name of the substituent and separated from it by a hyphen.

All isomeric molecules having the formula C_7H_{16} are shown along with their names (2.13a–2.13i).

- $CH_3-CH_2-CH_2-CH_2-CH_2-CH_2-CH_3$ (a) Heptane

  (Note: The prefix *n*- is not used in IUPAC nomenclature)

- $CH_3-CH-CH_2-CH_2-CH_2-CH_3$ (b) 2-Methylhexane
  $|$
  CH_3

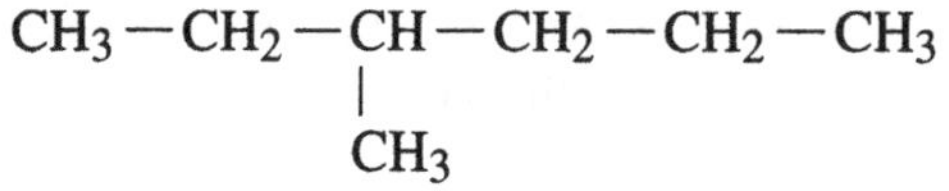

- $CH_3-CH_2-CH-CH_2-CH_2-CH_3$
  |
  CH_3

  (c) 3-Methylhexane

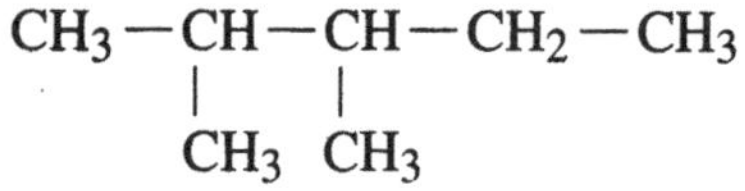

- $CH_3-CH-CH-CH_2-CH_3$
  | |
  CH_3 CH_3

  (d) 2, 3-Dimethylpentane

  (Note: Each substituent should have a numerical locant)

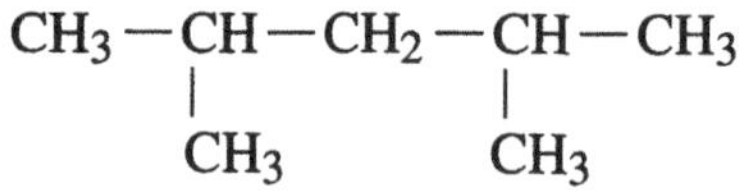

- $CH_3-CH-CH_2-CH-CH_3$
  | |
  CH_3 CH_3

  (e) 2, 4-Dimethylpentane

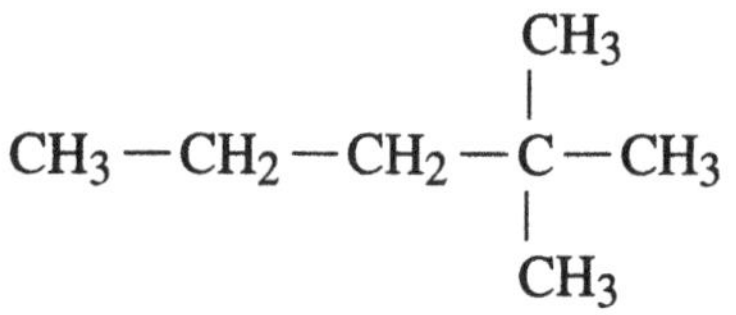

- $CH_3-CH_2-CH_2-\overset{\displaystyle CH_3}{\underset{\displaystyle CH_3}{C}}-CH_3$

  (f) 2, 2-Dimethylpentane

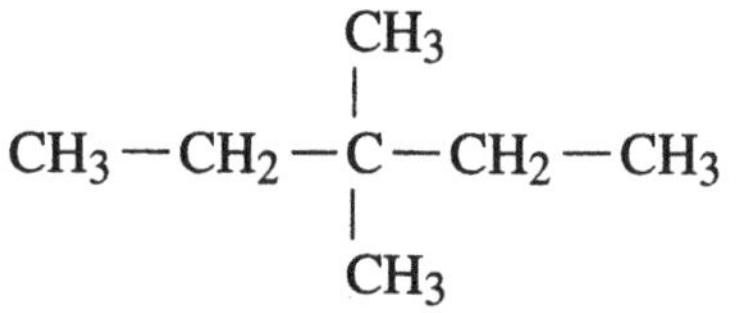

- $CH_3-CH_2-\overset{\displaystyle CH_3}{\underset{\displaystyle CH_3}{C}}-CH_2-CH_3$

  (g) 3, 3-Dimethylpentane

- $CH_3-\overset{\displaystyle CH_3}{C}-CH-CH_3$
  | |
  CH_3 CH_3

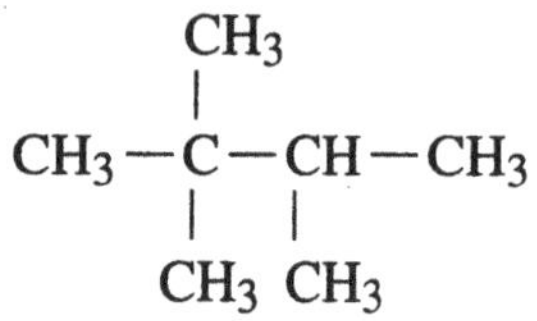

  (h) 2, 2, 3-Trimethylbutane
       (Not 2,3,3-Trimethylbutane)

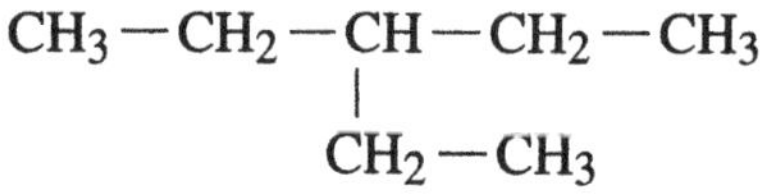

- $CH_3-CH_2-CH-CH_2-CH_3$
  |
  CH_2-CH_3

  (i) 3-Ethylpentane

  (2.13)

Some of the other rules of nomenclature are illustrated in the following examples (2.14).

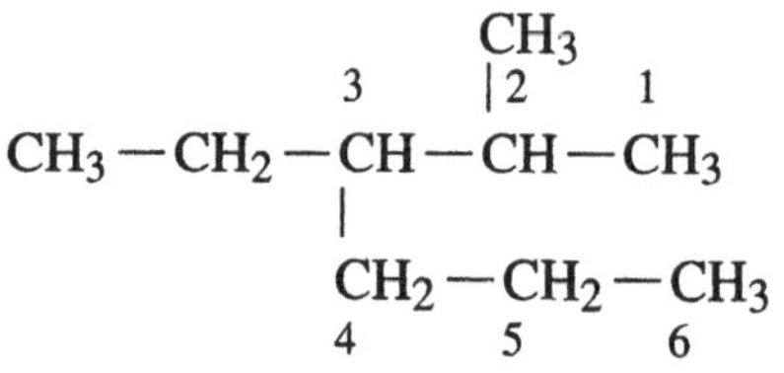

- $CH_3-CH_2-\overset{3}{C}H-\overset{2}{C}H-\overset{1}{C}H_3$ with CH_3 on C2
  |
  $\overset{}{C}H_2-\overset{}{C}H_2-\overset{}{C}H_3$
   4   5   6

  3-Ethyl-2-methylhexane
  (right name)

  (2.14a)

  (4-Ethyl-5 methylhexane is
  wrong. The sum of the locant
  numbers should be minimum.)

$$\underset{1}{CH_3}-\underset{2}{CH_2}-\underset{3}{CH}-\overset{\overset{\displaystyle CH_3}{|}}{CH}-CH_3$$

• 3-Isopropylhexane (wrong name) (2.14b)

$$\underset{4}{CH_2}-\underset{5}{CH_2}-\underset{6}{CH_3}$$

There are two different hexane chains in 2.14. The hexane chain in 2.14a has two branch points; that in 2.14b has only one branch point. The former is the right choice. In 2.14a, the substituents are arranged alphabetically, ethyl coming before methyl.

• $$CH_3-\overset{\overset{\displaystyle CH_3}{|}}{\underset{\underset{\displaystyle CH_3}{|}}{C}}-CH_2-CH_3$$ 2, 2-Dimethylbutane (not 2-Dimethybutane) (2.15)

• $$CH_3-\overset{\overset{\displaystyle H_3C}{|}}{\underset{\underset{\displaystyle H_3C}{|}}{C}}-\overset{\overset{\displaystyle CH_3}{|}}{\underset{\underset{\displaystyle H}{|}}{C}}-CH_3$$ 2, 2, 3-Trimethylbutane (2.16)

(**Note:** multipliers like di– and tri– are used to group together the same substituents, but there should be as many locant numbers as substituents.)

• $$\overset{1}{CH_3}-\overset{2}{CH}-\overset{3}{\underset{\underset{\displaystyle CH_3}{|}}{C}}-\overset{4}{\underset{\underset{\displaystyle CH_2-CH_3}{|}}{\overset{\overset{\displaystyle CH_3}{|}}{C}}}-\overset{5}{CH_2}-\overset{6}{CH_3}$$ 3-Ethyl -2,3-dimethylhexane (2.17)

(**Note:** The multiplier prefixes like di– and tri– are not taken into account when deciding the alphabetical order of substituents; ethyl comes before (di) methyl.)

2.2.1.3 Branched Alkyl Groups

When the substituent is an alkyl group which is itself branched, it is named in conformity with IUPAC rules, and included in brackets.

• $$\underset{1}{CH_3}-\underset{2}{CH_2}-\underset{3}{CH_2}-\underset{4}{\underset{\underset{\displaystyle CH_3-CH-CH_3}{|}}{CH}}-\underset{5}{CH_2}-\underset{6}{CH_2}-\underset{7}{CH_3}$$ (2.18)

The alkyl substituent at position 4 is given an IUPAC name as a derivative of ethyl.

$$\underset{1}{CH_3}-\overset{\displaystyle |}{\underset{2}{CH}}-CH_3$$ 1-methylethyl (2.19)

Thus 2.18 is named 4-(1-methylethyl)heptane.

Many alkyl groups have trivial names which are accepted in IUPAC nomenclature. Thus, 1-methylethyl (2.19), is more commonly known as isopropyl, which is accepted. It is correct to name 2.18 as 4-isopropylheptane. Both names are accepted. Some IUPAC accepted trivial names of alkyl groups are listed in Table 2.3.

Table 2. 3   Systematic names of alkyl groups and their IUPAC-accepted trivial names

Structure	Trivial name	Systematic name
$-CH-CH_3$ | CH_3	isopropyl	1-methylethyl
$-CH-CH_2-CH_3$ | CH_3	*sec*-butyl	1-methylpropyl
$-CH_2-CH-CH_3$ | CH_3	isobutyl	2-methylpropyl
CH_3 | $-C-CH_3$ | CH_3	*tert*-butyl	1,1-dimethylethyl
CH_3 | $-CH_2-C-CH_3$ | CH_3	neopentyl	2,2-dimethylpropyl
$-CH=CH_2$	vinyl	ethenyl
$-CH_2-CH=CH_2$	allyl	2-propenyl

2.2.2   NOMENCLATURE OF ALKENES AND ALKYNES

The basic rules laid down for alkanes can be applied for alkenes and alkynes. The double and triple bonds are functional groups. The name of the alkene or alkyne is obtained by replacing the suffix *–ane* of alkane by *–ene* or *–yne*. Thus, we have ethene, propene, butene and ethyne, propyne, butyne. The parent chain is the one containing the functional group—the double bond or triple bond.

$$CH_3 - CH_2 - \overset{2}{C} - \overset{3}{CH_2} - \overset{4}{CH_3} \qquad (a)$$
$$\overset{\|}{\underset{1\ CH_2}{}}$$

$$\text{not,} \quad \overset{1}{CH_3} - \overset{2}{CH_2} - \overset{3}{C} - \overset{4}{CH_2} - \overset{5}{CH_3} \qquad (b)$$
$$\overset{\|}{\underset{CH_2}{}}$$

(2.20)

The numbering of the chain is from the end nearer to the functional group. This molecule (2.20), is named 2-ethyl-1-butene. The features of the name are described below.

(a) Locant of the substituent, in this case, ethyl. This numeral is placed immediately before the name of the substituent, separated by a hyphen.

(b) Name of the substituent.

(c) Position of the functional group, in this case, the double bond. It is between carbons, 1 and 2. In cases where there is no ambiguity, only the first numeral, 1, is included in the name.

(d) Name of the parent carbon chain.

(e) Suffix *–ene*, for the functional group.

Note that the functional group is always incorporated in the name using the designated suffix. Its position is specified by the locant numeral which is placed before the name of the parent chain, after the substituents. The substituent(s) (which may also be a functional group in some situations) is included as prefix*. The following example (2.21) illustrates the relevant rules.

$$\overset{1}{C}H_3 - \overset{2}{C}H = \overset{3}{C}H - \overset{4}{C}H - \overset{5}{C}H_3$$
$$\underset{CH_3}{|}$$

(a) 4-Methyl-2-butene (Not 2-Methyl-3-butene. Numerbing begins with the end nearer to the double bond.)

$$\overset{4}{C}H_3 - \overset{3}{C}H_2 - \overset{2}{C} \equiv \overset{1}{C}H$$

(b) 1-Butyne

$$CH_2 = CH - CH = CH_2$$

(c) 1,3-Butadiene

$$CH_2 = C - CH = CH_2$$
$$\underset{CH_3}{|}$$

(d) 2-Methyl-1,3-butadiene

$$\overset{CH_3}{\underset{CH_3}{\overset{|}{C}H_3 - C - C \equiv C - H}}$$

(e) 3,3-Dimethyl-1-butyne   (2.21)

Some simple alkenes and alkynes have trivial names which are firmly entrenched. These names are accepted by IUPAC. The industry generally uses only these names. A partial list of such names is given in Table 2.4.

*See key points, nomenclature, for a more detailed picture.

Table 2. 4   Some commonly used trivial names of alkenes and alkynes

Structure	Trivial name	IUPAC name
$CH_2 = CH_2$	Ethylene	Ethene
$CH_2 = CH - CH_3$	Propylene	Propene
$CH_2 = \underset{\underset{CH_3}{\vert}}{C} - CH_3$	Isobutylene	2-Methylpropene
$CH \equiv CH$	Acetylene	Ethyne
$CH_2 = \underset{\underset{CH_3}{\vert}}{C} - CH = CH_2$	Isoprene	2-Methyl-1,3-butadiene

2.2.3 COMPOUNDS WITH OTHER FUNCTIONAL GROUPS

2.2.3.1 Alkyl Halides

The systematic name for alkyl halides is *haloalkanes* (2.22a–2.22e).

$$CH_3 - \underset{\underset{Cl}{\vert}}{CH} - CH_2 - CH_3 \qquad \text{2-Chlorobutane} \tag{a}$$

$$\underset{\underset{Br}{\vert}}{CH_2} - \underset{\underset{Cl}{\vert}}{CH} - CH_3 \qquad \text{1-Bromo-2-chloropropane} \tag{b}$$

$$CHCl_3 \qquad \text{Trichloromethane} \tag{c}$$

$$\overset{1}{CH_3} - \overset{2}{CH} - \overset{3}{CH} - \overset{4}{CH} - \overset{5}{CH_2} - \overset{6}{CH_3} \qquad \text{3-Bromo-4-iodo-2-methylhexane} \tag{d}$$

(with substituents CH_3, Br, I below positions 2, 3, 4)

$$\underset{6}{CH_3} - \underset{5}{CH_2} - \underset{4}{CH} - \overset{\overset{Br}{\vert}}{\underset{\underset{CH=CH_2}{\vert}}{CH}} - CH_2 - CH_2 - CH_3 \qquad \text{4-Bromo-3-propyl-1-hexene} \tag{e}$$

$$\tag{2.22}$$

(**Note:** The halo and the alkyl substituent are considered to be of the same category)

These compounds can also be named as alkyl halides (2.23a, 2.23b)

CH$_3$—I   Methyl iodide   (a)

(CH$_3$)C—Cl   *tert*-Butyl chloride   (b)   (2.23)

This is referred to as *functional class nomenclature*. In this nomenclature, the alkyl group is not a substituent. The two words in the name should be separated by a space.

2.2.3.2 Alcohols

The family of compounds known as alcohols have the hydroxyl group as the functional group (HO—). In IUPAC nomenclature, this group is denoted by the suffix –*ol*. This is introduced in the name of the alkane by replacing the ending –*e* by –*ol* (2.24).

ethane – ethan**ol**

propane – propan**ol**   (2.24)

The following examples illustrate the nomenclature of alcohols (2.25a–2.25e).

CH$_3$—CH$_2$—OH   Ethanol   (a)

CH$_3$—CH—CH$_3$ 2-Propanol   (b)
    |
    OH

$\quad\quad$ CH$_3$
$\quad\quad$ |
H$_3$C—C—CH$_3$ 2-Methyl-2-propanol   (c)
$\quad\quad$ |
$\quad\quad$ OH

CH$_2$—CH$_2$—CH$_2$—OH   3-Chloro-1-propanol (Not 1-Chloro-3-propanol.
    | The suffix, –*ol*, has priority over the prefixed
    Cl   substituents like halo or alkyl)   (d)

$\quad\quad$ OH
$\quad\quad$ |
CH$_2$—CH—CH=CH$_2$ 3-Buten-2-ol (Not 1-Buten-3–ol. –*ol* outranks –*ene*,
 1$\quad$ 2$\quad$ 3$\quad$ 4   and gets the lower number. Also note the
                                insertion of the locant for –ol by
                                splitting the name).   (e)   (2.25)

Simple alcohols are also named as *alkyl alcohol* based on their functional class or family. In such names, the two words should be separated by a space (2.26a–2.26f). Alcohols may also be named as derivatives of carbinol, a trivial name for methanol.

CH_3-OH Methanol   Methyl alcohol, carbinol   (a)

CH_3-CH_2-OH Ethanol   Ethyl alcohol, methylcarbinol   (b)

$CH_3-CH_2-CH_2-OH$ 1-Propanol   Propyl alcohol, ethylcarbinol   (c)

$$CH_3-\overset{\displaystyle OH}{\underset{\vphantom{x}}{C}}H-CH_3$$ 2-Propanol   Isopropyl alcohol, dimethylcarbinol   (d)

$$CH_3-\overset{\displaystyle OH}{\underset{\displaystyle CH_3}{C}}-CH_3$$ 2-Methyl-2-propanol   *tert*-Butyl alcohol, trimethylcarbinol   (e)

$CH_2=CH-CH_2-OH$ 2-Propen-1-ol   Allyl alcohol, vinylcarbinol   (f)

(2.26)

Names such as isopropanol and *tert*-butanol where two systems of nomenclature are mixed, are not accepted.

2.2.3.3 Ethers

Ethers are molecules where an oxygen has been inserted between two carbons in a chain. They can be seen as two alkyl or aryl groups connected by an oxygen. Their common names are based on this (2.27a, 2.27b).

CH_3-O-CH_3 Dimethyl ether   (a)

$$CH_3-\overset{\displaystyle}{\underset{\displaystyle O-CH_3}{C}}H-CH_2-CH_3$$

       sec-Butyl methyl ether   (b)
       (1-Methylpropyl) methyl ether   (2.27)

(**Note:** There is no space separation between methyl and propyl in 1-methylpropyl because methyl is a substituent. There is space separation between the three words. Ether is a family name which stands alone.)

The group $-OCH_3$ is also called methoxy. (In general $RO-$ is called alkoxy.) Using this name, the above two ethers are also named as in 2.28a, 2.28b. Both (2.27 and 2.28) are accepted IUPAC nomenclature. The former is the functional class nomenclature and the latter is the substitutive nomenclature.

$$CH_3-O-CH_3 \qquad\qquad \text{Methoxymethane \quad (a)}$$

$$CH_3-CH-CH_2-CH_3$$
$$\big|$$
$$O-CH_3 \qquad\qquad \text{2-Methoxybutane \quad (b)}$$

$$(2.28)$$

2.2.3.4 Aldehydes

Aldehydes contain the functional group –CHO, which can come only at the end of a chain and is denoted by the suffix *–al* (2.29a–2.29d). For example,

$$\overset{\displaystyle O}{\overset{\displaystyle \|}{H-C-H}} \qquad\qquad \text{Methanal \qquad (formaldehyde) \qquad (a)}$$

$$\overset{\displaystyle O}{\overset{\displaystyle \|}{CH_3-C-H}} \qquad\qquad \text{Ethanal \qquad (acetaldehyde) \qquad (b)}$$

$$CH_3-CH-CHO$$
$$\big|$$
$$CH_3 \qquad\qquad \text{2-Methylpropanal \quad (isobutyraldehyde) \ (c)}$$

$$CH_2=CH-CH_2-CHO \qquad \text{3-Butenal} \qquad\qquad (d)$$

Cyclopentanecarboxaldehyde   (e)

Benzenecarboxaldehyde (benzaldehyde) (f)

$$(2.29)$$

The group $-CH=O$, is called the formyl group. When the formyl group is attached to a ring, the resulting aldehyde is named as a carboxaldehyde as 2.29e and 2.29f.

2.2.3.5 Ketones

Ketones have the functional group $\big\rangle C=O$, which cannot be at the end of a chain. The suffix for the functional group is *–one*. (An alternate nomenclature where they are named as alkyl ketone, is also in vogue, 2.30.) For example,

$$CH_3-\overset{\overset{\displaystyle O}{\|}}{C}-CH_3$$

(a)   Propanone (acetone); (dimethyl ketone)

$$CH_3-\overset{\overset{\displaystyle O}{\|}}{C}-CH_2-CH_3$$

(b)   Butanone; (ethyl methyl ketone) (In these two cases, the position of the functional group is not specified because there is no ambiguity.)

$$CH_3-CH_2-\overset{\overset{\displaystyle CH_3}{|}}{CH}-\overset{\overset{\displaystyle O}{\|}}{C}-CH_3$$

(c)   3-Methyl-2-pentanone (*sec*-butyl methyl ketone)   (2.30)

(**Note:** When these compounds are named as ketones, there should be space separation between the names of the alkyl groups and the word ketone.)

2.2.3.6 Carboxylic Acids

Carboxylic acids have the functional group, *carboxy* —COOH, which can only be at the end of a chain. The suffix –*oic acid* is used and the numbering starts at the carboxy carbon. In the following examples, the names in brackets are well-accepted trivial names (2.31a–2.31e).

$$H-\overset{\overset{\displaystyle O}{\|}}{C}-OH$$

Methanoic acid   (formic acid)   (a)

$$CH_3-\overset{\overset{\displaystyle O}{\|}}{C}-OH$$

Ethanoic acid   (acetic acid)   (b)

$$CH_3-CH_2-\overset{\overset{\displaystyle O}{\|}}{C}-OH$$

Propanoic acid   (propionic acid)   (c)

$$CH_3-CH_2-CH_2-\overset{\overset{\displaystyle O}{\|}}{C}-OH$$

Butanoic acid   (butyric acid)   (d)

$$CH_3-\overset{\overset{\displaystyle O}{\|}}{\underset{\underset{\displaystyle CH_3}{|}}{CH}}-\overset{\overset{\displaystyle O}{\|}}{C}-OH$$

2-Methylpropanoic acid (isobutyric acid)   (e)

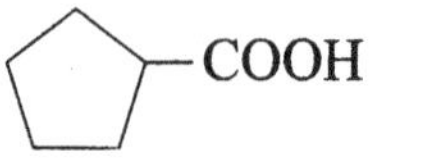

—COOH

Cyclopentanecarboxylic acid   (f)

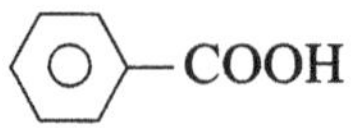

—COOH

Benzenecarboxylic acid (benzoic acid)   (g)

(2.31)

When the carboxyl group, —COOH, is attached to a ring, the suffix carboxylic acid is used for nomenclature as in cyclopentanecarboxylic acid and benzenecarboxylic acid (2.31f and 2.31g).

As can be seen, the numbering starts at the carboxyl carbon, even though that carbon can have no substituents, (recall that the same is the case with aldehydes).

An alternate practice of denoting the position of substituents using Greek letters is also very much in use. In this system, the carbon next to the carboxyl group is labelled α, the next one, β and so on. In this system, the trivial name of the carboxylic acid is used (2.32a–2.32c). For example,

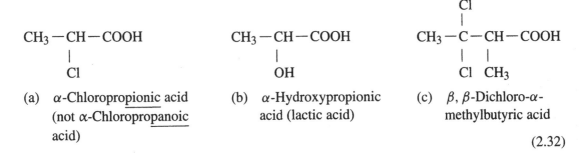

(a) α-Chloropropionic acid (not α-Chloropropanoic acid)

(b) α-Hydroxypropionic acid (lactic acid)

(c) β, β-Dichloro-α-methylbutyric acid

(2.32)

2.2.3.7 Acid Derivatives

Esters: Esters are formally divided into two parts for naming purposes (2.33).

$$CH_3 - \overset{\overset{\displaystyle O}{\|}}{C} - O - C_2H_5$$

(2.33)

portion from the acid (acetate)

alkyl group which has replaced the H of the carboxylic acid (ethyl)

This molecule is ethyl acetate (trivial name) or ethyl ethanoate (IUPAC name). The second part (derived from the acid) is obtained by replacing the *–ic* suffix of the carboxylic acid by *–ate*, regardless of whether the name is trivial or systematic.

$$CH_3 - (CH_2)_5 - \overset{\overset{\displaystyle O}{\|}}{C} - O - \underset{\underset{\displaystyle CH_3}{|}}{CH} - CH_3$$

(2.34)

benzoate (from benzoic acid) | methyl

Methyl benzoate (a)

Isopropyl heptanoate (b)

Acid halides   (Acyl halides): Acid halides are obtained when the $-OH$ of the carboxylic acid is replaced by a halogen. The portion of the acid, after the removal of the OH is called the *acyl group*. Its name is obtained by replacing the suffix *–ic* by *–yl* (2.35).

$$CH_3 - \overset{\overset{\displaystyle O}{\|}}{C} + OH$$

ethanoyl
from ethanoic acid

$$CH_3 - \overset{\overset{\displaystyle O}{\|}}{C} - Cl$$

Ethanoyl chloride

(trivial name: acetyl chloride)   (a)

$$CH_3 - \overset{\overset{\displaystyle CH_3}{|}}{CH} - \overset{\overset{\displaystyle O}{\|}}{C} - Cl$$

2-Methylpropanoyl chloride   (2.35)

(trivial name: isobutyryl chloride)   (b)

$$\underset{\bigcirc}{\overset{\overset{\displaystyle O}{\|}}{C} - Cl}$$

Benzoyl chloride   (c)

Amides: Amides are obtained by replacing the $-OH$ of carboxylic acid by $-NH_2$, $-NHR$ or $-NR_2$. They are named by replacing the *–oic* (or *–ic* in trivial names) by amide (2.36a, 2.36b).

$$CH_3 - \overset{\overset{\displaystyle O}{\|}}{C} - OH$$

Acetic acid
(ethanoic acid)

$$CH_3 - \overset{\overset{\displaystyle O}{\|}}{C} - NH_2$$ (a)

Acetamide
(ethanamide)

$$CH_3 - \overset{\overset{\displaystyle CH_3}{|}}{CH} - \overset{\overset{\displaystyle O}{\|}}{C} - \underset{\underset{\displaystyle CH_3}{|}}{NH}$$

N-Methyl-2-methylpropanamide
N-Methylisobutyramide (see nomenclature
of amines below for understanding
the usage, N-methyl)   (b)   (2.36)

Nitriles: Nitriles are compounds containing the cyano group, $-C\equiv N$. They are formally considered to be derivatives of carboxylic acids, since they can be hydrolysed to carboxylic acids (2.37).

$$R - C \equiv N \xrightarrow{\;H_2O\;} R - \overset{\overset{\displaystyle O}{\|}}{C} - OH + NH_3$$ (2.37)

Names of nitriles are based on the names of the related carboxylic acid (2.38a, 2.38b).

$$CH_3 - \overset{\overset{\textstyle O}{\|}}{C} - OH \qquad CH_3 - C \equiv N$$

Acetic acid   Acetonitrile (methyl cyanide)   (a)

(2.38)

COOH   C ≡ N

Benzoic acid   Benzonitrile (phenyl cyanide)   (b)

In the IUPAC system, nitriles are named by simply adding nitrile to the name of the parent hydrocarbon including the carbon of the cyano group (2.39).

$$CH_3 - C \equiv N \qquad \text{Ethanenitrile (one word)} \tag{2.39}$$

2.2.3.8 Amines

Amines are derivatives of ammonia obtained by replacing one, two or all three of the hydrogens or ammonia by alkyl or aryl groups. The groups could be the same or different. If only one hydrogen of ammonia has been replaced, the amine is primary; if two hydrogens are replaced, it is secondary; if all three have been replaced, it is tertiary. (This nomenclature is different from the primary, secondary and tertiary applied to alkyl groups or alcohols.)

Amines are named by prefixing the name(s) of the alkyl group(s) to the word amine (2.40a, 2.40b).

$$CH_3 - NH_2 \text{ methylamine (one word)} \qquad \text{(a)}$$

$$CH_3 - CH_2 - NH - CH \overset{\displaystyle \diagup CH_3}{\diagdown CH_3} \qquad \text{N-ethylisopropylamine} \qquad \text{(b)}$$

(2.40)

(The prefix N– is used to show that the ethyl group is attached to N of isopropylamine.)

Amines are also named by replacing the ending −e of the parent hydrocarbon by amine (2.41a, 2.41b). Both are accepted by IUPAC.

$$CH_3 NH_2 \text{ Methanamine (methylamine)} \qquad \text{(a)}$$

$$CH_3 - \underset{\underset{\underset{CH_3 \quad CH_3}{\diagup N \diagdown}}{|}}{CH} - CH_2 - CH_2 - CH_3 \qquad \text{(b)}$$

(2.41)

N, N-dimethyl-2-pentanamine
N, N-dimethyl(1-methylbutyl)amine

2.2.4 NOMENCLATURE OF ALICYCLIC COMPOUNDS

Cycloalkanes are hydrocarbons where the two ends of an alkane chain are joined. Cycloalkanes should have atleast 3 carbons. They are named by prefixing *cyclo-* to the name of the corresponding alkane.

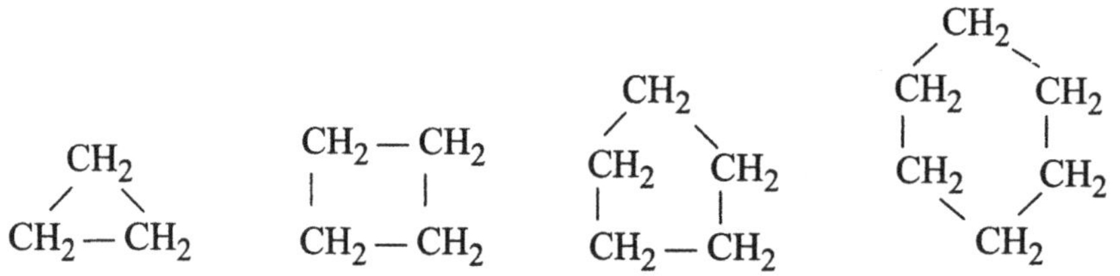

$$
\begin{array}{cccc}
\text{Cyclopropane} & \text{Cyclobutane} & \text{Cyclopentane} & \text{Cyclohexane}
\end{array}
\tag{2.42}
$$

Cycloalkanes are often drawn simply as framework structures, where only the polygon corresponding to the cyclic structure (ring) is shown (2.43).

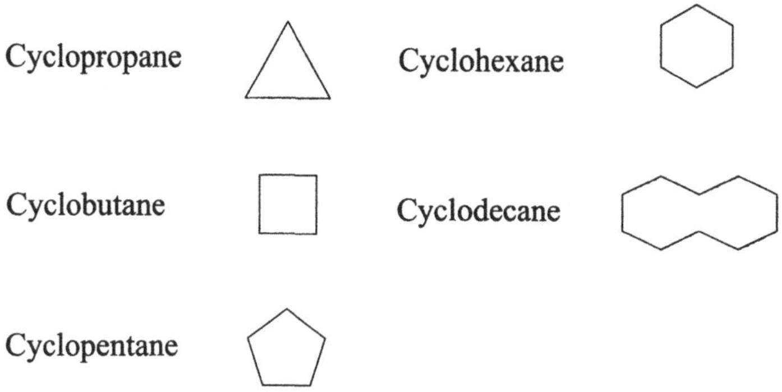

Cyclopropane △ Cyclohexane ⬡

Cyclobutane ☐ Cyclodecane

Cyclopentane ⬠

$$\tag{2.43}$$

Nomenclature, when substituents and functional groups are present, follow the same rules as for alkanes. These are illustrated in the following example (2.44a–2.44h).

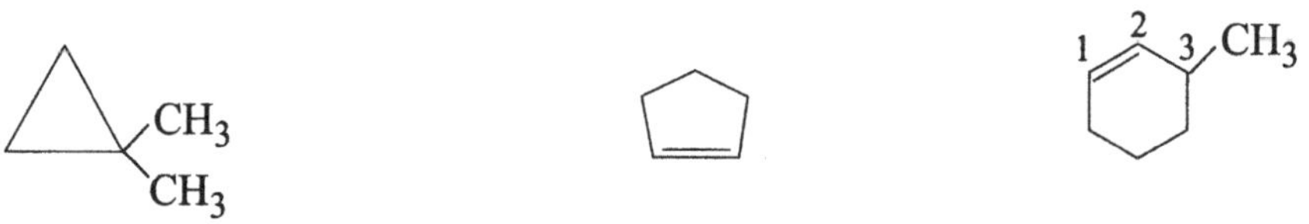

1, 1-Dimethylcyclopropane   (a)   Cyclopentene   (b)   3-Methylcyclohexene   (c)

(**Note:** Numbering starts with one end of the double bond and proceeds through the double bond. So the double bond is between carbons 1 and 2.)

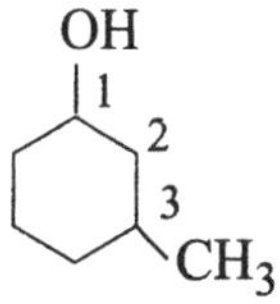

3-Methylcyclohexanol. (In
this case there is no need
to specify the location of
the hydroxyl group.)   (d)

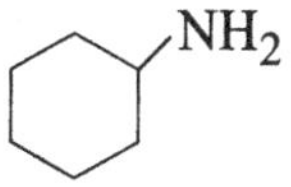

Cyclohexylamine or   (g)
Cyclohexanamine

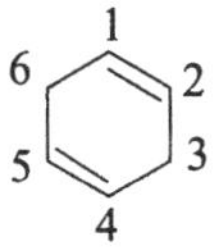

1, 4-Cyclohexadiene   (e)

(2.44)

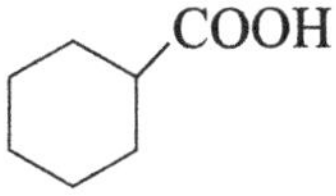

Cyclohexane carboxylic
acid (not cyclohexanoic   (h)
acid. There is no such
molecule)

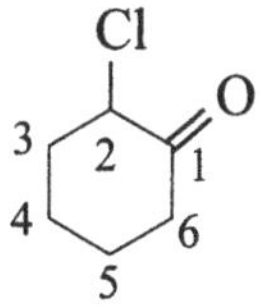

2-Chlorocyclohexanone   (f)

2.2.5 NOMENCLATURE OF AROMATIC COMPOUNDS

2.2.5.1 Monosubstituted Benzenes

We shall at this stage discuss only the derivatives of benzene.

The aryl group (corresponding to alkyl groups in aliphatic compounds) derived from benzene (C_6H_6) is C_6H_5 – and is named phenyl. [Not benzyl. Benzyl is the name given for the phenylmethyl group ($C_6H_5-CH_2-$); this could be confusing and should be borne in mind.]

Most of the simple derivatives of benzene have trivial names which are used in preference to IUPAC names which are rarely used. These are listed in Table 2.5.

2.2.5.2 Disubstituted Benzenes

There are two ways of naming disubstituted benzenes. One uses locant numbers as in the case of cycloalkanes. The other uses the prefixes *o–*, *m–*, and *p–* (for ortho, meta and para, respectively) for 1, 2–, 1, 3– and 1, 4– disubstituted benzene derivatives.

Table 2. 5 Names of anomatic compounds

Structure	Formula	(Trivial) Name	IUPAC name
	C_6H_5-	Phenyl	—
	$C_6H_5-CH_3$	Toluene	Methylbenzene
	C_6H_5-OH	Phenol	Hydroxybenzene
	C_6H_5-CHO	Benzaldehyde	Benzene carboxaldehyde
	C_6H_5-COOH	Benzoic acid	Benzene carboxylic acid
	$C_6H_5-NH_2$	Aniline	Phenylamine (benzenamine)
	$C_6H_5-\overset{\overset{O}{\|}}{C}-CH_3$	Acetophenone	
	$C_6H_5-NO_2$	Nitrobenzene	

Table 2. 5 (Continued)

Structure	Formula	(Trivial) Name	IUPAC name
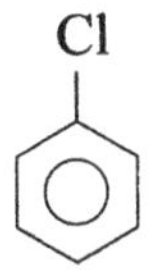 Cl	$C_6H_5 - Cl$	Chlorobenzene	Chlorobenzene
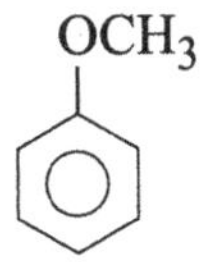 OCH_3	$C_6H_5 - OCH_3$	Anisole (methyl phenyl ether)	Methoxybenzene

The trivial name for hydroxytoluenes (or methylphenols) is cresol. There are three isomeric cresols (2.45a–2.45c).

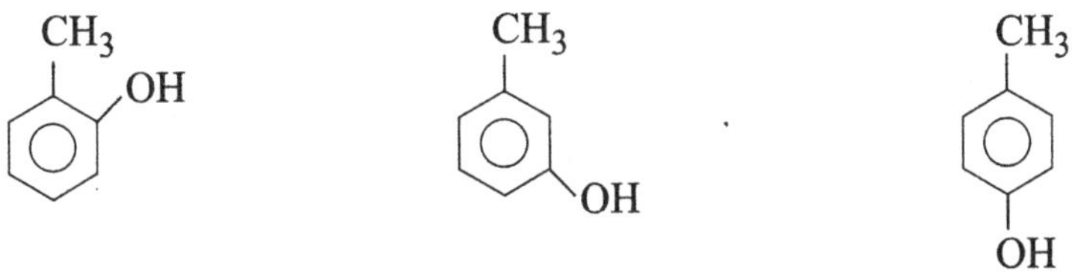

(a) *o*-Cresol
    2-Methylphenol

(b) *m*-Cresol
    3-Methylphenol

(c) *p*-Cresol
    4-Methylphenol

(2.45)

Dimelthylbenzenes are called xylenes (2.46a–2.46c).

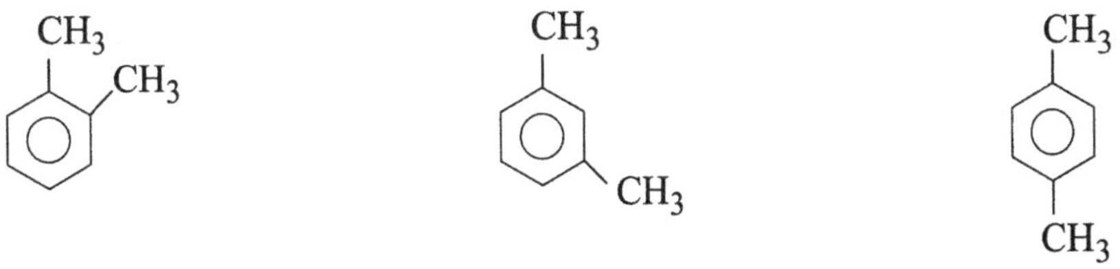

(a) *o*-Xylene
    1, 2-Dimethylbenzene

(b) *m*-Xylene
    1, 3-Dimethylbenzene

(c) *p*-Xylene
    1,4-Dimethylbenzene

(2.46)

Other disubstituted benzenes are also similarly named (2.47a–2.47c).

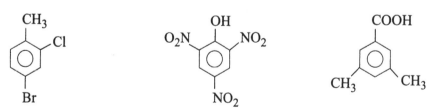

(2.47)

(a) *o*-Bromotoluene
2-Bromotoluene

(b) *m*-Chlorotoluene
3-Chlorotoluene

(c) *p*-Nitrotoluene
4-Nitrotoluene

(**Note:** In this name, the compound is treated as a derivative of toluene, the numbering starts with the carbon bearing the CH₃ group.) It is also acceptable to name it as 4-methylnitrobenzene.

Aminotoluenes (or methylanilines) are called *toluidines*. There are three of them (2.48a–2.48c).

(2.48)

(a) *o*-Toluidine
2-Methylaniline
2-Aminotoluene

(b) *m*-Toluidine
3-Methylaniline
3-Aminotoluene

(c) *p*-Toluidine
4-Methylaniline
4-Aminotoluene

2.2.5.3 Benzene with More than Two Substituents

o–, *m*– and *p*– prefixes are useless when there are more than two substituents. For such molecules, nomenclature based on numbering is resorted to. If one substituent is part of a trivial parent name – as with the methyl substituent in toluene or the hydroxy in phenol – numbering should start with the carbon bearing that substituent (2.49). For example,

(a) 4-Bromo-2-chlorotoluene
(position of CH₃ is not
specified)

(b) 2, 4, 6-Trinitrophenol
(trivial name: picric
acid)

(c) 3, 5-Dimethylbenzoic
acid

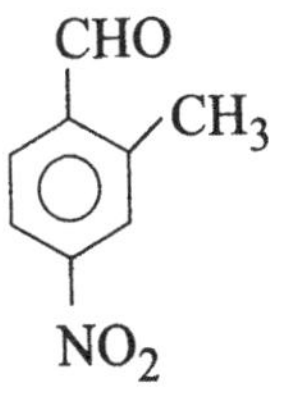

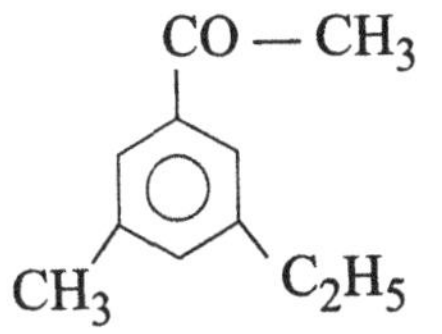

$$(2.49)$$

(d) 2-Methyl-4
nitrobenzaldehyde

(e) 3-Ethyl-5-methylacetophenone

When the compound is named as a derivative of benzene, all substituents should be specified by locant numerals (2.50a, 2.50b). For example,

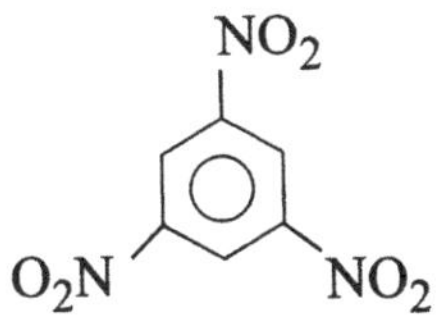

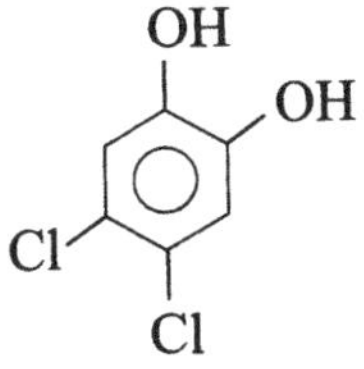

$$(2.50)$$

(a) 1, 3, 5-Trinitrobenzene

(b) 1, 2-Dihydroxy-4,
5-dichlorobenzene

2.2.5.4 Benzene with Side Chain

An alkyl group attached to a benzene ring is called a side chain (2.51).

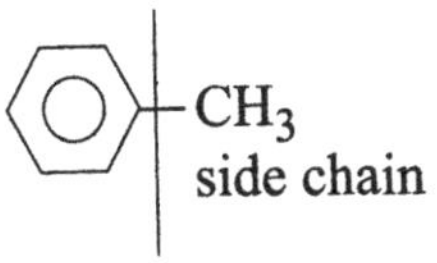

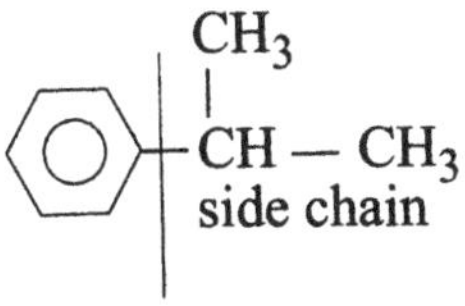

$$(2.51)$$

(a) Toluene

(b) Isopropylbenzene

The group, $C_6H_5-CH_2-$ is called benzyl. (As pointed out earlier, confusion between phenyl, C_6H_5-, and benzyl, $C_6H_5-CH_2-$ should be avoided.) The terms phenyl and benzyl can be used to denote substituents in aliphatic and alicyclic compounds (2.52a, 2.52b). For example,

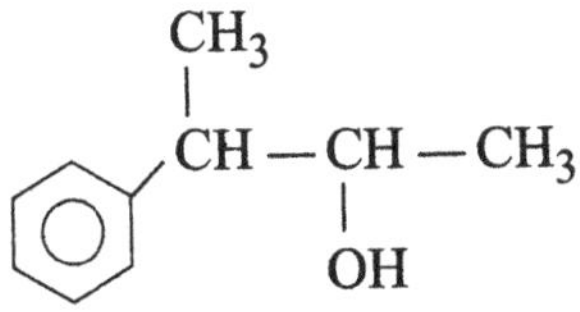

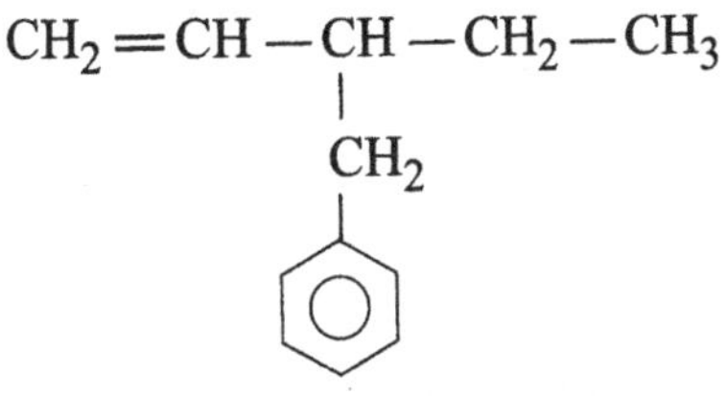

$$(2.52)$$

(a) 3-Phenyl-2-butanol

(b) 3-Benzyl-1-pentene

KEY POINTS

Classification; Functional Groups

Classification of organic compounds into families:

- aliphatic   : those derived from alkanes
- alicyclic   : similar to alkanes, but cyclic
- aromatic   : those derived from benzene (typically)
- heterocyclic   : cyclic compounds, similar to alicyclic or aromatic, with an atom other than carbon in the ring

Functional groups : groups of atoms or even single atoms, which have characteristic behaviour

Examples   : hydroxyl (OH) group in alcohols

                : double bond in alkenes

                : carbonyl group in aldehydes and ketones

Nomenclature

- Many trivial names are accepted in systematic nomenclature also.
- Names of alkanes to be remembered.
- Components of an IUPAC name:

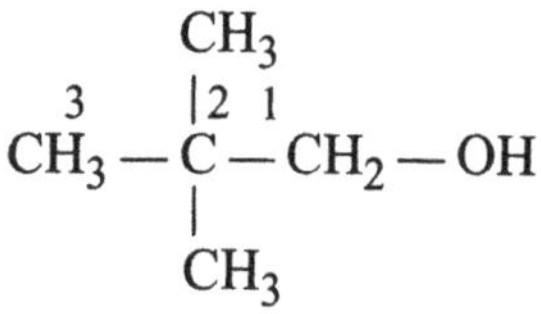

$$
\begin{array}{c}
\qquad\quad CH_3 \\
\overset{3}{CH_3} - \overset{2}{\underset{|}{C}} - \overset{1}{CH_2} - OH \quad \text{(parent chain propane; carbons} \\
\qquad\quad | \qquad\qquad\qquad \text{to be numbered)} \\
\qquad\quad CH_3
\end{array}
$$

$$
\underset{\text{a b a c d}}{2, 2 - \text{di methyl} - 1 - \text{propan ol}}
$$

2, 2 - di methyl - 1 - propan ol
a b a c d   e   c f c   g h

a : numeral (locant) to designate position of substituent

b : comma, to separate numerals

c : hyphen to separate numerals from names

d : multiplier prefix to denote two methyl groups

e : name of substituent

f : numeral (locant) to designate position of functional group

g : name of parent chain

h : suffix to denote functional group

EXERCISES

SECTION I

1. Draw the structures of the following compounds.

   (a) neopentyl alcohol

   (b) 2-chloro-1-propanol

   (c) 2-methyl-5-(1-methylpropyl)octane

   (d) 3, 3-dimethyl-1-butene

   (e) 2, 3-dimethyl-2-butanol

   (f) 2, 2-dimethylcyclopentanol

   (g) 4-hydroxypentanal

   (h) 2, 4-dichloro-3-hexanone

   (i) 3, 3-dimethylpentanoic acid

   (j) α, β-dichloropropionic acid

   (k) trichloroacetic acid

   (l) cyclohex-2-enone

   (m) 4-methylcyclohexylamine

   (n) 2-methyl-1, 3-cyclohexadiene

   (o) 1-phenylethanol

   (p) 2-phenylhexanenitrile

   (q) benzyl benzoate

   (r) 2, 4-di-*tert*-butylphenol

   (s) 3-(N, N-dimethylamino)benzamide

   (t) 2-bromo-3-chlorocyclopentanol

   (u) vinyl acetate

   (v) allyl alcohol

2. Give the IUPAC names of the following.

   (a) $CH_3 - CH - CH - CH_2 - CH_3$
   $\qquad\quad |\qquad |$
   $\qquad\quad CH_3 \;\; CH_3$

   (b) $CH_3 - CH - C = CH - CH_3$
   $\qquad\quad |\qquad |$
   $\qquad\quad CH_3 \;\; CH_3$

   (c) $CH_3 - CH - C - CH_3$
   $\qquad\qquad |\qquad ||$
   $\qquad\quad CH_3 \;\; O$

   (d)

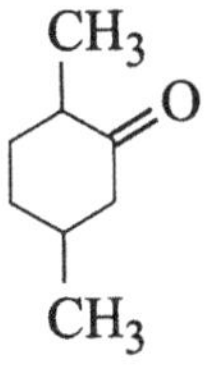

   (e) 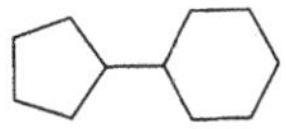
   (hint: name it as a substituted cyclohexane)

   (f) $CH_2 = CH - CH_2 - CH_2 - COOH$

   (g) 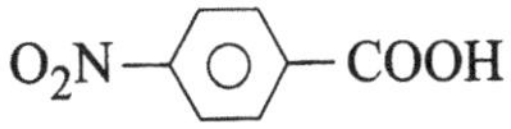
   $O_2N - \langle\!\!\;\bigcirc\!\!\;\rangle - COOH$

   (h)

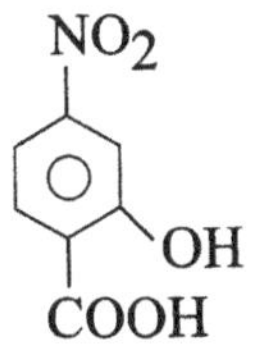

   (i)

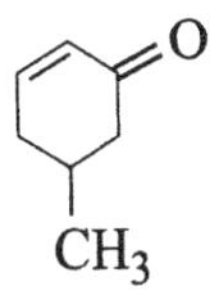

3.   Give one or more other acceptable names for the following molecules.

(a)

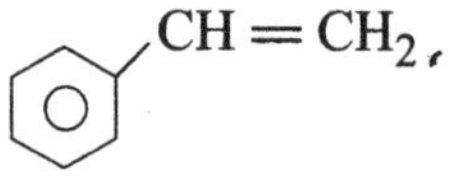

$CH=CH_2$,

Styrene

(b)

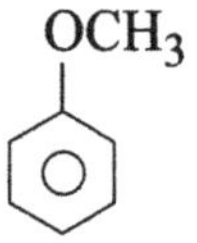

OCH_3

Anisole

(c)

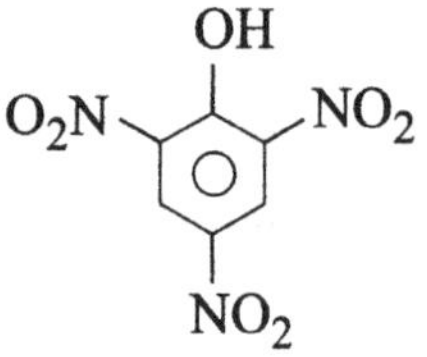

OH

O_2N NO_2

NO_2

Picric acid

(d)

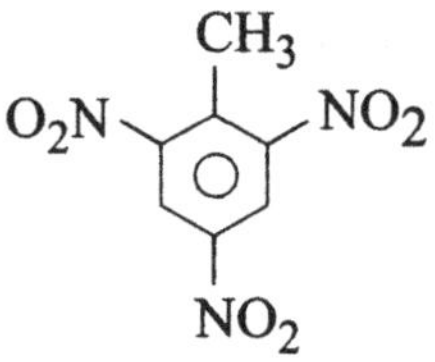

CH_3

O_2N NO_2

NO_2

TNT

(e)

Cl

Cl   Cl

Cl   Cl

Cl

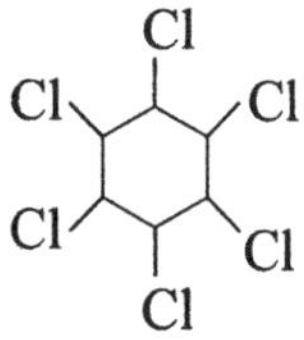

Gammexane (used
as an insecticide)

(f)

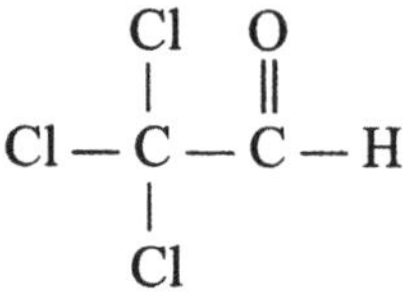

$$Cl-\overset{\displaystyle Cl}{\underset{\displaystyle Cl}{C}}-\overset{\displaystyle O}{C}-H$$

Chloral

(g)

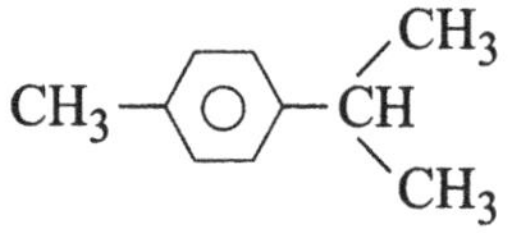

CH_3

$CH_3-\bigcirc-CH$

CH_3

p-Cymene

(h) 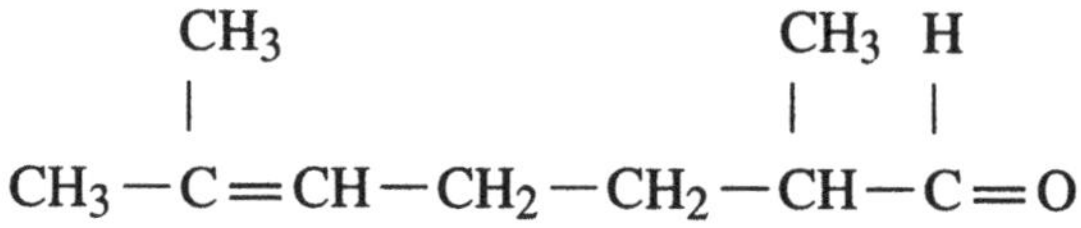 $CH_3-\overset{\displaystyle CH_3}{C}=CH-CH_2-CH_2-\overset{\displaystyle CH_3}{\underset{}{CH}}-\overset{\displaystyle H}{C}=O$

Citral (isolated from lemon grass oil)

(i)

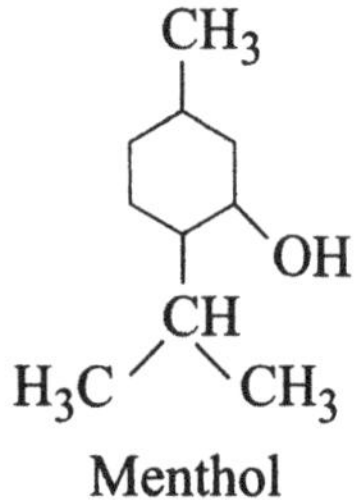

CH_3

OH

CH

H_3C CH_3

Menthol

(j) 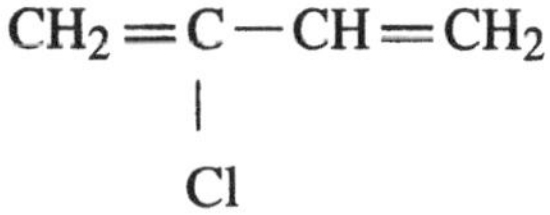 $CH_2=\overset{}{\underset{\displaystyle Cl}{C}}-CH=CH_2$

Chloroprene used in the manufacture
of synthetic rubber)

(k)

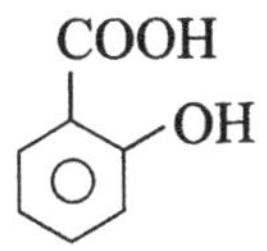

COOH

OH

Salicylic acid (used in the
manufacture of aspirin)

(l)

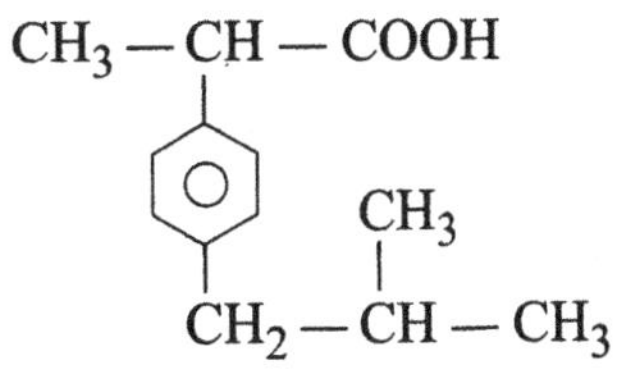

$CH_3 - CH - COOH$

CH_3

$CH_2 - CH - CH_3$

Ibuprofen (drug) (hint: name it as
a substituted propanoic acid)

(m)

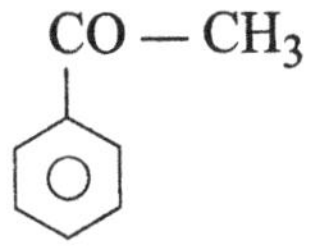

$CO - CH_3$

Acetophenone

4. Point out the errors in the following names.

(a) 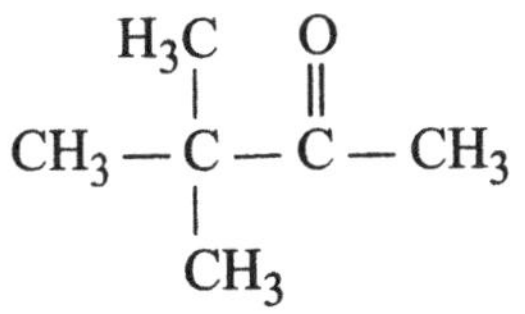 H_3C O

$CH_3 - C - C - CH_3$ 2, 2-Dimethyl-3-butanone

CH_3

(b)   2-Methyl propane (clue: a substituent and the parent should not be separated by space)
(c)   Methyl chloride (clue: no error, methyl is not a substituent)
(d)   Methylamine (no error)
(e)   N, N-Dimethyl aniline
(f)   Ethyl benzene
(g)   Ethylpropanoate
(h)   Ethylmethylketone
(i)   Diethylether

3 Alkanes

OBJECTIVES In this chapter, you will learn about,

- the sources of alkanes
- their physical properties and uses
- the methods of preparation
- the reactions of alkanes
- the mechanism of free radical chlorination of methane

3.1 INTRODUCTION

Alkane is the name given to the series of hydrocarbons, which are compounds containing only hydrogen and carbon, consisting of carbon chains connected by single bonds. They contain no double or triple bonds. They are saturated; this term is used to imply that they contain the maximum number of hydrogen atoms which can be attached to the carbon atoms of the chain. They conform to the molecular formula C_nH_{2n+2}. They are relatively unreactive, a property implied in the name *paraffin*. Alkanes (paraffins) are the major constituents of petroleum and natural gas. We have seen the nature of bonding in alkanes in Chapter 1 and their nomenclature in Chapter 2.

3.2 OCCURRENCE

Natural gas which is associated with petroleum deposits contains about 75% methane, along with ethane (about 10%) and propane (about 5%). The higher alkanes that are present in petroleum are mostly straight chain. Biogas which is formed by bacterial action on organic matter contains methane which is its combustible component. Many waxes of vegetable origin are alkanes containing 30 or more carbon atoms. Beeswax contains hentriacontane, $CH_3(CH_2)_{29}CH_3$. Alkanes are obtained on a commercial scale by the fractional distillation of petroleum.

3.3 PROPERTIES AND USES

The first few members of the series, C1 to C4, are gases at room temperature. They can be liquefied under pressure. LPG, liquefied petroleum gas, is mainly propane or butane.

The higher alkanes are liquids upto about C17, above which they are soft solids of waxy consistency. The boiling points of some alkanes are listed in Table 3.1. Two generalisations can be made from this.

- As the molecular weight increases, the boiling point increases.
- For molecules of the same molecular weight (compare the two C4 isomers, the three C5 isomers, and the three C8 isomers), straight chain alkanes are higher boiling than their branched isomers. This effect is more pronounced when the degree of branching is more.

Table 3. 1   Boiling points of some alkanes (°C)

Alkane	Boiling point	Alkane	Boiling point
methane	−160	2-methylbutane	28
ethane	−89	2,2-dimethylpropane	9
propane	−42	hexane	69
butane	0	octane	126
2-methylpropane	−9	2-methylheptane	116
pentane	36	2,2,3,3-tetramethylbutane	106

A simplified explanation of this phenomenon is that straight chain molecules are more thread-like and highly branched ones are more ball-like. The former provide for better contact between molecules causing them to 'stick together' better, due to short range intermolecular forces compared to the latter. In order to cause the liquid to boil, enough energy has to be supplied to disrupt the intermolecular forces and make the molecules escape into the gas phase.

3.3.1 POLARITY

Since alkanes contain no hetero atoms or polarisable groups such as double bonds, they are nonpolar. They readily mix with other hydrocarbons, but are immiscible with water and, to varying degrees, with other polar solvents. Their density (0.6–0.8 g ml^{-1}) is less than that of water. Thus, in a hydrocarbon–water mixture, hydrocarbon forms the upper layer.

3.3.2 USES

Alkanes find use as fuels, as feedstock for petrochemicals and as industrial and laboratory solvents. The petrochemical industries are those based on petroleum products. Alkenes and aromatics are the main raw materials, which in turn are obtained from alkanes.

The petroleum industry is a hydrocarbon industry. The components of petroleum, mainly alkanes, are crudely separated by fractionation, and the fractions are used as such or after modification, as industrial, automobile and domestic fuels and have various industrial applications.

Over-dependence on petroleum based fuels has associated risks. Petroleum, being a non-renewable fossil fuel, will eventually get exhausted and create an energy crisis. Hydrocarbon burning is also leading to severe environmental hazards. When plant and animal material is burnt, the CO_2 released is the same CO_2 which was taken from the atmosphere during photosynthesis. This does not disturb the carbon dioxide balance. Such is not the case with fossil fuels. When they burn, the CO_2 released increases the CO_2 content of the atmosphere and leads to increased green house effect and global warming.

3.4 PREPARATION OF ALKANES

3.4.1 FROM ALKYL HALIDES

Alkyl halides, RX, can be converted to alkanes, RH, by reduction, in which the X is replaced by H (Eq. 3.1) or by a coupling reaction (Eq. 3.2). These are illustrated below.

$$RX + (H) \rightarrow RH + (X) \tag{3.1}$$

$$2\,RX + reagent \rightarrow R - R + (2X\ reagent) \tag{3.2}$$

3.4.1.1 Reduction of Alkyl Halide by Catalytic Hydrogenolysis

Hydrogenolysis is the term used for the cleavage of a bond by hydrogen (cf. hydrolysis for cleavage by water, solvolysis for cleavage by solvent). This is done under catalytic hydrogenation conditions (see Chapter 4 for catalytic hydrogenation). Hydrogen reacts with alkyl halides in the presence of hydrogenation catalysts such as Pt or Pd as in Eq. 3.3, to give the alkane.

$$CH_3(CH_2)_6CH_2Cl + H_2 \rightarrow CH_3(CH_2)_6CH_3 + HCl \tag{3.3}$$

A variety of chemical reducing agents can also bring about such hyrogenolysis.

3.4.1.2 Reduction of Alkyl Halides by Magnesium Via Grignard Reagent

Alkyl halides, RX, react with magnesium in dry ether to form alkylmagnesium halides, called *Grignard Reagents* (Eq. 3.4).

Grignard reagents are named after the Nobel prize (1912) winning French chemist **Victor Grignard** (1871–1935), who discovered this reaction.

$$RX + Mg \rightarrow RMgX \tag{3.4}$$

When hydrolysed with water, they give the alkane, RH (Eq. 3.5).

$$RMgX + H_2O \rightarrow RH + Mg(OH)X \tag{3.5}$$

Note that in the nomenclature of the Grignard reagent, *alkylmagnesium* is written as one word. The hydrolysis can be understood in terms of the polarity of the $R-Mg$ bond, where the C of the alkyl group is partially negative and the more electropositive metal is partially positive. In water, or any other protic solvent, the alkyl groups get protonated. This is in contrast to the more familiar hydrolysis of alkyl halides, where the polarity of the C—Cl bond is such that the C is partially positive and reacts with the nucleophile, the hydroxy group.

3.4.1.3 Wurtz Reaction

RX reacts with certain metals like sodium in a *coupling reaction* to give the alkane corresponding to the joining of two alkyl groups (Eq. 3.6).

> This reaction is called the Wurtz Coupling, after the discoverer **Charles-Adolphe Wurtz** (1817–84), a French Chemist.

$$2RX + Na \rightarrow R-R + 2NaX \qquad (3.6)$$

Wurtz coupling of 1-bromobutane gives octane as the product. If an equimolar mixture of two different alkanes, say, propyl bromide and butyl bromide is used (hoping to get heptane), the result is a mixture of alkanes—hexane, heptane and octane.

3.4.2 FROM CARBOXYLIC ACIDS

Carboxylic acids, RCOOH, can be converted to the alkanes R—R, by *decarboxylative coupling* using electrolysis. When the salt of a carboxylic acid is electrolysed, the hydrocarbon, R—R is formed at the anode.

> This reaction is called *Kolbe electrolysis*, after Herman Kolbe (1818–84), a German Chemist.

$$RCOO^- \rightarrow R^\bullet + CO_2 + e^- \text{ (anode)} \qquad (3.7)$$

$$2R^\bullet \rightarrow R-R \qquad (3.8)$$

3.4.3 FROM ALKENES AND ALKYNES BY CATALYTIC HYDROGENATION

Hydrogen can be added to alkenes and alkynes in the presence of catalysts (Chapter 4, Section 4.4.4). Typical hydrogenation catalysts are based on transition metals such as Pt, Pd and Ni. Since alkenes themselves can be obtained from alkyl halides by dehydrohalogenation and from alcohols by dehydration (see Chapter 4 for both reactions), this is an indirect method of converting alcohols and alkyl halides to alkanes.

3.5 REACTIONS OF ALKANES

3.5.1 INTRODUCTION

Alkanes are relatively unreactive molecules. However they do react under appropriate conditions. Upon heating to more than 500°C or at lower temperatures in the presence of catalysts, they break down into smaller fragments, in a reaction called *cracking*. Alkanes undergo dehydrogenation in the presence of appropriate catalysts to form alkenes and polyenes. Cracking and dehydrogenation are important reactions in the petroleum industry and are discussed in Sections 3.5.5 and 3.5.4. Another reaction of great economic importance in today's world is combustion, which is the major source of energy for various human activities.

3.5.2 CHLORINATION

A reaction of relevance in the laboratory and also in the industry is *halogenation*.

3.5.2.1 Free Radical Substitution Reactions

Chlorine reacts with alkanes, in a *substitution reaction*. Reaction with methane results in successive replacement of hydrogens as in Eqs. 3.9a–3.9d.

$$CH_4 + Cl_2 \rightarrow CH_3Cl + HCl \qquad (a)$$

$$CH_3Cl + Cl_2 \rightarrow CH_2Cl_2 + HCl \qquad (b)$$

$$CH_2Cl_2 + Cl_2 \rightarrow CHCl_3 + HCl \qquad (c)$$

$$CHCl_3 + Cl_2 \rightarrow CCl_4 + HCl \qquad (d) \qquad\qquad (3.9)$$

The above reactions are correct from a stoichiometric point of view, but are more complex in mechanistic detail. Chlorine and other halogens react in two different ways. The difference lies in whether the Cl—Cl bond breaks homolytically or heterolytically. Homolytic dissociation gives chlorine atoms (free radicals) and heterolytic dissociation gives ions. Reactions involving the former type of dissociation are *free radical reactions*.

The latter type are *ionic* or *polar reactions*. We shall see an example of an ionic reaction of halogens, addition to alkenes, in Chapter 4. Chlorination of alkanes is a free radical reaction. This reaction takes place at relatively high temperatures (300–400°C) or under the influence of light (*photochemical reaction*). Use of high temperature or use of light is necessary to supply the required energy to bring about the dissociation of a chlorine molecule into chlorine atoms. In principle, the C—H and C—C bonds of alkanes can also dissociate homolytically, but the covalent bonds in halogens are weaker (see bond dissociation energies, Table 1.2, Chapter 1).

3.5.2.2 Mechanism of Chlorination of Methane

Under the influence of heat or light the following reactions take place.

$$Cl\!:\!Cl \xrightarrow{\Delta\ \text{or}\ h\nu} Cl^{\bullet} + {}^{\bullet}Cl \qquad\qquad (3.10)$$

$$Cl^{\bullet} + H\!:\!\overset{\displaystyle H}{\underset{\displaystyle H}{\overset{\bullet\bullet}{C}}}\!:\!H \longrightarrow Cl\!:\!H + {}^{\bullet}\overset{\displaystyle H}{\underset{\displaystyle H}{\overset{\bullet\bullet}{C}}}\!:\!H \qquad\qquad (3.11)$$

$$\text{Cl}:\text{Cl} + \cdot\overset{\displaystyle H}{\underset{\displaystyle H}{\ddot{\text{C}}}}:\text{H} \longrightarrow \text{Cl}^{\cdot} + \text{Cl}:\overset{\displaystyle H}{\underset{\displaystyle H}{\ddot{\text{C}}}}:\text{H} \tag{3.12}$$

Equation 3.10 represents the homolytic dissociation of a chlorine molecule. The symbols over the arrow, Δ and hν, are employed in chemistry to indicate heat and photochemical conditions respectively. The latter represents the energy of a photon. The chlorine atom is a reactive species, and in the reaction medium where it is generated, its chance of colliding with a methane molecule is high. Upon such a collision, it abstracts a hydrogen atom from methane, leaving behind a methyl radical, as in Eq. 3.11. The methyl radical so generated collides with other molecules present in the medium. The 'other molecules' referred to here are chlorine and methane. Collision with a chlorine molecule results in a reaction as in Eq. 3.12. The new chlorine atom produced in this step reacts with another methane molecule, a repetition of step 3.11. This kind of repetitive reaction sequence is called a *chain reaction*. It is obvious that all the chlorine molecules that we started with need not be dissociated to chlorine atoms. A few are enough. Each chlorine atom so generated, starts a chain which will use up a large number of methane and chlorine molecules (in principle, all the available methane and chlorine), till the chain stops (*terminates*). In fact, there are many ways in which such a chain can terminate, some of which are shown in Eqs. 3.13, 3.14 and 3.15.

$$\text{Cl}^{\cdot} + \cdot\overset{\displaystyle H}{\underset{\displaystyle H}{\ddot{\text{C}}}}:\text{H} \longrightarrow \text{Cl}:\overset{\displaystyle H}{\underset{\displaystyle H}{\ddot{\text{C}}}}:\text{H} \tag{3.13}$$

$$\text{Cl}^{\cdot} + \text{Cl}^{\cdot} \longrightarrow \text{Cl}:\text{Cl} \tag{3.14}$$

$$^{\cdot}\text{CH}_3 + {}^{\cdot}\text{CH}_3 \longrightarrow \text{H}_3\text{C}:\text{CH}_3 \tag{3.15}$$

All these termination steps are radical couplings, where two odd electron species (free radicals or atoms) join together to form non-radical products. Step 3.13 needs special mention. The product of this reaction is the desired final product, chloromethane. However, only a very small amount of the total chloromethane obtained can be formed by such a coupling. Radical coupling, though a facile reaction, requires two radicals to collide with each other. Concentration of the radicals in a medium of this type, is quite small and the chances of two of them encountering each other are slim. Each radical has a better chance of encountering one or the other of the abundantly available molecules of the starting materials, methane or chlorine.

Equation 3.10, is called the *initiating step*. Reactions represented by Eqs. 3.11 and 3.12 are called the *chain propagating steps*. Equations 3.13 to 3.15 represent the *chain terminating steps*. These three components are present in all free radical chain reactions.

We have seen that the first reaction of chain propagation (reaction of a chlorine atom with a methane molecule, Eq. 3.11), takes place readily because the probability of the chlorine atom colliding with a methane molecule is high. This is true only in the early stages of the reaction. As the reaction progresses, the concentration of chloromethane in the medium increases and the chances of the chlorine atom colliding with chloromethane molecules increase. This can lead to other reactions, Eqs. 3.16 and 3.17, and the product dichloromethane.

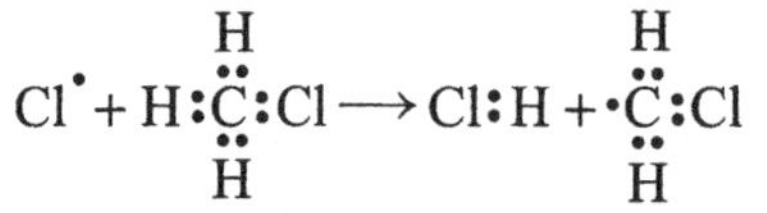

$$Cl^• + H:\overset{\overset{\displaystyle H}{\cdot\cdot}}{\underset{\underset{\displaystyle H}{\cdot\cdot}}{C}}:Cl \longrightarrow Cl:H + •\overset{\overset{\displaystyle H}{}}{\underset{\underset{\displaystyle H}{}}{C}}:Cl \tag{3.16}$$

$$Cl:Cl + •\overset{\overset{\displaystyle H}{\cdot\cdot}}{\underset{\underset{\displaystyle H}{\cdot\cdot}}{C}}:Cl \longrightarrow Cl^• + Cl:\overset{\overset{\displaystyle H}{\cdot\cdot}}{\underset{\underset{\displaystyle H}{\cdot\cdot}}{C}}:Cl \tag{3.17}$$

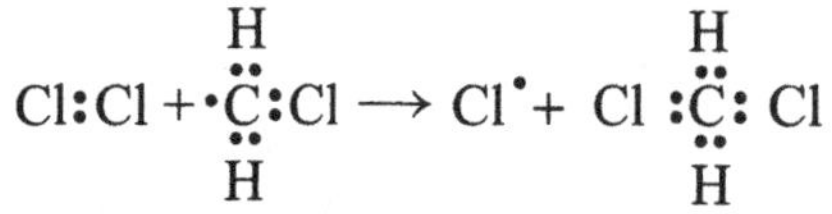

As a result of further such reactions, dichloromethane leads to trichloromethane and then to tetrachloromethane, as in Eq. 3.9.

3.5.3 HIGHER ALKANES

In methane, all the four hydrogens are equivalent. When chlorine substitutes one of the hydrogens, chloromethane is formed. It has no isomers. So also, chloroethane has no isomers. The situation is different with propane and the higher alkanes. There can be more than one *monochloroalkane*. In butane, there are four secondary hydrogens (highlighted in the structure below) and six primary hydrogens.

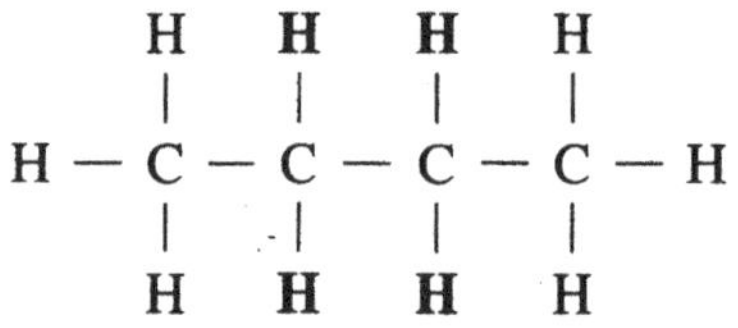

$$
\begin{array}{ccccc}
\mathbf{H} & \mathbf{H} & \mathbf{H} & H & \\
| & | & | & | & \\
H - C - C - C - C - H \\
| & | & | & | & \\
H & \mathbf{H} & \mathbf{H} & H &
\end{array}
$$

Two isomers of chlorobutane, 1-chlorobutane and 2-chlorobutane respectively, are possible, corresponding to the abstraction of the primary (Eqs. 3.18a and 3.18b) or the secondary (Eqs. 3.19a and 3.19b) hydrogens by the chlorine atom.

$$CH_3 - CH_2 - CH_2 - CH_3 + Cl^• \rightarrow {}^•CH_2 - CH_2 - CH_2 - CH_3 + HCl \tag{a}$$

$$^•CH_2 - CH_2 - CH_2 - CH_3 + Cl_2 \rightarrow Cl - CH_2 - CH_2 - CH_2 - CH_3 + Cl^• \tag{b}$$

$$\tag{3.18}$$

$$CH_3 - CH_2 - CH_2 - CH_3 + Cl^• \rightarrow CH_3 - {}^•CH - CH_2 - CH_3 + HCl \tag{a}$$

$$CH_3 - {}^•CH - CH_2 - CH_3 + Cl - Cl \rightarrow CH_3 - CHCl - CH_2 - CH_3 + Cl^• \tag{b}$$

$$\tag{3.19}$$

Statistically, 1-chlorobutane and 2-chlorobutane should have been formed in the ratio, 3:2. In actual experiments, in the photochemical chlorination of butane at 25°C, they are obtained in the approximate ratio 1:3. It should be noted that monochlorobutanes are not the only products. Dichloro-, trichloro- and other polychlorobutanes are also formed, and more of those are formed as the reaction progresses.

If the ratio of 1-clorobutane and 2-chlorobutane was 3:2, we would have said that the substitution takes place randomly and the reaction is not *selective*. Actually the reaction is selective, one particular isomer being selectively formed. This kind of selectivity, where the reaction takes place selectively at a particular position in a molecule, when more than one position is available for reaction, is called *regioselectivity*. The selectivity in the chlorination is due to the difference in the rates of the two competing hydrogen abstraction reactions (Eqs. 3.18a and 3.19a). This in turn, is due to the difference in the stability of the primary free radical formed in the former reaction and the secondary radical formed in the latter. As we have already seen in Chapter 2, secondary free radicals are more stable than primary free radicals.

In the case of 2-methylpropane, there are 9 primary hydrogens and only one tertiary hydrogen. The observed rate of formation (and hence, the yield) of 1-chloro-2-methylpropane (isobutyl chloride) and that of 2-chloro-2-methylpropane (*tert*-butyl chloride) in photochemical chlorination at 25°C is about 1:5 as against the statistically expected ratio of 9:1.

Bromine, which is less reactive than chlorine reacts with alkanes with even greater selectivity. In the photochemical bromination of 2-methylpropane, *tert*-butyl bromide is formed almost exclusively with only a negligible quantity of the other isomer.

3.5.4 DEHYDROGENATION

We have seen how alkanes are obtained by the hydrogenation of alkenes. The reverse of this reaction is dehydrogenation, which involves the elimination of a molecule of hydrogen. This can be achieved with the help of a catalyst, usually a transition metal like Pt, Pd or Ni, which are also hydrogenation catalysts. As we know, a catalyst increases the rate of both reverse and forward reactions, without affecting the equilibrium composition. By manipulating the conditions (temperature, concentration of reactants, removal of a product as and when formed), one can disturb the equilibrium and drive the reaction in the forward or reverse direction. Thus, under appropriate conditions, alkanes can be dehydrogenated to alkenes. Since there is no specific reactive centre in an alkane chain (as there is in the case of alkyl halides or alcohols) dehydrogenation tends to be random. Thus, when butane is passed over a catalyst based on nickel or chromium, different products are obtained (Eq. 3.20).

$$CH_3 - CH_2 - CH_2 - CH_3 \xrightarrow[-H_2]{\text{catalyst, }\Delta} \begin{array}{l} CH_2 = CH - CH_2 - CH_3 \\ CH_3 - CH = CH - CH_3 \\ \text{(butenes)} \end{array}$$

$$\Big\downarrow -H_2$$

$$CH_2 = CH - CH = CH_2$$
1,3-Butadiene

(3.20)

This is the method of manufacturing 1,3-butadiene which is a raw material for synthetic rubber. Aromatic hydrocarbons like benzene and toluene are present in petroleum, but can also be obtained by the dehydrogenation of the alkanes of petroleum (Eq. 3.21).

$$C_7H_{16} \xrightarrow[-H_2]{catalyst} \left. \begin{array}{l} CH_2=CH-C_5H_{11} \\ CH_3-CH=CH-C_4H_9 \\ C_2H_5-CH=CH-C_3H_7 \end{array} \right\} \xrightarrow{-nH_2} \begin{array}{l} Dienes \\ Trienes \end{array} \longrightarrow$$

C_7H_8

$$\underset{\text{(Heptane)}}{C_7H_{16}} \xrightarrow{-4H_2} \underset{\text{(Toluene)}}{C_7H_8}$$

(3.21)

Styrene, $C_6H_5-CH=CH_2$ (phenylethene) which is an important raw material for polymers and synthetic rubbers, is manufactured by the catalytic dehydrogenation of ethylbenzene (Eq. 3.22).

Ethylbenzene $\xrightarrow{-H_2}$ Styrene

(3.22)

3.5.5 CRACKING

Cracking in the context of petroleum refining, refers to the breakdown of long chain hydrocarbons (alkanes) into smaller molecules under the influence of heat, or heat and catalyst. This is done (i) to convert high boiling long chain hydrocarbons to low boiling shorter chain hydrocarbons and (ii) to obtain lower gaseous alkenes like ethene, propene and butenes which are the building blocks of a huge complex group of industries called the petrochemical industry.

All organic molecules including alkanes, when heated to high enough temperatures, acquire enough thermal energy to cause bond dissociation (homolytic). For this to happen in alkanes, temperatures should be above 500°C. A very simplified version of what happens then, is given in Eq. 3.23.

$$R-CH_2-CH_2-CH_2 \overset{\frown}{\vee} CH_2-CH_2-R'$$

$$\xrightarrow{\Delta} R-CH_2-CH_2-\overset{\bullet}{C}H_2 + \overset{\bullet}{C}H_2-CH_2-R'$$

(3.23)

Needless to say, the bond chosen for homolytic cleavage is purely random; any one of the bonds could have cleaved. Next, a typical reaction of free radicals favoured at such high temperatures occurs (Eq. 3.24).

$$R - CH_2 - CH_2 - CH_2 \xrightarrow{\Delta} R - \overset{\bullet}{C}H_2 + CH_2 = CH_2$$

$$R' - CH_2 - \overset{\bullet}{C}H_2 \xrightarrow{\Delta} \overset{\bullet}{R'} + CH_2 = CH_2 \tag{3.24}$$

$R - \overset{\bullet}{C}H_2$ and $\overset{\bullet}{R'}$ may continue to lose $CH_2 = CH_2$ in steps or they may become $R - CH_3$ and RH by picking up a H from another alkane. In this process, the small molecule obtained as the major product is the very important industrial raw material, ethene.

As seen, thermal cracking requires temperatures higher than 500°C. The use of catalysts can bring down this temperature to 400°C. The cracking catalysts usually used in industry are acidic solids such as silica–alumina and zeolites. The latter are aluminosilicates with 3-dimensional open framework structure – and relevant for our present purpose – having acidic nature. The acid catalyst generates a few initiating carbocations (Eq. 3.25).

$$CH_2 = CH - R + H^+ \text{ (catalyst)} \longrightarrow CH_3 - \overset{+}{C}H - R \tag{3.25}$$

The alkene (Eq. 3.25) initially present in the form of trace impurities in the alkane, becomes available in abundance as the reaction progresses. The initiating carbocation reacts with the alkane chain in a hydrogen exchange reaction (Eq. 3.26).

$$CH_3 - \overset{+}{C}H - R + CH_3 - \overset{\overset{\displaystyle H}{|}}{C}H - CH_2 - CH_2 \cdots CH_2 - R'$$

$$\longrightarrow CH_3 - \overset{\overset{\displaystyle H}{|}}{C}H - R + CH_3 - \overset{+}{C}H - CH_2 - CH_2 - CH_2 - R' \tag{3.26}$$

Any one of the hydrogens can be removed by the initiating cation in this step, so long as it is from a CH_2 and not from the CH_3 at the end of chain. This restriction is imposed because of the relative stabilities of the carbocations. H abstraction from a CH_2 gives a secondary carbocation; abstraction from a CH_3 gives a primary carbocation The fates of two possible secondary carbocations are given in Eq. 3.27.

$$CH_3 - \overset{+}{C}H - CH_2 - CH_2 - CH_2 - R'$$

$$\longrightarrow CH_3 - CH = CH_2 + \overset{+}{C}H_2 - CH_2 - R'$$

$$CH_3 - CH_2 - CH_2 - CH_2 - \overset{+}{C}H - CH_2 - R''$$

$$\longrightarrow CH_3 - CH_2 - CH_2 - CH_2 - CH = CH_2 + \overset{+}{R''}$$

$$R' - CH_2 - \overset{+}{CH_2} \text{ (or } \overset{+}{R}\text{'')} + CH_3 - \overset{\overset{\displaystyle H}{|}}{CH} - CH_2 \text{----}$$

$$\longrightarrow R' - CH_2 - \overset{\overset{\displaystyle H}{|}}{CH_2} \text{ (or } R'' - H) + CH_3 - \overset{+}{CH} - CH_2 \text{----} \qquad (3.27)$$

Alkenes – other than ethene – are the major products of catalytic cracking, in addition to lower molecular weight alkanes. Catalytic cracking is the main industrial source of propene, a major petrochemical building block.

KEY POINTS

- The main source of alkanes is petroleum.
- Alkanes are nonpolar molecules.
- As in other families of compounds, the boiling point increases with increase in molecular weight.
- Among isomeric alkanes, straight chain ones are higher boiling than branched ones.
- Uses: fuel, feedstock in petrochemical industries, solvents.
- Methods of preparation:
    — from alkyl halides, by hydrogenolysis, Wurtz coupling
    — from alkyl halides, via Grignard reagents
    — from carboxylic acids by decarboxylative coupling (Kolbe electrolysis)
    — by catalytic hydrogenation of alkenes and alkynes.
- Reactions:
    — cracking and dehydrogenation are important processes in the petroleum industry
    — halogenation, mechanism of chlorination of methane—free radical chain reaction, selectivity in the halogenation of higher alkanes related to the stability of alkyl free radicals.

EXERCISES

SECTION I

1. Draw the structures of the lowest boiling and highest boiling alkanes of molecular formula (a) C_5H_{12}, (b) C_8H_{18}.
2. Suggest a method of preparing each of the following alkanes starting from appropriate starting materials of 3 carbons or less: (a) methane (b) ethane (c) propane (d) butane (e) hexane.

3. Suggest a method of preparing each of the following from the starting material indicated.

    (a) 2-methylpropane from an alkene

    (b) 2,2,3,3-tetramethylbutane from a carboxylic acid containing less than eight carbon atoms

    (c) pentane from an alkyl halide

    (d) 2-methylbutane from an alkyne

    (e) *neo*-pentane from an alkyl halide

4. Write equations for the reactions that take place when,

    (a) 2-methylpropanoate is electrolysed in alkaline medium

    (b) 2-methylbutane reacts with bromine in the presence of light

    (c) butane is burned

    (d) hexylmagnesium bromide is treated with water

    (e) ethylmagnesium iodide is treated with acetic acid (hint: acetic acid is a source of protons)

    (f) 1-butene reacts with hydrogen in the presence of a platinum hydrogenation catalyst.

5. Identify the C_5H_{12} isomer which gives only one monochloro product upon chlorination.

SECTION II

1. Two isomeric alkyl halides, A and B having the molecular formula C_4H_9Cl, upon treatment with magnesium gave two isomeric Grignard reagents, both of which upon hydrolysis gave the same alkane, C. This alkane could also be obtained by a Wurtz coupling reaction from D. Identify A, B, C and D.

2. How many isomeric monochloro products are possible in the chlorination of isobutane? Which will be the major product? Explain the reason for the selectivity.

3. Explain the terms initiation, propagation and termination in free radical chain reactions (use chlorination of alkane as an example).

CHALLENGING QUESTION

1. Suggest a method of preparing $CH_3 — CHD — CH_3$. (This is propane with one deuterium instead of a hydrogen, specifically at the 2-position. Suggestion: You may use 2-bromopropane and D_2O.)

ASSIGNMENT

Prepare a report, collecting data from books and the internet, on *Fossil Fuels: Use and Misuse.*

4 Alkenes

OBJECTIVES   In this chapter, you will learn about,

- alkenes
- preparation of alkenes using a variety of reactions
- properties of alkenes
- prediction of their reaction products with different reagents

4.1   WHAT ARE ALKENES?

Alkenes, also called *olefins*, are hydrocarbons containing carbon–carbon double bonds. Compared to alkanes (C_nH_{2n+2}), the alkenes (C_nH_{2n}) have fewer number of hydrogens for the number of carbons present. In this sense they are *unsaturated* compounds—unsaturated with respect to the number of hydrogens that can be present. There is another class of compounds called the *cycloalkanes*, which also have the general formula C_nH_{2n}, but these do not contain double bonds, and are not considered unsaturated. *Cycloalkenes* can contain double bonds as in *cyclohexene*. Such cycloalkenes behave in the same way as alkenes. *Aromatic* compounds like benzene which do contain double bonds, are not considered unsaturated.

$$CH_3 - CH = CH_2 \qquad\qquad CH_3 - CH_2 - CH_3$$

C_3H_6 Propene (alkene) C_3H_8 Propane (alkane)

C_6H_{10} Cyclohexene (cycloalkene) C_6H_{12} Cyclohexane (cycloalkane)

C_6H_6 Benzene (aromatic)

4.2 PREPARATION OF ALKENES

A distinction is often made between laboratory methods of preparation of organic compounds and industrial preparation of organic compounds, even though the chemistry is the same. The main difference lies in the economics. The emphasis in industry is on cost (and time) saving, even though elaborate equipment and process conditions may be required which may not be readily obtained in the laboratory. However in those industries which need specialised chemicals (which may be required in small quantities) meant for end products which are of high value (low volume high value chemicals), laboratory methods are directly applicable. Prominent among such industries are the pharmaceutical industries.

4.2.1 ELIMINATION REACTIONS

Elimination reactions are discussed in detail in Chapter 12.

Consider the following reactions,

$$R - \underset{\underset{H}{|}}{CH} - \underset{\underset{H}{|}}{CH} - R' \quad \xrightarrow{\text{dehydrogenation}} \quad R - CH = CH - R' + H_2 \tag{4.1}$$

$$R - \underset{\underset{H}{|}}{CH} - \underset{\underset{Cl}{|}}{CH} - R' \quad \xrightarrow{\text{dehydrochlorination}} \quad R - CH = CH - R' + HCl \tag{4.2}$$

$$R - \underset{\underset{Br}{|}}{CH} - \underset{\underset{Br}{|}}{CH} - R' \quad \xrightarrow{\text{debromination}} \quad R - CH = CH - R' + Br_2 \tag{4.3}$$

$$R - \underset{\underset{H}{|}}{CH} - \underset{\underset{OH}{|}}{CH} - R' \quad \xrightarrow{\text{dehydration}} \quad R - CH = CH - R' + H_2O \tag{4.4}$$

In Eqs. 4.1 to 4.4, a molecule of H_2, HCl, Br_2 or H_2O respectively, have been removed or eliminated from a parent saturated molecule, to give an alkene. The atoms or groups that have been eliminated were on *vicinal* or adjacent carbon atoms. Taking one of the atoms or groups as reference, the elimination is called a 1,2-or α, β-elimination. The latter term is abbreviated to β-*elimination*.

$$\overset{2}{R} - \underset{\underset{H}{|}}{CH} - \overset{1}{\underset{\underset{Cl}{|}}{CH}} - R \qquad\qquad \overset{\beta}{R} - \underset{\underset{H}{|}}{CH} - \overset{\alpha}{\underset{\underset{Cl}{|}}{CH}} - R$$

β-eliminations constitute the most important group of reactions for the preparation of alkenes. We shall discuss Eqs. 4.2 to 4.4 now and take up Eq. 4.1 under industrial methods of preparation of alkenes.

4.2.1.1 Dehydrohalogenation

Alkyl halides on treatment with bases give alkenes. The usual procedure is to dissolve the alkyl halide in a solvent like ethanol along with alkali, usually KOH or NaOH, and heat.

$$CH_3-CH_2-CH_2Br + \xrightarrow{OH^-/ethanol, \Delta} CH_3-CH=CH_2 + H_2O + Br^- \qquad (4.5)$$

$$CH_3-\underset{\underset{Cl}{|}}{CH}-CH_2-CH_3 \xrightarrow{OH^-/ethanol, \Delta} CH_2=CH-CH_2-CH_3+$$

$$CH_3-CH=CH-CH_3 + H_2O + Cl^-$$

$$(4.6)$$

In the equations, the inscription over the arrow represents alkali (one of the reagents), ethanol (solvent), and the symbol Δ (delta) represents heating. In the molecule used in Eq. 4.5, 1-bromopropane, only one type of β-hydrogen is available, those on C2, and only one alkene can be formed. In Eq. 4.6, 2-chlorobutane, two types of β-hydrogens are present, one on C1 and the other on C3, and β-elimination of either one is possible. Two alkenes are expected, 1-butene and 2-butene, and both are formed. Usually in such reactions there is some degree of *selectivity*, namely, one of the alkenes is formed in greater quantity than the other. In this case, 2-butene is the major product. This is not what will be expected if the removal of the β-hydrogen is random. There are 3 hydrogens on C1 and 2 hydrogens on C2. If the hydrogens were removed in a random or in a statistical manner, 1-butene and 2-butene should have been formed in a 3:2 ratio, which is not the case. The preference for 2-butene is due to the fact that it is formed at a faster rate which is related to its greater stability (see Sections 1.4.3.5 and 4.4.4.1 for discussions on relative stabilities. Also see section 12.3.1 for a detailed discussion on regioselectivity of elimination).

In Eq. 4.6, two alkenes, 1-butene and 2-butene are shown as products. Actually the situation is a little more complex: three alkenes are formed. What is shown as 2-butene is a mixture of two *geometrical isomers*, *cis*-2-butene and *trans*-2-butene.

$$CH_3-\underset{\underset{Cl}{|}}{CH}-CH_2-CH_3 \xrightarrow{OH^-/ethanol, \Delta} CH_2=CH-CH_2-CH_3 \text{ (1-butene)}$$

$$(4.7)$$

          (*cis*-2-butene)   (*trans*-2-butene)

The relative abundance of the three alkenes is as expected from relative stabilities, namely, *trans*-2-butene > *cis*-2-butene > 1-butene (see Section 4.4.4).

It should be noticed that the two atoms or groups eliminated are not removed 'together'. Representations such as the following are not correct.

$$-\overset{|}{\underset{H}{C}}-\overset{|}{\underset{Cl}{C}}- \longrightarrow \overset{\diagdown}{\diagup}C=C\overset{\diagup}{\diagdown} + HCl \tag{4.8}$$

The proton of the β-carbon is removed by the OH^- ion to form water. A base like hydroxide or at least a *nucleophilic solvent* like H_2O is essential for this reaction. The reaction can be logically deduced using simple concepts that we are already familiar with. *Polarity of bonds* has an important role in such deductions.

In an alkane, $R-CH_2-CH_2-CH_2-CH_2-R$, the various CH_2 groups are not very different from each other and the $C-H$ bonds are more or less of the same strength. The $C-H$ bonds are non polar or nearly so, meaning that the electron density of the σ-bond is uniform between the two nuclei. In an alkyl halide, the situation is different.

$$R-CH_2-\overset{\overset{\displaystyle H}{|}}{\underset{\underset{\displaystyle Cl}{\curlyvee}}{C}}-R'$$

Because of the greater electronegativity of Cl compared to C, the $C-Cl$ bond is 'polar'. The bonding electrons between the C and Cl are displaced towards the Cl. In the extreme case this will appear as an ionic bond.

$$-\overset{|}{\underset{|}{C}}-\ddot{\underset{\cdot\cdot}{Cl}}\!: \longleftrightarrow -\overset{|}{\underset{|}{C}}{}^+ :\!\ddot{\underset{\cdot\cdot}{Cl}}\!:^- \tag{4.9}$$

In reality, there is only a partial charge separation and the covalent bond is intact. This situation is represented using the conventions of *resonance* as in Eq. 4.9. The hybrid of these two *canonical structures* is usually represented using the following symbolism.

$$-\overset{|}{\underset{|}{C}}{}^{\delta+}\!-Cl^{\delta-} \tag{4.10}$$

What is implied is that the carbon has a partial positive charge and the chlorine bears a partial negative charge. In those cases where the carbon can accommodate a positive charge comfortably – as in the case of *tert*-alkyl halides – actual ionisation can take place, albeit to a very amall extent, in solvents of high dielectric constant.

$$CH_3-\underset{\underset{\displaystyle CH_3}{|}}{\overset{\overset{\displaystyle CH_3}{|}}{C}}-Cl \;\;\underset{}{\overset{H_2O}{\rightleftharpoons}}\;\; CH_3-\underset{\underset{\displaystyle CH_3}{|}}{\overset{\overset{\displaystyle CH_3}{|}}{C^+}} \;\; + Cl^- \qquad (4.11)$$

The equilibrium is very much towards the left. The driving force is the stabilisation of the *carbocation* by *hyperconugation* and by *inductive effect*. In the case of secondary and, more so, primary alkyl halides, actual ionisation does not take place, only a partial charge separation occurs, as in (Eq. 4.10). In either case, whether in Eq. 4.10 or Eq. 4.11, the positive charge on the α-carbon induces a polarity in the C_β-H bond, making this hydrogen susceptible to abstraction by a base. In other words, the β-H is acidic enough to be removed by a base.

$$HO\!:^- \;+\; \overset{\overset{\displaystyle H}{|}}{CH_2}\!-\!\overset{+}{\underset{\underset{\displaystyle CH_3}{|}}{C}}\!-\!CH_3 \;\;\xrightarrow[\text{OH}^-\text{ or H}_2\text{O}]{\text{base, }\Delta}\;\; CH_2\!=\!\underset{\underset{\displaystyle CH_3}{|}}{C}\!-\!CH_3 + H_2O \qquad (4.12)$$

$$HO\!:^- \;+\; \overset{\overset{\displaystyle H}{|}}{CH_2}\!-\!\overset{\overset{\displaystyle H}{|}}{\underset{\underset{\displaystyle CH_3}{|}}{C^{\delta+}}}\!\!-\!Cl^{\delta-} \;\;\xrightarrow{\text{OH}^-,\,\Delta}\;\; CH_2\!=\!\underset{\underset{\displaystyle CH_3}{|}}{C}\!-\!H + H_2O + Cl^- \qquad (4.13)$$

Representation of the course of a reaction is referred to as the *mechanism* of the reaction. The curved arrow represents specific movement of a pair of electrons resulting in the formation or in the breaking of a bond heterolytically. As is evident, the driving force for the reaction is the polarity of the C—Cl bond. The group or atom which leaves in the reaction, carrying along with it the bonding pair of electrons —Cl in this case – is called the *leaving group*. It is the *leaving ability* of the leaving group that is crucial for the success of the reaction. A very useful rule of thumb for such situations is that *conjugate bases of strong acids are good leaving groups*. Halides – conjugate bases of the strong hydrohalic acids – are good leaving groups making dehydrohalogenations successful. Acetate, the conjugate base of the weak acid acetic acid, is a poor leaving group. Hydroxide, the conjugate base of the weak acid H_2O, is a poor leaving group. On the other hand H_2O, the conjugate base of the strong acid H_3O^+, is a good leaving group, as we shall see in our discussion of dehydration of alcohols.

4.2.1.2 Dehydration of Alcohols

In planning synthetic strategy for complex organic molecules, alcohols are important starting materials. Alkyl halides, which we discussed in the previous section, are often obtained from alcohols. Removal of the elements of H_2O, from an alcohol by β-elimination gives alkenes (Eq. 4.14). This is called dehydration.

$$-\overset{\displaystyle H}{\underset{\displaystyle OH}{\overset{|}{\underset{|}{C}}-\overset{|}{\underset{|}{C}}-}} \quad \xrightarrow{\text{–H, –OH}} \quad \overset{\diagdown}{\diagup}C=C\overset{\diagup}{\diagdown} \tag{4.14}$$

Using the terminology developed in the previous section, OH^- is the leaving group in this elimination. However, OH^- the conjugate base of the weak acid H_2O, is a poor leaving group. There have been many successful attempts to make the OH^- group a better leaving group by chemical modification. Converting an alcohol into an alkyl halide by suitable chemical conversions is one such methodology. There are other ways of doing this without completely changing the hydroxyl group. The simplest of these is to use an acid to activate the OH group. This is acid catalysed dehydration of alcohols.

$$CH_3-\underset{\underset{\displaystyle OH}{|}}{CH}-CH_3 \quad \xrightarrow{H^+,\,\Delta} \quad CH_2=CH-CH_3+H_2O \tag{4.15}$$

Heating is necessary in most cases. Thus 2-propanol when heated with sulphuric acid gives propene. Ethanol when heated with sulphuric acid up to 240°C gives ethene. At lower temperature (160°C), ethanol gives diethyl ether.

$$C_2H_5OH \xrightarrow{H_2SO_4,\,\Delta}
\begin{array}{l}
\xrightarrow{160°C} \quad C_2H_5-O-C_2H_5+H_2O \\
\qquad\qquad\quad \text{Diethylether} \\
\xrightarrow{240°C} \quad CH_2=CH_2+H_2O
\end{array} \tag{4.16}$$

Other acids, notably phosphoric acid, can also be used. Hydrohalic acids are not suitable because of a competing reaction which is the formation of the alkyl halide. Soild acids such as aluminium oxide also bring about dehydration at about 400°C. The first step in the dehydration reaction is the addition of a proton to the alcohol, to form an *oxonium ion*.

$$CH_3-\underset{\underset{\displaystyle OH}{|}}{CH}-CH_3+H^+ \quad \longrightarrow \quad CH_3-\underset{\underset{\displaystyle {}^+OH_2}{|}}{CH}-CH_3 \tag{4.17}$$

This is an acid base reaction and is similar to the addition of a proton to water to give the hydronium ion. In the cold, that is, when 2-propanol is treated with concentrated sulphuric acid, the former dissolves in the latter, forming the oxonium ion. Other oxygen containing compounds, like ethers, also behave similarly, that is, dissolve in sulphuric acid forming the oxonium ion. In the case of alcohols, the oxonium ion dissociates when heated.

$$CH_3-CH-CH_3 \xrightarrow{\Delta} CH_3-\overset{+}{C}H-CH_3 + H_2O$$
$$\underset{|}{\overset{|}{+OH_2}}$$

(4.18)

The OH_2 has behaved as a good leaving group. Thus, while the simple $C-OH$ bond in the alcohol will not undergo heterolysis readily, the $C-^+OH_2$ bond does, since H_2O is the conjugate base of the strong acid, H_3O^+. If the alcohol is heated with a base, as in the case of the dehydrohalogenation of alkyl halides, the expected dehydration will not happen.

$$CH_3-CH-CH_3 \xrightarrow{OH^-,\Delta} \text{No reaction}$$
$$\underset{OH}{|}$$

(4.19)

In the above representation, 'no reaction' is not strictly correct because the OH^- can and does abstract a proton from the alcoholic OH group, but this does not lead to any net reaction (Eq. 4.20).

$$CH_3-CH-CH_3 + OH^- \rightleftharpoons CH_3-CH-CH_3 + H_2O$$
$$\underset{OH}{|} \qquad\qquad \underset{O^-}{|}$$

(4.20)

Simple answers for the failure of the reaction (Eq. 4.19) are that OH^- is a poor leaving group or that the $C-OH$ bond is not sufficiently polar. Introduction of a positive charge on the oxygen (Eqs. 4.17 and 4.18) did the trick. Instead of a proton, a Lewis acid could also be used. In fact, the use of the solid acid alumina (Al_2O_3) is believed to belong to this category (active sites on the alumina surface are Lewis acids, shown as L in Eq. 4.21).

$$CH_3-CH-CH_3 \atop \underset{\overset{..}{O}H}{|} +L \rightleftharpoons CH_3-CH-CH_3 \atop \underset{\overset{+}{O}-H}{|} \atop \underset{L}{|} \xrightarrow{\Delta} CH_3-\overset{+}{C}H-CH_3 + LOH$$

(4.21)

Once the carbocation is formed from the alcohol, deprotonation (Eq. 4.23) is an option. In fact it is the preferred option at higher temperatures. Reaction with a *nucleophile* is the other option and this takes place at lower temperatures (Eq. 4.22).

$$CH_3-\overset{+}{C}H-CH_3 \underset{HSO_4^-}{\overset{HSO_4^-}{\rightleftarrows}} CH_3-CH-CH_3 \atop \overset{|}{OSO_3H}$$

(4.22)

$$\underset{HSO_4^-}{\rightleftarrows} CH_2=CH-CH_3 + H_2SO_4$$

(4.23)

Higher temperature favours the equilibrium leading to an alkene. Some points of difference between dehydration and dehydrohalogenation are as follows.

(i) Dehydrohalogenation usually requires a strong base like OH^- for the removal of the proton from the β-carbon, except from *tert*-alkyl halides where carbocations are intermediates. Dehydration requires acidic conditions and a strong base is incompatible under such conditions. In dehydration as generally depicted, the proton is removed by a water molecule, the conjugate base of an acid or any other nucleophile available in the medium.

(ii) In dehydrohalogenation, carbocations are involved as intermediates only in those cases where they are stable, as with *tert*-alkyl halides (Eq. 4.12). In other cases the partial positive charge on the carbon atom (Eq. 4.13) is enough to trigger the reaction. Acid catalysed dehydration reaction on the other hand involves a carbocation intermediate.

(iii) Dehydration usually requires heating. This is because all the steps in dehydration are reversible and heating helps to drive the reaction in the forward direction.

$$
\begin{array}{ccc}
\overset{\displaystyle H}{\underset{\displaystyle H}{-\overset{|}{C}-}}\overset{\displaystyle }{\underset{\displaystyle OH}{\overset{|}{C}-}} + H^+ \rightleftharpoons & \overset{\displaystyle H}{\underset{\displaystyle H}{-\overset{|}{C}-}}\overset{\displaystyle }{\underset{\displaystyle {}^+OH_2}{\overset{|}{C}-}} \rightleftharpoons & \overset{\displaystyle H}{\underset{\displaystyle }{-\overset{|}{C}-}}\overset{\displaystyle }{\underset{\displaystyle +}{\overset{|}{C}-}} \rightleftharpoons \diagdown C=C\diagup \\
& & + H_3O^+
\end{array}
\qquad (4.24)
$$

Dehydrohalogenation is also assisted by heating mainly for kinetic reasons. The rate of the reaction increases with temperature. The steps in base-induced dehydrohalogenation (Eqs. 4.12 and 4.13) are not reversible. A perceptive student will notice the difference in terminology. Dehydrations are acid-catalysed; dehydrohalogenations are base-induced. In the mechanism for the dehydrations described above (Eq. 4.24) the proton is not consumed, the reaction is catalysed by the proton. On the other hand dehydrohalogenation is not base-catalysed. A molecule of base is consumed for every molecule of alkyl halide reacting. The reaction is base-induced.

4.2.1.3 Dehalogenation

Elimination of a molecule of halogen from a vicinal dihalide gives an alkene. This elimination reaction is called dehalogenation. Zinc dust or iodide ion can bring about the reaction. Usually *vic*-dihalides are prepared by the addition of halogens to the double bond in alkenes. Hence this method cannot be considered to be useful for the preparation of alkenes. Unlike in the case dehydrohalogenation and dehydration, this reaction can give only one alkene since the position of the double bond is necessarily between the carbons bearing the halogen atoms.

$$
\underset{\underset{\displaystyle Br}{|}}{\overset{\overset{\displaystyle H}{|}}{CH_3-C}} - \underset{\underset{\displaystyle H}{|}}{\overset{\overset{\displaystyle Br}{|}}{C}} - CH_3 + Zn \longrightarrow CH_3 - CH = CH - CH_3 + Zn\,Br_2
$$

$$\text{(structure: bromocyclohexane with Br, H, H, Br)} + \text{Zn} \longrightarrow \text{(cyclohexene)} + \text{Zn Br}_2 \tag{4.25}$$

$$CH_3 - \underset{\underset{Br}{|}}{\overset{\overset{Br}{|}}{C}H} - CH - CH_3 + I^- \longrightarrow CH_3 - CH = CH - CH_3 + IBr + Br^- \tag{4.26}$$

4.2.1.4   Other β-Eliminations

Elimination of HZ is possible in all molecules where Z is a good leaving group. We have already seen that conjugate bases of strong acids are good leaving groups. We have also seen how the hydroxyl group can be changed into a good leaving group by protonation. Reaction of an alcohol with a sulphonic acid derivative can convert the OH to a sulphonate.

$$\begin{array}{c} H \quad OH \\ | \quad\; | \\ -C-C- \\ | \quad\; | \\ H \end{array} + Cl-SO_2-\!\!\!\bigcirc\!\!\!-CH_3 \longrightarrow \begin{array}{c} H \quad O-SO_2-\!\!\!\bigcirc\!\!\!-CH_3 \\ | \quad\; | \\ -C-C- \\ | \quad\; | \\ H \end{array} \tag{4.27}$$

p-Toluenesulphonyl
chloride
(tosyl chloride)   *p*-Toluenesulphonate
(tosylate)

This ester, *p*-toluenesulphonate, or tosylate for short, can undergo base-induced elimination in the same way as a halide.

An amino group can be converted by quaternisation to give a derivative with a good leaving group.

$$R-CH_2-CH_2-NH_2 + 3CH_3I \longrightarrow R-CH_2-CH_2-\overset{\overset{\displaystyle CH_3}{|}}{\underset{\underset{\displaystyle CH_3}{|}}{N}}{}^{+}\!\!-CH_3I^- \tag{4.28}$$

Such quaternary ammonium salts undergo base induced elimination to yield alkenes.

$$HO^- \quad\; \quad \quad \quad \quad \quad$$
$$R-\overset{\overset{\displaystyle H}{|}}{\underset{\underset{\displaystyle H}{|}}{C}}-CH_2-\overset{+}{N}(CH_3)_3 \longrightarrow H_2O + R-CH=CH_2 + N(CH_3)_3 \tag{4.29}$$

This reaction, known as *Hofmann elimination,* is very useful not so much for the preparation of alkenes as for *deaminating*, that is, removing the amino group from an amine (Section 12.3.2).

4.2.1.5 Eliminations Involving Cyclic Transition States

We have seen some indirect ways of dehydrating alcohols where the hydroxyl group has been converted to another, better leaving group. An interesting variation is the conversion of the hydroxyl group to an organic ester like acetate.

$$
\underset{\substack{|\quad|\\ H\ \ OH}}{\overset{\substack{H\ H\\ |\quad|}}{R-C-C-H}} + \underset{\substack{\|\\ O}}{CH_3-C-OH} \xrightarrow{H^+,\ \Delta} \underset{\substack{|\quad|\\ H\ \ O-C-CH_3\\ \quad\ \|\\ \quad\ O}}{\overset{\substack{H\ H\\ |\quad|}}{R-C-C-H}} + H_2O \tag{4.30}
$$

Acetic acid   Acetate

The acetate (CH_3COO^-), is not a particularly good leaving group, being the conjugate base of the weak acid, acetic acid. However acetates on heating to about $500°C$ undergo an elimination reaction giving the alkene and acetic acid. This is called *ester pyrolysis*. This reaction has been thoroughly studied and its mechanism is fully understood. This is an example of a reaction progressing through a cyclic transition state.

$$
\begin{array}{c}
\underset{\substack{H\quad\ O\\ \diagdown\ /\\ O=C\\ \diagdown\\ CH_3}}{\overset{\substack{H\quad\quad H\\ \diagdown\quad\ /\\ R-C-C-H\\ /\quad\ \diagdown}}{}}
\end{array}
\xrightarrow{500°C}
\begin{array}{c}
\underset{\substack{H\quad\quad O\\ \diagdown\quad\ \diagup\!\!\diagdown\\ O-C\\ \diagdown\\ CH_3}}{\overset{\substack{H\quad\quad H\\ \diagdown\quad /\\ C=C\\ \diagup\quad\ \diagdown\\ R\quad\quad H}}{}}
\end{array}
\tag{4.31}
$$

As can be seen, this is indeed a β-elimination. No base is added. The β-hydrogen is removed as a proton by a portion of the leaving group itself acting as the base. Esters other than acetate also behave similarly. A particularly useful ester is the xanthate. Useful because it can be easily prepared and it undergoes decomposition at lower temperatures ($200°C$) than acetates.

Xanthates are prepared by the reaction of an alcohol with carbon disulphide and alkali, followed by methylation with methyl iodide.

$$
\underset{\substack{|\quad|\\ H\ \ OH}}{\overset{\substack{H\ H\\ |\quad|}}{R-C-C-H}} + CS_2 \xrightarrow{KOH} \underset{\substack{|\quad|\\ H\ \ O-C-S^-K^+\\ \quad\ \|\\ \quad\ S}}{\overset{\substack{H\ H\\ |\quad|}}{R-C-C-H}} \xrightarrow{CH_3I} \underset{\substack{|\quad|\\ H\ \ O-C-SCH_3\\ \quad\ \|\\ \quad\ O}}{\overset{\substack{H\ H\\ |\quad|}}{R-C-C-H}}
$$

$$
\tag{4.32}
$$

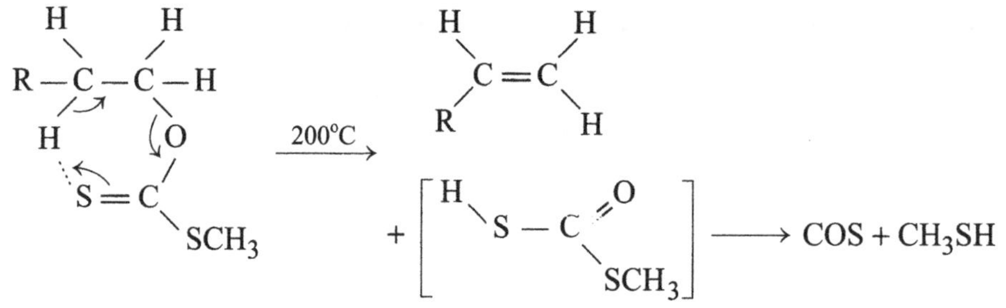

$$(4.33)$$

On pyrolysis at 200°C, they break down. The intermediate $CH_3-S-COSH$ (within square brackets) cannot be isolated, but undergoes further decomposition to COS and CH_3SH.

Xanthate pyrolysis is known as the **Chugaev reaction**. These pyrolytic reactions are indirect methods of dehydration of alcohols. As we shall see in the next section, direct dehydration of alcohols suffers from a serious drawback. Since they go through carbocation intermediates which are prone to rearrangement, the alkenes obtained are often isomers of the expected ones. Rearrangement is absent in ester pyrolysis because carbocations are not involved.

> Xanthate pyrolysis is named after Lev Chugaev (1873–1923), a Russian chemist.

4.2.1.6 Molecular Rearrangement during Dehydration

Acid-catalysed dehydration of alcohols involves intermediate formation of carbocations. (This and other molecular rearrangements are discussed in detail in Chapter 19.) The carbocations once formed, undergo one of the following reactions.

(i)   *Quenching* by combining with an anion or a neutral nueleophile.

(ii) Elimination of a positive entity, usually a proton to become a neutral product.

(iii) Rearrangement to a more stable carbocation.

If the possibility of rearrangement exists, reaction (iii) may happen before (i) or (ii). A carbocation is a high energy species and the three reactions described above are exothermic reactions and help to dissipate energy (and thus stabilise the molecule). In qualitative terms, the tendency is for an unstable (high energy) state to go to a more stable (lower energy) state. Rearrangement only partially remedies the situation since the product is still a carbocation though a more stable one. But the rearrangement which is an internal redistribution of atoms usually takes place at a faster rate than reactions (i) or (ii). These are illustrated in the dehydration of 3, 3-dimethyl-2-butanol (Eq. 4.34).

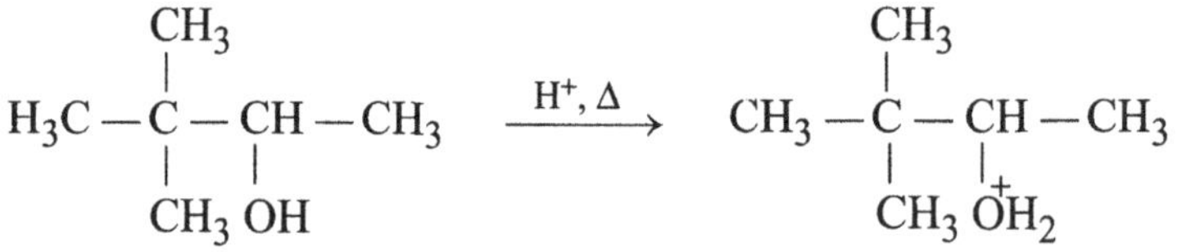

3, 3-Dimethyl-2-butanol

$$\longrightarrow \quad CH_3 - \overset{\overset{\displaystyle CH_3}{|}}{\underset{\underset{\displaystyle CH_3}{|}}{C}} - \overset{+}{C}H - CH_3 + H_2O$$

Secondary carbocation (less stable)

fast
rearrangement $\overset{-H^+}{\longrightarrow}$ $CH_3 - \overset{\overset{\displaystyle CH_3}{|}}{\underset{\underset{\displaystyle CH_3}{|}}{C}} - CH = CH_2$

3,3-Dimethyl-1-butene
unrearranged product

(4.34)

$$CH_3 - \overset{+}{\underset{\underset{\displaystyle CH_3}{|}}{C}} - \overset{\overset{}{}}{\underset{\underset{\displaystyle CH_3}{|}}{CH}} - CH_3$$

Tertiary carbocation
(more stable than secondary)

$\downarrow -H^+$

$$CH_3 - \overset{\overset{}{}}{\underset{\underset{\displaystyle CH_3}{|}}{C}} = \overset{\overset{}{}}{\underset{\underset{\displaystyle CH_3}{|}}{C}} - CH_3 \quad + \quad CH_2 = \overset{\overset{}{}}{\underset{\underset{\displaystyle CH_3}{|}}{C}} - \overset{\overset{}{}}{\underset{\underset{\displaystyle CH_3}{|}}{CH}} - CH_3$$

2,3-Dimethyl-2-butene
(major product)

2,3-Dimethyl-1-butene
(minor product)

When the alcohol, 3,3-dimethyl-2-butanol, is heated with sulphuric acid, the expected (unrearranged) alkene 3,3-dimethy-1-butene is formed only in very small quantity. The initially formed secondary carbocation rapidly rearranges to the more stable tertiary carbocation before deprotonation. This is known as the Wagner–Meerwein rearrangement.

The two alkenes 2,3-dimethyl-2-butene and 2,3 dimethyl-1-butene obtained by the deprotonation of the tertiary carbocation are the main products. Of these, the former is the major product compared to the latter. Statistically, by random deprotonotion, the former and the latter should have been in the ratio 1:6.

$$H \longleftarrow \textcircled{1H} \longrightarrow \text{2,3-Dimethyl-2-butene}$$

$$\textcircled{6H} \longrightarrow CH_3 - \overset{+}{\underset{\underset{\displaystyle CH_3}{|}}{C}} - \overset{\overset{}{}}{\underset{\underset{\displaystyle CH_3}{|}}{C}} - CH_3$$

2,3-Dimethyl-1-butene

The observed selectivity is because of the greater stability of 2,3-dimethyl-2-butene due to hyperconjugation. Such rearrangements are common in acid-catalysed dehydrations. Rearrangements are observed in other reactions also, including some dehydrohalogenations which involve carbocations. Some more examples of rearrangement are given in Eqs. 4.35a and 4.35b.

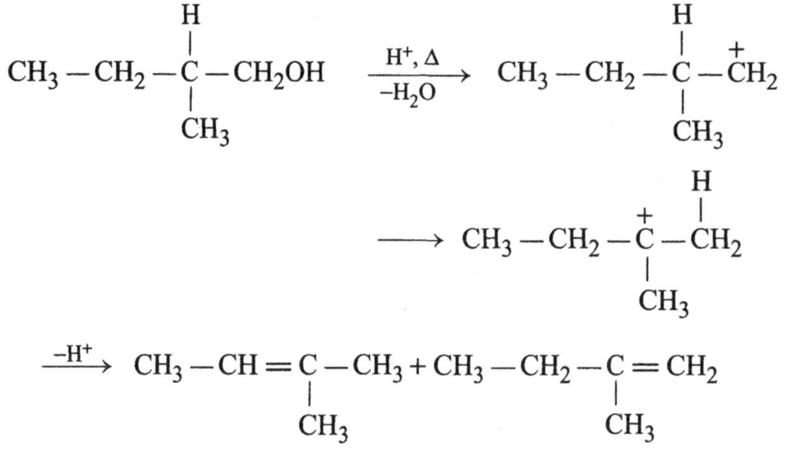

$$CH_3-CH_2-\underset{\underset{CH_3}{|}}{\overset{\overset{H}{|}}{C}}-CH_2OH \xrightarrow[-H_2O]{H^+,\,\Delta} CH_3-CH_2-\underset{\underset{CH_3}{|}}{\overset{\overset{H}{|}}{C}}-\overset{+}{C}H_2$$

$$\longrightarrow CH_3-CH_2-\overset{+}{\underset{\underset{CH_3}{|}}{C}}-\overset{\overset{H}{|}}{C}H_2$$

$$\xrightarrow{-H^+} CH_3-CH=\underset{\underset{CH_3}{|}}{C}-CH_3 + CH_3-CH_2-\underset{\underset{CH_3}{|}}{C}=CH_2 \qquad (4.35a)$$

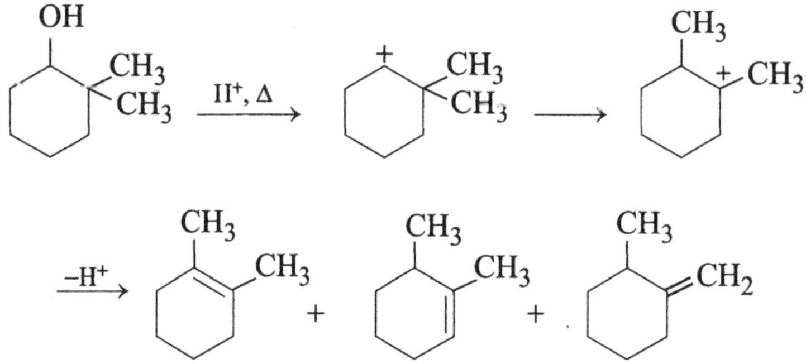

$$(4.35b)$$

The rearrangement of the secondary carbocation to the tertiary carbocation in Eq. 4.34 may be visualised as shown below,

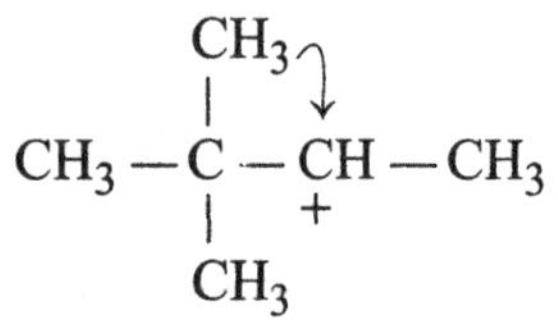

The methyl group migrates from its original point of attachment to the adjacent carbon carrying its bonding electron pair with it. In the transition state the migrating group is partially bonded to both carbons (Eq. 4.36).

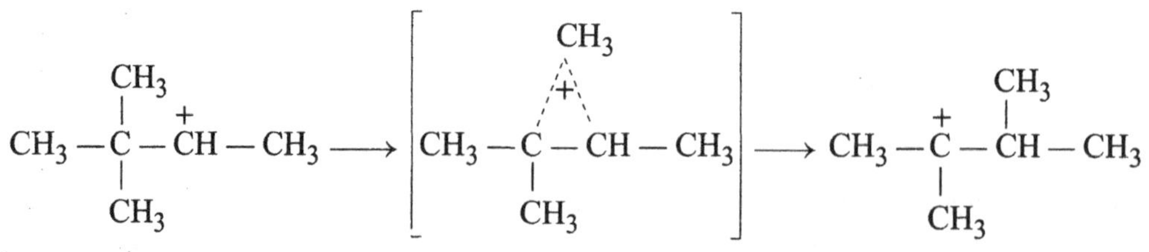

$$(4.36)$$

One of the dotted lines represents the partially broken bond and the other the partially formed bond. The migrating atom can also be hydrogen as in Eqs. 4.35a and 4.37.

$$CH_3-CH_2-\overset{\overset{\displaystyle H}{|}}{\underset{\underset{\displaystyle CH_3}{.|}}{C}}-\overset{+}{C}H_2 \longrightarrow CH_3-CH_2-\overset{\overset{\displaystyle H}{\wedge}}{\underset{\underset{\displaystyle CH_3}{|}}{C}}-CH_2 \longrightarrow CH_3-CH_2-\overset{+}{\underset{\underset{\displaystyle CH_3}{|}}{C}}-\overset{\overset{\displaystyle H}{|}}{C}H_2$$

$$(4.37)$$

4.2.2 FROM ALKYNES BY HYDROGENATION

The triple bond in an alkyne can be converted to a double bond by hydrogenation; involving one molecule of hydrogen. A noble metal catalyst is required (Eq. 4.38).

$$R-C\equiv C-R'+H_2 \xrightarrow[\substack{\text{Lindlar's}\\\text{catalyst}}]{Pd} \underset{H}{\overset{R}{>}}C=C\underset{H}{\overset{R'}{<}}$$

$$(4.38)$$

The product alkene can be further hydrogenated to the alkane (Eq. 4.39).

$$\underset{H}{\overset{R}{>}}C=C\underset{H}{\overset{R'}{<}} +H_2 \xrightarrow{Pd} H\overset{R}{\diagdown}C-C\overset{R}{\diagup}H$$

$$(4.39)$$

The two reactions differ in rates. It is possible to stop the reaction at the alkene stage, especially if a selective catalyst is used. Lindlar's catalyst which is a less active form of palladium catalyst (see Table 4.2) in comparison with ordinary palladium hydrogenation catalysts, is selective for the first step which is the conversion of alkyne to alkene (Eq. 4.38). The second step (Eq. 4.39) will not take place on this catalyst. The main interest in this method of preparing an alkene is that one particular geometrical isomer of the alkene – the *cis*-isomer – is selectively formed. The two hydrogens add from the same side of the alkyne. Stereochemically, this is called *cis-addition*.

Alkynes can also be converted to alkenes by chemical reduction using the alkali metal Li, and liquid ammonia. This reaction is called metal–ammonia reduction or the Birch reduction, named after A J Birch (1915–1995), an Australian chemist. This gives the other, the *trans* geometrical isomer. The hydrogen adds from opposite sides. This is called *trans-addition* (Eq. 4.40).

$$R-C\equiv C-R' \xrightarrow{\text{Li, liquid NH}_3} \underset{H}{\overset{R}{>}}C=C\underset{R'}{\overset{H}{<}}$$

$$(4.40)$$

The two methods are complementary and are useful for the preparation geometrically isomeric alkenes selectively (Eq. 4.41). Such reactions are examples of stereoselective reactions.

$$CH_3-C\equiv C-C_2H_5 \quad \begin{array}{c} \xrightarrow[]{\text{H}_2,\text{ Lindlar's catalyst}} \\ \\ \xrightarrow[]{\text{Li, liquid NH}_3} \end{array}$$

$$\begin{array}{cc} CH_3 & C_2H_5 \\ \diagdown & \diagup \\ C=C \\ \diagup & \diagdown \\ H & H \end{array}$$

$$\begin{array}{cc} CH_3 & H \\ \diagdown & \diagup \\ C=C \\ \diagup & \diagdown \\ H & C_2H_5 \end{array} \qquad (4.41)$$

4.2.3 INDUSTRIAL PREPARATION OF ALKENES

The laboratory methods of preparation discussed so far are applicable not only for the simple alkenes which are mentioned as examples but also for more complex molecules which may contain other functional groups and structural diversity. Such alkenes are often needed for industries which manufacture low volume, high value products like pharmaceuticals.

For bulk manufacture of simple alkenes like ethene, propene, butenes and their homologues, the laboratory methods discussed so far are not useful mainly because the alkyl halides and alcohols which are the starting materials, are themselves obtained industrially from alkenes. However there are some exceptions. Ethanol is obtained in bulk, not only from the petrochemical industry, but also as an agricultural product from molasses and grain. Dehydration of ethanol could be a method of manufacturing ethene in an agro-based economy. The petrochemical processes where alkenes are obtained along with other products are dehydrogentation and cracking (Chapter 3, Section 3.5.4 and 3.5.5).

4.3 PROPERTIES

In physical properties, alkenes are very similar to alkanes. The lower members upto C4 are gases at room temperature, C5 to C17 are liquids, after which they tend to be waxy solids. As in other series, the boiling and melting points increase with molecular weight. They are immiscible with water, but are miscible with other hydrocarbons and nonpolar solvents. They have relative densities in the liquid state in the range 0.65–0.75. The physical properties of some alkenes are given in Table 4.1. Isomerism appears with the C4 and higher alkenes. There can be four C4 alkenes (C_4H_8). They are,

$$\begin{array}{cc} CH_3 & CH_3 \\ \diagdown & \diagup \\ C=C \\ \diagup & \diagdown \\ H & H \end{array}$$

$$CH_2=CH-CH_2-CH_3$$

1-Butene   *cis*-2-Butene

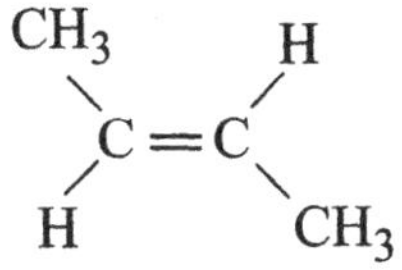

CH₃   H
   \   /
    C = C
   /   \
  H   CH₃

trans-2-Butene

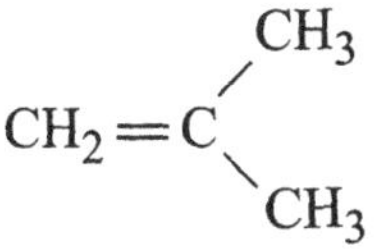

         CH₃
        /
CH₂ = C
        \
         CH₃

2-Methylpropene
(isobutylene)

Table 4. 1   Properties of alkenes

Name	Structure	b.p (°C)	m.p (°C)
Ethene (ethylene)	$CH_2 = CH_2$	−102.0	−169
Propene (propylene)	$CH_2 = CH - CH_3$	−48.0	−185
1-Butene	$CH_2 = CH - CH_2 - CH_3$	−6.5	–
Cis-2-butene	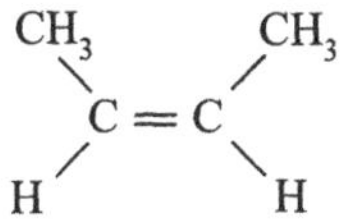 (structure)	4.0	−139
Trans-2-butene	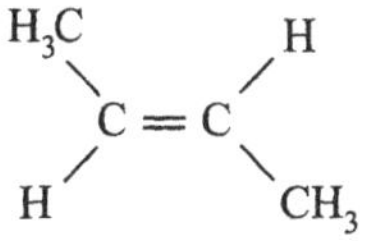 (structure)	1.0	−106
2-Methylpropene (isobutylene)	$CH_2 = C(CH_3)_2$	7.0	−141
1-Pentene	$CH_2 = CHCH_2 - CH_2 - CH_3$	30.0	–
1-Hexene	$CH_2 = CH - (CH_2)_3 - CH_3$	63.5	−138

The *cis*- and *trans*-2-butenes are called geometrical isomers (geometrical isomerism is discussed in detail in Chapter 22). These differ in the orientation of the atoms attached to the double bond in space. There is no free rotation around the double bond (unlike the case with a C—C single bond). Electron density in the C—C single bond (σ-bond) is symmetrical around σ-bond axis and rotation does not affect the degree of overlap. However with the double bond, rotation will involve the breaking of the π-bond of the double bond. Considering the original p-orbitals which overlapped to form the π-bond, the situation during rotation is as shown below (Fig. 4.1).

The middle structure corresponds to one where the π-bond has fully broken. The energy required for a reaction of this kind is at least as much as the bond energy of the π-bond which is 210–250 kJ mol⁻¹ (50–60 kcal mol⁻¹). This amount of energy is not available at ordinary temperatures. So such a rotation is not a 'free rotation'. The two isomers *cis*- and *trans*-2-butene are different molecules, stable indefinitely at room temperature. The rotation around the C—C single bond is also not exactly 'free'. For butane, rotation around the $C_2 - C_3$

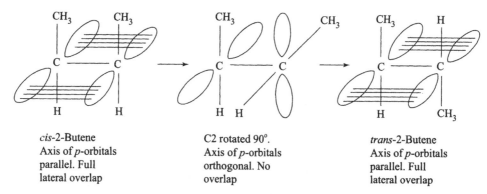

cis-2-Butene
Axis of *p*-orbitals
parallel. Full
lateral overlap

C2 rotated 90°.
Axis of *p*-orbitals
orthogonal. No
overlap

trans-2-Butene
Axis of *p*-orbitals
parallel. Full
lateral overlap

Fig. 4.1   Orientation of π-orbitals during rotation

bond involves crossing an energy barrier of about $25\,\text{kJ mol}^{-1}$ ($6\,\text{kcal mol}^{-1}$) due to what are kown as *non-bonded interactions*. But this amount of energy is available to the molecule at room temperature or even lower temperature, and rotation around the single bond is free for all practical purposes.

The nomenclature of geometrical isomers using prefixes *cis* and *trans* has been and still is, in wide use in organic chemistry. However this is not unambiguous in complex situations. A systematic nomenclature has been proposed and is in use today using the prefixes E and Z. According to this system, what has conventionally been named *cis*- and *trans*-2-butene are named Z-2-butene and E-2-butene. It should be emphasised that *cis* and Z are not synonymous, *trans* and E are not synonymous. This system of nomenclature is discussed in detail in section 22.1.1.

4.4   REACTIONS OF ALKENES

4.4.1   ELETROPHILIC ADDITION

The double bond in the alkene is a centre of reactivity because the electrons of the π-bond are more polarisable than the electrons of the σ-bond. The π-bond is an electron-rich centre. These electrons can enter into bond formation with electron-seeking reagents (electrophiles). In other words, the double bond is a nucleophilic centre. This is illustrated by the reaction with a proton which results in the opening of the double bond and the formation of a carbocation (Eq. 4.42).

$$\text{\textbackslash}C=C\text{/} + H^+ \longrightarrow \overset{+}{C}-\underset{}{C}-H$$

$$\text{C}-\text{C} + \left(H^+\right) \longrightarrow \overset{+}{C}-C \qquad (4.42)$$

vacant
s-orbital

vacant
p-orbital

Free radicals can also react with a double bond in a similar manner. Free radicals are electrophiles since they do not have fully occupied valence shells (Eq. 4.43).

$$\diagdown C = C \diagup + H^{\bullet} \longrightarrow \diagdown \overset{\bullet}{C} - \overset{\overset{\displaystyle H}{|}}{C} \diagup$$

$$(4.43)$$

Anions (nucleophiles) are repelled by the electron cloud of the double bond and do not react except in those molecules where the electron cloud has been distorted by polarisation to such an extent that one of the carbons of the double bond actually carries a partial positive charge and becomes electrophilic.

Double bonds undergo catalytic hydrogenation (addition of a molecule of hydrogen). Alkenes also participate in a class of reactions known as *cycloaddition*. The carbons next to the double bond *(allylic position)* are influenced by the π-bond so much that substitution reactions at that position are also possible *(allylic substitution)* under certain conditions.

$$R - CH_2 - CH = CH_2 + Br_2 \longrightarrow R - \underset{\underset{\displaystyle Br}{|}}{CH} - CH = CH_2 + HBr$$

$$(4.44)$$

4.4.1.1 Addition of Halogens

Alkenes react with halogens to form *vic*-dihalides (Eq. 4.45). Such reactions are called addition reactions and are the reverse of elimination reactions.

$$CH_3 - CH = CH - CH_3 + Br_2 \longrightarrow CH_3 - \underset{\underset{\displaystyle Br}{|}}{\overset{\overset{\displaystyle Br}{|}}{CH}} - CH - CH_3$$

$$(4.45)$$

Chlorine and iodine also add to double bonds. Fluorine reacts but with explosive violence, so the addition of fluorine is not of practical utility. The bromine addition reaction has been used as a convenient test for unsaturation, that is, to show that there is a carbon–carbon double bond in an unknown sample. A solution of bromine in carbon tetrachloride is red. If a few drops of an alkene are added to a dilute solution of bromine in carbon tetrachloride, the colour will disappear because of the consumption of bromine by the alkene. Some other classes of compounds may also answer this test. Thus, alkynes will react in the same way, except that a triple bond will consume two molecules of bromine. Some compounds which undergo easy bromination like phenol and aniline will also decolourise the bromine solution. Hence a positive test is indicative but does not confirm an alkene; a negative test (that is, no decolourisation) means that there is no unsaturation. Addition of Br_2, other halogens or interhalogens like ICl is used for the quantitative estimation of unsaturation. Addition of halogens is a true electrophilic addition. The adding species is the positive end of a polarised bromine molecule. Note that a bromine molecule is symmetric and has no polarity in the undisturbed state (Eq. 4.46).

$$Br-Br \longleftrightarrow \overset{+}{Br} : \overset{-}{Br} \longleftrightarrow \overset{-}{Br} : \overset{+}{Br} \tag{4.46}$$

An external polarising agent, that is, a molecule which can attract or repel electrons can induce the shift of electrons. When a halogen comes into proximity with the π-electron cloud of an alkene, it will act as shown below (Eq. 4.47) (see also Chapter 1, Section 1.3.3).

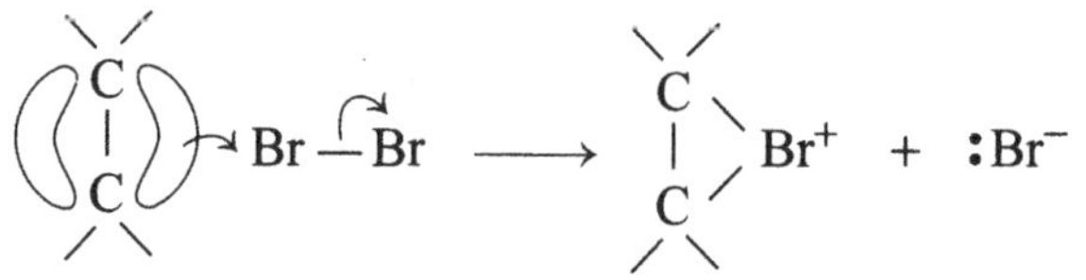

$$\tag{4.47}$$

In other words, the π-electron cloud acts as a nucleophile, displacing a bromide ion (leaving group) from the bromine molecule. The three membered ring structure is called the *bromonium ion*, a bromocation with an expanded valence shell. Such cations with a central atom having one valency more than the neutral atom are named with the suffix –onium. The bromonium ion this reaction is a real intermediate with a finite though short lived existence. Once formed, this reactive intermediate tends to react with any available nuclephile (for example, an anion). Br^- is available in the medium. When a solution of Br_2 in CCl_4 is used, Br^- is the only nucleophile available. If water was the solvent used, a water molecule also could have reacted as a nucleophile.

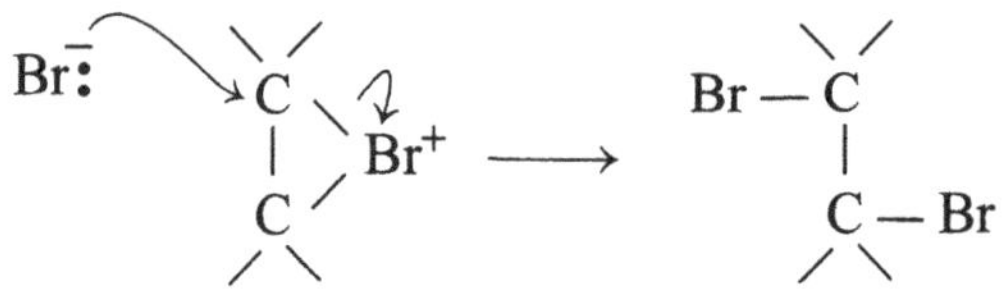

$$\tag{4.48}$$

The Br^- can attack any one of the two carbon atoms. The product will be the same. Since one face of the original alkene is obstructed by the 3-membered ring, Br^- attacks from the other face. Overall, the two Br atoms attach themselves to the opposite faces. This is called *trans*-addition. Why is a bromonium ion involved and not a carbocation as in Eq. 4.49?

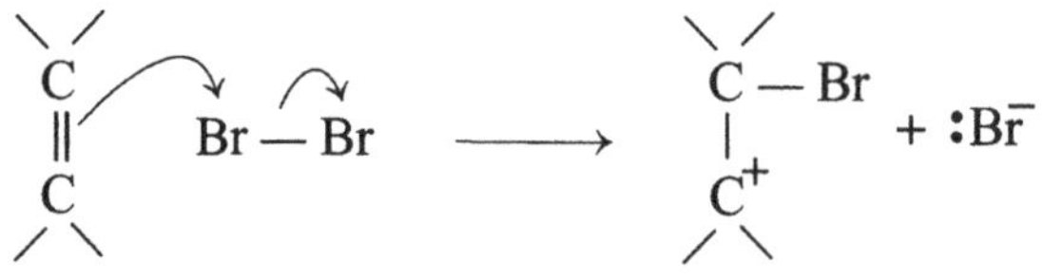

$$\tag{4.49}$$

The main evidence for the bromonium ion is based on *stereochemistry*. Geometrically isomeric alkenes like *cis*- and *trans*-2-butene give different steroisomers of 2,3-dibromobutane. This can be readily understood if the reaction goes through the bromonium ion and not through the open carbocation. If the medium contains other anions, products arising from those anions are also formed. For example, in a solution of bromine in a solvent containing the chloride ion, a chlorobromide is also formed in addition to the dibromide (Eq. 4.50).

$$\text{(4.50)}$$

If bromine in water (bromine water) is used instead of bromine in CCl_4, water can act as a nucleophile and attack the bromonium ion to give the *bromohydrin* (bromoalcohol) as one of the products (Eq. 4.51).

A Bromohydrin

$$\text{(4.51)}$$

4.4.1.2 Addition of Chlorine and Hypochlorous Acid

Chlorine adds to alkenes to give the dichloride in the same way as bromine does. When chlorine water, or better, a solution of chlorine in aqueous NaOH is used, the *chlorohydrin* is the major product (Eq. 4.52).

$$CH_2 = CH - CH_3 + Cl_2 + H_2O \xrightarrow{\text{NaOH}} \begin{array}{c} CH_2 - CH - CH_3 \\ | \quad\quad | \\ Cl \quad\; OH \end{array}$$

1-Chloropropan-2-ol
(propylene chlorohydrin)

$$\text{(4.52)}$$

The reaction is formally referred to as the addition of hypochlorous acid, HOCl but the sequence of steps is, (i) addition of Cl^+ followed by (ii) the addition of OH^- in the alkaline medium or of H_2O, if the medium is not alkaline.

4.4.1.3 Addition of Hydrogen Halide

Hydrogen halides like HCl, HBr and HI add themselves to the double bond when treated with alkenes (Eqs. 4.53 to 4.55).

$$CH_2 = CH_2 + HBr \longrightarrow \overset{\displaystyle \overset{H}{|}}{CH_2} - \overset{\displaystyle \overset{Br}{|}}{CH_2} \tag{4.53}$$

$$CH_2 = CH - CH_3 + HCl \longrightarrow \overset{\displaystyle \overset{H}{|}}{CH_2} - \overset{\displaystyle \overset{Cl}{|}}{CH} - CH_3 \tag{4.54}$$

$$CH_3 - CH = CH - CH_3 + HI \longrightarrow CH_3 - CH - \overset{\displaystyle \overset{H}{|}}{\underset{\displaystyle \underset{I}{|}}{C}} - CH_3 \tag{4.55}$$

These are reactions which involve the addition of a proton, H^+, to the double bond in an electrophilic reaction to form a carbocation (Eq. 4.42) followed by the addition of the anion (conjugate base) of the acid (Eq. 4.56).

$$CH_2 = CH_2 + H^+ \longrightarrow \overset{\displaystyle \overset{H}{|}}{CH_2} - \overset{+}{CH_2} \xrightarrow{Cl^-} \overset{\displaystyle \overset{H}{|}}{CH_2} - \overset{\displaystyle \overset{Cl}{|}}{CH_2} \tag{4.56}$$

4.4.1.4 Regioselectivity; Markowonikoff's Rule

In the example given in Eq. 4.54, one more product than what is shown is possible. Thus the reaction with propene (Eq. 4.54) can give two products (Eq. 4.57).

$$CH_2 = CH - CH_3 \xrightarrow{HCl} CH_3 - \underset{\displaystyle \underset{Cl}{|}}{CH} - CH_3 + CH_2 - \underset{\displaystyle \underset{Cl}{|}}{CH_2} - CH_3 \tag{4.57}$$

It is found experimentally that the first mentioned product 2-chloropropane is formed in greater quantity that the second one, 1-chloropropane. In other words there is selectivity in the addition of HCl to alkenes. This kind of selectivity where one of two possible *positional isomers* (*regioisomers*) is selectively formed is called *regioselectivity*. (If one of two possible steroisomers is selectively formed, it is called *stereoselectivity*). This kind of regioselectivity has been found to be general as in the following examples (Eqs. 4.58–4.60). Note that regioselectivity is relevant only when the groups on the two carbons of the double bond are not identical, that is, the alkene is unsymmetrical.

$$CH_2 = \underset{\displaystyle \underset{CH_3}{|}}{C} - CH_3 + HI \longrightarrow \overset{\displaystyle \overset{H}{|}}{CH_2} - \underset{\displaystyle \underset{CH_3}{|}}{\overset{\displaystyle \overset{I}{|}}{C}} - CH_3 \tag{4.58}$$

$$CH_3 - CH = \underset{\displaystyle \underset{CH_3}{|}}{C} - CH_3 + HCI \longrightarrow CH_3 - \overset{\displaystyle \overset{H}{|}}{CH} - \underset{\displaystyle \underset{CH_3}{|}}{\overset{\displaystyle \overset{Cl}{|}}{C}} - CH_3 \tag{4.59}$$

$$C_2H_5 - \underset{\underset{CH_3}{|}}{C} = CH - C_3H_7 + HBr \longrightarrow C_2H_5 - \underset{\underset{CH_3}{|}}{\overset{\overset{Br}{|}}{C}} - \overset{\overset{H}{|}}{CH} - C_3H_7 \qquad (4.60)$$

In the examples given (Eqs. 4.58–4.60) only the main products are shown. The other product is also formed but in lesser quantities. A certain pattern can be seen in this selectivity and the Russian chemist, Vladimir W Markownikoff (1838–1904) generalised these observations thus:

In the addition of an acid like HCl to an unsymmetrical alkene, the hydrogen of the acid is attached selectively to the carbon which already contains more number of hydrogens. This is known as Markownikoff's rule.

At that time, the modern concepts about carbocations and their relative stability did not exist. Today we can rationalise Markonwnikoff's rule in terms of carbocation stability.

The addition of hydrogen chloride to an alkene involves two steps (Eq. 4.56),

(i)   Addition of H^+ to produce the carbocation.

(ii) Addition of Cl^- to the carbocation (*quenching of the carbocation*) to produce the neutral product.

Carbocations and carbanions and other such reactive intermediates (also called transient intermediates) are high energy species and the driving force for their subsequent reactions is the desire to lower the energy of the system. The reaction with Cl^- is an exothermic reaction where heat is given out and the system 'cooled'; hence the expression 'quenching' of the carbocation. That the reaction $R^+ + Cl^- \longrightarrow RCl$ has to be an exothermic reaction can be qualitatively understood if one notes that a new covalent bond is formed as a result of the reaction. Covalent bond formations are exothermic reactions, the energy involved being equal to the bond-energy. In the case of the example under consideration, this statement has to be modified since the two species combining are ions, not atoms. However the conclusion is still valid.

In the addition to propene, the first step is (Eq. 4.61),

$$CH_2 = CH - CH_3 + H^+ \longrightarrow \underset{\underset{H}{|}}{CH_2} - \overset{+}{CH} - CH_3 \qquad \begin{array}{l} \text{Secondary} \\ \text{carbocation} \end{array}$$

$$\text{or } \overset{+}{CH_2} - \underset{\underset{H}{|}}{CH} - CH_3 \qquad \begin{array}{l} \text{Primary} \\ \text{carbocation} \end{array} \qquad (4.61)$$

It is this step that decides which product one gets eventually. Hence this is called the *product determining step*. (There is also another expression, the *rate determining step*. In a multi-step reaction, the overall rate of the reaction is usually the same as that of the slowest step, which is called the rate determining step. In the present example, the first step which is the product determining step is also the rate determining step. The addition of proton to the alkene is slower than the second step namely quenching).

Can we predict, of the two carbocations formed, which the preferred one should be? Yes, we can do that if we consider the stabilities of the carbocations. Of the two in Eq. 4.61, the first one, 2-propyl cation is more stable than the second one, the 1-propyl cation, because the former is secondary and latter is primary (see discussion on carbocations in Chapter 1, Sections 1.4.3.3 and 1.4.3.4). The major product of addition is the product arising from the more stable carbocation (Eqs. 4.62, 4.63).

$$CH_3 - \overset{+}{C}H - CH_3 + Cl^- \longrightarrow CH_3 - \overset{\overset{\displaystyle Cl}{|}}{C}H - CH_3 \qquad (4.62)$$
$$\text{Major product}$$

$$\overset{+}{C}H_2 - CH_2 - CH_3 + Cl^- \longrightarrow \underset{\text{Minor product}}{\overset{\overset{\displaystyle Cl}{|}}{C}H_2 - CH_2 - CH_3} \qquad (4.63)$$

In the light of modern ideas, Markownikoff's rule may be reworded as follows:

In ionic addition to unsymmetrical alkenes, the cation will add to the alkene in such a way that the resulting carbocation intermediate is the more stable of the two possibilities.

This generalised statement is applicable to all electrophilic additions. For the most common situation, addition of protonic acids, the rule may be more simply stated thus,

In the addition of protonic acids to unsymmetrical alkenes, the protons add to the double bond in such a way that the more stable carbocation is formed.

4.4.1.5   Addition of Halogens and Hypochlorous Acid (Revisited)

With halogens like Br_2 and Cl_2 – in the absence of any other nucleophile – the question of regiochemistry does not arise. When another nucleophile is involved, for example, Br^+ addition in the presence of Cl^-, regiochemistry does enter the picture (Eq. 4.64).

$$CH_3 - CH = CH_2 + Br_2 \longrightarrow CH_3 - \overset{\displaystyle \overset{Br^+}{\diagup \diagdown}}{C}H - CH_2$$

$$\xrightarrow{Cl^-} CH_3 - \underset{\underset{\displaystyle Cl}{|}}{\overset{\overset{\displaystyle Br}{|}}{C}H} - CH_2 \text{ and } CH_3 - \overset{\overset{\displaystyle Br}{|}}{C}H - \underset{\underset{\displaystyle Cl}{|}}{C}H_2 \qquad (4.64)$$

The preferred product is the first one, 1-bromo-2-chloropropane. This is because, in the bromonium ion, the two carbon atoms of the 3-membered ring do not have the same share of the total positive charge. Those carbons which can stabilise a positive charge better will be more positive. In this case, it is the secondary carbon. The Cl^- reacts with the carbon which is more positive. The charge distribution in the bromonium ion is more like (A) than like (B) (Eq. 4.65).

$$CH_3 - \overset{+}{C}H - CH_2 \qquad CH_3 - CH - \overset{+}{C}H_2$$
$$\text{(A)} \qquad \qquad \text{(B)}$$

(4.65)

$$Cl^- + CH_3 - CH - \overset{+}{C}H_2 \longrightarrow CH_3 - \underset{Cl}{\overset{Br}{CH}} - CH_2$$

When OH^- or H_2O is the nucleophile as in the case of halohydrin formation from propene, the secondary alcohol is formed in preference to the primary alcohol (Eq. 4.66).

$$CH_3 - CH = CH_2 + Cl^+ + OH^- \longrightarrow CH_3 - \underset{Cl}{\overset{OH}{CH}} - CH_2 + CH_3 - \underset{OH}{\overset{Cl}{CH}} - CH_2$$
$$\text{(Preferred product)}$$

(4.66)

4.4.1.6 Addition of HBr; Peroxide Effect

While the addition of HCl to alkenes strictly follows Markownikoff's rule, there have been many conflicting reports about the addition of HBr in the early chemical literature. In some studies the expected Markownikoff product was obtained and in others, the unexpected product (hence called the *anti-Markownikoff* product), was obtained (Eq. 4.67).

$$R - CH = CH_2 + HBr \longrightarrow \begin{cases} R - \underset{Br}{\overset{}{C}H} - CH_3 & \text{(Markownikoff product)} \\ R - CH_2 - \underset{Br}{\overset{}{C}H_2} & \text{(Anti-Markownikoff product)} \end{cases}$$

(4.67)

By careful experiments the American chemists Kharasch and Mayo, working at the University of Chicago, solved this problem in the 1930s. They showed that the culprits were certain impurities in the alkene, belonging to a class of compounds called peroxides. These are organic molecules containing the peroxy linkage, $R - O - O - R$. These are formed from all organic compounds by air oxidation in small quantities, when they are stored for a long time. Alkenes are particularly prone to air oxidation. When such peroxides are present, the addition of HBr to alkenes does not follow the ionic mechanism seen so far. The peroxides initiate another mechanism which involves free radicals. The weak O—O bounds in peroxides are readily cleaved homolytically by mild heat (Eq. 4.68).

$$R-O-O-R \xrightarrow{\Delta} 2R-O^{\bullet} \tag{4.68}$$

These *free radicals*, react with HBr to generate bromine atoms (Eq. 4.69).

$$RO^{\bullet} + HBr \longrightarrow ROH + Br^{\bullet} \tag{4.69}$$

Now the following sequence of reactions set in (Eqs. 4.70, 4.71),

$$Br^{\bullet} + CH_2 = CH - CH_3 \longrightarrow Br - CH_2 - \overset{\bullet}{C}H - CH_3 \tag{4.70}$$

$$Br - CH_2 - \overset{\bullet}{C}H - CH_3 + HBr \longrightarrow Br - CH_2 - CH_2 - CH_3 + Br^{\bullet} \tag{4.71}$$

The product, $^{\bullet}$Br in Eq. 4.71 now adds to another molecule of propene, that is, repetition of Eq. 4.70. Thus, a *chain reaction* takes place. A few molecules of RO$^{\bullet}$ are enough to initiate the chain. In principle, the chain will continue till all the propene or HBr is used up. In practice, the chain does not progress indefinitiely, but gets terminated by one of several termination steps (cf. chlorination of methane by free radical chain reaction Chapter 3, Section 3.5.2.2).

Thus, a small amount of peroxide present unintentionally triggers a free radical chain reaction. The selectivity in this free radical addition of HBr is not the same as in the electrophilic addition. Let us examine Eq. 4.70 which is the product determining step. Br$^{\bullet}$ could have added in one of two ways, only one of which is shown in Eq. 4.70. Both possible products are shown in Eq. 4.72.

$$Br^{\bullet} + CH_2 = CH - CH_3 \longrightarrow Br - CH_2 - \overset{\bullet}{C}H - CH_3 \text{ or } {^{\bullet}}CH_2 - \underset{|}{CH} - CH_3 \tag{4.72}$$
$$Br$$

The first one, the secondary free radical, is the preferred product. The reason is the greater stability of the secondary free radical compared to the primary (Chapter 1, Section 1.4.3.11). The overall preferred product is the anti-Markownikoff product 1-bromopropane, obtained in Eq. 4.71.

The anti-Markownikoff product can be obtained by deliberately adding a peroxide as an initiator or by using conditions favourable for homolysis like heating and light and avoiding the use of a polar solvent which may help the ionic reaction. The peroxide effect is unique for HBr addition. It is not

Kharasch and Mayo established that the anti-Markownikoff addition of HBr follows a free radical pathway. Since the presence of peroxides is the most common reason for this mechanism to be followed, they called the phenomenon the *peroxide effect*. Markownikoffs rule applies only to reactions which follow an ionic pathway via the carbocation.

seen in the addition of HCl or HI. This is because they do not add homolytically due to unfavourable energetics.

Since alkyl bromides, like other alkyl halides can be hydrolysed to alcohols, the two ways of adding HBr can be used for selectively converting an alkene to two types of alcohols as illustrated by the selective conversion of propene to 2-propanol and to 1-propanol (Eqs. 4.73, 4.74).

$$CH_2=CH-CH_3 + HCl \longrightarrow CH_3-\overset{\overset{\displaystyle Cl}{|}}{CH}-CH_3 \xrightarrow{H_2O} CH_3-\overset{\overset{\displaystyle OH}{|}}{CH}-CH_3$$

(or HBr in
the absence
of peroxides)

$$\longrightarrow CH_3-\overset{\overset{\displaystyle Br}{|}}{CH}-CH_3 \xrightarrow{H_2O} CH_3-\overset{\overset{\displaystyle OH}{|}}{CH}-CH_3 \qquad (4.73)$$

$$CH_2=CH-CH_3 + HBr \xrightarrow{peroxide} \overset{\overset{\displaystyle Br}{|}}{CH_2}-CH_2-CH_3 \xrightarrow{H_2O} \overset{\overset{\displaystyle OH}{|}}{CH_2}-CH_2-CH_3$$
$$(4.74)$$

As we shall see presently, the direct addition of water to propene by acid-catalysed hydration follows Markownikoff's rule and gives a 2-propanol. Note that the peroxide effect is applicable only to the addition of HBr and not to the addition of HCl or HI and certainly not to that of H_2O or H_2SO_4.

4.4.1.7 Addition of Water and Sulphuric Acid; Hydration

Water can be directly added to alkenes in the presence of strong acids like sulphuric acid to give alcohols, corresponding to the addition of H^+ and OH^- across the double bond. This is called hydration (Eq. 4.75).

$$CH_2=CH-CH_3 + H_2O \xrightarrow{H^+} \overset{\overset{\displaystyle H \quad OH}{|\quad\;\;|}}{CH_2}-CH-CH_3 \qquad (4.75)$$

The reaction follows Markownikoff's rule. Alkenes react with concentrated sulphuric acid in a reaction which can be readily understood (Eq. 4.76).

$$CH_2=CH-CH_3 + H^+ \longrightarrow CH_3-\overset{+}{CH}-CH_3 \xrightarrow{HSO_4^-} CH_3-\overset{\overset{\displaystyle O-SO_3H}{|}}{CH}-CH_3$$

2-Propyl
hydrogensulphate

$$(4.76)$$

This is the 2-propyl ester of sulphuric acid. If water is added to this product, it will be hydolysed to 2-propanol. Recall the hydrolysis of alkyl halides to alcohol (Eq. 4.74). Sulphates are hydrolysed even more easily than halides because the sulphate anion is a better leaving group than the halide. If aqueous sulphuric acid is used instead of concentrated sulphuric acid, 2-propanol is obtained without going through 2-propyl hydrogensulphate (Eq. 4.77).

$$CH_2 = CH - CH_3 + H_3O^+ \longrightarrow CH_2 - \overset{\overset{\displaystyle H}{|}}{\underset{\underset{\displaystyle H_2O}{+}}{C}}H - CH_3 \longrightarrow CH_2 - \overset{\overset{\displaystyle H}{|}}{C}H - CH_3 \tag{4.77}$$

The overall reaction is the addition of water (Eq. 4.75). Since neutral water does not have H_3O^+ concentration high enough to drive the first step – and hence the total reaction successfully in the forward direction – an acid like sulphuric acid is required in aqueous solution. As can been seen from Eq. 4.76, the role of the acid is strictly catalytic. The addition of water – hydration – is the reverse of the dehydration of alcohols to form alkenes, which is also acid catalysed. Industrially, alcohols are manufactured from alkenes by acid catalysed hydration. Alkenes are the bulk byproducts of *petroleum refining*.

4.4.2 HYDROBORATION

Hydroboration followed by oxidation of the boron-containing adduct, is an indirect method of getting the anti-Markownikoff alcohol from an alkene. As the name implies, hydroboration is the addition of hydrogen and boron across a double bond. The actual reagent is diborane, B_2H_6. But the effective reacting species in solution is the monomeric borane, BH_3 which does not exist free. The usual method of obtaining diborane is by the reaction of sodium borohydride with a Lewis acid like BF_3. Diborane is a very reactive gaseous molecule which is stored as a solution in an ether solvent like tetrahydrofuran or diglyme.

$$CH_3 - O - CH_2 - CH_2 - O - CH_2 - CH_2 - O - CH_3$$
Diethyleneglycol dimethyl ether
(diglyme)

Tetrahydrafuran (THF)

In the ether solvent, BH_3 is present as a complex with the ether molecule.

$$NaBH_4 \xrightarrow{BF_3} B_2H_6 \longrightarrow \overset{+}{O} - \overset{-}{B}H_3 \tag{4.78}$$
THF–BH_3 complex

When an alkene like propene is added to such a solution, hydroboration takes place.

$$CH_2=CH-CH_3 + \left[\begin{matrix}H \\ H\end{matrix}B-H\right] \longrightarrow CH_2-CH-CH_3 \xrightarrow{\ CH_2=CH-CH_3\ }$$

$$H-B\begin{matrix}CH_2-CH-CH_3 \\ CH_2-CH-CH_3\end{matrix} \xrightarrow{\ CH_2=CH-CH_3\ } CH_2-CH-CH_2-B\begin{matrix}CH_2-CH-CH_3 \\ CH_2-CH-CH_3\end{matrix}$$

$$(4.79)$$

As shown in Eq. 4.79, three molecules of alkene can react with one molecule of borane to give successively, mono- , di- and trialkylboranes. The regiochemistry of this addition is interesting, and is anti-Makownikoff. Even though hydrogen is one of the adding atoms, this addition does not come under the purview of Markownikoff's rule since the adding molecule, BH_3 is not a protonic acid. The hydrogen in borane is not a proton but a hydride, formally H^-.

BH_3 itself is a Lewis acid. The boron has only 6 valence electrons around it and it reacts with the π-bond of the double bond in a conventional elecrophilic addition, *a la* Markownikoff, at the terminal primary carbon, causing the development of a positive charge at the middle secondary carbon. There is a notable difference between this electrophilic addition and the addition of a proton. In the proton addition, the secondary carbocation is formed as a true intermediate which gets quenched by a nuclephile in a second step. In hydroboration, the carbocation is not fully formed; instead, the developing positive charge on the secondary carbon is quenched in a concerted reaction (Eq. 4.80).

$$(4.80)$$

The structure within square brackets is not an intermediate but is a representation of the 4-membered cyclic *transition state* of the reaction. The broken lines represent bonds which are either partially formed or partially broken. The regioselectivity is due to two factors, (i) the greater stability of the secondary carbocation (in this case a partially formed carbocation; this is called a polar effect) and (ii) the steric effect.

The steric effect, which we shall come across frequently in organic reactions, is simply speaking, due to crowding around the reaction centre. In propene, of the two alkene carbons, the CH_2 has only two hydrogens in its vicinity, the other carbon has one hydrogen and one carbon which occupy more space than the hydrogen. This carbon itself has three hydrogens on it. Altogether, the space surrounding the middle carbon is more crowded than the end carbon.

Less crowded   More crowded

This steric effect is felt more when the atom or molecule that is adding on is fairly large. The boron atom is not large, but the BH_3, the BHR and the BHR_2 molecules are much larger. The polar and the steric effects are together responsible for the regiochemistry. The cyclic transition state shown in (Eq. 4.87) also requires that the atoms, B and H form bonds with the alkene carbons at the same face of the double bond (*cis*-addition).

Hydroboration, discovered by the Nobel laureate chemist H C Brown, is a powerful tool in synthetic organic chemistry. Herbert C Brown (1912–2004), Purdue University, USA, won the Nobel Prize jointly with Georg Wittig in 1979.

4.4.2.1 Oxidation of the Alkylborane

The hydroboration reaction was developed and became a powerful tool for organic synthesis. The alkylboranes (mono-, di- and tri-) can be oxidised with alkaline hydrogen peroxide to give the alcohol (Eq. 4.81).

$$(CH_3 - CH_2 - CH_2)_3B \xrightarrow{\text{H}_2\text{O}_2/\text{NaOH}} CH_3 - CH_2 - CH_2 - OH + Na_3BO_3$$

$$(4.81)$$

When hydroboration is followed by oxidation using H_2O_2, the overall reaction constitutes a methodology for converting an alkene to an alcohol corresponding to the anti-Markownikoff addition of water (Eq. 4.82).

$$CH_2{=}C-CH_3 \xrightarrow[\text{(ii) H}_2\text{O}_2/\text{OH}^-]{\text{(i) B}_2\text{H}_6/\text{THF}} HO-CH_2-CH-CH_3$$

with CH_3 substituent groups below each middle carbon.

$$(4.82)$$

$$\text{(i) B}_2\text{H}_6/\text{THF}$$
$$\text{(ii) H}_2\text{O}_2, \text{OH}^-$$

4.4.3 OZONE ADDITION; OZONOLYSIS

Ozone, the triatomic oxygen allotrope is a very reactive molecule. It reacts with alkenes to form an addition compound called an *ozonide*. The kind of reaction where a cyclic compound is formed by addition to an alkene is called a *cycloadditon*. The ozonide formed in Eq. 4.83

> Much of our understanding of ozonolysis is due to Rudolf Crigee (1902–75), a German chemist.

is called the *molozonide*, the initial ozonide. It is a very unstable compound and quickly undergoes a molecular rearrangement to a more stable structure (stable in a relative sense) (Eq. 4.84). It is stable enough to be isolated but is still quite unstable and explosive.

$$\begin{array}{c} \underset{H}{\overset{H_3C}{>}}C=C\underset{H}{\overset{C_2H_5}{<}} \\[4pt] :\overset{+}{\underset{..}{O}}=\overset{..}{\underset{..}{O}}-\overset{..}{\underset{..}{O}}:^{-} \end{array} \longrightarrow \begin{array}{c} \overset{CH_3}{\underset{O}{}}\quad \overset{C_2H_5}{\underset{O}{}} \\ H-C-C-H \\ O \qquad O \\ \diagdown O \diagup \\ \text{(A molozonide)} \end{array} \tag{4.83}$$

$$\begin{array}{c} \overset{CH_3}{}\quad \overset{C_2H_5}{} \\ H-C-C-H \\ O \qquad O \\ \diagdown O \diagup \end{array} \longrightarrow \begin{array}{c} \overset{CH_3}{}\;\overset{O}{}\;\overset{C_2H_5}{} \\ H-C \qquad C-H \\ O-O \\ \text{Ozonide} \end{array} \tag{4.84}$$

For most applications, the ozonide need not be isolated. It is hydrolysed by water to form carbonyl compounds (Eq. 4.85).

$$\begin{array}{c} \overset{CH_3}{}\;\overset{O}{}\;\overset{C_2H_5}{} \\ C \qquad C \\ H \quad O-O \quad H \end{array} \xrightarrow{\;H_2O\;} \underset{H}{\overset{H_3C}{>}}C=O \;+\; O=C\underset{H}{\overset{C_2H_5}{<}} \;+\; H_2O_2 \tag{4.85}$$

Since the byproduct, hydrogen peroxide, causes oxidation of the aldehydes, the hydrolysis is carried out in the presence of a reducing agent – usually zinc dust in aqueous acetic acid – which reduces the hydrogen peroxide to water. The overall reaction is as follows (Eq. 4.86),

$$CH_3-CH=CH-C_2H_5 \xrightarrow[\text{(ii) Zn, CH}_3\text{COOH, H}_2\text{O}]{\text{(i) O}_3} CH_3-\overset{H}{\underset{|}{C}}=O+O=\overset{H}{\underset{|}{C}}-C_2H_5 \tag{4.86}$$

The ozonide formation followed by hydrolysis together is called *ozonolysis*. 'Lysis' is the suffix used to denote cleavage by a particular reagent as in hydrolysis or solvolysis. The alkene has been cleaved at the double bond into fragments, each fragment being an aldehyde or ketone obtained formally by inserting two oxygens (Eqs. 4.87, 4.88 and 4.89).

$$
\underset{\underset{CH_3}{|}}{CH_3-C}\!\!=\!\!\overset{O\ \ O}{CH-CH_3} \xrightarrow[\text{(ii) Zn, H}^+,\text{H}_2\text{O}]{\text{(i) O}_3} \underset{\underset{CH_3}{|}}{CH_3-C}\!\!=\!\!O \quad O\!\!=\!\!\underset{\underset{H}{|}}{C-CH_3} \tag{4.87}
$$

$$
\xrightarrow{\text{ozonolysis}} \tag{4.88}
$$

$$
\underset{\text{Isoprene}}{\underset{\underset{CH_3}{|}}{CH_2\!\!=\!\!CH-\overset{O\ \ O}{C}\!\!=\!\!\overset{O\ \ O}{CH_2}}} \xrightarrow{\text{ozonolysis}} \underset{H}{\overset{H}{\diagdown}}C\!\!=\!\!O \quad O\!\!=\!\!C\underset{\diagdown H}{\overset{\diagup H}{}}
$$

$$
O\!\!=\!\!\underset{\underset{CH_3}{|}}{C}-C\!\!=\!\!O \tag{4.89}
$$

Since the cleavage takes place exactly at the double bond, identification of the products of cleavage gives information about the structure of the alkene. The isomeric alkenes of C_4H_8 can be assigned structures by identifying the products of ozonolysis (Eq. 4.90).

$$
\underset{\text{1-Butene}}{CH_2=CH-CH_2-CH_3} \xrightarrow{\text{ozonolysis}} \underset{\substack{\text{Formaldehyde} \\ \text{(Methanal)} \\ \text{(1 molecule)}}}{CH_2O} + \underset{\substack{\text{Propanal} \\ \text{(1 molecule)}}}{CH_3-CH_2-CHO}
$$

$$
\underset{\substack{\text{2-Butene} \\ \text{(cis- or trans-)}}}{CH_3-CH=CH-CH_3} \xrightarrow{\hspace{2cm}} \underset{\substack{\text{Acetaldehyde (ethanal)} \\ \text{(2 molecules)}}}{2CH_3-CHO}
$$

$$
\underset{\underset{CH_3}{|}}{CH_2=C-CH_3} \xrightarrow{\hspace{2cm}} \underset{\substack{\text{Formaldehyde} \\ \text{(1 molecule)}}}{CH_2O} + \underset{\substack{\text{Acetone} \\ \text{(1 molecule)}}}{CH_3-\overset{\overset{\displaystyle O}{\|}}{C}-CH_3} \tag{4.90}
$$

4.4.4   HYDROGENATION

Addition of a molecule of hydrogen to a double bond to give an alkene is called *hydrogenation* (Eq. 4.91).

$$CH_2 = CH - CH_2 - CH_3 + H_2 \longrightarrow CH_3 - CH_2 - CH_2 - CH_3$$
1-Butene $\qquad\qquad\qquad\qquad\qquad$ Butane $\qquad\qquad\qquad$ (4.91)

Hydrogen molecule is nonpolar and is relatively unreactive in the absence of stimulation by a catalyst. Hydrogenation will not take place by just mixing the alkene with the hydrogen even on heating, though the reaction is quite exothermic (Eq. 4.92).

$$CH_2 = CH_2 + H_2 \longrightarrow CH_3 - CH_3 + 32\,kcal \qquad\qquad\qquad (4.92)$$

Recall that the reaction of H_2 and O_2 to form water is highly exothermic but requires a catalyst. Many transition metals like Pt, Pd and Ni and some oxides – like those of Cr and Mo – are effective hydrogenation catalysts. In the presence of one of the above metals in finely divided form, hydrogenation can be brought about. Thus, when a suspension of finely powdered platinum in 1-hexene (a liquid) taken in a vessel, is filled with hydrogen and agitated, the reaction takes place to give hexane. Agitation is necessary because the components of the reaction, hydrogen (gas), hexane (liquid) and platinum (solid) are in different phases. In some cases, heating is necessary. Hydrogen under pressure (that is, more than 1 atmosphere pressure) also helps the reaction. At the end of the reaction which can be followed by measuring the volume of hydrogen consumed, the catalyst is filtered off from the liquid. Some of the catalysts used in the laboratory are listed in Table 4.2.

Table 4. 2   Hydrogenation catalysts

Name	Description	Uses	Remarks
Adam's catalyst	Platinum oxide. The oxide gets reduced to the metal by the hydrogen	Alkene to alkane	Works at room temperature and 1 atm. pressure of H_2
Palladium on charcoal	Palladium metal dispersed on high surface area charcoal	Alkene to alkane	Room temperature 1 atm. pressure
Raney nickel	Obtained by treating powdered Ni-Al alloy with aq.NaOH. Al dissolves leaving behind nickel in a finely divided form	Aromatics to cycloalkanes	Requires hydrogen under pressure and heating
Lindlar's catalyst	Pd metal dispersed on $BaSO_4$ powder and partially poisoned by quinoline	Alkyne to alkene selective reduction	Room temp., 1 atm. pressure

Strong negative heat of reaction, in this case called heat of hydrogenation, implies that the reaction is feasible. However, without a catalyst, hydrogenations are impractically slow because of very high energies of activation. The catalyst makes the reaction faster by bringing down the activation energy.

The metal catalyst *adsorbs* both reactants – hydrogen and alkene – causing the dissociations of the H—H σ-bond and the C—C π-bond. The chemically adsorbed species then react on the surface of the catalyst and finally the product (alkane) *desorbs* from the surface. In the following schematic representation, the metal surface is symbolically shown as —M—M (*chemisorption*—chemical adsorption) (Eq. 4.93).

$$
H_2 + \ -M-M- \ \longrightarrow \
\begin{matrix} H & H \\ | & | \\ -M-M- \end{matrix}
\quad \text{(Chemisorption of hydrogen)}
$$

$$
\begin{matrix} \diagdown & \diagup \\ C=C \\ \diagup & \diagdown \end{matrix}
+ \ -M-M- \ \longrightarrow \
\begin{matrix} | & | \\ -C-C- \\ | & | \\ -M-M- \end{matrix}
\quad \text{(Chemisorption of alkene)}
$$

$$
\begin{matrix} | & | \\ -C-C- \\ | & | \\ -M-M- \end{matrix}
\quad
\begin{matrix} H & H \\ | & | \\ -M-M- \end{matrix}
\ \longrightarrow \
\begin{matrix} | & | \\ -C-C- \\ | \\ -M-M- \end{matrix}
\quad
\begin{matrix} H \\ | \\ -M-M- \end{matrix}
\quad \text{(Surface reaction)}
$$

$$
\begin{matrix} | & | \\ -C-C- \\ | \\ -M-M- \end{matrix}
\quad
\begin{matrix} H & H \\ | & | \\ -M-M \end{matrix}
\ \longrightarrow \
\begin{matrix} | & | \\ -C-C- \\ | & | \\ H & H \\ -M-M- \end{matrix}
\quad
\begin{matrix} H \\ | \\ -M-M \end{matrix}
\quad \text{(Surface reaction and desorption of product alkane)}
$$

$$(4.93)$$

Since both hydrogens come from the surface of the catalyst and react with the same face of the alkene even through the addition is step-wise, the overall reaction is *cis*-addition. The reaction takes place on the surface of a solid catalyst though the reactants are liquids or gases. Such catalysis is known as *heterogenous catalysis*. Adsorption of the reactants on the catalyst surface is a precondition for reaction. If the area of the catalyst surface is large, more number of reactant molecules can adsorb and react. Hence all good heterogeneous catalysts have large surface areas. This is achieved by (i) using finely powdered solid and (ii) dispersing the solid particles on the surface of another solid which already has a large surface area, like charcoal. Both concepts are seen in the description of the catalyst in Table 4.2.

Certain transition metal complexes which are soluble in appropriate solvents have been developed, which can catalyse hydrogenations. Since the catalysts and the reactant are in the liquid phase, this type of catalysis is called homogeneous catalysis. When hydrogenations are done at ordinary temperature where the alkene is a liquid, it is the hydrogen that is dissolved in the liquid that reacts, that is, the hydrogen is in the liquid phase.

4.4.4.1 Heats of Hydrogenation

The heats of hydrogenation of simple alkenes are all in the region of $126 \, \text{kJ mol}^{-1}$ ($30 \, \text{kcal mol}^{-1}$) with slight but significant differences (Table 4.3). Comparison of entries 3, 4, and 5 is instructive. All three alkenes, upon hydrogenation give the same product, butane. The differences in their heats of hydrogenation are due to the differences in the energy content of the alkene as can be seen from Fig. 4.2.

Table 4. 3 Heats of hydrogenation (kJ mol^{-1})

No.	Alkene	$-\Delta H$
1	$CH_2 = CH_2$	137
2	$CH_2 = CH - CH_3$	126
3	$CH_2 = CH - CH_2 - CH_3$	127
4	$\begin{array}{c} H_3C \qquad CH_3 \\ C = C \\ H \qquad\qquad H \end{array}$	120
5	$\begin{array}{c} H_3C \qquad H \\ C = C \\ H \qquad\qquad CH_3 \end{array}$	115

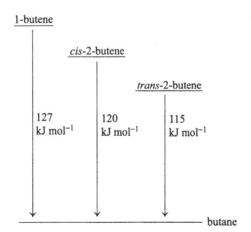

Fig. 4.2 Heats of hydrogenation and stability

1-Butene packs $11\,kJ\,mol^{-1}$ more than *trans*-2-butene. *Cis*-2-butene has about $5\,kJ\,mol^{-1}$ energy more than the *trans* isomer. The higher the energy content, the less stable is the molecule. Stability here refers to the thermodynamic stability. The differences seen in Fig. 4.2 are what have been qualitatively predicted on the basis of hyperconjugation (see Chapter 1, Section 1.4.2.8 and Fig. 1.2).

4.4.5 HYDROXYLATION; PERMANGANATE OXIDATION

A reaction which formally looks like an addition reaction is the conversion of an alkene to a *vic*-diol (Eq. 4.94).

$$
\underset{}{\overset{}{C}}{=}\underset{}{\overset{}{C}} \longrightarrow -\underset{OH}{\overset{|}{C}}-\underset{OH}{\overset{|}{C}}- \tag{4.94}
$$

This reaction has been called hydroxylation or *vic*-dihydroxylation. One of the simpler and more direct methods of hydroxylation is by the use of permanganate. Permanganate is a powerful oxidising agent and can oxidise many organic compounds. Alkenes can be oxidised under mild conditions using cold alkaline $KMnO_4$ to the *vic*-diol. This reaction is of importance for the synthesis of other molecules since the diols can be subjected to various reactions of utility for synthesis. The diols can be cleaved by oxidation with periodate or lead tetraacetate to the same products as those obtained by ozonolysis of the orginal alkene (Eq. 4.95).

$$
CH_3-CH{=}CH-CH_3 \xrightarrow{KMnO_4} CH_3-\underset{OH}{\overset{|}{CH}}-\underset{OH}{\overset{|}{CH}}-CH_3 \tag{4.95}
$$

$$
\xrightarrow{IO_4^- \,(or)\, Pb(OAc)_4} 2CH_3CHO
$$

$$
\tag{4.96}
$$

The reaction with permanganate, resulting in its decolourisation is used as a test for unsaturation (cf. the decolourisation of Br_2 in CCl_4 as a test for unsaturation). When mixed with an alkene, a dilute solution of $KMnO_4$, *Baeyer's reagent*, gets decolourised and a brown precipitate of MnO_2 is formed. This test also has the same limitation as the bromine test in the sense that it is not unique. Other easily oxidisable molecules like aldehydes also answer this test.

4.4.6 ALLYLIC BROMINATION BY N-BROMOSUCCINIMIDE (NBS)

There are some reactions of alkenes where hydrogen on the carbon next to the double bond, called the allylic position, is removed and another atom or group takes its place. Such a reaction is called allylic substitution reaction. One such reaction is allylic bromination which is best brought about by the reagent N-bromosuccinimide (NBS) (Eq. 4.97).

$$\text{CH}_2 = \text{CH} - \underset{\underset{\text{H}}{|}}{\text{CH}_2} + \underset{\underset{\text{CH}_2 - C}{|}}{\underset{||}{\overset{||}{\text{CH}_2 - \text{C}}}} \underset{\text{N--Br}}{\overset{\text{O}}{}} \xrightarrow[\text{or } h\nu,\, \Delta]{\text{peroxide}} \text{CH}_2 = \text{CH} - \underset{\underset{\text{Br}}{|}}{\text{CH}_2} + \underset{\underset{\text{CH}_2 - C}{|}}{\text{CH}_2 - \text{C}} \underset{\text{N--H}}{\overset{\text{O}}{}}$$

Allylic position · · · · · · · NBS · · · · · · · · · · · · · · · · Succinimide

(4.97)

The role of NBS is to release bromine into the reaction mixture in small concentrations by the oxidation of HBr which is a product of the reaction. (A trace of HBr is initially present in the NBS as an impurity).

$$\text{N--Br} + \text{HBr} \longrightarrow \text{N--H} + \text{Br}_2 \qquad (4.98)$$

The bromine so formed brominates by a free radical chain reaction. In practice, the reaction is carried out in the presence of a free radical initiator like peroxide or light (cf. addition of HBr by a free radical chain reaction—peroxide effect).

$$\text{Br}_2 \xrightarrow{h\nu} 2\,\text{Br}^\bullet \qquad (4.99)$$

$$\text{CH}_2 = \text{CH} - \underset{\underset{\text{H}}{|}}{\text{CH}_2} + \text{Br}^\bullet \longrightarrow \text{CH}_2 = \text{CH} - \overset{\bullet}{\text{CH}}_2 + \text{HBr}$$

(4.100)

(The HBr formed here goes to reaction Eq. 4.98 to continue the chain.)

$$\text{CH}_2 = \text{CH} - \overset{\bullet}{\text{CH}}_2 + \text{Br} - \text{Br} \longrightarrow \text{CH}_2 = \text{CH} - \underset{\underset{\text{Br}}{|}}{\text{CH}_2} + \text{Br}^\bullet$$

(4.101)

(The Br$^\bullet$ formed here goes to reaction Eq. 4.100 to continue the chain.)

Why does Br$_2$ not add to the alkene to form the *vic*-dibromide? The answer is that, the addition reaction also takes place, but since the first step is reversible under the conditions of the reaction (low concentration of bromine, high temperature), the equilibrium favours allylic substititution (Eqs. 4.102 and 4.103).

$$R-\overset{\bullet}{C}H-\underset{\underset{Br}{|}}{CH}-CH_3 \xrightarrow{Br_2} R-\underset{\underset{Br}{|}}{CH}-\underset{\underset{Br}{|}}{CH}-CH_3 + Br^{\bullet}$$

$$R-CH=CH-CH_3 + Br^{\bullet} \tag{4.102}$$

$$R-CH=CH-\overset{\bullet}{C}H_2 \xrightarrow{Br_2} R-CH=CH-\underset{\underset{Br}{|}}{CH_2}$$
$$+ HBr \qquad\qquad + Br^{\bullet} \tag{4.103}$$

If the Br_2 concentration is low, the first step of Eq. 4.102 reverses because the radical does not encounter a bromine molecule. The alternate path (Eq. 4.103) which goes through the more stable allyl radical, is followed. The allyl radical is relatively stable due to resonance and it stays long enough to meet a bromine molecule, even when present in low concentration.

The reasonance stabilisation of the allyl radical is shown in Eq. 4.104.

$$CH_3-CH=CH-\overset{\bullet}{C}H_2 \longleftrightarrow CH_3-\overset{\bullet}{C}H-CH=CH_2 \tag{4.104}$$

The odd electron of the allyl radical is not localised on one particular carbon atom. It also follows that the radical in Eq. 4.104 can react with Br_2 at either one of the two positions.

$$\left[CH_3-CH=CH-\overset{\bullet}{C}H_2 \longleftrightarrow CH_3-\overset{\bullet}{C}H-CH=CH_2\right]$$
$$\xrightarrow{Br_2} CH_3-CH=CH-\underset{\underset{Br}{|}}{CH_2} + CH_3-\underset{\underset{Br}{|}}{CH}-CH=CH_2 \tag{4.105}$$

Bromination of 1-butene or 2-butene can give both products of Eq. 4.105, 3-bromo-1-butene and 1-bromo-2-butene, not necessarily in the same quantity. 3-Bromo-1-butene is the expected product from 1-butene. 1-bromo-2-butene is a rearranged product. This kind or rearrangement is called *allylic rearrangement*.

4.5 USES

Alkenes are important starting materials or intermediates for laboratory synthesis and for raw materials in the chemical industry. Many of today's organic chemical industries are based on petroleum (petrochemical industries). The primary carbon source for most of these industries are ethylene, propylene, and to a lesser extent higher alkenes along with the aromatic hydrocarbons, benzene, toluene and xylenes (so-called BTX) which are also mainly from petroleum. The bulk of the alkenes go directly or after conversion to other monomers for the production of polymers. The rest are converted into basic chemicals such as ethanol, acetic acid, acetone and phenol. Table 4.4 summarises the major applications of alkenes. A minor use of ethylene is in speeding up the ripening of fruits where its role is that of a plant hormone.

Table 4. 4 Applications of alkenes

Alkene	Product	Applications/remarks
Ethylene	Polyethylene	
	Ethylene oxide → Ethylene glycol	antifreeze, polyester
	→ Polyethylene glycol	
	Vinyl chloride → PVC	
	Styrene → Polystyrene	
	Ethanol	
	Acetaldehyde	by direct oxidation, Whacker process
	Acetic acid	
	C6–C18 ethylene oligomers	
	(linear alpha olefins)	produced by Ziegler catalysis converted to aldehydes by hydroformylation (Oxo-process – reaction with CO and H_2 using cobalt octacarbonyl catalyst)
Propylene	Polypropylene	stereoregular (HDPP) by Ziegler–Natta polymerisation
	Propylene oxide → propylene glycol	
	Epichlorohydrin	epoxy resins
	Acrylonitrile	by ammoxidation
	→ polyacrylonitrile	
	Isopropylbenzene (cumene)	manufacture of phenol and acetone
Isobutylene	Polyisobutylenes	fuel and lube additives

KEY POINTS

Preparation of Alkenes

β-Elimination

$$\underset{\underset{X}{\underset{|}{\underset{\alpha}{\overset{\beta}{-C}}}}}{\overset{Y}{\overset{|}{-C}}}\!-\!C- \longrightarrow _{/}^{\backslash}C=C_{\backslash}^{/} + XY$$

Dehydrohalogenation: Elimination of HX, X = halogen

$$\underset{\underset{X}{\underset{|}{-C}}}{\overset{H}{\overset{|}{-C}}}\!-\!C- \xrightarrow{\text{OH}^-} _{/}^{\backslash}C=C_{\backslash}^{/} + H_2O + X^-$$

Dehydration: Acid catalysed elimination of water from alcohols

$$-\overset{\displaystyle H}{\underset{\displaystyle OH}{\overset{|}{\underset{|}{C}}}}-\overset{|}{\underset{|}{C}}- \quad \xrightarrow{H^+, \Delta} \quad \overset{\diagdown}{\diagup}C=C\overset{\diagup}{\diagdown} + H_2O$$

Dehalogenation: Debromination, elimination of Br_2

$$-\overset{\displaystyle Br}{\underset{\displaystyle Br}{\overset{|}{\underset{|}{C}}}}-\overset{|}{\underset{|}{C}} + Zn \quad \longrightarrow \quad \overset{\diagdown}{\diagup}C=C\overset{\diagup}{\diagdown} + Zn\,Br_2$$

$$-\overset{\displaystyle Br}{\underset{\displaystyle Br}{\overset{|}{\underset{|}{C}}}}-\overset{|}{\underset{|}{C}}- + I^- \quad \longrightarrow \quad \overset{\diagdown}{\diagup}C=C\overset{\diagup}{\diagdown} + I\,Br + Br^-$$

Hofmann elimination: Quaternary ammonium salts

$$-\overset{\displaystyle H}{\underset{\displaystyle {}^+N(CH_3)_3}{\overset{|}{\underset{|}{C}}}}-\overset{|}{\underset{|}{C}}- \quad + OH^- \quad \xrightarrow{\Delta} \quad \overset{\diagdown}{\diagup}C=C\overset{\diagup}{\diagdown} + H_2O + N(CH_3)_3$$

Ester pyrolysis: xanthate pyrolysis (Chugaev reaction):cyclic transition states

Carbocation rearrangements; governed by carbocation stability. Examples from dehydration reactions.

$$CH_3-\overset{\displaystyle CH_3}{\underset{\displaystyle CH_3}{\overset{|}{\underset{|}{C}}}}-\overset{|}{\underset{|}{\overset{CH}{OH}}}-CH_3 \quad \xrightarrow{H^+, \Delta} \quad CH_3-\overset{|}{\underset{CH_3}{C}}=\overset{|}{\underset{CH_3}{C}}-CH_3 + CH_2=\overset{|}{\underset{CH_3}{C}}-\overset{|}{\underset{CH_3}{CH}}-CH_3$$

Reactions of Alkenes

Electrophilic addition

$$\overset{\diagdown}{\diagup}C=C\overset{\diagup}{\diagdown} + E^+ \longrightarrow -\overset{\displaystyle E}{\overset{|}{C}}-\overset{+}{\underset{|}{C}}- \quad \xrightarrow{Nu\,\overset{..}{\underset{..}{\,}}^-} \quad -\overset{\displaystyle E}{\overset{|}{C}}-\overset{|}{\underset{\displaystyle Nu}{\underset{|}{C}}}-$$

E = electrophile;   Nu = nucleophile

Addition of halogen: Bromine

$$\text{C}=\text{C} + \text{Br}_2 \longrightarrow \overset{\overset{\text{Br}}{|}}{\underset{+}{\text{C}-\text{C}}} + \text{Br}^- \longrightarrow -\overset{\overset{\text{Br}}{|}}{\underset{|}{\text{C}}}-\overset{|}{\underset{\overset{|}{\text{Br}}}{\text{C}}}-$$

Halogen in the presence of other nucleophiles

$$\text{C}=\text{C} + \text{Br}_2 + \text{OH}^- \longrightarrow -\overset{\overset{\text{Br}}{|}}{\underset{|}{\text{C}}}-\overset{|}{\underset{\overset{|}{\text{OH}}}{\text{C}}}-$$

Protonic acids, HX

$$\text{C}=\text{C} + \text{H}^+ \longrightarrow -\overset{\overset{\text{H}}{|}}{\underset{|}{\text{C}}}-\overset{+}{\underset{|}{\text{C}}}- \xrightarrow{\text{X}^-} -\overset{\overset{\text{H}}{|}}{\underset{|}{\text{C}}}-\overset{\overset{\text{X}}{|}}{\underset{|}{\text{C}}}-$$

$$\text{CH}_2=\text{CH}-\text{CH}_3 + \text{HCl} \longrightarrow \text{CH}_3-\underset{\overset{|}{\text{Cl}}}{\text{CH}}-\text{CH}_3 + \text{CH}_2-\underset{\overset{|}{\text{Cl}}}{\text{CH}_2}-\text{CH}_3$$

Markownikoff's rule (major) (minor)

Regioselectvity is controlled by the stability of the carbocation.

Peroxide effect: Addition of HBr by free radical mechanism

$$\text{CH}_2=\text{CH}-\text{CH}_3 + \text{HBr} \xrightarrow{\text{peroxide}} \text{CH}_2-\underset{\overset{|}{\text{Br}}}{\text{CH}_2}-\text{CH}_3$$

Hydration: acid-catalysed addition of water

$$\text{CH}_2=\text{CH}-\text{CH}_3 + \text{H}_2\text{O} \xrightarrow{\text{H}^+} \text{CH}_3-\underset{\overset{|}{\text{OH}}}{\text{CH}}-\text{CH}_3$$

This follows Markownikoff's rule.

Other reactions of alkenes

Hydroboration: Addition of BH_3 (actual reagent B_2H_6) followed by oxidation

$$\text{CH}_2=\text{CH}-\text{CH}_3 + \text{BH}_3 \longrightarrow -\underset{|}{\text{B}}-\text{CH}_2-\text{CH}_2-\text{CH}_3$$

$$\xrightarrow{\text{H}_2\text{O}_2/\text{OH}^-} \text{HO}-\text{CH}_2-\text{CH}_2-\text{CH}_3$$

Indirect method for alkene to alcohol; product anti-Markownikoff.

Ozonolysis:

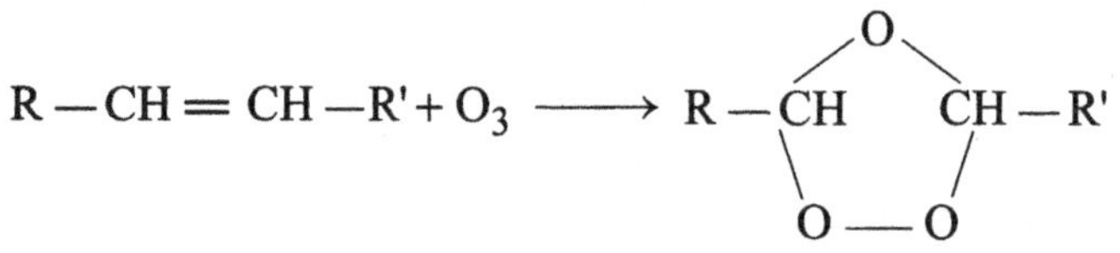

$$R - CH = CH - R' + O_3 \longrightarrow R - \overset{\displaystyle O}{\underset{\displaystyle O-O}{CH \quad CH}} - R'$$

$$\xrightarrow{\text{H}_2\text{O/Zn/H}^+} R - CHO + R' CHO$$

Useful for determining the position of the double bond.

Hydrogenation – cataytic/heterogenous:

Role of catalyst.

Heat of hydrogenation as a measure of the stability of the alkene.

Hydroxylation: By permanganate

$$R - CH = CH - R' \xrightarrow{\text{KMn O}_4} R - \underset{\underset{\displaystyle OH}{|}}{CH} - \underset{\underset{\displaystyle OH}{|}}{CH} - R'$$

Further oxidation by IO_4^- gives products similar to those in ozonolysis.

Allylic bromination: N-Bromosuccinimide(NBS), allylic rearrangement.

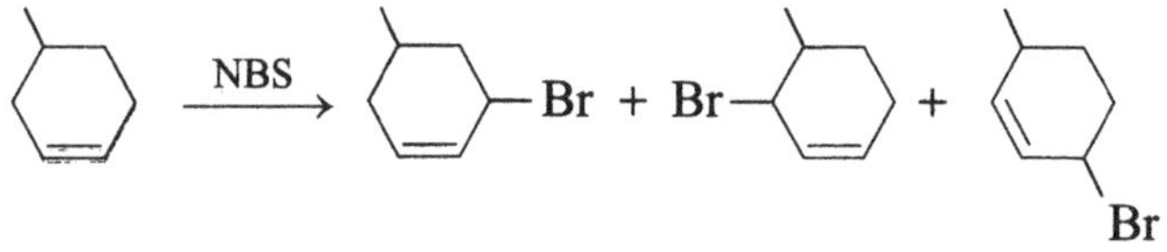

EXERCISES

SECTION I

1. Give names and structures of the alkenes formed in the following reactions,

    (a) $(CH_3)_2 CH - CH_2Cl \xrightarrow{\text{O}\overline{\text{H}}\text{/ethanol, } \Delta}$

    (b) $CH_3 - \overset{\displaystyle Br}{\underset{\displaystyle CH_3}{\overset{|}{\underset{|}{C}}}} - CH_2 - CH_3 \xrightarrow{\text{O}\overline{\text{H}}\text{/ethanol, } \Delta}$

(c)

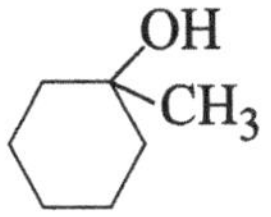

OH

CH₃

$$\xrightarrow{\text{H}^+ \ (\text{H}_2\text{SO}_4), \ \Delta}$$

(d) $CH_3 - CH_2 - CH_2 - CH_2 - OCOCH_3$ $\xrightarrow{\Delta, \ 500°C}$

(e)

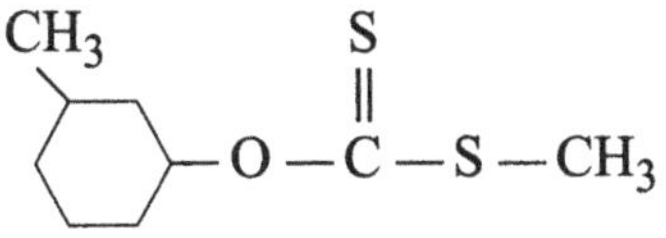

CH₃

$$\overset{\displaystyle S}{\underset{\displaystyle \parallel}{}}$$

$-O-C-S-CH_3$ $\xrightarrow{\Delta, \ 200°C}$

(f) $CH_3 - \underset{\underset{\displaystyle \overset{+}{N}(CH_3)_3 \ OH^-}{|}}{CH} - CH_2 - CH_3$ $\xrightarrow{\Delta}$

(g) $CH_3 - C \equiv C - CH_2 - CH_3$ $\xrightarrow{\text{H}_2, \ \text{Lindlar's catalyst}}$

(h) $CH_3 - C \equiv C - CH_3$ $\xrightarrow{\text{Li, liquid NH}_3}$

(i) $CH_3 - \underset{\underset{\displaystyle Br}{|}}{CH} - \overset{\overset{\displaystyle CH_3}{|}}{\underset{\underset{\displaystyle Br}{|}}{C}} - CH_3$ $\xrightarrow{\text{Zn dust}, \ \Delta}$

2. Give the names and structures of the major products of the following reactions

(a) $CH_2 = \underset{\underset{\displaystyle CH_3}{|}}{C} - CH_3$ (excess) + B_2H_6 (THF) $\longrightarrow$

(b) product of 2(a) + H_2O_2, NaOH $\longrightarrow$

(c) $CH_2 = \underset{\underset{\displaystyle CH_3}{|}}{C} - CH_3$ $\xrightarrow{\text{Cl}_2, \ \text{NaOH, H}_2\text{O}}$

(d) 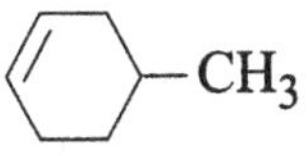 ⬡—CH₃ $\xrightarrow{\text{(i) O}_3 \ \text{(ii) Zn, CH}_3\text{COOH, H}_2\text{O}}$

(e) 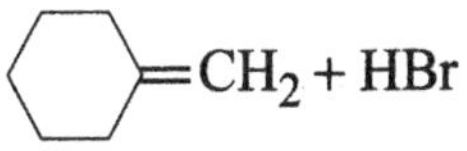 ⬡=CH₂ + HBr $\xrightarrow{\text{peroxide}}$

(f)   [cyclohexene]$-CH_3 + H_2O \xrightarrow{H^+}$

(g) CH_3-[cyclohexene] $+ NBS \longrightarrow$ (Give all possible products)

3. Which starting material will give the following as the **only product** by the type of reaction specified?

(a) $CH_2=CH-CH_2-CH_3$ by dehydrobromination

(b)   [cyclohexene]$-CH_3$ by dehydrochlorination

(c) $CH_2=CH-CH_2-CH_3$ by Chugaev reaction

(d) $CH_2=CH-CH_2-CH_3$ by Hofmann elimination

(e) $CH_3-\overset{\overset{\displaystyle O}{\|}}{C}-CH_3$ by ozonolysis

(f) $CH_3-\overset{\overset{\displaystyle OH}{|}}{CH}-CH_2-CH_3$ by hydroboration/oxidation

(g) $CH_2=\overset{\overset{\displaystyle }{|}}{\underset{\underset{\displaystyle CH_3}{|}}{C}}-CH_2-Br$ by bromination using NBS

4. (a) Rearrange the following alkenes in (i) the increasing order of stability and (ii) the decreasing order of heat of hydrogenation.

(i) $CH_2=CH-CH_2-CH_3$ (ii) $CH_2=\underset{\underset{\displaystyle CH_3}{|}}{C}-CH_2-CH_3$

(iii) $CH_3-\underset{\underset{\displaystyle CH_3}{|}}{C}=\underset{\underset{\displaystyle CH_3}{|}}{C}-CH_3$ (iv) $CH_3-\underset{\underset{\displaystyle CH_3}{|}}{C}=CH-CH_3$

(b) Rearrange the following in the increasing order of leaving group ability, as defined in section 4.2.1.1.

$$Cl^-, CH_3COO^-, OH^-, CF_3COO^-, HSO_4^-$$

(c) Rearrange the following alcohols in the increasing order of ease of acid catalysed dehydration.

$$(CH_3)_3C-OH, \quad CH_3CH_2CH_2OH, \quad CH_3-CH(OH)-CH_3$$

5. Show how the following conversion can be carried out selectively. (Carrying out a conversion selectively means, the expected product should be the only one or at least the major product).

(a) 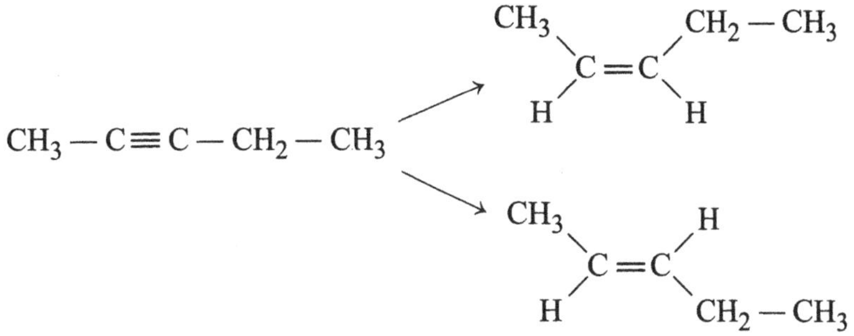 $CH_3-C\equiv C-CH_2-CH_3$

$$
\begin{array}{c}
CH_3 \\
\diagdown \\
C=C \\
H \diagup \qquad \diagdown H
\end{array}
\quad
\begin{array}{c}
CH_2-CH_3 \\
\diagup
\end{array}
$$

$$
\begin{array}{c}
CH_3 \\
\diagdown \\
C=C \\
H \diagup \qquad \diagdown CH_2-CH_3
\end{array}
\quad
\begin{array}{c}
H \\
\diagup
\end{array}
$$

(b)

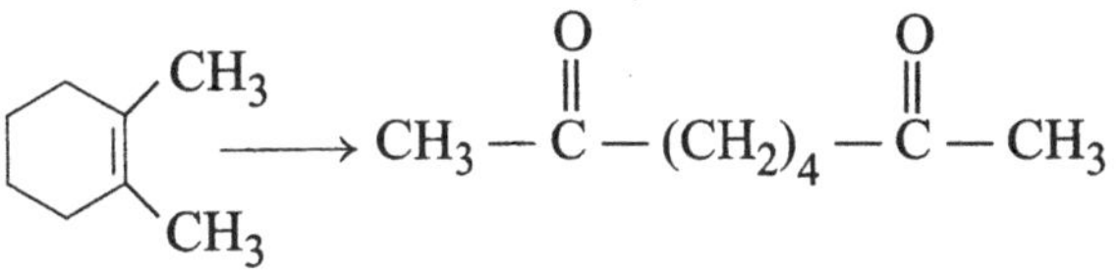

(c) 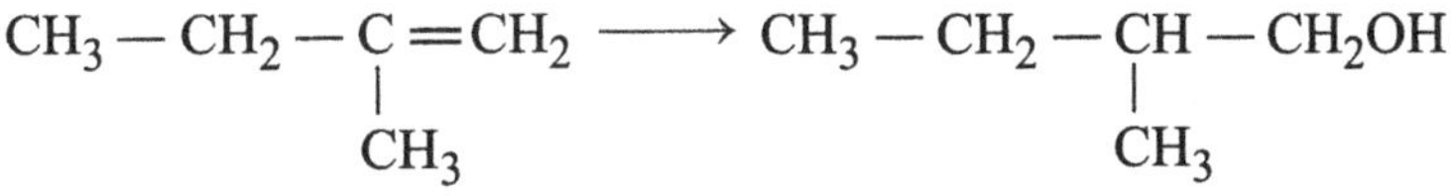 $CH_3-CH_2-\underset{\underset{CH_3}{|}}{C}=CH_2 \longrightarrow CH_3-CH_2-\underset{\underset{CH_3}{|}}{CH}-CH_2OH$

(d) 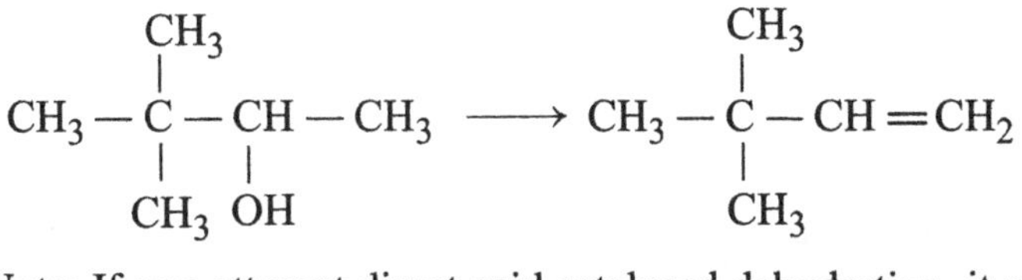 $CH_3-\underset{\underset{CH_3}{|}}{\overset{\overset{CH_3}{|}}{C}}-\underset{\underset{OH}{|}}{CH}-CH_3 \longrightarrow CH_3-\underset{\underset{CH_3}{|}}{\overset{\overset{CH_3}{|}}{C}}-CH=CH_2$

Note: If you attempt direct acid catalysed dehydration, it will result in rearranged products (see Eq. 4.34). You will have to resort to indirect methods.

(e) 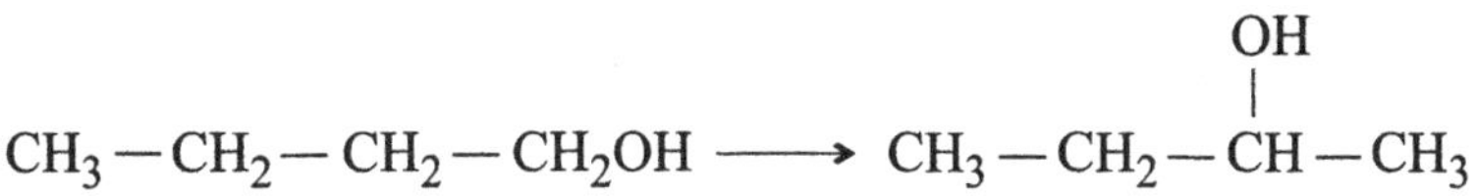 $CH_3-CH_2-CH_2-CH_2OH \longrightarrow CH_3-CH_2-\underset{\underset{OH}{|}}{CH}-CH_3$

SECTION II

1. Iodine monochloride readily adds to alkenes to form vicinal chloroiodides. Predict the regiochemistry of the addition of ICl to 1-hexene and justify your answer.

2. When 3-methyl-2-butanol, $CH_3-CH(CH_3)-CH(OH)-CH_3$, is dehydrated using aqueous acid, if the reaction is stopped before all the alcohol has reacted and then analysed, some

amount of 2-methyl-2-butanol, $(CH_3)_2C(OH)-CH_2-CH_3$ is found to be present. It is not present originally. Show how this could have formed during the reaction.

3. Alcohol (A) (also called neopentyl alcohol) does not contain β-hydrogens. Neverthless it undergoes dehydration upon heating with sulphuric acid. The products are 2-methyl-1-butene and 2-methyl-2 butene. Give the mechanism for their formation.

$$CH_3 - \underset{\underset{CH_3}{|}}{\overset{\overset{CH_3}{|}}{C}} - CH_2OH \qquad (A)$$

4. Give mechanisms for the following reactions.

(a) $CH_3 - CH_2 - \underset{\underset{CH_3}{|}}{CH} - CH_2OH \quad \xrightarrow{H^+, \Delta} \quad CH_3 - CH = \underset{\underset{CH_3}{|}}{C} - CH_3$

$$+ \; CH_3 - CH_2 - \underset{\underset{CH_3}{|}}{C} = CH_2$$

(b)

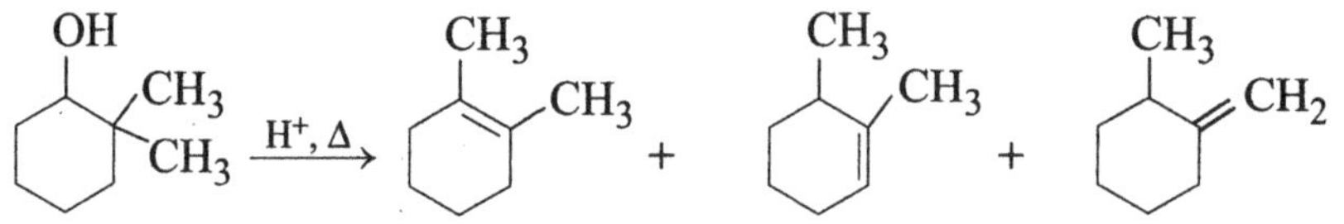

PROBLEMS

1. Identify A, B, C, D, E

(i) $C_5H_{11}Cl(A) \xrightarrow[\Delta]{OH^-/ethanol} B(C_5H_{10}, \text{ minor product}) + C(C_5H_{10}, \text{ major product})$

(ii) $C + HCl \longrightarrow C_5H_{11}Cl \; (D\text{—an isomer of A})$

(iii) $D \xrightarrow[\Delta]{OH^-/ethanol} C \text{ (major product)} + E(C_5H_{10}, \text{ minor product)} \; (B \neq E)$

(iv) $C \xrightarrow{ozonolysis} CH_3CHO + CH_3 - CO - CH_3$

2. An alkene, C_6H_{12} (A) reacts with HBr in the presence and in the absence of peroxides to give the same single product, $C_6H_{13}Br$,(B). What are the possible structures of (A)? Suggest a method to assign a definite structure.

3. An alkene C_5H_{10} (A) upon hydration using aqueous sulphuric acid gave an alcohol (B) (mainly one isomer, but with very little of another one, (C). (A) upon hydroboration followed by

oxidation with alkaline H_2O_2 gave the alcohol (C). Acetone is one of the products of ozonolysis of (A). Assign structures for (A)–(C).

CHALLENGING QUESTIONS

1. The following reactions **do not succeed**. Instead, rearranged products are formed. Predict the actual products and account for their formation.

(a)

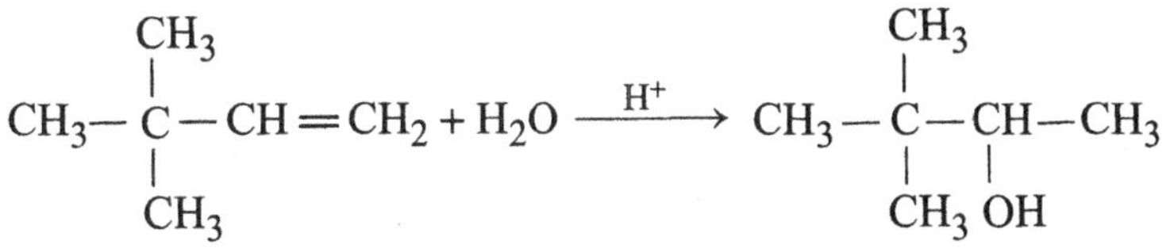

$$CH_3-\underset{\underset{CH_3}{|}}{\overset{\overset{CH_3}{|}}{C}}-CH=CH_2 + H_2O \xrightarrow{H^+} CH_3-\underset{\underset{CH_3}{|}}{\overset{\overset{CH_3}{|}}{C}}-\underset{\underset{OH}{|}}{CH}-CH_3$$

(b)

$$CH_3-\underset{\underset{CH_3}{|}}{\overset{\overset{CH_3}{|}}{C}}-CH=CH_2 + HCl \longrightarrow CH_3-\underset{\underset{CH_3}{|}}{\overset{\overset{CH_3}{|}}{C}}-\underset{\underset{Cl}{|}}{CH}-CH_3$$

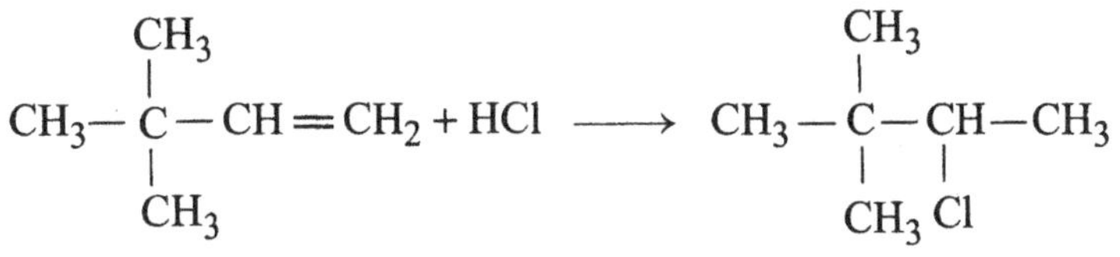

2. Consider reaction 4.13 in the text where it is mentioned in the discussion that the β-hydrogen is acidic enough to be removed by a base. The α-hydrogen should be even more acidic. Figure out what will happen if the α-hydrogen is removed by the base. Bear in mind that some of the steps are reversible.

ASSIGNMENT

From books and from the internet, prepare a report on chemical industries which use ethene (ethylene) and propene (propylene) as the principal raw materials.

5 Alkynes

OBJECTIVES In this chapter, you will learn about,

- acidity of acetylene and terminal alkynes
- methods of preparation of alkynes by
  - dehydrohalogenation of *vic-* and *gem-*dihalides
  - alkylation of acetylene and terminal alkynes
- addition reactions of alkynes
  - addition of water, how it obeys Markownikoff's rule
  - addition of hydrogen halides
  - addition of halogens
  - catalytic hydrogenation, reduction by alkali metal in liquid ammonia
  - hydroboration and further oxidation
- ozonolysis reaction

5.1 INTRODUCTION

Alkynes are hydrocarbons containing a carbon–carbon triple bond and have the molecular formula C_nH_{2n-2}. The first member of the series, ethyne, is better known as acetylene. Those alkynes with the triple bond at the end of a chain are called terminal alkynes and are monosubstituted acetylenes.

Acetylene was readily available to chemists by the end of the nineteenth century and it was prepared by the action of water on calcium carbide. Reactions of acetylene were studied in detail by the German chemist, Walter Reppe in the 1930s and 1940s. Many compounds

Acetylene was discovered by Davey in 1836.

containing triple bonds exist in nature, including the fatty acid tariric acid, the plant poison cicutoxin and the enediyne antibiotics like dynemicin A.

$$CH_3-(CH_2)_{10}-C\equiv C-(CH_2)_4-COOH \quad \text{(Tariric acid)}$$

$$CH_3-(CH_2)_2-CH(OH)-(CH=CH)_3-C\equiv C-C\equiv C-(CH_2)_3-OH \quad \text{(Cicutoxin)}$$

5.2 ACIDITY OF ACETYLENE AND THE TERMINAL ALKYNES

The striking property of the alkynes, unique among hydrocarbons, is the acidity of terminal alkynes. A hydrogen directly attached to the sp hybridised carbon atom is acidic, with a pK_a value of 26 (for acetylene). It is a very weak acid (Table 5.1).

Table 5. 1 Some acidity values

Acid	Formula	pK_a
Hydrochloric acid	$H-Cl$	-3.9
Acetic acid	CH_3COO-H	4.7
Phenol	C_6H_5O-H	10.0
Water	$HO-H$	15.7
Acetylene	$HC\equiv C-H$	26.0
Ammonia	NH_2-H	36.0
Ethylene	$H_2C=CH-H$	45.0
Ethane	H_3C-CH_2-H	62.0

The equilibrium, Eq. 5.1, will be favourable for the acetylide ion only when the base $\overline{B}$: is stronger than the acetylide; or using different terminology, when acetylene is a stronger acid than the conjugate acid of the base used, namely BH.

$$H-C\equiv C-H+\overline{B}: \ \rightleftharpoons \ H-C\equiv \overline{C}: +BH \tag{5.1}$$

Obviously hydroxide is not a strong enough base to generate the acetylide. If the acetylide is generated in the presence of water, it will be protonated by the reverse reaction corresponding to Eq. 5.1.

Examination of Table 5.1 tells us that the amide ion (conjugate base of ammonia) should be capable of generating the acetylide from acetylene in liquid ammonia medium. This is the method of choice to generate the anion from acetylene and other terminal alkynes (Eq. 5.2).

$$H-C\equiv C-H+ NaNH_2 \xrightarrow{NH_3(l)} H-C\equiv \overline{C}: \overset{+}{Na} +NH_3 \tag{5.2}$$

The sodium salt formed is called sodium acetylide. Even though acetylene contains two acidic hydrogens, only the monoanion is formed under these conditions. The negative charge on the acetylide makes the second hydrogen less acidic. The dianion can be generated by passing acetylene over heated sodium metal.

Calcium carbide, which is the calcium salt of acetylene dianion, is produced by heating limestone with coke in an electric furnace to around $2000°C$. Being a true acetylide, it reacts with water to regenerate acetylene. By passing acetylene gas through aqueous solutions of ammoniacal cuprous chloride or silver nitrate, copper and silver acetylides are precipitated out. This reaction can be used as a test for acetylene. Silver and copper acetylides are pressure sensitive explosives.

The reason for the acidity of acetylene is related to the hybridisation of the alkynyl carbon. The acidity of a molecule $A-H$ depends on the effective electronegativity of the atom to which the hydrogen is attached. The effective electronegativity of carbon, or any other atom is not the same in all molecules (see also Chapter 1, Section 1.4.2.1). Among other things, it depends upon the type of hybridisation. Electronegativity of carbon increases when the s character increases. The s character of hybrid orbitals is: sp^3—25%; sp^2—33% and sp—50%. This effect is reflected in the pK_a values of ethane, ethylene and acetylene listed in Table 5.1. Viewed in a different way, the lone pair of electrons of the acetylide anion are located in an sp orbital and the high degree of s character of this orbital makes the electrons less available for covalent bond formation (cf. low base strength of the nitrogen of nitriles, compared to amines).

5.3 PREPARATION OF ALKYNES

5.3.1 DEHYDROHALOGENATION OF GEM- AND VIC-DIHALIDES

Compounds where two halogens are attached to the same carbon are called geminal (abbreviated as *gem*-) dihalides and those where two halogens are attached to adjacent carbons are called vicinal (abbreviated as *vic*-, Section 4.2.1.3) dihalides. Both types of dihalides on treatment with strong bases undergo elimination of two molecules of hydrogen halide to give alkynes (see Chapter 4). This method is most often used for the preparation of terminal alkynes, though it is applicable to internal alkynes also. Preparation of 1-butyne is illustrated in Eqs. 5.3 and 5.4.

$$CH_3-CH_2-CH_2-CHCl_2 \xrightarrow{C_2H_5ONa/ethanol,\Delta} CH_3-CH_2-CH=CHCl$$

$$CH_3-CH_2-CH=CHCl \xrightarrow{C_2H_5ONa/ethanol,\Delta} CH_3-CH_2-C\equiv CH \quad (5.3)$$

$$CH_3-CH_2-\underset{\underset{Br}{|}}{CH}-\underset{\underset{Br}{|}}{CH_2} \xrightarrow{C_2H_5ONa/ethanol,\Delta} CH_3-CH_2-C\equiv CH \quad (5.4)$$

As shown, the reactions take place in two stages. A vinyl halide is the intermediate. For such reactions, a *nonaqueous solvent* (solvent other than water) is preferred because of the poor solubility of most organic substances in water. In this case, sodium ethoxde is used as the base and the corresponding alcohol, ethanol, as the solvent. With this base, heating is necessary to bring about the elimination reaction.

The more powerful base, sodamide in liquid ammonia, can bring about both steps of the elimination smoothly at low temperature (the temperature of liquid ammonia). Hence this is the base/solvent combination preferred for alkyne synthesis. Since the product is a terminal alkyne which will react with sodamide, three equivalents of base are consumed in this reaction, two for the two hydrogen halides and one for the acidic hydrogen of the alkyne. The sodium salt of the alkyne is the product. Treatment with water is required before the alkyne can be released (Eq. 5.5).

$$CH_3-CH_2-\underset{\underset{Br}{|}}{CH}-\underset{\underset{Br}{|}}{CH_2} \xrightarrow{NaNH_2/NH_3(l)} CH_3-CH_2-C\equiv CNa$$
$$\downarrow H_2O \qquad (5.5)$$
$$CH_3-CH_2-C\equiv CH$$

Since *vic*-dihalides are usually prepared by the addition of halogen to alkenes, this is an indirect method of converting an alkene to the alkyne.

5.3.2 ALKYLATION OF ACETYLENE OR TERMINAL ALKYNES

Introduction of an alkyl group into a molecule, which usually replaces a hydrogen, is called alkylation. The acidic hydrogen of acetylene or a terminal alkyne can be replaced by an alkyl group by treating the corresponding sodium acetylide with an alkyl halide. In acetylene itself, the two hydrogens can be successively replaced by the same or two different alkyl groups in two stages. These are illustrated in Eqs. 5.6 to 5.8, for the synthesis of propyne and 1-pentyne (terminal alkynes) and 3-heptyne (an internal alkyne).

$$CH\equiv CH \xrightarrow{NaNH_2/NH_3(l)} CH\equiv CNa \xrightarrow{CH_3I} CH\equiv C-CH_3 \qquad (5.6)$$

$$CH\equiv CNa + CH_3-CH_2-CH_2I \rightarrow CH\equiv C-CH_2-CH_2-CH_3 \qquad (5.7)$$

$$CH\equiv C-CH_2-CH_2-CH_3 \xrightarrow{NaNH_2/NH_3(l)} NaC\equiv C-CH_2-CH_2-CH_3$$
$$\downarrow CH_3-CH_2I$$
$$CH_3-CH_2-C\equiv C-CH_2-CH_2-CH_3$$
$$(5.8)$$

As can be seen, the equations are not balanced, which is often the practice in organic chemistry. In Eq. 5.8, two equations have been combined.

5.4 REACTIONS OF ALKYNES

Many of the reactions of alkynes are similar to the corresponding reactions of alkenes. A more detailed discussion of the reagents and reaction conditions is given in Chapter 4.

5.4.1 ELECTROPHILIC ADDITIONS

Like alkenes, alkynes are also electron rich molecules and are nucleophiles. Electrophilic reagents such as hydrogen halides, halogens and water add across the triple bond of alkynes.

5.4.1.1 Addition of Water (Hydration)

Recall that in the case of alkenes, addition of water requires catalysis by an acid. The same is the case with alkynes. Acidic conditions are required, but additionally mercuric ion is used as a catalyst. Typically, when an alkyne is treated with aqueous sulphuric acid containing mercuric sulphate, water adds across the triple bond. The product alcohol has a double bond and is called an *enol*. Aliphatic enols are in equilibrium with the corresponding aldehyde or ketone obtained by the shift of a hydrogen from the oxygen to the carbon. This kind of equilibrium is called *keto–enol tautomerism* (discussed in more detail in Chapter 17, Section 17.6.2). In the case of simple aliphatic enols, the equilibrium is very much towards the ketone or the aldehyde so that the only product that can be isolated is the ketone or the aldehyde. The two examples in Eqs. 5.9 and 5.10 illustrate the hydration of the terminal alkyne (propyne) and the internal alkyne (2-butyne) to form the corresponding ketones. Notice that in the case of propyne, where regioselectivity is relevant, the addition follows Markownikoff's rule (see Chapter 4, Section 4.4.1.7).

$$CH_3-C\equiv CH \xrightarrow{\text{H}_2\text{O/H}_2\text{SO}_4\text{/HgSO}_4} CH_3-\underset{\underset{\text{(enol)}}{\overset{|}{OH}}}{C}=CH_2 \rightleftharpoons CH_3-\underset{\underset{\text{(ketone)}}{\overset{||\quad|}{O\ \ H}}}{C}-CH_2 \tag{5.9}$$

$$CH_3-C\equiv C-CH_3 \xrightarrow{\text{H}_2\text{O/H}_2\text{SO}_4\text{/HgSO}_4} CH_3-\underset{\overset{|}{O\,H}}{C}=CH-CH_3 \rightleftharpoons CH_3-\underset{\overset{||\quad|}{O\ \ H}}{C}-CH-CH_3 \tag{5.10}$$

Acetylene is the only alkyne which gives an aldehyde (acetaldehyde) upon hydration (Eq. 5.11).

$$CH\equiv CH \xrightarrow{\text{H}_2\text{O/H}^+\text{/Hg}^{2+}} CH_2=\underset{\overset{|}{OH}}{CH} \rightleftharpoons CH_3-CHO \tag{5.11}$$

5.4.1.2 Addition of Hydrogen Halides

Hydrogen halides add to alkynes according to Markownikoff's rule. The triple bond can react with two molecules of hydrogen halide and the product is a *gem*-dihalide. The reaction takes place in two steps. Both steps obey Markownikoff's rule. Equations 5.12 and 5.13 illustrate the addition of HCl to 1-butyne and 2-butyne, respectively.

$$CH \equiv C - CH_2 - CH_3 + HCl \longrightarrow CH_2 = \underset{\underset{Cl}{|}}{C} - CH_2 - CH_3 \overset{HCl}{\longrightarrow} CH_3 - \underset{\underset{Cl}{|}}{\overset{\overset{Cl}{|}}{C}} - CH_2 - CH_3$$

$$(5.12)$$

$$CH_3 - C \equiv C - CH_3 + 2HCl \longrightarrow CH_3 - \underset{\underset{Cl}{|}}{\overset{\overset{Cl}{|}}{C}} - CH_2 - CH_3 \qquad (5.13)$$

As in the case of alkenes, peroxide effect is observed in the addition of HBr (Eq. 5.14).

$$CH \equiv C - CH_2 - CH_3 \xrightarrow{\text{2HBr, peroxide}} \underset{\underset{Br}{|}}{\overset{\overset{Br}{|}}{CH}} - CH_2 - CH_2 - CH_3 \qquad (5.14)$$

5.4.1.3 Addition of Halogens

Two equivalents of halogens add to the triple bond. The first stage gives the dihaloalkene by trans addition as in the case of alkenes. A tetrahalide is formed in the second stage (Eq. 5.15).

$$CH_3 - C \equiv C - CH_3 + Cl_2 \rightarrow \underset{Cl}{\overset{CH_3}{\underset{\diagdown}{}}} C = C \underset{CH_3}{\overset{Cl}{\diagup}} \xrightarrow{Cl_2} CH_3 - \underset{\underset{Cl}{|}}{\overset{\overset{Cl}{|}}{C}} - \underset{\underset{Cl}{|}}{\overset{\overset{Cl}{|}}{C}} - CH_3 \qquad (5.15)$$

5.4.2 HYDROBORATION

(See Chapter 4, Section 4.4.2 for hydroboration of alkenes.) Diborane in an ether solvent adds to alkynes. Even though two molecules of borane can add, the reaction of synthetic utility is the one corresponding to the addition of one molecule of borane. As seen from Eq. 5.16, the alkenylborane obtained can be oxidised by alkaline hydrogen peroxide to the enol which tautomerises to the aldehyde.

$$CH_3 - C \equiv CH + (BH_3) \; \rightarrow \; CH_3 - \underset{\underset{H}{|}}{C} = \underset{\underset{BH_2}{|}}{CH}$$

$$\xrightarrow{\;H_2O_2/OH^-\;} \; CH_3 - CH = \underset{\underset{OH}{|}}{CH} \; \rightleftharpoons \; CH_3 - CH_2 - \underset{\overset{||}{O}}{CH} \qquad (5.16)$$

For practical purposes, to stop the hydroboration when only one equivalent of B–H has added, a dialkylborane, R_2BH is used as the hydroborating agent instead of the usual diborane. Since dialkylboranes are fairly bulky, the addition stops when one equivalent has added, due to steric hindrance. Oxidation gives the aldehyde in addition to the alcohol corresponding to the R group of the dialkylborane. The usual dialkylborane employed for such purposes is di-*sec*-isoamylborane, abbreviated as disiamylborane, shown in Eq. 5.17.

$$CH_3 - C \equiv CH + HBR_2 \; \rightarrow \; CH_3 - \underset{\underset{H}{|}}{C} = \underset{\underset{BR_2}{|}}{CH} \; \xrightarrow{\;H_2O_2/OH^-\;} \; CH_3 - CH_2 - CHO$$

$$\left(R = CH_3 - CH - \underset{\underset{CH_3}{|}}{CH} - \text{sec-isoamyl} \atop \quad\;\; \underset{CH_3}{|} \right)$$

$$(5.17)$$

5.4.3   OXIDATION OF ALKYNES: OZONOLYSIS

Ozone reacts with alkynes and further hydrolysis without isolating the ozonide, causes the cleavage of the molecule at the triple bond to give two molecules of carboxylic acids as in Eq. 5.18.

$$CH_3 - CH_2 - C \equiv C - CH_3 \; \xrightarrow{\;(i)\,O_3;\;(ii)\,H_2O\;} \; CH_3 - CH_2 - COOH + CH_3 - COOH$$
$$(5.18)$$

When terminal alkynes are oxidised, the terminal carbon is released as carbon dioxide. Oxidation by permanganate also cleaves alkynes at the triple bond to give carboxylic acids.

5.4.4   CATALYTIC HYDROGENATION

5.4.4.1   Hydrogenation using Lindlar's Catalyst

In the presence of hydrogenation catalysts like Pt and Pd, alkynes take up two molecules of hydrogen to give the alkane. However, the reaction can be stopped to isolate the alkene – when one molecule of hydrogen has reacted – if a modified catalyst, which cannot hydrogenate alkenes, is used. Lindlar's catalyst which is specifically used for this purpose is palladium metal dispersed on barium sulphate and partially poisoned by quinoline (Chapter 4, Section 4.2.2 and Table 4.2). The addition of hydrogen is *cis*, that is, both hydrogen atoms add from the same side of the molecule (Eq. 5.19).

$$CH_3-C\equiv C-CH_3 \xrightarrow{\text{H}_2,\text{ Lindlar's catalyst}} + \underset{H}{\overset{CH_3}{\underset{\diagdown}{\diagdown}}}C=C\underset{H}{\overset{CH_3}{\diagup}} \qquad (5.19)$$

5.4.5 REDUCTION WITH LITHIUM IN LIQUID AMMONIA

Alkynes can also be reduced to alkenes by lithium in liquid ammonia. Lithium and other alkali metals dissolve in liquid ammonia to give a solution containing the metal as the cation and a solvated electron. This electron is a powerful reducing agent and can reduce alkynes to alkenes. This is known as *Birch reduction*, metal–ammonia reduction or *dissolving metal reduction*. The product of this reduction corresponds to *trans*-addition of hydrogen as in Eq. 5.20 (Chapter 4, Section 4.2.2).

$$CH_3-C\equiv C-CH_3 \xrightarrow{\text{Li, NH}_3(l)} \underset{H}{\overset{CH_3}{\underset{\diagdown}{\diagdown}}}C=C\underset{CH_3}{\overset{H}{\diagup}} \qquad (5.20)$$

KEY POINTS

- Acetylene is obtained by the action of water on calcium carbide.
- The striking property of acetylene and other terminal alkynes is their acidity, $pK_a \sim 26$.
- A strong base such as sodamide in liquid ammonia abstracts a hydrogen from acetylene to give the acetylide anion.
- Acidity of acetylene is due to the greater electronegativity of the *sp* carbon.
- Alkynes are prepared by the double dehydrohalogenation of *gem-* and *vic-* dihalides by a strong base.
- Higher alkynes can be prepared by the alkylation of acetylides using alkyl halides.
- Alkynes can enter into electrophilic addition reactions in the same way as alkenes.
- Addition of water (hydration) takes place in aqueous sulphuric acid in the presence of mercuric sulphate (catalyst). When one water molecule has added, the product is an enol which tautomerises to the ketone or aldehyde.
- Addition takes place according to Markownikoff's rule, so that all alkynes other than acetylene, give ketones. Acetylene alone gives acetaldehyde.
- Two molecules of HX or halogen add to give the *gem*-dihalide or the tetrahalide, respectively. The addition of each molecule of HX proceeds according to Markownikoff's rule.

> - Addition of HBr to terminal alkynes in the presence of peroxides gives the 1,1-dibromo compound (peroxide effect as in alkenes).
> - Hydroboration takes place with diborane or, preferably with the dialkylborane, di-*sec*-isoamylborane. The latter reagent is preferred because only one B–H unit adds to the triple bond leaving behind the double bond, due to steric hindrance. Oxidation by hydrogen peroxide/alkali gives the enol which tautomerises to the ketone. In the case of terminal alkynes, aldehyde is the product, (compare hydration using acid and mercuric ion, which gives the ketone). Ozonoylsis of alkynes cleave the molecule at the triple bond to give two carboxylic acids.
> - Catalytic hydrogenation to produce the alkene and further to the alkane, is possible. Hydrogenation can be stopped at the alkene stage by using Lindlar's catalyst. The addition of hydrogen is *cis*, 2-butyne giving *cis*-2-butene.
> - Alkyne can be reduced by lithium in liquid ammonia (Birch reduction) to alkene. In this case, the addition is *trans*. 2-Butyne gives *trans*-2-butene.

EXERCISES

SECTION I

1. Give the organic products of the following reactions.

(a) $CH\equiv C-\underset{\underset{\textstyle CH_3}{|}}{CH}-CH_3 + H_2O \xrightarrow{H_2SO_4,\ HgSO_4}$

(b) $CH_3-\underset{\underset{\textstyle CH_3}{|}}{CH}-C\equiv C-\underset{\underset{\textstyle CH_3}{|}}{CH}-CH_3 \xrightarrow{(i)\ O_3;\ (ii)\ H_2O}$

(c) $CH_2=CH-CH_2-C\equiv CH \xrightarrow{NaNH_2/NH_3(l)}$

(d) $CH_2=CH-CH_2-C\equiv CH + HCl(excess) \longrightarrow$

(e) $CH_3-C\equiv C-CH_2-\underset{\underset{\textstyle CH_3}{|}}{CH}-CH_3 \xrightarrow{H_2/Lindlar's/catalyst}$

(f) $CH_3-C\equiv C-CH_2-\underset{\underset{\textstyle CH_3}{|}}{CH}-CH_3 \xrightarrow{Li/NH_3(l)}$

(g) $CH_3-C\equiv CH + 2HBr \xrightarrow{\text{peroxide}}$

(h) $CH_3-CH_2-CH_2-C\equiv CH \xrightarrow{\text{(i) } R_2BH; \text{ (ii) } H_2O_2/OH^-}$
(R = disiamyl)

2. Show how the following can be prepared from acetylene.

  (a) $CH_3-C\equiv C-CH_3$ (c) $CH_3-CO-CH_3$

  (b) CH_3-CH_2-CHO (d) $CH_3-CCl_2CH_3$

3. What are the following? Give an application of each.

  (a) Lindlar's catalyst   (d) Sodamide

  (b) Birch reduction   (e) Solvated electron

  (c) Calcium carbide   (f) disiamylborane

4. Addition of which reagents (under controlled conditions so that only one molecule of the reagent adds) to acetylene will give the following molecules which are useful as monomers in polymer industry?

  (a) $CH_2=CHCl$ (vinyl chloride)   (b) $CH_2=CH-OCOCH_3$ (vinyl acetate)

SECTION II

1. Give reasons for the acidity of acetylene. Compare the acidity of ethylene and acetylene.
2. Show how 1-propanol can be converted to

  (a) propyne   (e) acetone

  (b) 1, 2-dichloropropane   (f) *cis*-2-butene

  (c) 2, 2-dichloropropane   (g) *trans*-butene

  (d) propanal

3. $R-X + CH\equiv C^- \overset{+}{Na} \rightarrow R-C\equiv CH$

  The above reaction is called ethynylation (cf. alkylation). Using ethynylation, how can the following transformations be brought about?

  (a) $RCl \rightarrow R-CO-CH_3$

  (b) $C_6H_5-CH_2Cl \rightarrow C_6H_5-CH_2-CO-CH_3$

  (c) $C_6H_5-CH_2Cl \rightarrow C_6H_5-CH_2-CH_2-CHO$

4. Explain the regio-selectivity in the addition of HCl to propyne. Also explain the peroxide effect in the addition of HBr to propyne.

5. Show how the following can be differentiated from each other by simple chemical tests. (A simple chemical test is one which can be done in the laboratory in test tubes using reagents readily available in the laboratory, and which will give a definite answer such as colour change, precipitate formation or gas evolution quickly).
   Ethane, ethylene and acetylene.

6. What will happen when the following are treated with methylmagnesium iodide?

       (a) Acetylene   (b) 1-Butyne   (c) 2-Butyne

   (Hint: $CH_3MgI + H - B \rightarrow CH_4 + BMgI$, when **BH** is an acid. See reaction of CH_3MgI with water in Chapter 3, Section 3.4.1.2, under preparation of alkanes by the hydrolysis of Grignard reagents).

CHALLENGING QUESTIONS

1. What are the possible products in the dehydrohalogenation by $NaNH_2$, of

$$\underset{\underset{\displaystyle Cl}{|}}{\overset{\overset{\displaystyle Cl}{|}}{CH_3-CH_2-CH_2-C-CH_3}}$$

   Do you expect any selectivity in the formation of the various possible products? (Hint: Stability of the product formed and the acidity of the hydrogen being abstracted by the base can be used as guide lines.)

2. Repeat the same exercise as in question 1 above, with the following compound.

$$\overset{\overset{\displaystyle Cl \quad Cl}{| \quad \; |}}{CH_3-CH_2-CH_2-CH-CH_2}$$

3. Your wish to prepare $C_6H_5-CH_2-\overset{\overset{\displaystyle O}{||}}{C}-CH_2-CH_3$ and plan the following route. Do you expect any complications?

$$C_6H_5-CH_2Cl + NaC\equiv C-CH_3 \rightarrow \text{1-Phenyl-2-butyne}$$

$$\xrightarrow{H_2O/H^+/Hg^{++}} C_6H_5-CH_2-\overset{\overset{\displaystyle \;}{\underset{\underset{\displaystyle O}{||}}{C}}}{}-CH_2-CH_3$$

6 Dienes

OBJECTIVES   In this chapter, you will learn about,

- classification of dienes into isolated, cumulated and conjugated
- prediction of stability of each class using their heats of hydrogenation
- prediction of chemical reactivity with particular reference to conjugated dienes, 1,2- and 1,4-addition; kinetic and thermodynamic control
- various aspects of the Diels–Alder reaction
- synthesis of dienes, specifically 1,3-butadiene, isoprene and chloroprene

6.1   INTRODUCTION

Alkenes containing only one double bond can be described as monoenes, to differentiate them from those that contain more than one double bond—the polyenes. Dienes are polyenes that contain two double bonds. A double bond is a functional group with characteristic properties and reactions, as we have seen in Chapter 4. Polyenes also exhibit these properties. However, the presence of more than one double bond – especially in conjugation – brings out some interesting additional features.

6.2   CLASSIFICATION

Dienes and polyenes in general, can be classified into three types—cumulated, conjugated and isolated. Cumulated dienes are those in which the double bonds are adjacent to each other so that the middle carbon is shared by both double bonds as in 1,2-propadiene (the trivial name of which is allene). The three carbons of allene are sp^2, sp and sp^2 hybridised respectively.

$$H_2C{=}C{=}CH_2 \quad \text{(1,2-Propadiene or allene)}$$

In conjugated dienes, the two double bonds are separated by a $C-C$ single bond as in 1,3-butadiene.

$$H_2C=CH-CH=CH_2 \quad \text{(1,3-Butadiene)}$$

Isolated dienes are those where the double bonds are separated by more than one $C-C$ single bond. That is, there is at least one sp^3 hybridised carbon between the double bonds. 1,4-Pentadiene is an example of an isolated diene.

$$H_2C=CH-CH_2-CH=CH_2 \quad \text{(1,4-Pentadiene)}$$

6.3 STABILITY AND STRUCTURE

The order of stability of the dienes is, conjugated (most stable), isolated (less stable) and cumulated (least stable). The stability referred to here is thermodynamic stability and should not be confused with reactivity which is related to reaction kinetics. As discussed in Chapters 1 and 4, heats of hydrogenation data are useful for comparing the stabilities of alkenes. The heats of hydrogenation of four hexadienes are compared with that of 1-hexene (Fig. 6.1).

The isolated diene, 1,5-hexadiene, has almost twice the heat of hydrogenation of the alkene, 1-hexene, demonstrating the truly 'isolated' nature of the two double bonds. The other dienes

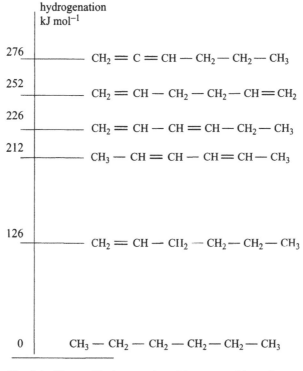

Fig. 6.1   Heats of hydrogenation of hexene and hexadienes

show interesting variations. The heat of hydrogenation of the cumulated diene, 1,2-hexadiene is 24 kJ mol^{-1} more than the isolated diene, showing that the former is the most unstable diene in the group. The conjugated dienes are more stable than the isolated dienes due to resonance brought about by the delocalisation of the π-electrons (as discussed in Chapter 1). The canonical structures shown below have charge separation, but they help to illustrate the delocalisation of the π-electrons.

$$\overset{+}{CH_2}-CH=CH-\overset{-}{CH_2} \longleftrightarrow \overset{-}{CH_2}-CH=CH-\overset{+}{CH_2}$$

2,4-hexadiene is more stable than 1,3-hexadiene due to the fact that both double bonds are internal in the former, giving rise the extra stablisation due to hyperconjugation.

The structure of 1,3-butadiene represented in Fig. 6.2, reveals certain interesting features. Each double bond has a planar arrangement of atoms. When the two double bonds are coplanar as in (a), delocalisation of electrons due to the overlap of the unhybridised p orbitals on C2 and C3 is possible. The condition that the two double bonds should be coplanar arises, because this overlap is possible only when the p orbitals on C2 and C3 are parallel to each other. The bond between C2 and C3 is a single bond, notwithstanding the canonical structures shown above, where it is a double bond. Free rotation is possible around this bond. However, when it is rotated, overlap between the p orbitals becomes less effective, and resonance energy is lost. Hence, the structure where the two double bonds are coplanar has the lowest energy among all the *conformations* of 1,3-butadiene obtained by rotation around the C2—C3 bond. Actually, there are two such conformations where resonance is possible—(a) and (b). (a) Is called the *s-trans* or *transoid* conformation, and (b) the *s-cis* or *cisoid* conformation. The *s-trans* conformation is more stable than the *s-cis* because of non-bonded interactions (steric interaction between the two hydrogens, circled in Fig. 6.2b) in the latter.

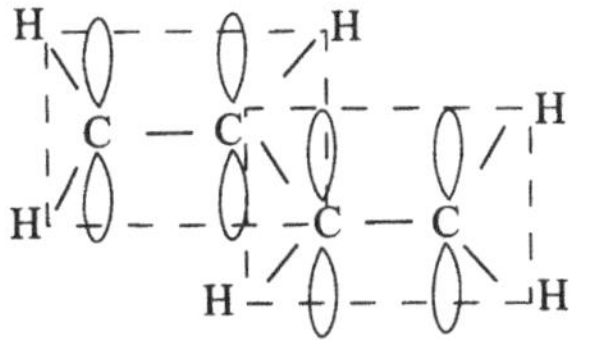

(a) s-*trans*-1, 3-Butadiene 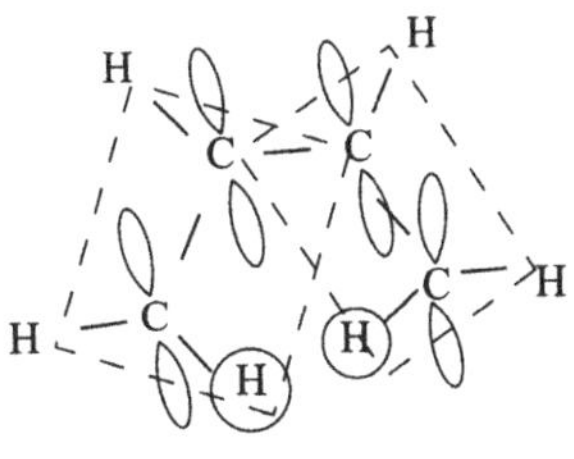   (b) s-*cis*-1, 3-Butadiene

Fig. 6.2 Structure of 1,3-butadiene

Cumulated dienes have an *sp* hybridised carbon in the middle. Since the two unhybridised *p* orbitals of the *sp* hybrised carbon have axes which are mutually perpendicular, the two double bonds which utilise these *p* orbitals have planes which are mutually perpendicular (Fig. 6.3). This shape makes 1,3-disubstituted allenes *asymmetric*, making such molecules optically active, more about which you will learn later.

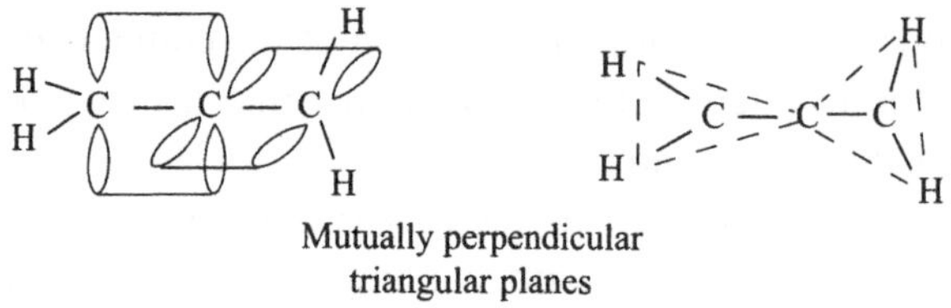

Mutually perpendicular
triangular planes

Fig. 6.3 Structure of allene

6.4   REACTIONS

Dienes take part in the characteristic reactions of alkenes such as electrophilic addition. However, electrophilic addition to conjugated dienes gives rise to products corresponding to the addition of the two fragments of the adding molecule to C1 and C2 (taking 1,3 - butadiene as the example) or to C1 and C4. These additions are referred to as *1,2-addition* or *direct addition* and *1,4-addition* or *conjugate addition*. The factors which influence the course of addition are discussed here.

6.4.1   ADDITION OF HBr AND Br$_2$

Addition of one equivalent of HBr to 1,3-butadiene gives two isomeric bromobutenes as in Eq. 6.1.

$$CH_2 = CH - CH = CH_2 + HBr \longrightarrow CH_3 - CH(Br) - CH = CH_2 \quad (a)$$

$$+ CH_3 - CH = CH - CH_2Br \quad (b) \quad (6.1)$$

The first product is the expected product arising from addition taking place according to Markownikoff's rule. The second one is an unexpected product. It is not a case of anti-Markownikoff addition as seen under conditions leading to the peroxide effect, since the position of the original double bond has changed. Similarly, in the addition of bromine to 1,3-butadiene, two products (Eq. 6.2) are obtained.

$$CH_2 = CH - CH = CH_2 + Br_2 \longrightarrow \underset{\displaystyle Br}{CH_2} - \underset{\displaystyle Br}{CH} - CH = CH_2 \quad (a)$$

$$+ \underset{\displaystyle Br}{CH_2} - CH = CH - \underset{\displaystyle Br}{CH_2} \quad (b) \qquad (6.2)$$

The product (a) arises from *1,2-addition* or *direct addition* and the product (b) from *1,4-addition* or *conjugate addition*. The ratio of 1,2- and 1,4-addition is temperature dependant.

6.4.2   MECHANISM OF ADDITION OF HBr—KINETIC VS THERMODYNAMIC CONTROL

The first step in the addition of HBr is the electrophilic addition of a proton, following Markownikoff's rule, to give the more stable carbocation (Eq. 6.3). Comparing this carbocation with the preferred carbocation formed by the addition of a proton to 1-butene (Eq. 6.4), it can be seen that while the latter is secondary, the former is not only secondary but is also a resonance stabilised allyl carbocation. The two canonical structures of the allyl carbocation are shown in Eq. 6.3.

$$\overset{1}{CH_2}=\overset{2}{CH}-\overset{3}{CH}=\overset{4}{CH_2} + H^+ \longrightarrow \begin{bmatrix} CH_3 - \overset{+}{CH}-CH=CH_2 & \text{(a)} \\ CH_3 - CH=CH - \overset{+}{CH_2} & \text{(b)} \end{bmatrix} \quad (6.3)$$

$$CH_2=CH-CH_2-CH_3 + H^+ \longrightarrow CH_3 - \overset{+}{CH}-CH_2-CH_3 \quad (6.4)$$

It is evident from the canonical structures that the positive charge of the allyl carbocation is distributed on C2 and C4. When the carbocation is quenched by the Br^- ion, either one of the products of Eq. 6.1 is possible, and that is exactly what happens. The two *regioisomers* are not formed in equal quantities. When the addition is carried out at $-80°C$, the 1,2-addition product (1,2-*adduct*) 3-bromo-1-butene (of Eq. 6.1a), is 81% of the mixture and the 1,4-adduct [1-bromo-2-butene (6.1b)] is 19%. At $0°C$, this changes to 70% and 30% and at $45°C$ to 15% and 85%, respectively. The composition of the product mixture is decided by the interplay of two factors—the rate at which the products are formed (*kinetic control*) and the thermodynamic stability of the two products (*thermodynamic control* or *equilibrium control*). Of the two haloalkenes, the 1,4-adduct [1-bromo-2-butene (Eq. 6.1b)], an internal alkene, is thermodynamically more stable than the 1,2-adduct, terminal alkene [3-bromo-1-butene (Eq. 6.1a)]. From the data on heats of hydrogenation, we have seen (Chapter 1, Section 1.4.3) that 2-butene is thermodynamically more stable than 1-butene, the reason being the greater stabilisation due to hyperconjugation in the former. The same reason applies to the greater stability of (b) (Eq. 6.1) compared to (a) (Eq. 6.1). The more stable product is predominantly formed at higher temperatures (why it is so is discussed below). Thermodynamic control operates under these conditions. At lower temperatures, addition is controlled by the rate of formation of the two products. The 1,2-adduct is formed at a faster rate than the 1,4-adduct. An examination of the canonical structures of the intermediate allyl cation in Eq. 6.3 shows that the positive charge is located on C2 and C4. The charge density at the two positions is not the same. Referring to the structures in Eq. 6.3, the first canonical structure (a), has the positive charge on a secondary carbon, and the second canonical structure (b), on a primary carbon. The charge density in the hybrid is more at C2 than at C4. Hence, the anion reacts with C2 leading to 1,2-addition at a faster rate than 1,4-addition. These ideas are best illustrated with the help of an energy diagram (Fig. 6.4).

For the following discussion, refer to Fig. 6.4 and Eq. 6.5. In this two-step reaction (Eq. 6.5), the overall rate of the reaction (rate of formation of both products taken together) is decided by the rate of proton addition, which is the first step. This is the rate determining step. The rate of the reaction is dependent on the activation energy for this step, E^*1, which depends upon factors which we shall not discuss at present. Protonation gives the resonance stabilised allylic carbocation (Eq. 6.5a), which undergoes two parallel reactions—quenching by the bromide ion either at C2 or at C4. Reaction at C2 has lower activation energy, $E^*(1, 2)$, because of the greater positive charge density at C2 than at C4, whose energy of activation, $E^*(1, 4)$, is more. Hence, bromide addition at C2 which is overall 1,2 addition of HBr, takes place faster than bromide addition at C4. The

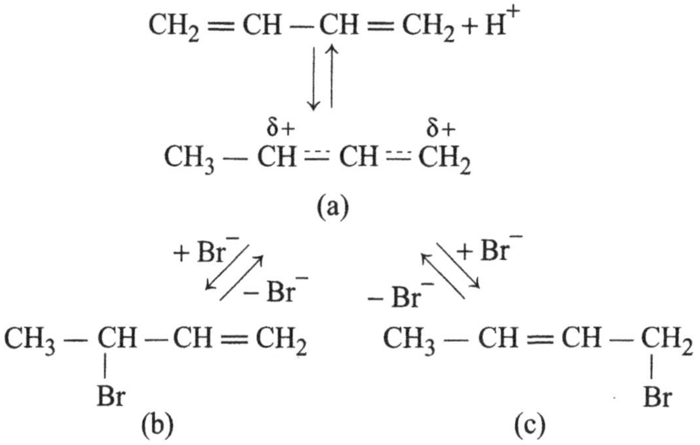

$$CH_2 = CH - CH = CH_2 + H^+$$

$$\updownarrow$$

$$CH_3 - \overset{\delta+}{CH} \cdots CH \cdots \overset{\delta+}{CH_2}$$

(a)

(6.5)

$$CH_3 - \underset{\underset{Br}{|}}{CH} - CH = CH_2 \qquad CH_3 - CH = CH - \underset{\underset{Br}{|}}{CH_2}$$

(b)   (c)

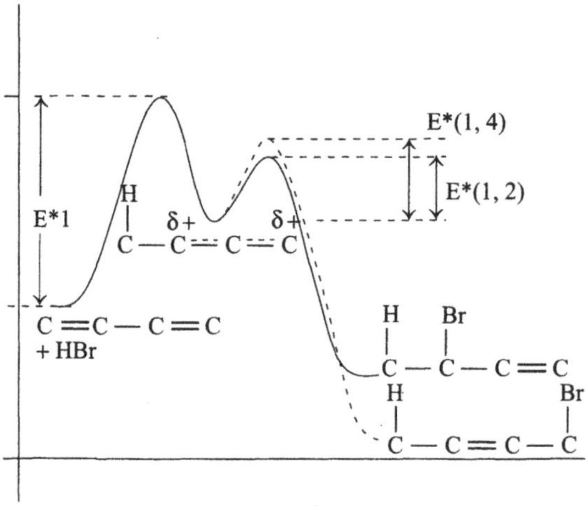

Fig. 6.4 Energy diagram for 1,2- and 1,4-addition (kinetic and thermodynamic control)

1,2-adduct (Eq. 6.5b), is the kinetically controlled product and predominates at low temperatures. However this step – the bromide addition – is reversible. The final products, 1-bromo-2-butene (c), and 3-bromo-1-butene (b), are both allylic bromides and they can undergo heterolytic dissociation (ionisation) to give back the carbocation and the bromide. Ionisation of alkyl halides is not a facile reaction. However, when the carbocation formed by ionisation is relatively stable (like tertiary or allyl), under appropriate conditions (polar solvent, high temperature), ionisation can take place (see Chapter 11 for a discussion of ionisation of alkyl halides in the dehydrohalogenation reaction). In the present case, the product of ionisation is the resonance stabilised allyl cation and that is the driving force for the reversal of the step. The consequence is that the two products can interconvert; that is, one can get converted to the other through the common intermediate, the cation. If enough time is given, equilibrium gets established between the two regioisomers. At the higher temperature, sufficient energy is available for the molecules to overcome the energy barrier for the reverse reactions and equilibrium is established rapidly. Under these conditions, the product composition does not correspond to the rate at which they are formed, but to their equilibrium composition. At equilibrium, the more stable isomer predominates.

Generally speaking, organic reactions are *kinetically controlled*. That is, if more than one product is possible in a reaction, the product composition is a reflection of the rates at which they are formed. When the products can intercovert and equilibrium gets established under the conditions of the reaction, the product composition is, or approaches, the equilibrium composition. Under these conditions, the reaction is *thermodynamically controlled*.

Coming back to the reaction under discussion, if the product mixture formed at low temperature – such as the one at −80°C with 81% of the 1,2-adduct – is heated in the presence of HBr to 45°C, the composition will change. It will become one with 15% of the 1,2-adduct and 85% of the 1,4-adduct, same as the composition if the reaction was initially carried out at 45°C.

This example also illustrates another important aspect of reactions which is often overlooked. The major product formed here at lower temperatures is not the more stable product, but the less stable one. Very often, we make statements such as, 'a particular product is the major product because it is the more stable one'. Such a statement is valid only if the more stable product is also the one which is formed at the faster rate, or if the reaction is thermodynamically controlled—which need not always be true.

6.4.3 DIELS–ALDER REACTION

An important reaction of dienes is their reaction with monoenes, to form 6-membered rings, as illustrated in the reaction of 1,3-butadiene and ethene to form cyclohexene (Eq. 6.6).

$$
\begin{array}{ccc}
\text{1,3-Butadiene} & + \ \| \ \text{Ethene} \longrightarrow & \text{Cyclohexene}
\end{array}
\tag{6.6}
$$

This example is of significance only for purpose of illustration because ethene itself does not readily undergo such a reaction.

This is the Diels–Alder reaction which is a *cycloaddition reaction*. Specifically, it is a $(4 + 2)$ cycloaddition, meaning it is a reaction between a 4π-electron molecule and a 2π-electron molecule. It is a member of a group of reactions called *pericyclic reactions*. (Some other members of this group are *sigmatropic rearrangements* like Claisen rearrangement (Chapter 19) and *electrocyclic reactions* such as interconversion of butadiene and cyclobutene). All these reactions have some things in common such as cyclic transition states, and that they take place in a single step without the formation of any intermediates.

Diels–Alder reaction is a very useful reaction discovered by the German chemists, Otto Diels (1876–1954) and Curt Alder (1902–1958), for which they were awarded the Nobel Prize in 1950.

The two reactants of the Diels–Alder reaction, 1,3-butadiene and ethene, in Eq. 6.6 are called the *diene* and the *dienophile*, respectively. The product cyclohexene, is the Diels–Alder *adduct*.

Ethene and simple alkenes are not particulary efficient dienolphiles. Molecules with electron withdrawing groups attached to one or both of the sp^2 hybridised carbons are good dienophiles. (See Chapter 1, Table 1.3, for a list of common electron-withdrawing groups). Carbonyl groups present in aldehydes, ketones, and carboxylic acids and their derivatives, when attached to the alkene carbon make such alkenes good dienophiles. Generally all dienes, preferably without electron-withdrawing substituents, react as the diene component provided it can assume the *s-cis* conformation. Some examples are listed in Eqs. 6.7 to 6.15.

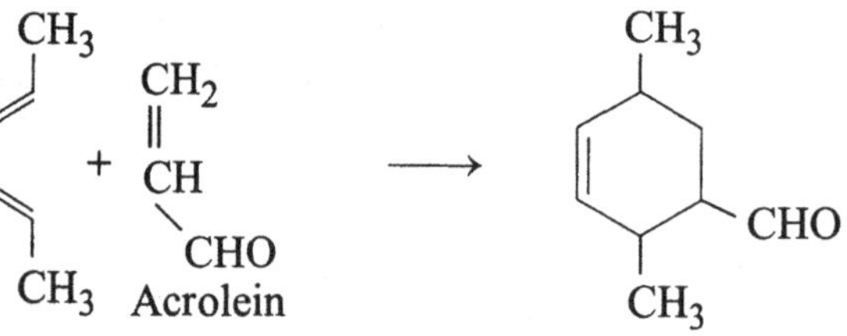

(6.7)

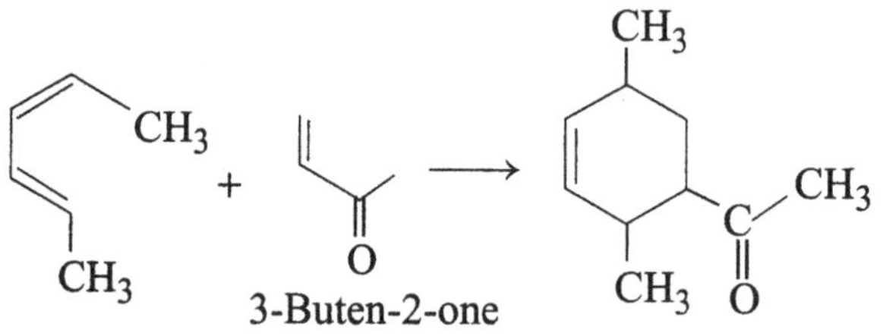

(6.8)

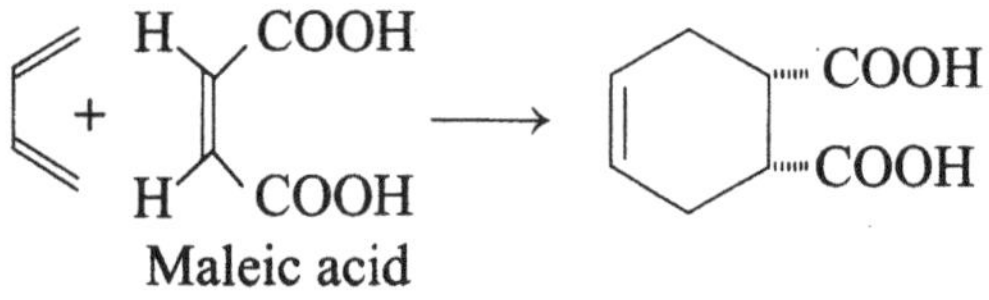

3-Buten-2-one

(6.9)

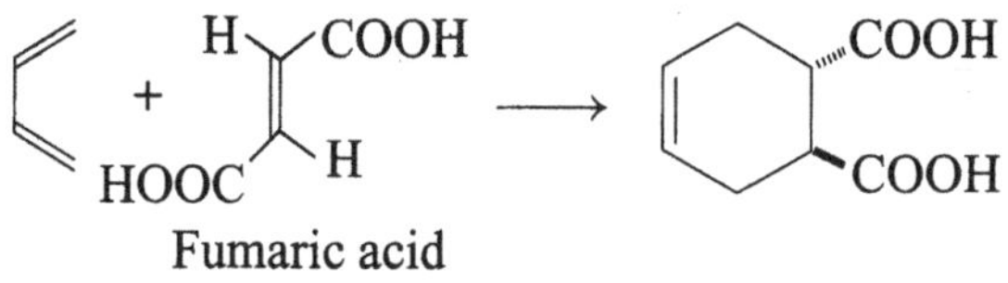

Maleic acid

(6.10)

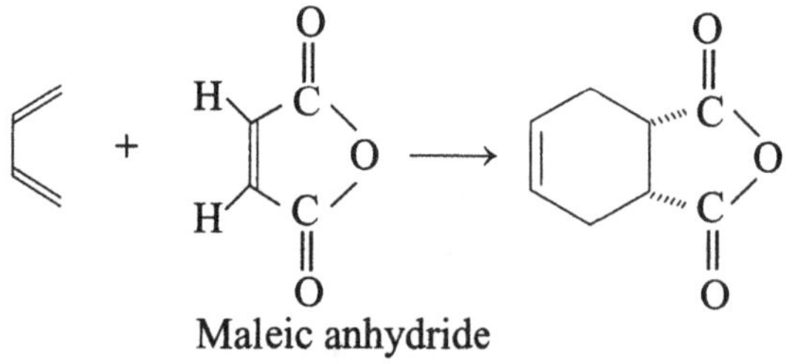

Fumaric acid

(6.11)

Maleic anhydride

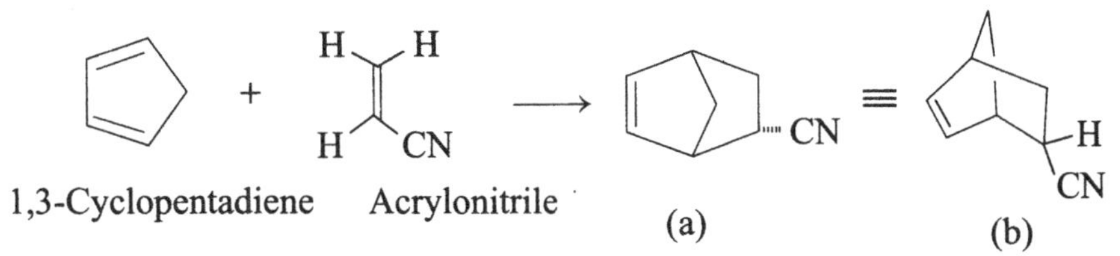

1,3-Cyclopentadiene   Acrylonitrile   (a)   (b)   (6.12)

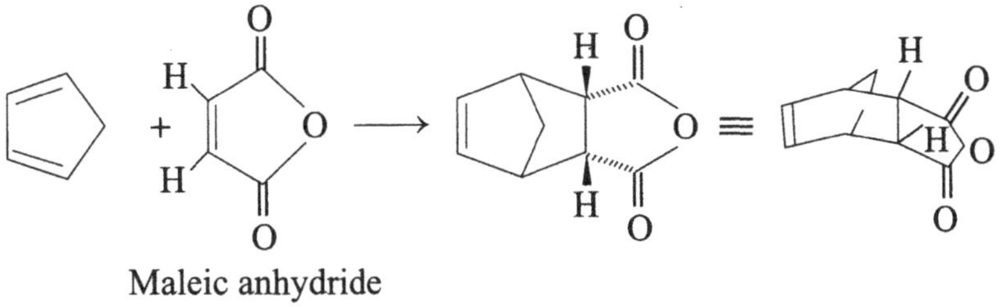

Maleic anhydride   (6.13)

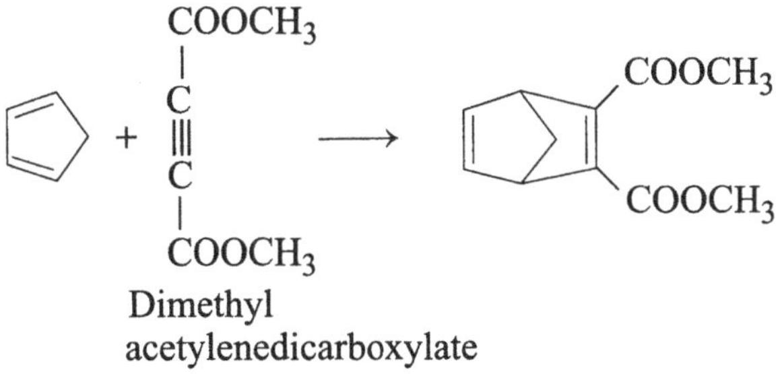

Dimethyl
acetylenedicarboxylate   (6.14)

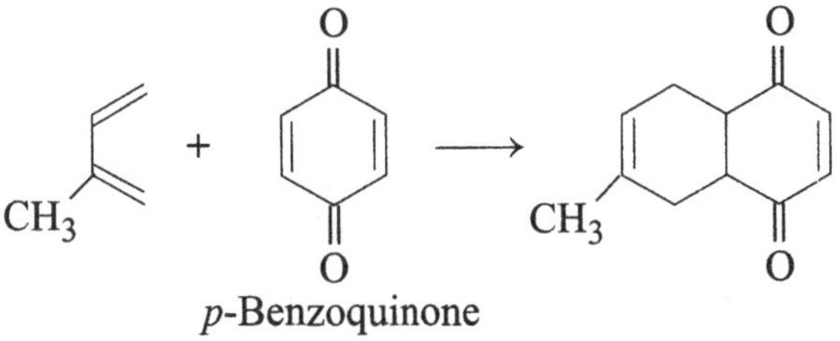

p-Benzoquinone   (6.15)

Since a cyclic compound is the final product, the diene should be oriented in the *s-cis* conformation, even though in most cases this is the less stable conformation. Equation 6.16 illustrates the need for the *s-cis* arrangement of the diene. The diene in this example – even though conjugated – is rigidly held in the *s-trans*-like conformation and hence does not react. In Eqs. 6.12, 6.13 and 6.14, the diene is cyclopentadiene which is rigid in the *s-cis* conformation and is a very reactive diene. The Diels–Alder reaction is a reversible reaction and many adducts which are thermodynamically unstable or highly sterically crowded, undergo the reverse reaction upon heating – the retro-Diels–Alder – to give back the starting materials. In some cases, the same adduct could have been formed from two different sets of diene–dienophile pairs. This is the case when the diene is a cyclohexadiene derivative as in Eq. 6.17. The dienophile component of the second pair is not a reactive one because it has no electron-withdrawing groups. So the reaction will go all the way to the right. (Section III, Challenging questions, question No. 4 gives an extreme example of this).

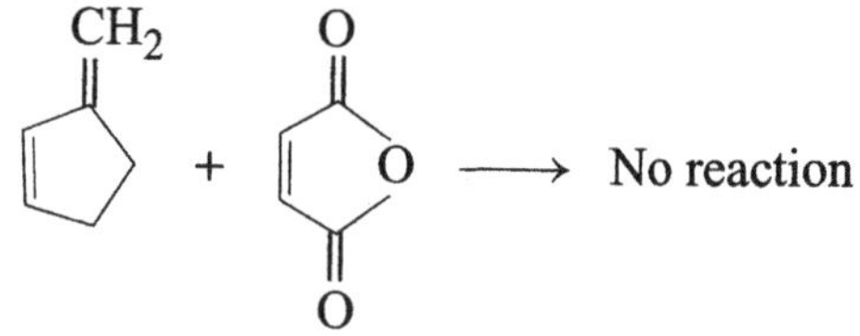

(6.16)

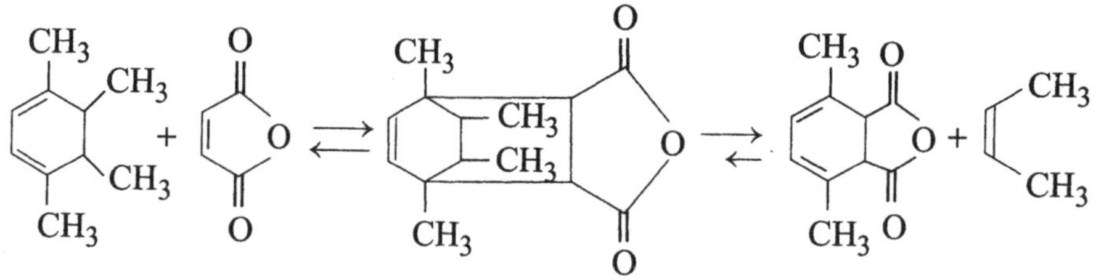

(6.17)

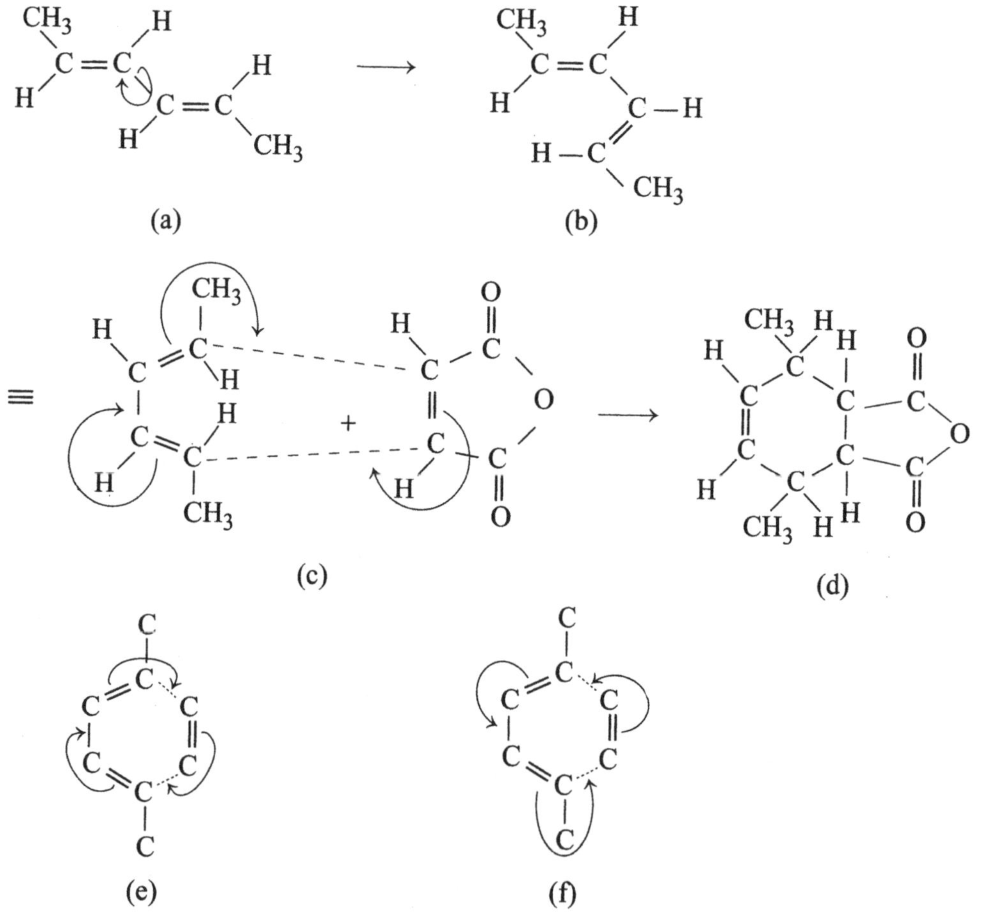

(a)   (b)

(6.18)

(c)   (d)

(e)   (f)

Equation 6.18 gives some guidelines for drawing the structure of the adduct. The diene is E,E- or (*trans, trans*-2,4-hexadiene), normally written in the *s-trans* conformation (a), as shown. The first step is to rotate a portion of the molecule around the C3—C4 single bond to arrive at the *s-cis* conformation (b). Next, redraw this so that the concave face of the cisoid molecule faces right as in (c). This need not be done if one has practice, but the beginner is advised to go through this step. Now, draw the dienophile on the right side of the diene, with the functional groups facing away from the diene (This is only to make the drawing less cluttered. Actually, as we shall see later, the functional groups may be facing the same direction as the double bonds of the diene). Draw broken lines as shown to remind us where the new bonds are to be formed. Bear in mind that the Diels–Alder reaction is a 1,4- or conjugate addition of the alkene to the diene. You may find it useful to draw curved arrows to keep track of the breaking and formation of bonds. In the equation, the arrows are shown to move clockwise, (e). This is arbitrary. They could as well have been shown to move counter-clockwise, (f). These curved arrows do not have the same significance as in conventional reaction mechanisms, where they imply the shifting of a pair of electrons. However, the curved arrows do help us to arrive at the correct arrangement of bonds in the final product. However complex the structures of the diene and the dienophile are, our job is made simple if the double bonds of the diene and dienophile are located and oriented face to face correctly, bearing in mind that as a result of the two new bonds formed, a six-membered ring is generated.

The above discussion might give a false idea that the diene and the dienophile have to come together in the same plane. This is far from the actual situation. The two molecules approach each other in two parallel planes (Fig. 6.5). It is enough to know at this stage – without going into the details of the role of orbital symmetry in pericyclic reactions – that the two new single bonds formed in the adduct are formed by the overlap of the concerned *p*-orbitals of the diene and dienophile along their axis, not laterally. The lobes of the *p*-orbitals utilised to form the new bonds are those on the same face of the diene (bottom face in the drawing) and the same face of the dienophile (top face of the dienophile in the drawing). In the terminology of pericyclic reactions, when *the lobes on the same face of a reactant* are used, the reaction is said to be *suprafacial* with respect to that reactant. The Diels–Alder reaction is a *suprafacial–suprafacial* or simply, a *supra–supra* addition. To use the terminology of other addition reactions, it is a *cis*-addition.

There are certain stereochemical aspects of the Diels–Alder reaction which follow from the mechanism discussed above. Groups which are *cis* in the dienophile remain *cis* in the adduct as in Eq. 6.9 with maleic acid. In Eq. 6.10 where fumaric acid is used, the carboxy groups in the adduct remain *trans*. In the dienophile, the terms *cis* and *trans* are used in relation to the geometrical isomerism in alkenes. In the adduct, the same terms refer to geometrical isomerism in cycloalkanes (discussed in Chapter 22). Wedge-shaped lines are used to indicate bonds pointing above the average plane of the molecule, while broken lines indicate bonds below the plane. Substituents on the dienophile usually end up below the plane in the adduct. In Eq. 6.11, the dienophile used is maleic anhydride. This is a commonly used dienophile because of its ready availability and high reactivity. It is a solid and the adducts are solids even when the dienes are liquid hydrocarbons. The Diels–Alder reaction generates a cyclohexene ring. If one of the reactants is already cyclic, the adduct will have one ring more than the number of rings started with. In Eq. 6.11, the dienophile is cyclic and in Eq. 6.12 the diene is cyclic. The product in both cases is bicyclic. In Eq. 6.13,

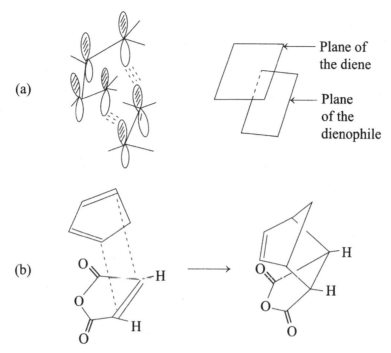

(a)

(b)

Plane of
the diene

Plane
of the
dienophile

Fig. 6.5 Diels–Alder reaction: diene and dienophile in parallel planes

both are cyclic and the product is thus tricyclic. In adducts such as those in Eqs. 6.12 and 6.13 where the CH_2 *bridge* is shown to be above the plane, the substituents of the dienophile, $-CN$ in Eq. 6.12 and the $-CO-O-CO-$ part of the anhydride in Eq. 6.13, are below the plane. This happens because substituents such as the anhydride portion of maleic anhydride are oriented in the transition state in such a way that they are close to the π-electrons of the diene (Fig. 6.5). Such an orientation of the substituent is said to be *endo*. Groups above the plane are said to be *exo*. In the Diels–Alder reaction, endo addition is preferred. (This is sometimes referred to as *Alder's endo rule*.)

The utility of the Diels–Alder reaction in organic synthesis cannot be over emphasised. It is a method for creating two $C-C$ bonds in one step and for the formation of six membererd rings. The stereochemistry is well defined. It is applicable to a wide range of dienes and dienophiles. The reaction is also used as a diagnostic tool for the detection of conjugated dienes and for their separation and estimation.

6.5 PREPARATION OF DIENES

Conjugated dienes are prepared in the laboratory by the dehydrohalogenation of alkyl halides of the type $R-CH=CH-CHX-CH_2-R'$ which are readily prepared from alkenes. Recall the allylic bromination of alkenes by N-bromosuccinimide (Chapter 4, Section 4.4.6). For example, 2-methyl-1,3-butadiene – whose trivial and more familiar name is isoprene – can be prepared as in Eq. 6.19.

$$CH_2 = CH - \underset{\underset{CH_3}{|}}{CH} - CH_3 \xrightarrow{NBS} CH_2 = CH - \underset{\underset{CH_3}{|}}{\overset{\overset{Br}{|}}{C}} - CH_3$$

$$\xrightarrow{ethanol/KOH/\Delta} CH_2 = CH - \underset{\underset{CH_3}{|}}{C} = CH_2$$

2-Methyl-1,3-butadiene
(isoprene)   (6.19)

A halogen on the carbon next to the allylic position – called the *homoallylic* position – upon dehydrohalogenation gives the conjugated diene regioselectively because of the greater stability of the conjugated diene (Eq. 6.20).

$$CH_2 = CH - CH_2 - \underset{\underset{Cl}{|}}{CH} - CH_3 \xrightarrow{ethanol/KOH/\Delta} CH_2 = CH - CH = CH - CH_3$$

(Not $CH_2 = CH - CH_2 - CH = CH_2$)   (6.20)

Allyl and homoallyl alcohols also give the corresponding conjugated dienes upon dehydration (Eq. 6.21).

$$CH_3 - \underset{\underset{CH_3}{|}}{C} = CH - \underset{\underset{OH}{|}}{CH} - CH_3 \xrightarrow{H^+, \Delta} CH_3 - \underset{\underset{CH_3}{|}}{C} = CH - CH = CH_2 \qquad (6.21)$$

These reactions can also be used for the preparation of isolated dienes, but regioselectivity cannot be expected.

Eneynes – molecules containing double and triple bonds – upon selective hydrogenation using Lindlar's catalyst, give dienes. Vinylacetylenes, $R - CH = CH - C \equiv C - R'$, give conjugated dienes (Eq. 6.22).

$$CH_2 = CH - C \equiv CH + H_2 \xrightarrow{Lindlar's\ catalyst} CH_2 = CH - CH = CH_2 \qquad (6.22)$$

Some specific dienes which are important industrially or in other ways, are discussed below.

6.5.1 1,3-BUTADIENE

This is used in the manufacture of synthetic rubbers such as butadiene–styrene copolymer (SBR). Industrially, it is obtained by the dehydrogenation of butane where the products are butenes and butadiene.

6.5.2 Isoprene

Isoprene is the trivial name for 2-methyl-1,3-butadiene. Many natural products including terpenes present in the *essential oils* of plants (the oils isolated from the *essences* of flowers, fruits and other plant materials) and natural rubber are formally polymers of isoprene. Equations 6.23 and 6.24 formally depict how the terpene citral, present in lemon grass oil and natural rubber are related to isoprene.

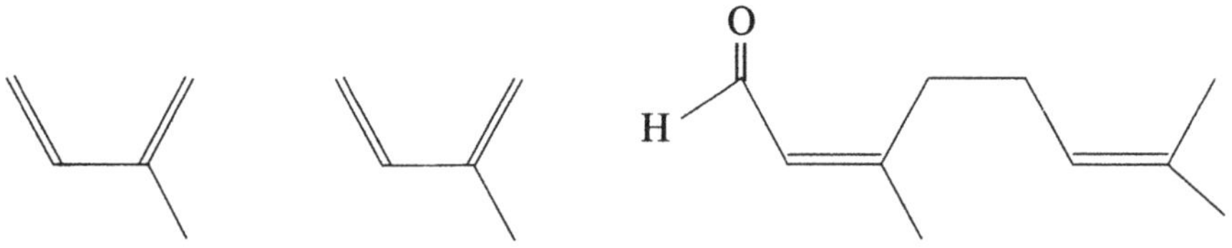

2 Isoprene units   Citral   (6.23)

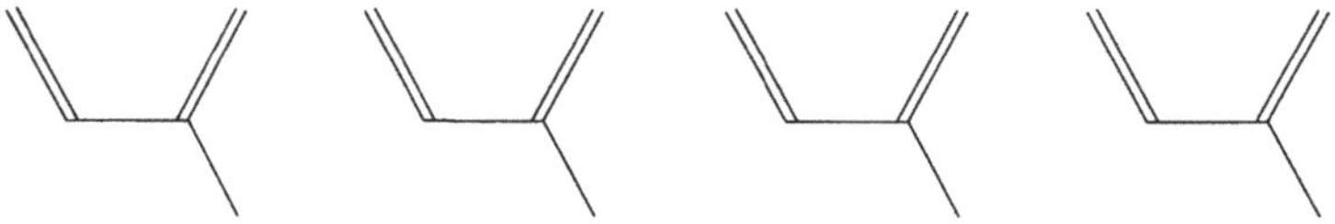

4 Isoprene units

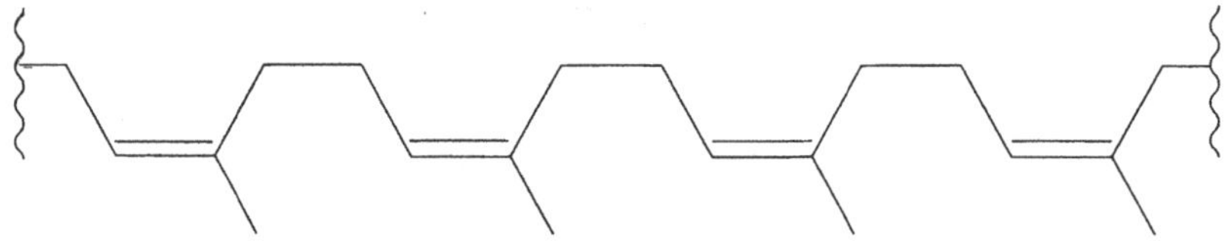

(6.24)

Chain of natural rubber (notice that all the double bonds are *cis*)

The actual molecule involved in the synthesis of these products in the plant (biosynthesis) is isopentenyl pyrophosphate, which is called the biological isoprene unit. This molecule is an intermediate in biosynthetic pathways.

$$CH_2=\overset{\overset{\large CH_3}{|}}{C}-CH_2-CH_2-O-\underset{\underset{\large OH}{|}}{\overset{\overset{\large O}{\|}}{P}}-O-\underset{\underset{\large OH}{|}}{\overset{\overset{\large O}{\|}}{P}}-OH$$

Isopentenyl pyrophosphate   (6.25)

Isoprene is industrially obtained from the C5 hydrocarbons cut from the products of thermal cracking of petroleum fractions.

6.5.3 CHLOROPRENE

Chloroprene is the trivial name of 2-chloro-1,3-butadiene. It is obtained by the addition of one equivalent of HCl to vinylacetylene, a dimer of acetylene (Eq. 6.26).

$$CH_2 = CH - C \equiv CH + HCl \longrightarrow CH_2 = CH - \underset{\underset{Cl}{|}}{C} = CH_2 \qquad (6.26)$$

A polymer of chloroprene (neoprene), is one of the first industrial *elastomers*, the name given to synthetic polymers having the property of elasticity. Neoprene has varied applications as a synthetic rubber.

KEY POINTS

- Polyenes are compounds containing more than one double bond. Those containing two double bonds are called dienes.
- Dienes are classified as cumulated, conjugated and isolated.
- Conjugated dienes are more stable than isolated dienes.The former are resonance stablised. Cumulated dienes are the least stable of the three types.
- Heats of hydrogenation help to compare stabilities.
- Conjugated dienes, in order to maintain conjugation, require all the *p*-orbitals of the double bonds to be parallel. This is possible only when the two alkene components are coplanar. This condition is met only in two conformations, the *s-trans* and the *s-cis*. The former is more stable than the latter.
- Dienes undergo all the reactions of simple alkenes. Of special interest is electrophilic addition to conjugated dienes. Two regioisomeric adducts correspoding to 1,2-addition and 1,4-addition are possible. The latter is also called conjugate addition. This reaction illustrates the phenomena of kinetic control and thermodynamic control.
- Conjugated dienes take part in cycloaddition reactions. The reaction of a conjugated diene with a monoene (called the dienophile) gives cyclohexene derivatives. This reaction is called the Diels–Alder reaction. Cycloadditions belong to a group of reactions called pericyclic reactions. Diels–Alder reaction is stereospecific. Substituents which are *cis* in the dienophile end up *cis* in the product. Those which are *trans* in the dienophile remain *trans* in the product.

EXERCISES

SECTION I

1. Locate the isolated and conjugated double bonds in the following.

   (a) $CH_2 = CH - CH = CH - CH = CH - CH_3$

(b) $CH_3-(CH_2)_6-CH_2-CH=C=CH-CH=CH-COOCH_3$ (This is a naturally occurring material—the sex attractant in the dried-bean beetle)

(c) 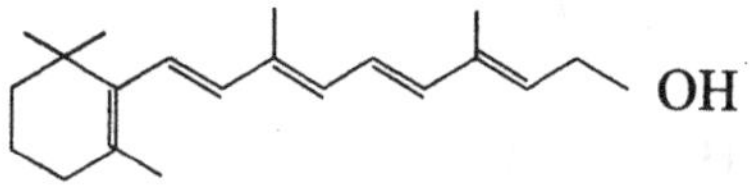   OH   (vit. A)

(d)

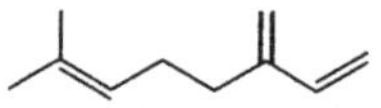

(myrcene, a terpene from bayberry oil)

(e)

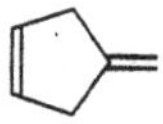

(f)

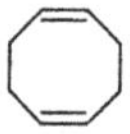

(g)

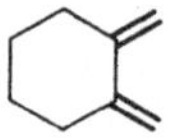

2.   Draw the *s-cis* conformations, if possible of the following.

(a)

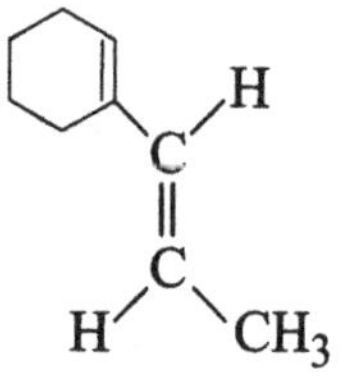

(b)

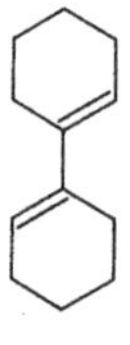

(c)

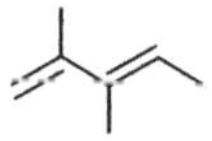

(d)

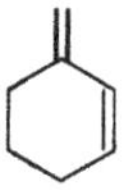

(e)

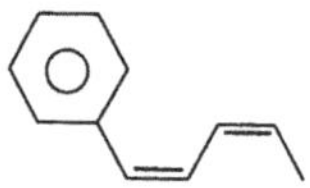

3.   Draw structure of the Diels–Alder adduct.

(a)

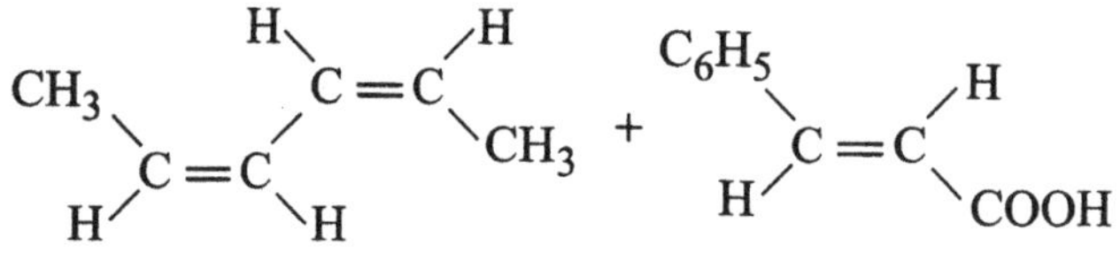

(b)

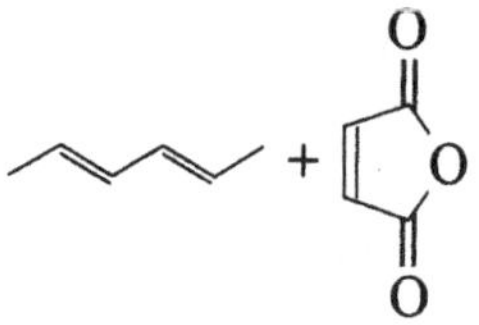

(c)

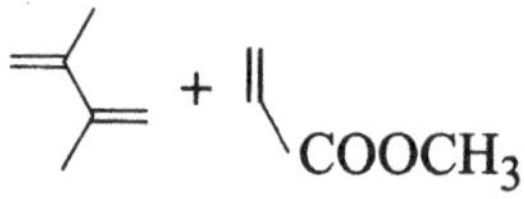

(d)

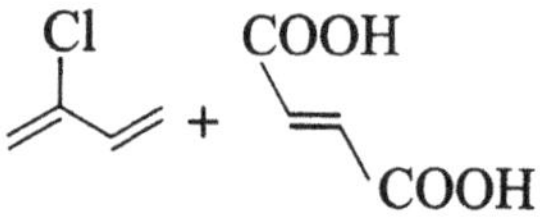

(e)

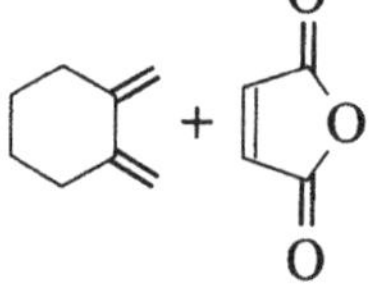

(f)

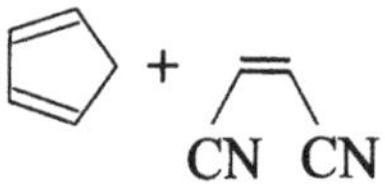

4. Draw the structures of the diene and the dienophile which upon reaction will give each of the following adducts.

(a)

(b)

(c)

(d)

(e)

SECTION II

1. 1, 3, 5, 7-Cyclooctatetraene (structure given below) is not a planar molecule. (It is described as tub-shaped.) Do you classify it as a conjugated or an isolated polyene? Comment on its resonance stabilisation.

2. Draw the products of 1, 2-addition and 1, 4-addition of HBr to the following.

(a)   (b)

Account for the fact that 1, 2- and 1, 4-addition give the same product.

3. Give an account of kinetic and thermodynamic control of reactions with examples.
4. Give an account of the Diels–Alder reaction.
5. In Section 6.4.2, it is stated that when the reaction product mixture obtained by addition of one equivalent of HBr to 1,3-butadiene at −85°C is heated with HBr to 40°C, the composition will change to one that will be obtained if the reaction is initially carried out at 40°C. Rationalise this observation.

CHALLENGING QUESTIONS

1.

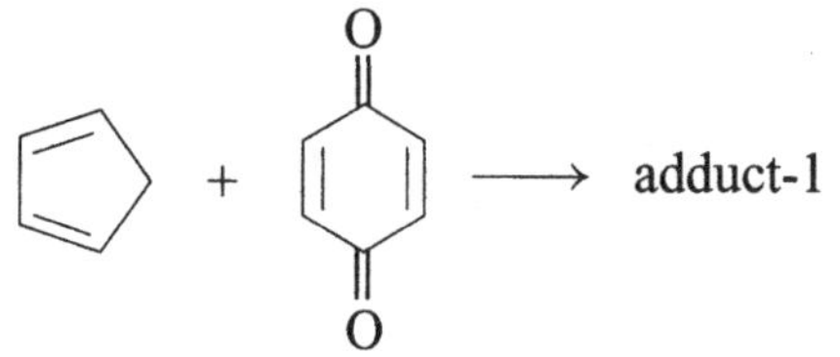

p-Benzoquinone

Draw the structures of both adducts.

adduct-1 + ⟶ adduct-2

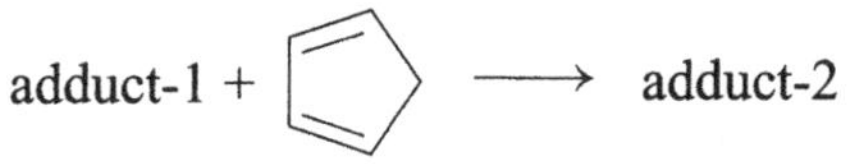

2. A mixture containing 1, 3-hexadiene, 2, 4-hexadiene, 1, 5-hexadiene and 1, 4-hexadiene was shaken with excess maleic anhydride and the solids removed by filtration. What are the liquid hydrocarbons in the filtrate? What are the solids present in the filtration residue?

(Note: maleic anhydride and its Diels–Alder adducts are solids insoluble in hydrocarbons.)

3. $CH_3 - CH = CH - CH_2 - \underset{\underset{Cl}{|}}{CH} - CH_3$

$\xrightarrow{\text{ethanol/KOH/}\Delta}$ $CH_3 - CH = CH - CH = CH - CH_3$ (Major product)

$+ CH_3 - CH = CH - CH_2 - CH = CH_2$ (Minor product)

Draw an energy profile diagram for the above reaction. The reaction is not reversible. The two products of the reaction cannot interconvert under the conditions of the reaction. Is the reaction kinetically or thermodynamically controlled? In the text (Section 6.6) it is mentioned that the conjugated diene is formed preferentially because it is more stable. Comment on the validity of this statement.

4. Rationalise the following transformations.

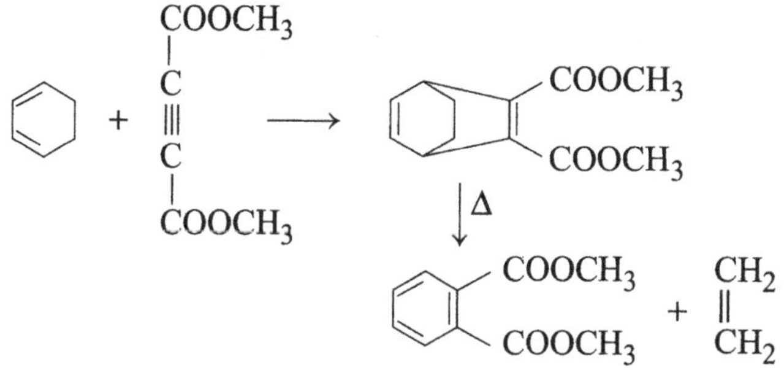

ASSIGNMENT

Write an essay on natural rubber explaining production, processing, structure, chemistry, properties and uses.

7 Polymerisation

OBJECTIVES In this chapter, you will learn about,

- definitions of various terms associated with polymers
- what macromolecules are
- what natural and synthetic polymers are
- how polymerisation based on addition reactions (vinyl polymers) and condensation reactions takes place
- vinyl polymerisation, mechanistic types—free radical, cationic, anionic and coordination
- mechanism of free radical initiated vinyl polymerisation
- what stereoregular polymers are
- polyesters and polyamides, phenol–formaldehyde resins
- some important synthetic polymers

7.1 INTRODUCTION TO MACROMOLECULES

Macromolecules are large molecules. Large can mean, in terms of molecular mass, anything from several hundreds to a few millions. The term is mainly used to refer to *polymers* which are macromolecules formed by the joining together of a large number of the same or similar molecules, called *monomers*. There are many naturally occurring polymers such as natural rubber (isoprene), carbohydrates like starch and cellulose (glucose), proteins (amino acids) and nucleic acids like DNA and RNA (nucleotides). The words in parenthesis, in the above listing, are the corresponding monomers. After studying the many useful properties of these natural polymers such as the elasticity of natural rubber, fibre quality of carbohydrates like cellulose and proteins like silk and wool, chemists have tried to create similar polymers in the laboratory (synthetic polymers). These attempts have been highly successful and the polymer industry is among the major chemical industries in the world today; the type and range of synthetic polymers available for various applications are indeed large. Synthetic polymer preparation is essentially based on two types of reactions, *addition* and *condensation*.

7.2 ADDITION POLYMERISATION

The monomers for addition polymerisation are alkenes or substituted alkenes. The grouping $CH_2=CH-$, is called the vinyl group and most of the important monomers have the structure $CH_2=CH-Z$, where Z can be H, alkyl, aryl, halogen, hydroxyl and its derivatives, carboxylic acid and its derivatives or nitrile. The group of atoms of the monomer appears in the polymer chain as repeating units (Eq. 7.1).

$$n CH_2=CHZ \rightarrow \cdots CH_2-\underset{\underset{Z}{|}}{CH}-CH_2-\underset{\underset{Z}{|}}{CH}-CH_2-\underset{\underset{Z}{|}}{CH}\cdots \text{ or, } \left(CH_2-\underset{\underset{Z}{|}}{CH}\cdots \right)_n$$

$$(7.1)$$

n Can be very large and determines the mass and size of the polymer. The grouping $-(CH_2-CHZ)-$ in the polymer structure (Eq. 7.1) is called the *repeating unit*. Such monomers are called vinyl monomers.

A list of the more important vinyl monomers, the polymers obtained from them and their uses is given in Table 7.1. The polymerisation reaction can involve one of four types of mechanisms, *free radical, cationic, anionic* or *coordination*. The basic steps involved in these mechanisms are outlined below.

7.2.1 FREE RADICAL VINYL POLYMERISATION

This is illustrated with the polymerisation of styrene (the trivial name for vinylbenzene) to polystyrene (Eqs. 7.2a–7.2d).

$$R-O-O-R \longrightarrow 2\,RO^{\bullet} \qquad \text{(a)}$$

$$R-O^{\bullet} + CH_2=CH \rightarrow R-O-CH_2-\overset{\bullet}{CH} \qquad \text{(b)}$$

$$R-O-CH_2-\overset{\bullet}{CH} + CH_2=CH \rightarrow R-O-CH_2-CH-CH_2-\overset{\bullet}{CH} \qquad \text{(c)}$$

$$\rightarrow\rightarrow R-O\left(CH_2-CH \right)_n CH_2-\overset{\bullet}{CH} \qquad \text{(d)}$$

$$(7.2)$$

Table 7. 1 Some important vinyl polymers

Monomer structure (name)	Polymer structure (name)	Applications
$CH_2 = CH_2$ (ethylene)	$-(CH_2 - CH_2)- n$ Polyethylene/polythene HDPE	Films, tubes, moulded articles
$CH_2 = CH - CH_3$ (propylene)	$\left(CH_2 - CH\right)_n$ $\quad\quad\quad\; CH_3$ Polypropylene, HDPP	Sheets, tubes, fibre, consumer articles
$CH_2 = CH-\bigcirc$ (styrene)	$\left(CH_2 - CH\right)_n$ $\quad\quad\quad\; \bigcirc$ Polystyrene	Packaging, consumer articles, elastomers
$CH_2 = CHCl$ (vinyl chloride)	$\left(CH_2 - CH\right)_n$ $\quad\quad\quad\; Cl$ Polyvinyl chloride (PVC)	Tubes, pipes, sheets, leather substitute
$CH_2 = CHCN$ (acrylonitrile)	$\left(CH_2 - CH\right)_n$ $\quad\quad\quad\; C \equiv N$ Polyacrylonitrile	Fibre (wool substitute)
$CF_2 = CF_2$ (tetrafluoro ethylene)	$-(CF_2 - CF_2)_n$ Teflon	Non-stick coating, bearings, gaskets, seals
$CH_2 = C - COO\ CH_3$ $\quad\quad\; CH_3$ methyl methacrylate	Polymethyl methacrylate (plexiglas)	Sheets, glass substitute

Simple alkenes other than ethylene are not suitable for free radical polymerisation. Other than propylene, all the monomers listed in Table 7.1 can be polymerised under free radical conditions. Styrene and other vinyl monomers require an *initiator* to start *(initiate)* the reaction. The initiator is usually a molecule which contains a weak covalent bond that can be cleaved homolytically under mild heating to give free radicals. Organic peroxides, $RO - OR$, whose $O - O$ bond is weak are good free radical initiators. Peroxides decompose readily to give two free radicals (alkoxy radicals) as in Eq. 7.2a. The alkoxy radical adds to the double bond of the vinyl monomer to give a new free radical (Eq. 7.2b). Recall the mechanisms of chlorination of methane (Chapter 3) and addition of HBr to alkenes in the presence of peroxides leading to the peroxide effect (Chapter 4). The addition is regiospecific, meaning that the addition product is the one shown and not the one with the alternate structure, $RO-CHZ-\bullet CH_2$. This is because, the former is the more stable secondary free radical and also in the case of styrene, it is a resonance stabilised benzyl free radical. This regiospecificity is maintained throughout, so that the group Z always appears on alternate carbons in the polymer backbone. This has been referred to as *head-to-tail linking* of the vinyl monomers. Head-to-tail linkage is the rule in all polymers, regardless of whether the mechanism is free radical, cationic, anionic or coordination. Various other Z groups of the vinyl monomer are also capable of stabilising the free radical. The free radical formed in step 7.2b adds to another monomer (step 7.2c) and the process keeps repeating. The addition in step 7.2c and all subsequent additions are regiospecific. The result is a free radical chain reaction which leads to the formation of a long chain of $-CH_2-CHZ-$ repeating units. The two 'chains' in the above sentence have two different meanings. The former refers to a repeating sequence of reactions, other examples of which were seen in Chapter 3 (chlorination of methane) and Chapter 4 (free radical addition of HBr to alkenes). The latter refers to the polymer chain. The above steps are called *chain propagation steps*. In principle, the chain can continue indefinitely. In practice, it will *terminate* when n, the number of monomer units in the chain reaches a certain value, decided by various factors which we shall not deal with here. Two possible termination steps are shown in Eqs. 7.3 and 7.4.

$$2RO-(CH_2-CH)_n CH_2-\overset{\bullet}{C}H \longrightarrow RO-(CH_2-CH)_n CH_2-CH_2$$

$$\qquad (7.3)$$

$$+ RO-(CH_2-CH)_n CH=CH$$

$$RO-(CH_2-CH)_n CH_2-\overset{\bullet}{C}H + R'H \longrightarrow R'\bullet + RO-(CH_2-CH)_n CH_2-CH_2$$

$$\qquad (7.4)$$

In the former, two growing chains react with each other, one abstracting a hydrogen atom from the penultimate carbon of the other chain, making itself an alkane and the other one an alkene. Since one chain gets an extra hydrogen and the other loses a hydrogen, it is called *termination by disproportionation*. In Eq. 7.4, the growing chain abstracts a hydrogen from another molecule – noncommittally shown as $R'H$ – making itself an alkane, and the chain terminates. However, the newly generated radical $R'\bullet$ can intiate a new chain. Hence, this process is called *chain transfer*. Earlier it was mentioned that propene and butene are not suitable for free radical polymerisation. This is because of the facile reaction (Eq. 7.5), which gives the resonance stabilised allyl radical which is poorly reactive.

$$RO^{\bullet} + CH_2\!=\!CH\!-\!CH_3 \;\rightarrow\; ROH + CH_2\!=\!CH\!-\!\overset{\bullet}{C}H_2 \tag{7.5}$$

From the mechanism as outlined here, it can be seen that the polymer chain grows linearly; there is no branching. However, a closer examination of Eq. 7.4 will show that the hydrogen donor, RH, can be the backbone of a growing chain or a completed chain. The new radical centre generated can add to a monomer and initiate a side chain (Eq. 7.6). Such reactions lead to branching of the polymer chain.

$$--- CH_2 - \underset{\underset{Z}{|}}{CH} - CH_2 - \underset{\underset{Z}{|}}{\overset{\bullet}{C}} - CH_2 - \underset{\underset{Z}{|}}{CH} --- + CH_2 = \underset{\underset{Z}{|}}{CH}$$

$$\tag{7.6}$$

$$\rightarrow \;\; --- CH_2 - \underset{\underset{Z}{|}}{CH} - CH_2 - \underset{\underset{Z}{|}}{\overset{\overset{\displaystyle CH_2 - \overset{\displaystyle Z}{\overset{|}{CH}}}{|}}{C}} - CH_2 - \underset{\underset{Z}{|}}{CH} ---$$

When a single monomer polymerises to give the polymer with identical repeating units as in polystyrene, the polymer is called a *homopolymer*. More than one monomer can be polymerised together, in which case the chain can contain both monomer units, either randomly or in ratios decided by various factors. Such polymers are called *copolymers*. A copolymer of styrene and butadiene was mentioned in Chapter 6 as an example of a synthetic elastomer, SBR.

7.2.2 CATIONIC POLYMERISATION

Just as the addition of a free radical to an alkene forms the basis of free radical vinyl polymerisation, the addition of a cation to an alkene forms the basis of cationic polymerisation. The cation of importance in this context is the proton. Reaction of 2-methylpropene (in industry, the preferred name is isobutylene) under acid (proton) catalysis is illustrated in Eqs. 7.7a–7.7d.

$$CH_2 = \underset{\underset{CH_3}{|}}{C} - CH_3 + H^+ \;\longrightarrow\; CH_3 - \underset{\underset{CH_3}{|}}{\overset{+}{C}} - CH_3 \tag{a}$$

$$CH_3 - \underset{\underset{CH_3}{|}}{\overset{+}{C}} - CH_3 + CH_2 = \underset{\underset{CH_3}{|}}{C} - CH_3 \longrightarrow CH_3 - \underset{\underset{CH_3}{|}}{\overset{\overset{CH_3}{|}}{C}} - CH_2 - \underset{\underset{CH_3}{|}}{\overset{+}{C}} - CH_3 \tag{b}$$

$$
\begin{array}{ccccc}
& \underset{|}{CH_3} & \underset{|}{CH_3} & & \underset{|}{CH_3} \\
CH_3-\underset{|}{C}-CH_2-\overset{+}{\underset{|}{C}} & & + CH_2= & \underset{|}{C} \\
& CH_3 & CH_3 & & CH_3
\end{array}
$$

$$
\longrightarrow CH_3 - \underset{\underset{CH_3}{|}}{\overset{\overset{CH_3}{|}}{C}} - CH_2 - \underset{\underset{CH_3}{|}}{\overset{\overset{CH_3}{|}}{C}} - CH_2 - \overset{+}{\underset{\underset{CH_3}{|}}{C}} - CH_3 \qquad (c)
$$

$$
CH_3 - \underset{\underset{CH_3}{|}}{\overset{\overset{CH_3}{|}}{C}} - CH_2 - \overset{+}{\underset{\underset{CH_3}{|}}{C}} + \overset{..}{B}{}^{-} \longrightarrow CH_3 - \underset{\underset{CH_3}{|}}{\overset{\overset{CH_3}{|}}{C}} - CH = \underset{\underset{CH_3}{|}}{C} - CH_3
$$

$$
+ CH_3 - \underset{\underset{CH_3}{|}}{\overset{\overset{CH_3}{|}}{C}} - CH_2 - \underset{\underset{CH_3}{|}}{C} = CH_2 + BH \qquad (d)
$$

$$(7.7)$$

This reaction takes place with isobutylene in the presence of fairly concentrated sulphuric acid. Recall (Chapter 4) that in the presence of aqueous sulphuric acid, alkenes undergo hydration, that is, addition of water. The carbocation formed in step 7.7a can undergo one of three possible reactions: (i) be quenched by an anion or any other nucleophile (like water, if available, leading to hydration); (ii) undergo deprotonation which in this case is the reverse reaction; (iii) react with another alkene molecule. This last option is really the same as option (i), because the alkene is a nucleophile. Option (iii) is shown in Eq. 7.7b. The product is a new carbocation, also tertiary like the one in 7.7a. Two fates that can befall this cation are shown in 7.7c and 7.7d. The former is the beginning of a carbocation chain reaction which can potentially lead to a polymer. The latter is deprotonation by any available base, like the bisulphate anion or another alkene molecule. The result of this reaction (7.7d) is a *dimer*. The product of the combination of two monomers is a dimer. We can have *trimers, tetramers;* when the number of monomers forming the polymer is not very large, the polymer is often called an *oligomer*. Notice that step 7.7d is actually a chain termination step. Cationic polymerisation seldom leads to large polymers. These reactions often stop at the dimer stage.

Dimerisation of isobutylene is an important process in the petroleum industry. The products 2,4,4-trimethyl-2-pentene and 2,4,4-trimethyl-1-pentene are liquid hydrocarbons obtained from the gaseous hydrocarbon isobutylene. The process is an example of the production of *polymer gasoline*. The two trimethylpentene isomers, upon catalytic hydrogenation (see Chapter 4) give 2,2,4-trimethylpentane, which is referred to in the industry as *isooctane* and is the standard employed to measure the octane number of gasoline. Cationic polymerisation is successful only with alkenes or other monomers which give fairly stable carbocations upon protonation. Ethylene is not amenable to cationic polymerisation.

7.2.3 ANIONIC POLYMERISATION

Just as polymerisation can be initiated by free radicals and cations, it can also be initiated by anions. As we have seen in earlier discussions, carbon–carbon double bonds are nucleophiles and addition of an anion to a double bond is normally not feasible. Nucleophilic addition (addition of anion) to a C—C double bond is possible under two conditions. The first situation arises if the double bond is sufficiently electron-depleted by the influence of a neighbouring electron-withdrawing group. An example is the addition of anions to α, β-unsaturated carbonyl compounds as in Michael addition (Chapter 16) (Eq. 7.8).

$$
B\ddot{:} + R-CH=CH-\underset{\underset{O}{\|}}{C}-Z \longrightarrow
\left[
\begin{array}{c}
\overset{R}{\underset{|}{}}\quad\quad \overset{\ddot{O}^-}{\underset{|}{}} \\
B-CH-CH=C-Z \\
\updownarrow \\
\overset{R}{\underset{|}{}}\quad\quad \overset{O}{\underset{\|}{}} \\
B-CH-\ddot{\underset{..}{C}}H-C-Z
\end{array}
\right]
\tag{7.8}
$$

As can be seen, the resultant anion is resonance stabilised. The other situation arises when the anion obtained is more stable (that is, a weaker base) than the adding base. An example is the addition of butyl carbanion obtained from butyllithium to styrene (Eq. 7.9).

$$
C_4H_9Li \rightleftarrows CH_3-(CH_2)_2-\overset{+}{\underset{..}{C}H_2}\,\overset{+}{Li}
$$

$$
\xrightarrow{\quad CH_2=CH-\langle O\rangle \quad} CH_3-(CH_2)_2-CH_2-CH_2-\overset{+}{\underset{\underset{\langle O\rangle}{|}}{\overset{..}{C}H}}\,\overset{+}{Li}
\tag{7.9}
$$

The resultant anion is a resonance stabilised benzyl carbanion, more stable than the unstabilised butyl anion.

Both types of situations can lead to anionic polymerisation. Acrylate esters and acrylonitrile are industrially important monomers which are polymerised under anionic conditions. Styrene is an example of a monomer which can be polymerised by anionic initiation. The first step of this is Eq. 7.9 which is followed by the propagation steps (Eq. 7.10). Termination of the chain occurs usually by reaction with a proton source like water. In the absence of water, the chain can continue till all the monomer is used up and what is left behind is a polymer chain with a carbanion end associated with a lithium cation, which is an alkyllithium similar to butyllithium.

$$CH_3 - (CH_2)_2 - CH_2 - CH_2 - \overset{..}{C}H \quad Li^+$$

$$\downarrow nCH_2 = CH$$

(7.10)

$$CH_3 - (CH_2)_2 - CH_2 \left(CH_2 - CH \right)_n CH_2 - \overset{..}{C}H \quad Li^+$$

These are stable molecules. An interesting property of such a molecule is that if more monomer is added at any later stage, the polymerisation reaction will continue and the chain will grow further. Such a polymer has been called a *living polymer*. The monomer added in the second stage could be a different one, in which case the new segment of the polymer has different repeating units (Eq. 7.11), where I^* is the initiating anion, A and B are monomers, and the asterisk indicates the reactive end of the chain. The product is a copolymer, but not a random copolymer; it is called a *block copolymer.*

$$I^* + nA \rightarrow I - (A)_{n-1} - A^* \xrightarrow{mB} I - (A)_n - (B)_{m-1} - B^*$$

(7.11)

7.2.4 COORDINATION POLYMERISATION

One of the major developments in the area of vinyl polymerisation is the discovery of Ziegler–Natta polymerisation. The original Ziegler–Natta catalysts were based on titanium tetrachloride in combination with diethylaluminium chloride. Polymerisation of propylene with this catalyst gave polypropylene with higher density and melting point and which could be drawn into fibres and had other desirable mechanical properties in comparison to the product of free radical polymerisation (which in any case – as we have seen earlier – is not particularly suitable for propylene). In the industry, this material is referred to as high density polypropylene (HDPP). Similarly high density polyethylene (HDPE) is obtained from ethylene. In later developments, this catalyst system has been replaced by those based on zirconium complexes. The main difference between conventional polypropylene and that made using Ziegler–Natta catalysts, is its stereochemistry. A polypropylene chain can have different stereochemical structures with respect

The discovery of coordination polymerisation by the German Chemist Karl Ziegler (1898–1973) and the Italian Chemist Giulio Natta (1903–1979) in the 1950's was a major milestone. They were awarded the Nobel Prize in Chemistry in 1963.

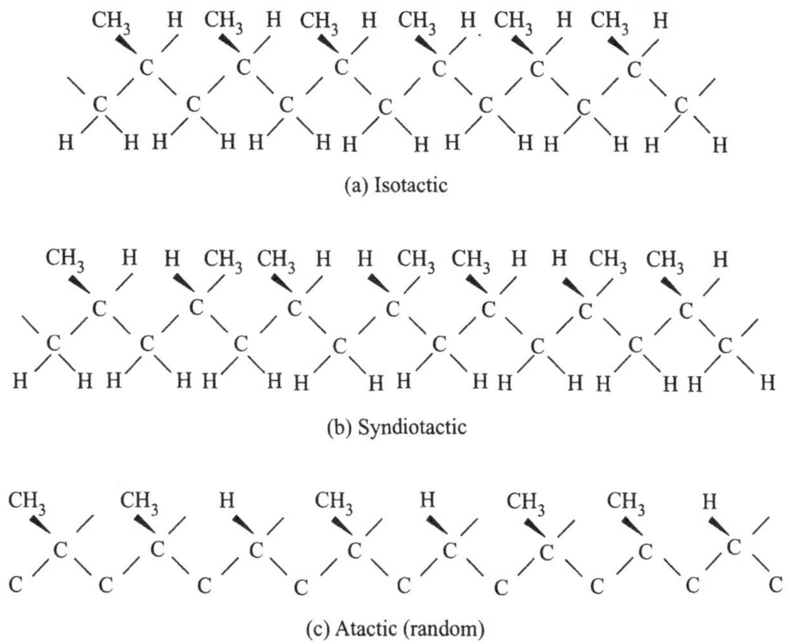

(a) Isotactic

(b) Syndiotactic

(c) Atactic (random)

Fig. 7.1 Tacticity in polypropylene

to the orientation of the methyl groups, as seen in Fig. 7.1. The particular sterechemical feature under consideration goes by the name *tacticity*.

In the figure, the linear chain is shown in a stretched conformation with all the carbons of the polymer chain (not of the methyl groups) in the plane of the paper. In structure (c), the methyl groups attached to alternate carbons of the polymer backbone are oriented randomly, some projecting to the front and some to the back of the plane of the paper. This called an *atactic polymer*. In structure (a), the methyl groups are all oriented in the same direction (pointing to the front). This is called *isotactic* and we say that the chain is *stereoregular*. The sterochemical arrangement where the methyl groups are arranged alternately to the front and to the back (7.1b) is called *syndiotactic*. The stereoregular polymers, isotactic or syndiotactic can be obtained by the use of appropriate coordination catalysts. The reaction involves coordination of the monomer to the central transition metal atom through its π-bond, and successive insertion of the monomer into the growing polymer chain, also anchored to the metal. The stereochemical requirements within the coordination sphere of the metal complex are stringent, and only one particular stereoisomer is formed.

The main difference between the stereoregular polymer and the atactic one is that the former has higher density. This is obtained by the ability of the regular polymer chains to pack more tightly than the irregular ones. Even in polyethylene, where tacticity is not relevant, Ziegler–Natta polymerisation gives a higher density polymer. We have seen earlier that vinyl polymers, though expected to be linear and unbranched, can develop branching due to chain transfer (Eq. 7.6). Branching is absent in coordination polymers. Branching is one of the reasons for looser packing of the chains and lower density in free radical polymers.

We have seen in Chapter 6 that natural rubber is polyisoprene and that the double bonds in it are all *cis*. Isoprene can be polymerised under Ziegler–Natta conditions to give all-*cis* polyisoprene which is similar in properties to natural rubber. 1,3-Butadiene is also polymerised by coordination catalysis to obtain polybutadiene which is an important elastomer.

7.2.5 CROSS-LINKING

Linear polymer chains, though very large, are still independent molecules. When such chains are joined together at intermediate points by bonds or even short chains, we have the phenomenon of cross-linking. Cross-linking gives rise to a three-dimensional network of interconnected chains. While linear polymers are soluble in appropriate solvents, cross-linked polymers may not dissolve. When a solid dissolves, the solute molecules get distributed uniformly among the solvent molecules. With cross-linked polymers, the reverse can happen. The solvent molecules occupy the space available within the network. The solid will *swell*. Swelling in compatible solvents is a characteristic property of cross-linked polymers. Vulcanised rubber is an example of a cross-linked polymer. These polymers, even when there is moderate cross-linking, melt or at least soften on heating. They are called thermoplastic polymers.

7.3 CONDENSATION POLYMERISATION

Another type of polymerisation where the bond formation between the monomer molecules takes place, not by an addition reaction but by a condensation reaction where a small molecule like water is eliminated, is called condensation polymerisation. The reaction between a carboxylic acid and an alcohol, with the elimination of water, to form an ester is an example of condensation (Eq. 7.12).

$$R-OH + HO-\overset{\overset{\displaystyle O}{\|}}{C}-R' \rightarrow RO-\overset{\overset{\displaystyle O}{\|}}{C}-R' + H_2O \qquad (7.12)$$

Amide formation (Eq. 7.13) from the reaction of a carboxylic acid and an amine is another example.

$$R-NH_2 + HO-\overset{\overset{\displaystyle O}{\|}}{C}-R' \rightarrow RNH-\overset{\overset{\displaystyle O}{\|}}{C}-R' + H_2O \qquad (7.13)$$

The above reactions do not result in polymers. If the reactants are bifunctional – having two functional groups like diols, dicarboxylic acids, hydroxy acids or amino acids – the condensation can continue to give a polymer (Eqs. 7.14, 7.15 and 7.16). These examples involve condensation by dehydration. The polymer in Eq. 7.14 is a well-known polyester called polyethylenegylcol terephthalate (PET, Terylene, Dacron) prepared from terephthalic acid and ethyleneglycol. Commercially, it is more often prepared from the methyl ester of terephthalic acid (Eq. 7.17), in which case the small molecule eliminated during condensation is methanol and the reaction involved is trans-esterification or ester exchange (a methyl ester is converted to another ester, in this case, a glycol ester).

$$HO-\overset{O}{\overset{\|}{C}}-\langle\bigcirc\rangle-\overset{O}{\overset{\|}{C}}-OH + HO-CH_2-CH_2-OH \longrightarrow$$

$$\cdots-O\left[\overset{O}{\overset{\|}{C}}-\langle\bigcirc\rangle-\overset{O}{\overset{\|}{C}}-O-CH_2-CH_2-O\right]_{n-1}\overset{O}{\overset{\|}{C}}-\langle\bigcirc\rangle-\overset{O}{\overset{\|}{C}}-O-CH_2-CH_2-O\cdots$$

$$(7.14)$$

$$H_2N-(CH_2)_6-NH_2 + nHO-\overset{O}{\overset{\|}{C}}-(CH_2)_4-\overset{O}{\overset{\|}{C}}-OH \longrightarrow$$

$$\cdots-HN-(CH_2)_6-NH\left[\overset{O}{\overset{\|}{C}}-(CH_2)_4-\overset{O}{\overset{\|}{C}}-NH-(CH_2)_6-NH\right]_{n-1}\overset{O}{\overset{\|}{C}}-(CH_2)_4-\overset{O}{\overset{\|}{C}}-\cdots$$

$$(7.15)$$

$$n\ \ H_2N-(CH_2)_5-COOH \longrightarrow$$

$$(7.16)$$

$$\cdots--HN\left[(CH_2)_5-\overset{O}{\overset{\|}{C}}-NH\right]_{n-1}(CH_2)_5-\overset{O}{\overset{\|}{C}}-NH---\cdots$$

$$n\ CH_3O-\overset{O}{\overset{\|}{C}}-\langle\bigcirc\rangle-\overset{O}{\overset{\|}{C}}-OCH_3 + n\ HO-CH_2-CH_2-OH \longrightarrow$$

$$(7.17)$$

$$\cdots-O\left[\overset{O}{\overset{\|}{C}}-\langle\bigcirc\rangle-\overset{O}{\overset{\|}{C}}-O-CH_2-CH_2-O\right]_{n}\cdots-- + CH_3OH$$

The polymers in Eqs. 7.15 and 7.16 are polyamides, known as nylons. The monomers used in Eq. 7.15 are 1,6-diaminohexane (also known as hexamethylene diamine) and the dicarboxylic acid, adipic acid. Both monomers contain 6 carbon atoms and the polyamide obtained is known as Nylon 66. The polyamide product of Eq. 7.16 is obtained from the monomer, 6-aminohexanoic acid (whose trivial name is ε-aminocaproic acid) which is an amino acid. This nylon has a repeating unit of 6 carbons and is known as Nylon 6. The nylons and polyesters are used in the manufacture of fibres for textiles, films, sheets, moulded objects and engineering plastics. Nylon 6 is industrially obtained not from the amino acid, but from the corresponding cyclic amide or lactam, called caprolactam. When caprolactam is used as the monomer, the reaction is not a condensation, but an addition. Its polymerisation is brought about by a base (which can be water) and is an example of ring-opening polymerisation (Eq. 7.18).

Caprolactam

(7.18)

(Nylon-6)

7.4 PHENOL–FORMALDEHYDE RESINS

A particular type of condensation polymer is obtained by the reaction of phenol and formaldehyde, and certain other similar pairs of reactants. The reaction is illustrated in Fig. 7.2.

The condensation can be either acid catalysed or base catalysed and results in cross-linking. The primary reaction between phenol and formaldehyde takes place as in Fig. 7.2a. This is the Lederer–Manasse reaction discussed in Chapter 15. A formaldehyde molecule reacts with the *para* or *ortho* positions of phenol to introduce a hydroxymethyl, $-CH_2OH$, group. Mono-, di- or trisubstitution at any or all active positions is possible. The hydroxymethylphenol can condense with other molecules or with itself as in *b*, *c*, and *d* to generate chains. One phenol molecule can act as the nucleus for three growing chains as in 7.2d. Continuation of this reaction results in a cross-linked polymer.

The partially formed resin can be moulded and on further heating (heat curing) undergoes further cross-linking and sets to a hard mass. Hence such polymers are called *thermosetting resins*. They do not melt upon heating but set to become hard solids. Strong heat can only char the material. *Bakelite* is an example of a thermosetting resin. A list of important condensation polymers is given in Table 7.2.

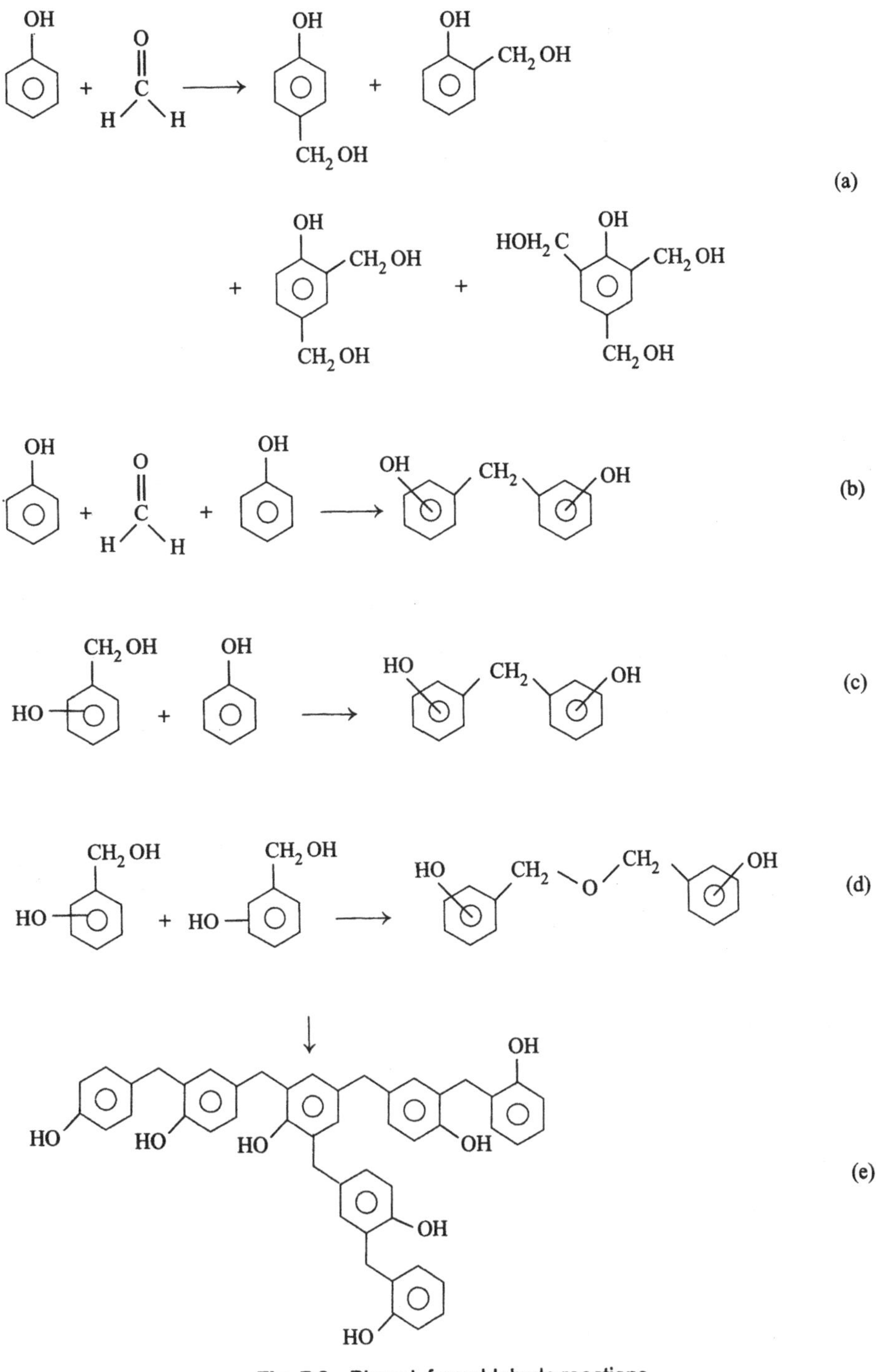

Fig. 7.2 Phenol–formaldehyde reactions

Table 7. 2   Condensation polymers

Monomer(s)	Polymer	Uses
Ethylene glycol + terephthalic acid or ester	$\left(\!\!\begin{array}{c} O-CH_2-CH_2-O-C(\!=\!\!O)-\!\!\bigcirc\!\!-C(\!=\!\!O)-O- \end{array}\!\!\right)_n$ Polyester (terylene, dacron, PET)	Fibre, sheets, utensils
Caprolactam	$\left[\!-NH-C(\!=\!\!O)-(CH_2)_5-\!\right]_n$ Polyamide (Nylon-6)	Fibre, sheets, tubes, objects
Hexamethylene diamine + adipic acid	$\left[\!-NH-(CH_2)_6-NH-C(\!=\!\!O)-(CH_2)_4-C(\!=\!\!O)-\!\right]_n$ Polyamide (Nylon-66)	Fibre, sheets, tubes, objects
Phenol + formaldehyde	Phenol–formaldehyde resins (Bakelite)	Moulded objects
Melamine + formaldehyde	Melamine resins	Utensils, crockery
p-Phenylene diamine + terephthalic acid	$\left(\!-NH-\!\!\bigcirc\!\!-NH-C(\!=\!\!O)-\!\!\bigcirc\!\!-C(\!=\!\!O)-\!\right)_n$ (Kevlar) (Polyamide, Aramid)	Fibre, cables, bullet-proof vests

---

KEY POINTS

- Polymers are macromolecules formed by the repeated linking of small molecules (called monomers) with each other.
- Two types of reactions, addition and condensation, can be used to achieve polymerisation.

- Addition reactions are essentially those of unsaturated compounds, containing the vinyl group, $CH_2 = CH-$, and hence are called vinyl polymerisations.
- These can involve four types of mechanisms, those initiated by free radicals, cations, anions and coordination complexes.
- All four types of mechanisms involve chain reactions, meaning that once initiated, the chain can grow by incorporating more and more monomer molecules.
- Where the polymer backbone has substituents (as in polypropylene), the orientation of the substituent in space has an influence on the properties of the polymer. This stereochemical aspect is referred to as the tacticity of the polymer.
- Polymerisation by conventional methods (free radical, cationic or anionic), leads to random orientation of the substituents giving rise to atactic polymers.
- Coordination polymerisation developed by Ziegler and Natta using transition metal complexes gives stereoregular polymers.
- Condensation polymerisation involves the reaction between bifunctional monomers containing groups such as carboxyl, hydroxyl or amino to give esters or amides.
- Important condensation polymers include polyesters and polyamides which are useful as synthetic fibres.
- These polymers are thermoplastic, in the sense that they become plastic or melt upon heating.
- The chains may be linear, branched or cross-linked. These have a bearing on the properties of the polymer.
- Condensation polymers obtained from phenol and formaldehyde are not thermoplastic, but set to a hard mass upon heating due to further cross-linking. These are thermosetting resins.

EXERCISES

SECTION I

1. Draw a section of the polymer chain obtained from each of the following monomers. Identify the repeating unit.

   (a) $CH_2 = CCl_2$ (1, 1-dichloroethene or vinylidene chloride). The polymer is used to make films, commercially known as saran—a wrapping material.

   (b) $CH_2 = CH - \overset{\overset{\displaystyle O}{\|}}{C} - OCH_3$ (methyl acrylate)

(c) $CH_2=CH-OC-CH_3$ (with O double-bonded to the C) (vinylacetate: the polymer upon hydrolysis, gives the water-soluble polymer, polyvinyl alcohol)

(d) $HO-CH_2-CH_2-OH$ [The polymer of this formed by condensation (ether formation), is water-soluble.]

(e) 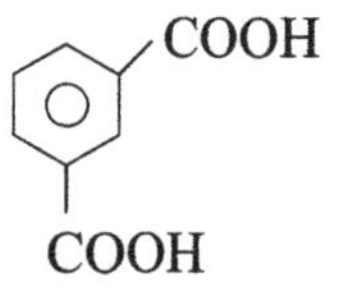 (isophthalic acid with COOH groups) $+ HO-CH_2-CH_2-OH \longrightarrow$ (a polyester)

2. Give the structure of the product of the first step in each of the following reactions.

(a) $C_6H_5-C-O-O-C-C_6H_5 \xrightarrow{\Delta}$ (with two O double bonds) (Benzoyl peroxide) (Hint: the weakest bond breaks)

(b) $CH_3-C(CH_3)(CH_3)-O-O-C(CH_3)(CH_3)-CH_3 \xrightarrow{\Delta}$ (*tert*-butyl peroxide)

(c) $R-O-Cl \xrightarrow{\Delta}$ (Alkylhypochlorite)

(d) $2 R-CH_2-\overset{\bullet}{C}H_2 \rightarrow$ Disproportionation products

(e) $2 R-CH_2-\overset{\bullet}{C}H \rightarrow$ Disproportionation products, with $COOCH_3$ substituent

(f) $R-(CH_2)_n-CH_2-\overset{\bullet}{C}H_2 + CH_2=CH-CH_3 \rightarrow$

3. Draw the structure of isotactic polystyrene.

4. Polyvinyl alcohol mentioned in question 1c, cannot be made from the monomer by direct polymerisation. Why?

5. Which conditions, free radical, cationic or anionic, are most suited for the polymerisation of the following monomers? In some cases, more than one condition may be suitable.

(a) Styrene   (e) Acrylonitile

(b) Propylene   (f) Tetrafluoroethylene

(c) Ethylene   (g) Vinyl chloride

(d) Isobutylene

SECTION II

1. Give the mechanism of anionic polymerisation of methyl acrylate,

$$CH_2=CH-\overset{\overset{\displaystyle O}{\displaystyle \|}}{C}-OCH_3,$$ initiated by an alkoxide. Show how head-to-tail linkage is exclusively obtained.

2. 1,3-Butadiene can be polymerised under free radical conditions. The chain propagation step can, in principle, involve either 1,2-addition or 1,4-addition. Outline the steps for both and arrive at the structures of the polymers formed exclusively by 1,2-addition and exclusively by 1,4-addition.

CHALLENGING QUESTIONS

1. The following addition reaction of an alcohol to an isocyanate, to form a carbamate or urethan, is the basis of the production of polyurethan resins.

$$R-OH + R'-N=C=O \longrightarrow R'-NH-\overset{\overset{\displaystyle O}{\displaystyle \|}}{C}-OR$$

Draw the structure of the polymer obtained from toluene-2,4-diisocyanate and ethylene glycol.

$$+ \ HO-CH_2-CH_2-OH \longrightarrow \text{Polyurethan}$$

2. The polymer, a polyether called polyethylene glycol, which can be considered to be the condensation polymer from ethylene glycol (see question 1d, under Section I) is in practice prepared by the ring opening polymerisation (see Eq. 7.18 in text) of ethylene oxide. Outline the mechanism of polymerisation of ethylene oxide initiated by OH^-.

$$\overset{\overset{\displaystyle O}{\diagup \ \diagdown}}{CH_2-CH_2} \longrightarrow \ ---CH_2-CH_2-O\left[CH_2-CH_2-O\right]_n---$$

3. If a small amount, say 1%, of 1,4-divinylbenzene is added to styrene in the polymerisation of the latter, it can be assumed that about 1 in every 100 of the monomer incorporated is divinylbenzene. Draw the structure of such a copolymer. The free vinyl group attached to the occasional phenyl ring which came from divinylbenzene can also be involved in polymerisation. What is the consequence of this on the structure of the polymer chain?

ASSIGNMENT

Environmental hazards associated with plastics (polymers) are very much in the news today. Prepare an essay on this topic. Touch upon the uses of polymers, their stability, how they cause pollution, their degradability including biodegradation and your suggestions for management of plastics pollution. 'Say No to Plastics' is easy to say but unrealistic. Give your views on how you will balance the benefits and the ill-effects of plastics.

8 Cycloalkanes

OBJECTIVES   In this chapter, you will learn about,

- strain, understand the significance of heat of combustion

- Baeyer's strain theory

- the causes of strain, understand the concept of strainless rings

- the shapes of cycloalkanes

- synthesis of cyclopropane and cyclobutane

- the general methods like Wurtz coupling, pyrolysis of salts of dicarboxylic acids, acyloin condensation and Dieckmann cyclisation to prepare cycloalkanes

- synthesis of cyclohexanes using specific methods like the Diels–Alder reaction and hydrogenation of aromatics

- the properties and reactions of cycloalkanes including the ring opening of cyclopropane

8.1   INTRODUCTION

Cycloalkanes are alkanes in which a bond exists between two carbons which are not adjacent to each other, giving rise to a *ring* or *cyclic structure*. They have the same general formula C_nH_{2n} as alkenes, but are not unsaturated. Cycloalkanes are named on the basis of the number of carbons forming the ring (which has to necessarily be three or more) and adding the prefix 'cyclo' to the name of the corresponding alkane. Thus, we have cyclopropane, cyclobutane, cyclohexane or cyclodecane. As shown in the following structures, the carbons and hydrogens are often omitted while drawing the rings.

CH₂
/ \
H₂C — CH₂

Cyclopropane

CH₂ — CH₂
| |
CH₂ — CH₂

Cyclobutane

Cyclohexane

Cyclodecane

Petroleum – the natural source of alkanes – does not contain a significant quantity of cycloalkanes. Some derivatives like naphthenic acids contain cyclopentanes. Cycloalkane derivatives of various ring sizes are present in essential oils from plants, as terpenoids. Among the various cycloalkanes, the most frequently encountered are the five and the six-membered compounds.

8.2   STABILITY OF CYCLOALKANES

8.2.1   BAEYER'S STRAIN THEORY

The above representations of cycloalkane structures may give an impression that the molecules are planar; in reality, this is not so. Other than cyclopropane, which has to necessarily be planar, all the higher cycloalkanes are nonplanar. Their shapes involve a *puckered* carbon framework. Till the end of the nineteenth century, this fact was not appreciated. *Baeyer's strain theory* is based

Adolf von Baeyer (1835–1917), made many contributions to organic chemistry, including the synthesis of indigo. He was awarded the Nobel prize in chemistry in 1905, for his contributions to the chemistry of organic dyes.

on the assumption that cycloalkanes are planar. The internal angles of regular polygons increase with size. The angle in a triangle and a square are 60° and 90° respectively. These are very different from the tetrahedral bond angle of an sp^3 hybridised carbon which is 109.5°. The angle in a pentagon (108°) is closest to the tetrahedral angle, the next closest being 120° in a hexagon. In higher polygons, the internal angle progressively increases, deviating more and more from the tetrahedral angle. Baeyer argued that deviation from the tetrahedral angle causes *angle strain* in the cycloalkanes. Those with large angle strain will be highly unstable and difficult to form. According to the theory, this is the reason for the observed fact (at that time) that cyclopentanes and cyclohexanes are readily formed and the others are difficult to synthesise. The ease of formation or synthesis is no longer considered to be a criterion for stability except in cases where the crucial bond forming step is reversible. (See discussion on kinetic and thermodynamic control of reactions, Section 6.4.2.) Other methods of assessing thermodynamic stability are available. Heat of combustion data are one such which will be discussed presently. According to these data, while cyclopropane and cyclobutane are undoubtedly strained and unstable thermodynamically, the higher cycloalkanes differ very little in stability from the open chain alkanes, and such differences that are observed are due more to other reasons than angle strain. Baeyer's strain theory in its totality is not valid, except for the small rings—cyclopropane and cyclobutane.

8.2.2 STRAIN AND HEATS OF COMBUSTION DATA

Heats of combustion of some cycloalkanes are listed in Table 8.1. Combustion of a cycloalkane gives carbon dioxide and water according to Eq. 8.1.

$$C_nH_{2n} + \frac{3}{2} nO_2 \rightarrow nCO_2 + nH_2O + heat \qquad (8.1)$$

The value $\Delta H^\circ/n$ is taken as heat of combustion (enthalpy) per CH_2 unit. The higher the ΔH° per CH_2 of a molecule, the greater is its energy content (Fig. 8.1) and lower is its stability. This value, 653 kJ mol^{-1} is remarkably constant for cycloalkanes larger than C11. This also happens to

Table 8. 1 Heats of combustion of cycloalkanes (kJ mol^{-1})

S. No.	Cycloalkane	Heat of combustion (ΔH°)	Heat of combustion per CH_2
1	Cyclopropane $(CH_2)_3$	2091	697
2	Cyclobutane $(CH_2)_4$	2697	687
3	Cyclopentane $(CH_2)_5$	3291	658
4	Cyclohexane $(CH_2)_6$	3920	653
5	Cycloheptane $(CH_2)_7$	4598	654
6	Cyclooctane $(CH_2)_8$	5268	658
7	Cyclononane $(CH_2)_9$	5933	659
8	Cyclodecane $(CH_2)_{10}$	6586	658
9	Cycloundecane $(CH_2)_{11}$	7238	658
10	Cyclododecane $(CH_2)_{12}$	7845	654
11	Cyclotetradecane $(CH_2)_{14}$	9138	653
12	(Very large rings)	—	653
13*	$CH_3(CH_2)_4CH_3$ (hexane)	4163	—
14*	$CH_3(CH_2)_5CH_3$ (heptane)	4817	654**
15*	$CH_3(CH_2)_6CH_3$ (octane)	5470	653**

*Alkanes
** ΔH° per CH_2 being the difference between two successive entries (13–15)

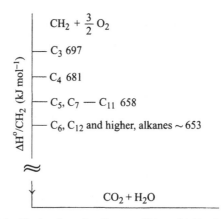

Fig. 8.1 Heats of combustion per CH_2 unit of cycloalkanes

be the difference in heat of combustion between two successive members of homologous straight chain alkanes, where again the increment is one CH_2 (entries 13–15).

In the open chain or acyclic alkanes, there is obviously no strain and 653 kJ mol^{-1} may be taken as ΔH° per CH_2 in a strainless molecule. Any higher value represents the magnitude of strain in that molecule as can be seen from Fig. 8.1. The lowest value, 653, is taken as that for a strain-free molecule. According to these data, cyclopropane and cyclobutane have strain, cyclohexane and the large cycloalkanes are similar to open-chain alkanes and are strain-free. Cyclopentane and the medium sized rings (C7 to C11) have some strain, but much less than for cyclopropane and cyclobutane.

8.2.3 SOURCE OF STRAIN: SHAPES OF CYCLOALKANES

Angle strain of the type envisaged by Baeyer is certainly present in cyclopropane and cyclobutane. The apparent $C-C-C$ angle in cyclopropane is 60°. Normal σ-bonds involving the overlap of the sp^3 hybrid orbitals of carbon along their axis is not possible here. Bonding in cyclopropane is represented with partial overlap resulting in a bent bond (Fig. 8.2a).

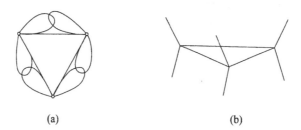

(a)   (b)

Fig. 8.2 Orbital overlap in cyclopropane and shape of cyclopropane

Partial overlap results in weak bonds and lower stability. This is what is described as *angle strain*. However, the strain in cyclopropane is only partially due to this angle strain. Another type of strain that exists in cyclopropane is *torsional strain*. Torsional strain is best understood by considering the conformations of ethane (seen briefly in Section 1.2.1 and in more detail in Chapter 21). Due to the free rotation around the C—C single bond, ethane can assume different, rapidly interconverting conformations. A very useful way of projecting the three-dimensional shape of the molecule on planar surface is the Newman projection, introduced by Melwin S Newman. In this projection, the molecule is viewed along the C—C axis and only one carbon, the one closer to the viewer, is visible. Let us call this the front carbon. The rear carbon is right behind the front carbon, hidden or eclipsed by it. The three C—H bonds on the front carbon and the three on the rear carbon will appear to be radiating out from the same point (Fig. 8.3a).

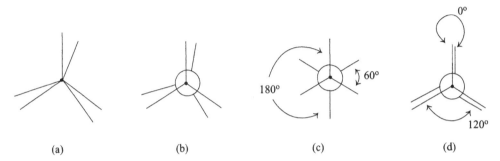

Fig. 8.3 Newman projections of ethane

To make the drawing easier to understand, Newman placed an imaginary disc between the carbons, shown in Fig. 8.3b as a circle. Now, there is no ambiguity between the bonds on the front carbon and those on the rear carbon. The dihedral angles between the bonds on the front and the rear carbons are clearly seen in these drawings. Some of the relevant dihedral angles are marked in 8.3c and 8.3d. Figure 8.3c shows the conformation where the bonds on the front carbon and those on the rear carbon are as far away from each other as possible. This is called the *staggered conformation* and the angles marked in the drawing are called the *torsional angles* or the *dihedral angles*. In conformation 8.3d, the torsional angle is 0° and the C—H bonds of the rear carbon are exactly behind those on the front carbon. They are eclipsed and will not be seen from the front. (The projection in the drawing is slightly tilted so that it is possible to view the rear bonds also.) This conformation is called the *eclipsed conformation*. The eclipsed conformation is less stable than the staggered conformation due to torsional strain. This is not steric in origin, because the distance between the eclipsed hydrogens is more than the sum of their van der Waals' radii. When the groups at the ends of the bonds are larger in size than hydrogen, steric strain may also come into play. Torsional strain occurs due to the repulsion between the two C—H bonds in the eclipsed conformation. It has also been suggested that there is more efficient delocalisation of electrons in the staggered conformation. Because of the torsional strain, the conformations do not possess the same potential energy. An energy diagram of the potential energy of ethane against the dihedral angle is shown in Fig. 8.4.

It is clear that the rotation is not 'free' and that energy has to be supplied to take the molecule above the barrier when going from one staggered conformation to another. The height of this barrier in terms of energy is 12 kJ mol^{-1}. Under normal conditions, the thermal energy present

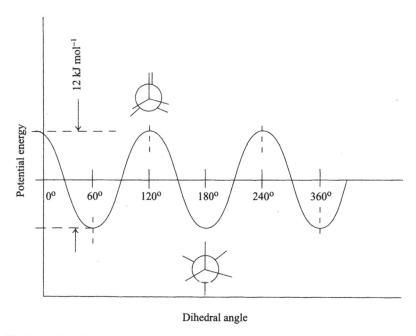

Fig. 8.4 Potential energy diagram for the rotation of ethane around the C—C bond

in molecules is enough to overcome this barrier and the rotation is free for all practical purposes. However, at any given temperature, most of the molecules of ethane will possess the staggered conformation. The value, $12\ \text{kJ mol}^{-1}$, is the magnitude of torsional strain in ethane for three pairs of C—H bonds. We can assign about $4.2\ \text{kJ mol}^{-1}$ for each pair of eclipsed C—H bonds.

Coming back to cyclopropane, it is obvious from Fig. 8.2b that all the C—H bonds are fixed in the eclipsed conformation. There are six pairs of eclipsing interactions, accounting for about $24\ \text{kJ mol}^{-1}$ of torsional strain. This and the angle strain, are together, responsible for the instability of cyclopropane. In cyclobutane (Fig. 8.5a), in the planar conformation, all the C—H bonds are eclipsed.

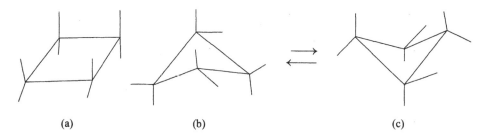

Fig. 8.5 Conformations of cyclobutane

This strain is somewhat relieved in 8.5b, where the molecule is twisted slightly to a nonplanar shape. Twisting to a nonplanar conformation causes an increase in angle strain, but overall the nonplanar cyclobutane has less energy than the planar one. For the same reasons, cyclopentane is also nonplanar (Fig. 8.6).

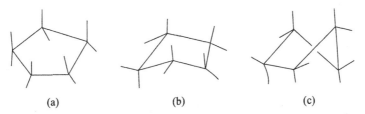

Fig. 8.6 Conformations of cyclopentane

Since there are more number of carbons in cyclopentane than in cyclobutane, the former is more flexible. The flexibility is illustrated by the equilibrating structures in Fig. 8.5a, b and c. The planar structure has less angle strain than the puckered ones, but has more torsional strain in this case as well. The conformation with the least energy is a compromise between the two types of strain.

In the case of cyclohexane, molecular models show that it can exist in nonplanar or puckered conformations without any angle strain (Fig. 8.7). Most of these have torsional strain in greater or lesser degrees. One particular set of conformations, 8.7a and 8.7b, referred to as *chair*

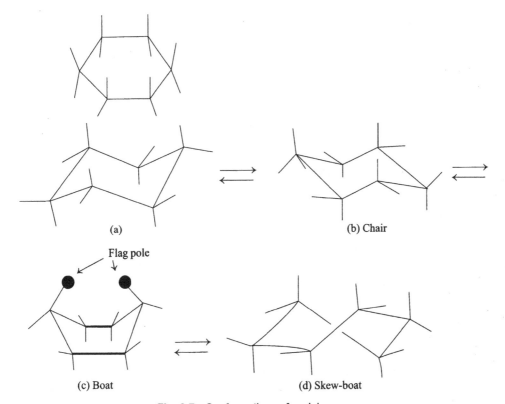

Fig. 8.7 Conformations of cyclohexane

conformations, has no torsional strain or angle strain. All the bonds on adjacent C—C pairs in this conformation are staggered. The other conformations in Fig. 8.7, the *twist boat* or *skew boat conformations* (8.7d) and the *boat conformations* (8.7c), are also free of angle strain but have torsional strain. Thus, the boat conformation has two pairs of eclipsed C—C units represented by thick lines as in 8.7c and also steric strain due to the proximity of the hydrogens on carbons 1 and 4 called the *flag pole hydrogens* – represented by solid circles. These are to some extent relieved in the skew boat conformations. The interactions – repulsive or attractive – between atoms which are not directly bonded to each other, are called *non-bonded interactions*. All the conformations of cyclohexane are interconvertible and are in equilibrium with each other. Of all the conformations, the chair form has the lowest energy and at any given time, maximum number of molecules will exist in this conformation.

The rings larger than cyclohexane also exist in torsion-free nonplanar or puckered conformations. The medium sized rings (C7—C11) in all their conformations have some degree of nonbonded repulsive interactions of the type seen in the boat form in cyclohexane. This is the reason why their heats of combustion per CH_2 are slightly more than for cyclohexane. In the larger rings these interactions are absent and they behave in a similar manner as the open chain alkanes. One of the many interconverting conformations of cycloheptane is shown in Fig. 8.8.

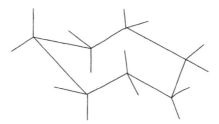

Fig. 8.8 Conformation of cycloheptane

8.3   SYNTHESIS OF CYCLOALKANES

8.3.1   GENERAL CONSIDERATIONS

A specific method applicable to the synthesis of cyclohexanes alone is the hydrogenation of aromatic compounds, namely, the derivatives of benzene. Because of this, cyclohexanes are sometimes called *hydroaromatic compounds*. Formation of a cyclic molecule from noncyclic (acyclic or open chain) starting materials is called *cyclisation*. There are two general approaches to the synthesis of cyclic molecules including cycloalkanes from acyclic starting materials. In the first, two open-chain molecules are joined together in a single step which involves the formation of two bonds. Such a reaction is called a cycloaddition and we have seen examples of this in the Diels–Alder reaction (Section 6.4.3) and ozonide formation (Section 4.4.3). We shall return to cycloaddition reactions presently. The other approach is to start with an open chain molecule with reactive centres at both ends (a head and a tail) and bring about a bond-forming reaction between the head and the tail as in Eq. 8.2.

$$
\begin{array}{ccc}
C-C-A & & C-C-A' \\
| & \rightarrow & | \qquad | \\
C-C-B & & C-C-B'
\end{array}
\qquad (8.2a)
$$

The risk in attempting such a cyclisation is that two molecules of $A-C-C-C-C-B$ may react head-to-tail to give a non-cyclic product as in Eq. 8.2b.

$$
A-C-C-C-C-B + A-C-C-C-C-B
$$
$$
\rightarrow A-C-C-C-C-B'-A'-C-C-C-C-B
\qquad (8.2b)
$$

This, as we know, is the beginning of a polymerisation rection. Cyclisation as in 8.2a is called an *intramolecular reaction,* that is, reaction within the same molecule. Reaction between two molecules (Eq. 8.2b) is an *intermolecular reaction.* In either case, the reaction takes place only when the reactive head A, comes into contact with the reactive tail B, of either the same molecule or another molecule. When the chain is long, an intramolecular reaction has less probability than an intermolecular reaction. This is especially so in concentrated solutions where the chance of A meeting B of another molecule is much greater than the chance of it meeting B of the same molecule. One of the means to partially overcome this, is to work with very dilute solutions. *High dilution conditions* are usually employed for the synthesis of large ring compounds. The situation described here is the reason why the formation of large ring molecules is difficult and not due to the reason originally ascribed—Baeyer strain.

In the case of smaller rings, the probability of intramolecular reaction is better. Their ease of formation is controlled by kinetic factors. The stability of the transition state in the cyclisation step is important. The transition state is already a cyclic structure, subject to the same kinds of strain that we have identified for the product cycloalkane. Thus, it is no surprise that the ease of formation of the cycloalkanes closely parallels ring strain for small and medium rings. The most readily formed are the six-membered rings whose transition states are strain-free, as are the cyclohexanes themselves. Seven to eleven membered rings are more difficult to form because of strain due to nonbonded interactions in the transition state and increasing difficulty of intramolecular reaction because of the probability factor. For C3–C5 rings, probability factor is in favour of cyclisation, but strain in the transition state is large for C3 and C4.

8.3.2 SYNTHESIS OF CYCLOPROPANES

8.3.2.1 Cycloaddition (Carbene Addition)

The transient intermediate carbene ($:CH_2$) adds to alkenes to form cyclopropanes. A carbene may be generated by the thermal decomposition (*thermolysis*) of diazomethane, CH_2N_2 (Eq. 8.3). This is not of much synthetic utility because of the occurrence of many side reactions. Dichlorocarbene ($:CCl_2$) generated by the action of alkali on chloroform, adds smoothly to alkenes to give dichloro cyclopropanes (Eq. 8.4).

$$
CH_2N_2 \xrightarrow{\Delta} :CH_2 + N_2
\qquad (8.3a)
$$

$$\ce{>C=C<} + \ce{:CH2} \longrightarrow \quad \text{(triangle structure)} \tag{8.3b}$$

$$\ce{CHCl3} \xrightarrow{\ce{OH^-}} \ce{:CCl2} \xrightarrow{\qquad} \quad \underset{\ce{Cl}\quad\ce{Cl}}{\text{(cyclopropane structure)}} \tag{8.4}$$

A useful method for the preparation of the hydrocarbon itself is the Simmons–Smith reaction (Eq. 8.5).

$$\ce{CH2I2} \xrightarrow{\ce{Zn|Cu}} \ce{CH2 \cdot ZnI2} \xrightarrow{\qquad} \quad \text{(dimethylcyclopropane structure)} \tag{8.5}$$

The reaction between $\ce{CH2I2}$ and zinc–copper couple generates $\ce{CH2}$ in the form of a complex with zinc iodide. This adds smoothly to alkenes to give cyclopropanes and the geometry of the alkene is maintained. In the example given, *cis*-2-butene gives *cis*-1,2-dimethylcyclopropane. These are cycloaddition reactions.

8.3.2.2 Dehalogenation of 1,3-dihalocomponds

A cyclisation reaction using an acyclic starting material is the dehalogenation of 1,3-dihaloalkanes by a metal like magnesium or zinc, (Eq. 8.6). (See the dehalogenation of 1,2- or vicinal dihaloalkanes to obtain alkenes, Section 4.2.1.3.) When applied to 1,3-dihalides, or dihalides other than vicinal, this can be considered to be an intramolecular Wurtz coupling (see Section 3.4.1.3 for Wurtz coupling).

$$\underset{\underset{\ce{Br}}{|}}{\ce{CH3-CH}} - \ce{CH2} - \underset{\underset{\ce{Br}}{|}}{\ce{CH-CH3}} + \ce{Zn} \longrightarrow \quad \underset{\ce{CH3}\quad\ce{CH3}}{\text{(triangle structure)}} + \ce{ZnBr2} \tag{8.6}$$

8.3.3 SYNTHESIS OF CYCLOBUTANES

Cyclobutanes can be obtained by cycloaddition reactions between two alkenes, under photochemical conditions. These are $2+2$ cycloadditions. A general reaction is shown in Eq. 8.7. The reaction is more successful with unsaturated carbonyl compounds such as acids, esters or ketones (Eq. 8.8).

$$2\,\ce{R-CH=CH-R} \xrightarrow{h\nu} \quad \underset{\ce{R}\quad\ce{R}}{\overset{\ce{R}\quad\ce{R}}{\text{(square structure)}}} \tag{8.7}$$

$$2C_6H_5-CH{=}CH-COOH \xrightarrow{h\nu}$$ 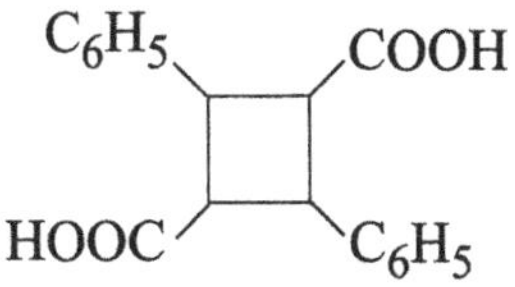 (8.8)

Cinnamic acid — Truxillic acid

Cyclobutanes can be obtained by the debromination (coupling) of 1,4-dibromo compounds in a cyclic analogue of the Wurtz reaction (Eq. 8.9).

$$R-\underset{\displaystyle Br}{\underset{|}{CH}}-CH_2-CH_2-\underset{\displaystyle Br}{\underset{|}{CH}}-R' \xrightarrow{Na}$$ 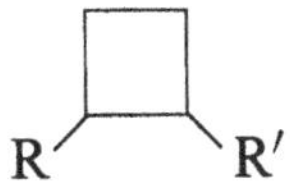 (8.9)

This is applicable to higher cycloalkanes also, as we shall see presently.

8.3.4 Higher Cycloalkanes

All the common methods used for C—C bond formation are in principle applicable to cyclisations leading to cycloalkanes. In all these reactions, yields are reasonably good for cyclopentanes and cyclohexanes, but decreases for larger rings, for reasons already discussed in Section 8.2.3. For the large ring compounds, high dilution techniques give better results. Some of these reactions are discussed below.

8.3.4.1 From Dihaloalkanes by Wurtz-Type Coupling

We have seen the application of this reaction to cyclopropane and cyclobutane. Metals like Na, Mg and more commonly Zn, have been employed.

8.3.4.2 From Dicarboxylic Acids by the Pyrolysis of Salts

This reaction is based on the thermal decomposition of the calcium salts of carboxylic acids to give ketones. For example, acetone from calcium acetate (Eq. 8.10).

$$(CH_3COO)_2^{2-}Ca^{2+} \xrightarrow{\Delta} CH_3-CO-CH_3 + CaCO_3 \quad (8.10)$$

When applied to dicarboxylic acids, cyclic ketones are obtained as products as illustrated in Eqs. 8.11 and 8.12 for the synthesis of cyclopentanone and cyclohexanone.

$$\begin{array}{l} H_2C-CH_2-\overset{\displaystyle O}{\overset{\|}{C}}-O^- \\ \;| \\ H_2C-CH_2-\underset{\displaystyle O}{\underset{\|}{C}}-O^- \end{array} Ca^{++} \xrightarrow{\Delta}$$ 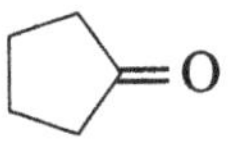 $+ CaCO_3$ (8.11)

Calcium adipate

$$\begin{array}{c} \text{CH}_2\text{---CH}_2\text{---COO}^- \\ \diagup \\ \text{CH}_2 \\ \diagdown \\ \text{CH}_2\text{---CH}_2\text{---COO}^- \end{array} \text{Ca}^{++} \xrightarrow{\Delta} \bigcirc=\text{O} + \text{CaCO}_3 \qquad (8.12)$$

Calcium pimelate

This reaction does not succeed when applied to lower dicarboxylic acids, succinic acid and glutaric acid. However, the reaction is successful to some extent with higher dicarboxylic acids. This reaction was applied by Ruzicka for the synthesis of several large ring ketones, including the naturally occurring perfumery chemicals, muscone and civetone.

Leopald Ruzicka (1887–1976), Swiss Federal Institute of Technology, Zurich, was the corecepient of the Nobel Prize in chemistry, in 1939.

Muscone $C_{16}H_{30}O$

Civetone $C_{17}H_{30}O$

Ruzicka found that cerium and thorium salts gave better results. The yields in the case of large ring ketones are low. Nevertheless, this was the only method to get those compounds. Ketones can be converted to the hydrocarbons by a variety of reactions, like Clemmensen reduction and Wolff–Kishner reduction.

8.3.4.3 Acyloin Condensation of the Esters of Dicarboxylic Acids

Acyclic acyloin condensation is shown in Eq. 8.13.

$$\begin{array}{cc} \text{O} & \text{O} \quad \text{OH} \\ \parallel & \parallel \quad \mid \\ \text{R---C---OEt} \xrightarrow{\text{Na/xylene}/\Delta} & \text{R---C---CH---R} \end{array} \qquad (8.13)$$

When an ester is treated with sodium metal in a hydrocarbon solvent like hot xylene, reductive coupling of two molecules of the ester takes place to give a ketoalcohol, also known as acyloin. When applied to the ester of a dicarboxylic acid like diethyl adipate (Eq. 8.14), the product is the cyclic acyloin, in this case, 2-hydroxycyclohexanone.

$$\text{(8.14)}$$

This reaction is the most useful one for the preparation of large ring compounds.

8.3.4.4 Dieckmann Cyclisation

Another cyclisation which also makes use of dicarboxylic acid esters is the Dieckmann cyclisation. The diester of a dicarboxylic acid is treated with a strong base like an alkoxide. An intramolecular Claisen condensation (Eqs. 8.15a, 8.15b) takes place (see Chapter 16 for Claisen condensation). The mechanism for the reaction of diethyl adipate is outlined in Fig. 8.9.

A hydrogen on the α-position of an ester is acidic enough to be abstracted by a base like the alkoxide. Action of the base on 1 gives the resonance stabilised enolate anion 2 whose negative charge is delocalised as shown. This carbanion attacks as a nucleophile at the carbonyl group of the second ester group. The resulting oxy-anion, 3, ejects an alkoxide to give the ketoester 4. The hydrogen (which is highlighted in structure 4) is made acidic as in acetoacetic ester, by the carbonyl group of the ester and the ketone. It reacts quantitatively with a base to form the resonance stabilised enolate anion 5. Acidification with a mineral acid in the working up will give back the ketoester 4, which is the desired final product. All the steps in this sequence are reversible except the last step 4 → 5. It is this irreversible last step which drives the entire multi-step reaction in the forward direction all the way to the final product. The actual cyclisation takes place in the step 2 → 3. Its success in this case is due to the fact that its transition state is a relatively strain free 5-membered ring. The reaction proceeds even better when the product is a 6-membered ring from diethyl pimelate (Eq. 8.15b). (In Fig. 8.7 and Eqs. 8.13–8.15, the ethyl group has been abbreviated as Et instead of C_2H_5. Other commonly used abbreviations are, Me for methyl, Pr for propyl, Ph for phenyl.)

$$\text{(a)}$$

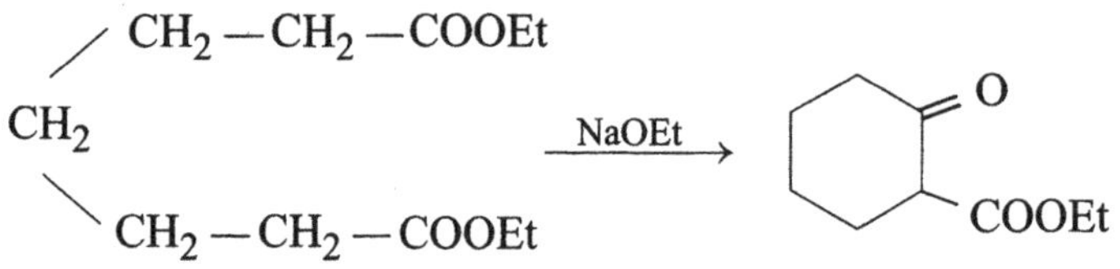

$$\text{(8.15)}$$

Diethyl pimelate

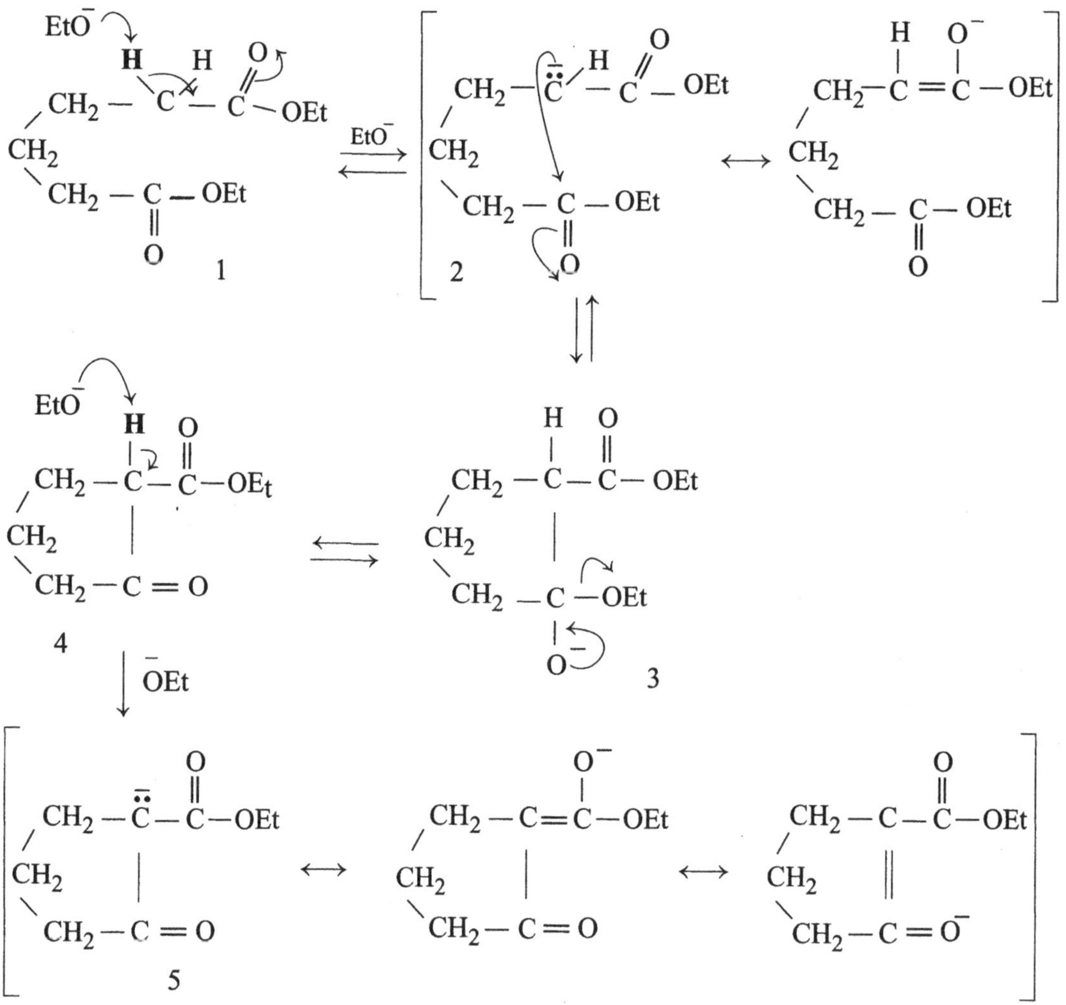

Fig. 8.9 Dieckmann cyclisation

The ketoesters obtained from Dieckmann cyclisation (called β-ketoesters) on hydrolysis give beta-ketoacids which readily lose CO_2 on heating (*decarboxylation*) to give the cyclic ketone, (Eq. 8.16), which can be converted to the hydrocarbon if necessary (as already mentioned). As expected, Dieckmann cyclisation is unsatisfactory with larger rings.

$$\text{(structure with COOEt and O)} \xrightarrow[\text{(ii) } H_3O^+]{\text{(i) } H_2O|OH^-} \text{(structure with COOH and O)} \xrightarrow{\Delta} \text{(structure with O)} + CO_2 \quad (8.16)$$

Notice that Dieckmann cyclisation gives a cyclic ketone with *one carbon less* than the parent diester. Acyloin condensation gives cyclic compounds with the *same number of carbons* as the parent diacid.

8.3.5 SYNTHETIC METHODS SPECIFICALLY APPLICABLE TO SIX-MEMBERED RINGS (DIELS–ALDER REACTION AND HYDROGENATION OF AROMATIC COMPOUNDS)

The Diels Alder reaction has already been discussed in detail (Section 6.4.3). The other method specifically applicable to cyclohexane and its derivatives is the hydrogenation of aromatics. Benzene and its derivatives cannot be hydrogenated under the mild conditions suitable for alkenes. Hydrogenation of benzene is carried out in the presence of a catalyst, usually Raney nickel, for laboratory applications under hydrogen pressures of about 100 bar and 100–200°C. Commercially, benzene is hydrogenated to cyclohexane for further conversion to chemicals like adipic acid, caprolactam and hexamethylene diamine, used for the manufacture of synthetic fibre (see Section 7.3).

Benzene and its derivatives can also be reduced by lithium and liquid ammonia (Birch reduction, see Section 4.2.2 for Birch reduction). Birch reduction of the benzene ring does not give cyclohexane, but a cyclohexadiene.

8.4 PROPERTIES

Cycloalkanes are similar to alkanes in physical and chemical properties, except for cyclopropane and cyclobutane. These two, because of strain, behave differently from the other cycloalkanes. Because of the cyclic structure, di- and higher substituted cycloalkanes exhibit geometrical isomerism. The geometrical isomers are really diastereomers. Conformations of cycloalkanes have a profound influence on stability and

> Otto Hassel (Norway, 1897–1981) and D H R Barton (England, 1918–1998) shared the Nobel Prize in chemistry in 1969, for their contributions in the area of conformational analysis.

reactivity. The study of conformations and their effects is called *conformational analysis*. These aspects are discussed in detail in Chapter 22.

8.5 REACTIONS

Cycloalkanes – other than 3- and 4-membered – take part in the same kind of reactions as alkanes. This is true of functional groups attached to the ring also, except for conformational effects. Cyclopropanes show unique reactivity mainly because of the facts that the $C—C$ bonds are not true sigma bonds and there is a large amount of strain in the ring. Because of the lack of full overlap, the electrons of the σ-bonds are more polarisable than those of the alkanes and they take part in certain reactions similar to those of alkenes. Typically, they take part in electrophilic additions resulting in ring opening. Reactions with HBr and bromine are illustrated in Eqs. 8.17 and 8.18.

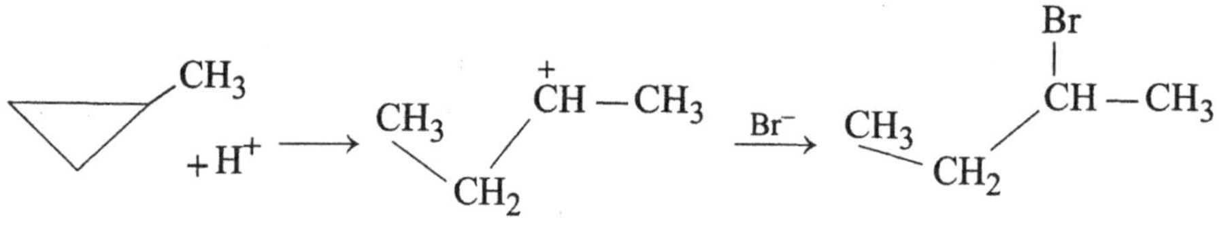

$$\text{(8.17)}$$

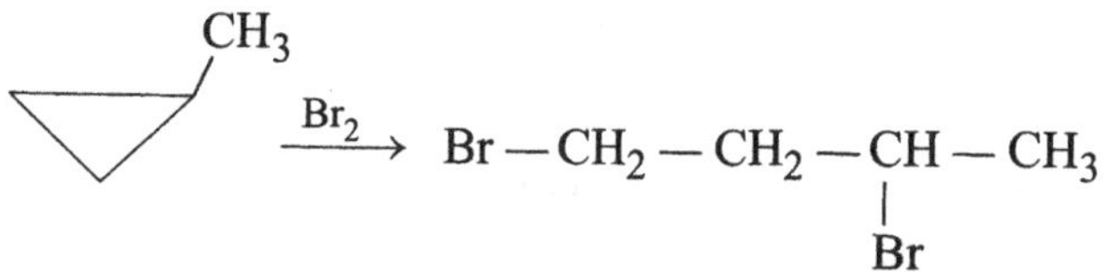

$$\text{(8.18)}$$

These reactions are electrophilic additions. The regioisomer obtained in the addition of HBr is as expected (cf. Markownikoff's rule, Section 4.4.1.4). Cyclopropanes can be catalytically hydrogenated under mild conditions as with alkenes (Section 4.4.4). This results in ring opening and the formation of an alkane (Eq. 8.19).

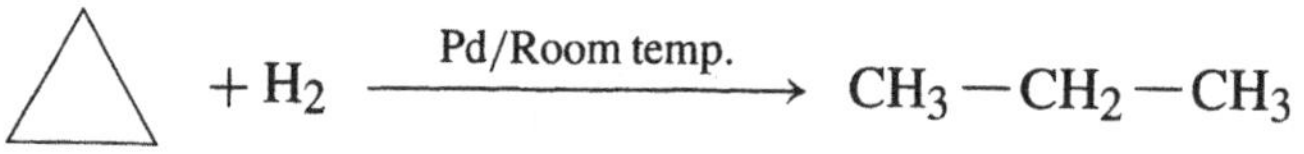

$$\triangle \;+ H_2 \xrightarrow{\text{Pd/Room temp.}} CH_3 - CH_2 - CH_3 \qquad \text{(8.19)}$$

Cyclobutanes do not undergo such addition reactions. For example, treatment with bromine gives bromocyclobutane rather than ring opening. Catalytic hydrogenation (hydrogenolysis) is possible with nickel at 200°C (Eqs. 8.20 and 8.21).

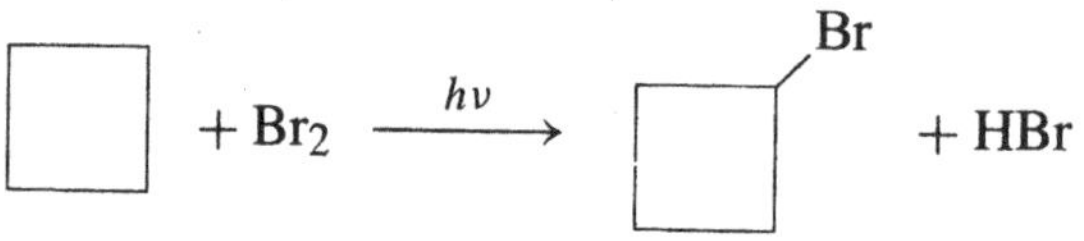

$$\square \;+ Br_2 \xrightarrow{h\nu} \square^{Br} \;+ HBr \qquad \text{(8.20)}$$

$$\text{(8.21)} \quad CH_3 - CH - CH - CH_3$$
$$\qquad\qquad\qquad\qquad |\quad\;\; |$$
$$\qquad\qquad\qquad CH_3\; CH_3$$

with Ni/200°C:

$$CH_3 \square CH_3 \;+ H_2 \xrightarrow{\text{Ni/200°C}} CH_3 - CH - CH - CH_3 \;(CH_3, CH_3) \qquad \text{(8.21)}$$

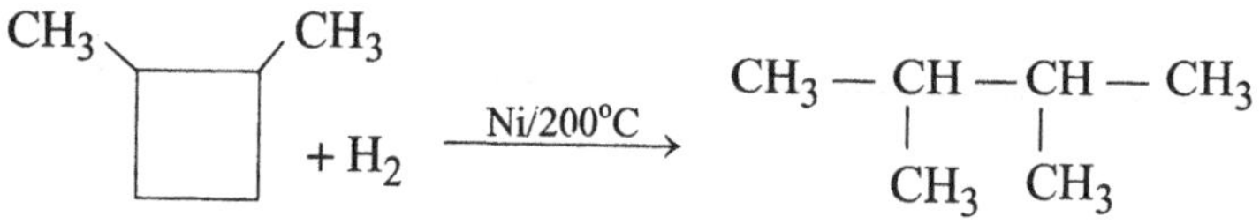

KEY POINTS

- Cycloalkanes are named by prefixing 'cyclo' to the name of the alkane with the same number of carbons.
- Baeyer's strain theory was based on the observation that cyclopentane and cyclohexane are readily formed; rings of other sizes are not easily synthesised.
- Baeyer's theory made the wrong assumption that all cycloalkanes are planar.

- Heats of combustion data give a method of quantitatively assessing strain. While C3 and C4 rings are undoubtedly strained, the higher ones are much less strained. Cyclopentane has some strain, cyclohexane none. As we go higher, cycloheptane to cycloundecane have some strain. The higher ones are strainless. The heat of combustion per CH_2 for cyclohexane and other cycloalkanes including cyclododecane and higher, is the same as for the acyclic alkanes and hence they are considered strainless.

- Angle strain and torsional strain are the main causes of strain in C3 to C5 rings. Cyclobutane and cyclopentane have nonplanar (puckered) shapes which relieves torsional strain to some extent at the cost of slightly increasing angle strain. The optimum conformation of the molecule is the one where the sum of the two strains and of any other strain, if present, is minimum.

- Torsional strain can be understood by looking at the conformations of ethane.

- Cyclohexane can have several conformations with no angle strain. One particular conformation – referred to as the chair form – has no torsional strain either, all the bonds on adjacent carbons being staggered. The other conformations referred to as the boat, the skew or the twist boat, have varying degrees of torsional strain and strain due to other nonbonded interactions due to proximity and hence, steric crowding, of hydrogens on 1 and 4 positions. Most of the molecules of cyclohexane in a sample at any given temperature will exist in the chair coformation.

- In C7—C11 cycloalkanes, the main source of strain is due to nonbonded interactions between hydrogens on non-adjacent carbons. The higher cycloalkanes are large enough to assume conformations which are essentially strain-free as in the open chain alkanes.

- Early chemists had observed that rings other than cyclohexane and cyclopentane are not easily formed or are not at all formed (in the case of the higher rings). The reason for this is now identified as due to strain in the transition state leading to cyclisation for the small rings. In the case of large rings, the reason is the low probability of the two reactive ends (head and tail) of a long chain meeting each other. The greater probability is for the head of one chain reacting with the tail of another chain leading to polymerisation. Working at high dilution partially remedies this problem.

- Cycloalkanes are synthesised by adopting the same reactions which are useful for carbon–carbon bond formation in open chain molecules. Some of these are discussed.

  — From dihalides by Wurtz-type coupling.

  — By heating the salts of dicarboxylic acids.

  — From esters of dicarboxylic acids by acyloin condensation.

  — From esters of dicarboxylic acids by Dieckmann cyclisation.

- Cycloaddition reactions are applicable to specific cycloalkanes. They include
  - Carbene addition and the related Simmons–Smith reaction for cyclopropanes.
  - Photochemical $2 + 2$ cycloaddition of alkenes, for cyclobutanes.
  - Diels–Alder reaction for cyclohexanes.
- Another method, specific for cyclohexanes, is the reduction of aromatic compounds. Reduction can be either by catalytic hydrogenation or by Birch reduction.
- Cycloalkanes are similar to alkanes in properties.
- The C—C bonds in cyclopropane, due to incomplete overlap of the sp^3 orbitals, are more polarisable than in alkanes and they take part in addition reactions like the alkenes. These additions lead to ring opening. Typical addition reactions of cyclopropanes are those of HBr, Br_2 and H_2.

EXERCISES

SECTION I

1. Draw the structures of the following compounds.

   (a) 1,1-dimethylcyclopropane

   (b) 2-methylcyclopropanone

   (c) *cis*-cyclobutane-1,2-dicarboxylic acid

   (d) 1,3-cyclopentadiene

   (e) 2-chloro-4-methylcyclohexanol

   (f) 1,3,5,7-cyclooctatetraene

   (g) 1,6-dimethylcyclodecane

2. Rearrange the following in the decreasing order of strain: cyclopropane, cyclobutane, cyclohexane, cycloheptane.

3. Mark the following as planar/nonplanar (with respect to the carbon skeleton)—cyclopropane, cyclobutane, cyclopentane, cyclohexane, benzene.

4. Mark the following as true or false.

   (a) Propene and cyclopropane are isomers.

   (b) The boat and chair forms of cyclohexane are geometrical isomers.

   (c) Cyclopentane is the most stable of all cycloalkanes because its C—C—C bond angle is closest to the tetrahedral angle.

   (d) Large ring cycloalkanes cannot be synthesised because they are highly strained.

   (e) Large cycloalkanes with ring sizes such as 50 or 100 cannot exist.

5. Show how methylcyclohexane can be obtained from (a) an aromatic compound by hydrogenation and (b) a dibromide by Wurtz-type coupling.

6. Show how 1,2-cyclohexanedicarboxylic acid can be obtained from (a) an aromatic compound and (b) by a Diels–Alder reaction.

7. Show how ethylcyclopropane can be synthesised (a) by debromination of a dibromide and (b) by the Simmons–Smith reaction.

8. Give the structure of the product formed when,

   (a) calcium adipate is heated
   (b) diethyl adipate is heated with sodium in dry xylene (acyloin condensation)
   (c) diethyl adipate is treated with sodium ethoxide in ethanol (Dieckmann cyclisation)
       [Adipic acid is $HOCO-(CH_2)_4-COOH$].

9. Identify the dicarboxylic acid, whose calcium salt upon heating will give dihydrocivetone. (Structure of civetone is given in the text, Section 8.3.4.2. Dihydrocivetone is civetone without the double bond.)

SECTION II

1. Draw the structure of all hydrocarbons of molecular formula C_4H_8.

2. Take the example of Wurtz-type cyclisation of dihaloalkanes of different chain lengths and discuss the factors influencing cyclisation.

3. Identify the compounds $A-C$. Benzene upon hydrogenation with Raney nickel gave hydrocarbon, A, C_6H_{12}. A upon partial catalytic dehydrogenation gave hydrocarbon, B, C_6H_{10} which upon ozonolysis and work-up without using zinc dust gave a diacid, C.

4. How can adipic acid be converted to cyclopentanone and to cyclohexanone?

CHALLENGING QUESTION

1. What type of strain(s) can you identify in cyclopropene and in cyclopropanone?

9 Aromatic Hydrocarbons and Aromaticity

OBJECTIVES   In this chapter, you will learn about,

- the definition of aromaticity
- applications of Hückel's rule
- concept of antiaromaticity
- general reactivity of aromatic compounds
- electrophilic aromatic substitution, general mechanism
- specific study of nitration, sulphonation, chlorination, Friedel–Crafts alkylation and acylation
- basic aspects of aromatic nucleophilic substitution including benzyne mechanism
- orienting effect and activation/deactivation effects of groups
- kinetic and thermodynamic control of aromatic substitution
- addition reactions of arenes
- side-chain halogenation of toluene

9.1   INTRODUCTION

We have learnt about the unusual structural features of benzene which are representative of arenes or aromatic hydrocarbons in Chapter 1 (Section 1.2.5). The term 'aromatic' came to be used because the earlier known members were isolated from aromatic or fragrant plant-extrudates like balsams and resins. The molecular formula of benzene, C_6H_6, suggests that it is highly 'unsaturated'. But, benzene does not have the expected chemical reactivity of alkenes. While alkenes characteristically undergo *addition reactions* to get converted to saturated molecules, benzene and other arenes undergo *substitution* reactions, preserving their unsaturation. The distinctive structural features and reactions of benzene and its derivatives lead to the concept of *aromaticity*.

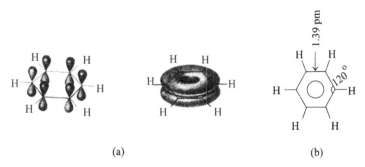

(a)   (b)

Fig. 9.1 Structure of benzene

9.2 STRUCTURE OF BENZENE

The typical features of the structure of benzene were discussed in Chapter 1. Let us recapitulate.

(i) Benzene is a fully conjugated cyclic molecule.

(ii) The six carbon atoms of benzene are sp^2 hybridised and the six unhybridised p orbitals overlap to form rings of π-electron clouds above and below the plane of the molecule (Fig. 9.1a).

(iii) It is a planar molecule, a necessary condition for the delocalisation described above.

(iv) All the C — C bonds are equivalent, with the bond length being intermediate between C — C single and double bonds. The bond lengths and bond angles are as shown in Fig. 9.1b.

(v) Benzene is more stable than the hypothetical cyclohexatriene as evaluated by the heats of hydrogenation data (Fig. 1.1, Chapter 1).

(vi) The structure of benzene is not that of cyclohexatriene as proposed by Kekulé. It is a resonance hybrid.

9.3 AROMATICITY

Cyclic molecules which have the characteristics listed above, posses the property called *aromaticity*. A majority of such molecules are benzene-like or benzenoid. What about other cyclic conjugated molecules like cyclobutadiene and cyclooctatetraene?

Richard Willstatter (1872–1942), a German chemist, made important contributions towards the determination of the structure of chlorophyll. He was awarded the Nobel Prize in chemistry in 1915.

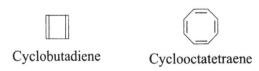

Cyclobutadiene   Cyclooctatetraene

Attempts to prepare these molecules to check for their aromaticity or otherwise were frustrating to the early organic chemists till 1911 when Richard Willstatter prepared cyclooctatetraene by the degradation of an alkaloid. It turned out to be a stable molecule – a typical polyene – with none of the characteristics expected of an arene. Resonance stabilisation of the type seen in benzene, is absent in cyclooctatetraene, as evident from the heats of hydrogenation data (Fig. 9.2). The heat of hydrogenation of the tetraene is more than four times that of *cis*-cyclooctene.

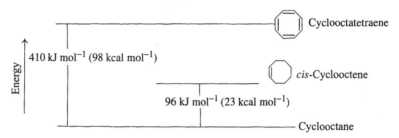

Fig. 9.2   Heat of hydrogenation of cyclooctatetraene

Cyclobutadiene, on the other hand, could not be prepared. It could only be detected in recent years using sophisticated techniques as a highly reactive, transient intermediate. Why is it not stable and apparently not aromatic? Molecular orbital calculations have shown that cyclic conjugation is not sufficient for aromaticity. The number of π-electrons in the delocalised system should be 2, 6, 10, 14 and so on, with the numbers conforming to the formula $4n+2$, where n is 0 or a positive integer. This is in accordance with *Hückel's rule*, which states that, *in planar, monocyclic, fully conjugated polyenes, only those which possess $(4n + 2)$ π-electrons will have aromaticity and the associated special stability.* Though the rule is meant for monocyclic molecules, it is valid for polycyclic arenes like naphthalene and anthracene as well. A corrollary of Hückel's rule is that planar, closed, conjugated, cyclic polyenes which do not conform to Hückel's rule not only do not have aromatic stability, but are more unstable than what the corresponding valence bond structures predict. Such molecules are *antiaromatic*. Cyclobutadiene is an antiaromatic compound. Cyclooctatetraene, first prepared by Willstatter, and now also available by other synthetic methods, is not a planar molecule. It is *tub-shaped* and does not come under the purview of Hückel's rule. It is not antiaromatic. It is *nonaromatic*.

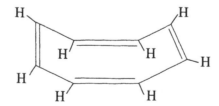

Cyclooctatetraene (tub-shaped)

The modern criterion for aromaticity is based on the nuclear magnetic resonance (NMR) spectra of the compounds (discussed in Chapter 13).

Benzene and benzenoid molecules conform to Hückel's rule where $n = 1$. Larger rings with 10 electrons ($n = 2$), 14 electrons ($n = 3$), 18 electrons ($n = 4$) and more have been synthesised. These have been shown to be aromatic by the NMR spectroscopy criterion.

Several molecules where $n = 1$, but which do not possess the benzene ring (nonbenzenoid) are known. These include ions like the cyclopentadienyl anion, cycloheptatrienyl cation, and heterocyclics like furan, pyrrole and thiophene. Polycyclic aromatics containing the benzene ring include naphthalene, anthracene, phenanthrene and many others.

| Cyclopentadienyl anion | Cycloheptatrienyl cation | Furan | Pyrrole | Thiophene |

Naphthalene   Anthracene   Phenanthrene

The Frost's circle is a convenient device to follow the distribution of electrons in the molecular orbitals of conjugated cyclic polyenes. The polyene skeleton is inscribed in the circle with one apex pointing down. The horizontal line separates the bonding and anti-bonding orbitals. Orbitals falling on the line are non-bonding. The π-electrons of the polyene are filled in pairs, starting with the lowest energy level at the bottom. This is represented for cyclobutadiene, benzene and cyclooctatetraene. Benzene has one bonding molecular orbital (lowest energy), two bonding orbitals of equal energy, two anti-bonding orbitals of equal energy and one anti-bonding orbital of even higher energy. The six π-electrons of benzene are placed in the three bonding orbitals in pairs, giving a closed shell configuration (all the bonding orbitals filled) with all the electrons being paired. In cyclobutadiene, after filling the lowest bonding orbital with an electron pair, the next two electrons go to the two degenerate non-bonding orbitals, one in each, according to Hund's rule. Two electrons are unpaired and what we have is a diradical which would be very reactive and unstable. Following the same reasoning, planar cyclooctatetraene would also be a diradical.

Cyclobutadiene →   Benzene   ← Cyclooctatetraene

(Frost's circles)

9.4   PROPERTIES OF BENZENE

Benzene – boiling point 80.1°C, melting point 5.4°C, density 0.879/20°C – is a liquid hydrocarbon with a characteristic smell, partially miscible with water but readily miscible with most organic solvents. It is toxic, inflammable and carcinogenic. It used to be manufactured from coal tar, but today, another major source of benzene, toluene and xylenes is petroleum (directly or from products of reforming).

9.5   REACTIONS

Reactions of benzene and substituted benzenes can be classified into three categories. These are:

(i)   Reactions in which a hydrogen atom or another group is replaced, leaving the benzene structure intact. These are *substitution reactions*. A typical example is chlorination of benzene to obtain chlorobenzene.

(ii)   Reactions where the arene structure is destroyed, that is, aromaticity is lost. These are essentially *addition reactions*, a typical example of which is the hydrogenation of benzene to cyclohexane.

(iii)   Reactions of the side chain, illustrated by the conversion of toluene to benzoic acid or benzyl chloride.

9.5.1   AROMATIC ELECTROPHILIC SUBSTITUTION

The two faces of the planar benzene ring are electron rich regions (Fig. 9.1a). Reagents which can readily attach themselves to such a molecule are called *electrophiles*. So, the predominant substitution reactions of benzene are electrophilic substitutions. This does not mean that *nucleophiles* cannot react with an aromatic ring. They can, if the electron cloud is distorted by the influence of groups present on the ring in such a way that certain regions of the ring are relatively electron deficient. In such situations, nucleophilic substitution is possible. Some of these reactions are discussed later on in this chapter.

9.5.1.1   Basic Aspects of Electrophilic Substitution

When an electrophile collides with a benzene ring, a weak bond is established between the π-electron cloud and the electrophile. This is described as a π-complex. This is illustrated for the situation where the electrophile is a proton (Eq. 9.1a). The bonding here is very weak and the aromaticity of the ring is not seriously disturbed. However, in media of high proton concentration, some of these get converted to species with a covalent bond between the proton and one of the carbons, called a σ-complex (Eq. 9.1b). This is a high energy process because the aromaticity of the ring is destroyed. The only redeeming feature of this reaction is that the non-aromatic carbocation – the cyclohexadienyl cation (σ-complex) – has some stability due to the delocalisation of the positive charge as seen from the canonical structures of the hybrid

(Eq. 9.1b). Recall that the situation here is similar to what happens when a protonic acid, HX adds to 1,3-butadiene, an alkadiene (Eq. 9.2). The product of protonation of the diene (Eq. 9.2a), is a resonance-stabilised cation which can also be called a σ-complex. The difference between the σ-complexes of the arene and the alkadiene is in their subsequent reactions. The carbocation from the alkadiene can either revert to the alkadiene by deprotonation (reversal of the protonation step) or be quenched by reaction with any available nucleophile, X⁻ in this case (Eq. 9.2b). The observed reaction is the latter, the addition of HX to 1,3-butadiene. If we had taken an alkene like ethene instead of the diene, the course of the reaction would have been the same, namely, the addition of HX.

The carbocation (σ-complex) from the arene also has these options (eq. 9.1c), but with one major difference. Quenching by reaction with X⁻ which is the addition of HX, will result in a product (I of Eq. 9.1c), which has lost the aromaticity of the starting material. Hence, reversal is the predominant reaction here. Now, let us consider the situation where D^+ from DX is used as the electrophile (Eq. 9.1d). Reversal can give either the original benzene or C_6H_5D, deuterated benzene. The latter is the product of electrophilic substitution of H of benzene by D.

$$(a)$$

$$(b)$$

$$(c) \qquad (9.1)$$

$$(d)$$

The situation that arises with any (generalised) electrophile E^+, is shown in Eq. 9.3. Here, Nu⁻ is the nucleophile associated with the electrophile. The π-complex is omitted in this and in future representations. The σ-complex can (i) revert to the starting material, (ii) undergo deprotonation to give the substitution product, or (iii) react with the nucleophile Nu⁻, to give the addition product. The last option does not occur because this will result in loss of aromaticity. Only the substitution product is formed.

$$CH_2=CH-CH=CH_2+H^+ \rightleftharpoons \left[CH_3-\overset{+}{C}H-CH=CH_2 \leftrightarrow CH_3-CH=CH-\overset{+}{C}H_2\right]$$

$$\downarrow X^-$$

$$CH_3-CH=CH-CH_2X$$

(9.2a and 9.2b)

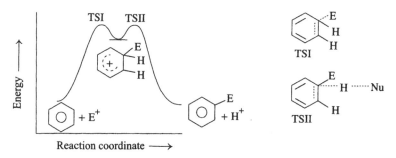

(σ-Complex)

(Substitution)   (9.3)

(Addition)

This is the basic mechanism of electrophilic substitution in arenes – with individual differences – when different electrophiles are involved. The behaviour of the arene is dictated by the tendency to retain aromaticity. This fact was recognised by chemists from the very early days as the characteristic behaviour of aromatic compounds. The course of the reaction is represented in an energy diagram (Fig. 9.3), omitting the π-complex.

Fig. 9.3 Aromatic electrophilic substitution

A variety of electrophiles can be introduced in the aromatic ring by substitution. The reactivity of different aromatic substrates and electrophiles varies greatly. Some useful electrophilic substitutions are:

- nitration—introduction of NO_2 (the nitro group) usually using a mixture of HNO_3 and H_2SO_4,
- halogenation—introduction of a halogen,
- sulphonation—introduction of the SO_3H group,
- Friedel–Crafts alkylation—introduction of an alkyl group, R, using the alkyl halide RX and a Lewis acid like $AlCl_3$ as catalyst, and

- Friedel–Crafts acylation—introduction of an acyl group, RCO, using RCOX and AlCl₃ as catalyst.

These reactions are discussed in the following sections.

9.5.1.2 Nitration

Introduction of a nitro group, NO_2, is one of the most useful substitution reactions of benzene. There are several species with their composition corresponding to NO_2. These include the cation, NO_2^+ called the nitronium ion, the nitrite anion NO_2^-, and the neutral molecules NO_2 and N_2O_4. It is the nitronium ion that is of interest to us. It has a structure with the positive charge mainly concentrated on the nitrogen.

$$\overset{+}{O=N=O}$$

There are several ways of generating this cation. The most important and relevant one for us in the context of nitration of arenes, is that it is formed by the ionisation of nitric acid in a solution of sulphuric acid (Eq. 9.4).

$$HNO_3 + 2H_2SO_4 \;\rightarrow\; NO_2^+ + H_3O^+ + 2HSO_4^- \tag{9.4}$$

Nitration of a variety of aromatic compounds is brought about by a mixture of concentrated nitric and sulphuric acids, called the nitrating mixture. Benzene itself is nitrated by such a mixture at about 50°C to yield nitrobenzene. At a higher temperature of about 100°C, with fuming nitric acid instead of the usual concentrated nitric acid, *m*-dinitrobenzene is obtained (Eq. 9.5).

$$\tag{9.5}$$

The facts that (i) the introduction of a second nitro group requires a higher temperature showing that it is not as easy as the introduction of the first nitro group, and (ii) of the three possible dinitrobenzenes, *o*-, *m*- and *p*-, the meta isomer is selectively formed, are both significant. The significance and explanation of these aspects are dealt with in subsequent sections.

Following the general mechanism outlined above, the mechanism of nitration of benzene can be depicted as follows (Eq. 9.6).

$$\tag{9.6}$$

Since the attachment of the electrophile to one of the ring-carbons is the crucial step, electron density at the ring has a great bearing on the rate of reaction. Those benzene derivatives containing substituents which are electron releasing (Chapter 1, Table 1.3) are more reactive in electrophilic substitutions than benzene itself (for example, nitration). Thus, toluene undergoes nitration even at 30°C to give a mixture of 58% *o*-nitrotoluene and 38% *p*-nitrotoluene (Eq. 9.7).

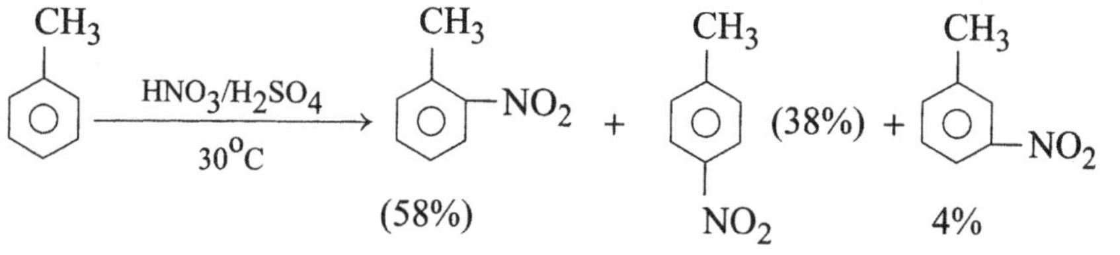

The electron-releasing resonance effect (hyperconjugation) of the methyl group is illustrated here. In contrast to the methyl group, the presence of the electron-withdrawing nitro group in nitrobenzene makes the ring less reactive and further nitration of nitrobenzene to dinitrobenzene requires more drastic conditions than for the first nitration (Eq. 9.5).

Nitration is a useful reaction because (i) the nitronium ion is a powerful electrophile and most aromatic compounds can be readily nitrated, and (ii) the nitro group once introduced, can be converted to a variety of other functional groups and thus opens up the entry to many substituted benzene derivatives.

9.5.1.3 Halogenation

Chlorine and bromine can be introduced by the halogenation reaction using the respective halogen. Recall the addition of halogen to alkenes (Chapter 4, Section 4.4.1.1). As seen from the generalised mechanism, the first step in the electrophilic reaction which is the formation of the σ-complex, is a high energy (endothermic) step. For this step to take place readily, either the ring should have high electron density at the reaction site, or the electrophile should be highly reactive and in high concentration. The nitronium ion is a very reactive electrophile. Still, as we have seen, mild heating is required for the reaction with benzene. With the more reactive (more electron rich) toluene, heating is not required. With the less electron rich (than benzene) nitrobenzene, heating to a higher temperature and the use of fuming nitric acid (to provide higher concentration of the nitronium ion) is required.

Benzene – which we shall use as a standard benchmark for arene reactivity – does not react with chlorine or bromine in the absence of additional inducement, which in this case is done by increasing the concentration of the electrophile, X^+. Recall that in the addition of halogen to alkenes, the halogen molecule without any catalyst, acted as the electrophile, because of the polarity induced by the alkene itself. The benzene molecule also brings about such an induced polarity in the halogen, but it is not good enough to overcome the unfavourable energy of the first step. The use of the Lewis acid catalyst $FeCl_3$, for chlorination is the strategy adopted for the halogenation of benzene. In practice, instead of $FeCl_3$, metallic iron in the form of iron filings is used. Iron reacts with chlorine to form $FeCl_3$ and the substitution reaction proceeds as in Eq. 9.8. Iron metal is often referred to as a 'halogen carrier'.

$$2Fe + 3Cl_2 \longrightarrow 2FeCl_3$$

$$FeCl_3 + Cl_2 \longrightarrow FeCl_4^- \ Cl^+$$

(9.8)

With more reactive (electron-rich) benzene derivatives, the Lewis acid catalyst or halogen carrier is not required. The most common electron-rich benzene derivatives which are used as examples to illustrate such effects are phenol, aniline and phenol ethers like anisole. The first step in the reaction of phenol with bromine, formation of the σ-complex, and subsequent deprotonation to yield *p*-bromophenol are depicted in Eq. 9.9a. The σ-complex, I, in this equation is one of the canonical structures of the delocalised cation. Some of the other canonical structures of this cation are shown in Eq. 9.9b. The canonical structure, I, in Eq. 9.9a shows how the OH group provides additional stabilisation to the cation by helping to delocalise the positive charge better. Examination of the resonance structures of phenol (Eq. 9.9c) will show that there are three positions in the ring, two *ortho* and one *para*, marked by arrows, which have greater negative charges than the other positions. The electrophile can react at these positios also to give *o*-bromophenol, Eq. 9.9d. In actual practice, the phenol molecule is so reactive that bromination takes place, rapidly one after the other, at all three positions and the product isolated is 2,4,6-tribromophenol (Eq. 9.9e). In fact, it is not even necessary to use pure bromine, but an aqueous solution of bromine (bromine water) will suffice. A simple laboratory test to detect phenol is to add bromine water to the sample under investigation. The colour (yellowish red) of bromine water will immediately disappear due to the formation of tribromophenol. Recall that a similar test was also mentioned as useful to detect the presence of alkenes (Chapter 4). A useful observation to differentiate between the phenol test and the alkene test is that in the former, a white precipitate will appear in the test tube due to the formation of tribromophenol. The dibromide formed from the alkene will usually not appear as a solid precipitate. Aniline (Chapter 1, Eq. 1.40) behaves in the same way as phenol.

(I)

(a)

(b)

(c)

(d)

(9.9)

$$H-O-\bigcirc + 3\ Br_2 \longrightarrow H-O-\overset{Br}{\underset{Br}{\bigcirc}}-Br + 3\ HBr \qquad (e)$$

Fluorination and iodination reactions are usually not done. Fluorine reacts very violently and cannot be controlled. Iodination does not succeed because the reaction is highly reversible (Eq. 9.10). If special conditions are employed, such as use of an oxidising agent or base to remove the HI as soon as it is formed, iodination can be carried out.

$$\bigcirc + I_2 \rightleftarrows \bigcirc^I + HI \qquad (9.10)$$

9.5.1.4 Sulphonation

Sulphonation is the reaction used to introduce the $-SO_3H$ group, to prepare sulphonic acid. Benzene is converted to benzene sulphonic acid upon heating with concentrated sulphuric acid or better still, with oleum. Sulphur trioxide, SO_3, is the electrophile. It is a neutral molecule having a formal positive charge on sulphur. Sulphonation is reversible (Eq. 9.11). The substituent can be removed by heating with steam.

(9.11)

9.5.1.5 Friedel–Crafts Alkylation

An important aromatic electrophilic substitution reaction, alkylation, was discovered by Charles Friedel (France) and James M Crafts (USA) in 1877 using alkyl halides as alkylating agents. Alkyl halides do not ionise to carbocations spontaneously. Friedel and Crafts used anhydrous aluminium chloride as the catalyst for this reaction. $AlCl_3$ – a powerful Lewis acid – complexes with alkyl halides (Eq. 9.12). When R is a primary alkyl group, complete charge separation may not take place but the polarised molecule is reactive enough. Alkylation of benzene occurs when benzene reacts with alkyl chloride and anhydrous aluminium chloride, in the presence of excess benzene as solvent, (Eq. 9.13). $AlCl_3$ is required only in catalytic quantities.

$$R-Cl + AlCl_3 \rightarrow \overset{\delta+}{R} - \overset{\delta-}{Cl} - AlCl_3 \rightarrow R^+ + AlCl_4^- \tag{9.12}$$

$$\tag{9.13}$$

The alkylation reaction has two major limitations.

(i) Since carbocations are involved, carbocation rearrangements take place (Chapter 4).

(ii) Since the initial product of alkylation (monoalkylation product) is more reactive than the parent arene, dialkylation or polyalkylation cannot usually be avoided.

When alkylation is carried out with isobutyl chloride, no isobutylbenzene is obtained as the monoalkylation product, *tert*-butylbenzene is the only product. Similarly, propyl chloride gives only isopropylbenzene (Eqs. 19.14a, b, c and d).

$$
\begin{array}{ll}
CH_3-CH-CH_2-Cl + \bigcirc \xrightarrow{AlCl_3} & (a) \\
\quad\ |\ \\
\quad CH_3 &
\end{array}
$$

$$
\begin{array}{ll}
 & (b) \\
 & \tag{9.14}
\end{array}
$$

$$
\begin{array}{ll}
CH_3-CH_2-CH_2-Cl + \bigcirc \xrightarrow{AlCl_3} & (c)
\end{array}
$$

$$
\begin{array}{ll}
 & (d)
\end{array}
$$

The introduction of more than one alkyl group (polyalkylation) limits the synthetic utility of the Friedel–Crafts alkylation. Thus, Friedel–Crafts alkylation with ethyl chloride or any other ethyl halide is not the method of choice for the laboratory synthesis of ethylbenzene. However, a variation of this reaction, involving Friedel–Crafts acylation (*vide infra*) comes in handy for this synthesis.

Carbocations generated by other routes – under appropriate conditions – can alkylate arenes. An industrial method of preparing ethylbenzene is the catalytic ethylation of benzene by ethene. The ethylbenzene so manufactured is catalytically dehydrogenated to styrene which is a raw material for the polymer industry (Eq. 9.15). Vinyl chloride does not react under Friedel–Crafts conditions;

hence direct synthesis of styrene is not possible. To manufacture linear alkylbenzenes (LAB, in industrial terminology) for the detetergent industry, benzene is alkylated with a mixture of linear alkenes, using anhydrous HF as catalyst (Eq. 9.16).

$$CH_2 = CH_2 + \text{Catalyst . } H^+ \longrightarrow CH_3 \overset{+}{C}H_2 \text{ . Catalyst} \xrightarrow{} \text{(benzene)} \overset{CH_2-CH_3}{\diagup}$$

(9.15)

$$\text{(benzene)}-CH_2 CH_3 \xrightarrow[\text{catalyst}]{\text{dehydrogenation}} \text{(benzene)}-CH=CH_2$$

$$\text{(benzene)} + R - CH = CH - R' \xrightarrow{\text{HF}} \text{(benzene)} \begin{matrix} CH - CH_2 - R' \\ | \\ R \end{matrix}$$

(9.16)

9.5.1.6 Friedel–Crafts Acylation

When an acyl chloride is used instead of an alkyl chloride in the Friedel–Crafts reaction, Friedel–Crafts acylation takes place. For example, reaction of acetyl chloride (ethanoyl chloride) with toluene in the presence of anhydrous aluminium chloride gives *p*-methylacetophenone (Eq. 9.17).

$$H_3C-\text{(benzene)} + CH_3 COCl \xrightarrow{AlCl_3} H_3C-\text{(benzene)}-CO-CH_3 \qquad \text{(a)}$$

$$CH_3-\text{(benzene)} \overset{O}{\underset{CH_3}{\overset{||}{C}}} \overset{+}{\ldots} \overset{-}{Cl}.AlCl_3 \longrightarrow CH_3-\text{(benzene)} \overset{O}{\underset{H}{\overset{||}{\underset{\nwarrow AlCl_4^-}{C-CH_3}}}}$$

$$\longrightarrow CH_3-\text{(benzene)}-\overset{O}{\overset{||}{C}}-CH_3 + HCl + AlCl_3 \qquad \text{(b)}$$

(9.17)

Acid anhydrides can be used instead of acid chlorides for acylation. Thus, benzene can be converted to acetophenone using acetic anhydride and anhydrous aluminium chloride.

Acylation, which gives an alkyl aryl ketone, is a more useful reaction than alkylation on at least two counts.

(i) Rearrangements of the type seen in alkylation are absent. This is because the acyl carbocation is not prone to rearrangement.

(ii) Polyacylation is absent. The product of monoacylation, the ketone, is less reactive in electrophilic substitution than the unsubtituted arene. The acyl group, –COR, is an electron-withdrawing group and deactivates the ring.

The utility of the reaction is further enhanced by the fact that the aryl ketones can be converted to the alkyl benzenes by several well established reactions such as Clemmensen reduction and Wolff–Kishner reduction. Thus, isobutylbenzene which cannot be prepared by alkylation, reaction between isobutyl chloride and benzene, (Eq. 9.14), can be prepared indirectly by acylation, followed by reduction (Eq. 9.18).

$$
\langle\bigcirc\rangle + CH_3-CH-C-Cl \xrightarrow{AlCl_3} \langle\bigcirc\rangle-\overset{O}{\overset{\|}{C}}-CH-CH_3
$$

(9.18)

$$
\xrightarrow[\substack{\text{(Clemmensen} \\ \text{reduction)}}]{\text{Zn(Hg), HCl}} \langle\bigcirc\rangle-CH_2-CH-CH_3
$$

Friedel–Crafts reactions, both alkylation and acylation, cannot be carried out with deactivated aromatic compounds; namely, those bearing electron-withdrawing groups (Table 9.1, category 4). In fact, nitrobenzene is often used as a solvent for Friedel–Crafts reaction of other arenes.

A significant difference between alkylation and acylation is that while the former requires only catalytic quantities of $AlCl_3$, the latter requires the 'catalyst' in more than stoichiometric quantity. This is because the product (ketone) ties up an equivalent quantity of $AlCl_3$ as a complex, $Ar(R) CO.AlCl_3$, making it unavailable for catalysis.

9.5.2 EFFECT OF SUBSTITUENTS

If the molecule which is subjected to electrophilic reaction already contains one or more substituents, these can affect the reaction in two ways. (i) The group(s) can activate or deactivate the ring, making it more reactive or less reactive than unsubstituted benzene and (ii) the group(s) can direct the incoming group to one or more specific positions, like the ortho, meta or para positions. The latter effect is called the *orienting effect* of the existing group.

9.5.2.1 Activation/Deactivation

Electron-releasing groups (Chapter 1, Table 1.3) increase the electron density in the ring making it more reactive towards electrophilic reactions. Conversely, electron-withdrawing groups deactivate the ring (Table 9.1). Groups such as hydroxy, alkoxy and amino are strongly activating groups even though they have electron-withdrawing inductive effects. Recall that bromination of phenol and aniline takes place under very mild conditions and yields the tribromo derivatives. These groups have a lone pair of electrons on the atom which is directly attached to the ring. The orbital containing this lone pair overlaps with orbitals containing the π-electrons of the ring, increasing the electron density. Alkyl groups are also activating, but not as strongly as those mentioned above. Among the groups in Table 9.1, category 2, –OCOR and –NHCOR are only moderately activating even though they carry lone pairs of electrons on the atom directly attached to the ring, O and N respectively. This is because of the presence of the electron withdrawing –COR group. Halogens also have lone pairs of electrons which have an electron-releasing resonance effect due to the overlap with the π-electrons of the ring. However, their strong electron-withdrawing inductive effect is more important. Thus, chlorobenzene is less reactive than benzene. Groups such as nitro, carbonyl (which includes aldehyde, ketone, carboxylic acid and derivatives), and cyano which are electron-withdrawing by both resonance and inductive effects, are deactivating.

9.5.2.2   Orientation

We have seen several examples of the orienting effect of groups in the reactions discussed so far. Thus, nitration of nitrobenzene gives *m*-dinitrobenzene (Eq. 9.5), nitration of toluene gives a mixture of *o*- and *p*-nitrotoluenes (Eq. 9.7) and bromination of phenol gives 2,4,6-tribromophenol (Eq. 9.9e). Resonance structures of the corresponding starting materials (for example, Eq. 9.9c for phenol) give an idea of the relative electron densities at the different carbon atoms of the ring.

Thus, it is not surprising that the bromination of phenol gives a product with bromines at the *ortho* and *para* positions. Based on such observations, groups are classified as *ortho/para orienting* and *meta orienting*. The groups (Chapter 1, Table 1.3) which are electron-releasing by resonance, regardless of their inductive effect, are ortho/para orienting. These include hydroxy, amino, alkoxy and alkyl groups which are activating groups. A halogen on the benzene ring acts as an *ortho/para* orienting but deactivating group. Electron-withdrawing groups are *meta* orienting. These include nitro, carbonyl, carboxyl and cyano groups, all of which are electron-withdrawing by resonance and also by inductive effect. These groups deactivate the benzene ring. A list of common groups and their effect on electrophilic substitution is given in Table 9.1.

Table 9. 1   Orienting and activating effect of groups in aromatic electrophilic substitution

Category	Groups
1. *Ortho/para* orienting and strongly activating	$-OH$, $-OR$, $-NH_2$, $-NHR$, $-NR_2$
2. *Ortho/para* orienting and moderately activating	alkyl, aryl, $-OCOR$, $-NHCOR$
3. *Ortho/para* orienting and deactivating	halogen
4. *Meta* orienting and deactivating	$-NO_2$, $-CN$, $-COR$, COAr, COH, COOH, COOR, $-CONH_2$, $-SO_3H$, $-\overset{+}{N}R_3$ (where R = H or alkyl) $-CX_3$ (where X = halogen)

A scientific way of looking at orientation is in terms of the rates of competing reactions. These reactions, barring certain notable exceptions, are kinetically controlled (See Chapter 6 for a discussion of kinetic and thermodynamic control of reactions). The experimentally observed composition of products, such as the percentages of the *o*-, *m*-, and *p*-isomers is a consequence of the rates of formation of these products. These can be understood qualitatively by comparing the activation energies of the formation of the products. A qualitative comparison of the activation energies can be made by comparing the stabilities of the σ-complex corresponding to each isomer, making the reasonable assumption that the transition state is very similar in structure and hence stability, to the σ-complex in the electrophilic substitution on phenol (Fig. 9.4).

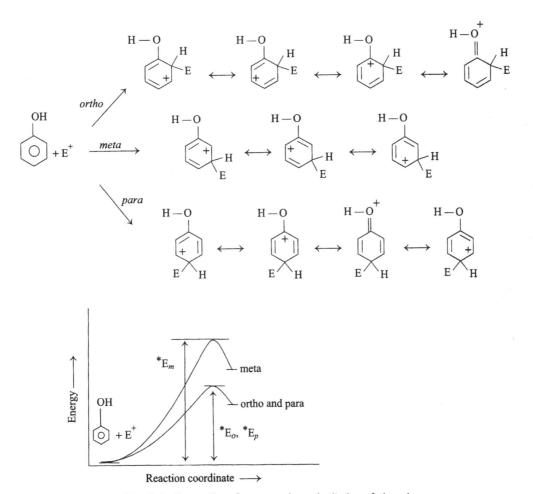

Fig. 9.4 Energetics of *o*-, *m*- and *p*-substitution of phenol

The σ-complexes for *o*- and *p*- have four 'good' canonical structures each whereas *m*- has only three. Since the former two are more stable than the latter – which is reflected in the energy diagram – we can predict that the *o*- and *p*-products will be formed at faster rates than the *m*-isomer. This is exactly what is observed. These analyses predict that the activation energy for the formation of the *o*- and *p*-isomers is the same. Since there are two *ortho* positions as against one *para* position, the rate of formation of the *ortho* isomer should be twice that of *para* and the ratio of the yield of *ortho* and *para* isomers should be 2:1. In actual practice, this is rarely, if at all observed. The *o-/p*-ratio is controlled by other factors which will be discussed in the next section.

The *ortho/para* orienting effect of the other groups in categories 1, 2 and 3 in Table 9.1 can be rationalised in a similar manner. The situation with respect to electrophilic substitution on nitrobenzene is represented in Fig. 9.5.

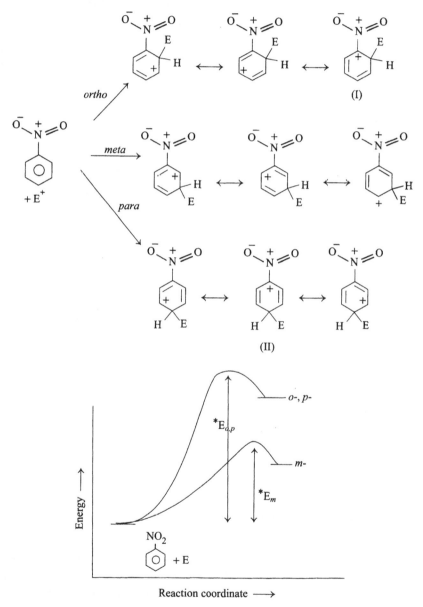

Fig. 9.5   Energetics of electrophilic substitution of nitrobenzene

The σ-complexes for all three isomers have three canonical structures each but with a notable difference. Of the three structures for *o-* and *p-*, two for each are 'good' while one for each (marked I and II) is 'bad' since it bears a positive charge on the carbon attached to the nitro group. This is a relatively unstable structure because there are two positive charges on adjacent atoms. The σ-complex for the *m-*isomer does not have this drawback, since all the three canonical structures are 'good'. Hence, substitution takes place at a greater rate at the meta position. Since the ring is electron deficient overall, its reactivity is less than that of benzene.

The *meta* orienting effect of the other groups in Table 9.1, category 4, can be rationalised by a similar comparison of the relative stabilities of the σ-complexes.

9.5.2.3 Kinetic vs Thermodynamic Control of Aromatic Substitution

The statement made in the previous paragraph that aromatic substitution reactions are kinetically controlled, needs further comment. The product formation steps in nitration and chlorination are irreversible and the ratio of the isomers formed is controlled by their rates of formation. This is not true of sulphonation and Friedel–Crafts alkylation. Under certain conditions such as higher temperatures in the case of sulphonation, and the presence of excess catalyst in the case of Friedel–Crafts alkylation, the reactions are reversible. If sufficient time is given for the product mixture to equilibrate, the product composition will not be dependent on the rates of formation, but will be the same as, or will approach, the equilibrium composition of the isomer mixture. At equilibrium, the more stable isomer will predominate. An often quoted example of the illustration of this effect is the sulphonation of naphthalene. The reactions of naphthalene are discussed in detail in Chapter 10, but this reaction is mentioned here to illustrate kinetic vs thermodynamic control (Eq. 9.19). When sulphonation of naphthalene (I), is carried out at 80°C, the only product is naphthalene-1-sulphonic acid (II). When the reaction is carried out at 160°C, a mixture of naphthalene-1-sulphonic acid (II) and naphthalene-2-sulphonic acid (III), in the ratio 20:80, is obtained. As mentioned, the reaction is reversible. At 80°C, the rate of the reverse reaction is very slow and the kinetically controlled product, II, is formed. At 160°C the reverse reaction also takes place and as shown in Eq. 9.19b, equilibrium is established. At equilibrium, the thermodynamically more stable product (III) predominates. The lower stability of II is believed to be due the steric crowding present in the molecule (Eq. 9.19c).

$$III \xrightleftharpoons{} I \xrightleftharpoons{} II \qquad (b)$$

(9.19)

(c)

Another example is the Friedel–Crafts methylation of toluene to produce the xylenes (Eq. 9.20). The reaction is not clean because a mixture of tri-, tetra- and up to hexamethylbenzene is obtained. The xylene isomers formed at the lower temperature, especially at the beginning of the reaction, are mainly *ortho* and *para* as expected from the orienting effect of CH_3. These are the kinetically controlled products. At higher temperatures, equilibrium is established between the xylenes because of reversibility. Of the three isomers, the *meta* isomer is the most stable. This is the main product under conditions of equilibrium or thermodynamic control.

$$\underset{\text{AlCl}_3}{\overset{\text{CH}_3\text{Cl}}{\longrightarrow}}$$

(9.20)

Major products at 0°C

Major product at 80°C

9.5.2.4 Ortho/Para Ratio

In the case of those substituents which are *ortho/para* orienting, structures such as in Fig. 9.4 will predict an *ortho/para* ration of 2:1, which is never observed in practice. The reasons for this are manifold.

(i)   The *positive charge density in a cyclohexadienyl cation* (the σ-complex in Eq. 9.1b) is not equal at the different positions, but is as shown in Eq. 9.21a. In the σ-complexes, Eqs. 9.21b (*ortho*) and 9.21c (*para*), the stabilisation of the charge by the electron-donating group Z is expected to be more in the *para* complex than in the ortho complex.

+0.26 +0.26
+0.09 +0.09
+0.30

(9.21)

(a)   (b)   (c)

(ii) *Steric effect*, especially when both Z and E are large, hinders *ortho* substitution. Such steric hindrance is absent for *para* substitution (Table 9.2). Comparing entries 1 and 2, we see that in the *ortho/para* distribution in the nitration of toluene and *tert*-butylbenzene, there is a drastic reduction in the *ortho* isomer in the latter which can be attributed to the

Table 9. 2 *o/p* Ratio in some reactions

	Reactant	Reaction	% Ortho	% Para
1.	CH$_3$	Nitration	58	37
2.	C(CH$_3$)$_3$	Nitration	16	73
3.	Cl	Nitration	30	69
4.	Cl	Sulphonation	1	99
5.	F	Nitration	12	88

steric hindrance by the larger *tert*-butyl group. Comparing entries 3 and 4, nitration and suphonation of chlorobenzene, very little *ortho* isomer is formed in sulphonation. This is because the electrophile SO$_3$, is much bulkier than NO$_2^+$. Again, steric hindrance is, at least partly, the explanation. Data in entries 3 and 5 are not compatible with this reasoning based on steric effect. F is a smaller group than Cl. The former should have produced more *ortho* substitution than the latter, which is contrary to what is observed. This brings us to the third factor which controls the *ortho/para* ratio, namely the inductive effect of the group.

(iii) *Inductive effect of the group Z* affects the ratio of the products. Fluorine has a powerful electron-withdrawing inductive effect which decreases the electron density at the neighbouring atoms. The ortho position is more affected than the para position. The substitution takes place preferentially at the more electron-rich para position.

9.5.3 AROMATIC NUCLEOPHILIC SUBSTITITION

In the aliphatic series, nucleophilic substitution is a well known type of reaction. Thus, alkyl halides, RX, are readily converted to other alkyl derivatives, RY, by the displacement of the halide by a nucleophile, which may be the anion, Y$^-$ (Eq. 9.22a). This is illustrated by the hydrolysis of chloropropane to propanol (Eq. 9.22b).

$$R-X + Y:^- \rightarrow R-Y + X:^- \tag{a}$$

$$CH_3-CH_2-CH_2-Cl + OH^- \rightarrow CH_3-CH_2-OH + Cl^- \quad (b) \tag{9.22}$$

These reactions are discussed in Chapter 11. These reactions are controlled by the polarity of the C−X bond of the haloalkane.

$$C-X \longleftrightarrow C^+ + X^- \tag{9.23}$$

Apart from the electronegativity of X, which we shall consider as a constant factor, the polarity is decided by the electronegativity of the carbon and the ability of the carbon to sustain a positive charge. An sp^2 hybridised carbon is more electronegative than an sp^3 carbon (Chapter 1, 1.4.3). The carbon–halogen bond in aryl and vinyl halides is less polar than in haloalkanes. (Recall that vinyl chloride is not reactive in Friedel–Crafts reactions, Section 9.5.1.) In haloalkanes, the residual positive charge of the carbon of the C−X bond is stabilised by the alkyl group. The presence of this positive charge makes the attack by the nucleophile Y, one of the steps of nucleophilic substitution, facile. These conditions are not satisfied in aryl halides. The aryl–X bond is relatively less polar than the alkyl–X bond. It is also seen that the nucleophile Y^-, will encounter an electron rich region (the π-electron cloud, which sandwiches the carbon framework) when it collides with the aryl halide.

This is not to say that aromatic nucleopilic substitution is impossible. If the unfavourable conditions described above are mitigated, such reactions are possible. There are two classes of aromatic nucleophilic substitutions. One, where the presence of electron-withdrawing substituents decreases the electron density of the ring, making certain positions of the ring amenable to nucleophilic attack. This is an addition–elimination mechanism. The other, where a powerful base causes dehydrohalogenation of the aryl halide to generate a highly reactive molecule called the *aryne*, which further reacts to give products corresponding formally to nucleophilic substitution products. This is the aryne mechanism or eliminaton–addition mechanism.

9.5.3.1 Nucleophilic Substitution by Addition–Elimination: Reaction of Halonitroarenes

Chlorobenzene cannot be hydrolysed to phenol by heating with aqueous alkali under laboratory conditions. (There is an old industrial process for the manufacture of phenol, the *Dow's process*, where chlorobenzene is heated with aqueous sodium hydroxide to a high temperature under pressure, which proceeds through a benzyne mechanism.) In sharp contrast, *p*-nitrochlorobenzene is hydrolysed to *p*-nitrophenol by refluxing with aqueous sodium hydroxide. 2,4-Dinitrochlorobenzene and 2,4,6-trinitrochlorobenzene are progressively more readily hydrolysed (Eq. 9.24).

$$\tag{9.24}$$

The mechanism for *p*-nitrochlorobenzene is outlined in Eq. 9.25.

$$ \text{(a)} $$

$$ \text{(9.25)} $$

$$ \text{(b)} $$

The canonical structures (in square brackets) of the intermediate σ-complex (called a *Meisenheimer complex*) show how the nitro group stabilises the anion. Other nucleophiles like alkoxide and ammonia can displace the halogen from such 'activated' aryl halides. Notice that the expression 'activation by the nitro group' in this context and description of the nitro group as a deactivating group in electrophilic aromatic substitution have meaning only when the reaction type is also specified.

9.5.3.2 The Benzyne Mechanism

In contrast to the poor reactivity of chlorobenzene in nucleophilic substitution under normal conditions is its reaction with sodium amide in liquid ammonia. Under the latter conditions, the chlorine is displaced by an amino group and the product is aniline (Eq. 9.26).

$$ \text{(9.26)} $$

It was known early that this reaction is not as simple as it appears. *p*-Chlorotoluene under these conditions gave not only *p*-aminotoluene but also *m*-aminotoluene (Eq. 9.27).

$$ \text{(9.27)} $$

Detailed studies have established that this reaction involves the elimination of HCl from chlorobenzene brought about by the strong base (Eq. 9.28). The intermediate, formally represented with a triple bond is called benzyne (aryne in general). This intermediate, a transient one, has been isolated and is highly reactive. It reacts as shown in Eq. 9.28.

$$(9.28)$$

A six membered ring with triple bond is expected to be highly strained. The molecule is represented (9.29) with the aromatic structure intact.

$$(9.29)$$

9.5.4 ADDITION REACTIONS OF BENZENE

Addition to the double bonds of benzene destroys its aromaticity. However, there are some reactions where addition takes place. Noteworthy among these are, addition of hydrogen (hydrogenation) and addition of chlorine.

9.5.4.1 Hydrogenation

This reaction requires a catalyst. For laboratory purposes, nickel catalysts (especially Raney nickel) are useful, at high temperature and hydrogen pressure (Chapter 4, Table 4.2). Cyclohexane and substituted cyclohexanes can be prepared readily by the hydrogenation of the appropriate aromatic compounds (Eq. 9.30).

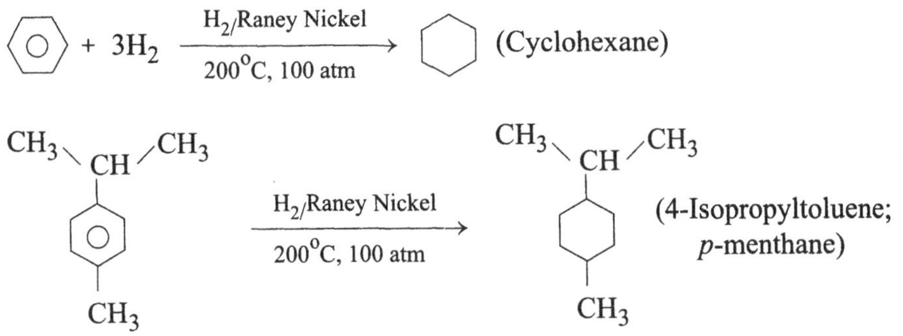

$$\text{benzene} + 3H_2 \xrightarrow[\text{200°C, 100 atm}]{\text{H}_2\text{/Raney Nickel}} \text{cyclohexane} \quad \text{(Cyclohexane)}$$

(9.30)

(4-Isopropyltoluene; *p*-menthane)

The ring can be reduced by chemical means as well; the most important of which is the Birch reduction (Eq. 9.31, See also Chapter 4, Section 4.2.2). An alkali metal like lithium or sodium dissolved in liquid ammonia is the reagent. 'Solvated electrons' present in such a solution constitute the actual reducing agent. The product of Birch reduction is the unconjugated cyclohexadiene, which does not get further reduced by this reagent.

$$Li + NH_3(l) \longrightarrow Li^+ + e^-(NH_3) \text{ (Solvated electron)}$$

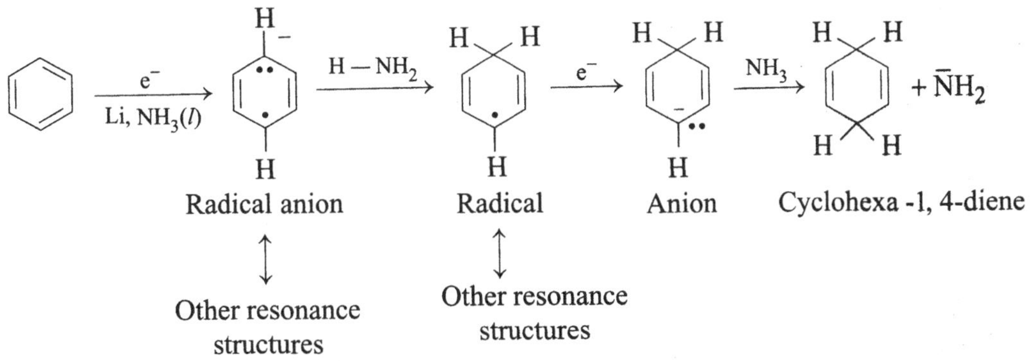

| Radical anion | Radical | Anion | Cyclohexa -1, 4-diene |

Other resonance structures

Other resonance structures

(9.31)

9.5.4.2 Addition of Chlorine

Recall that the familiar reaction of chlorine with benzene is electrophilic substitution which is an ionic reaction. This requires conditions favourable for the generation of the electrophile, formally Cl^+, or its precursor. This is the role of the catalyst (halogen carrier).

Addition of chlorine to the benzene ring can be brought about under free radical conditions. When chlorination is carried out under conditions that are favourable for free radical reaction, addition takes place instead of substitution. The conditions required in this context are light (photochemical halogenation) and heat. Also, the conditions favourable for the ionic reaction, that is, those favourable for the generation of the Cl^+ ion, should be avoided. Lewis acids and polar solvents should be avoided. Under photochemical conditions the following reaction (Eq. 9.32) takes place. The fully chlorinated product, hexachlorocyclohexane (benzene hexachloride) is a mixture of various stereoisomers. One of these is the γ-isomer, which is used as an insecticide (gammexane).

$$Cl_2 \xrightarrow{H_3} 2Cl^\bullet$$

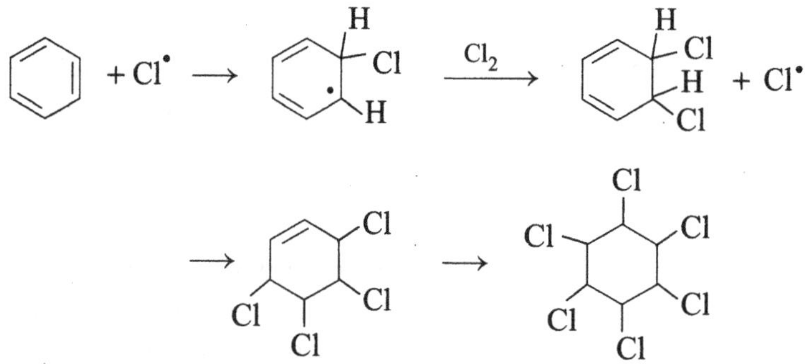

(9.32)

9.5.5 Reactions of the Aromatic Side Chain

An alkyl or substituted alkyl group attached to the benzene ring is called a side chain. While in most reactions, the side chain and functional groups attached to them behave in the same way as in aliphatic compounds, in some reactions the aromatic ring exerts a notable influence on its reactivity. Thus, the alkyl side chain can be easily oxidised as in the conversion of toluene to benzoic acid (Eq. 9.33).

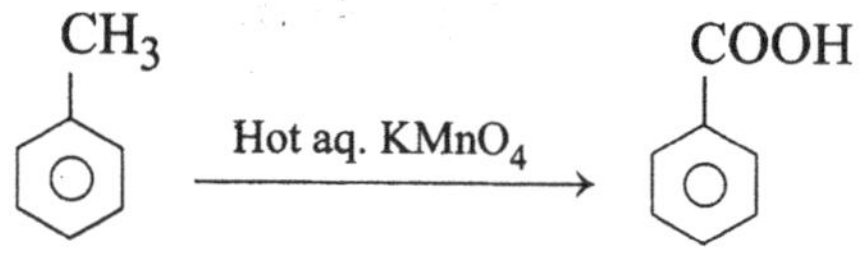

(9.33)

Another reaction of particular interest is side-chain halogenation.

9.5.5.1 Side-chain Chlorination of Toluene

Chlorine reacts with alkylbenzenes under free radical conditions to give products of side chain halogenation. Thus, toluene gives products with one, two or all three hydrogens of the methyl group replaced by chlorine (Eq. 9.34).

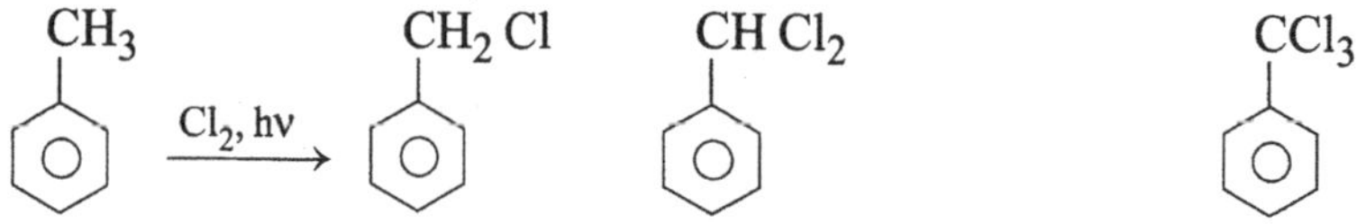

Benzyl   (Dichloromethyl)benzene   (Trichloromethyl)benzene
chloride   (benzal chloride)   (benzotrichloride)

(9.34)

This is a free radical chain reaction similar to the halogenation of alkanes (Chapter 3). The mechanism for the introduction of the first chlorine is outlined in Eq. 9.35. Subsequent chlorination leading to the di- and trichloro products takes place in the same manner.

$$Cl_2 \xrightarrow{h\nu} 2Cl^{\bullet} \qquad (a)$$

$+ Cl^{\bullet} \longrightarrow$ $+ HCl \qquad (b)$

$+ Cl_2 \longrightarrow$ $+ Cl^{\bullet} \qquad (c)$

$$(9.35)$$

In the second step (9.35b) hydrogen abstraction by the chlorine atom takes place. The hydrogens available in the toluene molecule are of two types, the ones on the benzene ring and the ones on the side chain. The bond dissociation energy of the former is about $469\,\text{kJ mol}^{-1}$ ($112\,\text{kcal mol}^{-1}$) and that of the latter, about $360\,\text{kJ mol}^{-1}$ ($86\,\text{kcal mol}^{-1}$). The chlorine atom selectively abstracts the latter hydrogen. The benzyl radical formed is resonance stabilised (9.36).

$$(9.36)$$

Recall that when chlorine atoms are introduced into benzene which does not have a side chain, the reaction is not hydrogen abstraction, but addition to the double bond (see Section 9.5.2). When ethylbenzene is chlorinated under free radical conditions, the major monochloro product is 1-chloroethylbenzene. 2-Chloroethylbenzene is also formed, but is only a minor product (Eq. 9.37).

$$(9.37)$$

Chlorination of toluene is an indirect way of oxidising toluene; the chloro derivatives can be hydrolysed to the oxygenated compounds (Eq. 9.38).

$$(9.38)$$

KEY POINTS

- *Aromaticity*: Fully conjugated, planar, cyclic compounds have a delocalised π-electron cloud and are aromatic, provided the number of π-electrons conform to the formula $4n + 2$, where n is a positive integer. This is Hückel's rule. The Hückel numbers are $2(n = 0)$, $6(n = 1)$, $10(n = 2)$ and so on.

- Aromatic structures possess extra stability (resonance stability).

- Planar structures, not conforming to Hückels's rule not only do not possess aromaticity, but are destabilised. They are antiaromatic. Cyclobutadiene (4π-electrons) is the prominent example of such a molecule. The next non-Hückel polyene which has (8π-electrons), cyclooctatetraene, is a stable compound, but non-planar. It is not antiaromatic, but is non-aromatic.

- Benzene undergoes substitution reactions, under similar conditions where alkenes undergo addition reactions. The preference for substitution is because the products of substitution retain the aromaticity (stability).

- The most common substitution reactions of benzene are electrophilic. The high electron density on the benzene ring attracts electrophiles and repels nucleophiles. Electrophilic substitution mechanism essentially involves two steps. The first step is the addition of the electrophile across the double bond to form a σ-complex, which is the same as the first step in the addition of electrophiles to alkenes. The second step is the deprotonation of the σ-complex to regain the aromaticity lost in the first step. It is described as an addition–elimination mechanism.

- Some of the important electrophilic substitutions are nitration, sulphonation, halogenation, Friedel–Crafts alkylation and Friedel–Crafts acylation.

- When the reaction is carried out on a benzene derivative which already contains a substituent, the existing group directs or orients the incoming group selectively to either the *meta* position or to the *ortho* and *para* positions. This is the orienting effect of groups. The orienting effect can be understood by drawing the canonical structures of the *ortho, meta* and *para* σ-complexes and comparing their relative stabilities. This is valid because the product formation in aromatic substitution is in most cases, kinetically controlled.

- In a few reactions which are reversible (notably alkylation and sulphonation) under conditions of equilibration, thermodynamic control may take over and the preferred product will be the more stable one, rather than the one which is formed at a faster rate.

- Groups attached to the benzene ring may be classified as strongly activating, moderately activating or deactivating, with respect to electrophilic substitution.

- When strong electron-withdrawing groups like one or more nitro groups are present, due of the reduction of electron density in the ring, nucleophilic aromatic substitution is possible. An illustrative example is the ready hydrolysis of 2,4-dinitrochlorobenzene to 2,4-dinitrophenol. This reaction also proceeds through a σ-complex, addition–elimination mechanism.

- In the presence of very strong bases, like sodamide in liquid ammonia, a formal nucleophilic diaplcement reaction, illustrated by the conversion of chlorobenzene to aniline takes place. This reaction involves an elimination–addition mechanism and the highly reactive molecule benzyne is the intermediate.
- Addition reactions are also known, illustrated by catalytic hydrogenation, Birch reduction and free radical addition of chlorine to benzene giving hexachlorocyclohexane.
- Side chains, like the methyl group in toluene can be chlorinated step-wise under free radical conditions to give benzyl chloride, (dichloromethyl)benzene and (trichloromethyl)benzene.

EXERCISES

SECTION I

1. Classify the following as aromatic, antiaromatic or non-aromatic. You may apply Hückel's rule even to the polycyclic molecules.

(i) (The carbon skeleton is nearly planar)

2. Classify the following groups as either *meta* orienting or *ortho/para* orienting.

(a) $-F$ (b) $-O-\langle\bigcirc\rangle$ (c) $-\overset{O}{\underset{\|}{C}}-Cl$ (d) $-\overset{O}{\underset{\|}{C}}-H$

(e) $-O-\overset{\overset{\displaystyle O}{\|}}{C}-\langle O \rangle$ (f) $-CF_3$ (g) $-\overset{+}{N}(CH_3)_3$ (h) $-N=O$

(i) $-CH=CH_2$ (j) $-\overset{\overset{\displaystyle O}{\|}}{\underset{\underset{\displaystyle O}{\|}}{S}}-R$

3. Predict the major product or products of mono-nitration of the following.

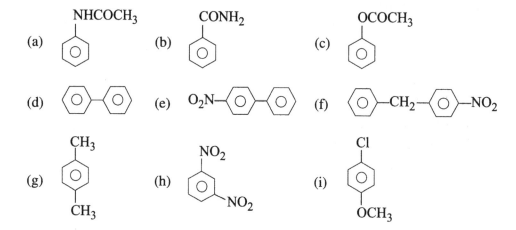

4. Suggest suitable reagents and reaction conditions to bring about the following conversions.

5. Identify the products A to F.

(a) [benzene ring] + $CH_3 CH_2 CH_2 Cl$ $\xrightarrow{AlCl_3}$ A (monoalkylation)

(b) [toluene with CH_3] + $H_2 SO_4$ $\longrightarrow$ B

(c) [4-chlorobiphenyl with Cl] + $NaNH_2$ $\xrightarrow{NH_3(l)}$ C + D

(d) [benzene with CH_3 top and NO_2 bottom] $\xrightarrow{KMnO_4/H_2O, \Delta}$ E

(e) [benzene with $COCH_3$] $\xrightarrow{HNO_3/H_2SO_4}$ F

6. Show how the following conversions can be brought about. More than one step may be involved.

(a) [toluene with CH_3] $\longrightarrow$ [benzene with CH_3 top and $CH_2 CH (CH_3)_2$ bottom]

(b) [benzene] $\longrightarrow$ [benzene with NO_2 and $COCH_3$]

(c) [toluene with CH_3] $\longrightarrow$ [benzene with $COOH$ top and NO_2 bottom]

(d) [toluene with CH_3] $\longrightarrow$ [benzene with $COOH$ and NO_2]

SECTION II

1. Draw resonance structures of the σ-complexes for *o*-, *m*- and *p*- substitution, compare stabilities and explain the orienting effects in nitration, for the following molecules. Draw energy profiles diagrams in each case.

(a) [benzene with CF_3]

(b) [benzene with $\overset{+}{N}(CH_3)_3 NO_3^-$]

2. (a) Outline the mechanism of the acetylation of benzene by acetic anhydride in the presence of anhydrous $AlCl_3$.

(b) In continuation of (a) above, suggest the mechanism of the following reaction.

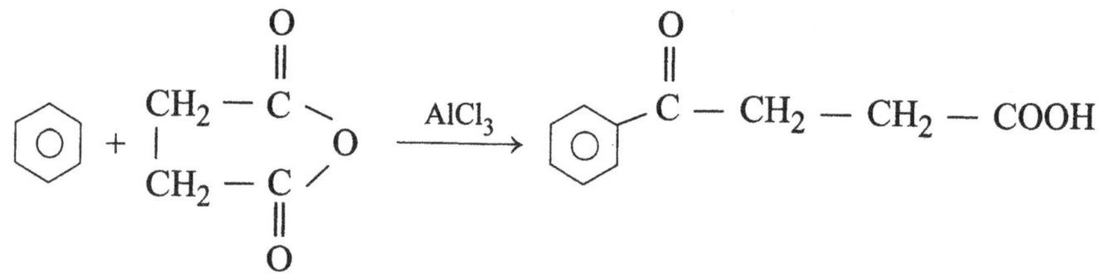

3. Account for the following observations.

(a) Photochemical conditions are not suitable for the bromination of toluene to obtain *o*- and *p*-bromotoluene.

(b) Bromination of aniline using bromine in aqueous HBr, gives 2,4,6-tirbromo anililine, even though in the acid medium, the ion (I) is present which is expected to give *meta* bromination.

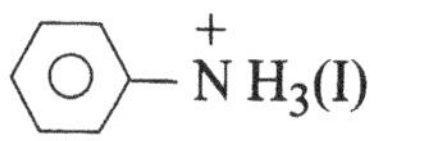

CHALLENGING QUESTIONS

1. One of the experiments (J D Roberts, 1953) which demonstrated the presence of benzyne involved the use of ^{14}C-labelled chlorobenzene.

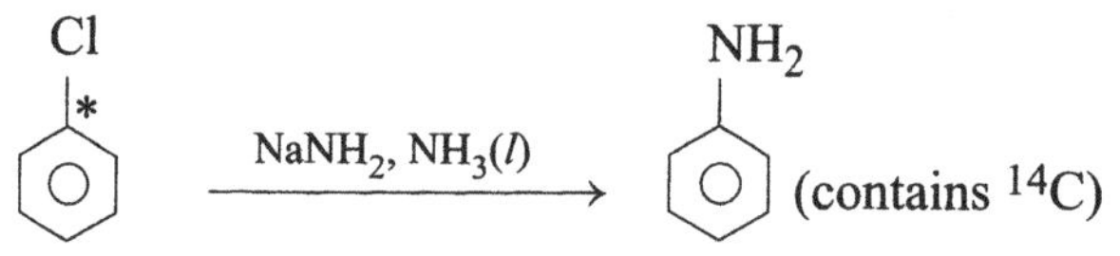

(* denotes ^{14}C)

If benzyne is the intermediate, predict the distribution of ^{14}C in the product.

2. Benzyne is a powerful dienophile in Diels–Alder reactions. Give the structure of the adduct in the following reaction.

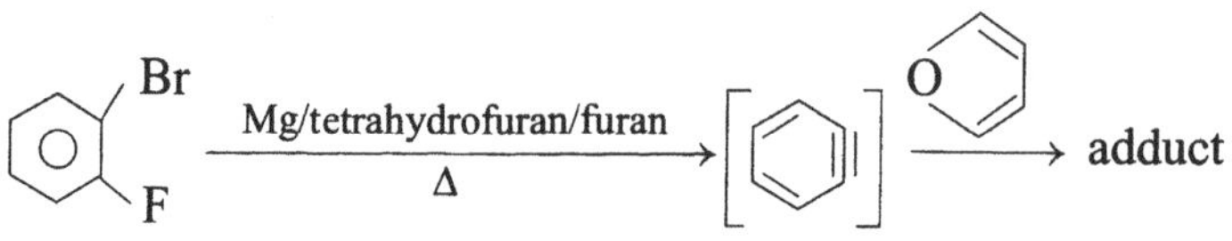

Notes:

(i) The first step is the generation of benzyne by the elimination of Br and F as MgBrF.

(ii) Tetrahydrofuran (THF) is an ether solvent used when nonaqueous conditions are needed.

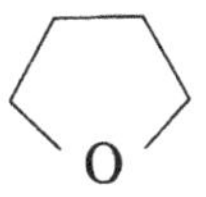

(THF)

(iii) Furan, though aromatic reacts as a diene when powerful dienophiles are present.

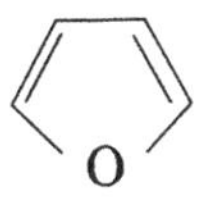

(Furan)

3. Suggest mechanisms for the following reactions.

(a)

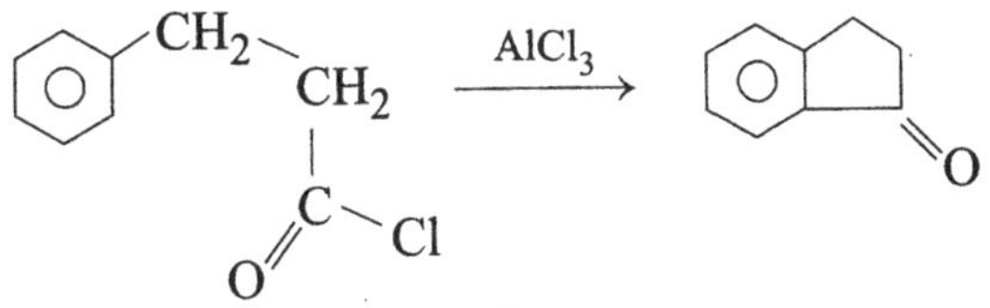

(Hint: It is an intramolecular Friedel–Crafts acylation)

(b)

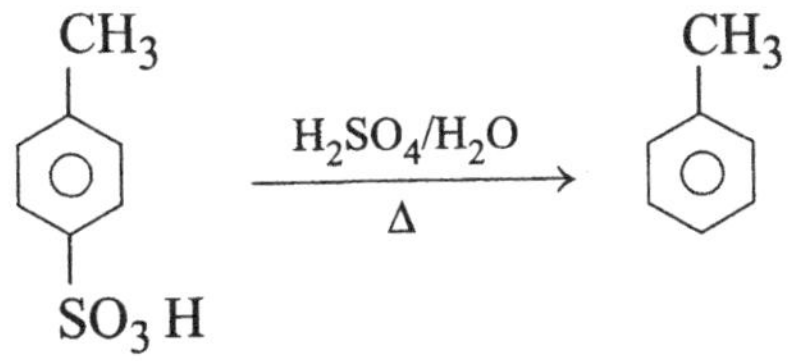

(Hint: This is the reverse of sulphonation. Sulphonation is a reversible reaction)

(c)

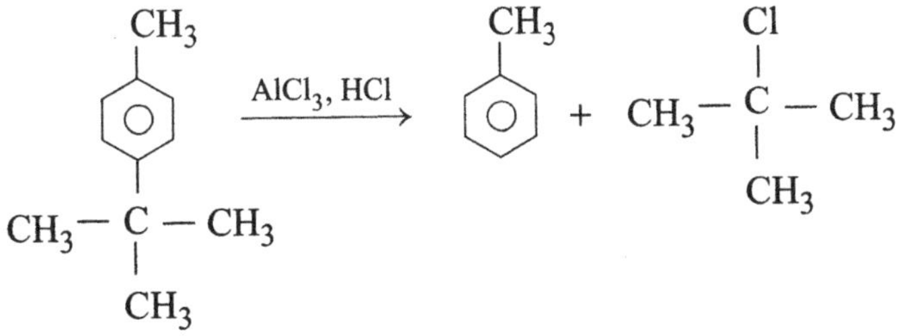

(Hint: Friedel–Crafts, alkylation is reversible)

4. Explain the methodology of the following synthetic schemes. (Why is the synthesis done by these indirect routes?)

(a)

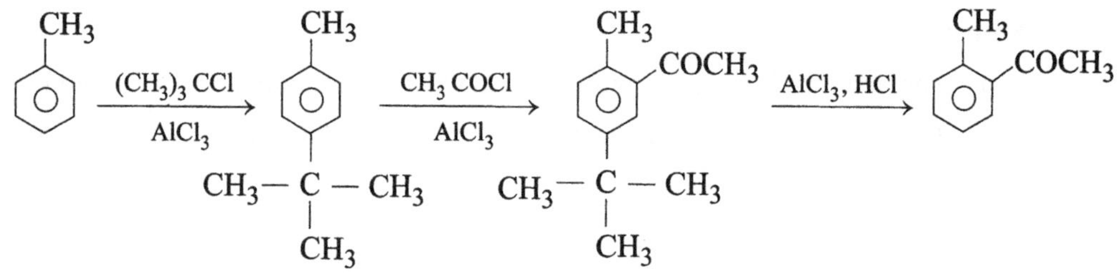

$$\text{Toluene} \xrightarrow[\text{AlCl}_3]{(CH_3)_3 CCl} \xrightarrow[\text{AlCl}_3]{CH_3 COCl} \xrightarrow{\text{AlCl}_3, \text{HCl}}$$

with substituent $CH_3-\overset{\displaystyle CH_3}{\underset{\displaystyle CH_3}{C}}-CH_3$

and $CH_3-\overset{\displaystyle CH_3}{\underset{\displaystyle CH_3}{C}}-CH_3$

(b)

$$\text{Phenol} \xrightarrow{H_2SO_4} \left(\text{para-} SO_3H + \text{ortho-} SO_3H \right) \xrightarrow{HNO_3} \text{Picric acid}$$

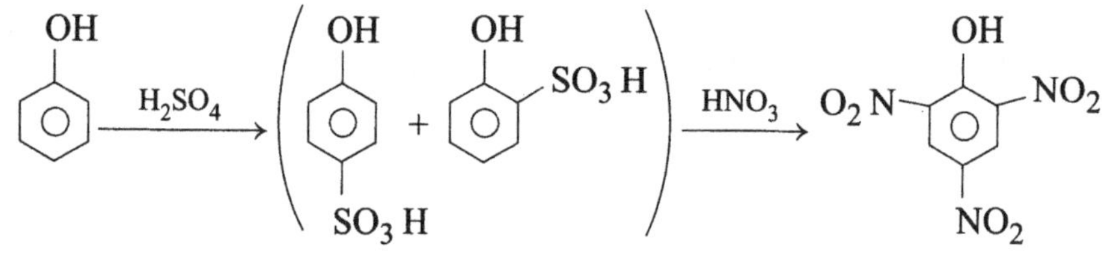

Picric acid

(Note: Phenol itself is oxidised by nitric acid, sulphonated phenol is not oxidised.)

(c)

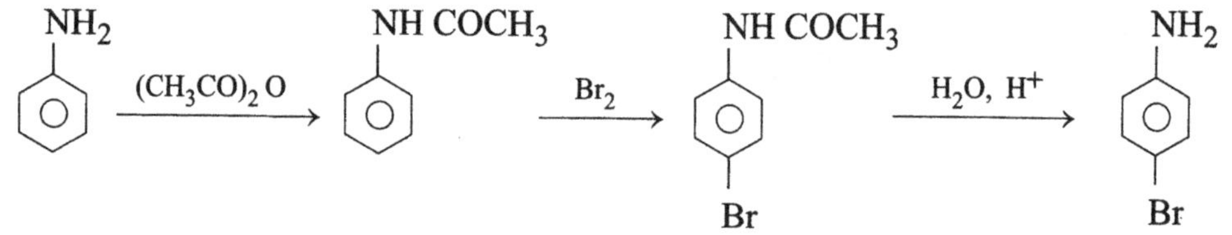

$$\xrightarrow[\text{(CH}_3\text{CO)}_2 O]{} \xrightarrow[\text{Br}_2]{} \xrightarrow[\text{H}_2\text{O, H}^+]{}$$

PROJECT

Prepare a project report on annulenes, their synthesis, properties and aromaticity (Annulenes are cyclic polyenes).

10 Polynuclear Aromatic Hydrocarbons

OBJECTIVES   In this chapter, you will learn about,

- polynuclear or polycyclic aromatics
- structure and aromaticity of naphthalene, anthracene and phenanthrene
- their occurrence, properties and uses
- reactions of the above: addition, oxidation
- why they are more reactive than benzene
- electrophilic substitution, mechanism, site selectivity
- nitration, halogenation, sulphonation, Friedel–Crafts acylation
- kinetic vs thermodynamic control
- synthesis of naphthalene, anthracene and phenanthrene ring systems

10.1 INTRODUCTION

A large number of aromatic hydrocarbons which contain benzene rings fused to each other have been identified. The structures and names of some of the simpler ones are given below. These are referred to as polynuclear or polycyclic aromatic hydrocarbons (PAHs), also as fused ring or condensed ring aromatic compounds. Each ring shares two or more carbon atoms with other rings and all the rings have one common aromatic π-electron cloud.

         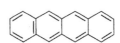

Naphthalene   Anthracene   Phenanthrene   Naphthacene

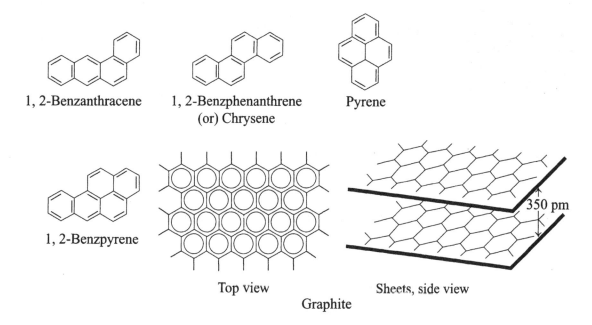

1, 2-Benzanthracene   1, 2-Benzphenanthrene   Pyrene
(or) Chrysene

1, 2-Benzpyrene

Top view   Sheets, side view

Graphite

Graphite is the ultimate polynuclear aromatic structure. The number of benzene rings in graphite is very large and its sheet structure confers the characteristic property of slipperiness. The extended π-cloud containing mobile electrons makes it a good conductor of electricity.

Most of the simpler polynuclear aromatic hydrocarbons are isolated from coal tar and their industrial importance is based on their use as starting materials for dyes. The higher ones are present in soot, cigarette tar and in the combustion gases from the burning of fossil fuels. More than a hundred such compounds have been identified. They have acquired well-deserved notoriety as potent carcinogens. Benzpyrene and benzanthracene are two well known carcinogens.

In this chapter we shall study the chemistry of three simple polycyclic aromatic hydrocarbons—naphthalene, anthracene and phenanthrene.

10.2 NAPHTHALENE, ANTHRACENE AND PHENANTHRENE

10.2.1 KEKULÉ STRUCTURES, RESONANCE AND AROMATICITY

These are planar conjugated polyenes, satisfying Hückel's rule. Hückel's rule is applicable to polycyclic systems also when the sp^2 carbons are on the periphery of the extended ring system. Their resonance energies obtained from heats of hydrogenation data are,

Benzene   : 151 kJ mol^{-1} (36 kcal mol^{-1})

Naphthalene : 255 kJ mol^{-1} (61 kcal mol^{-1})

Anthracene   : 351 kJ mol^{-1} (84 kcal mol^{-1})

Phenanthrene : 385 kJ mol^{-1} (92 kcal mol^{-1})

The stabilisation energy of naphthalene ($255\,kJ\,mol^{-1}$, $61\,kcal\,mol^{-1}$) is less by $47\,kJ\,mol^{-1}$ than what is estimated for two benzene rings ($2 \times 151 = 302\,kJ\,mol^{-1}$). Those for anthracene and phenanthrene are also less than the expected value for three benzene rings. We can draw three Kekulé structures for naphthalene (Eq. 10.1).

$$(10.1)$$

The numbering in naphthalene is as shown. It can be clearly seen that the three structures are not identical. For example, the bond between C1 and C2 is a double bond in two structures and a single bond in one; that between C2 and C3 is single in two and double in one. Such differences are present for the other bonds also. We expect that the C1—C2 bond length may not be the same as the C2—C3 bond length. This is what is actually found.

Naphthalene, bond lengths

Such differences are found in the other polycyclic molecules also.

10.2.2 OCCURRENCE, ISOLATION, PROPERTIES AND USES

Most of the polycyclic aromatic hydrocarbons and several of their derivatives occur in coal tar, the tarry liquid obtained by the distillation of coal. Before the advent of petroleum-based industries, coal was the major source of not only these, but also aromatics like benzene and other hydrocarbons, phenols and heterocyclic compounds. Later, it was discovered that polycyclic aromatics are present in petroleum distillation residue, are also formed during the pyrolysis and burning of most organic matter, and are present in soot of diverse origin. Cigarette smoke and cigarette tar, automobile exhaust and smoke from other sources contain these compounds. PAHs are included among the major carcinogenic components of atmospheric pollutants.

Naphthalene, anthracene and phenanthrene are isolated from the appropriate fractions of coal tar distillation. Naphthalene is also manufactured by the catalytic conversion of petroleum hydrocarbons. All the three compounds are solids, insoluble in water and soluble in hydrocarbon solvents, as expected.

Compound	Melting point (°C)	Boiling point (°C)
Naphthalene	80	218
Anthracene	217	354
Phenanthrene	101	340

Naphthalene has limited use as a moth repellant. Its major industrial use is in the manufacture of phthalic anhydride by oxidation (Eq. 10.2c). Naphthalene is also converted to derivatives bearing the hydroxyl, amino and sulphonic acid groups which can further be converted to azo dyes.

The most useful derivatives of anthracene and phenanthrene are the quinones, 9,10-anthraquinone and 9,10-phenanthraquinone, which are used as raw materials in the dye industry. Pure anthracene is used in scintillation counters.

The numbering of these compounds is shown below. For positions 1 and 2 of naphthalene, the Greek letters α and β are also commonly used.

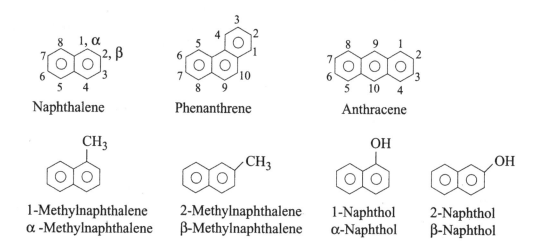

| Naphthalene | Phenanthrene | Anthracene |

CH₃		OH	
1-Methylnaphthalene	2-Methylnaphthalene	1-Naphthol	2-Naphthol
α -Methylnaphthalene	β-Methylnaphthalene	α-Naphthol	β-Naphthol

10.2.3 REACTIONS

10.2.3.1 General

As aromatic compounds, these compounds undergo electrophilic substitution reactions in which they are generally more reactive than benzene. Such reactions proceed through σ-complexes, as in the case of benzene. The greater reactivity can be understood when we consider the nature of the σ-complexes. The σ-complexes from polynuclear aromatics still have one or more benzene rings intact and the cations have positive charges which are more effectively delocalised than those in benzene (Eqs. 10.4 and 10.5). Polynuclear aromatics also undergo addition reactions such as catalytic hydrogenation, chemical reduction and oxidation where one ring gets affected more readily than benzene. Some examples of such reactions are given Eqs. 10.2a to 10.2e. Equation 10.2a illustrates the ease of reduction of one ring in naphthalene. The product, tetrahydronaphthalene (tetralin), which is in effect a dialkylbenzene, can be further reduced to decahydronaphthalene (decalin) only under more drastic conditions, reminiscent of the reduction of benzene to cyclohexane (Chapter 9, Eq. 9.30). In Eq. 10.2b, the chemical reduction of naphthalene by sodium and ethanol (Bouveault–Blanc reduction) is shown. Benzene cannot be reduced under these conditions. Naphthalene can be oxidised to phthalic acid (Eqs. 10.2c and 10.2d) and to naphthaquinone (Eq. 10.2e) by a variety of reagents. Anthracene and phenanthrene can be selectively reduced and oxidised at the middle ring, leaving the other two rings as intact benzene rings (Eq. 10.3).

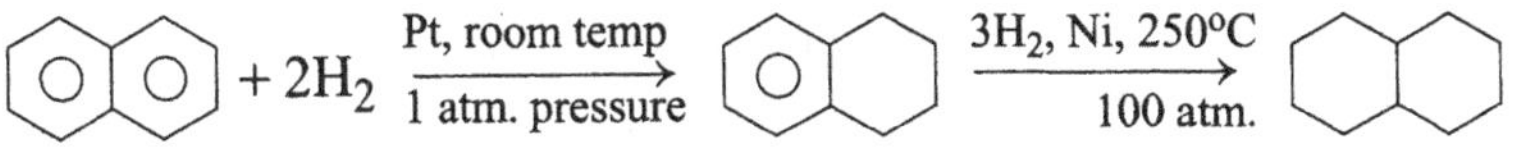

Tetrahydronaphthalene (tetralin) Decahydronaphthalene (decalin) (a)

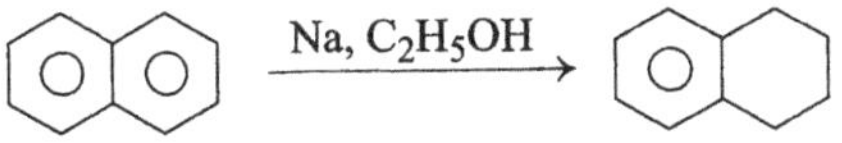

(b)

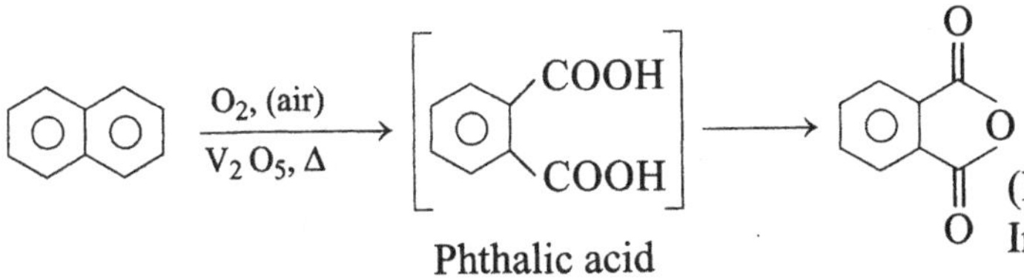

Phthalic acid (Phthalic anhydride, Industrial method) (c)

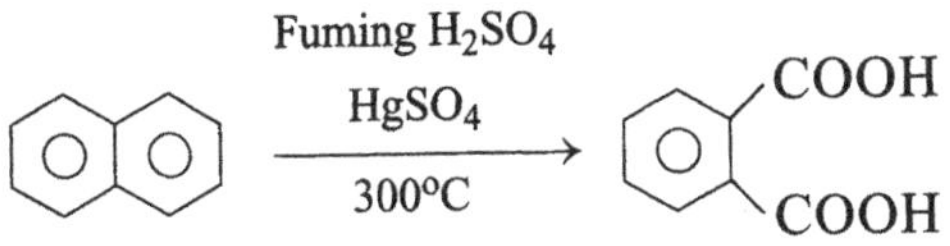

(d)

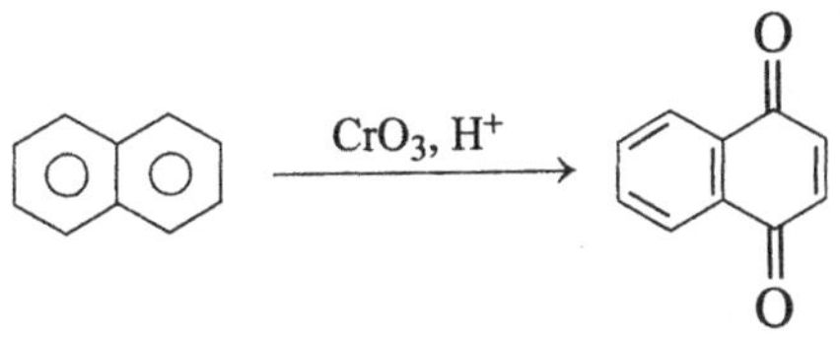

1, 4-Naphthaquinone (e)

(10.2)

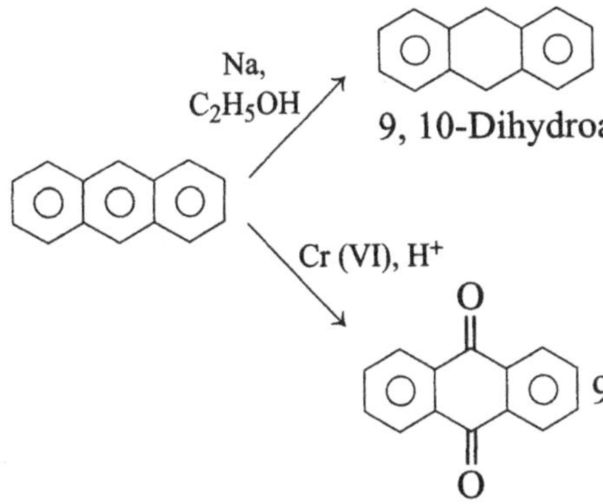

Na, C$_2$H$_5$OH → 9, 10-Dihydroanthracene

Cr (VI), H$^+$ → 9, 10-Anthraquinone (a)

9, 10-Dihydrophenanthrene

(10.3)

9, 10-Phenanthraquinone   (b)

10.2.3.2 Electrophilic Substitution Reactions of Naphthalene: General Considerations

Unlike benzene, the different positions in naphthalene are not identical. Two mono-substituted naphthalenes are possible, corresponding to substitution at the 1-position or at the 2-position. The 1-position is more reactive for electrophilic substitution. The kinetically controlled products of mono-substitution on naphthalene are the 1-substituted products. This selectivity can be understood by comparing the rates of formation of the two isomers, using the method that we have already employed for comparing the rates of formation of the *ortho*, *meta* and *para* isomers from mono-substituted benzenes when a second substituent is introduced (Chapter 9, Section 9.5.2.2).

Using the generalised electrophile E^+, the σ-complexes and their canonical structures for 1- and 2-substitution are given in Eqs. 10.4 and 10.5 respectively. Both σ-complexes have 5 canonical structures and in each case the positive charge is delocalised over the extended conjugated systems. However, of the canonical structures in Eq. 10.4 (1-substitution), two (labelled I and II) have intact benzene rings. In Eq. 10.5 (2-substitution) only one canonical structure (labelled VI), has an intact benzene ring. Assuming that more the number of intact benzene rings (fully aromatic rings) among the canonical structures, the greater the resonance stability of that particular hybrid, we conclude that the σ-complex (and the transition state leading to it) for 1-substitution is more stable. The activation energy for that reaction is less and so it will take place at a faster rate (Fig. 10.1). The resonance hybrids of structures I–V (1-substitution) are more stable than those of structures VI–X (2-substitution). Comparing the respective activation energies for their formation, $^*E_1$ is less than $^*E_2$. 1-Substitution takes place at a faster rate than 2-substitution. The former is the kinetically controlled product.

(10.4)

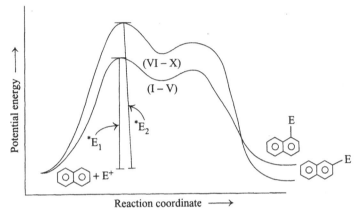

$$(10.5)$$

Fig. 10.1 Potential energy diagram for 1- vs 2-substitution on naphthalene

Comparing the energy levels of the final products, 1-substituted naphthalene and 2-substituted naphthalene in Fig. 10.1, the former is *thermodynamically* less stable than the latter, which is usually the case. The cause and consequence of this are discussed in Section 10.2.3.4.

10.2.3.3 Typical Electrophilic Substitutions on Naphthalene

Naphthalene undergoes all the usual electrophilic aromatic substitutions, generally under milder conditions than benzene. 1-Substituted naphthalene is the preferred product. The mechanism is similar to that for benzene, through the σ-complex (Eq. 10.4). Some of the reactions are listed below.

Nitration: HNO_3/H_2SO_4, warm → 1-Nitronaphthalene, (10.6)

Bromination: Br_2, no catalyst → 1-Bromonaphthalene, (10.7)

Sulphonation: con. H_2SO_4, 80°C → Naphthalene-1-sulphonic acid,   (10.8a)

Sulphonation: con. H_2SO_4, 160°C → Naphthalene-2-sulphonic acid,   (10.8b)

Friedel–Crafts acylation: RCOCl, AlCl₃ → 1-Acylnaphthalene,   (10.9)

10.2.3.4   Kinetic vs Thermodynamic Control

The change in specificity of 1-substitution vs 2-substitution, (site specificity or regiospecificity) in sulphonation at different temperatures, merits comment. Sulphonation is a reversible reaction. The reverse reaction is slow at the lower temperature. Also, the reaction leading to the formation of 2-sulphonic acid is slower than the one leading to the 1-sulphonic acid, for reasons discussed in Section 10.2.3.2 and the product obtained selectively is the expected kinetically controlled isomer, 1-sulphonic acid. At higher temperatures, the rates of both 1-substitution and 2-substitution increase; as do the rates of the reverse reactions. Equilibrium is rapidly established between the two isomeric sulphonic acids (Eq. 10.10). At equilibrium, the thermodynamically more stable 2-sulphonic acid predominates (Fig. 10.1). The reason for the greater thermodynamic stability of the 2-isomer is essentially steric in nature (see discussion in Chapter 9, Section 9.5.2.3 and Eq. 9.19c). Generally, 1-substituted naphthalenes are thermodynamically less stable than 2-substituted naphthalenes. This is because of steric crowding involving the H or any other group at the 8-position and the group, G, at the 1-position (10.10b). This effect is particularly evident in the sulphonic acid since $-SO_3H$ is a bulky group.

(10.10a)

(10.10b)

Friedel–Crafts acylation of naphthalene with simple acyl halides like acetyl chloride give mainly 1-acylnaphthalenes. When bulkier acyl groups are introduced, a mixture of the 1- and 2-acyl derivatives are obtained. Acylation with succinic anhydride and aluminium chloride gives a mixture of products (Eq. 10.11).

(10.11)

Succinic anhydride   4-(1-Naphthyl)-4-
                            oxobutanoic acid

4-(2-Naphthyl)-
4-oxobutanoic acid

Friedel–Crafts alkylations of naphthalene are of limited utility because the high reactivity of naphthalene causes polyalkylations.

10.2.3.5 Reactions of Anthracene and Phenanthrene

Anthracene and phenanthrene are more reactive than naphthalene. The 9 and 10 positions are the most reactive. These positions are vulnerable to addition and substitution reactions as illustrated by bromination (Eqs. 10.12 and 10.13).

(a)

I   II

(10.12)

(b)

$$(10.13)$$

Both products (I and II of Eq. 10.12a) are obtained, depending upon the temperature of the reaction. The carbocation intermediate (σ-complex) in 10.12b can undergo either addition of Br^- (overall addition of Br_2) or elimination of H^+ (substitution). This is the situation with the σ-complex from benzene also (see Chapter 9, Eq. 9.3). In the case of benzene, the choice is predominantly in favour of substitution because the alternate path will result in the total loss of aromaticity. In the case of anthracene and phenanthrene, the two choices are more evenly matched. In these cases, though addition will result in loss of aromaticity, the loss of stabilisation energy is relatively small. The resonance energy of athracene is $351\,kJ\,mol^{-1}$; that of a structure like I of Eq. 10.12a can be estimated to be that of two benzene rings, $2 \times 151 = 302\,kJ\,mol^{-1}$.

10.2.4 SYNTHESIS

10.2.4.1 Synthesis of the Naphthalene Ring System

The usual methodology is to start with a derivative of benzene and build the second ring up on it. Friedel–Crafts reactions are especially useful for this purpose. Haworth synthesis of naphthalene is outlined in Eq. 10.14. Friedel–Crafts acylation of benzene with succinic anhydride (Eq. 10.14, step 1) gives 3-benzoylpropanoic acid. This is reduced by Clemmensen reduction (Chapter 9, Section 9.5.1.6, Eq. 9.18) to 4-phenylbutanoic acid (step 2). This is converted to 1-tetralone (also called α-tetralone; the names are based on the trivial name tetralin, for tetrahydronaphthalene) either directly by treating with concentrated sulphuric acid or anhydrous HF, or better with polyphosphoric acid (step 3). It can also be carried out by converting the acid to the acid chloride (step 4) and then carrying out Friedel–Crafts acylation using aluminium chloride (step 5). Step 3 is a variation of Friedel–Crafts acylation, applicable for intramolecular acylation. Steps 6 and 7 are Clemmensen reduction and catalytic dehydrogenation (reverse of catalytic hydrogenation) respectively.

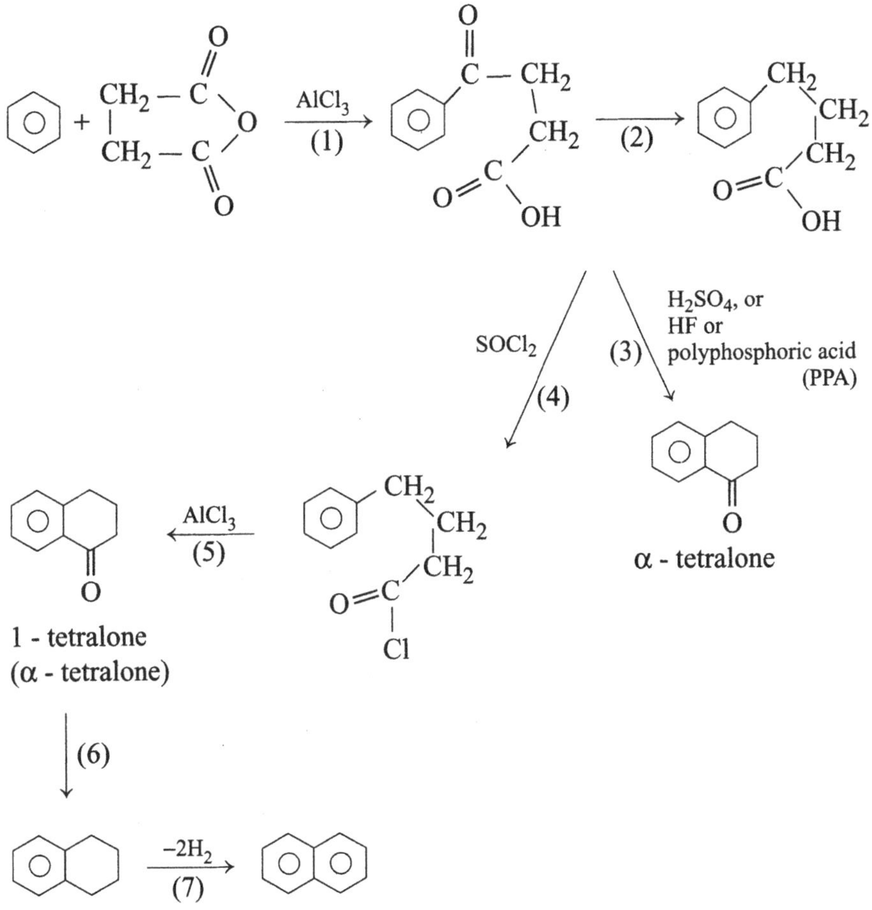

(10.14)

α - tetralone

1 - tetralone
(α - tetralone)

Some aspects of this scheme deserve comment. In step 2, reduction of the ketone function to CH_2 is necessary because the keto group deactivates the benzene ring and Friedel–Crafts acylation of this ring (step 3 or step 5) becomes difficult. We have seen in Chapter 9, Section 9.5.1.6, that Friedel–Crafts reactions do not succeed when electron-withdrawing groups like $-CO.R$ are attached to the ring. The situation is not so bad for intramolecular acylation where both reactants are present as parts of the same molecule. Nevertheless, prior conversion of $-CO-$ to $-CH_2-$ is advisable. Steps 3 and 5 come under the category of *intramolecular acylation* and the reaction is a *cyclisation* or *ring closure*. Intramolecular cyclisations proceed better when 5 and 6 membered rings are formed (See discussion on ring formation in Chapter 8). In step 3, the the free carboxylic acid is directly cyclised, instead of the acid chloride. This generally works only when the reaction is intramolecular. Polyphosphoric acid (PPA), essentially a solution of P_2O_5 in phosphoric acid, is the reagent of choice for such cyclisations. Dehydrogenation (step 7) is the reverse of catalytic hydrogenation and is especially successful when a cyclohexane ring is to be converted to an aromatic ring. This process is also called *aromatisation*. Hydrogenation catalysts (Pd/C, Pt) can bring about dehydrogenation also, if the conditions are adjusted so that the equilibrium shifts to the desired direction. Other reagents (not catalysts) for aromatisation are sulphur and selenium, which combine with hydrogen to form H_2S or H_2Se respectively, upon heating. Equation 10.14 outlines the synthesis of naphthalene from benzene and is useful only for academic discussion. Naphthalene itself is obtained from coal tar. However the basic scheme can be modified to prepare a variety of naphthalene derivatives, (Eq. 10.15). Steps 2 and 3 indicated with multiple arrows, are sequences of more than one step involving well established reactions. (*What are these reactions?*)

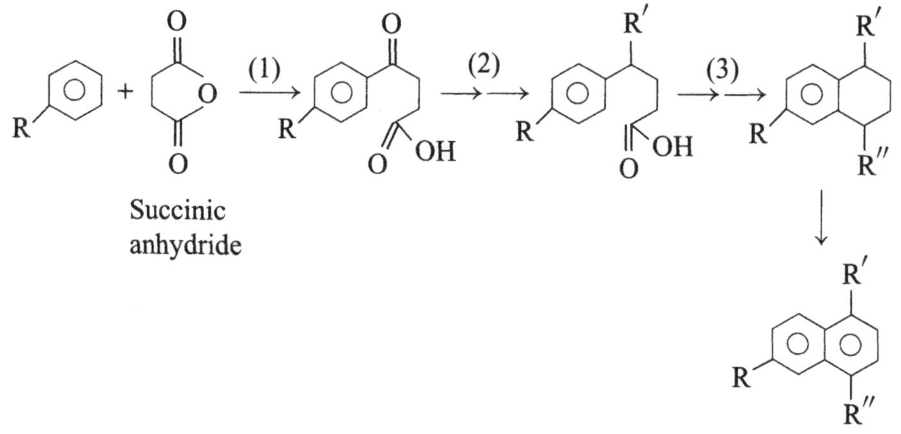

Succinic
anhydride

(10.15)

10.2.4.2 Synthesis of Phenanthrene and Anthracene

Friedel–Crafts reactions are useful for the synthesis of the higher polycyclic aromatics also. *Anthracene* can be obtained from *o*-benzoylbenzoic acid (Eq. 10.16). In this case, intramolecular acylation (step 2) succeeds even though the ring is deactivated, mainly because it is intramolecular. The conversion representd by step 3 can be achieved by one of several known reactions. This conversion is usually not necessary because anthraquinone is the most useful athracene derivative, used in the manufacture of dyes.

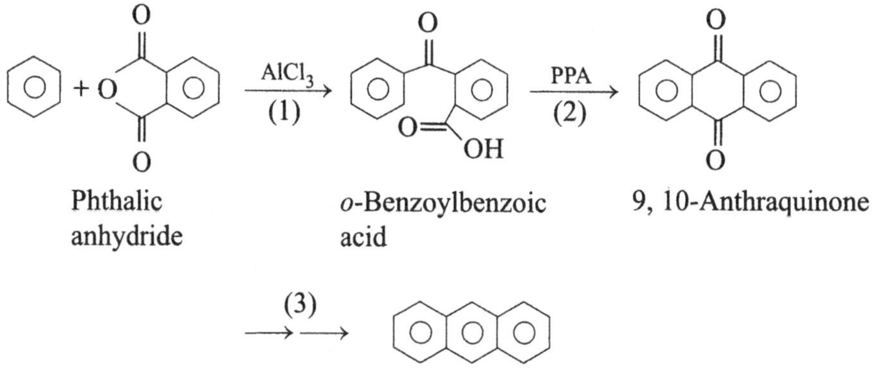

Phthalic
anhydride

o-Benzoylbenzoic
acid

9, 10-Anthraquinone   (10.16)

The phenanthrene ring system, usually in the fully or partially hydrogenated form, appears in many natural products – like steroids – of biological importance. Structures of some examples of steroids and of abietic acid, a terpenoid, are given below.

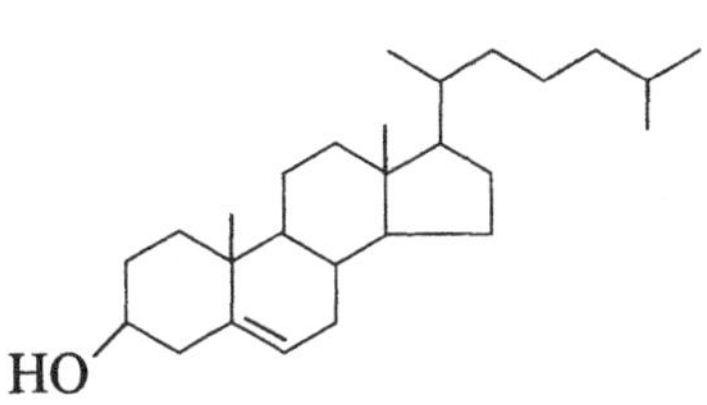

Cholesterol

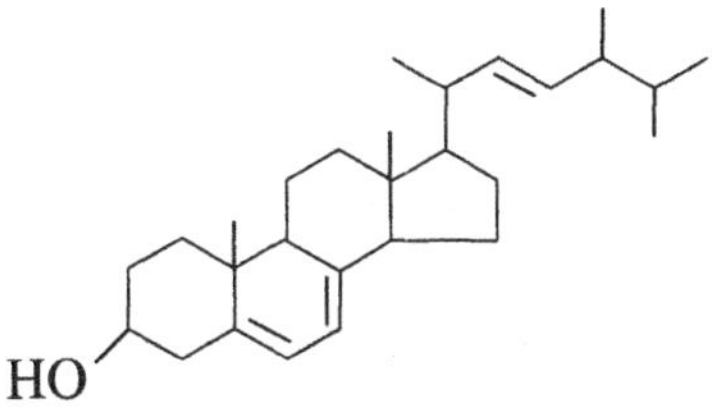

Ergosterol
(precursor of vitamin D)

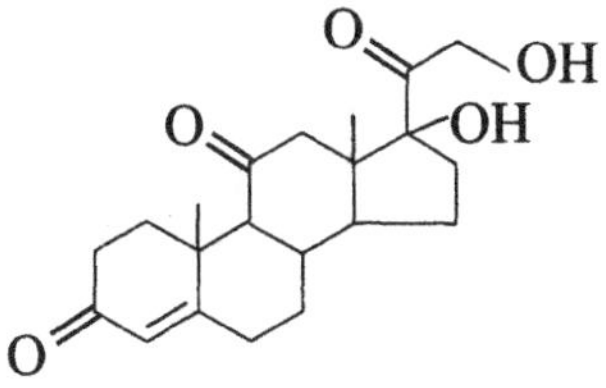

Cortisone

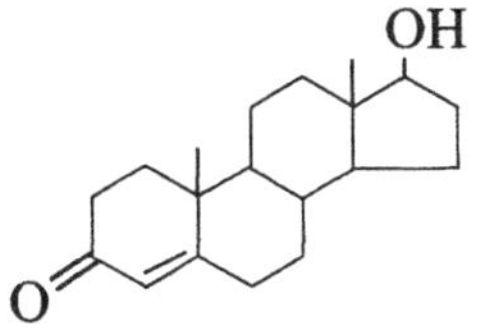

Testosterone

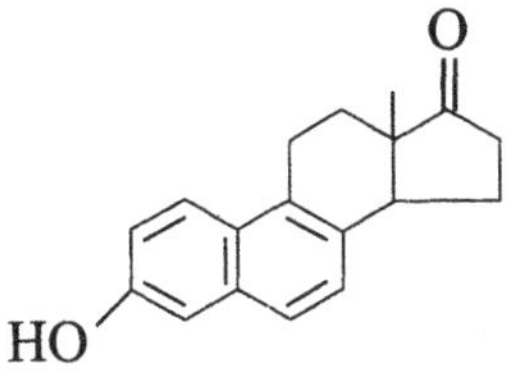

Equilenin

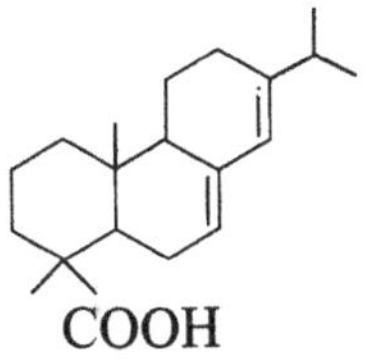

Abietic acid

Since these compounds and their structural variations have immense pharmaceutical importance, the synthesis of such molecules has attracted the attention of chemists. Of the many routes to obtain the phenanthrene nucleus, we shall discuss only the Friedel–Crafts acylation of naphthalene involving the Haworth synthesis (Eq. 10.17).

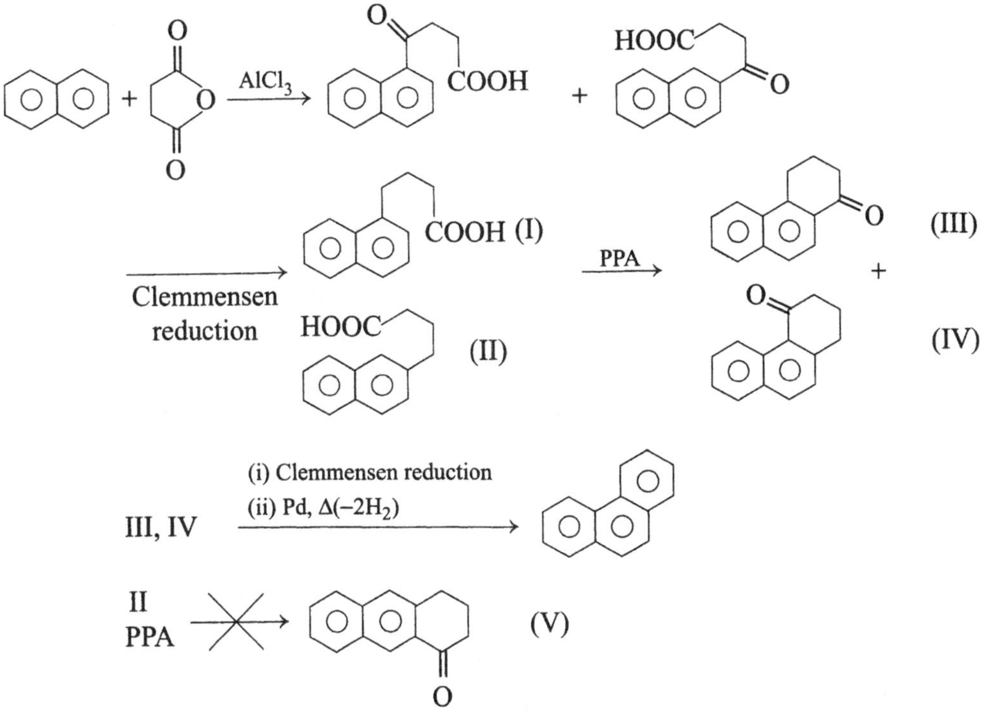

$$(10.17)$$

Acylation of naphthalene with succinic anhydride gives a mixture of the 1- and 2-acyl derivatives (Eq. 10.11). They can either be separated into the two isomers, or the mixture can be directly reduced by Clemmensen reduction to a mixture of acids, I and II. Cyclisation of the mixture with polyphosphoric acid gives the mixture of ketones III and IV, both containing the phenanthrene skeleton. They can be converted to phenanthrene itself in the usual way. Isomer II can, in principle, cyclise to give the anthracene skeleton V. This does not happen because cyclisation takes place at the more reactive 1-position of naphthalene. As already mentioned in connection with naphthalene, this basic scheme can be modified to synthesise a variety of phenanthrene derivatives.

KEY POINTS

- Polynuclear aromatic hydrocarbons, PAHs, contain two or more benzene rings fused together with at least two rings sharing more than one carbon atom.
- Naphthalene, anthracene and phenanthrene have two, three and three rings, respectively. There are hydrocarbons containing an even larger number of rings. Graphite is the ultimate in this series.
- Polycyclic aromatics are present and are isolated from coal tar. Other tars and soot also contain these. Most of these are carcinogenic.
- They have delocalised π-electron systems and can be represented by Kekulé structures. Unlike benzene, their Kekule structures are not identical. C–C bond lengths are also not the same.
- Resonance energies calculated using heats of hydrogenation data show that naphthalene has less than twice the stabilisation of benzene; anthracene and phenanthrene have less than three times the stabilisation of benzene.
- Naphthalene, anthracene and phenanthrene are solids, insoluble in water, soluble in hydrocarbon solvents. Their major industrial uses are in the manufacture of dyes. Naphthalene is also used to manufacture phthalic anhydride.
- One ring in naphthalene can be reduced (hydrogenated) or oxidised under milder conditions than benzene. The middle ring in anthracene and phenanthrene can also be readily reduced and oxidised. This reactivity is attributed to the fact that the loss of aromaticity (loss of resonance energy) when one ring is lost or made nonaromatic, is smaller than when benzene loses aromaticity.
- In anthracene and phenanthrene, the addition of H_2 or halogen, or oxidation takes place at the 9 and 10 positions.
- They undergo typical electrophilic substitution reactions like nitration, halogenation, sulphonation and Friedel–Crafts acylation more readily than benzene. The mechanism is similar to that with benzene, involving σ-complexes.
- The greater reactivity is attributed to the fact that the energy cost of forming the σ-complex which involves partial loss of aromaticity, is less than that with benzene.

- In naphthalene, substitution takes place preferentially at position 1 or the α-position, rather than at position 2 or the β-position. The 1-substituted product is the kinetically controlled one. This can be understood by a comparison of the stabilities of the σ-complexes for substitution at the two places by drawing resonance structures in the same way as is done to rationalise *ortho*, *meta* and *para* orientation in monosubstituted benzenes.

  Upon sulphonation of naphthalene, which is a reversible reaction, 1-naphthalene-sulphonic acid, the kinetically controlled product, is formed at low temperatures. At higher temperatures more of the more stable 2-sulphonic acid is formed under thermodynamic control. The greater stability of the 2-sulphonic acid is due to steric reasons. The 1-substituent and the H at position 8 crowd each other causing steric strain.

- Friedel–Crafts acylation is very useful for the synthesis of all three hydrocarbons. Acylation of benzene with succinic anhydride introduces a 4-carbon side chain with a keto group and a carboxyl group. Conversion of the carbonyl to CH_2 by Clemmensen reduction followed by intramolecular acylation, (cyclisation, ring closure) using H_2SO_4, HF or polyphosphoric acid gives terahydronaphthalene-1-one (α-tetralone). After Clemmensen reduction, this can be aromatised (dehydrogenated) by heating with S, Se, or catalytically using Pd or Pt. Similar methodologies are applicable for the synthesis of anthracene and phenanthrene. These reactions can be appropriately modified to synthesise substituted naphthalenes, substituted anthracenes and substituted phenanthrenes.

EXERCISES

SECTION I

1. Name the following compounds.

2. Draw the structures of the following compounds.

    (a) β-naphthylamine
    (b) α-tetralone

(c) 1,4-dimethyldecalin

(d) 9,10-dibromo-9,10-dihydroanthracene

(e) 2,6-diacetylnaphthalene

(f) 1-benzoylphenanthrene

3. Which of the following are fully aromatic? (That is, π-electron cloud covering all the carbon atoms.)

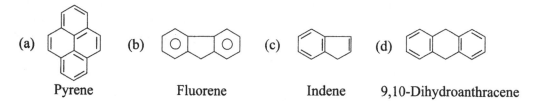

(a) Pyrene   (b) Fluorene   (c) Indene   (d) 9,10-Dihydroanthracene

4. Predict the major organic products of the following reactions.

(a) [naphthalene] $+ C_6H_5COCl \xrightarrow{\text{AlCl}_3}$?   (b) [tetralin] $\xrightarrow{\text{Se, }\Delta}$?

(c) [CH$_3$-naphthalene] $\xrightarrow{\text{HNO}_3/\text{H}_2\text{SO}_4}$?   (d) [CH$_3$-naphthalene] $\xrightarrow{\text{CH}_3\text{COCl, AlCl}_3}$?

5. Suggest the reagents which can bring about the following reactions.

(a) Naphthalene to tetralin

(b) Phenanthrene to 9,10-phenanthraquinone

(c) Tetralin to naphthalene

(d) 4-Phenylbutanoic acid to α-tetralone

(e) β-Tetralone to tetralin

(f) Benzene to *o*-benzoylbenzoic acid

SECTION II

1. Explain why naphthalene undergoes sulphonation faster than benzene.

2. Anthracene on reaction with bromine gives only 9,10-dibromo-9,10-dihydroanthracene, not any of the other possible dibromo derivatives. For example, not 1,4-dibromo-1,4-dihydroanthracene. Account for this.

3. Outline the mechanism of the cyclisation of 4-phenylbutanoic acid to α-tetralone catalysed by protonic acid.

4. Refer to Eq. 10.17. Explain in terms of mechanism, why the cyclisation of II gives IV and not V.

5. Diels–Alder reactions are useful for the synthesis of 6-membered rings. Show how 1,4-dihydronaphthalene can be synthesised by a Diels–Alder reaction. (Hint: the triple bond of benzyne is a powerful dienophile.)
6. Anthracene is formed when benzyl chloride is treated with anhydrous $AlCl_3$ or when benzyl alcohol is heated in the vapour phase with certain zeolites (crystalline aluminosilicates, which act as proptonic acids). Outline the mechanisms of both reactions.

CHALLENGING QUESTION

1. Anthracene takes part in Diels–Alder reactions as a diene. The following compounds I and II, are adducts of anthracene and the corresponding dienophile.
   (i)   Identify the dienophiles.
   (ii) Graphically represent the Diels–Alder reaction, paying attention to stereochemistry. (Refer to Chapter 6 for Diels–Alder reaction.)
   (iii) Discuss the behaviour of anthracene, why it acts as a diene, why there is selective reactivity at the 9 and10 positions.

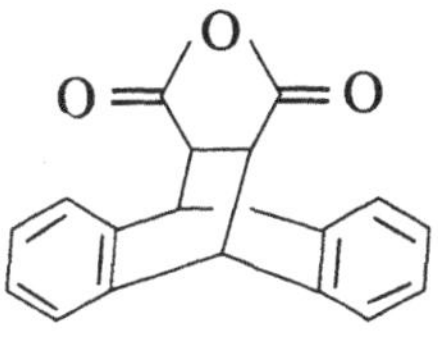

I 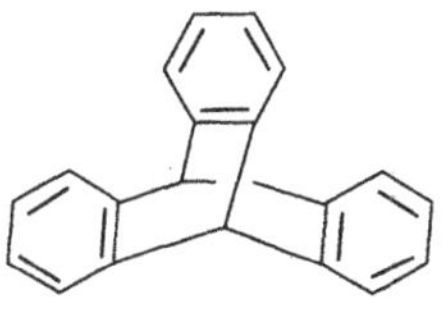   II

PROJECT

1. Prepare a report on carcinogenic hydrocarbons.
2. An important discovery in the 1980s was that of the carbon allotrope C_{60}, named buckminster fullerene. Prepare a report on fullerenes and related materials such as carbon nanotubes (focussing on their discovery, structure, properties and reactions).

11 Aliphatic Nucleophilic Substitution Reactions

OBJECTIVES   In this chapter, you will learn about,

- the types of substitution reactions
- definitions of terms such as leaving group, nucleophile and substrate
- S_N1, S_N2 reactions and kinetic differentiation
- factors favouring a particular mechanism
- stereochemical consequences
- influence of leaving group, nucleophile, substrate structure and solvent
- nucleophilicity and base strength
- S_Ni mechanism—retention of configuration
- hydroxyl as a leaving group
- some applications of nucleophilic substitution reactions

11.1   INTRODUCTION

Reactions of the type illustrated in Eq. 11.1 are called substitution reactions and are very useful for introducing desired functional groups or for converting one functional group to another.

$$CH_3-CH_2-Cl + OH^- \longrightarrow CH_3-CH_2-OH + Cl^- \qquad (a)$$

$$(CH_3)_3C-OH + HCl \longrightarrow (CH_3)_3C-Cl + H_2O \qquad (b)$$

$$\underset{\displaystyle CH_3-\overset{\textstyle O}{\overset{\textstyle \|}{C}}-OCH_3}{} + OH^- \longrightarrow \underset{\displaystyle CH_3-\overset{\textstyle O}{\overset{\textstyle \|}{C}}-OH}{} + CH_3O^- \qquad (c)$$

$$\text{C}_6\text{H}_5\overset{\text{H}}{\diagup} + \text{NO}_2^+ \longrightarrow \text{C}_6\text{H}_5\overset{\text{NO}_2}{\diagup} + \text{H}^+ \qquad \text{(d)}$$

$$\text{CH}_3-\textbf{MgBr} + \text{H}_2\text{O} \longrightarrow \underset{(\text{CH}_4)}{\text{CH}_3-\textbf{H}} + \text{Mg(Br)OH} \quad \text{(e)} \qquad (11.1)$$

In all these examples, one group in the reactant has been replaced by another group (highlighted). We are already familiar (Chapter 9) with reaction 11.1d, where one $-\text{H}$ of benzene is replaced by $-\text{NO}_2$ which enters as the electrophile NO_2^+, and the hydrogen which is replaced leaves as H^+. This is (aromatic) electrophilic substitution. In Eq. 11.1e, the group $-\text{MgBr}$ is replaced by the electrophile H^+, and is an example of aliphatic electrophilic substitution. In Eqs. 11.1a, 11.1b and 11.1c, the groups $-\text{Cl}^-$, $-\text{OH}^-$ and $-\text{OCH}_3^-$ are replaced by $-\text{OH}^-$, $-\text{Cl}^-$ and $-\text{OH}^-$, respectively, all nucleophiles. These are examples of aliphatic nucleophilic substitution reactions. The site of substitution in the first two examples is the sp^3 hybridised carbon and that in the third example is sp^2 hybridised.

In this chapter, we shall discuss aliphatic nucleophilic substitutions at sp^3 carbons – reactions of the type illustrated by 11.1a and 11.1b – also described as nucleophilic substitutions at a saturated carbon. Consider reaction 11.1a. OH^- is called the *nucleophile*, Cl^- the *leaving group*, $\text{CH}_3\text{CH}_2\text{Cl}$, the *substrate*, and C1 of this molecule is called the *reaction site*. Some more examples of nucleophilic substitution reactions are given in Eq. 11.2.

$$\text{CH}_3-\text{CH}_2-\text{Cl} + \text{H}_2\text{O} \longrightarrow \text{CH}_3-\text{CH}_2-\overset{+}{\underset{\text{H}}{\text{O}}}-\text{H} + \text{Cl}^-$$

$$\longrightarrow \text{CH}_3-\text{CH}_2-\text{OH} + \text{HCl} \qquad \text{(a)}$$

$$\text{C}_6\text{H}_5-\text{CH}_2-\text{Cl} + \text{CH}_3-\text{S}^- \longrightarrow \text{C}_6\text{H}_5-\text{CH}_2-\text{SCH}_3 + \text{Cl}^- \qquad \text{(b)}$$

$$\text{CH}_3-\text{CH}_2-\text{Br} + \text{NH}_3 \longrightarrow \text{CH}_3-\overset{\cdot}{\text{CH}}_2-\overset{+}{\text{NH}}_3 + \text{Br}^- \qquad \text{(c)}$$

$$\text{CH}_3-\text{I} + \text{CH}_3-\text{NH}-\text{CH}_3 \longrightarrow \text{CH}_3-\overset{+}{\underset{\text{CH}_3}{\text{NH}}}-\text{CH}_3 + \text{I}^- \qquad \text{(d)}$$

$$\text{C}_6\text{H}_5-\text{CH}_2-\text{Br} + \text{CH}_3-\text{COO}^- \longrightarrow \text{C}_6\text{H}_5-\text{CH}_2-\text{O}-\text{COCH}_3 + \text{Br}^- \qquad \text{(e)}$$

$$\text{C}_2\text{H}_5-\text{OH} + \text{C}_2\text{H}_5-\text{OH} \xrightarrow{\text{H}^+} \text{C}_2\text{H}_5-\overset{+}{\underset{\text{H}}{\text{O}}}-\text{C}_2\text{H}_5 + \text{H}_2\text{O} \qquad \text{(f)}$$

$$(11.2)$$

The nucleophile may be an anion like OH^-, CH_3O^-, CH_3S^- or CH_3COO^- (Eqs. 11.1a, 11.2b, 11.2c, 11.2e and 11.2f) or a neutral molecule like H_2O, NH_3 or an amine (Eqs. 11.2a, 11.2c and 11.2d). In all the above examples, the leaving group is a halide. In 11.1b and 11.2f, the leaving group is OH^-, but only under acidic conditions where the substrate is effectively $R—OH_2^+$ and the leaving group is H_2O.

11.2   KINETICS AND REACTION MECHANISMS

11.2.1   THE S_N1 AND S_N2 MECHANISMS

Kinetics studies are very useful for understanding reaction mechanisms. Such studies have been especially useful in the case of nucleophilic substitution reactions. Rate expressions for reactions of the type Eq. 11.3, have been found to fall between the extremes represented by Eqs. 11.4 and 11.5.

$$R—X + Y\mathbf{:} \longrightarrow R—Y + X\mathbf{:} \tag{11.3}$$

$$\text{Rate} = k_1[R—X] \tag{11.4}$$

$$\text{Rate} = k_2[R—X][Y\mathbf{:}] \tag{11.5}$$

Equation 11.4 describes a true unimolecular reaction and 11.5, a true bimolecular reaction. These are extreme cases. The actual reactions may follow intermediate kinetics. Sir Christopher Ingold – a British physical organic chemist – who studied these reactions in detail, named the mechanisms as S_N1 (substitution, nucleophilic, unimolecular) and S_N2 (substitution, nucleophilic, bimolecular). Whether the given reaction under investigation follows S_N1 or S_N2 kinetics is decided by several factors.

11.2.1.1   General Considerations

One of the primary requirements for such reactions to occur is that the molecule, $R—X$ should be 'reactive'. We have seen in Chapter 9 that even among halides, aryl halides are not normally 'reactive' in nucleophilic substitution. One of the reasons for this is the relatively low polarity of the C–halogen bond in aryl halides. For a $C—X$ bond to break heterolytically, (i) X should be highly electronegative, should be able to depart with the bonding pair of electrons of the covalent bond which is broken, and (ii) the C should be able to accommodate a positive charge (Eq. 11.6).

$$C—X \longrightarrow C^+ + \mathbf{:}X^- \tag{11.6}$$

These conditions have to be satisfied since even if an actual carbocation is not formed during the reaction, this carbon will have a partial positive charge in the transition state.

11.2.1.2 S_N1 Mechanism

This mechanism is best illustrated with two examples. Reactions 11.7 and 11.8 follow S_N1 kinetics. The overall mechanism for 11.7a is given in Eqs. 11.7b–11.7d.

$$(CH_3)_3C-Cl + OH^-/H_2O \longrightarrow (CH_3)_3C-OH + Cl^- + H^+ \quad \text{(a)}$$

$$(CH_3)_3C-Cl \underset{}{\overset{\rightarrow}{\longleftarrow}} (CH_3)_3C^+ + Cl^- \quad \text{(b)}$$

$$(CH_3)_3C^+ + H_2O \longrightarrow (CH_3)_3C-\overset{+}{\underset{H}{O}}-H \quad \text{(c)}$$

$$(CH_3)_3C-\overset{+}{O}H_2 + H_2O \longrightarrow (CH_3)_3C-OH + H_3O^+ \quad \text{(d)}$$

$$(CH_3)_3C^+ + OH^- \longrightarrow (CH_3)_3C-OH \quad \text{(e)} \quad (11.7)$$

$$\underset{}{\overset{CH_2-Cl}{\bigcirc}} + H_2O \longrightarrow \underset{}{\overset{CH_2-OH}{\bigcirc}} + Cl^- + H^+ \quad (11.8)$$

Cl is a strongly electronegative atom and leaves readily as $\overline{C}l\text{:}$. The cation left behind is relatively stable. The *tert*-butyl carbocation (Eq. 11.7b) is stablised by hyperconjugation and by inductive effect of the alkyl groups (see Chapter 1, Section 1.4.3). Another reason for the heterolysis to occur readily is due to steric reasons (Chapter 1, Section 1.5.1). The benzyl cation (Eq. 11.8) is resonance stabilised (Chapter 1, Section 1.4.3). Therefore, the dissociation (ionisation) takes place, albeit to a small extent (Eq. 11.7b). Even though the carbocation formed by ionisation is stable as carbocations go, it is an extremely reactive species. The preferred reaction here is *quenching* by the nucleophilic solvent water (Eq. 11.7c, also see Chapter 4, Section 4.4.1.4 for a discussion of the term *quenching*). The product of 11.7c is deprotonated by water (Eq. 11.7d). If the medium is alkaline, that is, if OH^- is present, this will also quench the carbocation and the alcohol is obtained directly (11.7e). The significance of S_N1 kinetics in this case is that the rate of the reaction is unaffected by the concentration of the nucleophile (OH^-) or, in its absence, water, which is usually present in excess. Whether OH^- is present or not, the reaction will take place at the same rate. When the solvent is the nucleophile, such reactions are called solvolysis reactions. Thus, the reaction is hydrolysis in this case and is pseudo-unimolecular. The rate of the overall reaction is the same as the rate of the ionisation (Eq. 11.7b) which is the first step. This is the slow step, and is called the *rate-determining step*. Aqueous medium, which has a high dielectric constant, helps ionisation further favouring the S_N1 mechanism.

The factors which favour the S_N1 mechanism are those which favour ionisation of the substrate. They are:

- *Nature of the leaving group:* The *leaving group* should be a 'good' one, which means that $X\text{:}$ should be a 'stable' anion. (It can also be a neutral molecule, as we shall see presently.)

What are stable anions? A working definition is that the conjugate base of a strong acid is a stable anion. Since the relative acidities of acids are well understood, this is a useful definition. Cl^-, the conjugate base of the strong acid HCl, is a stable anion and a good leaving group. Similarly, the other halides are also good leaving groups. OH^- which is the conjugate base of the weak acid H_2O, is a poor leaving group. The leaving ability of the acetate anion lies in between. Sulphate, bisulphate and sulphonates are good leaving groups. Alkoxides and amino groups are poor leaving groups. The neutral molecule H_2O, which appears as a leaving group in the reactions of protonated alcohols, is a good leaving group because it is the conjugate base of the strong acid H_3O^+ (Eq. 11.9a). The nucleophile Cl^- alone, in the absence of acid, is incapable of bringing about this reaction (Eq. 11.9b).

$$(CH_3)_3C-OH + HCl \longrightarrow (CH_3)_3C- \overset{+}{\underset{\underset{H}{|}}{O}}-H$$

$$\longrightarrow (CH_3)_3C^+ + H_2O \xrightarrow{Cl^-} (CH_3)_3C-Cl \quad (a)$$

$$(CH_3)_3C-OH + KCl \longrightarrow \text{No reaction} \qquad (b) \qquad (11.9)$$

- *Nature of the alkyl group in the substrate:* If the carbocation corresponding to the alkyl group is stable, S_N1 reaction can take place. The preference for S_N1 is maximum for tertiary alkyl halides, less for secondary and least for primary, in accordance with the stabilities of the corresponding carbocations. Benzyl and allyl halides though primary, react through the S_N1 mechanism, because the carbocations are resonance stabilised.

- *Steric effect:* tert-Alkyl halides ionise not only because of the stability of the carbocations but also because ionisation results in relief of steric strain. The reaction site in the substrate is sp^3 hybridised, with bond angles close to $109°$, whereas the carbocations are sp^2 hybridised with bond angles of $120°$ (Eq. 11.7b). When the substituents at the reaction site are large in size, like the methyl group in *tert*-butyl halides, there is steric crowding and strain, which can be relieved by ionisation. This is an additional factor which drives the *tert*-alkyl halides to react by the S_N1 mechanism (Chapter 1, Section 1.5.1).

- *Solvent:* Solvents like water which have a high dielectric constant help ionisation and hence favour S_N1 reactions.

It should be pointed out that molecules undergoing nucleophilic substitution can also undergo elimination reactions, under similar conditions. Factors which affect the competition between substitution and elimination are discussed in detail in Chapter 12. At this stage it is enough to mention that lower temperatures favour substitution.

The energy diagram (Fig. 11.1) emphasises the fact that S_N1 reactions are two-step reactions. Though the intermediate carbocation is a high energy species, it has a finite life time, though short-lived.

Since the S_N1 mechanism involves carbocations as intermediates, two features characteristic of carbocation reactions are found in these reactions also. (i) Carbocations are prone to rearrangement (Chapter 4). S_N1 reactions may also involve rearrangement. (ii) Carbocations are

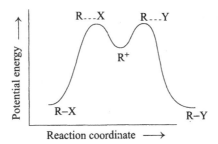

Fig. 11.1 S_N1 reactions: energy profile

planar. This has got stereochemical consequences (Eq. 11.10) which can be seen in the hydrolysis of 2-chloro-2-phenylbutane by the S_N1 mechanism.

$$ \tag{11.10} $$

R-2-Phenyl-2-butanol   S-2-Phenyl-2-butanol

(I)   (II)   (III)   (IV)

This molecule exhibits *optical isomerism*. In a polarimeter, it shows a measurable *optical rotation*. It has no plane of symmetry and is asymmetric. (the presence of four different gropups, chlorine, ethyl, methyl and phenyl on the central carbon atom is the reason for the asymmetry of this molecule). The *configuration* (I) shown in Eq. 11.10 represents one of the isomers. The other isomer which is the mirror image, has the same structure but differs from the first in the orientation of the groups in space. From the shape of the planar carbocation (II, Eq. 11.10), it can be seen that the two faces are equally accessible to the incoming nucleophile. Corresponding to the attack of the nucleophile on the two faces, two stereoisomeric products are possible. They are identical structurally but differ in the arrangement of the groups in space. They are mirror images of each other, and are optical isomers. In the example given, even though we started with one particular optical isomer of 2-chloro-2-phenylbutane, the produt is a 50:50 mixture of the two optical isomers of 2-phenyl-2-butanol. An equal mixture of the two optical isomers is called a *racemic mixture*. Experimentally, such a mixture will show no optical acivity when measured in a polarimeter. The starting material – which is a single optical isomer – will show its characteristic optical activity. (Optical isomers are named according to the Cahn–Ingold–Prelog rules, using the prefixes R and S for the two opposite configurations. The structures in Eq. 11.10 represent R-2-phenyl-2-butanol and S-2-phenyl-2-butanol. Chapter 20 gives a detailed disussion of the rules of nomenclature of optical isomers.)

If the starting material is optically active and if the reaction proceeds by S_N1 mechanism, the product will be racemic or almost racemic. This fact can be used to verify the S_N1 mechanism, in addition to kinetic studies.

11.2.1.3 S_N2 Mechanism

Nucleophilic substitution reactions of primary alkyl halides generally show bimolecular kinetics (Eq. 11.5). Hydrolysis of chloroethane is very slow in neutral water, but takes place faster in aqueous alkali. The rate of the reaction increases as the concentration of the alkali increases. Concentrations of both the alkyl halide (substrate) and the OH$^-$ (nucleophile) appear in the rate expression. Based on kinetics and some other details (mainly stetreochemical), a bimolecular mechanism has been proposed where the breaking of the C—X bond and the formation of the C—Y bond take place simultaneously (Eq. 11.11a). This is called a *concerted reaction* as opposed to the S_N1 mechanism which occurs in a step-wise manner. The potential energy diagram for the S_N2 reaction is shown in Fig. 11.2.

$$
\text{HO}^- + \underset{\underset{H}{|}}{\overset{\overset{CH_3}{|}}{C}}-\text{Cl} \longrightarrow \left[\text{HO} ---- \underset{\underset{H\ \ H}{}}{\overset{\overset{CH_3}{|}}{C}} ---- \text{Cl} \right] \longrightarrow \text{HO}-\overset{CH_3}{C} + \text{Cl}^- \qquad (11.11a)
$$

Using curved arrows, the S_N2 reaction can be represented as follows (Eq. 11.11b).

$$
\text{HO}^- + \overset{\overset{CH_3}{|}}{\text{CH}_2}-\text{Cl} \longrightarrow \text{HO}-\overset{\overset{CH_3}{|}}{\text{CH}_2} + \text{Cl}^- \qquad (11.11b)
$$

It needs to be emphasised that the structure in square brackets (Eq. 11.11a) is not an intermediate, unlike the carbocation in the S_N1 reactions. The structure is the representation of the *transition state*. In the transition state, the leaving group and the nucleophile (the incoming and the outgoing groups) are bonded to the reaction site by partial bonds. These two groups are attached to the opposite faces of the planar or nearly planar central carbon atom. It is planar because it is almost like a carbocation. The transition state has five groups attached to the central carbon (5-coordinated), even though two of the groups are attached by partial bonds which are longer than normal covalent bonds. When the product is formed by the collapse of the 5-coordinated

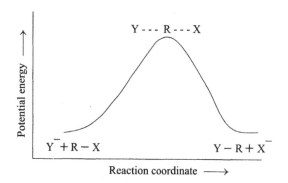

Fig. 11.2 S_N2 reaction: energy profile

transition state, the new (incoming) group Y, is attached to the central carbon at the opposite side to that originally occupied by the leaving group. The three other groups on the central carbon also undergo a change in their orientation in space. The configuration of the substrate is said to have undergone an *inversion* (also known as *Walden inversion,* after the chemist who first reported it). Walden inversion can be experimentally verified if the starting material is optically active. The product will not be racemic, but will be a single optical isomer with the inverted configuration. The optically active R-2-chlorobutane upon hydrolysis under S_N2 conditions gives S-2-butanol with a configuration opposite to that of the original R-2-chlorobutane (Eq. 11.12).

$$(11.12)$$

R-2-Chlorobutane   S-2-Butanol

11.2.1.4 Conditions for S_N2 Mechanism

Structure of the substrate: Two aspects of the S_N2 mechanism are of importance in this context. (i) Carbocations are not intermediates, hence carbocation stability is not a deciding factor though it is not unimportant (see discussion of benzyl halides below). (ii) The transition state in the S_N2 mechanism has a 5-coordinated central carbon where there is more steric crowding than in the starting material. If the groups on the central carbon are bulky, as in *tert*-alkyl halides, the steric strain due to this crowding will be high. Hence, *tert*-alkyl halides are the least likely to react by S_N2 mechanism, and primary alkyl halides the most likely. Secondary alkyl halides represent a border line situation. The reactivity of alkyl halides is represented thus:

S_N1 reactivity: primary < secondary < tertiary

S_N2 reactivity: primary > secondary > tertiary

It was pointed out earlier (Section 11.2.1.2) that benzyl and allyl halides react readily following the S_N1 mechanism because of the stability of the corresponding carbocations. Interestingly, they also react readily by the S_N2 mechanism. This is becaue the transition state for S_N2 reaction (Eq. 11.11a) has a partial carbocation character and resonance stabilisation plays a part here as well.

Nature of the leaving group: Good leaving groups tend to leave early leading to the S_N1 mechanism. Poorer leaving groups do not ionise and leave readily, but can be made to leave with help from the incoming nucleophile, leading to the S_N2 mechanism.

Nature and concentration of nucleophile: Concentration of the nucleophile does not appear in the rate expression for the S_N1 mechanism, but does appear in the S_N2 mechanism. In a borderline case, if the concentration of the nucleophile is increased, the mechanism may shift to true S_N2. A more active nucleophile, one which can give a more effective 'push', will also favour the S_N2 pathway.

Solvent: We have seen that solvents with a high dielectric constant favour the S_N1 mechanism because ionisation and charge separation is involved. The reverse is true for the S_N2 mechanism. In the transition state, the unit negative charge of the nucleophile (if it is an anion) is not destroyed, but gets spread out. Solvents of low dielectric constant favour this.

Nucleophilicity and leaving group efficiency: Earlier, we have seen that stable anions – conjugate bases of strong acids – are good leaving groups. They are also weaker bases. Does any relationship between base strength and nucleophilicity exist?

Terms like 'more active nucleophile' or 'more efficient nucleophile' have been used. What do they mean? Nucleophiles are also bases. Though the converse is also true, the two terms are not synonymous. Base strength is related to the equilibrium of an acid–base reaction, usually protonation. In this context, nucleophilicity refers to the rate of the reaction of the nucleophile with the carbon at the reaction site. It is a kinetic property. The polarisability of the electron cloud on the nucleophile which goes to form the partial bond in the transition state is the deciding factor. Large ions or species with diffuse electron clouds are better nucleophiles than smaller ions. Thus, iodide is a better nucleophile than chloride even though HI is a stronger acid than HCl. The latter fact predicts that I^- (conjugate base of the stronger acid) is a better leaving group than Cl^- and has been experimentally proven. Thus, iodide is a better leaving group and a better nucleophile than chloride. This is contrary to the usual observation that a good leaving group is a poor nucleophile and vice versa. Some other good nucleophiles are RS^- (in contrast to RO^-) due to the larger size of the S, and azide, N_3^- due to the diffuse electron cloud.

11.3 S_Ni MECHANISM

There are some nucleophilic displacement reactions where a portion of the substrate itself acts as the nucleophile, effecting an internal or intramolecular nucleophilic displacement. The best known example of this is the reaction of alcohols with thionyl chloride to give alkyl chlorides (Eq. 11.13).

$$R-OH + SOCl_2 \rightarrow R-Cl + SO_2 + HCl \tag{11.13}$$

This is a useful reaction for such conversions. The reaction is also mechanistically interesting. When the substrate is asymmetric, like one of the optical isomers of 2-butanol, the product 2-chlorobutane has the same configuration as the starting material. Racemisation as in S_N1 reactions and inversion as in S_N2 reactions do not occur. The stetreochemistry in such reactions is *retention of configuration.*

Alkyl chlorosulphite (ROSOCl) is the intermediate and has been isolated under mild conditions. This has the same configuration as the alcohol because the $C-O$ bond is not broken during its formation. This intermediate dissociates into an ion pair (I) which then breaks down to another ion pair (II) and recombines to form the product, the alkyl halide. Even though a carbocation is present in the ion pair, the Cl^- combines with the carbocation only at one face because that is where the Cl^- is present (Eq. 11.14). Reactions of this type have been called S_Ni, for substitution nucleophilic internal.

$$\begin{array}{c}H\\ \diagdown\\ C-OH + SOCl_2\\ \diagup\quad\vdots\\ CH_3\ C_2H_5\end{array} \longrightarrow \begin{array}{c}H\qquad O\\ \diagdown\qquad\parallel\\ C-O-S-Cl + HCl\\ \diagup\quad\vdots\\ CH_3\ C_2H_5\end{array}$$

(11.14)

$$\longrightarrow \left[\begin{array}{c}H\qquad O\\ \diagdown\quad -\parallel\\ C^+\ OSCl\\ \diagup\quad\vdots\\ CH_3\ C_2H_5\end{array}\right] \longrightarrow \left[\begin{array}{c}H\\ \diagdown\\ C^+\ \bar{Cl}\\ \diagup\quad\vdots\\ CH_3\ C_2H_5\end{array}\right] + SO_2 \longrightarrow \begin{array}{c}H\\ \diagdown\\ C-Cl\\ \diagup\quad\vdots\\ CH_3\ C_2H_5\end{array}$$

$$\qquad\qquad\qquad I \qquad\qquad\qquad\qquad II$$

Notice that the OH has been made a good leaving group by converting it to a chlorosulphite ($ClSO_2^-$). It is the conjugate base of the unstable but strong chlorosulphurous acid, $ClSO_2H$.

11.4 OTHER NUCLEOPHILIC SUBSTITUTION REACTIONS

11.4.1 HYDROXYL AS A LEAVING GROUP

Alcohols are readily available starting materials, and the need for the conversion of the hydroxyl group to other functional groups often arises. Direct displacement of the hydroxyl is usually not possible because the hydroxyl is a poor leaving group since it is the conjugate base of the weak acid, H_2O. Methods are available to modify the hydroxyl so that it can be replaced by other nucleophiles. *The easiest is protonation.*

Secondary alcohols and more readily tertiary alcohols, on treatment with hydrohalic acids, are converted to the halides. Thus, 2-butanol, on treatment with con. HBr gives 2-bromobutane. 2-Methyl-2-propanol (*tert*-butyl alcohol) upon treatment with hydrochloric acid gives 2-chloro-2-methylpropane (*tert*-butyl chloride) (Eq. 11.15).

$$\begin{array}{c}CH_3-CH-CH_2-CH_3 + HBr\\ \mid\\ OH\end{array} \longrightarrow \begin{array}{c}CH_3-CH-CH_2-CH_3 + Br^-\\ \mid +\\ OH_2\end{array}$$

$$\longrightarrow \begin{array}{c}+\\ CH_3-CH-CH_2-CH_3\\ + H_2O + Br^-\end{array} \longrightarrow \begin{array}{c}CH_3-CH-CH_2-CH_3\\ \mid\\ Br\end{array} \qquad (a)$$

$$(CH_3)_3C-OH + HCl \longrightarrow (CH_3)_3C-\overset{+}{O}H_2 + Cl^- \longrightarrow$$

$$(CH_3)_3C^+ + H_2O + Cl^- \longrightarrow (CH_3)_3C-Cl \qquad \text{(b)}$$

$$(11.15)$$

In these reactions, the hydroxyl has been modified by protonation, so that the leaving group is H_2O and not OH^-. The former, the conjugate base of the strong acid H_3O^+, is a good leaving group.

Preparation of ethers from alcohols by heating with sulphuric acid is another example of the application of this principle. Ethanol, when heated with con. H_2SO_4 at 140°C undergoes *intermolecular dehydration* to give diethyl ether. Other primary alcohols also behave in a similar manner. When heated to higher temperatures, *intramolecular dehydration* to give ethene (elimination reaction) predominates. Secondary and tertiary alcohols, even at lower temperatures, give the alkenes. The mechanism of diethyl ether formation (Eq. 11.16) shows it is a nucleophilic displacement reaction.

$$(11.16)$$

Modification of the hydroxyl by *conversion to esters of inorganic acids* is another useful strategy. Reaction of alcohols with a sulphonyl chloride, like *p*-toluenesuphonyl chloride (Eq. 11.17) gives an alkyl suphonate. The sulphonate in Eq. 11.17 is propyl *p*-toluenesulphonate, abbreviated as propyl tosylate. The tosylate anion, $-OSO_2Ar$ (Ar stands for 4-methylphenyl or *p*-tolyl) is the conjugate base of a sulphonic acid which is a strong acid comparable in strength to sulphuric acid. It is a good leaving group. Tosylates can be subjected to various nucleophilic substitution reactions (Eq. 11.18). The overall reaction – conversion of an alcohol to the nitrile – cannot be carried out by a direct reaction of CN^- with the alcohol.

$$CH_3-CH_2-CH_2-OH + Cl-\overset{\overset{O}{\|}}{\underset{\underset{O}{\|}}{S}}-\langle\bigcirc\rangle-CH_3$$

p-Toluenesulphonyl chloride
(tosyl chloride)

(11.17)

$$\longrightarrow CH_3-CH_2-CH_2-O-\overset{\overset{O}{\|}}{\underset{\underset{O}{\|}}{S}}-\langle\bigcirc\rangle-CH_3 + HCl$$

Propyl *p*-toluenesulphonate (propyl tosylate)

$$CH_3-CH_2-CH_2-O-SO_2-Ar + NaCN$$
$$\longrightarrow CH_3-CH_2-CH_2-CN + Ar-SO_3^- + Na^+$$

(11.18)

The conversion of alcohols to chlorides using $SOCl_2$ follows the same strategy.

Some useful nucleophilic substitution reactions are listed in Eqs. 11.19a–11.19f. Equation 11.19a illustrates the cleavage of an ether by a strong acid. This reaction is useful for removing the alkoxy group from alkyl aryl ethers. The nitriles prepared in reactions like 11.19b can be hydrolysed to the corresponding carboxylic acids, or reduced using the powerful reducing agent lithium aluminium hydride to a primary amine, containing one carbon more than the starting alkyl halide. The nitrile upon hydrolysis by aqueous acid yields 2-methylbutanoic acid and upon reduction gives 2-methylbutylamine (2-methylbutanamine). Azides are good nucleophiles. The azide prepared as in Eq. 11.19c can be reduced by lithium aluminium hydride to cyclopentylamine, (cyclopentanamine). Equation 11.19d illustrates the successive alkylation (in this case ethylation) of ammonia to give primary, secondary and tertiary amines and finally the quaternary ammonium salt. Synthesis of mercaptans is illustrated in Eq. 11.19e. The very useful Williamson's synthesis of ethers is illustrated in Eq. 11.19f.

$$C_6H_5-O-CH_3 + HI \longrightarrow C_6H_5-OH + CH_3I \qquad\text{(a)}$$

$$CH_3-\underset{\underset{Br}{|}}{CH}-CH_2-CH_3 + NaCN \longrightarrow CH_3-\underset{\underset{CN}{|}}{CH}-CH_2-CH_3$$

$$\overset{H^+, H_2O}{\underset{\Delta}{\swarrow}} \qquad \overset{LiAlH_4}{\underset{(reduction)}{\searrow}}$$

$$CH_3-\underset{\underset{COOH}{|}}{CH}-CH_2-CH_3 \qquad CH_3-\underset{\underset{CH_2NH_2}{|}}{CH}-CH_2-CH_3 \qquad\text{(b)}$$

$$\text{cyclopentyl—Br} + \text{Na N}_3 \longrightarrow \text{cyclopentyl—N}_3 \xrightarrow{\text{LiAlH}_4} \text{cyclopentyl—NH}_2 \qquad \text{(c)}$$

Sodium azide   Cyclopentyl
                         azide

$$C_2 H_5 - I + NH_3 \rightarrow C_2 H_5 - NH_2 \xrightarrow{C_2 H_5 I} (C_2 H_5)_2 NH \xrightarrow{C_2 H_5 I} (C_2 H_5)_3 N$$

$$\Big\downarrow C_2 H_5 I$$

$$(C_2 H_5)_4 N^+ I^- \qquad \text{(d)}$$

$$\text{cyclohexyl—Cl} + \text{KSH} \longrightarrow \text{cyclohexyl—SH} \qquad \text{(e)}$$

$$CH_3 - CH_2 - CH_2 - O \, Na + CH_3 - CH_2 - I$$

$$\longrightarrow CH_3 - CH_2 - CH_2 - O - CH_2 - CH_3 \qquad \text{(f)}$$

Ethyl propyl ether
(Williamson synthesis)

$$(11.19)$$

KEY POINTS

- A typical nucleophilic substitution reaction at a saturated carbon is illustrated by the hydrolysis of ethyl chloride, namely replacement of the chlorine in ethyl chloride by the hydroxyl group.
- In this reaction, the chloride is the leaving group, the hydroxide or water is the nucleophile, and ethyl chloride is the substrate.
- The kinetic rate expressions for such reactions fall between two extremes, unimolecular (the rate depends upon the concentration of the substrate alone) and bimolecular (the rate depends upon the concentration of the substrate and the nucleophile).
- The former is called the S_N1 mechanism and the latter, the S_N2 mechanism.
- The rate-determining step in the S_N1 mechanism involves ionisation (carbocation formation).
- The S_N1 mechanism is favoured by alkyl halides whose alkyl group is tertiary and bulky (sterically crowded). The presence of a good leaving group and a solvent of high dielectric constant favour this mechanism. Concentration of the nucleophile does not appear in the rate expression. Hence, the nature and concentration of the nucleophile have little effect.

- S_N2 is a one-step concerted reaction. The presence of good nucleophiles in high concentration and solvents of low dielectric constant favour this mechanism.
- S_N2 reactivity decreases in the order, primary > secondary > tertiary. S_N1 reactivity is in the reverse order.
- The stereochemistry of the S_N1 reaction is racemisation; that of S_N2 is inversion of configuration. These can be demonstrated using optically active substrates.
- Nucleophilicity is not the same as base strength. The latter is a thermodynamic propery while the former is a kinetic property. Iodide is a better nucleophile and a better leaving group than chloride.
- Reaction of alcohols with thionyl chloride proceed with retention of configuration. This is an example of a S_Ni reaction.
- Poor leaving groups like hydroxyl can be made better leaving groups by methods such as protonation.
- Many nucleophilic substitution reactions have important applications.

EXERCISES

SECTION I

1. Give the structures of the main organic products of the following reactions.

   (a) $CH_3 - CHCl - CH_3 + C_2H_5SNa \rightarrow$
   (b) *p*-Chlorobenzyl chloride + NaCN $\rightarrow$
   (c) $(CH_3)_3C - OK$ (potassium *tert*-butoxide) $+ CH_3 - I \rightarrow$
   (d) $(CH_3)_2CH - CH_2 - CH_2 - I + CH_3 - COOK$ (potassium acetate) $\rightarrow$
   (e) Iodocyclopentane + NaN_3 (sodium azide) $\rightarrow$

2. Suggest the reactants (alkyl halide and nucleophile) for preparing the following compounds.

   (a) Benzyl mercaptan, $C_6H_5 - CH_2 - SH$
   (b) Butyl ethyl ether, $CH_3 - CH_2 - O - CH_2 - CH_2 - CH_2 - CH_3$
   (c) Allyl cyanide, $CH_2 = CH - CH_2 - CN$
   (d) N-Ethyl-N,N-dimethylamine, $CH_3 - CH_2 - N(CH_3)_2$
   (e) Anisole, $C_6H_5 - O - CH_3$
   (f) 1-Heptyne (from acetylene)
   (g) 1-Naphthyl ethyl ether

3. In the following pairs, which will be more reactive in the reaction specified?

   (a) $CH_3 - CH_2 - I$ and $CH_3 - CH_2 - C(CH_3)_2 - I$ (in hydrolysis)
   (b) $(CH_3)_3C - Br$ and $CH_3 - Br$ (in S_N2 reaction with KSH)
   (c) $CH_3 - I$ and $CH_3 - Br$ (in S_N2 reaction with NaCN)

(d) Benzyl chloride and 2-phenylchloroethane, $C_6H_5-CH_2-CH_2-Cl$, (in hydrolysis)

(e) *tert*-Butyl alcohol and sec-butyl alcohol (in reaction with HCl to form the corresponding alkyl chloride).

4. *tert*-Butyl chloride is hydrolysed in a mixture of water and methanol.

(a) In which solvent would the hydrolysis be faster, 1:1 methanol-water or 1:3 methanol-water? (Water has a higher dielectric constant than methanol).

(b) If the solution is made slightly alkaline by adding KOH, how will the rate of the reaction be affected?

SECTION II

1. Give the mechanism of the following reaction.

   2-Butanol + HBr → 2-Bromobutane

2. Give the structures of the compounds designated by letters.

(a) Propanol + *p*-toluenesulphonyl chloride → A
A + KCN → B

(b) 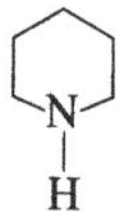 + CH$_3$I (1 equivalent) → C

C + CH$_3$I (1 equivalent) → D

(c) Benzyl chloride, $C_6H_5-CH_2-Cl$ + E → F $\xrightarrow{H^+, H_2O, \text{hydrolysis}}$ $C_6H_5-CH_2-COOH$

(d) $CH_3-CHBr-CH_3 + G \rightarrow H \xrightarrow{\text{LiAlH}_4 \text{ reduction}} (CH_3)_2CH-CH_2-NH_2$

(e) $CH_3-CHBr-CH_3 + I \rightarrow J \xrightarrow{\text{LiAlH}_4 \text{ reduction}} (CH_3)_2CH-NH_2$

(f) $CH_3-CH(OH)-CH_2-CH_3$ + *p*-toluenesulphonyl chloride → K $\xrightarrow{\text{KI}}$ L

(g) $M + NaNH_2 \rightarrow N \xrightarrow{O}$ 3-phenylpropyne $\xrightarrow{\text{NANH}_2} P \xrightarrow{\text{CH}_3\text{I}} Q$

3. Show how the following conversions can be brought about.

(a) R-2-Butanol → R-2-chlorobutane (retention of configuration)
(b) R-2-Butanol → S-2-iodobutane (Hint: refer question 2 f, above)
(c) R-2-Butanol → R-2-methoxybutane (Hint: Williamson synthesis)
(d) R-2-Butanol → S-2-methoxybutane

4. Comment on the statement, 'S$_N$2 reactions proceed through a transition state; there is no transition state, only an intermediate in S$_N$1 reactions'.

5. By examining the transition states, show why benzyl halides react faster than ethyl halides, in both S_N1 and S_N2 reactions.
6. In Section 11.2.1.2, 2-chloro-2-phenylbutane is described as 'a suitable candidate for S_N1 reaction'. Justify this description.

CHALLENGING QUESTIONS

1. Consider the following reaction:

    $CH_3 — O — C_6H_5$ (anisole) $+ HI \rightarrow CH_3 — I + C_6H_5OH$

    (a) Suggest a mechanism for the reaction.
    (b) The reaction does not give $CH_3 — OH + C_6H_5 — I$. Why?
    (c) The reaction (cleavage of the ether) does not take place if KI is used as the source of iodide ion, instead of HI, even in a highly concentrated solution. Ethers are also not cleaved by alkaline hydrolysis. Explain.
    (d) Among the hydrohalic acids, the efficiency of the cleavage of ether decreases in the order, HI > HBr > HCl. Explain.

2. Account for the observation that optically active 1-phenyl-1-propanol, kept in contact with aqueous sulphuric acid, slowly racemises. (Hint: in aqueous sulphuric acid, protonation of the OH, and then dissociation to the carbocation can take place, both of which are reversible reactions.)

3. Account for the observation that a solution of optically active 2-iodooctane in acetone, containing dissolved iodide ions (from some other dissolved iodide salt) racemises upon keeping (these are S_N2 reaction conditions). (Hint: Iodide is a good nucleophile and a good leaving group. I^- can displace I^- from alkyl iodides by S_N2 mechanism.)

4. Many S_N1 reactions of optically active alkyl halides give products of partial inversion of configuration, rather than total racemisation. Rationalise this. (Hint: Carbocations are formed, but the incoming nucleophile may not be free to approach it with equal ease from both faces. Elaborate.)

12 Elimination Reactions

OBJECTIVES   In this chapter, you will learn about,

- mechanisms and kinetic aspects of 1,2-eliminations
- characteristics of E1 and E2 mechanisms
- competition between elimination and substitution
- E1cB mechanism
- orientation of the double bond: Saytzeff and Hofmann rules
- stereochemistry of elimination
- rationalisation of stereochemistry

12.1 INTRODUCTION

You have already been introduced to elimination reactions in Chapter 4, as a class of reactions useful for the preparation of alkenes. Two atoms or groups attached to adjacent carbons are eliminated – in effect as a small molecule – not necessarily together and not necessarily in one step, to create a double bond between the two carbons concerned. This is called a β-elimination or a 1,2-elimination and is illustrated by the examples in Eq. 12.1.

$$CH_3-\underset{\underset{H}{|}}{CH}-\underset{\underset{Cl}{|}}{CH_2} \longrightarrow CH_3-CH=CH_2 + HCl \qquad \text{(a)}$$
$$\text{(Dehydrohalogenation)}$$

$$CH_3-\underset{\underset{H}{|}}{CH}-\underset{\underset{OH}{|}}{CH}-CH_3 \longrightarrow CH_3-CH=CH-CH_3 + H_2O \qquad \text{(b)}$$
$$\text{(Dehydration)}$$

$$CH_3-\underset{\underset{Br}{|}}{CH}-\underset{\underset{Br}{|}}{CH}-CH_3 \longrightarrow CH_3-CH=CH-CH_2 + Br_2 \qquad \text{(c)}$$
$$\text{(Dehalogenation)}$$

$$CH_3-CH_2-CH_2-CH_2-O-\overset{\underset{\displaystyle \|}{S}}{C}-SCH_3 \xrightarrow{\Delta} \begin{array}{l} CH_3-CH_2-CH=CH_2 \quad (d) \\ + COS + CH_3SH \end{array}$$

(Chugaev reaction – ester pyrolysis)   (12.1)

In this chapter, we shall discuss the mechanisms of these reactions in greater detail. We shall see how two mechanistic extremes – one unimolecular and the other bimolecular – designated E1 and E2, parallel to the S_N1 and S_N2 mechanisms, exist. Conditions that favour one or the other of these mechanisms and the competition between substitution and elimination will be discussed. We shall also see the stereochemistry and regiochemistry (orientation of the double bond) of elimination reactions, in greater detail.

12.2 MECHANISMS: KINETIC STUDIES

As in the case of substitution reactions, we shall take alkyl halides as representative substrates. 2-Chloropropane on treatment with alkali eliminates HCl and gives propene. Under these conditions 2-chloropropane can also give 2-propanol by substitution reaction (Eq. 12.2).

$$CH_3-\underset{\underset{\displaystyle Cl}{|}}{CH}-CH_3 + OH^- \rightarrow CH_3-CH=CH_2 + CH_3-\underset{\underset{\displaystyle OH}{|}}{CH}-CH_3 \quad (12.2)$$

These two are competing reactions. Not surprisingly, there are many similarities between the mechanisms of substitution and elimination reactions. Kinetically, elimination reactions may fall between two extreme rate laws, first order or second order kinetics, corresponding to unimolecular and bimolecular reactions (Eqs. 12.3a and 12.3b).

$$Rate = k_1[RX] \qquad (a)$$

$$Rate = k_2[RX][OH^-] \quad (b) \qquad\qquad (12.3)$$

They are called E1 (elimination, unimolecular) and E2 (elimination, bimolecular) mechanisms.

12.2.1 E1 MECHANISM

E1 mechanism is observed in alkyl halides whose alkyl groups correspond to stable carbocations, like tertiary carbocations. The first step (12.4a) is ionisation which is the rate-determining step, as in S_N1 reactions. The second step (b or c), the *product-determining step*, is different for substitution and elimination (Eq. 12.4).

$$CH_3-\underset{\underset{\displaystyle Cl}{|}}{CH}-CH_3 \rightarrow CH_3-\overset{\overset{\displaystyle +}{|}}{\underset{\underset{\displaystyle H}{|}}{C}}-CH_3 + Cl^- \qquad (a)$$

$$CH_3 - \overset{+}{\underset{\underset{H}{|}}{C}} - CH_3 + OH^- \rightarrow CH_3 - \overset{\overset{OH}{|}}{\underset{\underset{H}{|}}{C}} - CH_3 \qquad (b)$$

(Substitution)

$$CH_3 - \overset{+}{\underset{\underset{H}{|}}{C}} - CH_3 + OH^- \rightarrow CH_3 - CH = CH_2 + H_2O \qquad (c) \qquad (12.4)$$

(Elimination)

E1 elimination is a two-step reaction, proceeding through a carbocation. The mechanism can be described using the potential energy diagram (Fig. 12.1).

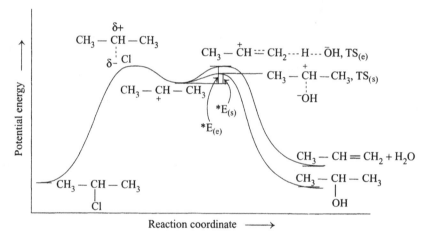

Fig. 12.1. Energy profile diagram for E1 and S_N1 reactions ($^*E_{(s)}$: Activation energy for substitution, $^*E_{(e)}$: Activation energy for elimination, $TS_{(s)}$: Transition state for substitution, $TS_{(e)}$: Transition state for elimination)

The factors which we listed as favouring the S_N1 reaction will also favour the E1 reaction, as far as kinetically measured reaction rates are concerned. This is usually the rate of the first step. The competition between the formation of the actual products – alcohol or alkene – is controlled by several factors. One of them is the activation energy. From Fig. 12.1, we can see that the competing reactions are controlled by the two activation energies, $^*E_{(s)}$ and $^*E_{(e)}$. The respective transition states for the substitution (Eq. 12.4b) and the elimination (Eq. 12.4c) steps are as shown in Eqs. 12.5a, 12.5b.

$$CH_3 - \overset{\overset{H}{|}}{\underset{\underset{OH}{|}}{\overset{\delta+}{C}}} \overset{\delta+}{-} CH_3, TS_{(s)}, \text{ for quenching by } OH^- \qquad (a)$$

(substitution)

$$CH_3 - \overset{\overset{\displaystyle H}{|}}{\underset{\underset{\displaystyle H}{|}}{C}} \overset{\delta+}{=} \overset{\delta+ \quad \delta-}{\underset{\underset{\displaystyle H}{|}}{C}} \cdots H \cdots OH, \ TS_{(e)}, \ \text{for deprotonation} \qquad (b) \qquad (12.5)$$
$$\text{(elimination)}$$

Steric factors: We have already seen that relief of steric strain has a role in the rate of ionisation (the first step). It has a role also in the second step, competition between substitution and elimination. The carbocation is three-coordinated and planar with a bond angle of 120°. The transition state for substitution, $TS_{(s)}$, is not planar. It is four-coordinated and moving towards tetrahedral. This increases the steric strain, more so if the carbocation has bulky groups attached. The transition state for elimination, $TS_{(e)}$, is moving towards the alkene which has the same planar geometry around the central carbons as the carbocation and there is no increase in steric strain. Hence, bulky R groups (tertiary included) favour elimination over substitution in general, 'other conditions' being equal. Now let us look at the other conditions that can affect the reaction.

Nature and concentration of the nucleophile/base: Recall our efforts to make a distinction between nucleophilicity and base strength in our discussion of substitution reactions (Chapter 11, Section 11.2.1.4). In the deprotonation of the carbocation leading to elimination (Eq. 12.4c), the role of the OH^- is strictly that of a base, namely to abstract a proton. While high concentration of the base/nucleophile will help both reactions, a strong base will favour elimination.

Solvent: The transition states for both reactions are neutral overall but with charge separation. The positive and negative charges are more spread out in the one for elimination (see structures in Fig. 12.1 and Eqs. 12.5a and 12.5b). This is favoured by solvents of low dielectric constant (less polar solvents). Thus, ethanol is a better solvent for elimination than water, other things being equal.

Temperature: Both reactions will go at faster rates when the temperature is raised. The increase in rate for the reaction with higher activation energy is more than for the reaction with the lower activation energy. Generally, elimination has higher activation energy than substitution and the former is favoured at higher temperatures.

12.2.2 E2 MECHANISM

Elimination reactions which do not meet the required conditions for E1 mechanism, usually follow second order kinetics conforming to bimolecular reactions (Eq. 12.3b). These are concerted reactions where the departure of the leaving group and the abstraction of the β-hydrogen by a base take place almost simultaneously. They are called E2 reactions and the mechanism is represented in Eq. 12.6.

$$HO^- \quad H - CH_2 - CH_2 - Cl \longrightarrow HOH + CH_2 = CH_2 + Cl^- \qquad (a)$$

Transition state:
$$\delta- \quad \delta+ \qquad \delta-$$
$$HO \text{---} H \text{---} C = C \text{---} Cl \qquad \qquad (b)$$

(with H H above and H H below the two carbons)

Stereochemistry

(12.6)

(c)

Some of the features of this mechanism are as follows.

(i) A fully formed carbocation is not involved. Hence, carbocation stability is less important than in E1 reactions. A C—C double bond develops between the α- and β-carbons, even as the leaving group is departing and the hydrogen from the β-carbon is being abstracted by the base. All these changes may not be taking place in a perfectly synchronous manner. Some of the changes may take place ahead of the others. An extreme case of such a situation is, if the leaving group departs well before the rest of the changes take place. This is what occurs in the E1 mechanism.

(ii) The bond between the α- and β-carbons is acquiring a double bond character in the transition state. It is said to be a partial double bond. Factors which stablise the double bond, such as resonance and hyperconjugation will lower the energy of the transition state and make the reaction proceed faster.

(iii) Both carbons undergo rehybridisation in the transition state. They change from sp^3 to sp^2 hybridisation. The portion of the molecule around the two carbons becomes planar and the bond angles undergo a change from about 109° to about 120°. Steric crowding of groups attached to both carbons decreases in the process. This is in contrast to the situation in the S_N2 reactions where bulky groups on the reaction site will hinder substitution. Bulky groups on C_α or C_β will facilitate elimination at the expense of substitution. This is true for both E1 and E2 eliminations.

(iv) In the E2 transition state, the negative charge of the incoming base is spread out over many atoms. This is favoured in low dielectric solvents, as in the case of E1 reactions.

(v) The role of the base in the E2 transition state is to abstract a proton from the β-carbon, and not to act as a nucleophile. Strong bases are required for elimination. Comparing OH⁻ and SH⁻, we see that the former is a stronger base but a poorer nucleophile than the latter. With a given substrate, the OH⁻ is likely to preferentially cause elimination and SH⁻, substitution. High concentration of base will also be conducive to elimination. These are true, whether the elimination is E1 or E2.

(vi) Higher temperatures favour elimination over substitution as discussed under E1 reactions.

12.2.3 COMPETITION BETWEEN ELIMINATION AND SUBSTITUTION

The factors discussed in the preceding sections may be summarised as given below.

- Good leaving groups are necessary for both substitution and elimination.
- Sterically crowded substrates with bulky substituents on the α- and/or β-carbon tend to undergo elimination rather than substitution.
- Groups that stablilise the developing double bond in the transition state like aryl, vinyl and alkyl, if present on the α- or β- carbon will help elimination.
- A strong base in high concentration favours elimination. A weak base but good nucleophile will favour substitution.
- High temperature favours elimination.
- Solvents of low polarity (low dielectric constant) favour elimination.

12.2.4 E1cB MECHANISM

In the discussion on the E2 transition state (Section 12.2.2), it was mentioned that some of the bond changes could take place ahead of others. The E1 mechanism is an extreme situation where the heterolytic dissociation of the $C-X$ bond (ionisation and departure of the leaving group) takes place ahead of the other changes. The other extreme is the possibility that the abstraction of the β-hydrogen as a proton by the base takes place first and the other changes follow (Eq. 12.7).

$$B\overset{..}{:} + H-\overset{|}{\underset{|}{C}}-\overset{|}{\underset{|}{C}}-X \xrightarrow{\text{fast}} BH + \overset{..}{:}\overset{|}{\underset{|}{C}}-\overset{|}{\underset{|}{C}}-X$$

Carbanion   (a)
(conjugate base
of substrate)

(12.7)

$$\overset{..}{:}\overset{|}{\underset{|}{C}}\curvearrowright\overset{|}{\underset{|}{C}}\curvearrowleft Cl \longrightarrow \overset{\backslash}{\underset{/}{C}}=\overset{/}{\underset{\backslash}{C}} + Cl^-$$

(b)

This is unimolecular elimination from the conjugate base (cB) of the substrate and has been designated as the E1cB mechanism. The first step (Eq. 12.7a) is a fast acid–base equilibrium. The next step, loss of the leaving group, is the slow step. A substrate in which the hydrogen on the β-carbon is quite acidic is expected to react by this mechanism. The best known example is the dehydration of aldol under alkaline conditions (Eq. 12.8).

$$
\begin{array}{c}
\underset{\substack{| \\ H}}{\overset{\substack{OH \\ |}}{CH_3-C}}-\underset{\substack{| \\ H}}{\overset{\substack{\mathbf{H} \\ |}}{C}}-\underset{\substack{\| \\ O}}{C}-H \quad \underset{\longleftarrow}{\overset{OH^-}{\longrightarrow}} \quad CH_3-\overset{OH}{C}\cdots\overset{..}{C}-C-H \quad \longrightarrow \quad \underset{H}{\overset{CH_3}{\diagdown}}C=C\underset{H}{\overset{C-H}{\diagup}}^{O}
\end{array}
$$

(Aldol)

$$\updownarrow$$

$$
CH_3-\underset{\substack{| \\ H}}{\overset{\substack{OH \\ |}}{C}}-\underset{\substack{| \\ H}}{C}=\underset{\substack{| \\ O^-}}{C}-H
$$

(12.8)

The carbanion is resonance-stabilised as shown. Hydrogens, such as the ones highlighted in the aldol, are relatively acidic and many of the reactions of aldehydes and ketones are due to this property (Chapter 15).

The E1cB mechanism can be said to operate only if it can be proved experimentally that a carbanion is involved and that it has a finite life time. One way of doing this is by deuterium exchange. The reaction should be carried out in a solvent containing D_2O. If the unreacted substrate is recovered before the reaction is complete and analysed for deuterium content, D will be present at the β-position if a carbanion is involved (Eq. 12.9). There are very few simple aliphatic systems where this has been established.

$$
\begin{array}{ccc}
\underset{\substack{| \\ H \\ X \\ +\, OH^-}}{\overset{\substack{H \quad H \\ | \quad |}}{R-C-C-R'}} & \underset{\longleftarrow}{\rightleftharpoons} \overset{\substack{H \\ |}}{R-\overset{..}{C}-C-R'} & \overset{D_2O}{\underset{\rightleftharpoons}{\longrightarrow}} \underset{\substack{| \\ H \\ X}}{\overset{\substack{D \quad H \\ | \quad |}}{R-C-C-R'}}
\end{array}
$$

(12.9)

12.3 ORIENTATION OF THE DOUBLE BOND

12.3.1 SAYTZEFF'S RULE

When HBr is eliminated from 2-bromobutane (Eq. 12.10), 1-butene and 2-butene are possible as products; the latter composed of *cis-* and *trans-*geometrical isomers. The selectivity between 1- and 2-butene is referred to as *regioselectivity* or selectivity in the orientation of the double bond. In the dehydrohalogenation of 2-halobutanes and in other similar situations where more than one regioisomer is possible, it has been observed that the more heavily substituted alkene is preferentially formed. 1-Butene is ethylethene, a monosubstituted ethene and 2-butene is dimethylethene, a disubstituted ethene. More of the second compound is obtained. Dehydrochlorination of 2-chloro-2,3-dimethylbutane (Eq. 12.11), can give 2,3-dimethyl-1-butene and 2,3-dimethyl-2-butene. The former is a disubstituted alkene and the latter tetrasubstituted. Both are formed, but the mixture contains more of the tetrasubstituted alkene.

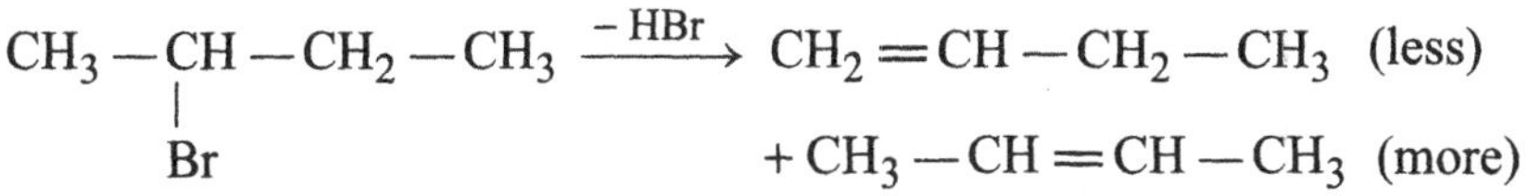

$$CH_3-CH-CH_2-CH_3 \xrightarrow{-HBr} CH_2=CH-CH_2-CH_3 \text{ (less)}$$
$$\underset{\underset{Br}{|}}{} \qquad\qquad + CH_3-CH=CH-CH_3 \text{ (more)} \tag{12.10}$$

$$CH_3-\underset{\underset{CH_3}{|}}{\overset{\overset{Cl}{|}}{C}}-\overset{\overset{CH_3}{|}}{CH}-CH_3 \xrightarrow{-HCl} CH_2=\underset{\underset{CH_3}{|}}{C}-CH-CH_3 \text{ (less)}$$

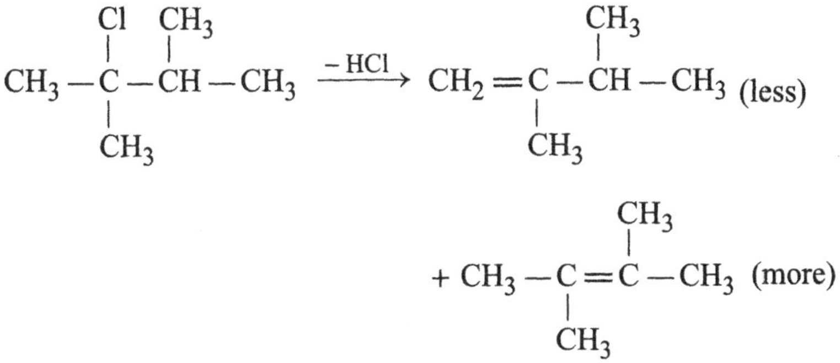

(12.11)

$$+ CH_3-\underset{\underset{CH_3}{|}}{C}=\overset{\overset{CH_3}{|}}{C}-CH_3 \text{ (more)}$$

It is a general observation, that *more of the more heavily substituted alkene is formed when more than one isomeric alkene is possible*. It is applicable to a variety of elimination reactions like dehydrohalogenation and dehydration. This rule is readily rationalised by examining the transition states for elimination, either the E1 transition state or the E2 transition state leading to 1-butene and 2-butene (Eq. 12.12).

> The Saytzeff's rule has been called so after Alexander M Saytzeff (Zaitsev), the Russian Chemist who formulated it (1875).

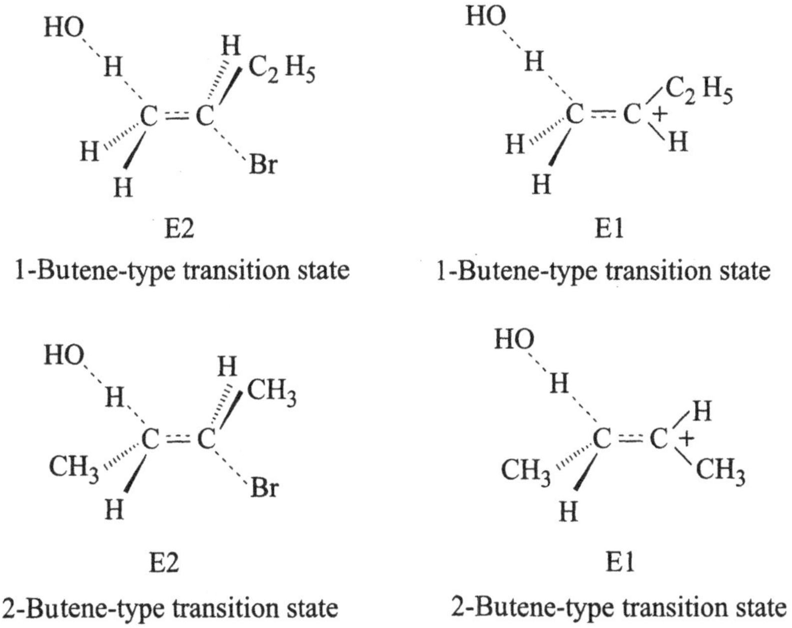

E2	E1
1-Butene-type transition state	1-Butene-type transition state

(12.12)

E2	E1
2-Butene-type transition state	2-Butene-type transition state

Both transition states have partially formed double bonds. Alkyl substituents on the α- or β-carbon will stabilise this partial double bond in much the same way as a real alkene is stablised by alkyl groups through hyperconjugation. 2-Butene-type transition states are better stabilised and have lower energy of activation than 1-butene-type transition states. 2-Butene is formed at a faster rate than 1-butene. The greater thermodynamic stability of 2-butene compared to 1-butene due to hyperconjugation, is discussed in grater detail in Chapter 1, Section 1.4.3.5 (also see Fig. 1.3). The reason for the preference for the more substituted double bond in the product is the relative stabilities of the partially formed double bonds in the transition state. It is inaccurate to say that more of 2-butene is formed because it is the more stable isomer. What we are looking at is a kinetically controlled reaction, not a thermodynamically controlled one.

The ratio of 2-butene to 1-butene in the reactions of various 2-butyl substrates under different conditions is not constant. It may vary over a wide range, so much so that there are reactions where 1-butene is the preferred product and not 2-butene, contrary to Saytzeff's rule. This again emphasises the statement that the product distribution is kinetically controlled. Hofmann elimination is the best known example of this apparent anomaly. A lesser known example is that when 2-butanol is dehydrated under heterogeneous catalysis over hot thorium oxide, 1-butene is selectively formed.

12.3.2 HOFMANN ELIMINATION: HOFMANN RULE

A W Hofmann, a German chemist, studied the decomposition of quaternary ammonium hydroxides by heat. Such reactions have been named *Hofmann elimination* of amines. He observed that in reactions of the type Eq. 12.13, the less heavily substituted alkene is the preferred product. This is referred to as the *Hofmann rule*.

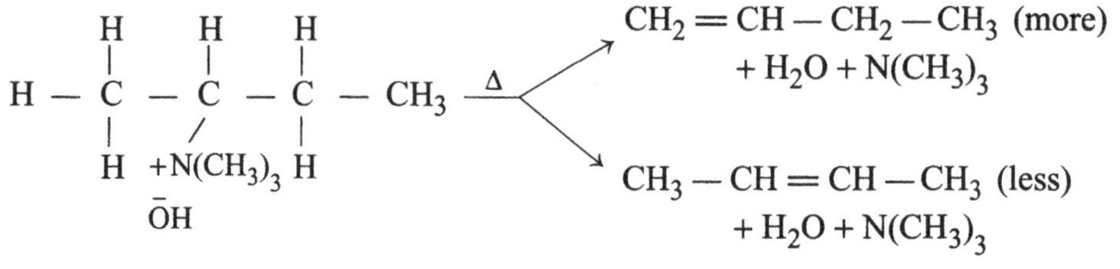

$$CH_2=CH-CH_2-CH_3 \text{ (more)} + H_2O + N(CH_3)_3$$

$$H-\underset{\underset{\overset{+N(CH_3)_3}{|}}{\overset{H}{|}}{C}-\underset{/}{\overset{H}{|}}{C}-\underset{\overset{H}{|}}{\overset{H}{|}}{C}-CH_3 \xrightarrow{\Delta}$$

$$CH_3-CH=CH-CH_3 \text{ (less)} + H_2O + N(CH_3)_3 \qquad \text{(a)}$$

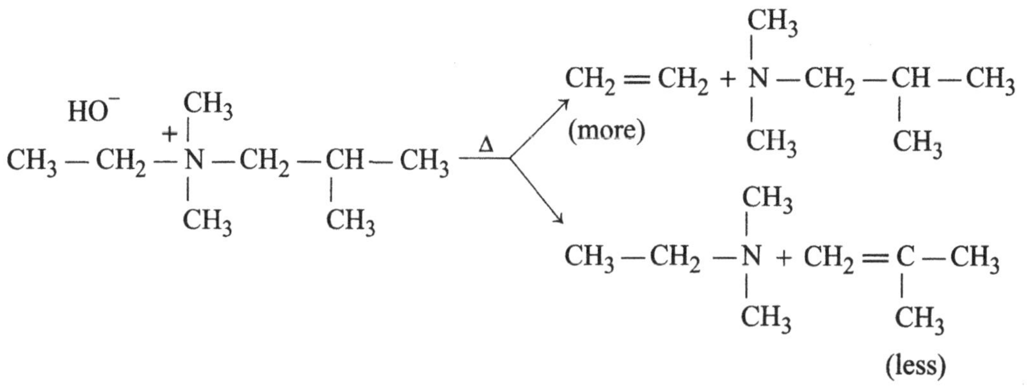

$$CH_3-CH_2-\underset{\underset{CH_3}{|}}{\overset{+\overset{CH_3}{|}}{N}}-CH_2-CH-CH_3 \xrightarrow{\Delta}$$

$$CH_2=CH_2 + \underset{\underset{CH_3}{|}}{\overset{\overset{CH_3}{|}}{N}}-CH_2-\underset{\overset{|}{CH_3}}{CH}-CH_3 \text{ (more)}$$

$$CH_3-CH_2-\underset{\underset{CH_3}{|}}{\overset{\overset{CH_3}{|}}{N}} + CH_2=\underset{\overset{|}{CH_3}}{C}-CH_3 \qquad \text{(b)}$$

(less)

(12.13)

Hofmann eliminations are useful reactions for converting amines to alkenes, but their greater utility has been in the study of alkaloids. These are natural products containing amine functions and *Hofmann elimination* has been extensively used for unravelling their structure by Hofmann himself and others.

$$
\text{H}_2\text{O} + \text{CH}_2 = \text{CH} - \text{CH}_2 - \text{CH}_3 \qquad \text{(a)}
$$
$$
+ \text{N}(\text{CH}_3)_3
$$

$$
\text{H}_2\text{O} + \text{CH}_3 - \text{CH} = \text{CH} - \text{CH}_3 \qquad \text{(b)}
$$
$$
+ \text{N}(\text{CH}_3)_3
$$

$$(12.14)$$

The mechanism of the reaction is given in Eq. 12.14. The apparent contradiction between the Saytzeff and Hofmann rules needs to be explained. The transition states for the two products of Hofmann elimination (Eq. 12.14) appear to be very similar to the ones in dehydrohalogenation reactions that obey Saytzeff rule (Eq. 12.12). There are two explanations for the Hofmann selectivity, one electronic and the other steric. Probably both effects operate simultaneously and in varying degrees in different instances.

Electronic effect: An assumption was made in accounting for Saytzeff rule, that the transition state has a double bond character, and the relative stabilities of the partially formed double bonds leading to the two products control the regioselectivity. This is not the whole picture. There are four changes that occur in the transition state, two bonds are broken (C—X and C—H) and two bonds are formed (H—OH and the second bond in the double bond). As already stated, these changes may not be perfectly synchronised. The breaking of the bond between C and H is an acid–base reaction. The acidity of this hydrogen has a strong influence on this. The acidity of a C—H is dependent on the stability of the conjugate base, the carbanion. As we have seen in Chapter 1, Section 1.4.3.7, a primary carbanion is more stable than a secondary carbanion. A hydrogen on the CH_3 of the 2-butyl substrate is more acidic than one on the CH_2. In the quarternary ammonium ion, which undergoes Hofmann elimination, because of the electron-withdrawing effect of the $-\text{NR}_3^+$ group, the C—H bond-breaking progresses faster than the double bond formation. The deciding factor for regioselctivity in this case is not the stability of the partial double bond but the stability of the partially formed carbanion. Proton abstraction from the CH_3 (primary) takes place at a faster rate than from the CH_2 (secondary).

Steric effect: Influence of steric factors in the regiochemistry of elimination arises from the fact that the terminal position of a chain is more accessible to an attacking reagent than an internal position. OH^- can approach and abstract a proton from the CH_3 of a butane chain more readily than from a CH_2, purely because the latter is more sterically hindered. In small molecules this difference is not seriously felt. This is not the case when a bulky group like $-NR_3$ is present. In Hofmann elimination, steric hindrance to proton abstraction from the CH_2 is quite large and is reflected in the reversal of 1-butene: 2-butene ratio compared to the reaction of the less bulky haloalkanes.

In summary, the two effects electronic and steric, are represented in Eq. 12.15.

$$
\begin{array}{l}
\text{(structure)} \longrightarrow \\
\end{array}
$$

$RCH = C(R) - CH(R) - CH_3$
(I, Hofmann product)

$RCH_2 - C(R) = C(R) - CH_3$
(II, Saytzeff product)

(12.15)

Sterically more accessible, more acidic

Sterically hindered, less accessible, less acidic

A shift from Saytzeff to Hofmann product is observed,

* when the leaving group is large and positively charged as in Hofmann degradation,
* even in dehydrohalogenations, when the R groups are large in size, increasing the steric factor,
* also in dehydrohalogenations, when the base used for hydrogen abstraction is large as in the series (in terms of size) $OH^- < OCH_3^- < OCH(CH_3)_2^- < OC(CH_3)_3^-$, increasing the steric factor. (See question 1, under *Challenging Questions,* for some data illustrating these.)

It has been suggested that the transition state for elimination is a variable one with the carbocation (E1) at one end of the spectrum and the carbanion (E1cB) at the other end, both being extremes. In between are the various E2 transition states with different degrees of progress of the different bond-breaking and bond-formation events.

E1cB . . . .   E2 . . . . E1

$\leftarrow$ region where $\rightarrow$ $\leftarrow$ region of Saytzeff rule $\rightarrow$
Hofmann rule operates

(Variable elimination reaction transition state)

The Hofmann reaction is somewhere to the right of the truly synchronous E2 transition state. It should be emphasised that Hofmann elimination is *not* an E1cB reaction. Kinetically, it is an E2 reaction.

12.4   OTHER ELIMINATION REACTIONS

Dehydration of alcohols and *dehalogenation* of *gem*-dihalides were discussed in Chapter 4 (Sections 4.2.1.2 and 4.2.1.3). To recapitulate, the former is necessarily acid catalysed because OH^- is a poor leaving group and needs activation. (However, see the base-induced dehydration of aldol, Eq. 12.8. Why does this succeed?) Dehydration generally requires heating and in the majority of cases, follows E1 mechanism. Since carbocations are involved, rearrangement is observed in most cases (Chapter 4, Section 4.2.1.6). *Pyrolytic eliminations* of methyl xanthates (Chugayev reaction) and pyrolysis of esters like acetates have also been discussed. These pyrolytic eliminations are *cis*-eliminations that proceed through cyclic transition states (Chapter 4, Section 4.2.1.5).

12.5   STEREOCHEMISTRY

Sterochemistry of elimination refers to the orientation of the leaving group and the β-hydrogen in the transition state of the product determining step. There are two extremes referred to as *anti*-elimination and *syn*-elimination (Fig. 12.2).

In examples (a) and (b), the starting materials I and III, are the same, one particular stereoisomer (diastereomer) of 2-chloro- 3-methylpentane. The two rotational forms (rotamers, conformers) of the same molecule are shown as I and III. They are not isomers. (a) Shows *anti*- or *trans*-elimination, where the two groups —Cl and H are *anti*- or *trans*- to each other – are oriented with a dihedral angle of 180° between them. In (b), *syn*- or *cis*-elimination is shown. The two groups eliminated are oriented *syn*- or *cis*- to each other with a dihedral angle of 0°. The two modes of elimination give two geometrically isomeric alkenes, *Z*-3-methyl-2-pentene (II), and *E*-3-methyl-2-pentene (IV). In (c) and (d), examples of cyclic molecules are given. Here, V and VII are two geometrical isomers of 2-methylchlorocyclohexane. The products shown are those of *anti*-elimination. V gives only one alkene, VI, by *anti*-elimination. VII gives two products, VI and VIII, with more of the latter (operation of Saytzeff rule). *Syn*-elimination (not shown) would have given VI and VIII from V, and only VI from VII.

By studying the product selectivity in the elimination reactions of model compounds of this type, which give different alkenes by *syn*- and *anti*-elimination, it has been generalised that E2 eliminations selectively give the *anti*-elimination products. E1 reactions give a mixture of products corresponding to *syn*- and *anti*-elimination, often with some selectivity for the latter. By contrast, pyrolytic eliminations like the Chugaev reaction give only *syn*-elimination products (Chapter 4, Section 4.2.1.5). It may be recalled that the addition of bromine to alkenes proceeds by *anti*-addition and that debromination is *anti*-elimination.

We have seen how 2-butene is formed preferentially over 1-butene in Saytzeff elimination. 2-Butene is composed of two geometrical isomers, *cis* and *trans*, of which more of the

anti-(*trans-*) Elimination

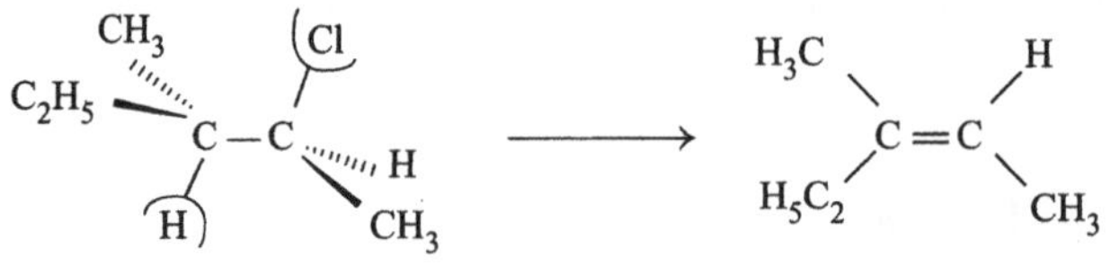

(a)

(I) (II) Z-3-Methyl-2-pentene

syn-(*cis-*) Elimination

(b)

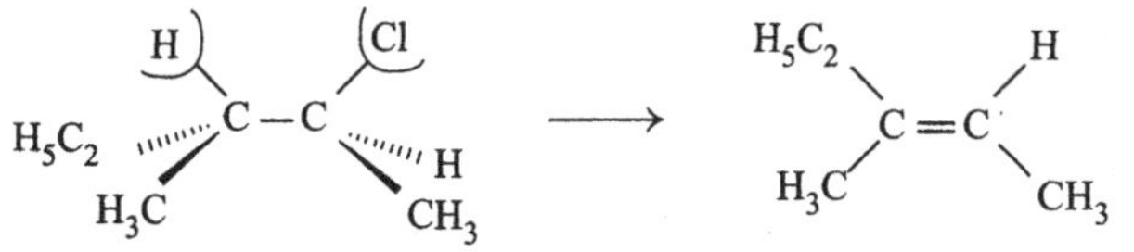

(III) (IV) E-3-Methyl-2-pentene

anti-elimination

(c)

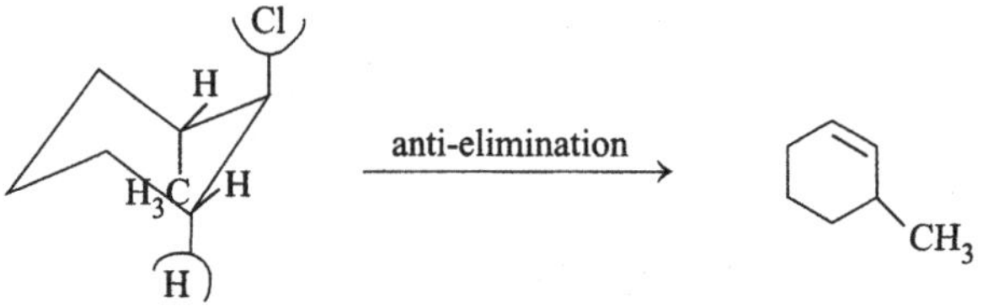

(V) *trans-*2-Methylchlorocyclohexane (VI) 3-Methylcyclohexene

anti-elimination

(VIII) 2-Methylcyclohexene (more)

+

(d)

(VII) *cis-*2-Methylchlorocyclohexane

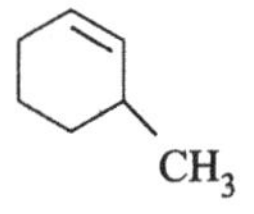

(VI) 3-Methylcyclohexene (less)

Fig. 12.2 Sterochemistry of elimination

latter is formed. Why? It is too facile to say that it is because *trans*-2-butene is the more stable isomer. Now we know that this is not an adequate reason as to why it is preferred, knowing that its formation is kinetically controlled. Let us examine the transition states (Eq. 12.16).

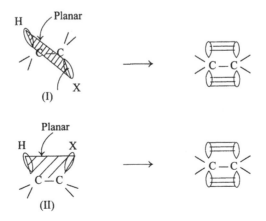

As the transition state I leading to *cis*-2-butene develops, the non-bonded interaction between the methyl groups (which are moving towards an eclipsed conformation with a dihedral angle of 0° due to the planarity of the double bond) keeps increasing. In the other transisition state, II, the CH_3 groups are moving farther apart to an *anti* conformation with a dihedral angle of 180°. The latter is formed at a faster rate.

12.5.1 RATIONALISATION OF TRANS ELIMINATION

In the E2 transition state, sp^3 orbitals on the α- and β-carbons are being transformed to unhybridised *p*-orbitals which form the developing double bond. Maximum overlap between these two *p*-orbitals is achieved when they are parallel to each other. For this to happen, the parent sp^3 orbitals should be coplanar (Fig. 12.3). This condition is satisfied when the two orbitals concerned are *anti* to each other in the staggered conformation (I) or when the two are *syn* to each other in the eclipsed conformation (II). This is the *stereoelectronic requirement*. Of the two, the staggered conformation

Fig. 12.3 Stereoelectronic requirement for E2 elimination

is of lower energy than the eclipsed conformation, due to less non-bonded interactions in the former. This is the *steric constraint*. The two factors, stereoelectronic and steric, contribute to make *anti*-elimination the least energy route for elimination.

The stereochemical terms used in this chapter such as staggered, eclipsed, *syn* and *anti* will be discussed in detail in Chapters 19 and 20.

KEY POINTS

- 1,2-Eliminations or β-eliminations conform to either unimolecular or bimolecular mechanisms. These have been designated E1 and E2 mechanisms.

- E1 eliminations are two-step reactions. The first step is ionisation to form a carbocation, same as for S_N1 reactions. The carbocation is converted to the alkene by hydrogen abstraction by a base. The factors which favour S_N1 reactions also favour E1 reactions.

- E2 elimination proceeds by a concerted mechanism where all the bond-breaking and bond-formation events are nearly synchronous.

- Tertiary and sterically crowded substrates favour elimination over substitution.

- Strong bases in high concentration, less polar solvents and higher temperatures favour elimination over substitution.

- Two seemingly contradictory rules, Saytzeff rule and Hofmann rule, predict the orientation of the double bond in elimination.

- In dehydrohalogenation and dehydration reactions, the more heavily substituted alkene is preferentially formed (Saytzeff rule).

- In the degradation of quaternary ammonium hydroxides, the less heavily substituted alkene is preferentially formed (Hofmann rule).

- Saytzeff rule can be rationalised on the basis of the stability of the developing double bond in the transition state. The more heavily substituted double bond is better stabilised by hyperconjugation.

- In the transition state for Hofmann elimination, abstraction of the β-hydrogen takes place ahead of the rest of the changes, mainly due to the strong electron withdrawing effect of the leaving group. Greater acidity of the β-hydrogen at the CH_3 position and greater steric accessibility of the CH_3 hydrogens are the reasons for Hofmann selectivity.

- When highly acidic β-hydrogens are present, the reaction may proceed through a carbanion and is called the E1cB mechanism. Alkaline dehydration of aldol is an example of this.

- E2 reactions are *anti*-eliminations. This is controlled by the stereoelectronic requirements of the transition state.

EXERCISES

SECTION I

1. Give the major organic products of the following reactions. If more than one alkene is possible predict relative abundance.

(a) $(CH_3)_3C-Cl + NaOCH_3 \xrightarrow[\Delta]{CH_3OH}$

(b) $C_6H_5-CH_2-\underset{\underset{Cl}{|}}{CH}-CH_3 + NaOC_2H_5 \xrightarrow[\Delta]{C_2H_5OH}$

(c) $C_6H_5-CH_2-\underset{\underset{Cl}{|}}{CH}-CH_3 + NaCN \xrightarrow{C_2H_5OH, \Delta}$

(d) $C_6H_5-CH_2-\underset{\underset{Cl}{|}}{CH}-CH_3 + NaSH \xrightarrow{C_2H_5OH, \text{ cold}}$

(e) $CH_3 - CH_2 - \overset{\overset{\displaystyle CH_3}{|}}{\underset{\underset{\displaystyle CH_3}{|}}{\underset{\overset{\displaystyle CH-CH_3}{|}}{{}^+N}}} - CH_2 - CH_2 - CH_3 \quad {}^-OH \xrightarrow{\Delta}$

(f) 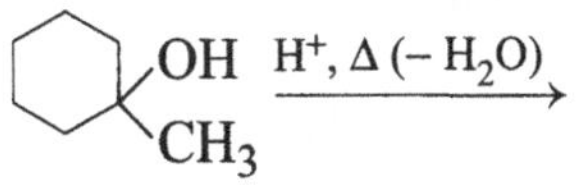 $\overset{OH}{\underset{CH_3}{\diagdown}} \xrightarrow{H^+, \Delta\,(-H_2O)}$

(g) 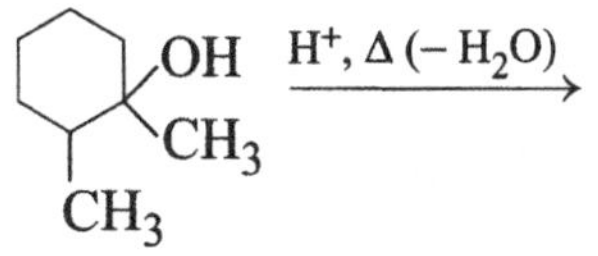 $\xrightarrow{H^+, \Delta\,(-H_2O)}$

2. Give the better conditions to bring about elimination in each of the following.

(a) $CH_3-CH(Br)-CH_2-CH_3 \rightarrow CH_3-CH=CH-CH_3$
$+ CH_2=CH-CH_2-CH_3$
(Heating with $NaOCH_3$ in methanol, or treating in the cold with aqueous methanol and NaOH.)

(b) *tert*-butyl alcohol → isobutene

(Heating with aqueous H_2SO_4, or stirring in the cold with aqueous HCl.)

3. Rearrange the following in the increasing order of reactivity (least reactive, first).

(a) 1-butanol; 2-methyl-2propanol; 2-butanol (acid-catalysed dehydration)

(b) 2-phenyl-2-propanol; 2-phenyl-1-propanol; 1-phenyl-2-propanol (acid-catalysed dehydration)

(c) 2-methyl-1-propanol; 2-phenyl-1-propanol; 2-phenyl-2-butanol (acid-catalysed dehydration)

(d) 2-chloropentane; 1-chloropentane; 2-methyl-2-chlorobutane
(base induced dehydrochlorination)

(e) OH^-, SH^-, CH_3COO^- (reactivity as a base for E2 elimination)

4. Identify the compounds A, B and C.

(a) $A + Br_2 \longrightarrow B \xrightarrow{\text{Zinc dust}}$

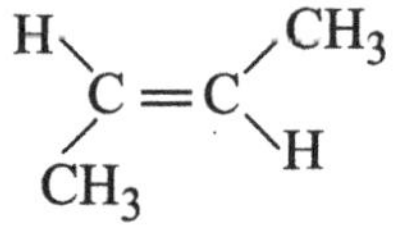

$$\underset{CH_3}{\overset{H}{\diagdown}} C = C \underset{H}{\overset{CH_3}{\diagup}}$$

(b) $C \xrightarrow{\Delta} CH_3{-}CH_2{-}\overset{\overset{\displaystyle CH_3}{\displaystyle |}}{C} = CH_2$ (major product) $+ N(CH)_3 + H_2O$

SECTION II

1. Explain/rationalise the following observations.

(a) Reaction (i) gives the ether (Williamson synthesis). While reaction (ii) gives no ether, but only an alkene.

(i) $(CH_3)_3C{-}OK + CH_3I \rightarrow (CH_3)_3C{-}OCH_3$

(ii) $(CH_3)_3C{-}Br + CH_3ONa \rightarrow H_2C = \overset{\overset{\displaystyle |}{\displaystyle C}}{\underset{\displaystyle CH_3}{|}} {-}CH_3$

(b) Reaction (i) is faster than (ii) under similar conditions.

(i) $CH_2{=}CH{-}CH_2{-}CH_2Cl \xrightarrow[C_2H_5OH, \Delta]{NaOC_2H_5} CH_2{=}CH{-}CH{=}CH_2$

(ii) $CH_3{-}CH_2{-}CH_2{-}CH_2{-}Cl \xrightarrow[C_2H_5OH, \Delta]{NaOC_2H_5} CH_3{-}CH_2{-}CH{=}CH_2$

(c) In the following reaction, I is the major product, II is the minor product.

$$CH_2=CH-CH_2-\underset{\underset{Cl}{|}}{CH}-CH_2-CH_3 \xrightarrow[C_2H_5OH, \Delta]{NaOC_2H_5}$$

$$CH_2=CH-CH=CH-CH_2-CH_3 + CH_2=CH-CH_2-CH=CH-CH_3$$
$$\qquad\qquad\quad (I) \qquad\qquad\qquad\qquad\qquad\qquad\qquad (II)$$

(d) Reaction (i) is faster than (ii) under similar conditions.

(i) $CH_3-\underset{\underset{CH_3}{|}}{CH}-CH_2Cl \xrightarrow[C_2H_5OH, \Delta]{NaOC_2H_5} CH_3-\underset{\underset{CH_3}{|}}{C}=CH_2$

(ii) $CH_3-CH_2-CH_2-CH_2Cl \xrightarrow[C_2H_5OH, \Delta]{NaOC_2H_5} CH_3-CH_2-CH=CH_2$

(e) For the dehydration of alcohols, H_2SO_4 is a better reagent than HCl.

(f) Reaction (i) takes places faster than (ii) under similar conditions.

(i) $CH_3-\overset{\overset{O}{||}}{C}-CH_2-CH_2Br \xrightarrow[C_2H_5OH]{NaOC_2H_5} CH_3-\overset{\overset{O}{||}}{C}-CH=CH_2$

(ii) $CH_3-\overset{\overset{O}{||}}{C}-\underset{\underset{Br}{|}}{CH}-CH_3 \xrightarrow[C_2H_5OH]{NaOC_2H_5} CH_3-\overset{\overset{O}{||}}{C}-CH=CH_2$

(g) Reaction (i) takes place readely, reaction (ii) does not take place.

(i) $CH_3-\underset{\underset{OH}{|}}{CH}-CH_2-\overset{\overset{O}{||}}{C}-H \xrightarrow{OH^-} CH_3-CH=CH-\overset{\overset{O}{||}}{C}-H$

(ii) $CH_2-CH_2-CH_2-\overset{\overset{O}{||}}{C}-H \xrightarrow{OH^-} CH_2=CH-CH_2-\overset{\overset{O}{||}}{C}-H$
$\quad\;\;\overset{|}{OH}$

2. With suitable examples, show how E1 eliminations can give a mixture of products corresponding to syn- and anti-eliminations.

3. Recommended experimental conditions for preparing alkenes from alkyl halides is: "reflux with alcoholic potash (solution of potassium hydroxide in ethanol)". All the words in this sentence namely 'reflux', 'alcoholic' and 'potash', are important. Explain.

CHALLENGING QUESTIONS

1. Rationalise the following observations. (Changes in the Saytzeff: Hofmann ratio when different kinds of changes are made).

    (a) $CH_3-CH-CH_2-CH_2-CH_3 + OH^-$
             $\,\,\,\,\,\,\,\,\,\,\,|$
             $\,\,\,\,\,\,\,\,\,\,\,X$

    $\rightarrow CH_2=CH-CH_2-CH_2-CH_3 + CH_3-CH=CH-CH_2-CH_3$
    $\,(I)\,(II)$

X		(I)	(II)
$-Br$	....	31%	69%
$-\overset{+}{S}(CH_3)_2$	....	87%	13%
$-\overset{+}{N}(CH_3)_2$	....	98%	2%

    (b)
    $\,CH_3$
    $\,|$
    $CH_3-C-CH_2-CH_3+Base \rightarrow CH_2=C-CH_2-CH_3+CH_3-C=CH-CH_3$
    $\,\,\,\,\,\,\,\,\,\,\,\,\,\,\,|\,|\,\,\,\,\,\,\,(III)\,|\,\,\,\,\,\,\,\,(IV)$
    $\,\,\,\,\,\,\,\,\,\,\,\,\,\,\,Br\,CH_3\,CH_3$

Base		(III)	(IV)
$C_2H_5O^-$	....	30%	70%
$(CH_3)_3C-O^-$	....	72%	28%
$(C_2H_5)_3C-O^-$	....	78%	22%

    (c) $CH_3-CH_2-CH_2-CH-CH_3$
             $\,|$
             $\,X$

    $\xrightarrow{C_2H_5O^-} CH_3-CH_2-CH_2-CH=CH_2 + CH_3-CH_2-CH=CH-CH_3$
    $\,(V)\,(VI)$

X		(V)	(VI)
F	....	85%	15%
Br	....	31%	69%

(d) $CH_3 - \underset{\underset{Br}{|}}{\overset{\overset{R}{|}}{C}} - CH_2 - CH_3 \xrightarrow{C_2H_5O^-} CH_2 = \underset{}{\overset{\overset{R}{|}}{C}} - CH_2 - CH_3 + CH_3 - \overset{\overset{R}{|}}{C} = CH - CH_3$

$\qquad\qquad\qquad\qquad\qquad\qquad\qquad\qquad$ (VII) $\qquad\qquad\qquad$ (VIII)

R		(VII)	(VIII)
H	....	19%	81%
CH₃	....	30%	70%

2. Refer to Fig. 12.2. Write down the mechanism of the Chugaev reaction of the S-methylxanthate of 3-methyl-2-pentanol, corresponding to the structure in this figure (structure I or III, with OH instead of Cl). Draw the structure of the transition states and arrive at the structure of alkenes formed.

13 Determination of Structure Using Spectroscopy

OBJECTIVES   In this chapter, you will learn about,

- the electromagnetic spectrum
- an electronic spectrum
- types of electronic excitations
- applications of UV spectroscopy
- infrared region, presentation of infrared spectra
- sampling techniques
- characteristic IR frequencies
- applications of IR spectroscopy
- principles of NMR spectroscopy
- chemical shift, proton counting and spin–spin coupling
- characteristic chemical shifts
- aromaticity and ring current effect
- applications of NMR spectroscopy
- principle of mass spectrometry
- applications of mass spectrometry

13.1 INTRODUCTION

Determination of the structure of unknown compounds, whether of natural origin or prepared in the laboratory, is a challenging task for the organic chemist. The classical approach may be summarised as follows:

- (i)   determination of elemental composition, molecular weight and molecular formula,
- (ii) detection of functional groups and a study of their chemical properties and reactions,
- (iii) degradation to simpler, known molecules,
- (iv) proposing a structure, and
- (v) confirmation of the structure by unambiguous synthesis.

Many elements of this scheme have now been replaced by instrumental methods, especially spectroscopic methods. Spectroscopy can provide information on molecular weight, chemical composition, molecular formula, structural features like conjugation and aromaticity, presence or absence of functional groups and to a large extent, the total structure. In this chapter, an introduction to the application of spectroscopic techniques to structure determination will be provided. Ultraviolet (UV), Infrared (IR), Nuclear Magnetic Resonance (NMR) spectroscopy and Mass Spectrometry (MS) will be discussed.

13.2 THE ELECTROMAGNETIC SPECTRUM

The three spectroscopic techniques of immediate interest to us – UV, IR and NMR – are based on absorption of radiation of appropriate wavelength from the electromagnetic spectrum (Fig. 13.1).

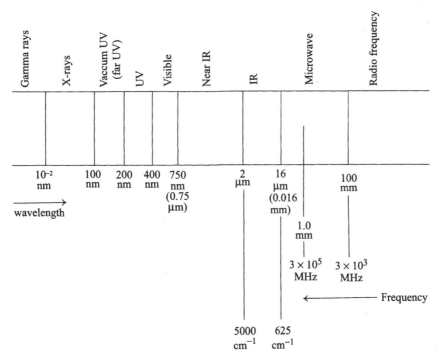

Fig. 13.1 Electromagnetic spectrum

Absorption of radiation in the UV and visible region, corresponding to wavelengths between 200 and 750 nm (2×10^{-7} to 7.5×10^{-7} m) causes electronic excitation. UV and visible spectra are electronic spectra. The region, 2.5×10^{-6} m to 16×10^{-6} m (2.5–16 micrometer, also called micron, represented as μ m or simply μ), corresponds to infrared radiation. This higher wavelength has lower energy than UV radiation and cannot cause electronic excitation. Its energy corresponds to that of the vibrational levels and thus causes vibrational excitation. IR spectroscopy is vibrational spectroscopy. The NMR spectrum arises due to excitation from one spin state to another caused by radiation in the radiofrequency region.

Frequency is related to wave length by the expression,

$$\nu = c/\lambda$$

where ν = frequency in cycles $\sec^{-1}$, whose SI units are reciprocal seconds (s^{-1}) and named hertz (Hz), λ = wavelength in m and c = velocity of light, 3.0×10^8 m s^{-1}. Another frequency unit in common usage in IR spectroscopy is wave number, expressed as reciprocal centimeter (cm^{-1}), which is the number of waves per centimeter. The IR region 2.5 μm to 16 μm corresponds to 4000 to 625 cm^{-1}. It should be remembered that higher the wave length, lower the energy. Energy is proportional to the frequency ($E = h\nu$).

13.3   UV SPECTROSCOPY

When a beam of light composed of the full UV-visible spectrum of wavelengths is passed through a substance (usually a solution of the substance), not all wavelengths are absorbed. Only photons of frequencies which correspond in energy to specific allowed excitations of electrons from a lower energy level to a higher level will be absorbed because these energy levels are quantised, not continuous.

In a spectrometer, light is allowed to fall on the substance. The sample is not subjected to all frequencies of light together, but it is scanned from one end of the spectrum to the other. Only frequencies compatible in energy to electronic excitation will be absorbed, the others will pass through unhindered. A comparison of the intensity of the incident radiation and the transmitted radiation can be made in terms of absorbance (absorption) or transmittance (transmission). The instrument measures absorbance or transmittance of different frequencies and plots the absorbance or transmittance against frequency or wavelength. This plot is called the spectrum.

Electronic excitation occurs in molecules when electrons in stable, filled orbitals absorb energy and get excited to orbitals of higher energy, usually into the anti-bonding orbitals. In organic molecules that are in the ground state, electrons can be in σ-bonds or in π-bonds, or can be present as non-bonded or unshared electrons, like the lone pairs on oxygen and nitrogen. They can all be excited to the anti-bonding orbitals when energy is supplied in the form of photons of appropriate frequency. The energy required for the excitation of electrons from σ-bonds is large and corresponds to wavelengths lower than 200 nm (called the far-UV region). These are not accessible in common UV spectrometers. This is because, in the far-UV region, absorption due to container material, air and solvents make instrumentation complicated. Only UV-visible absorption in the 200–750 nm range is useful for routine studies. In the case of simple alkenes, an electron from the bonding π-orbital, which is the highest occupied molecular orbital (HOMO), can be excited to the anti-bonding π-orbital (π^* orbital) which is the lowest unoccupied molecular orbital (LUMO) in the ground state (Fig. 13.2a).

The wavelength corresponding to this excitation is 170 nm which is not available in a laboratory UV instrument. In 1,3-butadiene, the energy gap between the HOMO and LUMO is smaller and corresponds to 217 nm which is available (Fig. 13.2b). In polyenes, as the number of double bonds in conjugation increases, the energy gap between the HOMO and LUMO decreases so much so

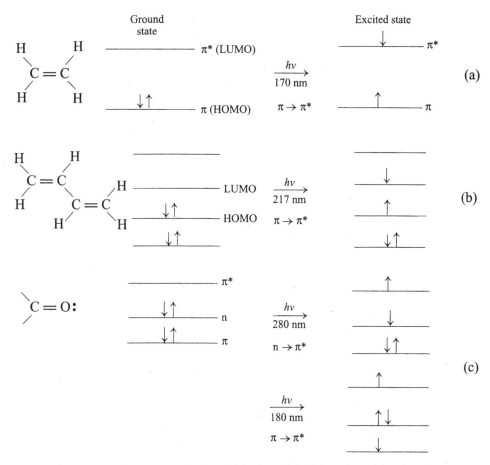

Fig. 13.2 Electronic excitation of (a) ethene (b) 1,3-butadiene (c) carbonyl

that visible light is absorbed, making the molecules coloured. Carotenoids – which are naturally occurring plant pigments – are polyenes (see the structure of β-carotene, the pigment isolated from carrots).

β-carotene
λ_{max} 451 nm

In the case of simple ketones, there are two UV excitations which are of importance. These are the $\pi \to \pi^*$ excitation and the $n \to \pi^*$ excitation (Fig. 13.2c). The latter requires lower energy than the former and takes place at higher wavelengths. The absorption maximum for $n \to \pi^*$ excitation falls in the region 270–300 nm. It is a weak absorption with low molar absorptivity, ε.

ε is a measure of the fraction of the molecules that are exposed to the radiation which actually absorb the radiation and get excited. The $\pi \rightarrow \pi^*$ is an intense band but appears in the far-UV region. However, when the carbonyl is conjugated with a carbon–carbon double bond, this gets shifted to 215–250 nm. Representative UV spectra, those of isoprene (a) and a bicyclic conjugated ketone (b) are given in Fig. 13.3.

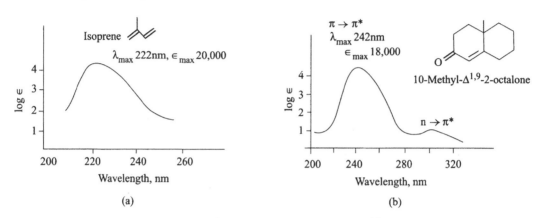

Fig. 13.3 UV spectra of (a) isoprene (b) 10-methyl-$\Delta^{1,9}$-2-octalone

The maxima are not sharp and are generally referred to as bands. UV spectra are rarely reproduced as such, but are reported by giving λ_{max} and ε_{max}, thus:

Isoprene, λ_{max} 222 nm (ε_{max} 20, 000)

10-Methyl-$\Delta^{1,9}$-octalone, λ_{max} 242 nm (ε_{max} 18,000);

λ_{max} 300 nm (ε_{max} 100)

UV spectroscopy has been very useful in the study of terpenes and steroids. The structures of testosterone (a conjugated ketone) and abietic acid (a conjugated diene) are given in Chapter 10, Section 10.2.4.2. Their UV maxima are at 240 and 238 nm respectively, with log ε_{max} of more than 4 for both.

Aromatic compounds show distinctive UV spectra. Aliphatic and alicyclic compounds without conjugated unsaturation, largely do not show any maxima in the UV region. As is evident, the main use of UV is in the study of conjugation in unsaturated molecules. This is illustrated by the UV spectra of *cis*- and *trans*-stilbenes.

trans-Stilbene

λ_{max} 295.5 nm

ε 29,000

cis-Stilbene

λ_{max} 280 nm

ε 10,500

Extended conjugation between the two benzene rings through the intervening double bond is possible only when the benzene rings and the double bond are all planar. This condition is satisfied in *trans*-stilbene. In *cis*-stilbene, due of steric crowding, one ring is forced out of plane thus curtailing extended conjugation. Electronic excitation in *cis*-stilbene requires more energy than in *trans*-stilbene. So, the absorption maximum of the former is lower than that of the latter. Some further aspects of electronic excitation are discussed in Chapter 26.

13.4 IR SPECTROSCOPY

Electromagnetic radiation in the infrared region has energies appropriate to bring about excitations in the vibrational levels. Vibration is a fundamental property of bonds. A bond, like a stretched spring, can vibrate in different ways. These are known as different modes of vibration, such as different kinds of stretching and bending. Since we are looking at vibrations of specific bonds in IR spectroscopy, the information gained is about functional groups. Each functional group like carbonyl, hydroxyl, $C=C$ double bond, $C-H$ bond in different molecules where the C is sp^3, sp^2 or sp hybridised, or $N-H$ bonds in amines, gives rise to peaks in IR spectra at specific frequencies. These frequencies are characteristic of the bond (group) with small, but significant variations. These characteristic frequencies help to identify the groups and the variations give valuable information about the environment of the groups in the molecule. The total IR spectrum is uniquely characteristic of a particular molecule and can be used to establish its identity. No two molecules have their IR spectra identical in all respects. An IR spectrum is often used as a 'finger print' of the molecule for identification purposes.

13.4.1 RECORDING AND PRESENTATION OF IR SPECTRA

The sample, for which the IR spectrum has to be recorded, is taken in a container (IR cell) whose walls are transparent to the IR beam. Cells are made of IR-transparent salts like NaCl. Glass or quartz cannot be used. If the sample is a liquid, it can be directly taken in the cell. If it is a solid, it should be dissolved in a solvent like CCl_4. Other techniques are to grind the solid with a mineral oil (Nujol mull) or grind the solid with an IR-transparent salt like KBr and press the mixture into a pellet or wafer (KBr pellet). Spectra recorded using different sampling techniques may differ in detail.

IR spectra are presented with % transmittance on the *y*-axis and wavelength (μm) and/or frequency (cm^{-1}) on the *x*-axis. Since it is a transmittance spectrum, absorption maxima appear as negative peaks. The IR spectrum of methyl acetate is shown in Fig. 13.4.

The salient features of this spectrum are the peaks at 2900, 1740 and 1250 cm^{-1}. Their assignments are as follows: 2900 cm^{-1} (3.4 μm) (s) assigned to $C-H$ stretching vibration, 1740 cm^{-1} (5.75 μm) (s) – $C=O$ stretching, 1250 cm^{-1} (8.0 μm) (s) and 1050 cm^{-1} (9.5 μm) (s) – $C-O$ stretching. Other peaks are present at 1450, 1380 and 850 cm^{-1}, which are vibrations of the molecular skeleton, and are not very useful for the assignment of structure. The notations, s, m and w are used to denote strong, medium and weak, with reference to the intensity of the peak concerned.

When it is not convenient to reproduce the spectrum in total, it is customary to list the prominent peaks alone, as done above. The region 1600–4000 cm^{-1} is usually given particular attention to because the frequencies of most functional groups appear here and hence is useful for

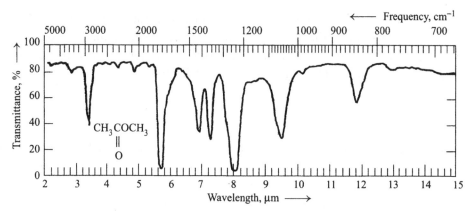

Fig. 13.4 IR spectrum of methyl acetate

structure determination. The region 625–1500 cm^{-1} has many small peaks which are of limited use for structural assignment, but are characteristic of the molecule. This region is referred to as the *finger-print region* of the spectrum. The characteristic frequencies of the common groups in organic molecules are listed in tables. Such tables are very useful in interpreting IR spectra. A limited version of such a listing is given in Table 13.1.

Table 13. 1 Characteristic infrared frequencies

Bond	Frequency, cm^{-1}
Single bonds, stretching	
O—H (alcohols, phenols monomeric)	3610–3640
O—H (alcohols, phenols, hydrogen boded)	3200–3600 (broad)
N—H (amines)	3300–3500
O—H (carboxylic acids)	2500–3000
C—H (alkynes)	3300
C—H (aomatic)	3000–3100
C—H (alkenes)	3020–3080
C—H (alkanes)	2850–2960
C—O (alcohols, phenols, esters, ethers)	1080–1300
Multiple bonds, stretching	
C=O (aldehydes, ketones, acids, their derivatives)	1690–1760
C=C (alkenes)	1640–1680
C=C (aromatic)	1500, 1600
Substituted benzenes	
monosubstituted	730–770 and 690–710
o-disubstituted	730–770
m-disubstituted	750–810 and 680–730
p-disubstituted	790–840

The problems given below illustrate how IR spectra are useful in structure identification.

Problem 13.1

Molecular formula: C_4H_8O

IR spectrum: $1710\,cm^{-1}$ (s)

$2900\,cm^{-1}$ (m-s)

Solution: The molecular formula shows that it is aliphatic or alicyclic containing one double bond or ring. [For this calculation, for every monovalent atom (like halogen) present, add one H to the molecular formula. Discard the halogen. For every trivalent atom like N, subtract 1 from the number of hydrogens in the molecular formula and discard the nitrogen. Disregard divalent atoms like O and S. Examine the resultant molecular formula, C_4H_8 in this case. If the number of carbons is n and number of hydrogens m, then, $2n + 2 - m = d$, where d is the hydrogen deficit compared to the parent alkane. $d/2$ is the *index of unsaturation*, or *index of hydrogen deficiency*, namely, the number of double bonds or rings. In the present case, $(10 - 8)/2 = 1$ is the number of double bonds or rings. A carbonyl group accounts for one double bond]. Let us assign the IR frequencies with the help of Table 13.1.

$$2900\,cm^{-1} : C—H \text{ stretching, } 1710\,cm^{-1} : C{=}O \text{ stretching.}$$

The compound may be 2-butanone, butanal or 2-methylpropanal. It cannot be an unsaturated alcohol like 3-butene-1-ol or any of its isomers (which will also have the same molecular formula and index of unsaturation). Aldehydes should have a characteristic $C—H$ stretching peak at about $2720\,cm^{-1}$. Since it is not mentioned, it is presumed to be absent. The compound can be tentatively identified as 2-butanone. Independent evidence is desirable to rule out the aldehyde structure. Since 2-butanone is a known compound, confirmation can be obtained by taking the spectrum of 2-butanone and the unknown sample under identical conditions and comparing them peak for peak (finger print comparison). If they are identical, the identification is complete.

When such IR frequencies are listed, it is not implied that other peaks are absent. Those frequencies which are presumed to be relevant are listed. Admittedly, this is not a safe presumption. The total IR spectrum of 2-butanone is given in Fig. 13.5.

Problem 13.2

Molecular formula: C_7H_8O

IR Spectrum: $3300\,cm^{-1}$ (s, broad)

$3000\,cm^{-1}$ (w)

$1230\,cm^{-1}$ (s)

$820\,cm^{-1}$ (s)

Solution: The index of unsaturation is 4 which in the first instance, suggests a benzene ring (3 double bonds and one ring). The strong broad band at $3300\,cm^{-1}$ is due to the OH

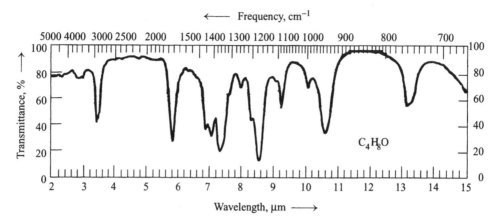

Fig. 13.5 IR spectrum (Problem 13.1)

group—phenolic or alcoholic. The band at 1230 cm^{-1} strongly suggests C—O in phenols. The corresponding band in alcohols is usually between 1050 and 1200 cm^{-1}. The region below 850 cm^{-1} gives information on the number and position of substituents on benzene (Table 13.1). The peak at 820 cm^{-1} suggests *p*-disubstituted benzene. A tentative identification of the unknown compound is that it is *p*-cresol. This needs confirmation. Structures like benzyl alcohol and *o*- or *m*-cresol cannot be ruled out. The full IR spectrum of *p*-cresol is given in Fig. 13.6.

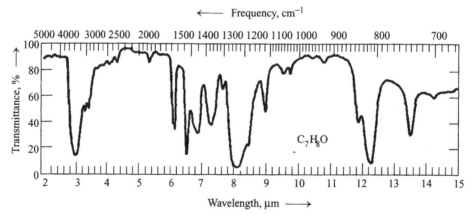

Fig. 13.6 IR spectrum (Problem 13.2)

13.5   NUCLEAR MAGNETIC RESONANCE (NMR) SPECTROSCOPY

13.5.1   INTRODUCTION

The nucleus of the proton, ^{1}H, has two spin states, $+1/2$ and $-1/2$. Both the states are of equal energy and are equally populated under normal conditions. They do not absorb energy in the form of electromagnetic radiation. Absorption can take place only if the two spin states have unequal energies. This condition is achieved by placing the nuclei in a strong magnetic field. In the magnetic field, some of the nuclei (which can be considered as tiny magnets) align themselves in the direction parallel to the applied field, and some others are aligned antiparallel to the applied field. Now, the two types of nuclei differ in energy, the ones parallel to the magnetic field being of lower energy. Very high magnetic fields are required to cause a difference in energy between the two spin states, large enough to create a signal corresponding to the absorption of radiation, which can be instrumentally detected. In a magnetic field of 4.7 T (Tesla) the two spin states of the proton, ^1H, have an energy difference of 8×10^{-5} kJ mol^{-1} which corresponds to the frequency 200 MHz. This is in the radiofrequency (rf) region. The energy difference between the two spin states is proportional to the applied magnetic field. When a sample containing protons is placed in a magnetic field of magnitude of the above range and the appropriate radiofrequency is applied, nuclei in the lower energy spin state can absorb the radiation and 'flip' to the higher energy state, which in turn lose the energy by emitting rf and flip back to the ground state. Under these conditions the nuclei are said to be in a state of resonance. In order to record an NMR spectrum, one of the two variables, magnetic field (which in a modern high resolution instrument may be 4.7 T) or the frequency [for this magnetic field the frequency may be 200 megahertz (MHz)] is kept constant and the other varied. When the NMR spectrum is presented, it is the frequency that is shown as being varied.

The statement that the difference in energy of the two spin states is proportional to the applied magnetic field has to be modified to read, 'poroportional to the effective magnetic field felt by the nucleus'. This is because, the actual or effective magnetic field in the environment of the nucleus is slightly lower than the applied field, because of opposing magnetic fields created by the spinning of electrons in the molecule.

Here, NMR spectroscopy of the proton, ^1H, is discussed and is referred to as ^1H NMR, proton-NMR or PMR spectroscopy. This tells us about the environment of the hydrogens in a molecule. All nuclei do not have the necessary spin properties to give rise to NMR spectra. The natural isotope of carbon with mass number 13 is another nucleus amenable to NMR spectroscopy. ^{13}C NMR spectra tell us about the environment of the carbon atoms present in a molecule. The conditions to obtain the ^1H spectrum are different from those required for the ^{13}C spectrum or the spectra of any other nuclei. The NMR spectrum of one nucleus does not suffer from interference due to other NMR-active nuclei. The solvent most commonly used for NMR is deuterated chloroform, $CDCl_3$. Other solvents which do not contain hydrogen can also be used.

13.5.2 CHEMICAL SHIFT

The hydrogens in a molecule, referred to in the NMR context as protons, are all not equivalent in terms of their electronic environment. Each nonequivalent proton in a molecule absorbs at a characteristic frequency and the difference between each is not more than a few hundred hertz (Hz) while the incident frequency is in the megahertz (MHz) range. To what extent the NMR absorption of each proton in a molecule differs from that of a bare proton, is a measure of the extent to which the neighbouring electrons affect the magnitude of the applied field. Since the 'bare proton' is a hypothetical entity, protons in a real molecule are taken as reference. The molecule used as the reference is tetramethylsilane $(CH_3)_4Si$, TMS. The difference in absorption frequency between TMS and the proton in a molecule is called the *chemical shift* of that proton. The chemical shift, expressed in Hz, depends upon the frequency of the instrument. If the chemical shift is 400 Hz in a 200 MHz instrument, it will be 800 Hz in a 400 MHz instrument. In order to have the same chemical shift in different instruments, working at different frequencies, the ratio between the chemical shift in Hz and the instrument frequency in MHz, expressed as parts per million (ppm) is used as the unit of chemical shift designated δ (delta). The chemical shift of the protons in TMS is taken to be δ 0. Chemical shift of 400 Hz in a 200 MHz instrument can be calculated as,

$$\delta = (400/200 \times 10^6) \times 10^6 = 2\,\text{ppm}$$

An alternate unit, τ (tau) where $\tau = 10 - \delta$, is also in use.

The NMR spectra of neopentane, acetone and dimethyl ether (each compound contains only one kind of proton) are shown in Fig. 13.7. The x-axis is marked in the δ or the τ scale. The x-axis, from left to right, represents decreasing frequency/increasing field strength. The right side of the spectrum constitutes the 'up-field' region. The frequency may be shown on the x-axis, usually on the top. Its value is significant only if the instrument frequency is also specified. δ or τ value – which is independent of the instrument frequency – is usually marked at the bottom.

One of the factors which affect the value of the chemical shift is inductive effect. This is evident from the spectra. In the three molecules, the electron density around the hydrogen of the CH_3 progressively decreases in the order, (a) > (b) > (c) because of the inductive effect of the atom/group attached to the carbon. There is greater electron density around the protons in neopentane than in dimethyl ether. Therefore, the effective field experienced by the proton in neopentane is less than in dimethyl ether. The protons in neopentane are said to be more *shielded* than those in dimethyl ether. The reverse expression is also in use, that is, the protons in dimethyl ether are more *deshielded* than in neopentane. In order to get the same effective magnetic field, slightly more external field has to be applied on neopentane than on dimethyl ether. The neopentane signal appears more *upfield* compared to that of dimethyl ether.

Two generalisations can be made at this stage. (i) Each magnetically non-equivalent proton gives rise to an NMR signal with a characteristic chemical shift. (ii) More shielded protons appear upfield and have smaller δ values. The greater the electron density around the proton, the more shielded it is.

Chemical shift information obtained from an NMR spectrum is useful for locating the position of the hydrogens in the molecule. Some useful chemical shift values are listed in Table 13.2.

13.5.3 MULTIPLICITY: SPIN–SPIN COUPLING

The NMR spectrum of 2-butanone is given in Fig. 13.8.

2-Butanone has three sets of magnetically non-equivalent protons—the protons on CH_2, and those on the two types of CH_3 groups. The CH_3 of the ethyl group, being further away from the

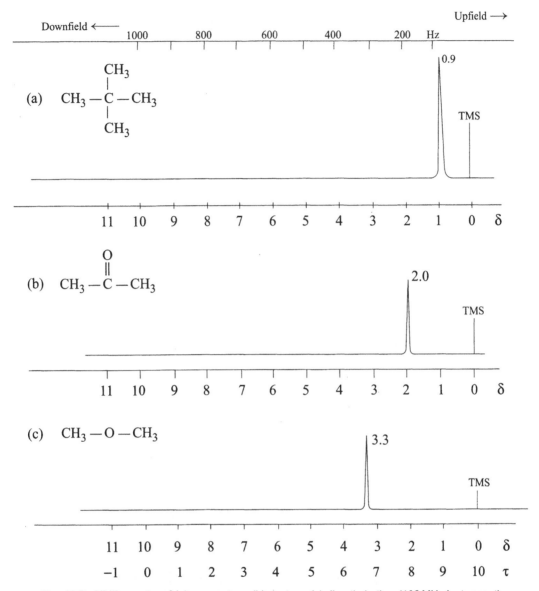

Fig. 13.7 NMR spectra of (a) neopentane (b) acetone (c) dimethyl ether (100 MHz instrument)

carbonyl than the other two sets, is more shielded. Hence, we assign the signals at δ 1.0 to the CH_3 of the ethyl and δ 2.0 to the CH_3 attached to the CO. The other 2H signal, also deshielded, at δ 2.3 is assigned to the CH_2. The ethyl signals are not single peaks as in Fig. 13.7, but show *splitting* or *multiplicity*.

Proton counting: An integrator in the instrument gives the areas under the peaks. The three signals or clusters of signals (Fig. 13.8), going from downfield to upfield are, in the ratio of their areas, 2 : 3 : 3. This is the ratio of the number of hydrogens responsible for the signals. Since the

Table 13. 2   Chemical shifts of protons

Type of proton	Chemical shift (δ)
H—C—R (as in alkanes)	0.9–1.0 (primary)
	1–1.2 (secondary)
	0.3–1.5 (tertiary)
H—C=C (vinyl)	4.5–6
H—Ar (aromatic)	6–8.5
H—C—Ar (benzyl)	2.2–3
H—C—C=C (allyl)	1.7–2
H—C—X (halides)	2–4
H—C—CO— (carbonyl compound acids, esters)	2–2.7
H—C—O (alcohols, ethers, esters)	3.3–4
H—C=O (aldehydes)	9–10
H—O (alcohols)	1–5.5*
II—O (phenols)	4–12*
H—O—CO (carboxyl)	10.5–12*
H—N (amines)	1–5*

Chemical shifts marked with an asterisk (*) are variable and depend upon the temperature and concentration.

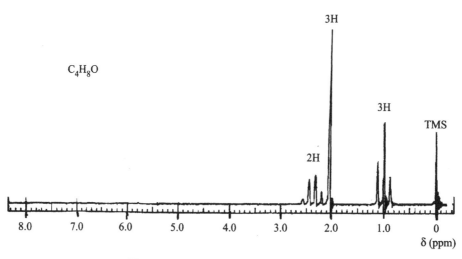

Fig. 13.8   NMR spectrum (Problem 13.3)

molecular formula is known to be C_4H_8O, we can conclude that there are two CH_3 groups and one CH_2 group.

Splitting of peaks: The CH_3 signal at δ 1.0 is a composite of three signals (a triplet) and the CH_2 signal at δ 2.3, a quartet. This is due to *spin-spin coupling*. A working rule is that the number of lines in the multiplet (multiplicity) of a particular set of protons is equal to one more than the number of protons on adjacent carbons. For the CH_3 of the ethyl group, the multiplicity is $2 + 1 = 3$ and for the CH_2, it is $3 + 1 = 4$. The source of the splitting is that the spin of the proton under consideration is influenced by the spins of the neighbouring protons. The CH_3 attached to the CO has no adjacent protons. It is therefore not split and appears as a singlet.

It is not always convenient to reproduce the entire NMR spectrum. Often NMR data are presented in abbreviated form as shown below for 2-butanone.

NMR spectrum, 2-butanone, δ 1.0, triplet, 3H

2.0, singlet, 3H

2.3, quartet, 2H

3H and 2H refer to the number of hydrogens corresponding to those signals.

Problem 13.3

(This unknown compound is the same as the one in Problem 13.1)

Molecular formula: C_4H_8O

IR spectrum: see problem 13.1

NMR spectrum: 1.0, triplet, 3H

2.0, singlet, 3H

2.3, quartet, 2H

Solution: There is one double bond or one ring present (see problem 13.1). The signals at δ 1.1 and 2.0 are likely to be CH_3 groups and the one at δ 2.3 due to CH_2. The splitting patterns for the first and third signals suggest the presence of an ethyl group. The chemical shifts suggest the presence of a $-CH_2-CO-CH_3$ grouping. The compound is identified to be $CH_3-CO-CH_2-CH_3$, 2-butanone. This is consistent with the IR spectrum. From the IR spectrum alone, the aldehyde structure cannot not be ruled out, but can definitely be ruled out on the basis of the NMR spectrum, (Fig. 13.8).

Problem 13.4

(This unknown compound is the same as the one in Problem 13.2)

Molecular formula: C_7H_8O

IR spectrum: see problem 13.2

NMR spectrum: δ 2.25, singlet, 3H

5.1, singlet, 1H

6.7, doublet, 2H

7.0, doublet, 2H

Solution: The compound is likely to be a benzene derivative, based on the index of unsaturation (see problem 13.2). The signal at δ 2.5 is likely to be a CH_3 attached to the benzene ring. That accounts for all 7 carbons. The oxygen should be phenolic, which accounts for the 1H singlet at δ 5.0. The two doublets accounting for a of total 4H in the aromatic region is characteristic of *p*-disubstitution. Hence, the compound is *p*-cresol, confirming the identification by IR spectroscopy (problem 13.2)

The total NMR spectrum is given in Fig. 13.9.

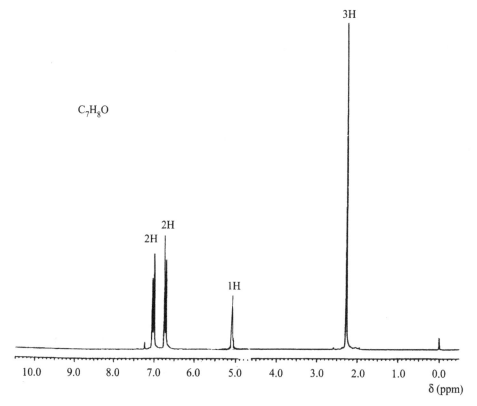

Fig. 13.9 NMR spectrum (Problem 13.4)

Some features of the NMR spectrum of *p*-cresol deserve comments. The position of the hydrogen of the phenolic OH, like those of alcohols and carboxylic acids and the H of amines, can vary and depends upon temperature and concentration. They often appear as broad peaks. The hydrogens of *p*-disubstituted benzenes (when the substituents are different) are of two kinds, and usually appear as a quartet of the type seen in Fig. 13.9. This kind of quartet is seen in the NMR spectra of other *p*-disubstituted benzenes also, as illustrated by the apectrum of *p*-methoxybenzaldehyde (Fig. 13.10). The protons on the benzene ring are of two types labelled H_a and H_b in the structure, which we shall refer to as the A hydrogen and the B hydrogen.

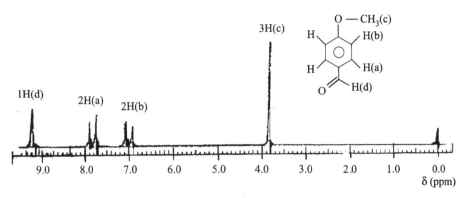

Fig. 13.10   NMR spectrum of *p*-methoxybenzaldehyde

The 'A' hydrogen splits the 'B' hydrogen into a doublet and the 'B' hydrogen splits 'A' into a doublet and together they appear as a pair of skewed doublets, referred to as an AB quartet. This is distinctive for *p*-disubstitution. Note the characteristic downfield location of the aldehyde proton H_d.

13.5.4   RING CURRENT EFFECT AND AROMATICITY

Vinylic hydrogens – those attached to the sp^2 hybridised carbon of an alkene – are deshielded and appear downfield in the region δ 4.5–6.5 (in contrast to the Hs in simple alkanes). This is partly due to the electronegativity of the sp^2 carbon. The chemical shift is also influenced by the anisotropic effect of the π-electrons of the double bond. The field induced by the π-electrons have directional properties (anisotropic). The protons attached to the double bond are in a strongly deshielding region and hence appear downfield. The anisotropy is more pronounced in the case of the carbonyl group. The very downfield location of the aldehyde proton (Fig. 13.10) is due to this effect. The Hs attached to the benzene ring appear even more downfield than in alkenes at δ 6.5–8.5. These protons are subjected to a strong deshielding effect, called the ring current effect (Fig. 13.11).

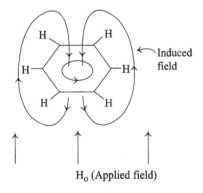

Fig. 13.11   Ring current effect

The circulating electrons of the π-system are responsible for an induced magnetic field which reinforces the applied magnetic field in the peripheral regions of the ring and causes deshielding of a greater magnitude than in an alkene. This is called the *ring current effect* and is taken as the definitive evidence for aromaticity.

13.5.5 HYDROGEN ATTACHED TO OTHER ELECTRONEGATIVE ATOMS/GROUPS

The following generalisations can be made on the effect of the environment on chemical shifts. Consider the system $H-C-Z$ where the C is sp^3 hybridised and the deshielding effect of various groups/atoms designated as Z is in the order,

$$Z = \text{alkyl} < \text{vinyl} < \text{carbonyl} < \text{aryl} < \text{amino} < \text{Cl, Br} < \text{O}$$

(in the increasing order of deshielding).

When the C of $H-C$ is sp^2 hybridised, the order is,

$$H-C{=}C < H-Ar < H-C{=}O \text{ (aldehyde) in the increasing order of deshielding.}$$

13.6 MASS SPECTROMETRY

Mass spectrometry is not based on electromagnetic radiation. A molecule is bombarded by electrons of high energy – typically 70 ev – under high vacuum. The energy gets transferred to the molecule and one or more electrons are removed in the process. When a neutral molecule loses an electron, the product is a cation radical. Since the initially formed cation radical itself packs a lot of energy, it gets further fragmented into a cation and a neutral radical, or in some other way (Eq. 13.1).

$$A - B + e^- \rightarrow \overset{+}{AB^\bullet} + 2e^-$$

$$\overset{+}{AB^\bullet} \rightarrow A^+ + B^\bullet \tag{13.1}$$

The positively charged species are accelerated in an electric field, separated by applying a magnetic field and analysed for their mass and their relative abundance. The mass spectrum thus obtained is a plot, in the form of a bar chart, of the relative abundance of the various positively charged fragments, including the cation of the parent molecule (the molecular ion), against their masses. Actually, it is their mass/charge (m/z) ratios that are obtained. Since most of the fragments are unipositive, m/z = m. The mass spectrum gives information about (i) the molecular mass, which is the mass of the molecular ion, $\overset{+}{M^\bullet}$, and (ii) the masses of the cationic fragments obtained by the breaking up of the molecule. In a high resolution instrument, the exact mass is obtained. The exact mass will not be a whole number because, the masses of particular isotopes of most atoms – other than ^{12}C – are not whole numbers. Thus, the accurate masses of the *isobaric molecules*, C_2H_2O and C_3H_6 are 42.015 and 42.048 respectively. This is because the accurate masses of the isotopes ^1H, ^{12}C, ^{14}N and ^{16}O are respectively, 1.008, 12.000, 14.007 and 15.999. The molecular ions C_2H_2O and C_3H_6 can be separated in a high resolution mass spectrometer and thus help to establish the identity. Identification of the major fragments help in structure determination, because fragmentation patterns of a given structure can largely be predicted. This is illustrated in the mass spectra of acetophenone (C_8H_8O) and *n*-propylbenzene (C_9H_{12}), both having a molecular mass of 120 (Fig. 13.12). In high resolution MS, they can be identified by the difference in their molecular mass.

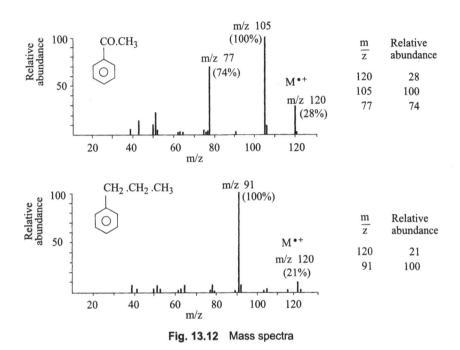

$\frac{m}{z}$	Relative abundance
120	28
105	100
77	74

$\frac{m}{z}$	Relative abundance
120	21
91	100

Fig. 13.12 Mass spectra

The data is also tabulated (Fig. 13.12) and this is resorted to when the full mass spectrum need not be reproduced. Small peaks next to large peaks are mostly due to isotopes, mainly deuterium. The most abundant mass in the mass spectrum of the former is at m/z = 105 and in the latter it is at m/z = 91. The most abundant peak is called the *base peak* in the spectrum. The fragmentation patterns of the two molecules are shown in Eq. 13.2.

$$CH_2-CH_2-CH_3 \xrightarrow[-e^-]{} \left[CH_2-CH_2-CH_3 \right]^{\overset{+}{\bullet}} \longrightarrow \left[\overset{+}{C}H_2 \cdots \longrightarrow \underset{\text{Tropylium cation}}{\oplus} \right] \quad (13.2)$$

$$\overset{\bullet}{CH_3-CH_2} +$$

$$m/z = 91$$

II

The major fragmentation of acetophenone (I), is the cleavage of the carbonyl–CH_3 bond, leaving behind a resonane stabilised $C_6H_5-CO^+$ (benzoyl cation) at m/z $= 105$. This is typical of ketones, where cleavage of the bond next to the carbonyl gives an acyl cation. Mono-substituted benzenes also give a phenyl cation of m/z 77. There are many other minor peaks due to various other less favoured fragmentations. In the case of *n*-propylbenzene (II), cleavage to give the resonance stabilised benzyl cation, m/z $= 91$, is the preferred fragmentation pattern. It has been shown that the benzyl cation rearranges into the more stable tropylium cation, which is fully aromatic. This fragmentation into m/z 91 dominates over all other fragmentations such that there are no other prominent peaks in the mass spectrum. Even m/z $= 77$ due to the phenyl cation is not seen.

KEY POINTS

- Absorption of electromagnetic radiation by molecules is quantised.
- Radiation in the UV–Vis region brings about electronic excitation.
- UV absorption in the 200–750 nm range corresponds to $n \to \pi^*$ or $\pi \to \pi^*$ excitations. These are most useful for the study of conjugated systems.
- Steric inhibition of conjugation affects UV absorption.
- Radiations in the IR region bring about vibrational excitations.
- IR spectroscopy gives information about functional groups.
- Characteristic IR frequencies of different groups are listed and these are useful for identifying functional groups.
- The proton ^1H, has two spin states, $+1/2$ and $-1/2$. In a magnetic field, they are aligned with or against the applied field. Nuclei in the lower energy state can absorb energy in the radio frequency region and get excited to the higher spin state.
- The energy difference between the spin states depends upon the applied magnetic field.
- The effective magnetic fields felt by different protons in a molecule differ because of the differences in their electronic environment. This gives rise to differences in chemical shift. This property is used in NMR spectroscopy to identify non-equivalent protons in a molecule and thus get structural information.

- The area under the peaks (signals) in the NMR spectrum is proportional to the number of protons responsible for the signal. This is used for counting the number of protons of different types.
- The signal of a particular proton is split by coupling with the spins of neighbouring protons. This spin-spin coupling gives rise to multiplicity, which is useful to calculate how many neighbouring protons there are.
- Chemical shifts can be understood in terms of electronegativity of the atoms/groups attached to the carbon bearing the proton under study.
- Ring current effect is seen in aromatic compounds. This is taken as criterion for aromaticity.
- In mass spectrometry, a molecule is bombarded by high energy electrons, when it gets ionised by electron loss and gets fragmented.
- The cationic species so obtained are analysed for their mass and abundance.
- The mass spectrum is a plot of the relative abundance against the mass of the fragments.
- High resolution mass spectrometry gives information about the exact mass of a species and helps in arriving at the molecular formula.
- Identification of the masses of the fragments and their relative abundances helps to arrive at the structure of the molecule.

EXERCISES

SECTION I

1. Assign the structure, (A) or (B), on the basis of the spectroscopic data provided.

    (a) UV max = 237 nm

    (A) 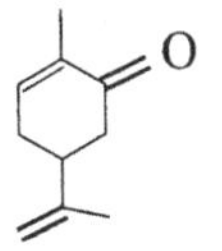 (B)

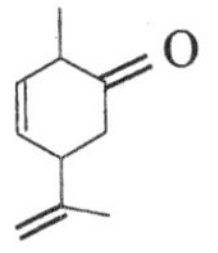

    (b) UV max = 232 nm

    (A) 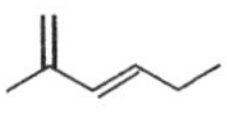 (B)

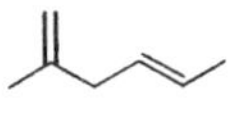

    (c) IR band 1700 cm^{-1}

    (A) 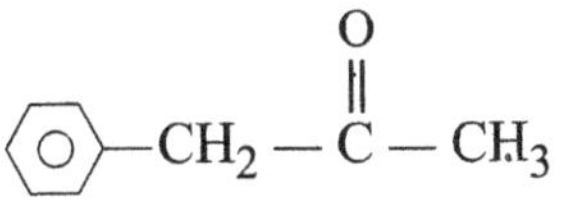 $\langle O \rangle$—CH_2—$\overset{\overset{\displaystyle O}{\|}}{C}$—$CH_3$ (B) 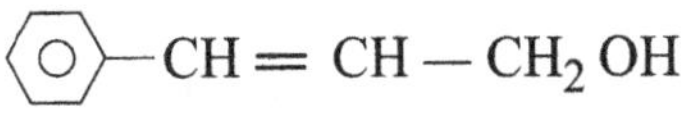 $\langle O \rangle$—CH = CH—$CH_2\,OH$

(d)   IR band $3300 \, cm^{-1}$

(A) OH (cyclohexane ring)   (B) OCH_3 (cyclohexane ring)

(e)   2 singlets of equal area in NMR

(A) $CH_3 - O - CO - CH_3$ (B) $CH_3 - CH_2 - O - CO - CH_3$

(f)   Three signals of unequal area in the NMR

(A) $CH_3 - CO - CH_3$ (B) $CH_3 - CH_2 - CHO$

(g)   Base peak at m/z 105 in MS

(A) $C_6H_5 - CH_2 - CH_2 - CH_3$ (B) $C_6H_5 - CH(CH_3)_2$

(h)   2 doublets in NMR of equal area in the region around δ 7, in addition to other signals.

(A) benzyl chloride   (B) *p*-chlorotoluene

(i)   Singlet at δ 3.6 in NMR in addition to other signals

(A) allyl alcohol   (B) methyl vinyl ether

2.   Identify the unknown compounds.

(a) $C_5H_{12}O$, two singlets in NMR at δ 1.0 and 3.3, in the area ratio 3 : 1.

(Solution: There are no double bonds or rings as seen from the index of unsaturation which is 0—see solved problem 13.1. From the area ratio we conclude that there are 9 hydrogens corresponding to the singlet at δ 1.0, and 3 hydrogens corresponding to δ 3.3 (total 12 hydrogens). Since the 9 high field hydrogens appear as one singlet, they are three equivalent sets of three as in *t*-butyl. From the chemical shift value of the other three hydrogens, we conclude that it is a CH_3 group attached to an electronegative atom, O in this case. Hence, the structure is *t*-butyl methyl ether.)

(b) C_5H_{12}, single signal, singlet, in the NMR at δ 0.9.

(c) $C_9H_{18}O$, single signal, singlet in the NMR at δ 1.1.

(d) $C_3H_6O_2$, two singlets of equal area in NMR.

(e) C_7H_8, two singlets in NMR at δ 2.3 and 6.8, in the area ratio, 3 : 5.

(f) C_7H_8O, two singlets in the NMR at δ 3.5 and 6.7, in the ratio, 3: 5.

(g) C_8H_8O, two singlets in the NMR at δ 2.5 and 7.2.

(h) C_3H_6O, broad IR band at $3300 \, cm^{-1}$, decolourises bromine water.

(i) C_3H_6O, no IR band in the region 1680–$1720 \, cm^{-1}$ and 3000–$3500 \, cm^{-1}$, decolourises bromine water.

(j) C_8H_{10}, base peak in MS at m/z 91.

SECTION II

1. Explain why the highlighted protons appear at very low field in NMR spectra (refer to table 13.2 for their chemical shifts).

$$RCOO-H, \quad R-CO-H(aldehyde)$$

2. Explain why the proton signals of OH and COOH disappear when D_2O is added to the solution.
3. The proton signal of TMS is assigned $\delta = 0$. Is it justified?
4. Refer to Problem 13.3. The structure, 2-butanone, was assigned on the basis of NMR even though the IR spectrum could not definitely rule out aldehyde structures. Sketch the expected NMR spectra of the isomeric aldehydes, C_4H_8O and show how they can be differentiated from one another and also from 2-butanone.
5. Show how NMR spectra can differentiate between methyl vinyl ether, $CH_2=CH-O-CH_3$, and allyl alcohol, $CH_2=CH-CH_2-OH$.

CHALLENGING QUESTIONS

Assign structures and interpret the spectra.

1. $C_6H_{10}O_4$
   IR band at 1720 cm^{-1}
   NMR: Singlet in the region δ 2.3 and 3.5 in the area ratio, 2 : 3

2. No signal in IR in the carbonyl region.
   NMR δ 1.1, doublet, 6H

          3.5, singlet, 3H

          3.8, septuplet (7 lines), 1H

   (Molecular formula is not given.)

14 Alcohols

14.1 INTRODUCTION: CLASSIFICATION

Alcohols are characterised by the hydroxyl functional group—OH. Depending on the type of carbon to which the OH is attached, alcohols are classified as primary, secondary and tertiary. Typical examples and their names along with boiling points, are given in Table 14.1.

14.2 PHYSICAL PROPERTIES

14.2.1 BOILING POINTS

The lower members are liquids at room temperature, having boiling points noticeably higher than other classes of compounds of similar molecular weights, but which do not contain the hydroxyl group. The boiling point of butanol (mol. wt. 74) is 118°C as compared to pentane (mol. wt. 72) b.p 36°C, diethyl ether (mol. wt. 74) b.p 35°C and 2-butanone (mol. wt. 72) b.p 80°C. The higher boiling points of alcohols are due to intermolecular hydrogen bonding. Higher temperatures are required to break the hydrogen bonds between molecules and release them into the gas phase. Higher alcohols containing more than 12 carbon atoms are waxy solids.

14.2.2 MISCIBILITY WITH WATER

The lower members up to C3 are fully miscible with water. Higher members are also partially miscible and the solubility in water decreases as the molecular weight increases. Hydrogen bonding between water and the alcohol molecule is the main reason for solubility.

Table 14. 1   Structures, names and boiling points of simple alcohols

Structure	Classification	IUPAC name	Name as alcohol	Name as carbinol	B.P(°C)
CH_3—OH	Primary	Methanol	Methyl alcohol	Carbinol	64.5
CH_3CH_2—OH	Primary	Ethanol	Ethyl alcohol	Methylcarbinol	78
$CH_3CH_2CH_2$—OH	Primary	1-Propanol	*n*-Propyl alcohol	Ethylcarbinol	97
$(CH_3)_2CH$—OH	Secondary	2-Propanol	Isopropyl alcohol	Dimethylcarbinol	82.5
$CH_3(CH_2)_2CH_2$—OH	Primary	1-Butanol	*n*-Butyl alcohol	*n*-Propylcarbinol	118
$(CH_3)_2CHCH_2$—OH	Primary	2-Methylpropanol	Isobutyl alcohol	Isopropylcarbinol	108
CH_3—CH(OH) —CH_2CH_3	Secondary	2-Butanol	*sec*-Butyl alcohol	Ethylmethylcarbinol	99.5
$(CH_3)_3C$—OH	Tertiary	2-Methyl-2-propanol	*tert*-Butyl alcohol	Trimethylcarbinol	83
CH_2=CH—CH_2—OH	Primary	2-Propene-1-ol	Allyl alcohol	Vinylcarbinol	97
$C_6H_5CH_2$—OH	Primary	Phenylmethanol	Benzyl alcohol	Phenylcarbinol	205
(cyclo) C_6H_{11}—OH	Secondary	Cyclohexanol	Cyclohexyl acohol	—	161.5

The fifth column in the table gives the carbinol nomenclature of alcohols. All alcohols are named as substituted carbinol, another name for methanol.

14.3 SOURCES: METHODS OF PREPARATION

14.3.1 METHANOL

Methanol is a component of the volatile fraction obtained when wood is heated for the production of charcoal. This process – destructive distillation of wood – yields a liquid called pyroligneous acid which contains methanol, acetone and acetic acid. Methanol is called wood alcohol since this used to be the major source from which methanol was obtained. Today, methanol is manufactured mainly by the catalytic hydrogenation of carbon monoxide. Carbon monoxide is obtained by the action of steam on heated coal which gives water gas, a mixture of CO and H_2 in the ratio 1:1. This is mixed with half its volume of hydrogen to give synthesis gas, which is CO + H_2 in the ratio, 1:2. Synthesis gas, over a mixture of zinc oxide–chromium oxide catalyst at 400°C, gives methanol. A mixture of CO and H_2, in the ratio 1:3, is obtained by the reaction of methane (natural gas) with steam at about 850°C over a nickel catalyst. This mixture can also be used for the synthesis of methanol. The excess hydrogen, after the methanol synthesis goes to the fertiliser industry for the manufacture of ammonia.

Methanol is toxic. It is an important industrial chemical used in the manufacture of formaldehyde (by dehydrogenation), methyl esters and ethers like *tert*-butyl methyl ether (methyl *tert*-butyl ether, MTBE) used as a gasolene additive to improve octane number, as an industrial solvent and for the denaturation of ethyl alcohol.

14.3.2 ETHANOL

Ethanol is one of the earliest known organic chemicals, and manufacture of ethanol by fermentation is one of the earliest chemical processes practiced by man. Ethanol is still manufactured by fermentation of various kinds of vegetable matter containing sugars and/or other carbohydrates (sugarcane, various grains). Ethanol is also produced as a petrochemical by the acid catalysed hydration of ethene, a product of petroleum cracking. As an industrial chemical, ethanol is used as a starting material in the preparation of other chemicals. One of the recent applications of ethanol is as a *biofuel*, blended with hydrocarbon fuels, for automobiles and combustion engines. If the source of ethanol is agricultural, that is, by fermentation of vegetable matter, this use is considered to be eco-friendly for two reasons. One, it is a renewable resource as against fossil fuels. Second, the carbon dioxide emitted by the burning of agricultural ethanol is the same carbon dioxide that was absorbed by the plant during photosynthesis. Hence, this does not increase the carbon dioxide level of the atmosphere and has little impact on greenhouse effect and global warming. Such a fuel is *carbon-neutral*. Another type of biofuel is *biodiesel* which is a methyl or ethyl ester of long chain fatty acids present in vegetable oils. Vegetable oils are esters of fatty acids with glycerol. They can be *trans-esterified* by monohydric alcohols like methanol or ethanol to give the methyl or ethyl esters. Representative components of such esters are ethyl stearate, $C_2H_5-O-CO-C_{17}H_{35}$, and ethyl oleate, $C_2H_5-OCO-C_{17}H_{33}$. The methyl or ethyl ester mixture obtained from vegetable oils can be used in diesel engines, either neat or blended with hydrocarbon diesel.

Ethanol used for beverages is sold at a higher price than industrial alcohol. In order to prevent the diversion of industrial ethanol for use as liquor, it is *denatured* by the addition of noxious or toxic chemicals like methanol. Denatured spirit is meant only for industrial and non-edible uses and is toxic.

Ethanol forms an azeotrope with water and when distilled, the azeotropic mixture containing about 95% ethanol and the rest water – called *rectified spirit* – is obtained. *Absolute alcohol* of 99.5% purity is obtained by further distillation with benzene, which removes most of the water as a ternary azeotrope. The last traces of water can be removed by distillation with quick lime or with magnesium metal.

Other alcohols are manufactured from alkenes (from petroleum). Many laboratory methods for the preparation of alcohols use alkenes as starting materials. Some of these reactions have been discussed in Chapter 4, as addition reactions of alkenes. These methods are summarised here. For detailed discussion, refer to the relevant sections in Chapter 4.

14.3.3 HYDRATION OF ALKENES

Addition of water to alkenes gives alcohols (Chapter 4, Section 4.4.1.7). The reaction follows Markownikoff's rule. Carbocations are intermediates and molecular rearrangements characteristic of carbocations are encountered. Hydration is industrially important. The petrochemical route for ethanol is the hydration of ethene.

14.3.4 HYDROBORATION FOLLOWED BY OXIDATION OF THE ALKYLBORANES

In hydroboration, a solution of diborane, B_2H_6, in an ether solvent like tetrahydrofuran (THF) or diethyleneglycol dimethyl ether (diglyme), reacts with the alkene (Chapter 4, Section 4.4.2). A molecule of monomeric BH_3 adds across the double bond to give an alkylborane. This addition is anti-Markownikoff, in the sense that the H gets attached to the more substituted carbon of the double bond. Oxidation by alkaline hydrogen peroxide converts the $C-B$ bond to a $C-OH$ bond (Chapter 4, Section 4.4.2.1). Effectively, the two steps of the reaction are together equivalent to the addition of water to the double bond in the anti-Markownikoff manner. Thus, the hydroboration route is complementary to the acid-catalysed hydration route (Eq. 14.1).

$$CH_2 = CH - CH_2 + H_2O \; \rightarrow \; CH_3 - \overset{\overset{\displaystyle OH}{|}}{CH} - CH_3$$

(Acid catalysed hydration)   (a)

$$CH_2 = CH - CH_2 + (H_2O) \; \rightarrow \; HO - CH_2 - CH_2 - CH_3$$

(Indirect hydration via hydroboration followed by oxidation)   (b)   (14.1)

The regiochemistry of hydroboration is different from that of acid-catalysed hydration. The stereochemistry of the former is also different. Hydroboration is a syn-addition and the alcohol corresponds to the syn-addition of H and OH across the double bond.

14.3.5 OXYMERCURATION OF ALKENES

Another indirect method of hydration of alkenes is via oxymercuration (Eq. 14.2).

$$
CH_2 = CH - R + Hg(O - CO - CH_3)_2 \longrightarrow
\begin{array}{c}
\overset{\displaystyle OH}{\overset{\displaystyle |}{CH_2 - CH - CH_3}} \\
| \\
Hg - OCOCH_3
\end{array}
\qquad (a)
$$

$$
CH_2 = CH - R + \overset{+}{Hg} - OCOCH_3 \longrightarrow
\begin{array}{c}
CH_2 - \overset{+}{CH} - CH_3 \\
\diagdown \quad \diagup \\
Hg - OCOCH_3
\end{array}
\qquad (14.2)
$$

$$
\downarrow H_2O, -H^+
$$

$$
\begin{array}{c}
\overset{\displaystyle OH}{\overset{\displaystyle |}{CH_2 - CH - CH_3}} \\
| \\
Hg - OCOCH_3
\end{array}
\qquad (b)
$$

Mercuric acetate in aqueous THF, adds to alkenes in an electrophilic reaction (Eq. 14.2a), where CH_3COOHg^+ formed by the ionisation of $(CH_3COO)_2Hg$ is the electrophile. The carbocation is quenched by water to give the adduct (Eq. 14.2b). The addition is according to Markownikoff's rule with the OH adding on to the secondary carbon. This reaction is called oxymercuration. In a subsequent reaction (Eq. 14.3), the oxymercurated adduct is demercurated by treatment with the reducing agent $NaBH_4$ (sodium borohydride).

$$
\begin{array}{c}
\overset{\displaystyle OH}{\overset{\displaystyle |}{CH_2 - CH - R}} \\
| \\
Hg - OCOCH_3
\end{array}
+ NaBH_4 \rightarrow
\begin{array}{c}
\overset{\displaystyle OH}{\overset{\displaystyle |}{CH_3 - CH - R}}
\end{array}
+ Hg + CH_3COOH
\qquad (14.3)
$$

The product obtained is the same as the one obtained by direct acid catalysed hydration. However, there is a significant difference between the two reactions. While the former is prone to rearrangement, the oxymercuration–demercuration route gives the alkene without rearrangement (Eq. 14.4).

$$
CH_3-\underset{\underset{CH_3}{|}}{\overset{\overset{CH_3}{|}}{C}}-\overset{+}{CH}-CH_2 \longrightarrow CH_3-\overset{+}{\underset{\underset{CH_3}{|}}{C}}-\underset{\underset{CH_3}{|}}{CH}-CH_3
$$

$$
CH_3-\underset{\underset{CH_3}{|}}{\overset{\overset{OH}{|}}{C}}-\underset{\underset{CH_3}{|}}{CH}-CH_3
$$

$$
CH_3-\underset{\underset{CH_3}{|}}{\overset{\overset{CH_3}{|}}{C}}-CH=CH_2 \qquad H_3O^+
$$

(i) Hg(OCOCH$_3$)$_2$
(ii) NaBH$_4$

$$
CH_3-\underset{\underset{CH_3}{|}}{\overset{\overset{CH_3}{|}}{C}}-\underset{\underset{OH}{|}}{CH}-CH_3
$$

$$(14.4)$$

The absence of rearrangement is because the intermediate in Eq. 14.2b is not a true carbocation but a 'bridged' one, as shown.

14.3.6 GRIGNARD SYNTHESIS OF ALCOHOLS

An important laboratory method for the synthesis of alcohols is by the use of Grignard reagents. We came across Grignard reagents in connection with the preparation of alkanes (Chapter 3, Section 3.4.1). The Grignard synthesis of alcohols is very versatile. The starting materials are (i) alkyl halide or aryl halide and magnesium to prepare the Grignard reagent, and (ii) a carbonyl compound—aldehyde, ketone or ester. Basically, the reaction involves the addition of the Grignard reagent, R–Mg–X, to the carbonyl group of an aldehyde, ketone or ester, followed by hydrolysis of the resultant complex (Eq. 14.5). R and R$'$ can be H, alkyl or aryl. R$''$MgX is obtained by the reaction of R$''$–X with clean magnesium metal in dry diethyl ether. R$''$ can be alkyl or aryl. When it is alkyl, X can be Cl, Br or I. When it is aryl, X can be Br or I. Chlorobenzene and other chloroarenes do not react under normal conditions to give Grignard reagents. Anhydrous ether (diethyl ether) is the solvent of choice. Water or other protic solvents like alcohols should not be present even in traces, because (i) these will foul up the surface of magnesium metal and prevent its reaction with the alkyl halide and (ii) Grignard reagents are destroyed by protic reagents, by protonation of the alkyl group to give the corresponding alkane. Organomagnesium halides are soluble in ether due to complexation of the metal with the oxygen of the ether. The initial product of the addition of RMgX to the aldehyde or ketone is the Mg salt of the alcohol, $-$OMgX, which is hydrolysed by adding water to release the free alcohol and Mg(OH)X. To facilitate this hydrolysis, aqueous acids like HCl or H_2SO_4 can be used at this stage. By appropriate variations of R, R' and R'', any primary, secondary or tertiary alcohol can be prepared as can be seen from the examples (Eq. 14.6). Notice that formaldehyde (methanal) is the only aldehyde – when used as carbonyl compound in the Grignard synthesis – that gives primary alcohols. All other aldehydes give secondary alcohols and ketones give tertiary alcohols. By appropriately choosing R and R' of the aldehyde or ketone, and R''–X, any alcohol can be synthesised.

$$
\underset{\displaystyle \overset{\displaystyle O}{\parallel}}{R-C-R'} + R''-Mg-X \rightarrow R-\underset{\underset{R''}{|}}{\overset{\overset{OMgX}{|}}{C}}-R' \xrightarrow{H_2O} R-\underset{\underset{R''}{|}}{\overset{\overset{OH}{|}}{C}}-R' + Mg(OH)X
$$

$$\text{where} \quad R, R' = H, \text{alkyl or aryl}; \quad R'' = \text{alkyl or aryl} \qquad (14.5)$$

$$
\underset{\displaystyle \overset{\displaystyle O}{\parallel}}{H-C-H} + CH_3-CH_2-CH_2-MgI \longrightarrow CH_3-CH_2-CH_2-\underset{\underset{H}{|}}{\overset{\overset{H}{|}}{C}}-OMgI
$$

$$
\xrightarrow{H_2O} CH_3-CH_2-CH_2-CH_2-OH \qquad (a)
$$

$$
\underset{\displaystyle \overset{\displaystyle O}{\parallel}}{CH_3-C-H} + \langle\bigcirc\rangle-MgBr \longrightarrow CH_3-\overset{\overset{O-MgBr}{|}}{\underset{|}{C}}-H \xrightarrow{H_2O} CH_3-\overset{\overset{OH}{|}}{\underset{|}{C}}-H \qquad (b)
$$

$$
\underset{\displaystyle \overset{\displaystyle O}{\parallel}}{CH_3-C-CH_3} + C_2H_5-MgBr \longrightarrow CH_3-\underset{\underset{C_2H_5}{|}}{\overset{\overset{O-MgBr}{|}}{C}}-CH_3 \xrightarrow{H_2O} CH_3-\underset{\underset{C_2H_5}{|}}{\overset{\overset{OH}{|}}{C}}-CH_3 \qquad (c)
$$

$$
\underset{\displaystyle \overset{\displaystyle O}{\parallel}}{\langle\bigcirc\rangle-C-H} + CH_3-MgI \longrightarrow \langle\bigcirc\rangle-\overset{\overset{O-MgI}{|}}{CH}-CH_3 \xrightarrow{H_2O} \langle\bigcirc\rangle-\overset{\overset{OH}{|}}{CH}-CH_3 \qquad (d)
$$

$$\text{cyclohexanone} = O + C_4H_9 - MgCl \longrightarrow \underset{C_4H_9}{\overset{O-MgCl}{\diagdown}} \xrightarrow{H_2O} \underset{C_4H_9}{\overset{OH}{\diagdown}} \qquad (e)$$

$$(14.6)$$

Esters can also be used as the carbonyl compound in Grignard synthesis (Eq. 14.7) to yield tertiary alcohols.

$$\underset{\displaystyle R - \overset{\textstyle O}{\overset{\|}{C}} - OR' + 2R'' - MgX}{} \longrightarrow \underset{\displaystyle R - \overset{OMgX}{\underset{R''}{\overset{|}{C}}} - R'' + R'O - Mg - X}{}$$

$$\xrightarrow{H_2O} \underset{\displaystyle R - \overset{OH}{\underset{R''}{\overset{|}{C}}} - R'' + R'OH} \qquad (14.7)$$

Note that the R' group of the ester does not appear in the final product and has been eliminated as R'OMgX. Two equivalents of the Grignard reagent are used up. In the product, the tertiary alcohol, two of the alkyl groups attached to the tertiary carbon are the same (R'') and come from the Grignard reagent, that is, from the alkyl halide R''X. In that sense, the tertiary alcohols obtained when esters are used, are different from those obtained when ketones are used. In the case of ketones, all three alkyl groups attached to the tertiary carbon can be different.

Free acids cannot be used as the carbonyl component of the Grignard synthesis since the acid will destroy one equivalent of the Grignard reagent, converting it to the alkane (Eq. 14.8). Also see Chapter 3, Section 3.4.1.

$$R - COOH + R''MgX \rightarrow R - COOMgX + R'' - H \qquad (14.8)$$

RCOO—MgX formed in this reaction is a salt and does not react even if more Grignard reagent is added.

Addition of the Grignard reagent to the carbonyl group can be understood in terms of the polarity of the two reagents. The carbonyl group has an inherent polarity, with the carbon being positive and the oxygen being negative as seen in the resonance structures (Eq. 14.9).

$$\underset{\diagup}{\diagdown} C = O \quad \leftrightarrow \quad \underset{\diagup}{\diagdown} \overset{+}{C} - \bar{O} \qquad \underset{\diagup}{\diagdown} \overset{\delta+}{C} \overset{\delta-}{=} O \qquad (14.9)$$

The polarity of the metal–carbon bond in the Grignard reagent (as in all organometallic molecules) is with the metal being positive and the carbon being negative (Eq. 14.10).

$$R-Mg-X \longleftrightarrow R^- \overset{+}{Mg}-X \qquad \overset{\delta-}{R}\cdots\overset{\delta+}{Mg}-X \tag{14.10}$$

The R group in the Grignard reagent is not a true carbanion but can react as a carbanion, and as a nucleophile. It reacts with the carbonyl carbon as follows (Eq. 14.11).

$$R-\overset{\overset{\displaystyle O}{\|}}{C}-R' \qquad R''-MgX \longrightarrow R-\overset{\overset{\displaystyle O-MgX}{|}}{\underset{\underset{\displaystyle R''}{|}}{C}}-R' \tag{14.11}$$

The reaction of the Grignard reagent with the ester involves the following steps (Eq. 14.12). The intermediate ketone cannot be isolated. It reacts with the Grignard reagent as soon as it is formed in the medium.

$$R-\overset{\overset{\displaystyle O}{\|}}{C}-OR' + R''-MgX \longrightarrow R-\overset{\overset{\displaystyle O-MgX}{|}}{\underset{\underset{\displaystyle R''}{|}}{C}}-OR' \longrightarrow R-\overset{\overset{\displaystyle O}{\|}}{C}-R'' + Mg\overset{X}{\underset{OR'}{\big\langle}}$$

$$\downarrow R''-MgX$$

$$R-\overset{\overset{\displaystyle O-MgX}{|}}{\underset{\underset{\displaystyle R''}{|}}{C}}-R'' \tag{14.12}$$

An extension of the Grignard synthesis is the use of ethylene oxide instead of formaldehyde for the preparation of primary alcohols (Eq. 14.13).

$$R-MgX + \overset{\displaystyle O}{CH_2-CH_2} \longrightarrow R-\overset{\overset{\displaystyle OMgX}{|}}{CH_2}-CH_2$$

$$\xrightarrow{H_2O} R-CH_2-CH_2-OH + Mg\overset{X}{\underset{OH}{\big\langle}} \tag{14.13}$$

Ethylene oxide gives a primary alcohol with one CH_2 more than what is obtained if formaldehyde is used. Thus, phenylmagnesium bromide and formaldehyde will give benzyl alcohol; the same Grignard reagent with ethylene oxide will give 2-phenylethanol (Eq. 14.14). (Note on nomenclature: 1-phenylethanol and 2-phenylethanol are also called α-phenylethanol and β-phenylethanol, respectively.)

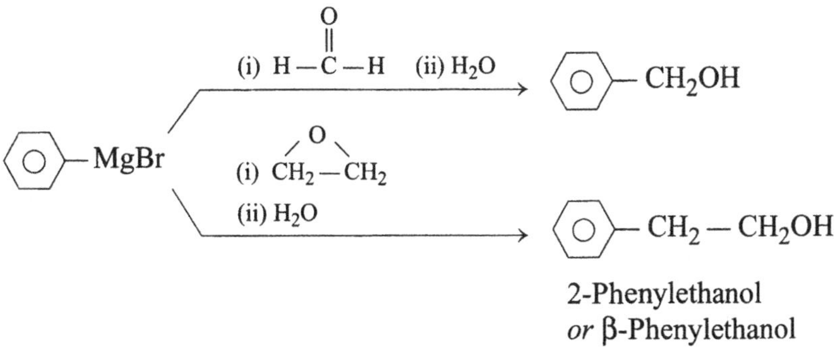

$$
\langle O \rangle\!-\!MgBr
\begin{cases}
\xrightarrow[\text{(i) } H-\overset{\overset{O}{\|}}{C}-H \quad \text{(ii) } H_2O]{} \quad \langle O \rangle\!-\!CH_2OH \\[2em]
\xrightarrow[\begin{array}{l}\text{(i) } CH_2\!-\!CH_2 \text{ (epoxide)}\\ \text{(ii) } H_2O\end{array}]{} \quad \langle O \rangle\!-\!CH_2-CH_2OH
\end{cases}
$$

(14.14)

2-Phenylethanol
or β-Phenylethanol

Grignard syntheses of alcohols are summarised in Fig. 14.1

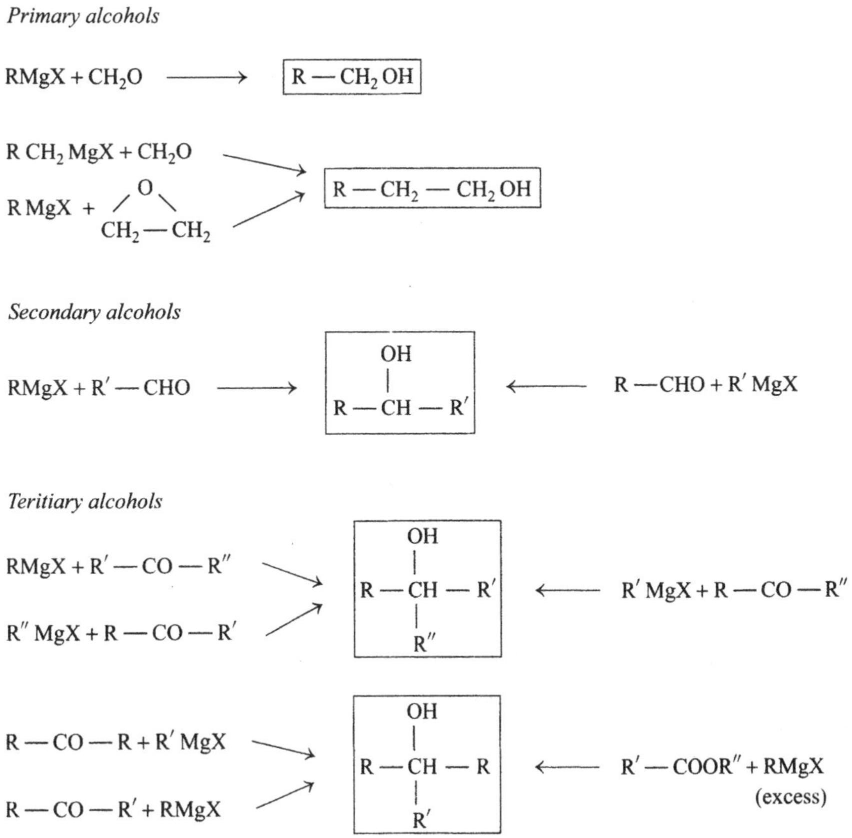

Primary alcohols

$$RMgX + CH_2O \longrightarrow \boxed{R-CH_2\,OH}$$

$$R\,CH_2\,MgX + CH_2O$$

$$R\,MgX + \;CH_2-CH_2\;(\text{epoxide}) \longrightarrow \boxed{R-CH_2-CH_2\,OH}$$

Secondary alcohols

$$RMgX + R'-CHO \longrightarrow \boxed{\begin{array}{c} OH \\ | \\ R-CH-R' \end{array}} \longleftarrow R-CHO + R'\,MgX$$

Teritiary alcohols

$$RMgX + R'-CO-R''$$

$$R''\,MgX + R-CO-R' \longrightarrow \boxed{\begin{array}{c} OH \\ | \\ R-CH-R' \\ | \\ R'' \end{array}} \longleftarrow R'\,MgX + R-CO-R''$$

$$R-CO-R + R'\,MgX$$

$$R-CO-R' + RMgX \longrightarrow \boxed{\begin{array}{c} OH \\ | \\ R-CH-R \\ | \\ R' \end{array}} \longleftarrow \begin{array}{c} R'-COOR'' + RMgX \\ (\text{excess}) \end{array}$$

Fig. 14.1 Grignard syntheses of alcohols

14.3.7 SYNTHESIS BY THE REDUCTION OF CARBONYL COMPOUNDS

Aldehydes and ketones can be reduced to primary and secondary alcohols respectively, by a variety of reducing agents. The complex metal hydrides – sodium borohydride ($NaBH_4$) and lithium aluminium hydride (LiAlH$_4$, abbreviated as LAH) – are the most important of these (Eq. 14.15). Reduction with sodium borohydride can be carried out in methanol or ethanol solution. Lithium aluminium hydride is decomposed by hydroxylic solvents and therefore, LAH reductions are done in anhydrous diethyl ether. Esters and free carboxylic acids can be reduced to alcohols by LAH but not by sodium borohydride. Stoichiometrically, one molecule of LAH or sodium borohydride is equivalent to four molecules of carbonyl compound. In the reduction of the free COOH group, one extra equivalent of LAH should be used since the acid consumes one equivalent of the hydride.

$$O_2N-\langle\bigcirc\rangle-CHO \xrightarrow{\text{(i) NaBH}_4,\ \text{(ii) H}_3O^+} O_2N-\langle\bigcirc\rangle-CH_2OH \qquad (a)$$

$$\langle\bigcirc\rangle{=}O \xrightarrow{\text{(i) LiAlH}_4,\ \text{(ii) H}_3O^+} \langle\bigcirc\rangle-OH \qquad (b)$$

$$CH_3-(CH_2)_4-CH_2-COOC_2H_5$$

$$\xrightarrow[\text{(ii) H}_3O^+]{\text{(i) LiAlH}_4,} CH_3-(CH_2)_4-CH_2-CH_2OH + C_2H_5OH \qquad (c)$$

$$(14.15)$$

Sodium borohydride is a mild reducing agent. The carbonyl group of aldehydes and ketones can be reduced by this reagent even in the presence of other reducible functional groups like the nitro group and the ester group. LAH is a more powerful reducing agent and can reduce the above mentioned functional groups as well. In the equations 14.15a–14.15c, the entries over the arrow refer to, (i) reduction by the reagent and (ii) working up using aqueous acid. The latter step is necessary because the immediate product of reduction is the salt of the alcohol from which the free alcohol is released by acidification.

Esters can be reduced to the same products as obtained by LAH reduction, by employing the Bouveault–Blanc reduction (treating with sodium metal and ethanol) or by catalytic hydrogenation over copper chromite catalyst at high temperature under pressure. Aldehydes and ketones can also be reduced by these methods.

14.4 REACTIONS

Alcohols are *protic* molecules. The hydrogen of the hydroxyl group can be replaced by metals. Alkali metals react to give alkoxides which are strong bases. Reaction of alcohols with sodium metal with the liberation of hydrogen can be used as a test to differentiate them from other neutral molecules like ethers or hydrocarbons.

14.4.1 REPLACEMENT OF THE HYDROXYL GROUP

We are already familiar with the difficulty in replacing the hydroxy group in nucleophilic substitution because of its poor leaving ability. Some of the indirect methods adopted have been mentioned in Chapter 11, Section 11.4.1. These may be summarised as follows.

Replacement of OH by halogen can be carried out by treatment with hydrohalic acid in which case the protonated form of the alcohol is the substrate and H_2O is the leaving group. Tertiary alcohols are readily converted to the halides by treatment with the corresponding hydrohalic acid. Secondary alcohols react less readily. Primary alcohols do not react under these conditions. The reaction is faster if a Lewis acid, zinc chloride, is used along with HCl. This mixture of zinc chloride and HCl is called Lucas reagent and is used to differentiate between primary, secondary and tertiary alcohols (*Lucas test*). The alcohol being tested, is shaken with Lucas reagent. If it is a tertiary alcohol, a turbidity develops immediately, due to the formation of the insoluble alkyl chloride. If it is secondary, the turbidity still appears, but over a period of several minutes. Primary alcohols may also develop the turbidity, but only after prolonged reaction time. The test relies on the fact that alcohols, in small concentration, dissolve in the reagent initially due to the formation of oxonium ions; chlorides do not. The role of the Lewis acid is the same as that of protons, that is, to activate the OH.

Alcohols are converted to chlorides by treatment with $SOCl_2$, in a $S_N i$ reaction (Chapter 11, Section 11.3). The OH is converted to chlorosulphite ($-OSOCl$), which is a more effective leaving group and gets replaced by the chloride ion present in the medium. Other reagents for the conversion of OH to halogen are phosphorous halides such as PCl_3 and PCl_5. These reagents convert the alcohol to chlorine-containing phosphite or phosphate esters which undergo replacement by the chloride (Eq. 14.16). These reagents yield the chloride with inversion of configuration. Red phosphorus and bromine, or red phosphorus and iodine can be used instead of PBr_3 and PI_3 for preparing the alkyl bromide and alkyl iodide respectively.

$$R-OH + PCl_3 \longrightarrow R-O-P\begin{smallmatrix}Cl \\ \\ Cl\end{smallmatrix} + HCl$$

(14.16)

$$Cl^- + \quad R-O-P\begin{smallmatrix}Cl \\ \\ Cl\end{smallmatrix} \longrightarrow Cl-R + POCl_2^-$$

Reaction with a sulphonyl chloride, usually in pyridine solution, gives the sulphonate. The sulphonate being a good leaving group, can be replaced by other nucleophiles with overall inversion of configuration (Eqs. 11.17 and 11.18, Chapter 11).

Dehydration of alcohols as a method of preparing alkenes and its mechanism, have been disucssed in Chapter 4 (Section 4.2.1.2) and Chapter 12 (Section 12.4). Dehydration is an acid-catalysed reaction. Since carbocations are involved, dehydration is often accompanied by rearrangement. Rearrangement can be avoided by indirect dehydration by ester pyrolysis, typically by Chugaev reaction (Chapter 4, Section 4.2.1.5).

14.4.2 OXIDATION OF ALCOHOLS

Primary alcohols can be oxidised to aldehydes and secondary alcohols to ketones, by several oxidising agents. A typical reagent is chromic acid which readily oxidies secondary alcohols to ketones. Primary alcohols are first oxidised to the aldehydes which are further oxidised to the carboxylic acids. A selective reagent, chromic oxide–pyridine complex, is useful to prepare the aldehyde from the primary alcohol (Eq. 14.17).

$$\text{[cyclohexanol]}-OH \xrightarrow{H_2CrO_4} \text{[cyclohexanone]}=O$$

$$CH_3-(CH_2)_4-CH_2OH \xrightarrow{H_2CrO_4} CH_3-(CH_2)_4-COOH$$

$$\searrow CrO_3, \text{ pyridine}$$

$$CH_3-(CH_2)_4-CHO$$

(14.17)

Tertiary alcohols do not get oxidised under the mild conditions which are suitable for primary and secondary alcohols.

14.4.3 ETHER FORMATION

Simple alcohols like ethanol undergo intermolecular dehydration to the symmetrical ethers upon heating with sulphuric acid. This reaction occurs at lower temperatures than that required for alkene formation. Its mechanism as an S_N2 reaction has been discussed in Chapter 11 (Eq. 11.16). This technique does not work for higher ethers or for mixed ethers ($R-O-R'$, where R and R' are different). *Williamson synthesis* is the method of choice in such cases (Eq. 14.18).

$$CH_3-CH-CH_2-CH_3 \longrightarrow CH_3-CH-CH_2-CH_3 + NaI$$

$$\underset{ONa}{|} \qquad\qquad \underset{O-CH_3}{|}$$

$$\qquad +CH_3-I$$

(14.18)

The sodium salt of the alcohol (sodium alkoxide) is required for Williamson synthesis and is prepared by reacting sodium metal with the alcohol. Its role here is that of a nucleophile.

14.4.4 ESTERIFICATION

Alcohols react with carboxylic acids in an acid-catalysed reaction to form esters (Eq. 14.19a). The oxygen of the hydroxyl group of the alcohol adds on as a nucleophile to the carbonyl group of the protonated acid (Eq. 14.19c).

$$R-COOH + C_2H_5-OH \xrightleftharpoons{H^+} R-\overset{\displaystyle O}{\overset{\|}{C}}-OC_2H_5 + H_2O \qquad \text{(a)}$$

Acetic anhydride

$$R-\overset{\displaystyle O}{\overset{\|}{C}}-OH + H^+ \rightleftharpoons R-\overset{\displaystyle \overset{+}{O}-H}{\overset{\|}{C}}-OH$$

$$\rightleftharpoons R-\overset{\displaystyle O}{\overset{\|}{C}}-OR' + H_2O + H^+ \qquad \text{(c)}$$

$$\longrightarrow R-\overset{\displaystyle H}{\overset{|}{\underset{+}{O}}}-\overset{\displaystyle O}{\overset{\|}{C}}-CH_3 + CH_3COO^- \longrightarrow R-O-\overset{\displaystyle O}{\overset{\|}{C}}-CH_3 + CH_3COOH \qquad \text{(d)}$$

$$(14.19)$$

Acid catalysed esterification (Eqs. 14.19a and 14.19c) is an equilibrium reaction. In order to drive the reaction in the forward direction, either an excess of one of the reagents – acid or alcohol – can be used, or the product – water – can be removed as and when formed. Esterification can also be carried out by heating with an acid anhydride. Thus, cyclohexanol upon heating with acetic anhydride gives cyclohexyl acetate (Eq. 14.19b). Acid anhydrides are more reactive than free carboxylic acids. They react according to the mechanism given in Eq. 14.19d.

14.5   UNSATURATED ALCOHOLS: ALLYL ALCOHOL

Vinyl alcohol $CH_2 = CH - OH$, an *enol*, does not exist. It is a *tautomer* of acetaldehyde. (See *keto–enol tautomerism* in Chapter 16.) Allyl alcohol, 2-propen-1-ol is a stable molecule. It can be prepared by the hydrolysis of allyl halides. Allyl alcohol reacts with HBr readily to form allyl bromide. Its reactivity is similar to that of tertiary alcohols and is due to the stability of the allyl cation by delocalisation. Allyl halides also have the same high reactivity as benzyl and tertiary halides. They readily undergo both S_N1 and S_N2 reactions. The higher homologue of allyl alcohol is crotyl alcohol, $CH_3 - CH = CH - CH_2 - OH$, 2-buten-1-ol. This alcohol reacts with HBr to yield a mixture of 3-bromo-1-butene and 1-bromo-2-butene; the former being the major product.

$$CH_3 - CH = CH - CH_2 - OH + HBr \longrightarrow \left[\begin{array}{c} CH_3 - CH = CH - \overset{+}{CH_2} \\ \updownarrow \\ CH_3 - \overset{+}{CH} - CH = CH_2 \end{array} \right] + H_2O + Br^-$$

$$CH_3 - CH = CH - CH_2 - Br + CH_3 - \overset{\overset{\displaystyle Br}{|}}{CH} - CH = CH_2 \tag{14.20}$$

The positive charge on the cation is delocalised with greater charge density at the secondary position. The cation is quenched by the bromide ion to give more of the rearranged product. This kind of rearrangement is called *allylic rearrangement*. Substitution reactions of crotyl halides also show this rearrangement.

14.6   POLYHYDRIC ALCOHOLS

Alcohols containing only one hydroxyl group are called monohydric alcohols. Those containing two hydroxyl group are called glycols. Molecules containing more than one hydroxyl group are called polyhydric alcohols or *polyols*. Some examples of polyols are given in Table 14.2.

Ethylene glycol and propylene glycol are 1,2- or vicinal glycols (*vic*-glycols). If the two hydroxyl groups are on the same carbon atom, it is called a geminal or *gem*-glycol. *gem*-Glycols are really the products of addition of water to a carbonyl group (Eq. 14.21).

$$R - \overset{\overset{\displaystyle O}{\|}}{C} - R' + H_2O \xrightleftharpoons{\longrightarrow} R - \overset{\overset{\displaystyle OH}{|}}{\underset{\underset{\displaystyle OH}{|}}{C}} - R' \tag{14.21}$$

This equilibrium is very much to the left for simple aldehydes and ketones and therefore, *gem*-glycols cannot be isolated.

Table 14. 2 Polyhydric alcohols

Name	Structure	Boiling point
Ethylene glycol (1,2-ethanediol)	$HO-CH_2-CH_2-OH$	197
Propylene glycol (1,2-propanediol)	$HO-CH_2-CH(OH)-CH_3$	187
Glycerol (glycerine) (1,2,3-propanetriol)	$HO-CH_2-CH(OH)-CH_2-OH$	290
Pentaerythitol	$C(CH_2OH)_4$	260 (melting point)
Sorbitol	$HOCH_2-(CHOH)_4-CH_2OH$	

14.6.1 PREPARATION OF VIC-GLYCOLS

Formation of vic-glycols by the permanganate oxidation of alkenes was discussed in Chapter 4 (Section 4.4.5). Another reagent which oxidises an alkene to a glycol is osmium tetroxide.

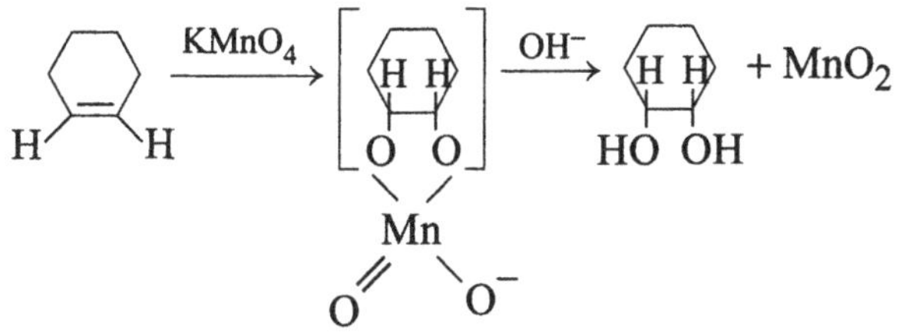

(a)

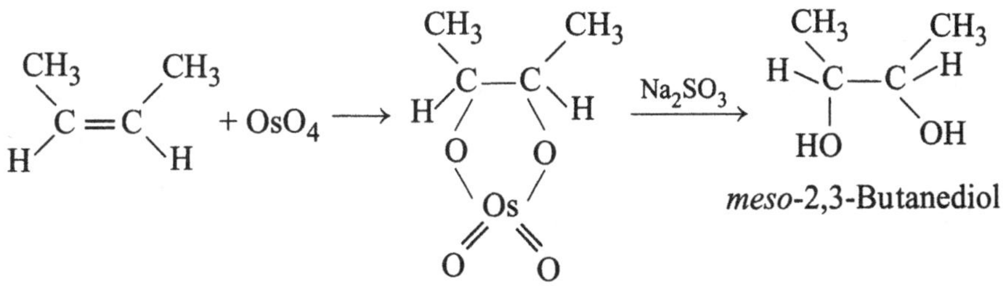

(b)

meso-2,3-Butanediol

(14.22)

Both reactions proceed through cyclic intermediates and the final products correspond to *cis*-addition of two OH groups (*cis*-dihydroxylation). Cyclohexene upon oxidation by either reagent gives *cis*-1,2-cyclohexanediol. *cis*-2-Butene gives meso-2,3-butanediol. The latter is one of the two possible *diasterostereomers* of 2,3-butanediol (Chapter 21).

Glycols can also be prepared indirectly by converting the alkene to the epoxide by oxidation with a peracid and then opening the epoxide ring by hydrolysis (Eq. 14.23).

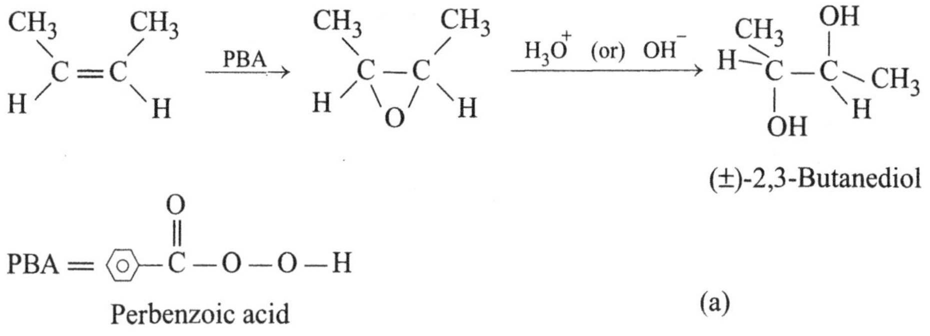

(±)-2,3-Butanediol

(a)

Acid hydrolysis:

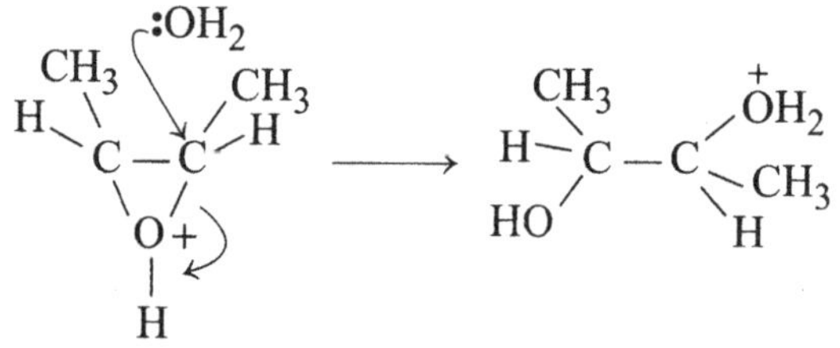

(b)

Alkaline hydrolysis:

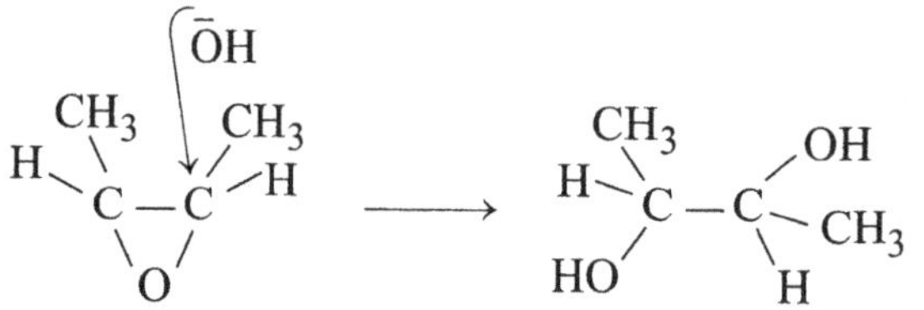

(c)

(14.23)

The reagent used for *epoxidation* is perbenzoic acid. Other peracids can also be used. The epoxide is a cyclic ether, which is a strained three-membered ring. It can be opened to the glycol by a nucleophilic displacement reaction using a base (hydroxide ion, Eq. 14.23c) or under acidic conditions (Eq. 14.23b). Epoxidation involves *cis*-addition of oxygen across the double bond, so that *cis*-2-butene gives *cis*-2-butene oxide. The ring opening is a neucleophilic displacement of the S_N2 category with the nuclcophile, OH^- or H_2O, coming from the direction opposite to the side occupied by the epoxide ring. The final outcome is *trans*-addition of two hydroxyl groups across the double bond (*trans*-dihydroxylation). The product, (±)-2,3-butanediol is the diastereomer other than the one obtained by *cis*-dihydroxylation.

14.6.1.1 Ethylene Glycol and Ethylene Oxide

Ethylene oxide or epoxyethane or ethane epoxide is manufactured by the catalytic oxidation of ethylene by oxygen (Eq. 14.24a). The IUPAC name for the three-membered ring containing one oxygen is *oxirane*. An earlier method of preparation of ethylene oxide is conversion of ethylene to the chlorohydrin by chlorine and water (Chapter 4, Section 4.4.1.2) followed by treatment with alkali (Eq. 14.24b, 14.24c).

$$CH_2 = CH_2 + O_2 \xrightarrow{\text{Ag catalyst}} H_2C\underset{\diagdown O \diagup}{-}CH_2 \qquad \text{(a)}$$

Ethylene oxide

$$CH_2 = CH_2 + Cl_2 + H_2O \longrightarrow \overset{\overset{\displaystyle H-O}{|}}{H_2C}-\overset{\overset{\displaystyle}{|}}{\underset{Cl}{CH_2}} \qquad \text{(b)}$$

$$(14.24)$$

$$HO \overset{\frown}{} H-O \qquad H_2O \quad O \\ \overset{|}{H_2C}-CH_2 \longrightarrow H_2C-CH_2 \qquad \text{(c)} \\ \underset{Cl}{} \qquad\qquad Cl^-$$

Polymerisation of ethylene oxide gives polyethylene glycols, $OH-CH_2-CH_2-(O-CH_2-CH_2)_n-O-H$. Ethylene glycol is manufactured by the hydrolysis of either ethylene oxide or ethylene chlorohydrin. Ethylene glycol is used in the manufacture of polyester resin and also as antifreeze.

14.6.2 OXIDATION OF GLYCOLS

Glycols can be selectively oxidised to cleave the molecule at the $C-C$ bond between the OH groups. This can be done either using periodic acid, HIO_4, or lead tetraacetate, $Pb(OCOCH_3)_4$ to give carbonyl compounds (Eq. 14.25).

$$\underset{\underset{HO}{|}}{R}-\underset{\underset{OH}{|}}{CH}-CH-R' \xrightarrow[\text{(or) } Pb(OCOCH_3)_4]{HIO_4} R-CHO + R'CHO \qquad \text{(a)}$$

$$\text{(b)}$$

$$\underset{HO\quad OH}{\bighexagon} \xrightarrow{HIO_4 \text{ (or) } Pb(OCOCH_3)_4} \underset{\underset{H\ H}{|\ |}}{\bigwedge} O=C\ C=O$$

$$\underset{\underset{HO}{|}}{R}-\underset{\underset{OH}{|}}{CH}-CH-R' + HIO_4 \longrightarrow R-CH-CH-R' \longrightarrow \underset{\underset{O}{||}}{R}-CH\ \ HC-\underset{\underset{O}{||}}{R'} + HIO_3 \qquad \text{(c)}$$

$$(14.25)$$

This sequence of reactions, alkene → glycol → products of oxidative cleavage, gives the same products as those obtained by ozonolysis of the alkene (Chapter 4, Section 4.4.5). Oxidative cleavage by periodic acid proceeds through a cyclic intermediate (Eq. 14.25c). This reaction, also called the *Malaprade reaction*, was discovered by the French chemist, Malaprade and has been very useful for the study of carbohydrates.

14.6.3 GLYCEROL

Glycerol (glycerine) is the most important trihydric alcohol. It is present in vegetable and animal oils and fats as triglycerides, which are the glyceryl triesters of long chain fatty acids. Glycerol is released when these are hydrolysed (saponified) as is done in the soap industry which is the major source of glycerol.

$$
\begin{array}{ccc}
& O & \\
& \parallel & \\
CH_2-O-C-R & & \\
| & O & CH_2-OH \\
& \parallel & | \\
CH-O-C-R \ + \ 3NaOH \ \longrightarrow & CH-OH \ + \ 3RCOONa \\
| & O & | \\
& \parallel & CH_2-OH \\
CH_2-O-C-R & & (Soap)
\end{array}
\qquad (14.26)
$$

The salt of the fatty acid shown as RCOONa (Eq. 14.26), is actually a mixture in which the R group may vary from C11 to C17 or more, and can be with or without double bonds, depending upon the composition of the fat. Regardless of the composition of the oil, glycerol is the byproduct. Glycerol on treatment with nitric acid gives a triester, glyceryl trinitrate, which is known by the trivial, incorrect, name trinitroglycerine (TNT). It is an explosive and is the main ingredient of dynamite which was developed by Alfred Nobel.

$$
\begin{array}{ccc}
CH_2-OH & & CH_2-O-NO_2 \\
| & \xrightarrow{\ H_2SO_4\ } & | \\
CH-OH \ + \ 3HNO_3 & & CH-O-NO_2 \ + \ 3H_2O \\
| & & | \\
CH_2-OH & & CH_2-O-NO_2
\end{array}
\qquad (14.27)
$$

Glyceryl trinitiate
('Trinitroglycerine', TNT)

Glycerol is used in the manufacture of alkyd resins. It is also used in the manufacture of pharmaceuticals and cosmetics.

KEY POINTS

- Alcohols are classified into primary, secondary and tertiary.
- Alcohols have higher boiling points than other classes of organic compounds with comparable molecular weights due to intermolecular hydrogen bonding.
- Methanol (wood alcohol) used to be obtained by the destructive distillation of wood. Today, the major industrial source of methanol is by the hydrogenation of carbon monoxide.
- Ethanol is manufactured by the fermentation of plant carbohydrates and also by the hydration of ethene. It is a major industrial chemical. A recent use of ethanol is as a biofuel substitute for petroleum-based fuel. It is also converted to biodiesel by reaction with oils and fats. Ordinary distilled ethanol is of 95% purity and is called rectified spirit. Absolute alcohol is made by azeotropic distillation with benzene. Ethanol meant for industrial use is denatured (made unfit for drinking) by mixing it with methanol and/or other chemicals.
- Alcohols can be prepared from alkenes by direct acid catalysed hydration or by indirect hydration via hydroboration or oxymercuration.
- They can also be made by the reduction of carbonyl compounds using reducing agents like sodium borohydride and lithium aluminium hydride.
- The most versatile method for the synthesis of alcohols is by the use of Grignard reagents.
- Grignard reagents, R–Mg–X, are prepared by the reaction between alkyl or aryl halides with metallic magnesium in dry diethyl ether.
- Grignard reagents react with

    —formaldehyde or ethylene oxide to give primary alcohols
    —other aldehydes to give secondary alcohols
    —ketones and esters to give tertiary alcohols.

- Unsaturated alcohols like allyl alcohol are similar in reactivity to tertiary and benzyl alcohols. Crotyl alcohol reacts with HBr to give mainly 3-bromo-1-butene, due to allylic rearrangement.
- vic-Dihydric alcohols are called glycols. Glycols are prepared from alkenes by direct oxidation using permanganate or osmium tetroxide. Both the reagents yield products corresponding to the addition of two OH groups to the double bond in a *cis*-manner (*cis*-dihydroxylation). Glycols can also be prepared by conversion of the alkene to the epoxide, followed by the hydrolytic opening of the epoxide ring by aqueous acid or alkali. The epoxide can be made by oxidation of the alkene using a peracid or by conversion of the alkene to the chlorohydrin followed by treatment with alkali. The epoxide route gives the glycol corresponding to *trans*-dihydroxylation.

- Glycols can be oxidised with $C-C$ bond cleavage by periodic acid or lead tetraacetate. 2,3-Butanediol is oxidised to two molecules of acetaldehyde.
- Glycerol, a trihydric alcohol, occurs as triglycerides in oils and fats. It is a byproduct of the soap industry. Reaction of glycerol with nitric acid gives the trinitrate, which is the explosive TNT.

EXERCISES

SECTION I

1. Give the products of the following reactions

    (a) $C_6H_5-CO-CH_2-COOC_2H_5$ $\xrightarrow{\text{NaBH}_4}$

    (b) $C_6H_5-CH_2-COOC_2H_5$ $\xrightarrow{\text{LiAlH}_4}$

    (c) 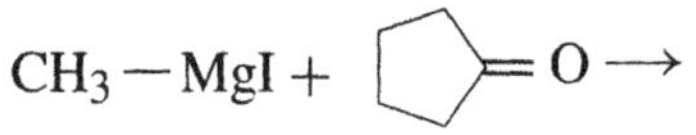 $CH_3-MgI +$ $= O \longrightarrow$

    (d)

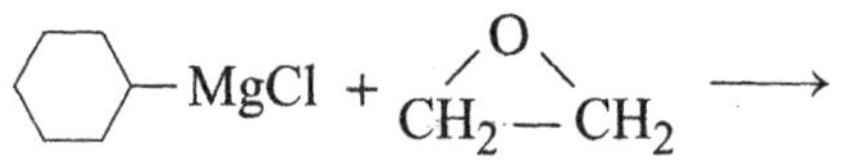

    (e) 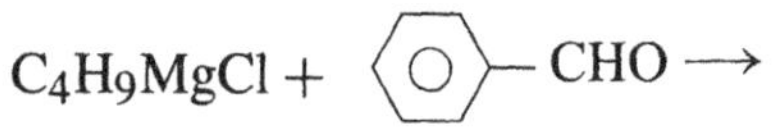 $C_4H_9MgCl +$ $-CHO \longrightarrow$

    (f) R-2-Butanol $+ PCl_3 \longrightarrow$

    (g) 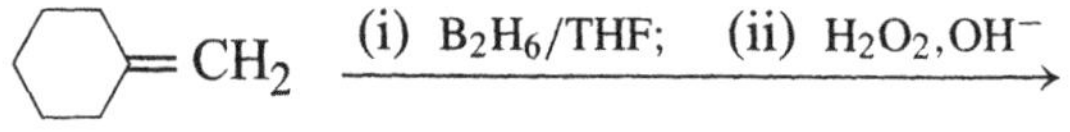 $= CH_2$ $\xrightarrow{\text{(i) } B_2H_6/\text{THF};\quad \text{(ii) } H_2O_2, OH^-}$

    (h) $\xrightarrow{H^+}$

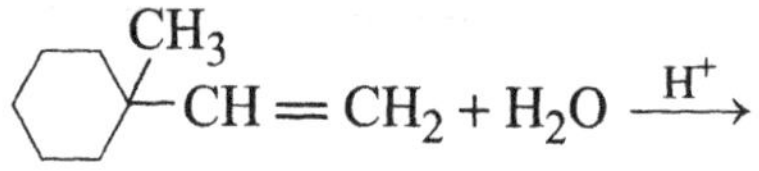

    (i) 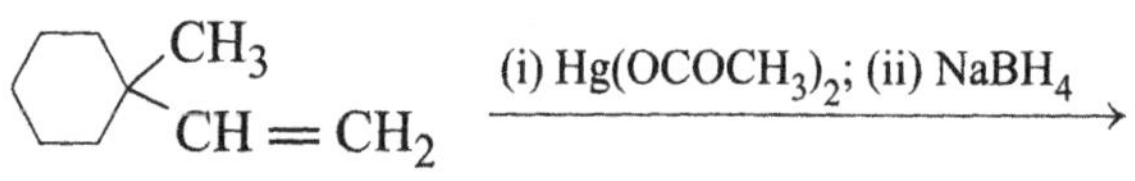 $\xrightarrow{\text{(i) Hg(OCOCH}_3)_2;\ \text{(ii) NaBH}_4}$

    (j) 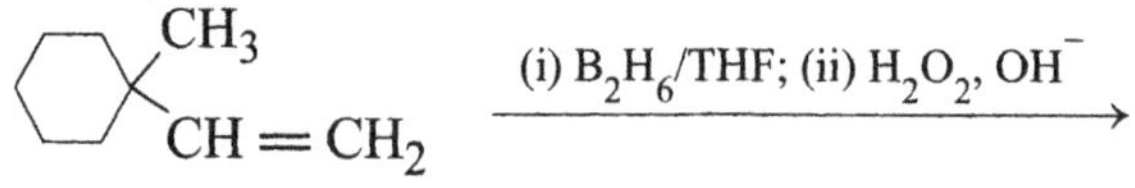 $\xrightarrow{\text{(i) } B_2H_6/\text{THF};\ \text{(ii) } H_2O_2,\ OH^-}$

    (k) 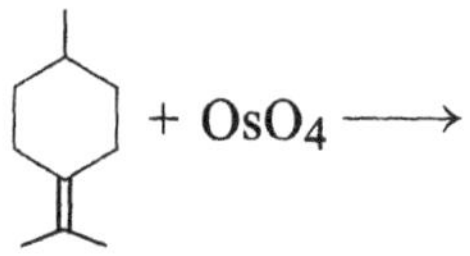 $+ OsO_4 \longrightarrow$

(l) [cyclohexane with OH and CH_2OH] $\xrightarrow{\text{HIO}_4}$

(m) [diphenyl substituted alkene] $C=C$ $+ OsO_4 \longrightarrow$

(n) [phenyl substituted alkene] $C=C$ $\quad$ (i) [phenyl]$-\overset{\overset{\displaystyle O}{\|}}{C}-O-OH$; (ii) OH^- $\xrightarrow{\hspace{3cm}}$

(o) [phenyl]$-Br + CH_3ONa \longrightarrow$

(p) [cyclohexanone with isopropyl substituent] $+ NaBH_4 \longrightarrow$

2. Show how the following conversions can be brought about. Other necessary reagents may be used.

(a) $CH_3-CH_2-CH_2-CH=CH_2 \rightarrow CH_3-CH_2-CH_2-CH_2-CH_2OH$

(b) [phenyl]$-CH=CH-CH_3 \longrightarrow$ [phenyl]$-\underset{\underset{\displaystyle OH}{|}}{CH}-CH_2-CH_3$

(c) $CH_3MgI \rightarrow CH_3-CH_2-CH_2OH$

(d) [phenyl]$-MgBr \longrightarrow$ [phenyl]$-\underset{\underset{\displaystyle \text{[phenyl]}}{|}}{\overset{\overset{\displaystyle OH}{|}}{C}}-$[phenyl]

(e) [phenyl]$-\overset{\overset{\displaystyle O}{\|}}{C}-OC_2H_5 \longrightarrow \left(\text{[phenyl]} \right)_3 C-OH$

(f)

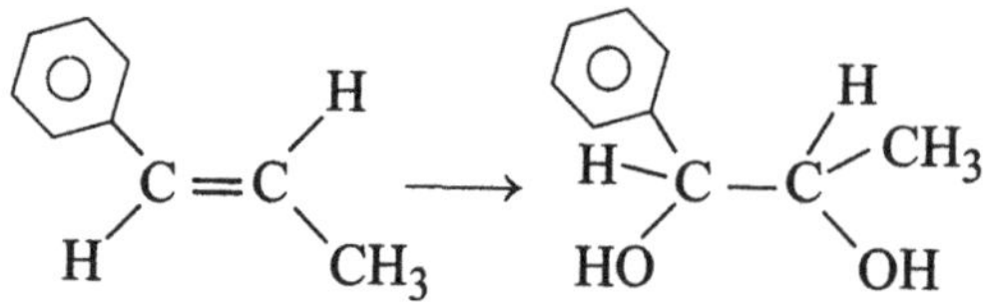

(g)

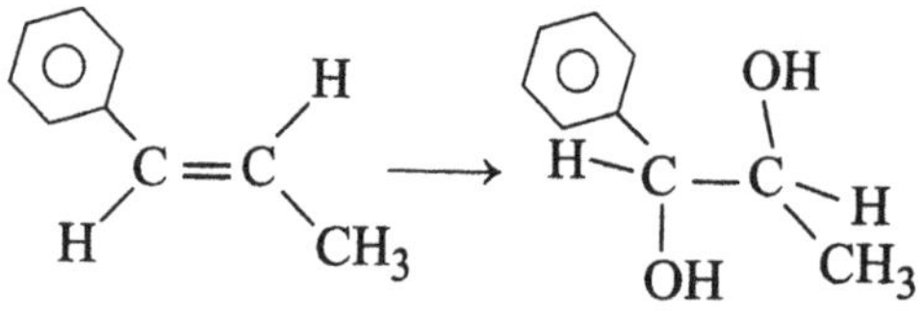

(h)

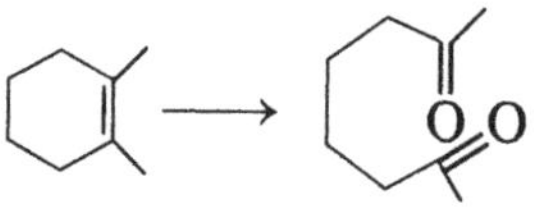

(i)

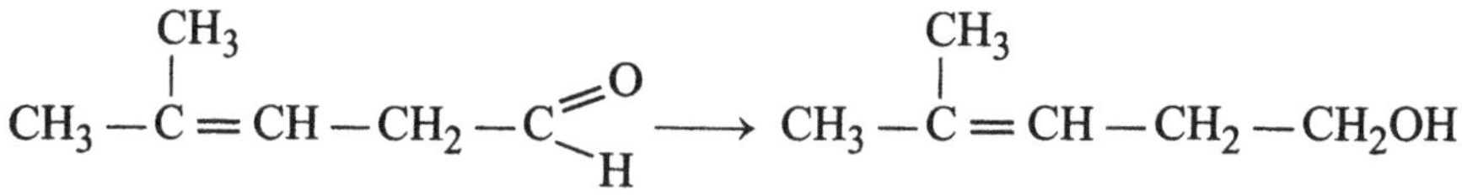

$$CH_3 - \underset{\underset{CH_3}{|}}{C} = CH - CH_2 - C \overset{O}{\underset{H}{\diagup}} \longrightarrow CH_3 - \underset{\underset{CH_3}{|}}{C} = CH - CH_2 - CH_2OH$$

(j)

$$CH_3 - \overset{O}{\overset{||}{C}} - CH_2 - CH_2 - COOCH_3 \longrightarrow CH_3 - \overset{OH}{\overset{|}{CH}} - CH_2 - CH_2 - CH_2OH$$

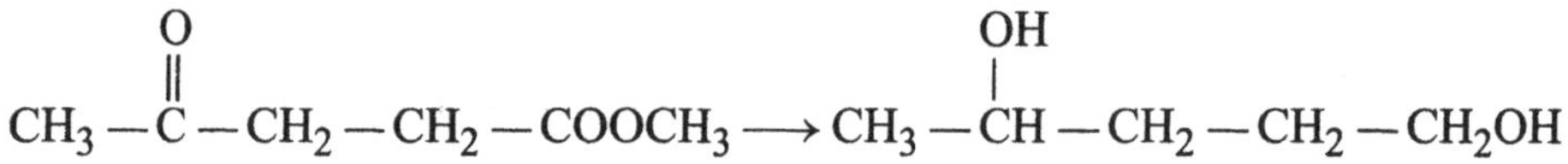

(k)

3. Give the structures of the following.

   (a) Triphenylcarbinol
   (b) 2-Phenyl-1-propanol
   (c) Epoxycyclohexane
   (d) *cis*-Stilbene oxide

   (e) Methoxycyclohexane
   (f) α-Phenylethanol
   (g) 1,4-Butanediol

SECTION II

1. Show how the following compounds can be synthesised using Grignard reagents.

   (a) 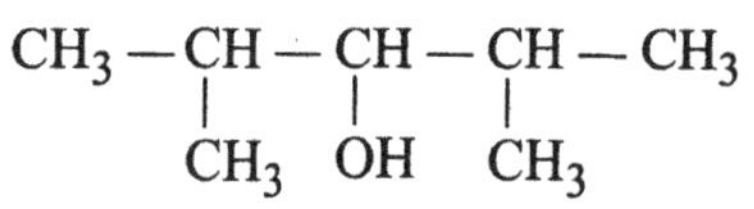 $CH_3 - \underset{\underset{CH_3}{|}}{CH} - \underset{\underset{OH}{|}}{CH} - \underset{\underset{CH_3}{|}}{CH} - CH_3$

   (b)

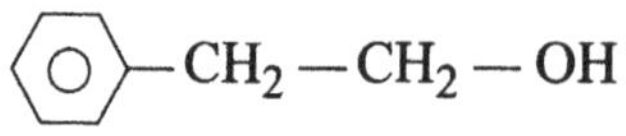

(c)

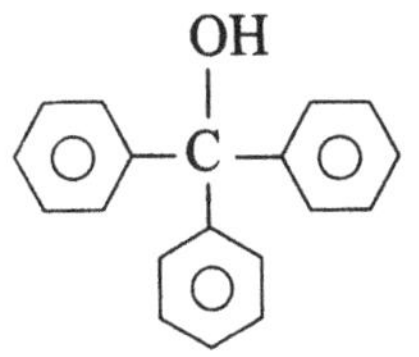

(d)

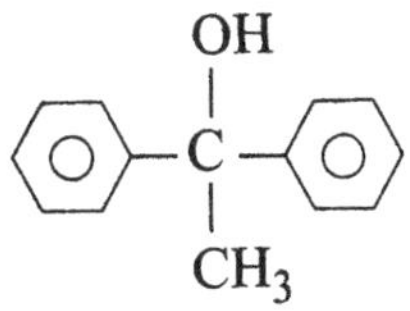

(e)

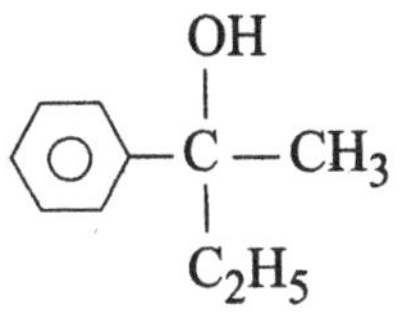

(f)

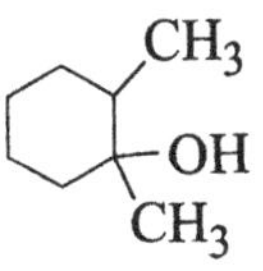

2. How can the following conversions be brought about? They may involve more than one step. You may use any other required reagents.

(a)

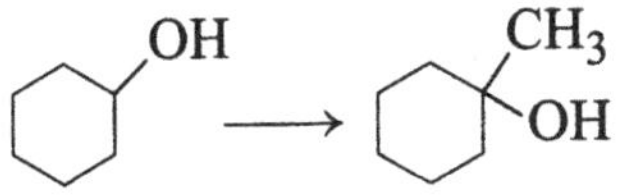

(Hint: It is a tertiary alcohol. You may need a ketone, which can be obtained from the given starting material.)

(b)

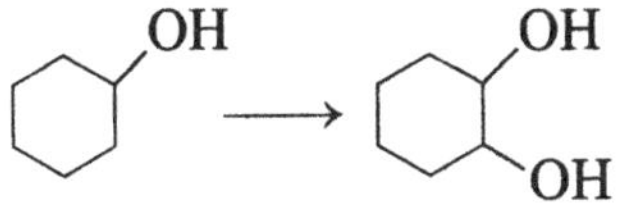

(c)

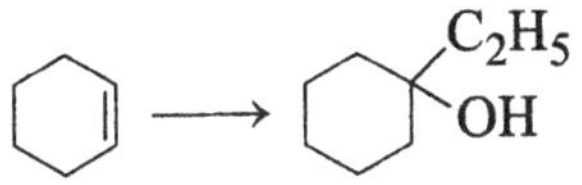

(d) $CH_3-(CH_2)_2-COOH \rightarrow CH_3-(CH_2)_2-CH_2OH$

(e)

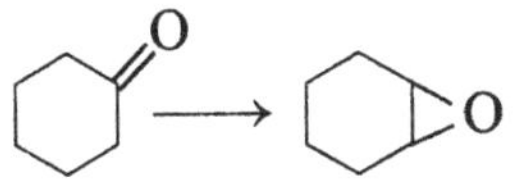

(f) $CH_3-CH_2-CH=CH_2 \rightarrow CI'_3-CH_2-CH_2-CHO$

(g) $CH_3-CH_2-CH=CH_2 \rightarrow CH_3-CH_2- \underset{\underset{O}{\|}}{C} -CH_3$

3. Give 3 different Grignard syntheses for

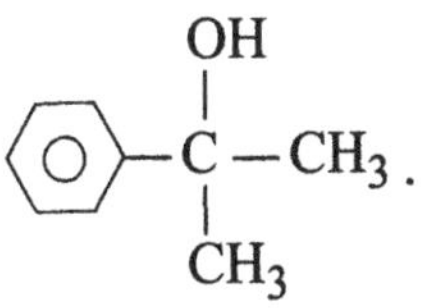

4. How does the following reaction take place?

$$R-CH=CH_2 + Hg(O\overset{\bullet}{C}OCH_3) \xrightarrow{\text{acetic acid}} R-\underset{\underset{OCOCH_3}{|}}{CH}-CH_2-HgOCOCH_3$$

5. Explain the role of $ZnCl_2$ in the Lucas test.

6. 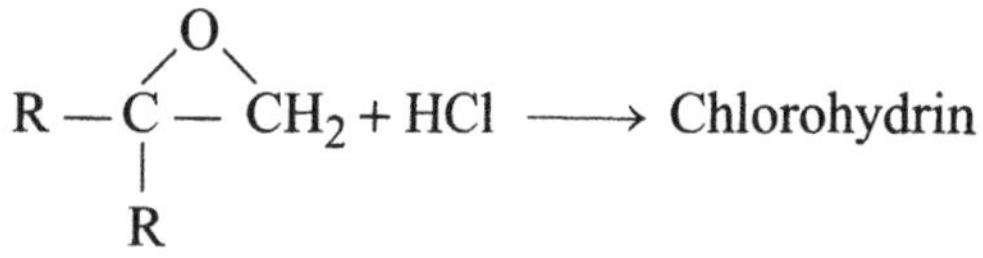 $R-\overset{O}{\overset{\diagup\diagdown}{\underset{\underset{R}{|}}{C}}}-CH_2 + HCl \longrightarrow$ Chlorohydrin

Give the structure of the chlorohydrin and the mechanism of the reaction.

7. Explain how the following reaction takes place.

$$R-OH + D_2O \rightleftharpoons R-OD + H-O-D$$

8. Identify the compounds A, B, C ...

(a) $C_4H_8(A) + H_2O \xrightarrow{H^+} (B)$

   (B) gives an immediate turbidity with Lucas reagent.

(b) $C_4H_8(C) + \text{perbenzoic acid} \rightarrow C_4H_8O$ (D)

   $(D) \xrightarrow{H_3O^+} C_4H_{10}O_2(E) \xrightarrow{HIO_4} 2CH_3CHO$

(c) $(F) \xrightarrow{\text{Red P}+I_2} (G) \xrightarrow{\text{Mg}} (H)$

   Excess of $(H) + (I) \xrightarrow{\text{after hydrolysis}} CH_3-\underset{\underset{CH_3}{|}}{\overset{\overset{OH}{|}}{C}}-CH_3$

   (Give two possible structures for I.)

CHALLENGING QUESTIONS

1. Design synthesis for the following compounds, using starting materials containing less number of carbon atoms than the final product.

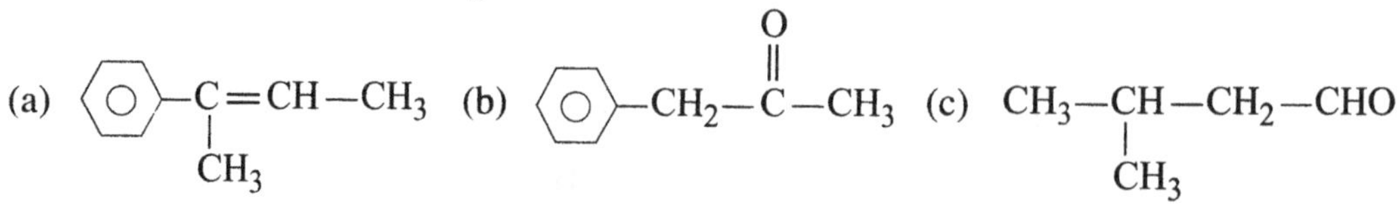

(a) $\langle\bigcirc\rangle-\underset{\underset{CH_3}{|}}{C}=CH-CH_3$ (b) $\langle\bigcirc\rangle-CH_2-\overset{\overset{O}{\|}}{C}-CH_3$ (c) $CH_3-\underset{\underset{CH_3}{|}}{CH}-CH_2-CHO$

2. $CH_3-O-C(CH_3)_3$ is manufactured by the reaction of methanol with 2-methylpropene under pressure. What kind of catalyst, acid or base, is required? Give the mechanism of the reaction.

3. How can ethyl acetate by prepared using methanol alone as the organic starting material. Solvents and other inorganic reagents can be used.

4. Identify C_3H_6O which after oxymercuration and demercuration gives $C_3H_8O_2$ which gets oxidised by HIO_4.

5. Structures of the type

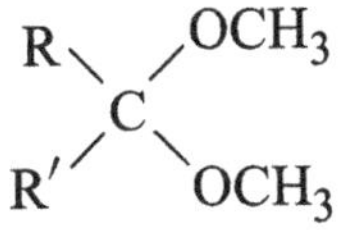

$$\begin{array}{c} R \\ \diagdown \quad \diagup OCH_3 \\ C \\ \diagup \quad \diagdown \\ R' \quad OCH_3 \end{array}$$

, called ketals are ethers of *vic*-glycols. They are stable in alkaline media, but are readily hydrolysed by aqueous acids. Explain. Give the mechanism of acid catalysed hydrolysis.

6. Structures of the type

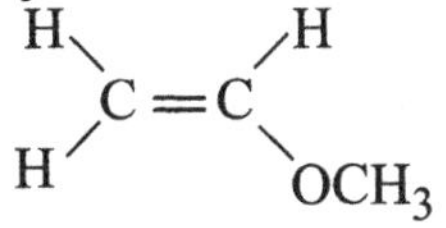

$$\begin{array}{c} H \\ \diagdown \quad \diagup H \\ C=C \\ \diagup \quad \diagdown \\ H \quad OCH_3 \end{array}$$

are enol ethers. They are stable in alkali, but are hydrolysed by aqueous acids. Explain. The acid catalysed hydrolysis may involve protonation as the first step. Protonation may be at the oxygen or at the double bond. Write down both mechanisms and decide which one is more likely.

PROJECT

Prepare a project report on **biofuels**. The field is very wide. Fire-wood and charcoal are biofuels; so are biogas and various fuel gases obtained from wood. Your report should look at not only the merits of biofuels, but also at their demerits.

15 Phenols

OBJECTIVES In this chapter, you will learn about,

- types of phenols
- the acidity of phenols and the factors affecting their acidity
- general methods of synthesis of phenols
- cumene hydroperoxide route for the production of phenol and acetone
- reactions of phenol—electrophilic substitution, Lederer–Manasse, Reimer–Tiemann and

  Kolbe–Schmitt reaction, Gattermann synthesis and Houben–Hoesch synthesis, coupling with diazonium salts, esterification
- polyhydric phenols, names and structures
- some representative phenols and cresols
- nitro and aminophenols
- chelation and physical properties
- preparation and properties of naphthols
- diazo coupling reaction
- phenol ethers
- Fries rearrangement
- hindered phenols as antioxidants
- spectroscopy of phenols and alcohols

15.1 INTRODUCTION

Phenols are aromatic compounds containing the hydroxyl group but are very different from alcohols in their properties and reactions. In phenols, the OH group is directly attached to the aromatic ring. Some of the simpler members are listed in Table 15.1.

Table 15. 1 Phenols

Name	Structure	Boiling point (b.p)/ Melting point (m.p)	pK$_a$ value
Phenol		182°C (b.p)	10.0
o-Cresol		191°C (b.p)	10.3
m-Cresol		201°C (b.p)	10.1
p-Cresol		202°C (b.p)	10.3
p-Chlorophenol		220°C (b.p)	9.4
o-Chlorophenol		173°C (b.p)	8.6
p-Nitrophenol		114°C (m.p)	7.2
m-nitrophenol		96°C (m.p)	8.4
1-Naphthol		96°C (m.p)	9.2
2-Naphthol		122°C (m.p)	9.5
2,4,6-Trinitrophenol (picric acid)		122°C (m.p)	0.4

The phenolic OH differs from that in alcohols in many ways. It cannot be replaced in nucleophilic substitution reactions for the same reasons that halobenzenes are not reactive in such reactions. Of course, they cannot be dehydrated or oxidised to ketones. On the other hand, the H is protic and can be replaced by metals and they can be esterified; both reactions involve the O—H bond. The most important difference is their higher acidity.

15.1.1 ACIDITY OF PHENOLS

The pK_a values of some phenols are listed in Table 15.1. Recall that $pK_a = -\log K_a$. A lower pK_a value denotes a stronger acid. All phenols are stronger acids than water whose $pK_a \sim 15.7$, and alcohols whose pK_a values are nearly the same as that of water. Among the phenols, there is a wide variation. Acidity is measured by the acid dissociation constant, K_a, which is the equilibrium constant of the reaction in Eq. 15.1.

$$Ar-OH + H_2O \rightleftharpoons Ar-\overset{-}{O} + H_3O^+ \qquad (15.1)$$

Phenols are more acidic compared to alcohols due to the fact that the phenoxide ion is stabilised by resonance and the negative charge is effectively delocalised (Eq. 15.2a).

$$\qquad (15.2)$$

The electron-withdrawing inductive effect of the aromatic ring also helps to dissipate the negative charge (Eq. 15.2b). The alkoxide ion has no resonance stabilisation and the electron-releasing inductive effect helps to further localise the negative charge. Substituents on the aromatic ring may help or hinder the delocalisation of the charge and thereby increase or decrease acidity. In general, electron-withdrawing groups anywhere in the ring cause increase in acidity. Compare the pK_a value of phenol with those of *p*- and *m*-chlorophenols. Both are more acidic than phenol. This is due to the electron-withdrawing effect of the chlorine. The effect is more pronounced in the *ortho* isomer than in the *para*. In the former, the inductive effect is stronger due to the proximity of the Cl to the anionic centre.

Now let us compare phenol with the nitrophenols. Both *m*- and *p*-nitrophenols are stronger acids than phenol. Between the two, the *para* isomer is stronger than the *meta*. In the nitrophenols, the resonance effect operates in addition to the inductive effect of the nitro group. The *p*-nitrophenoxide ion is stabilised by resonance as follows (Eq. 15.3).

(I)   (15.3)

Structure I, in Eq. 15.3 shows the involvement of the nitro group in delocalising the charge. The *meta* nitro group also helps to delocalise the charge but only by inductive effect. A resonance structure similar to I in Eq. 15.3 is not possible for the *meta* isomer. When three nitro groups are present – as in 2,4,6-trinitrophenol – all the three groups contribute to strengthening the acid by both inductive and resonance effects. This molecule is comparable to mineral acids in strength. The name picric acid – with the acid tag – is justified though it is not a carboxylic acid.

15.2   SYNTHESIS OF PHENOLS

Phenol and cresols are present in coal tar. Many substituted phenols are prepared from phenol by standard reactions like nitration. An important method for introducing the OH group in the benzene ring is via diazotisation (Eq. 15.4). This reaction is discussed in detail in Chapter 18.

$$Ar-H \xrightarrow{HNO_3/H_2SO_4} Ar-NO_2 \xrightarrow{Sn/HCl} Ar-NH_2$$

$$\xrightarrow[0°C]{HNO_2/H_2SO_4} Ar-N_2^+ \; HSO_4^- \xrightarrow[\Delta]{H_2O} Ar-OH \qquad (15.4)$$

Another method – especially applicable to the synthesis of naphthols – is the hydrolysis of the corresponding sulphonic acids (Eq. 15.5).

An early method for the manufacture of phenol, the *Dow's process*, involved the hydrolysis of chlorobenzene by alkali at high temperature and pressure, a reaction which involves the benzyne intermediate (Chapter 9, Section 9.5.3.1). The more recent industrial process which gives both phenol and acetone is illustrated in Eq. 15.6.

$$(15.6)$$

In this reaction sequence, the first step is a Friedel–Crafts-like isopropylation of benzene by propene. Isopropylbenzene (trivial name is cumene) obtained in this reaction is oxidised by bubbling air through the boiling hydrocarbon. A reaction called autoxidation, which is a free radical chain reaction, takes place to give cumene hydroperoxide. The hydroperoxide undergoes an acid-catalysed rearrangement when treated with aqueous sulphuric acid. Phenol and acetone are the products.

15.3 POLYHYDRIC PHENOLS

As in the case of alcohols, phenols containing more than one OH group are called polyhydric phenols. There are three isomeric dihydric phenols. These are catechol (*o*-dihydroxybenzene), resorcinol (*m*-dihydroxybenzene) and hydroquinone or quinol (*p*-dihydroxybenzene). The three trihydric phenols are pyrogallol (1,2,3-trihydroxybenzene), hydroxyquinol (1,2,4-trihydroxybenzene) and phloroglucinol (1,3,5-trihydroxybenzene).

All these are solids and are partly soluble in water.

15.4 REACTIONS

Phenols are weak acids, weaker than carboxylic acids. Phenols dissolve in aqueous alkali forming the phenoxide ion. This reaction can be used to distinguish them from alcohols and other nonacidic compounds, provided they are themselves insoluble in water. Carboxylic acids also dissolve in aqueous alkali. Carboxylic acids dissolve also in aqueous sodium bicarbonate solution, with the liberation of carbon dioxide. Phenols do not react with sodium bicarbonate. This property can be used to distinguish between phenols and carboxylic acids.

15.4.1 Electrophilic Substitution

Phenols are very reactive in electrophilic substitution. The OH group is highly activating and *ortho/para* directing (Chapter 9, Section 9.5.2.1 and Table 9.1). Treatment of phenol with bromine water at room temperature gives 2,4,6-tribromophenol. Under controlled conditions at low temperature, *p*-bromophenol can be obtained. Other electrophilic substitutions like nitration, sulphonation and Friedel–Crafts acylation take place readily. Low temperature and controlled conditions are necessary to get mono-substitution products. These reactions follow the general mechanism outlined in Chapter 9, Fig. 9.4. More interesting are reactions involving electrophiles which are much less reactive than the ones mentioned above. Some of these which are unique for phenols, are discussed below.

15.4.2 Lederer–Manasse Reaction

Phenol reacts with aldehydes or ketones, under acid or base catalysis, leading to the introduction of a hydroxymethyl or substituted hydroxymethyl group in the *para* and *ortho* positions. The reaction with formaldehyde is illustrated in Eqs. 15.7 and 15.8.

Acid catalysis

$$\tag{15.7}$$

Base catalysis

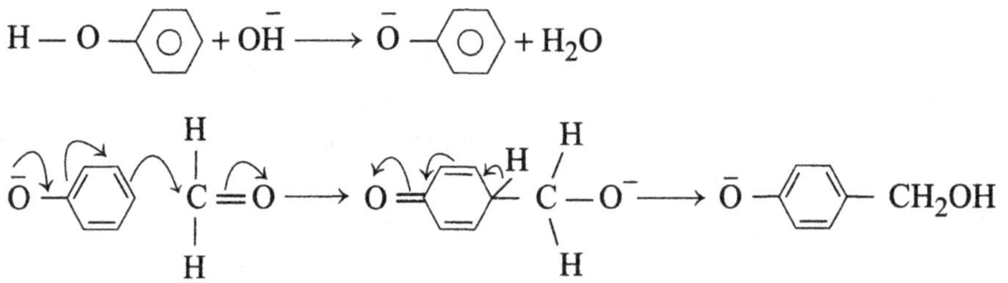

$$(15.8)$$

This reaction is an electrophilic substitution by the weak electrophile (the carbon of the carbonyl group). This is called the *Lederer–Manasse reaction*. With excess formaldehyde, the reaction can take place at all the active positions to give 2,4-di(hydroxymethyl)phenol and 2,4,6-tri(hydroxymethyl)phenol. Hydroxymethylphenol can condense with another phenol molecule in a Friedel–Crafts type reaction to give a diphenylmethane structure (Eq. 15.9).

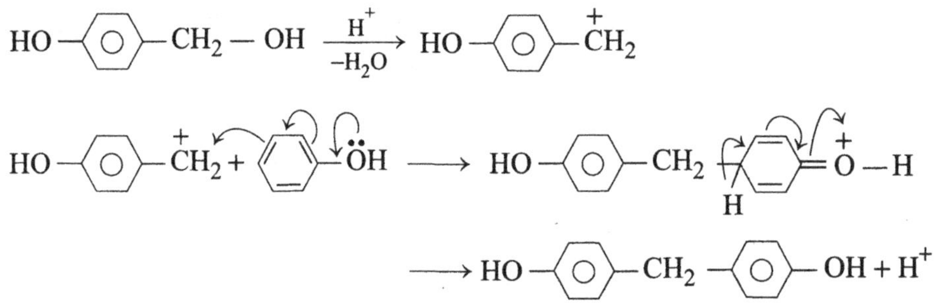

$$(15.9)$$

This is the basis of condensation polymerisation leading to phenol–formaldehyde resins (Chapter 7, Section 7.4).

15.4.3 REIMER–TIEMANN REACTION

Phenol, when heated with chloroform and alkali gives *o*- and *p*-hydroxybenzaldehyde. The former is the main product and is also called salicylaldehyde (Eq. 15.10).

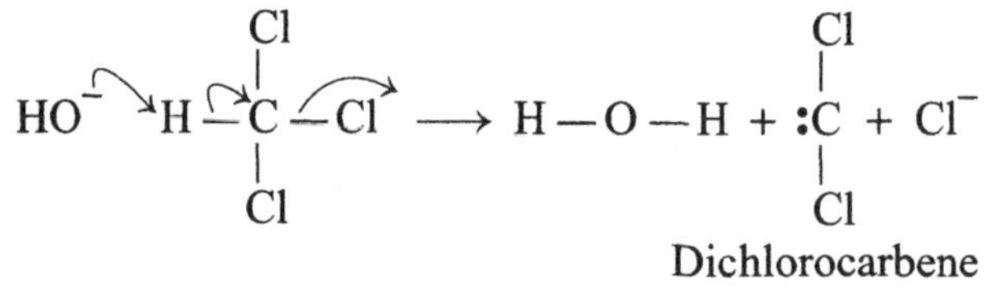

Dichlorocarbene

$$(a)$$

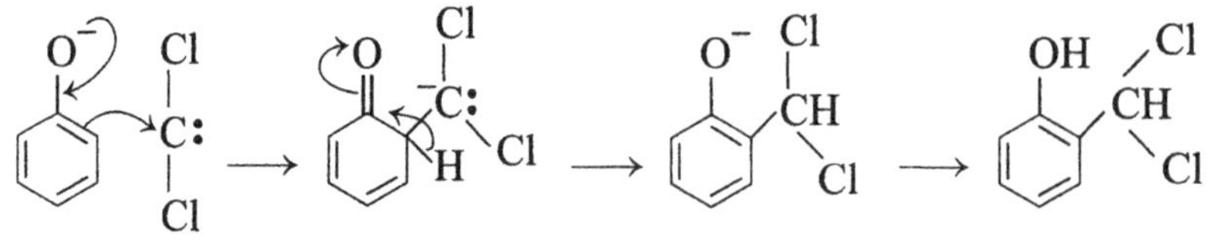

$$(b)$$

$$(c)$$
$$(15.10)$$

Chloroform in the presence of a strong alkali undergoes a 1,1-elimination (α-elimination) to give dichlorocarbene (Eq. 15.10a). The driving force for this reaction is the relatively high acidity of the hydrogen in chloroform. Hydroxide (OH^-), abstracts a proton from chloroform and the resulting carbanion ejects a Cl^- to give the molecule $:CCl_2$, in which carbon is divalent. The parent molecule $:CH_2$, is called carbene or methylene. It has a lone pair of electrons on the carbon, is neutral and highly reactive. There are many reactions in which carbenes are generated as intermediates.

Dichlorocarbene, like other carbenes, is a neutral molecule. It is a short-lived, highly reactive intermediate. Even though the carbon has a lone pair of electrons, there are only six electrons in the valence shell and it behaves as an electropihle. Its reaction with the electron-rich phenol molecule yields a benzal chloride, which hydrolyses to the aldehyde (see Chapter 9, Section 9.3.5.1).

15.4.4 KOLBE–SCHMITT REACTION (KOLBE SYNTHESIS)

When sodium phenoxide is heated with carbon dioxide under pressure, sodium salicylate is formed (Eq. 15.11).

$$(15.11)$$

Carbon dioxide behaves as the electrophile here. *o*-Hydroxybenzoic acid (salicylic acid) is the preferred product over the *para* isomer. It is an equilibrium controlled reaction. The former is thermodynamically more stable due to intramolecular hydrogen bonding.

Acetylation of salicylic acid gives acetyl salicylic acid, otherwise called aspirin (Eq. 15.12).

$$(15.12)$$

15.4.5 GATTERMANN ALDEHYDE SYNTHESIS AND HOUBEN–HOESCH SYNTHESIS OF PHENOLIC KETONES

Gattermann aldehyde synthesis works best with phenols for the synthesis of phenolic aldehydes. It is successful with phenolic ethers like anisole as well, but less successful with other arenes. The reagent that brings about the reaction is, hydrogen cyanide + HCl in the presence of aluminium chloride (Eq. 15.13a).

$$H-C \equiv N + HCl \longrightarrow ClCH = NH$$

$$ClCH = NH + AlCl_3 \longrightarrow AlCl_4^- \; [\overset{+}{C}H = NH \longleftrightarrow CH \equiv \overset{+}{N}H]$$

$$HO-\langle O \rangle + ClCH = NH \xrightarrow{AlCl_3} HO-\langle O \rangle - CH = NH \cdot AlCl_3$$

$$\downarrow H_3O^+$$

$$HO-\langle O \rangle - CHO \qquad (a)$$

Resacetophenone   (b)

(15.13)

Addition of HCl to HCN gives the imide of chloroformaldehyde, $Cl-CH=NH$, which reacts with phenol in the presence of $AlCl_3$, in a Friedel–Crafts reaction. The carbocation intermediate is stabilised by resonance. The product is the imide of the phenolic aldehyde as a complex with $AlCl_3$, which gets hydrolysed by aqueous acid to liberate the aldehyde. Instead of free HCN, zinc cyanide + HCl can be used.

When polyhydric phenols like phloroglucinol and resorcinol are treated with alkyl cyanides +HCl + $AlCl_3$, aryl ketones are obtained. This is the Houben–Hoesch synthesis of phenolic ketones. Thus, resorcinol reacts with acetonitrile and HCl in the presence of $AlCl_3$ to give resacetophenone (Eq. 15.13b). $CH_3-C(Cl)=NH$ is the intermediate.

15.4.6 AZO COUPLING

When an aromatic primary amine like aniline in dilute hydrochloric acid solution is treated with sodium nitrite at $0°C$, a diazonium salt is formed. The salt obtained from aniline is called benzene diazonium chloride (Eq. 15.14a). This reaction is called diazotisation and is the gateway to a number of aromatic compounds because the $-N_2^+$ group can be replaced or modified to introduce a variety of groups. We have already seen how it can be hydrolysed to obtain phenols (Eq. 15.4). The diazonium cation can react as an electrophile and enter into electrophilic substitution reactions with highly active aromatic molecules like phenols. Primary aromatic amines are diazotised in the cold by adding $NaNO_2$ solution to the solution of the amine in dilute mineral acid. A solution of the phenol in alkali is added to this. A coloured precipitate is formed. This reaction is called azo coupling (Eq. 15.14b). The products are azo compounds. The product from phenol and aniline is *p*-hydroxyazobenzene. The azo compounds are coloured substances. The coupling reaction of aniline with 2-naphthol yields a deep red precipitate (Eq. 15.14c). These azo compounds are called azo dyes. Formation of the intensely coloured azo dye is used as a test for primary aromatic amines and also for phenols, in qualitative analysis. Azo coupling is a typical aromatic electrophilic substitution (Eq. 15.14d).

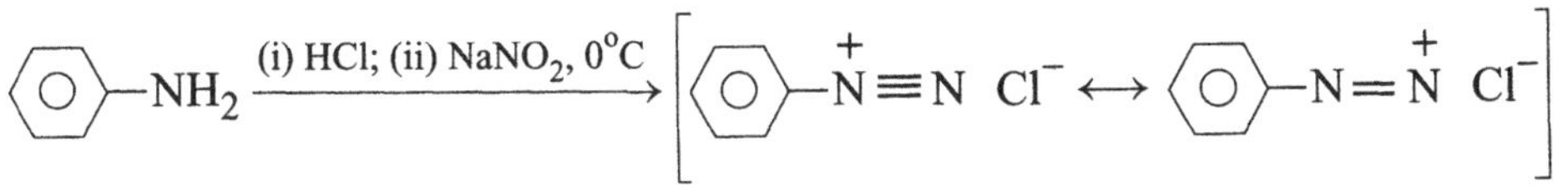

Benzenediazonium chloride   (a)

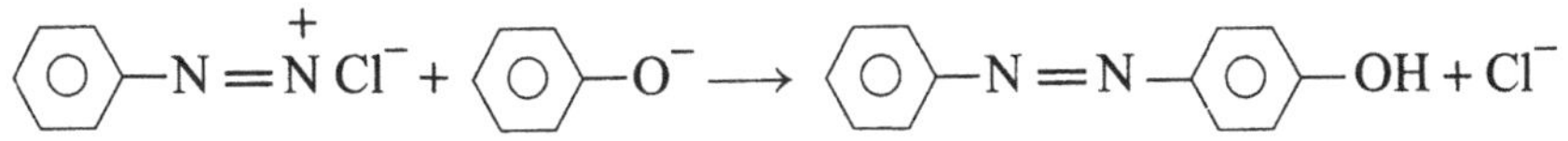

p-Hydroxyazobenzene   (b)

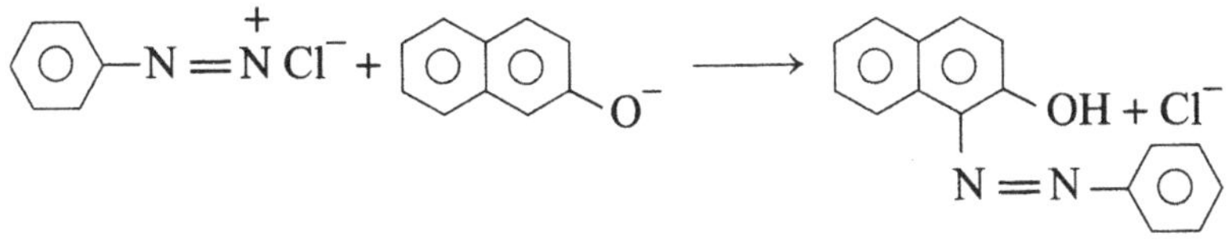

(c)

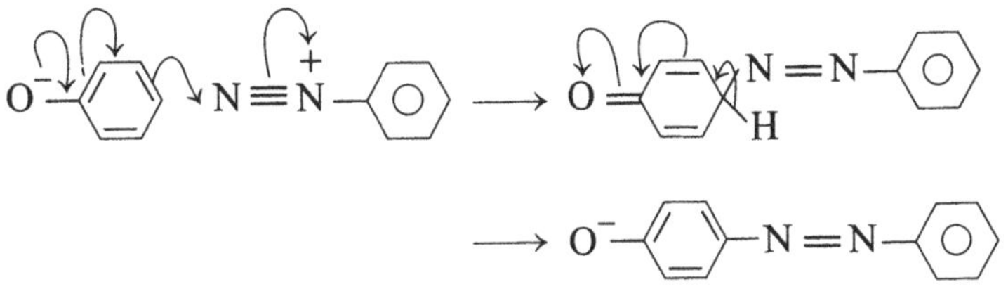

(d)

(15.14)

Other applications of the diazotisation reaction are discussed in Chapter 18.

15.4.7 ESTERIFICATION

The hydroxyl group of phenol, like that of alcohols, can be esterified. In the case of alcohols, the esterification is done by refluxing the alcohol with an acid in the presence of a mineral acid catalyst (Chapter 14, Eq. 14.19). This procedure does not succeed with phenol because of the lower nucleophilicity of the oxygen of phenol. Acid anhydrides or acyl halides are more reactive and can be used to esterify phenols. Esterificaton is also referred to as acylation, because an acyl group is introduced in the place of hydrogen. Recall acetylation of salicylic acid by heating with acetic anhydride to prepare aspirin (Eq. 15.12). Phenols react with benzoyl chloride in the presence of alkali to give the benzoyl derivative. Phenol itself gives phenyl benzoate (Eq. 15.15). This reaction is known as the *Schotten–Baumann reaction.*

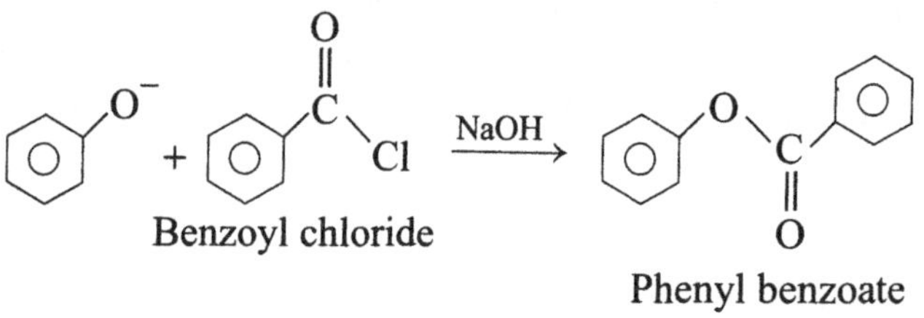

Benzoyl chloride

Phenyl benzoate

(15.15)

15.5   SOME REPRESENTATIVE PHENOLS

15.5.1   PHENOL AND CRESOLS

Phenol and the cresols are present in coal tar. The three cresols may be prepared from the corresponding aminotoluenes (toluidines) as illustrated for *p*-cresol (Eq. 15.16).

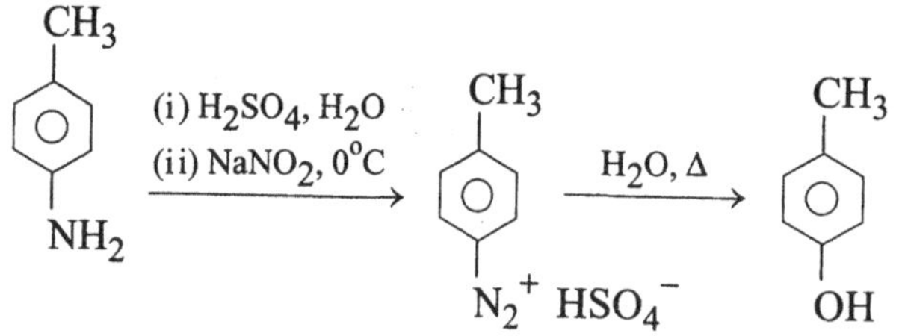

(15.16)

15.5.2   NITROPHENOLS

Some of the physical properties of the three nitrophenols are given in Table 15.2.

Table 15. 2   Physical properties of nitrophenols

Nitrophenol	Melting point (°C)	Boiling point (°C)	Solubility in water (g/100 g)	Steam volatility
ortho	45	217	0.2	Volatile
meta	96	Very high	1.35	Nonvolatile
para	115	Very high	1.69	Nonvolatile

Nitration of phenol at low temperature using dilute nitric acid yields a mixture of *o*- and *p*-nitrophenols. The yield is low because phenol gets oxidised by nitric acid. The *ortho* and *para* isomers can be separated by steam distillation. Only *o*-nitrophenol is steam volatile. Steam volatility and lower boiling and melting points of *o*-nitrophenol are due to *intramolecular hydrogen bonding (chelation)* in this isomer (Eq. 15.17a). In the *para* and *meta* isomers, such intramolecular hydrogen bonding is not possible. However, intermolecular hydrogen bonding is possible (Eq. 15.17b).

(a)   (b)   (15.17)

Volatility including steam volatility, requires intermolecular forces to be overcome by heat before the molecules can escape into the gas phase. This requires a higher temperature for the *para* isomer, where intermolecular hydrogen bonds have to be broken, than for the *o*-isomer whose molecules are monomeric. Solubility in water is also affected by hydrogen bonding. Solubility of phenols in water is assisted by hydrogen bonding with water molecules. This is possible in the *m*- and *p*-isomers but is hindered due to chelation in the *o*-isomer (whose hydrogen bonding tendency is satisfied internally).

The preparation of *p*-nitrophenol can be carried out efficiently in the laboratory using an indirect route (Eq. 15.18).

(15.18)

Direct nitration of phenol suffers from the following drawbacks.

- The hydroxyl group of phenol is prone to oxidation by nitric acid.
- Phenol is very reactive and therefore, polynitration cannot be avoided.

Acetylation of phenol gives phenyl acetate which is less reactive than phenol, but is still *ortho/para* directing (Chapter 9, Table 9.1). Nitration of phenyl acetate gives mainly the *para* product. *p*-Nitrophenyl acetate upon hydrolysis gives *p*-nitrophenol. In this strategy, the OH group is acetylated, the required reaction carried out, and then deacetylated. This process is called *protection/deprotection*.

15.5.3 AMINOPHENOLS

Aminophenols are prepared by reduction of nitrophenols, using metals such as tin, zinc or iron in the presence of hydrochloric acid. The amino group of aminophenols can be esterified more readily than the hydroxyl group. By heating *p*-aminophenol with acetic anhydride, *p*-acetamidophenol (paracetamol, acetaminophen), a drug used for controlling fever (antipyretic), is obtained.

$$HO-\underset{\bigcirc}{\bigcirc}-NH-\underset{\underset{O}{\|}}{C}-CH_3 \quad \text{\textit{p}-Acetamidophenol}$$

15.5.4 NAPHTHOLS

There are two isomeric naphthols, 1- and 2-naphthol or α- and β-naphthol. Naphthols are prepared from the corresponding sulphonic acids by hydrolysis with sodium hydroxide (Eq. 15.5). The alkyl ethers of 2-naphthol are used in perfumery. Methyl 2-naphthyl ether is known as nerolin. Naphthols are intermediates in the dye industry.

15.5.5 PHENOL ETHERS

Phenol ethers are prepared by treating sodium phenoxide with alkyl halide (Williamson synthesis). Dialkyl sulphates, like dimethyl and diethyl sulphate in alkali, often give better yields than the alkyl halides. Methyl phenyl ether (anisole) and ethyl phenyl ether (phenetole) are important perfumery chemicals.

15.5.6 FRIES REARRANGEMENT

Phenyl esters, upon heating with anhydrous aluminium chloride rearrange to phenolic ketones. This is illustrated for phenyl acetate (Eq. 15.19) (see also Chapter 19).

$$\underset{\bigcirc}{\overset{\overset{O}{\|}}{O-C-CH_3}} \xrightarrow{AlCl_3, \Delta} \underset{\bigcirc}{\overset{OH}{}} COCH_3 \quad + \quad \underset{\underset{COCH_3}{\bigcirc}}{\overset{OH}{}} \tag{15.19}$$

This is a better method for preparing such ketones, than direct Friedel–Crafts acylation of phenols.

15.6 SPECTROSCOPY OF ALCOHOLS AND PHENOLS

15.6.1 IR SPECTROSCOPY

The O—H stretching frequency in phenols and alcohols appears as a broad band in the region 3200–3600 cm^{-1}. The band is broad because of intermolecular hydrogen bonding. In molecules where such hydrogen bonding is absent, it appears as a sharp peak at 3610–3640 cm^{-1}. In very dilute solutions, intermolecular hydrogen bonding is relatively less. As a result, the OH band of alcohols and phenols becomes sharper and shifts to about 3600 cm^{-1} from about 3300 cm^{-1} in very dilute solutions. In molecules where intramolecular hydrogen bonding due to chelation is present, as in *o*-nitrophenol, the OH band appears as a sharp peak at about $3200\,cm^{-1}$ and is unaffected by dilution. Phenols and alcohols show the $C-O$ stretching band(s), one or two, at $1000-1250\,cm^{-1}$. These bands are also seen in ethers, carboxylic acids and esters.

15.6.2 NMR SPECTROSCOPY

The chemical shift of the proton of the OH group is sensitive to hydrogen bonding. Hydrogen bonding is a deshielding effect and causes a downfield shift of the OH signal. Since the degree of intermolecular hydrogen bonding depends on concentration, temperature and solvent, all these variables affect the chemical shift. In simple alcohols, it may appear in the range 1–5 ppm and in phenols, in the range 4–7 ppm. If intermolecular hydrogen bonding is present, the OH signal is usually broad. Addition of a drop of D_2O causes exchange of H for D and the signal may diminish or completely disappear. This technique is employed in NMR spectroscopy to identify such exchangeable protons.

15.7 PHENOLS AS ANTIOXIDANTS

Sterically hindered phenols like 2,6-di-*tert*-butyl-*p*-cresol, are employed as antioxidants. For example, the deterioration of lubricating oils by autoxidation during storage and use can be slowed down by the addition of such antioxidants in small quantities. Autoxidation is a free radical chain reaction. Any molecule which stops or slows down chain propagation can function as an antioxidant. Typically, such a molecule reacts with a free radical which is involved in propagating the chain (Eq. 15.20).

$$R^\bullet + H-O-Ph \rightarrow RH + Ph-O^\bullet \tag{15.20}$$

$R^\bullet$ is the chain propagating free radical and $Ph-O-H$ is the phenolic antioxidant. Normally, the new free radical $PhO^\bullet$ will continue propagating the chain. However, this will be slowed down if the radical $PhO^\bullet$ is very unreactive, either due to stability or due to steric effect. The phenoxy radical, $Ph-O^\bullet$ is resonance stabilised. 2,6-Di-*tert*-butyl-4-methylphenoxy radical (Eq. 15.21) is highly unreactive due to resonance and additionally because of steric hindrance.

$$\tag{15.21}$$

Hindered phenols and hindered anilines are one class of antioxidants. The role of many naturally occurring phenols like flavanoids and vitamin E in biological systems is now recognised to be that of antioxidants.

KEY POINTS

- Phenols are acidic unlike alcohols. The acidity of phenols is explained in terms of the resonance stabilistation of the phenoxide anion. Electron-withdrawing groups increase acidity. Groups like nitro in the *para* position which withdraw electrons by both resonance and inductive effects, cause a large increase in acidity.
- One of the methods of synthesising phenols is the hydrolysis of diazonium salts.
- Phenol itself is synthesised industrially by the following process:
  Benzene + propene → cumene → cumene hydroperoxide → phenol + acetone
- Being weak acids, phenols dissolve in aqueous alkali forming salts called phenoxides.
- The OH group of phenol activates the ring towards electrophilic substitution. All the common electrophilic substitutions like nitration and halogenation take place readily, to give *ortho/para* products.
- There are electrophilic substitution reactions which are shown only by phenols. Some of these are discussed.
- Lederer–Manasse reaction is illustrated by the reaction of phenol with formaldehyde under acid or base catalysis to give *o-/p*-hydroxymethylphenol. This is the basis for the production of phenol–formaldehyde resins.
- Phenol reacts with chloroform in alkali to give *o*-hydroxybenzaldehyde (salicylaldehyde). Dichlorocarbene formed from chloroform and alkali is the electrophile.
- The phenoxide ion condenses with carbon dioxide under pressure and high temperature to give *o*-hydroxybenzoic acid (salicylic acid). This is the Kolbe–Schmitt reaction. Acetylation of salicylic acid gives aspirin.
- Gattermann aldehyde synthesis involves the reaction of phenol or anisole with HCN in the presence of HCl and $AlCl_3$ to give phenolic aldehydes. $ClCH=NH$ formed by the addition of HCl to HCN is the reactive species.
- Houben–Hoesch synthesis of phenolic ketones is an extension of this reaction, where a nitrile, like acetonitrile +HCl condenses in the presence of $AlCl_3$ to give ketones.
- Phenols couple with diazonium salts to give azo-dyes. The diazonium cation is the electrophile.
- Acylation of phenols (esterification) can be done using acid anhydride or acyl halide.
- Phenyl acetate, on heating with aluminium chloride, rearranges to hydroxyacetophenone. This is the Fries rearrangement.
- Alkyl halides or alkyl sulphates react with phenoxides to give phenol ethers.
- *p*-Nitrophenol is prepared by nitrating phenyl acetate and then removing the acetyl group by hydrolysis. This is an example of protection/deprotection.
- *o*- And *p*-nitrophenols can be separated by steam distillation. The former is steam volatile, not the latter. Chelation in the former is the reason for its lower boiling point and melting point.

- The naphthols are prepared from the sulphonic acids by hydrolysis with alkali.
- Hindered phenols like 2,6-di-*tert*-butyl-*p*-cresol, are used as antioxidants.
- The hydroxyl group of phenols and alcohols can be studied by both IR and NMR spectroscopy.

EXERCISES

SECTION I

1. Rearrange the following in the increasing order of acid strength.

   (a) Phenol, *p*-cresol, *p*-nitrophenol
   (b) Phenol, cyclohexanol, acetic acid
   (c) *p*-Methoxyphenol, *p*-hydroxybenzaldehyde, *p*-chlorophenol
   (d) *o*-Chlorophenol, *o*-fluorophenol, phenol
   (e) 2, 6-Dichlorophenol, 3, 4-dichlorophenol, phenol

2. Which has a higher boiling point? Why?

   (a) Salicylaldehyde or *p*-hydroxybenzaldehyde
   (b) *o*-Nitrophenol or *m*-nitrophenol
   (c) *o*-Nitroaniline or *p*-nitroaniline

3. How can the following conversions be brought about?

   (a)

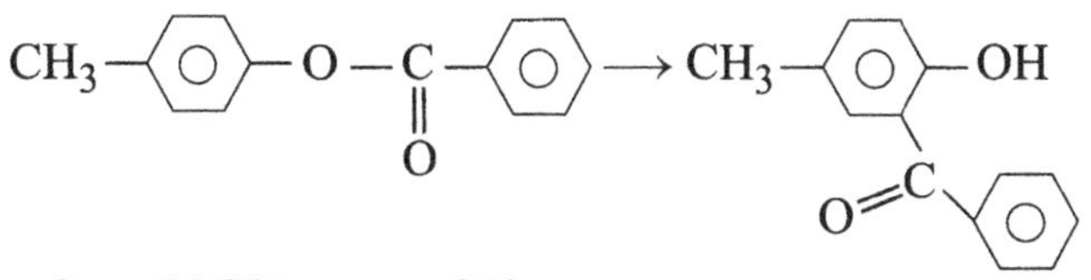

   (b)

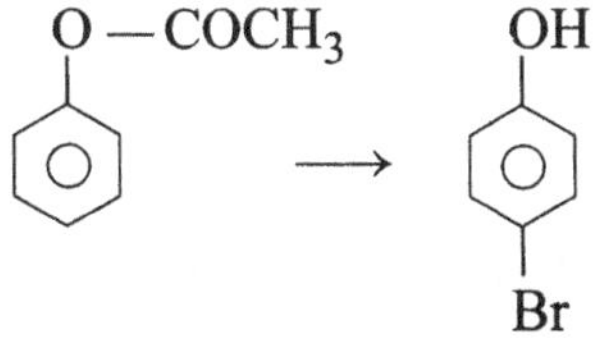

   (c)

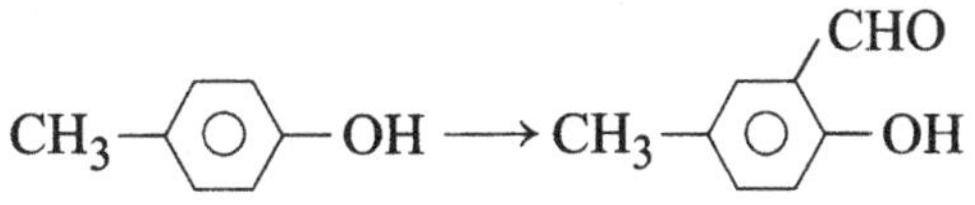

   (d)

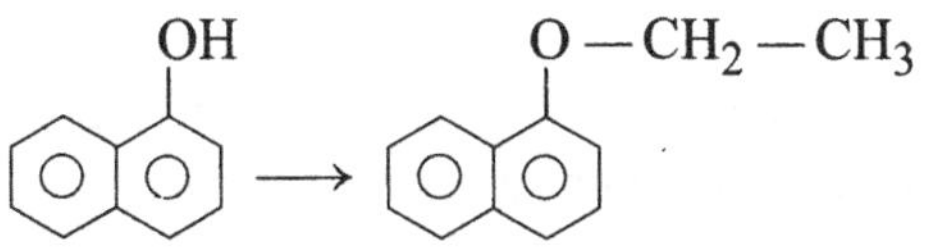

(e)

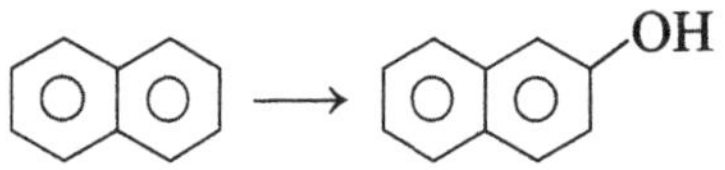

(f)

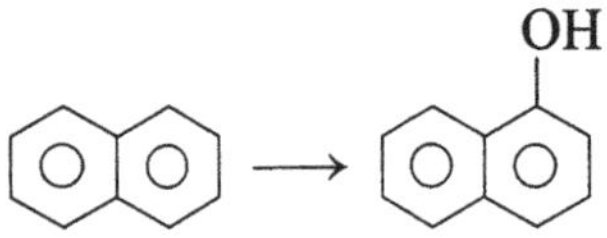

4. Give the products of the following reactions.

(a) 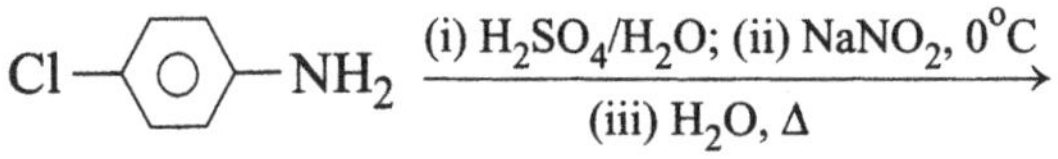 Cl—⬡—NH_2 $\xrightarrow[\text{(iii) } H_2O, \Delta]{\text{(i) } H_2SO_4/H_2O;\ \text{(ii) } NaNO_2,\ 0°C}$

(b) 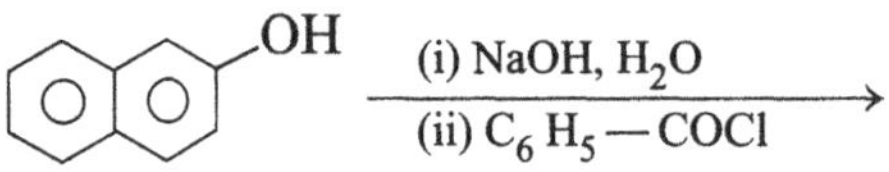 $\xrightarrow[\text{(ii) } C_6H_5-COCl]{\text{(i) NaOH, } H_2O}$

(c) $+ CHCl_3 + NaOH \xrightarrow{\Delta}$

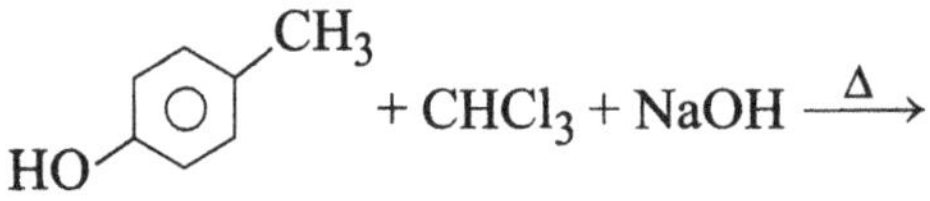

(d) 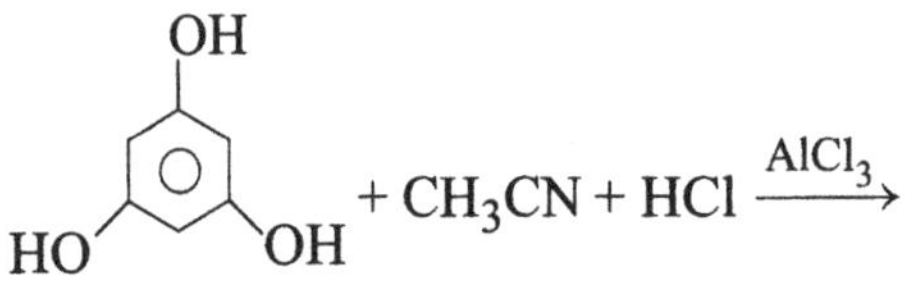 $+ CH_3CN + HCl \xrightarrow{AlCl_3}$

(e) 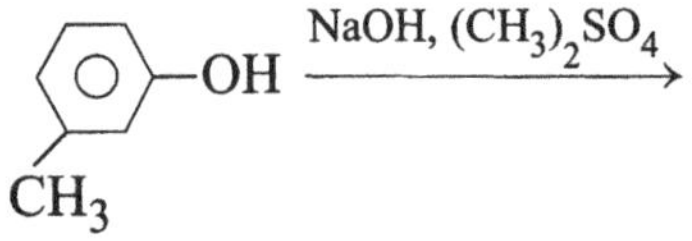 $\xrightarrow{NaOH,\ (CH_3)_2SO_4}$

(f) 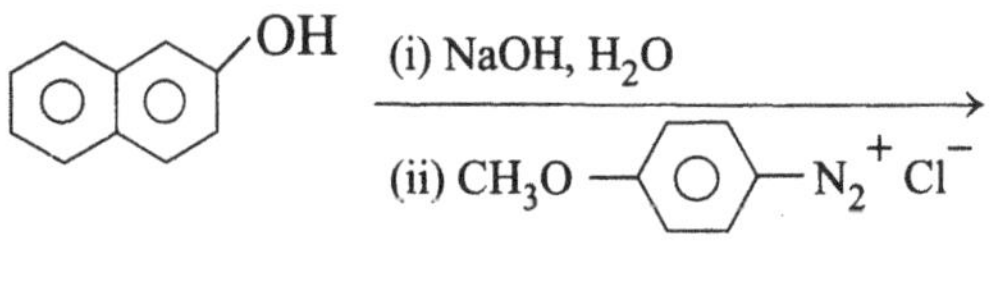 $\xrightarrow[\text{(ii) } CH_3O-⬡-N_2^+\ Cl^-]{\text{(i) NaOH, } H_2O}$

SECTION II

1. OH stretching band of *p*-cresol appears at $3300\ cm^{-1}$ as a broad band. That of the hindered phenol, 2, 6-di-*tert*-butyl-*p*-cresol (see Eq. 15.21) appears as a sharp band at about $3600\ cm^{-1}$ even in concentrated solution. Explain.

2. Show how *o*-nitrophenol can be differentiated from *p*-nitrophenol by IR spectroscopy.

3. Show how *p*-cresol, benzyl alcohol and anisole can be differentiated by (i) chemical methods; (ii) spectroscopic methods.

4. In Section 15.4.7 it is stated that phenols cannot be esterified by refluxing with alcohol and H^+ because of the low nuclcophilicity of the oxygen in phenol. Why is it less nuclcophilic than in alcohols? Refer to the mechanism of esterification, Eq. 14.19, Chapter 14, and explain if this statement is valid.

5. Refer to Eq. 5.10a. Explain why the H of chloroform is acidic.
6. Outline the mechanism of the Schotten–Baumann reaction. Explain why alkali is necessary for the reaction.
7. Apply the methodology of protection of the OH group and suggest a method of synthesis of *p*-bromophenol.
8. Refer to the NMR spectrum of *p*-cresol, Chapter 13, Fig. 13.9. One of the signals disappeared when the spectrum was run after the addition of a drop of D_2O. Which signal? Why?

CHALLENGING QUESTIONS

1. Autoxidation of cumene (Eq. 15.6) is a free radical chain reaction. The first two steps are given below. Write down the subsequent steps. Show how an antioxidant like a hindered phenol can slow down the oxidation.
   (Notes: (i) In• is an initiating free radical of unspecified origin. (ii) Molecular oxygen in the ground state is a diradical, represented as $\overset{\bullet}{O} - \overset{\bullet}{O}$.)

$$In^{\bullet} + C_6H_5-\underset{CH_3}{\overset{CH_3}{C}}-H \longrightarrow InH + C_6H_5-\underset{CH_3}{\overset{CH_3}{C^{\bullet}}}$$

$$C_6H_5-\underset{CH_3}{\overset{CH_3}{C^{\bullet}}} + \overset{\bullet}{O}-\overset{\bullet}{O} \longrightarrow C_6H_5-\underset{CH_3}{\overset{CH_3}{C}}-O-\overset{\bullet}{O}$$

2. Fries rearrangement of phenyl acetate gives a mixture of *o*-hydroxyacetophenone and *p*-hydroxyacetophenone. They can be separated by steam distillation.

   (a) Which isomer is steam volatile? Why?
   (b) They can be identified by IR spectroscopy, by studying the O—H stretching frequency. How?
   (c) They can be identified by their NMR spectra, from the splitting pattern of the aryl protons. How?
   (d) They can be identified by NMR by the chemical shift difference of the O—H proton. How?
   (e) After the Fries rearrangement, some unreacted ester will also be present. How can it be removed from the phenolic products by chemical separation?

16 Carbonyl Chemistry

OBJECTIVES   In this chapter, you will learn about,

- the polarity of the carbonyl group
- the methods of preparation, properties and uses of aldehydes and ketones
- the reason for the acidity of the α-hydrogens
- enolate anion formation
- acetal formation, addition of HCN, addition of bisulphite, condensation with amine reagents, aldol condensation, Claisen reaction, Knoevenagel reaction, Reformatsky reaction, Cannizzaro reaction, benzoin condensation, Michael addition and Wittig reaction
- reduction of the carbonyl group by complex metal hydrides, Meerwein–Ponndorf–Verley reduction, Wolff–Kishner reduction and bimolecular reduction to pinacol
- haloform reaction
- photolysis, Norrish type I and type II reactions
- qualitative tests for alehydes and ketones
- IR and NMR spectroscopy of carbonyl compounds

16.1   NATURE OF THE CARBONYL GROUP

The carbonyl group, $-C=O$, with hydrogen and/or alkyl or aryl groups attached to the carbon, is present in aldehydes and ketones. Much of the chemistry of this class of compounds is dictated by two aspects of the carbonyl group. These are, (i) the polarity of the $C=O$ bond and (ii) the ability of the carbonyl group to delocalise and thereby stabilise a negative charge on the adjacent carbon atom (α-carbon). The latter property makes the hydrogens on the α-carbon relatively acidic.

16.1.1 CARBONYL POLARITY

The inherent polarity of the carbonyl group is denoted by the resonance structures (Eq. 16.1a). The carbonyl carbon is an electrophile.

$$\left[\begin{array}{c} \diagup \\ C = O \end{array} \longleftrightarrow \begin{array}{c} \diagup \\ \overset{+}{C} - \bar{O} \end{array} \quad \begin{array}{c} \overset{\delta^+}{\diagup} \quad \overset{\delta^-}{} \\ C \cdots O \end{array} \right] \qquad (a)$$

$$\bar{Nu:} + \diagup C \overset{\frown}{=} \bar{O} \longrightarrow Nu - \underset{|}{\overset{|}{C}} - \bar{O} \xrightarrow{\;H^+\;} Nu - \underset{|}{\overset{|}{C}} - OH \qquad (b) \qquad\qquad (16.1)$$

$$\diagup C = O + H^+ \longrightarrow \left[\diagup C = \overset{+}{O} - H \longleftrightarrow \diagup \overset{+}{C} - O - H \right] \qquad (c)$$

As a result of this polarity, the carbonyl carbon is susceptible to nucleophilic attack (Eq. 16.1b). When the anion is quenched by a positive species like H^+, the overall reaction is the nucleophilic addition of Nu–H to the carbonyl group. This is illustrated by the addition of HCN (Eq. 16.2a). Addition of Grignard reagents is already familiar to us (Eq. 16.2b).

$$\diagup C = O + \bar{CN} \longrightarrow \begin{array}{c} CN \\ | \\ - \underset{|}{C} - \bar{O} \end{array} \xrightarrow{\;H^+\;} \begin{array}{c} CN \\ | \\ - \underset{|}{C} - OH \end{array} \qquad (a)$$

Cyanohydrin

$$R - MgX + \diagup C = O \longrightarrow R - \underset{|}{\overset{|}{C}} - O - MgX \qquad (b) \qquad\qquad (16.2)$$

$$\underset{X}{\diagup} C = O + \bar{OH} \longrightarrow HO - \underset{X}{\overset{O}{\underset{|}{C}}} - \longrightarrow H - O - \overset{O}{\underset{}{\overset{\|}{C}}} - + X^- \qquad (c)$$

$$\underset{R'O}{\overset{R}{\diagdown}} C = O + \bar{OH} \longrightarrow R - \underset{HO}{\overset{O}{\underset{|}{C}}} - OR' \longrightarrow R - \overset{O}{\underset{HO}{\overset{\|}{C}}} + R'\bar{O} \qquad (d)$$

Many reactions of carbonyl compounds fall under this category. Hydrolysis of esters and other acid derivatives involves the nucleophilic addition of OH^- followed by elimination of a leaving group (Eqs. 16.2c, 16.2d). The oxygen of the carbonyl group is a nucleophile. It can bond with a proton and other electrophiles including Lewis acids. Protonation of the carbonyl oxygen increases the reactivity of the carbonyl group in nucleophilic addition (Eq. 16.1c).

When the carbonyl group is conjugated with a double bond, we have α, β-unsaturated carbonyl compounds. In such molecules, a partial positive charge is present on the β-carbon as well (Eq. 16.3) and nucleophilic addition can also take place at the β-carbon. Recall conjugate addition to conjugated dienes (Chapter 6).

$$\text{C}=\text{C}-\text{C}=\text{O} \longleftrightarrow \text{C}=\text{C}-\overset{+}{\text{C}}-\overset{-}{\text{O}} \longleftrightarrow \overset{+}{\text{C}}-\text{C}=\text{C}-\overset{-}{\text{O}} \qquad (16.3)$$

16.1.2 ACIDITY OF HYDROGEN AT THE α-POSITION

A hydrogen at the α-position to the carbonyl group is relatively acidic. A hydrogen bound to a carbon, called a carbon acid, is a weak acid compared to those attached to more electronegative atoms. Among carbon acids, the acidity can vary over a wide range. We have seen in Chapter 5 how alkynes are more acidic than other hydrocarbons. The pK_a values of some weak acids are listed in Table 16.1.

Table 16. 1   Acid strengths of some weak acids

Structure	Name	Approximate pK_a
CH_3-COOH	Acetic acid	4.75
C_6H_5-OH	Phenol	10
$HO-H$	Water	15.7
CH_3-OH	Methanol	16
$(CH_3-CO)_2CH-H$	2,4-Pentanedione	9
$CH_3-CO-\overset{\overset{\textstyle H}{\textstyle \mid}}{\underset{\underset{\textstyle COOC_2H_5}{\textstyle \mid}}{CH}}$	Ethyl acetoacetate	11
$(C_2H_5O-CO)_2CH-H$	Diethyl malonate	13
$C_2H_5O-CO-CH_2-H$	Ethyl acetate	25
$CH_3-CO-CH_2-H$	Acetone	20
$H-C\equiv C-H$	Acetylene	26
H_2N-H	Ammonia	35

Acetone is a weak acid, weaker than all the oxygen acids listed here, but stronger than acetylene. Carbonyl compounds with two carbonyl groups flanking a CH_2 – illustrated here by 2,4-pentanedione, ethyl acetoacetate and diethyl malonate – are stronger acids than acetone. They are in fact stronger than water and methanol. Resonance stabilisation and delocalisation of the charge, which is the cause of the acidity, is more for the anion of 2,4-pentanedione than that for acetone (Eq. 16.4).

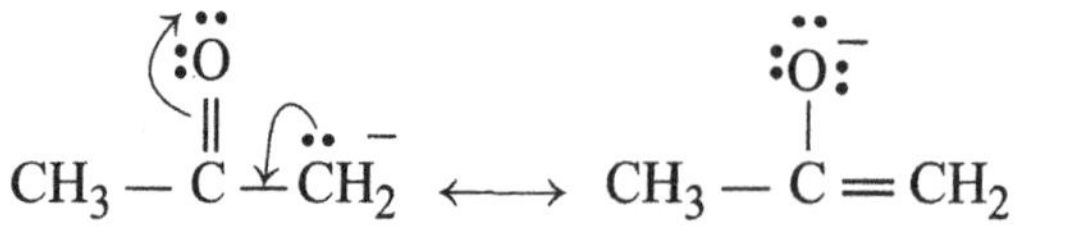

$$CH_3-\overset{\overset{\textstyle :\ddot{O}:}{\textstyle \|}}{C}-\ddot{C}H_2^- \longleftrightarrow CH_3-\overset{\overset{\textstyle :\ddot{O}:^-}{\textstyle \mid}}{C}=CH_2 \qquad (a)$$

$$\underset{\substack{\text{O} \\ \parallel}}{CH_3-C} \xleftarrow{} \underset{\substack{\cdot\cdot \\ -}}{CH} - \underset{\substack{\text{O} \\ \parallel}}{C} - CH_3 \longleftrightarrow CH_3 - \underset{\substack{\text{O}^- \\ \mid}}{C} = CH - \underset{\substack{\text{O} \\ \parallel}}{C} - CH_3$$

$$\tag{16.4}$$

$$\longleftrightarrow CH_3 - \underset{\substack{\text{O} \\ \parallel}}{C} - CH = \underset{\substack{\text{O}^- \\ \mid}}{C} - CH_3 \quad (b)$$

For the acid–base reaction (Eq. 16.5a), the equilibrium will be very much to the right if BH is a weaker acid than AH. In other words, B^- is a stronger base than A^-. Treatment of ethyl acetoacetate with aqueous alkali or with sodium methoxide in methanol, will convert it almost entirely to the sodium salt (Eq. 16.5b). Treatment of acetone with sodium methoxide will also result in the formation of the anion – but in very small concentration – with the equilibrium very much to the left (Eq. 16.5c).

$$B^- + AH \rightleftharpoons BH + A^- \tag{a}$$

$$\underset{\substack{\text{O} \\ \parallel}}{CH_3 - C} - CH_2 - \underset{\substack{\text{O} \\ \parallel}}{C} - OC_2H_5 + Na^+ \ \overset{-}{OCH_3}$$

$$\longrightarrow \underset{\substack{\text{O} \\ \parallel}}{CH_3 - C} - \underset{\substack{- \\ Na^+}}{CH} - \underset{\substack{\text{O} \\ \parallel}}{C} - OC_2H_5 + CH_3OH \quad (b) \tag{16.5}$$

$$\underset{\substack{\text{O} \\ \parallel}}{CH_3 - C} - CH_3 + NaOCH_3 \xrightleftharpoons{} \underset{\substack{\text{O} \\ \parallel}}{CH_3 - C} - \overset{-}{CH_2} + CH_3OH \quad (c)$$

The significant fact is that the carbanion with the negative charge on the α-carbon is formed, even from monocarbonyl compounds like acetone, though in very low concentration. Since carbanions are very reactive, many reactions arise from this intermediate when acetone is treated with alkali. Anions of the type seen in Eq. 16.4a and b, can be considered as the conjugate bases of the enol forms of the ketones. They are called *enolate anions*.

16.2 PROPERTIES OF ALDEHYDES AND KETONES

Structures, names and boiling points of simple aldehydes and ketones are listed in Table 16.2.

Formaldehyde, acetaldehyde and acetone are miscible with water in all proportions. The solubility of the others decreases with increasing molecular weight. Formaldehyde, a gas at room temperature, dissolves in water to give a solution known as formalin. Commercial formalin is 37% formaldehyde in water along with some methanol. Formaldehyde exists as two solid polymers. Trioxane with a m.p of 62°C is a cyclic trimer while paraformaldehyde, which decomposes on heating, is a linear polymer (see structures). Pure gaseous formaldehyde which may be required

Table 16. 2 Boiling points of simple aldehydes and ketones

IUPAC name	Trivial name	Structure	Boiling point (°C)
Methanal	Formaldehyde	H_2CO	−21
Ethanal	Acetaldehyde	$CH_3—CHO$	20
Propanal	Propionaldehyde	$CH_3—CH_2—CHO$	49
Benzaldehyde	Benzaldehyde	$C_6H_5—CHO$	178
Propanone	Acetone	$CH_3—CO—CH_3$	56
Butanone	Ethyl methyl ketone	$CH_3—CO—CH_2—CH_3$	80
Cyclohexanone	Cyclohexanone	$(cyclo)C_5H_{10}C=O$	157
Acetophenone	Acetophenone	$CH_3—CO—C_6H_5$	202
Benzophenone	Benzophenone	$C_6H_5—CO—C_6H_5$	306

for reactions such as Grignard synthesis, is obtained by heating paraformaldehyde or trioxane. Acetaldehyde also forms a trimer called paraldehyde.

Trioxane   Paraformaldehyde   Paraldehyde

16.3 PREPARATION

16.3.1 INDUSTRIAL PREPARATION

Formaldehyde is manufactured by the catalytic oxidation of methanol over silver, or iron/molybdenum oxide catalysts (Eq. 16.6).

$$CH_3OH + \frac{1}{2}O_2 \xrightarrow{\text{Ag, }500^\circ C} CH_2O + H_2O \tag{16.6}$$

Acetaldehyde and propionaldehyde are also made by this method from the alcohols. Catalytic dehydrogenation using Pd catalysts can also be used. The name aldehyde is derived from the words alcohol and dehydrogenation. Formaldehyde is used in the manufacture of phenol–formaldehyde and other resins. The aqueous solution, formalin, is used as a preservative.

Hydroformylation: Terminal alkenes, $R—CH=CH_2$, condense with carbon monoxide and hydrogen in the presence of metal carbonyls to form aldehydes. This reaction is known as *hydroformylation* and, in the industry as the *oxo-process* (Eq. 16.7).

$$R-CH=CH_2+CO+H_2 \xrightarrow{Co_2(CO)_8} R-CH_2-CH_2-\overset{\overset{\displaystyle O}{\|}}{C}-H \qquad (16.7)$$

The aldehydes can be further hydrogenated to alcohols. Ketones, other than acetone, are prepared by the oxidation of secondary alcohols which in turn are manufactured by the hydration of alkenes. Acetone can also be obtained using this route by the oxidation of isopropyl alcohol. However, as already seen, the important method for manufacturing acetone is via cumene hydroperoxide (Chapter 15, Eq. 15.6).

16.3.2 LABORATORY METHODS OF PREPARATION

16.3.2.1 From Alkenes by Ozonolysis, Via Glycol, and Via Hydration and Oxidation

Most of the laboratory methods of preparation of aldehydes and ketones employ alkenes as starting materials. Alkenes are directly converted to carbonyl compounds containing fewer carbon atoms by ozonolysis. The same carbonyl compounds are obtained if the alkene is first oxidised to the glycol and then oxidised by periodic acid or lead tetraacetate (Eqs. 16.8a and 16.8b). Hydration of alkenes either directly using aqueous acid or indirectly via hydroboration or oxymercuration gives alcohols which can be oxidised to the carbonyl compounds using chromic acid. If terminal alkenes containing the $=CH_2$ grouping are hydrated via hydroboration (Eq. 16.8c), primary alcohols are obtained which can be oxidised using CrO_3–pyridine complex to the aldehydes. Direct acid catalysed hydration or indirect hydration through oxymercuration of terminal mono-substituted alkenes of the type $R-CH=CH_2$, yields secondary alcohols $[R-CH(OH)CH_3]$ which can be oxidised to the ketones (Eq. 16.8d). Symmetrically di-substituted alkenes of the type $R-CH=CH-R$ can be hydrated and oxidised to $R-CO-CH_2-R$ (Eq. 16.8e). With di-substituted alkenes of the type $R-CH=CH-R'$, regioselectivity is likely to be poor if this method is followed (Eq. 16.8f). If the molecule is prone to rearrangement during acid catalysed hydration (Eq. 16.8g), indirect hydration using the oxymercuration reaction has to be resorted to (Eq. 16.8h).

$$\underset{R'}{\overset{R}{>}}C=C\underset{R'''}{\overset{R''}{<}} \xrightarrow{O_3} \underset{R'}{\overset{R}{>}}C=O+O=C\underset{R'''}{\overset{R''}{<}} \qquad (a)$$

$$\underset{R'}{\overset{R}{>}}\underset{\underset{HO}{|}}{C}-\underset{\underset{OH}{|}}{C}\underset{R'''}{\overset{R''}{<}} \xrightarrow[\text{Pb(OCOCH}_3)_4]{HIO_4 \text{(or)}} \underset{R'}{\overset{R}{>}}C=O+O=C\underset{R'''}{\overset{R''}{<}} \qquad (b)$$

$$\underset{R}{\overset{R'}{>}}C=CH_2 \xrightarrow{\text{(i) B}_2\text{H}_6;\text{ (ii) H}_2\text{O}_2/\text{OH}^-} \underset{R}{\overset{R'}{>}}CH-CH_2OH \longrightarrow \underset{R}{\overset{R'}{>}}CH-CHO \qquad \text{(c)}$$

$$\underset{R}{\overset{H}{>}}C=CH_2 \xrightarrow[\text{oxymercuration}]{\text{H}_3\text{O}^+\text{(or)}} \underset{R}{\overset{H}{>}}\underset{\underset{OH}{|}}{C}-CH_3 \xrightarrow{[o]} R-\overset{\overset{O}{\parallel}}{C}-CH_3 \qquad \text{(d)}$$

$$R-CH=CH-R \xrightarrow{\text{H}_3\text{O}^+} R-\underset{\underset{OH}{|}}{C}H-CH_2-R \xrightarrow{[o]} R-\overset{\overset{}{\underset{\underset{O}{\parallel}}{}}}{C}-CH_2-R \qquad \text{(e)}$$

$$R-CH=CH-R' \xrightarrow[\text{hydroboration}]{\text{H}_3\text{O}^+\text{(or)}} R-\underset{\underset{OH}{|}}{C}H-CH_2-R' + R\,CH_2-\underset{\underset{OH}{|}}{C}H-R'$$

$$R-\underset{\underset{O}{\parallel}}{C}-CH_2-R' \qquad R-CH_2-\underset{\underset{O}{\parallel}}{C}-R' \qquad \text{(f)}$$

(16.8)

$$CH_3-\underset{\underset{H}{|}}{\overset{\overset{CH_3}{|}}{C}}-CH=CH_2 \xrightarrow{\text{H}_3\text{O}^+} CH_3-\underset{\underset{HO}{|}}{\overset{\overset{CH_3}{|}}{C}}-CH_2-CH_3 \qquad \text{(g)}$$

$$CH_3-\underset{\underset{H}{|}}{\overset{\overset{CH_3}{|}}{C}}-CH=CH_2 \xrightarrow{\text{oxymercuration}} CH_3-\overset{\overset{CH_3}{|}}{CH}-\underset{\underset{OH}{|}}{CH}-CH_3$$

$$\longrightarrow CH_3-\overset{\overset{CH_3}{|}}{CH}-\underset{\underset{O}{\parallel}}{C}-CH_3 \qquad \text{(h)}$$

Alkynes can be hydrated (disussion in Chapter 5) using aqueous H_2SO_4 and $HgSO_4$ to ketones (Eq. 16.9a). This is especially suitable for the preparation of methyl ketones from terminal alkynes. If internal alkynes are used (Eq. 16.9b), a mixture of ketones will result.

$$R - C \equiv CH + H_2O \xrightarrow{H_2SO_4, HgSO_4} R - \underset{\underset{OH}{|}}{C} = CH_2 \rightarrow R - \underset{\underset{O}{\|}}{C} - CH_3 \quad \text{(a)}$$

$$R - C \equiv C - R' + H_2O \rightarrow R - \underset{\underset{O}{\|}}{C} - CH_2 - R' + R - CH_2 - \underset{\underset{O}{\|}}{C} - R' \quad \text{(b)}$$

(16.9)

We are already familiar with the conversion of $R-X$ to a terminal alkyne, which can then be converted to $R-CO-CH_3$. Hence this sequence of reactions forms an indirect route for the conversion of $R-X$ to $R-CO-CH_3$ (Eq. 16.10).

$$R - X + HC \equiv C^- Na^+ \rightarrow R - C \equiv CH \xrightarrow{H_2O} R - \underset{\underset{O}{\|}}{C} - CH_3 \quad (16.10)$$

An aldehyde can be converted to a ketone by first converting it to a secondary alcohol by Grignard reaction and then oxidising it (Eq. 16.11).

$$R - CHO + R'MgX \rightarrow R - \underset{\underset{OH}{|}}{CH} - R' \xrightarrow{Cr(VI)} R - \underset{\underset{O}{\|}}{C} - R' \quad (16.11)$$

16.3.2.2 Aryl Ketones by Friedel–Crafts Acylation

Reaction of an arene with an acyl halide or acid anhydride in the presence of anhydrous aluminium chloride produces the ketone (Chapter 9, Section 9.5.1.6). This reaction is controlled by the orienting effect of the groups (if any) which are already present. Thus, toluene can be acetylated to *p*-methylacetophenone. Ethylbenzene can be benzoylated to give *p*-ethylbenzophenone (Eq. 16.12).

$$CH_3 - \langle O \rangle + (CH_3CO)_2O \xrightarrow{AlCl_3} CH_3 - \langle O \rangle - COCH_3 \quad \text{(a)}$$

$$C_2H_5 - \langle O \rangle + \langle O \rangle - COCl \xrightarrow{AlCl_3} C_2H_5 - \langle O \rangle - \overset{\overset{O}{\|}}{C} - \langle O \rangle \quad \text{(b)}$$

(16.12)

Recall the acylation of naphthalene by succinic anhydride (Chapter 10, Section 10.2.3.4, Eq. 10.11). Friedel–Crafts acylation is not successful if deactivating groups are present.

16.3.2.3 Phenolic Aldehydes and Ketones

Reimer–Tiemann reaction and Gattermann aldehyde synthesis for phenolic aldehydes have already been discussed (Chapter 15, Sections 15.4.3 and 15.4.5). Phenolic ketones can be synthesised by Houben–Hoesch reaction and by Fries rearrangement (Chapter 15, Sections 15.4.5 and 15.5.6).

16.4   REACTIONS

Reactions of aldehydes and ketones can be broadly grouped into two types. (i) Nucleophilic addition to the carbonyl group and (ii) reactions due to the acidity of the α-hydrogens; which can also be called reactions of the enolate⁻ anion. The reactivity in nucleophilic addition depends upon the partial positive charge on the carbonyl carbon. As expected aldehydes – with only one alkyl group – are more reactive than ketones. Formaldehyde is the most reactive among the aldehydes. Aryl aldehydes and ketones are less reactive than aliphatic ones, because the positive charge density on the carbonyl carbon is less in the former because of delocalisation as a result of resonance.

16.4.1   NUCLEOPHILIC ADDITIONS

16.4.1.1   Addition of Water and Alcohols; Formation of Acetals and Ketals

Addition of water is the simplest example of nucleophilic addition. It can give an adduct, called a hydrate, which is a *gem*-glycol. As we know (Chapter 14, Section 14.2), these are unstable and exist in equilibrium with the parent carbonyl compound, with the latter predominating. Acid catalysed addition of water can be visualised as in Eq. 16.13.

$$(16.13)$$

If the reaction is carried out in methanol instead of water, a methyl acetal (VI) is formed (Eq. 16.14).

$$(I)$$

$$\xrightarrow[]{(3)} \quad \underset{\underset{+H^+ \;\;(II)}{\overset{\mid}{\underset{O-CH_3}{}}}}{CH_3-CH-OH} \quad \xrightarrow[]{(4)} \quad \underset{\underset{(III)}{\overset{\mid}{\underset{C:O-CH_3}{}}}}{CH_3-CH \overset{H}{\underset{+.}{-O}}-H}$$

(Hemiacetal)

$$\xrightarrow[]{(5)} \quad \underset{\underset{(IV)}{\overset{+}{\underset{:O-CH_3}{}}}}{CH_3-CH} \quad \longleftrightarrow \quad CH_3-CH \quad :O-CH_3$$

$$\xrightarrow[]{(6)} \quad \underset{\underset{(V)}{\overset{H}{\underset{+}{\underset{O-CH_3}{}}}}}{CH_3-CH-O-CH_3} \quad \xrightarrow[]{(7)} \quad \underset{(VI)}{CH_3-CH \overset{O-CH_3}{\underset{O-CH_3}{}}} \quad +H^+ \qquad (16.14)$$

Steps (1) and (2) show the acid catalysed addition of methanol to form structure (I). This can, in principle, undergo deprotonation as in step (3) to give (II). It has an —OH group and an —OCH$_3$ group attached to the same carbon atom and is called a hemiacetal. The hemiacetal, like a *gem*-glycol, is unstable and reacts further by the loss of a water molecule as in steps (4) and (5) to give a resonance stabilised cation, (IV). This cation undergoes one more nucleophilic addition of methanol and then undergoes deprotonation [steps (6) and (7)] to give the final product (VI), which has two —OCH$_3$ groups attached to the original carbonyl carbon. It is 1,1-dimethoxyethane, ether of the *gem*-glycol. Such 1,1- or *gem*-dialkoxy compounds when prepared from aldehyes, are called acetals. Similar addition of two molecules of methanol to a ketone like acetone gives 2,2-dimethoxypropane. *gem*-Dialkoxy compounds obtained from ketones are called ketals. Acetals and ketals, unlike the parent *gem*-glycols, are stable molecules. They are quite stable and resistant to hydrolysis in aqueous alkali, but can be hydrolysed to the parent carbonyl compound by aqueous acid. The acid catalysed hydrolysis is the reverse of all the steps involved in their formation.

16.4.1.2 Addition of HCN and Bisulphite

HCN adds to aldehydes and ketones to form cyanohydrins. Nucleophilic addition of CN$^-$ to the carbonyl group takes place, followed by protonotation (Eq. 16.15a). Cyanohydrins can be hydrolysed to hydroxyacids (Eq. 16.15b).

$$\text{Ph}-\overset{\underset{\displaystyle \overset{\cdot\cdot}{C}\equiv N}{\shortparallel}}{C}=\overset{\curvearrowleft}{O} \rightarrow \text{Ph}-\overset{\underset{\displaystyle C\equiv N}{|}}{\overset{\displaystyle H}{C}}-O^- \rightarrow \text{Ph}-\overset{\displaystyle H}{\underset{}{C}}\overset{\displaystyle \diagup OH}{\diagdown CN}$$

Cyanohydrin   (a)

$$\text{Ph}-\overset{\diagup OH}{\underset{\diagdown CN}{CH}} \xrightarrow{\text{H}^+,\ \text{H}_2\text{O}} \text{Ph}-\overset{\diagup OH}{\underset{\diagdown COOH}{CH}}$$

Mandelic acid   (b)   (16.15)

Aldehydes and ketones react with sodium hydrogen sulphite, to form bisulphite addition compounds (Eq. 16.16).

$$R-\overset{\diagup O}{\underset{\diagdown H}{C}} + \text{NaHSO}_3 \longrightarrow R-\overset{\overset{\displaystyle OH}{|}}{CH}-\text{SO}_3^-\ \text{Na}^+ \qquad (16.16)$$

These are crystalline solids and can be used to recover aldehydes or ketones from mixtures. After purification, acidification of the solid gives back the original aldehyde or ketone.

16.4.1.3 Addition of Organometallic Reagents: Reformatsky Reaction

Grignard reagents come to mind when organometallic reagents are discussed. Addition of Grignard reagents has already been dealt with (Chapter 14, Section 14.1.3.4). A related reaction is the Reformatsky reaction: α-bromo esters like ethyl α-bromoacetate react with aldehydes or ketones in the presence of zinc dust to give β-hydroxyesters (Eq. 16.17). In practice, zinc is added to a mixture of the bromoester and the carbonyl compound and then warmed. After the reaction, dilute acid is added to hydrolyse the complex.

$$\text{Br}-\text{CH}_2-\text{COOC}_2\text{H}_5 + \text{Zn} \longrightarrow \text{Br}-\text{Zn}-\text{CH}_2\text{COOC}_2\text{H}_5$$

$$\xrightarrow{\text{RCHO}} R-\overset{\overset{\displaystyle |}{CH}-\text{OZn Br}}{\underset{\displaystyle CH_2-\text{COOC}_2\text{H}_5}{}} \xrightarrow{\text{H}_3\text{O}^+} R-\overset{\diagup OH}{\underset{\diagdown CH_2-\text{COOC}_2\text{H}_5}{CH}}$$

β-Hydroxyester   (a)

$$R-\overset{\diagup OH}{CH}-\text{CH}_2-\text{COOC}_2\text{H}_5 \xrightarrow{-\text{H}_2\text{O}} R-\text{CH}=\text{CH}-\text{COOC}_2\text{H}_5$$

α, β-Unsaturated ester   (b)

(16.17)

The mechanism is similar to that of the Grignard reaction. The Reformatsky reaction is not as versatile as the Grignard reaction but has got specific applications. A Grignard reagent cannot be prepared from a bromoester since such a Grignard reagent will react with another molecule of the ester. β-Hydroxy esters can be dehydrated to α, β-unsaturated esters (Eq. 16.17b).

16.4.1.4 Addition of Reagents Containing the NH₂ Group

Amines and reagents containing the NH_2 group add to the carbonyl group with the elimination of a molecule of water according to the following mechanism (Eq. 16.18).

$$R-CH=\overset{+}{\underset{H}{N}}-R' \xrightarrow{-H^+} R-CH=N-R'$$

(16.18)

In the first step, the amino group adds as a nucleophile to the carbonyl group through the nitrogen. In some cases, if the reaction is carried out in acid medium, this may be preceded by protonation of the oxygen. The initial adduct is a 1-hydroxy-1-amino compound, (I). This generally undergoes dehydration spontaneously to give the final product with a $C=N$ double bond. The overall reaction is condensation rather than addition. Table 16.3 gives a list of the amino reagents that will react in this manner, along with the structure of the products and their names.

Table 16. 3 Condensation of NH_2 reagents with aldehydes or ketones, $R(R')C=O$

Name of reagent	Structure of reagent	Name of product	Structure of product
Amine	NH_2-R''	Schiff base (imine)	$R(R')C=NR''$
Hydroxylamine	NH_2-OH	Oxime	$R(R')C=NOH$
Semicarbazide	$NH_2-NH-CO-NH_2$	Semicarbazone	$R(R')C=N-NH-CO-NH_2$
Hydrazine	NH_2-NH_2	Hydrazone	$R(R')C=N-NH_2$
Phenylhydrazine	$NH_2-NH-C_6H_5$	Phenylhydrazone	$R(R')C=N-NH-C_6H_5$
2,4-Dinitro phenylhydrazine	$NH_2-NH-C_6H_3(NO_2)_2$	2,4-Dinitrophenyl-hydrazone	$R(R')C=N-NH-C_6H_3(NO_2)_2$

Condensation with simple amines gives imines which are called Schiff bases. The $C=N$ double bond can be reduced either by catalytic hydrogenation or by using LAH to obtain secondary amines. This process is called *reductive amination*. Oximes obtained by condensation with hydroxylamine can also be reduced to amines by LAH or by catalytic hydrogenation. During this reduction, the OH group is removed by hydrogenolysis. Oximes of ketones are converted to amides when treated with PCl_5 or sulphuric acid (Eq. 16.19a).

$$\underset{\substack{\text{Oxime}}}{\underset{R'}{\overset{R}{\diagdown}}C=N\diagdown_{OH}} \xrightarrow[\substack{H_2SO_4}]{PCl_5\ (or)} R'-\overset{\overset{\displaystyle O}{\|}}{C}-NH-R \qquad \text{(a)}$$

$$\underset{\substack{}}{R'-\overset{\overset{\displaystyle O}{\|}}{C}-NHR} \xrightarrow{H_3O^+} R'-\overset{\overset{\displaystyle O}{\|}}{C}-OH+R-NH_2 \qquad \begin{matrix}\text{(b)}\\ \\ (16.19)\end{matrix}$$

This reaction, known as Beckmann rearrangement, is discussed in detail in Chapter 19. The amides can be hydrolysed to amines and carboxylic acids (Eq. 16.19b). Hydrazones are involved in Wolff–Kishner reduction (Section 16.4.6). All these condensation products are solids with characteristic melting points. They are useful in identifying individual carbonyl compounds in qualitative analysis. 2,4-Dinitrophenylhydrazine in the presence of HCl condenses with carbonyl compounds to give sparingly soluble, high melting, intensely coloured (orange or red) precipitates of 2,4-dinitrophenylhydrazone. This reagent, called *Borsche's reagent*, is used as a sensitive test for carbonyl compounds.

16.4.2 REACTIONS OF THE ENOLATE ION

The anion obtained by the abstraction of a proton from the α-position is resonance stabilised, and is called the enolate anion. As discussed in Section 16.1.2, simple carbonyl compounds containing only one carbonyl group, are very weak acids. However, in the presence of bases like HO^- and RO^-, the enolate anion can be generated and can react with electrophiles available in the medium. One such electrophile is another molecule of the carbonyl compound itself. There are several reactions belonging to this category.

16.4.2.1 Aldol Condensation

The simplest example of aldol condensation is the condensation of two molecules of acetaldehyde to give 3-hydroxybutanal (Eq. 16.20). The name aldol is derived from the aldehyde and alcohol functions present in the molecule.

$$2\ CH_3-CHO \xrightarrow{H^+\ (or)\ OH^-} \underset{\substack{\text{Aldol}}}{CH_3-\overset{\overset{\displaystyle H}{|}}{\underset{\underset{\displaystyle OH}{|}}{C}}-CH_2-\overset{\overset{\displaystyle H}{|}}{C}=O} \qquad (16.20)$$

The reaction can be either base catalysed or acid catalysed (Eqs. 16.21a and 16.21b).

Base catalysed reaction

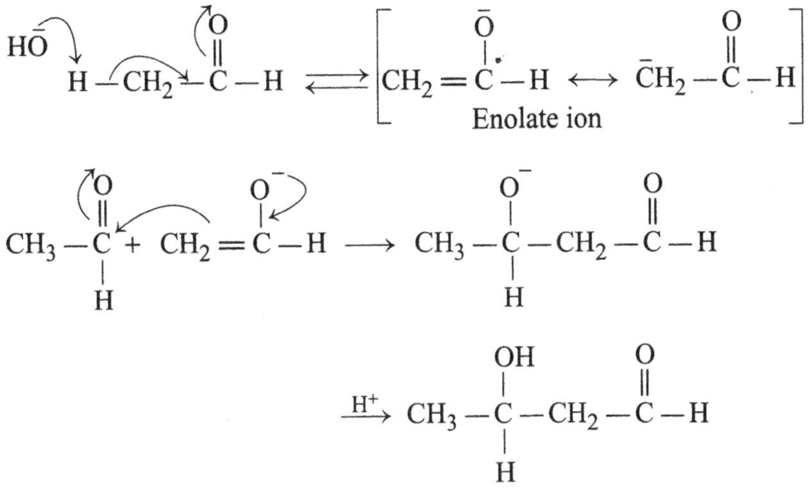

Enolate ion

$$CH_3 - \overset{O}{\overset{\|}{C}} + CH_2 = \overset{O^-}{\overset{|}{C}} - H \longrightarrow CH_3 - \overset{O^-}{\overset{|}{\underset{H}{C}}} - CH_2 - \overset{O}{\overset{\|}{C}} - H$$

$$\xrightarrow{\ H^+\ } CH_3 - \overset{OH}{\underset{H}{\overset{|}{C}}} - CH_2 - \overset{O}{\overset{\|}{C}} - H \qquad \text{(a)}$$

Acid catalysed reaction

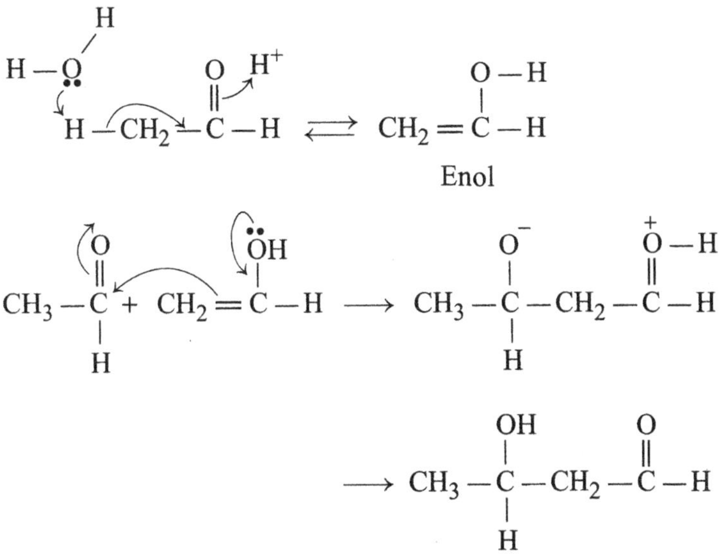

Enol

$$CH_3 - \overset{O}{\overset{\|}{C}} + CH_2 = \overset{\ddot{O}H}{\overset{|}{C}} - H \longrightarrow CH_3 - \overset{O^-}{\underset{H}{\overset{|}{C}}} - CH_2 - \overset{\overset{+}{O} - H}{\overset{\|}{C}} - H$$

$$\longrightarrow CH_3 - \overset{OH}{\underset{H}{\overset{|}{C}}} - CH_2 - \overset{O}{\overset{\|}{C}} - H \qquad \text{(b)}$$

$$(16.21)$$

In the base-catalysed reaction, the enolate anion is first formed. Nucleophilic addition of this anion to the carbonyl carbon of another acetaldehyde molecule, followed by protonation gives the enol. In the acid-catalysed reaction, the first step is not the formation of an anion but acid catalysed enolisation. This is an equilibrium process, called *tautomerism*. The enol adds to the double bond in much the same way as the enolate anion does. The similarity can be appreciated by comparing this step with the corresponding step in the base catalysed reaction (Eq. 16.21a). The aldol readily undergoes dehydration to the unsaturated aldehyde, crotonaldehyde (Eq. 16.22a). We have encountered this reaction in Chapter 12, Section 12.2.4, as an example of E1cB reaction. If the reaction is carried out under controlled conditions, the aldol can be isolated.

Ketones also undergo similar reactions. Ketone condensation also goes by the general name, aldol condensation. From acetone, the ketoalcohol 4-methyl-4-hydroxy-2-pentanone (diacetone alcohol) can be isolated under controlled conditions using barium hydroxide as the base (Eq. 16.22b). Its dehydration product mesityl oxide, is obtained if it is treated with acids even in the cold (Eq. 16.22c). Treatment of acetone with hydrochloric acid gives phorone (Eq. 16.22d), via the intermediate formation of mesityl oxide.

$$CH_3-\overset{\overset{\displaystyle OH}{|}}{CH}-\overset{\overset{\displaystyle |}{CH}}{\underset{\underset{\displaystyle OH}{H}}{|}}-\overset{\overset{\displaystyle O}{\|}}{C}-H \longrightarrow CH_3-CH=CH-\overset{\overset{\displaystyle O}{\|}}{C}-H \qquad \text{(a)}$$

Crotonaldehyde

$$CH_3-\overset{\overset{\displaystyle O}{\|}}{\underset{\underset{\displaystyle CH_3}{|}}{C}}+CH_3-\overset{\overset{\displaystyle O}{\|}}{C}-CH_3 \longrightarrow CH_3-\overset{\overset{\displaystyle OH}{|}}{\underset{\underset{\displaystyle CH_3}{|}}{C}}-CH_2-\overset{\overset{\displaystyle O}{\|}}{C}-CH_3 \qquad \text{(b)}$$

4-Methyl-4-hydroxy-2-pentanone
(diacetone alcohol)

$$\underset{\text{alcohol}}{\text{Diacetone}} \xrightarrow{-H_2O} CH_3-\overset{\overset{\displaystyle }{}}{\underset{\underset{\displaystyle CH_3}{|}}{C}}=CH-\overset{\overset{\displaystyle O}{\|}}{C}-CH_3 \qquad \text{(c)}$$

Mesityl oxide

(16.22)

$$CH_3-\overset{}{\underset{\underset{\displaystyle CH_3}{|}}{C}}=CH-\overset{\overset{\displaystyle O}{\|}}{C}-CH_3 \; + \; O=\overset{}{\underset{\underset{\displaystyle CH_3}{|}}{C}}-CH_3$$

$$\longrightarrow CH_3-\overset{}{\underset{\underset{\displaystyle CH_3}{|}}{C}}=C-\overset{\overset{\displaystyle O}{\|}}{C}-C=\overset{}{\underset{\underset{\displaystyle CH_3}{|}}{C}}-CH_3 \qquad \text{(d)}$$

Phorone

If two carbonyl compounds, for example ethanal and propanal, or ethanal and acetone are taken, mixtures will result due to self-condensation and cross-condensation. Cross-condensation can be of two types, enolate of ethanal adding to the carbonyl group of propanal and vice versa.

There is one set of reactants which gives better selectivity. Mixtures of aliphatic aldehydes and aromatic aldehydes give products corresponding to the addition of the anion of the aliphatic compound to the carbonyl of the aromatic compound. Of course, benzaldehyde which has no α-hydrogens cannot provide the enolate ion. Thus, benzaldehyde and acetaldehyde condense in the presence of dilute alkali to give only cinnamaldehyde (Eq. 16.23).

$$\text{C}_6\text{H}_5-\overset{\text{H}}{\underset{}{\text{C}}}=\text{O} + \text{CH}_3-\overset{\text{O}}{\underset{}{\text{C}}}-\text{H} \xrightarrow{\text{OH}^-} \text{C}_6\text{H}_5-\overset{\text{H}}{\underset{}{\text{C}}}=\text{CH}-\overset{\text{H}}{\underset{}{\text{C}}}=\text{O} \qquad (16.23)$$

Cinnamaldehyde

This reaction goes by the name *Claisen reaction* or *Claisen–Schmidt* reaction, not to be confused with another reaction called *Claisen condensation* which is an ester condensation (Chapter 17).

16.4.2.2 Knoevenagel Reaction

This reaction is very similar to aldol condensation. It is a mixed condensation between an aldehyde (the carbonyl component) and an *active methylene compound* which provides the enolate ion.

Referring to Table 16.1, it can be seen that molecules containing the $-\text{CO}-\text{CH}_2-\text{CO}$ grouping, that is, where a CH_2 is flanked by two CO groups, are much stronger acids than simple carbonyl compounds. The table lists the acidity of 2,4-pentanedione, ethyl acetoacetate and ethyl malonate. The CH_2 (methylene) groups in these and similar molecules are called *active methylenes*. That is, they are acidic and the H can be removed by bases like OH^-, OR^- and amines. Diethyl malonate reacts with benzaldehyde in the presence of an organic base like pyridine to give a condensation product (Eq. 16.24a).

$$\text{C}_6\text{H}_5-\overset{\text{H}}{\underset{}{\text{C}}}=\text{O} + \text{CH}_2\overset{\text{COOC}_2\text{H}_5}{\underset{\text{COOC}_2\text{H}_5}{\big\langle}} \xrightarrow{\text{(pyridine)}} \text{C}_6\text{H}_5-\overset{\text{H}}{\underset{}{\text{C}}}=\text{C}\overset{\text{COOC}_2\text{H}_5}{\underset{\text{COOC}_2\text{H}_5}{\big\langle}} \qquad (a)$$

Diethyl malonate

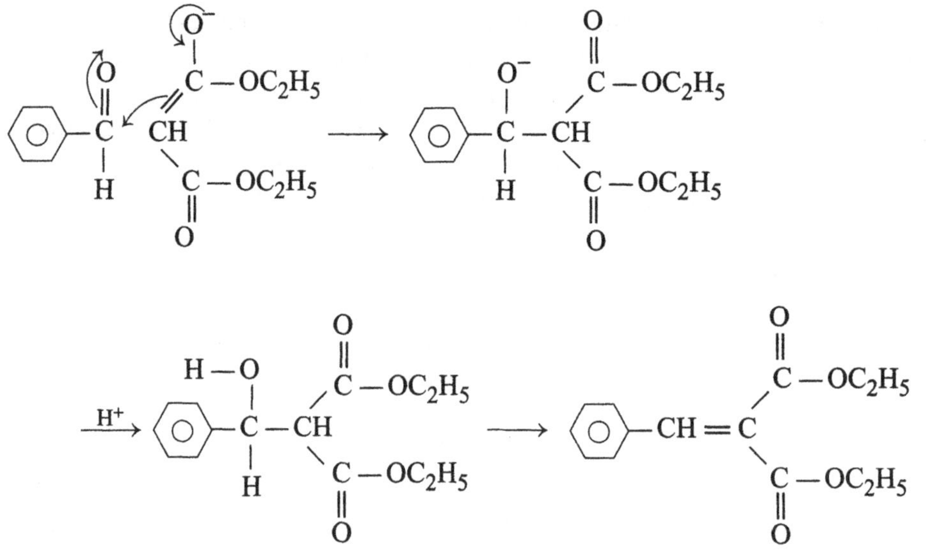

(b)

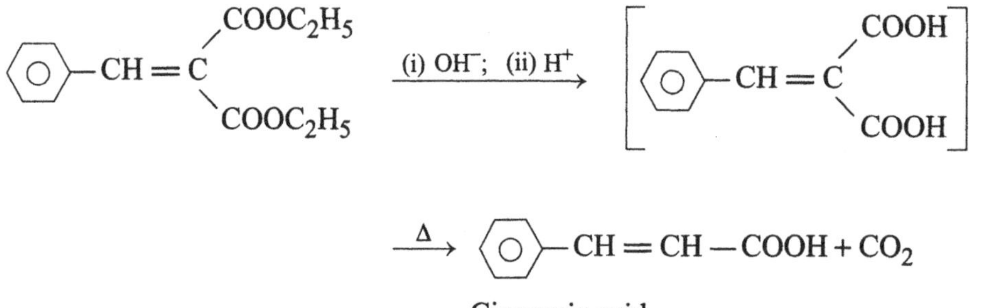

(c)

Cinnamic acid

(16.24)

The mechanism involves the addition of the carbanion formed from the active methylene compound to the aldehyde (Eq. 16.24b). The alcohol formed as the intermediate cannot be isolated. It undergoes dehydration to give the condensation product. One of the applications of this reaction is that the diester upon hydrolysis gives the diacid which on heating loses a molecule of CO_2 and yields, in this case, the unsaturated monocarboxylic acid, 3-phenylpropenoic acid, better known as cinnamic acid (Eq. 16.24c). This reaction in which there is loss of carbon dioxide from a carboxylic acid, is called *decarboxylation*. Acids like malonic acid – possessing two carboxyl groups attached to the same carbon – readily undergo deacarboxylation by heat. Recall that α, β-unsaturated carboxylic acid esters can be synthesised by the Reformatsky reaction also.

16.4.2.3 Perkin Reaction

Cinnamic acid can be synthesised by yet another carbonyl condensation reaction, the Perkin condensation. In this reaction, benzaldehyde is heated with acetic anhydride and sodium acetate. Acetic anhydride is the source of the enolate anion and sodium acetate is the base. The reaction requires heating upto 140°C. Even though acetate is a weak base, the reaction takes place facilitated by the second acetyl group of the anhydride taking part in the reaction (Eq. 16.25).

$$CH_3 - COO^- + H - CH_2 - \overset{\overset{\displaystyle O}{\|}}{C} - O - \overset{\overset{\displaystyle O}{\|}}{C} - CH_3$$

Acetate   Acetic anhydride

$$\xrightarrow{(1)} CH_3 - COOH + \overset{-}{C}H_2 - \overset{\overset{\displaystyle O}{\|}}{C} - O - \overset{\overset{\displaystyle O}{\|}}{C} - CH_3$$

(16.25)

The base CH_3COO^-, abstracts a proton from acetic anhydride, step (1). This is not a very favourable reaction because both the base and acid are very weak. However, this is driven forward by the subsequent steps. As the anion adds to the carbonyl group (step 2), the O^- is stabilised by interaction with the other carbonyl group of the anhydride, structure (I) and (II). The cyclic structure (II) then opens to structure (III) (step 3). Structure (III) eliminates a molecule of acetic acid (step 4) in much the same way as an alcohol will undergo dehydration. (Substitute OH^-

instead of CH_3COO^-, you will see the analogy.) Cinnamate anion, (IV) is converted to cinnamic acid, (V) by treating the reaction mixture with mineral acid, step (5).

Perkin condensation is applicable to aromatic aldehydes and anhydrides of simple carboxylic acids like acetic and propionic anhydride.

16.4.2.4 Michael Addition

Michael addition can be seen as an extension of the Knoevenagel reaction. It is a conjugate addition of an active methylene compound to an α, β-unsaturated carbonyl compound like an unsaturated aldehyde, ketone or ester. For example, the addition of diethyl malonate to 3-phenyl-2-propenal (cinnamaldehyde) in the presence of base to give the product as in Eq. 16.26.

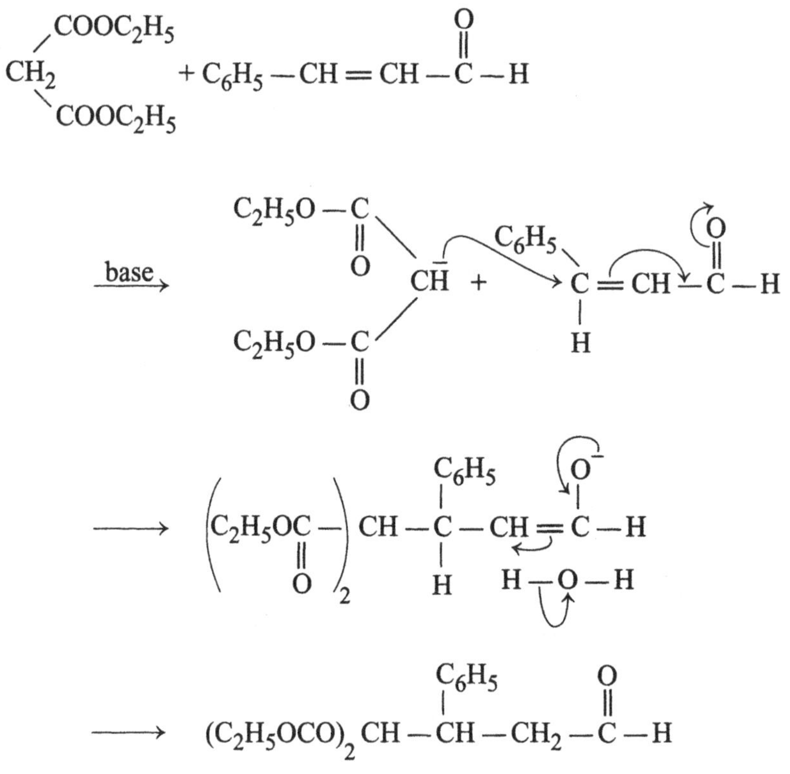

$$
\begin{array}{c}
\nearrow COOC_2H_5 \\
CH_2 \qquad\qquad + C_6H_5-CH=CH-\overset{\displaystyle O}{\overset{\|}{C}}-H \\
\searrow COOC_2H_5
\end{array}
$$

$$\xrightarrow{\text{base}}$$

(16.26)

Other enolate ions also react, but the reaction is most successful with active methylene compounds.

16.4.3 OTHER REACTIONS INVOLVING CARBONYL ADDITION

16.4.3.1 Benzoin Condensation

Aromatic aldehydes condense with themselves in the presence of cyanide ion, to give hydroxyketones, called acyloins. The condensation product of benzaldehyde is benzoin (V). The mechanism is outlined below (Eq. 16.27).

$$(16.27)$$

The first step is the addition of CN^- to the carbonyl to give (I). The hydrogen attached to the middle carbon in (I) is acidic due to the electron-withdrawing effect of the CN. This undergoes a rearrangement, step (2). The resultant carbanion (II), is resonance stabilised which is the driving force for the hydrogen shift. This carbanion adds as a nucleophile to another molecule of benzaldehyde, step (3). The product (III) rearranges to (IV) which then ejects a cyanide ion, step (5), to give the final product, benzoin (V). The cyanide ion is a specific catalyst for this reaction. Its role is to increase the acidity of the H in structure (I) and to stabilise the carbanion, (II).

16.4.3.2 Wittig Reaction

The Wittig reaction is used to convert the $C=O$ group of aldehydes and ketones to $C=CR_2$.

For example, acetophenone can be converted to 2-phenylpropene by this reaction. The reagent which will do this is a phosphonium ylide, also called a phosphorane. Preparation of a ylide is illustrated by the following example (Eq. 16.28).

This reaction was discovered by Georg Wittig, a German chemist, who was awarded the Nobel prize for this contribution.

$$(C_6H_5)_3P + CH_3I \longrightarrow (C_6H_5)_3\overset{+}{P} - CH_3 \; I^-$$

$$\xrightarrow{C_6H_5Li} \left[(C_6H_5)_3\overset{+}{P} - \overset{-}{C}H_2 \longleftrightarrow (C_6H_5)_3P = CH_2\right] + C_6H_6 + LiI \qquad (16.28)$$

Triphenylphosphonium
methylide

Reaction of triphenylphosphine or any triarylphosphine with an alkyl halide gives the quaternary phosphonium halide. This is deprotonated using a strong base like C_6H_5Li (phenyllithium) to generate a negative charge on the alkyl carbon. The zwitterion so obtained is called the ylide. Instead of CH_3I, any alkyl halide of the type $R(R')CH-X$ can be used to obtain the ylide.

$$Ar_3\overset{+}{P} - \overset{-}{C}(R')R \leftrightarrow Ar_3P = C(R')R$$

Reaction of an aldehyde or ketone with the ylide gives the alkene (Eq. 16.29).

Oxaphosphetane

Triphenylphosphine
oxide

(16.29)

A 4-membered ring containing O and P, called an oxaphosphetane, is the intermediate. This further dissociates to the alkene and triphenylphosphine oxide. P—O bond formation is a highly exothermic reaction and this is the driving force for the reaction. By choosing an appropriate carbonyl compound and an appropriate alkyl halide for the formation of the ylide, a wide variety of alkenes can be synthesised. Another example is given in Eq. 16.30.

(16.30)

16.4.4 Reduction of the Carbonyl Group

16.4.4.1 Sodium Borohydride and Lithium Aluminium Hydride

Reduction of the carbonyl group to the alcohol can be achieved in a number of ways. We have seen in Chapter 14, Section 14.1.3.5, that the complex metal hydrides, sodium borohydride and lithium aluminium hydride are very effective for this reduction. In these reagents, H is present as the hydride ion H^-, and the reduction can be viewed as a nucleophilic addition of the H^- to the carbonyl group. The complex metal hydride ions, BH_4^- and AlH_4^- are the hydride donors. Each of them can donate successively four hydride ions, and can reduce four equivalents of the aldehyde or ketone. The final product is the boron or aluminium salt of the alcohol which has to be decomposed during working up by adding aqueous mineral acid to liberate the alcohol (Eq. 16.31).

$$R_2C = O + NaBH_4 \longrightarrow R_2CHOH$$

$$\longrightarrow \longrightarrow (R_2CH-O-)_4B^- \xrightarrow{H_3O^+} 4\,R_2CHOH + B(OH)_4^- \qquad (a)$$

$$\longrightarrow \longrightarrow (R_2CH-O)_4Al^- \xrightarrow{H_3\overset{+}{O}} 4\,R_2CHOH \qquad (b)$$

$$(16.31)$$

Esters and free carboxylic acids can also be reduced using lithium aluminium hydride.

$$(16.32)$$

$$\xrightarrow{AlH_4^-} R-CH_2O-Al^-H_3 \xrightarrow{H_3O^+} RCH_2OH$$

16.4.4.2 Cannizzaro Reaction

Metal hydride reductions are reactions involving hydride transfer. There are some other hydride-transfer reactions where the source of the hydride is not a metal hydride, but the $C-H$

bond of an organic molecule. The Cannizzaro reaction is one such. This reaction is applicable only to aldehydes which do not contain α-hydrogens. This group includes all aromatic aldehydes, aliphatic aldehydes of the type R, R′R″C—CHO (trisubstituted acetaldehyde) and formaldehyde. Benzaldehyde, when treated with fairly concentrated sodium or potassium hydroxide solution gives one equivalent each of benzyl alcohol and benzoic acid (Eq. 16.33). Formaldehyde gives methanol and formic acid. One molecule of the aldehyde acts as the reducing agent, getting itself oxidised in the process by another molecule of the aldehyde. The reaction can be described as a disproportionation.

$$C_6H_5 - \overset{\overset{\text{O}}{\|}}{\underset{\text{HO}^-}{C}} - H \longrightarrow C_6H_5 - \overset{\overset{\text{O}^-}{|}}{\underset{\underset{\text{(I)}}{H-O}}{C}} - H + \overset{\overset{\text{O}}{\|}}{\underset{H}{C}} - C_6H_5$$

$$\longrightarrow C_6H_5 - \overset{\overset{\text{O}}{\|}}{\underset{H-O}{C}} + H - \overset{\overset{\text{O}^-}{|}}{\underset{H}{C}} - C_6H_5 \longrightarrow C_6H_5 - \overset{\overset{\text{O}}{\|}}{C} - O^- + \overset{\overset{\text{OH}}{|}}{CH_2} - C_6H_5$$

$$(16.33)$$

The crucial step is the hydride-transfer, (I). Heterolytic dissociation of the C–H bond to generate a hydride is quite rare. It takes place only if it results in a stable molecule and there is a hydride-acceptor available. In this reaction, a second molecule of the aldehyde acts as the hydride-acceptor. When two different eligible aldehydes are mixed and treated with alkali, crossed Cannizzaro reaction takes place. Each aldehyde can react with itself or react with the other aldehyde, either as the hydride-donor or as the hydride-acceptor. There are altogether four combinations, all of which may take place. If formaldehyde and an aromatic aldehyde are used, the reaction is more selective. Formaldehyde functions as the reducing agent, namely, as the hydride-donor. This reaction takes place at a faster rate than all the other alternatives. The products are benzyl alcohol and formic acid. This can be used as a method of reducing a benzaldehyde to the corresponding benzyl alcohol.

16.4.4.3 Meerwein–Ponndorf–Verley Reduction and Oppenauer Oxidation

Meerwein–Ponndorf reduction (also known as Meerwein–Ponndorf–Verley reduction) is the hydride transfer reduction of aldehydes and ketones, especially of the latter, where the hydride-donor is a secondary alcohol like isopropyl alcohol, in the form of the aluminium alkoxide. Aluminium isopropoxide is prepared by the reaction of isopropyl alcohol with aluminium metal. When acetophenone is heated with aluminium isopropoxide dissolved in isopropyl alcohol, 1-phenylethanol and acetone are formed in a reversible reaction (Eq. 16.34).

$$3CH_3-\underset{\underset{CH_3}{|}}{CH}-OH + Al \longrightarrow \left(CH_3-\underset{\underset{CH_3}{|}}{CH}-O-\right)_3 Al + 1\tfrac{1}{2}H_2 \qquad (a)$$

Aluminium
isopropoxide (I)

$$I + C_6H_5-\overset{\overset{O}{\|}}{C}-CH_3 \longrightarrow CH_3-\overset{\overset{O}{\|}}{C}-CH_3 + \left(C_6H_5-\underset{\underset{CH_3}{|}}{CH}-O-\right)_3 Al$$

$$\xrightarrow{H_3O^+} C_6H_5-\underset{\underset{CH_3}{|}}{CH}-OH \qquad (b)$$

(c)

$$(16.34)$$

In practice, acetone – which is low boiling – is continuously distilled out from the reaction mixture, to drive the equilibrium in the forward direction. 1-Phenylethanol is present in the reaction mixture as its aluminium salt. The free alcohol is released by acidification of the mixture. The crucial step in the reaction is the transfer of a hydride from the α-position of isopropyl alcohol to the carbonyl group of acetophenone. The electrophilicity of the carbonyl carbon is increased because the carbonyl oxygen is coordinated to aluminum—a Lewis acid. The hydride transfer step involves a cyclic transition state (Eq. 16.34c) and is reversible. This approach can be used for the reverse reaction as well. Thus, if aluminium 1-phenylethoxide is heated with acetone or any other ketone, the former can be oxidised to acetophenone. Suitable modifications in the experimental conditions have to be made in order to drive the reaction in the desired direction. When used as a method of oxidising a secondary alcohol to the ketone, it is known as *Oppenaur oxidation*.

16.4.5 BIMOLECULAR REDUCTION: REDUCTION OF ACETONE TO PINACOL

When acetone is treated with metallic magnesium, electron transfer from the metal to acetone occurs. A complex composed of two molecules of acetone and one atom of magnesium is formed. This complex upon decomposition with mineral acid gives 2,3-dimethyl-2,3-butanediol, also known as pinacol (Eq. 16.35).

$$CH_3 - \underset{\underset{\underset{\overset{\cdots}{Mg}}{O}}{\overset{CH_3}{\|}}}{C} \;\; \underset{\overset{O}{\|}}{\overset{CH_3}{C}} - CH_3 \longrightarrow CH_3 - \underset{\underset{O}{|}}{\overset{CH_3}{C}} \;\; \underset{\underset{O}{|}}{\overset{CH_3}{C}} - CH_3$$

$$\underset{Mg}{}$$

(16.35)

$$\xrightarrow{\text{HCl, H}_2\text{O}} \quad CH_3 - \underset{\underset{OH}{|}}{\overset{CH_3}{C}} - \underset{\underset{OH}{|}}{\overset{CH_3}{C}} - CH_3 + MgCl_2$$

Pinacol

16.4.6 REDUCTION OF KETONE TO HYDROCARBON: WOLFF–KISHNER REDUCTION AND CLEMMENSEN REDUCTION

This transformation of the carbonyl group to an alkyl group is illustrated by the conversion of acetophenone to ethylbenzene. Such reductions can be carried out in a number of ways. We have seen how alkyl aryl ketones prepared by Friedel–Crafts acylation can be converted to the hydrocarbons by such reactions. Wolff–Kishner reduction involves heating the hydrazone of the ketone with alkali (Eq. 16.36a). The mechanism of the reaction is outlined in Eq. 16.36b.

$$\underset{R}{\overset{R}{>}}C = N - NH_2 \xrightarrow[\Delta]{OH^- \text{ (or) OR}'} \underset{R}{\overset{R}{>}}CH_2 + N_2 \qquad \text{(a)}$$

$$\underset{R}{\overset{R}{>}}C = N - NH_2 + OH^- \longrightarrow \left[\underset{R}{\overset{R}{>}}C = N - \overset{-}{N}H \longleftrightarrow \underset{R}{\overset{R}{>}}\overset{-}{C} - N = NH \right]$$

$$\xrightarrow{H_2O} \underset{R}{\overset{R}{>}}\overset{\overset{H}{|}}{C} - N = NH \xrightarrow{OH^-} \underset{R}{\overset{R}{>}}\overset{\overset{H}{|}}{C} - N = \overset{-}{N} \longrightarrow \underset{R}{\overset{R}{>}}C\overset{\nearrow H}{\underset{\searrow H}{}} + N \equiv N \qquad \text{(b)}$$

$$\underset{H - OH}{}$$

(16.36)

Clemmensen reduction brings about the same conversion, by heating the ketone with zinc amalgam and hydrochloric acid (Eq. 16.37).

$$\underset{R}{\overset{R}{>}}C = O \xrightarrow{\text{Zn/Hg, HCl}} R_2CH_2 \qquad (16.37)$$

16.4.7 HALOGENATION OF CARBONYL COMPOUNDS: HALOFORM REACTION

Aldehydes and ketones can be halogenated at the α-position. Thus, cyclohexanone can be converted · to 2-bromocyclohexanone by treating with bromine. The reaction can take place in the presence of acid or alkali. The enol or the enolate is the intermediate reactive species. The two types of reactions are illustrated in Eqs. 16.38a and 16.38b. In the reaction in alkali, hypohalite (NaOX) which is a source of X^+, is the halogenating agent.

$$(16.38)$$

The reaction can continue to introduce more number of halogens. The second halogen atom is introduced at the same carbon where the first halogen is present, since that hydrogen is more acidic due to the inductive effect of the halogen, than hydrogens on the other α-position (Eq. 16.38c). In the case of methyl ketones, that is, those containing the $-CO-CH_3$ (acetyl) group, this leads to an interesting $C-C$ bond cleavage, called the *haloform reaction*. When a methyl ketone like acetone is treated with iodine and alkali, as a result of successive iodinations (Eq. 16.39a), 1,1,1-triiodo-2-propanone is formed. The base now induces a cleavage of this molecule (Eq. 16.39b).

$$CH_3-\overset{\overset{\displaystyle O}{\|}}{\underset{\underset{\displaystyle HO^-}{}}{C}}-CI_3 \longrightarrow CH_3-\overset{\overset{\displaystyle O^-}{|}}{\underset{\underset{\displaystyle HO}{|}}{C}}-CI_3 \longrightarrow CH_3-\overset{\overset{\displaystyle O}{\|}}{\underset{\underset{\displaystyle OH}{|}}{C}}+CI_3^-$$

(I)
(16.39)

$$\longrightarrow CH_3-\overset{\overset{\displaystyle O}{\|}}{C}-O^- + HCI_3 \qquad\qquad \text{(b)}$$

The crucial step in this reaction is the elimination of the carbanion, CI_3C- from the intermediate, (I). Such an elimination of the anion, other than a carbanion, is common in the reactions of carboxylic acids and their derivatives (Eq. 16.40).

$$CH_3-\overset{\overset{\displaystyle O}{\|}}{\underset{\underset{\displaystyle Nu:}{}}{C}}-X \longrightarrow CH_3-\overset{\overset{\displaystyle O^-}{|}}{\underset{\underset{\displaystyle Nu}{|}}{C}} \overset{}{\diagup} X \longrightarrow CH_3-\overset{\overset{\displaystyle O}{\|}}{\underset{\underset{\displaystyle Nu}{|}}{C}}+X^-$$

(16.40)

This occurs in ester hydrolysis, where $Nu = OH^-$ and $X = OR$; or in esterification, where $Nu = RO^-$ and $X = OH$, Cl or OCOR. In these examples, the leaving groups – oxygen anions or halides – are considered to be 'good leaving groups'. Carbanions are not good leaving groups. What makes the CI_3C^- a good leaving group? Obviously, it is the presence of the three halogen atoms, which by electron withdrawal stabilise the anion. $H-CX_3$ are stronger acids than simple hydrocarbons.

When acetone or other methyl ketones ($R-CO-CH_3$, where R is any alkyl or aryl group) are treated with chlorine, bromine or iodine in alkaline solution, the haloform reaction takes place to produce $R-COO^-$ and HCX_3. When iodine is used, iodoform, a yellow solid with a characteristic smell, precipitates out. This is the iodoform test which is useful for detecting methyl ketones in qualitative analysis. The only aldehyde that answers the iodoform test is acetaldehyde. Alcohols which are oxidised to methyl ketones in situ by the halogen, also undergo the haloform reaction. Thus, ethanol which gets oxidised to acetaldehyde, and secondary alcohols like 2-propanol and 2-butanol with the general structure $CH_3-CH(OH)-R$, undergo the iodoform reaction. Three halogens on a carbon are necessary for the ejection of the carbon as a carbanion. Molecules like $-CO-CH_2-R$ do get halogenated but do not cleave. The haloform reaction is useful for the oxidation of methyl ketones to the carboxylic acids (Eq. 16.41).

$$\text{C}_6\text{H}_5-\overset{\overset{\displaystyle O}{\|}}{C}-CH_3 \quad\xrightarrow{\text{NaOCl, NaOH}}\quad \text{C}_6\text{H}_5-\overset{\overset{\displaystyle O}{\|}}{C}-O^- + CHCl_3 \qquad (16.41)$$

16.4.8 PHOTOLYSIS OF KETONES

We have seen in Chapter 13 that electronic excitations of organic molecules take place when irradiated by light in the UV region. The energy of this radiation of wavelength 200–400 nm is of the order of 628–239 kJ mol^{-1} (150–70 kcal mol^{-1}). This is in the same range as the bond energies in molecules. As a result, the electronically excited molecules can undergo chemical reactions. Such reactions, brought about by the supply of energy in the form of light, are known as *photochemical reactions*. The symbol, hν is used to indicate that energy is supplied in the form of photons (cf., Δ for heat energy). Ketones undergo electronic excitation when irradiated with UV light (Chapter 13, Section 13.3). Ketones in the electronically excited states can undergo different types of reactions. When the reaction involves a cleavge, it is called *photolysis*. Here, we shall see two types of photolytic cleavages, called Norrish type I and type II cleavages.

Acetone in the gas phase on irradiation with UV light undergoes cleavage generating an acetyl and a methyl free radical (Eq. 16.42).

$$CH_3-\overset{\overset{O}{\|}}{C}-CH_3 \xrightarrow{h\nu} \left[CH_3-\overset{\overset{O}{\|}}{C}-CH_3\right]^* \longrightarrow CH_3-\overset{\overset{O}{\|}}{C}\cdot + \dot{C}H_3$$

$$CH_3-\overset{\overset{O}{\|}}{C}\cdot \longrightarrow \dot{C}H_3 + CO \qquad (16.42)$$

$$2\,\dot{C}H_3 \longrightarrow C_2H_6$$

The first step, which is the photochemical step, is a Norrish type I cleavage. The structure within square brackets and with an asterisk, denotes the electronically excited molecule. The Type I cleavage of cyclohexanone and the product arising from this are shown in Eq. 16.43.

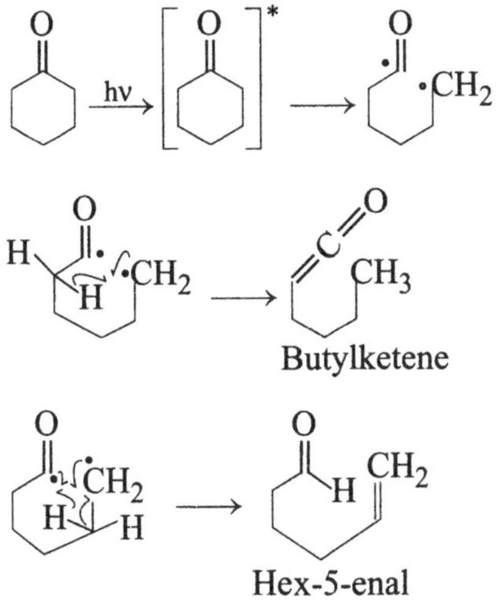
hν
•CH₂
H
CH₂
CH₃
Butylketene
CH₂
H
Hex-5-enal

(16.43)

Ketones containing a hydrogen at the 4-position (γ-position) undergo a different type of cleavage as a result of intramolecular hydrogen transfer (γ-hydrogen abstraction). This is the Type II cleavage (Eq. 16.44).

$$\overset{\gamma}{C}H_3-\overset{\beta}{C}H_2-\overset{}{C}H_2-\overset{\alpha}{C}H_2-\overset{O}{\overset{\|}{C}}-CH_3 \xrightarrow{h\nu} CH_3-CH=CH_2+CH_3-\overset{O}{\overset{\|}{C}}-CH_3$$

(16.44)

The products are an alkene and an enol. The enol immediately tautomerises to the ketone.

16.5   TESTS FOR ALDEHYDES AND KETONES

Aldehydes are reducing agents. When a solution of ammoniacal silver nitrate (Tollen's reagent) is warmed with an aldehyde, Ag^+ gets reduced to silver metal. If the glass container is clean, the metal will appear coated on the glass, forming a mirror. This is the Tollen's test – or the silver mirror test – for aldehydes. Cupric ion in alkaline medium, kept in solution by complexation with tatarate, is reduced to the cuprous ion when warmed with an aldehyde. The cuprous ion precipitates out as red cuprous oxide. The reagents used for this test are Fehling's solution or Benedict's solution.

Aldehydes and ketones give precipitates with the reagents listed in Table 16.3. Borsche's reagent is especially useful for qualitative testing.

16.6   SPECTROSCOPY

16.6.1   IR SPECTROSCOPY

The carbonyl stretching frequency in the IR spectrum appears in the region $1680–1800\,cm^{-1}$. Different types of aldehydes and ketones show characteristic variations in the IR frequency, which are very useful for identification. Some of these are listed in Table 16.4.

It may be noted that conjugated unsaturation shifts the frequency to lower values. In cyclic ketones, strained rings (4 and 5-membered) have higher frequency than open chain ketones and cyclohexanone.

Table 16. 4 IR frequencies of carbonyl groups

Type	Approximate IR frequency (cm^{-1})
RCHO	1725
(Aldehydes also show characteristic C—H band at $2720\,cm^{-1}$)	
R_2CO	1710
ArCHO	1700
ArCOR	1690
$C{=}C{-}C{=}O$ (α, β-unsaturated)	1690
Cyclohexanones	1710
Cyclopentanones	1740
Cyclobutanones	1780

16.6.2 NMR SPECTROSCOPY

The large downfield shift of the aldehyde hydrogen to the region δ 9–10, is diagnostic for aldehydes. See this signal in the NMR spectrum of *p*-methoxybenzaldehyde (Chapter 13, Fig. 13.10).

KEY POINTS

- The chemistry of aldehydes and ketones is dominated by the polarity of the carbonyl group where the carbon is partially positive and the oxygen is partially negative.
- Most of the methods of preparation of carbonyl compounds involve the oxidation of primary or secondary alcohols.
- Other methods include ozonolysis of alkenes, glycol oxidation, Friedel–Crafts acylation of arenes, and the specific reactions applicable to phenolic aldehydes.
- Reactions of aldeydes and ketones are mainly of two types. (i) Those involving nucleophilic addition to the carbonyl group and (ii) those involving the enolate anion generated by the removal of a hydrogen from the α-position as a proton.
- Acid catalysed addition of alcohols gives acetals and ketals from aldehydes and ketones respectively.
- HCN adds to aldehydes to form cyanohydrins, which can be hydrolysed to hydroxy acids.
- Sodium bisulphite adds to give bisulphite addition compounds.

- Amine reagents like primary amines, hydroxylamine, semicarbazide, hydrazine and phenylhydrazine add with the elimination of water to give derivatives containing the $C=N$ bond. The products of condensation of primary amines can be hydrogenated to obtain secondary amines. This is called reductive amination. Oximes obtained by the condensation with hydroxylamine undergo Beckman rearrangement to amides.

- Aldol condensation is the reaction where the enolate ion or the enol obtained from an aldehyde or ketone adds as the nucleophile to the carbonyl group of another similar molecule. The product is a hydroxy aldehyde or hydroxy ketone. The hydroxy compound readily undergoes dehydration to the α, β-unsaturated aldehyde or ketone. The reaction can be acid or base catalysed.

- Aromatic aldehydes undergo self-condensation catalysed by the cyanide ion to yield benzoins.

- Aldehydes which do not contain α-hydrogens (aromatic aldehydes, formaldehyde, trisubsituted acetaldehydes) undergo an oxidation–reduction reaction in the presence of alkali. One molecule gets reduced to the primary alcohol and another gets oxidised to the carboxylic acid. This is known as Cannizzaro reaction.

- The condensation of an aliphatic aldehyde and an aromatic aldehyde in the presence of alkali is called Claisen reaction or Claisen–Schmidt reaction.

- α-Bromoesters and zinc metal (through the organozinc compound) add to aldehydes to give α, β-unsaturated esters. This is the Reformatsky reaction. Ethyl cinnamate can be synthesised by Reformatsky reaction from benzaldehyde, ethyl bromoacetate and zinc.

- Active methylene compounds like diethyl malonate, in the presence of organic bases like pyridine, condense with aldehydes, in the Knoevenagel reaction. The product of condensation with malonic ester gives the unsaturated diester which can be hydrolysed to the unsaturated diacid which in turn can be decarboxylated by heat to an α, β-unsaturated acid. Thus, cinnamic acid can be synthesised from benzaldehyde and diethyl malonate by Knoevenagel reaction followed by hydrolysis and decarboxylation.

- Another reaction for synthesising cinnamic acid is the Perkin reaction. Benzaldehyde is heated with acetic anhydride and sodium acetate. The product is cinnamic acid, after acid-work up.

- When active methylene compounds react in the presence of base with α, β-unsaturated carbonyl compounds, conjugate addition takes place. This is the Michael addition.

- Aldehydes and ketones react with triphenylphosphonium ylides (phosphoranes) to give alkenes in which the $C=O$ of the original carbonyl compound gets converted to $C=CR_2$. This is the Wittig reaction.

- Aldehydes and ketones are reduced to the alcohols by sodium borohydride or by lithium aluminium hydride.

- They can also be reduced to the alcohols by refluxing with aluminium isopropoxide. This reaction is known as Meerwein–Ponndorff–Verley reduction.

- By heating the hydrazone with alkali, ketones can be converted to the hydrocarbons, $R-CO-R'$ to $R-CH_2-R'$. This is the Wolff–Kishner reduction. The same conversion can be carried out by Clemmensen reduction, which involves treating the ketone with zinc amalgam and HCl.

- Ketones containing α-hydrogens can be halogenated by the halogens under acidic or basic conditions. When methyl ketones are treated with halogen in alkali, they cleave to give haloform and the carboxylate. 2-Butanone with iodine in aqueous NaOH, gives sodium propanoate and iodoform. This is the haloform reaction. It is characteristic of compounds containing the CH_3CO- group or the $CH_3CH(OH)-$ group (the latter since it can be oxidised in situ by the halogen to CH_3CO-).

- Ketones upon irradiation with UV light may undergo photolysis. The $C-C$ bond between the carbonyl and the α-carbon can dissociate homolytically to two free radicals (such as $CH_3\cdot$ and $CH_3CO\cdot$ from acetone) which can undergo further reactions. This cleavage is called Norrish type I cleavage. Those ketones which have a γ-hydrogen, cleave after the abstraction of the γ-hydrogen by the oxygen of the electronically excited ketone. This is Norrish type II cleavage.

- Aldehydes are reducing agents. They reduce Ag^+ to metallic silver in the Tollen's test, and cupric to cuprous in the Fehling's test.

- The carbonyl stretching frequency in the IR spectrum appears in the region 1680–$1800\ cm^{-1}$. Structural variations cause variations in the stretching frequency. These are useful for structural identification.

EXERCISES

SECTION I

1. Give the structure(s) of the main organic product(s).

    (a) $(C_6H_5)_3P{=}CHCH_3 + CH_3-CO-CH_2-CH_2-COO^- \quad \rightarrow$

    (b) $CH_2{=}CH-CO-CH_3 + CH_2(COOC_2H_5)_2 \xrightarrow{\text{pyridine}}$

    (c) $p\text{-}CH_3-C_6H_4-CHO+(CH_3-CH_2-CO)_2O+CH_3-CH_2-COONa \xrightarrow{\Delta}$

    (d) $p\text{-}Cl-C_6H_4-CHO + CH_2O + KOH \xrightarrow{\Delta}$

    (e) $p\text{-}Cl-C_6H_4-CHO + CH_3-CH_2-CHO + OH^- \quad \rightarrow$

    (f) $CH_3-CH_2-CH(OH)-CH_3 + Br_2 + NaOH \quad \rightarrow$

    (g) $p\text{-}Cl-C_6H_4-CHO + KCN \quad \rightarrow$

    (h) $(CH_3)_3C-CHO + \xrightarrow[\Delta]{KOH/H_2O}$

2. How can the following conversions be brought about?

   (a) Benzene → $C_6H_5-CO-CH_2CH_3$

   (b) $CH_3-CH_2-CH_2-CH=CH_2$ → $CH_3-CH_2-CH_2-CH_2-CHO$

   (c) $CH_3-CH_2-CH_2-CH=CH_2$ → $CH_3-CH_2-CH_2-CO-CH_3$

   (d) $C_6H_5-CH_2-COOC_2H_5$ → $C_6H_5-CH_2-CHO$

   (e) $C_6H_5-CH_2Cl$ → $C_6H_5-CH_2-CO-CH_3$

   (f) $C_6H_5-CO-CH_3$ → $C_6H_5-C(CH_3)=CH-COOH$

   (g) Cyclohexanone → methylenecyclohexane $\left(\langle\bigcirc\rangle{=}CH_2\right)$

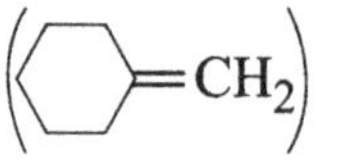

   (h) Phenol → *o*-hydroxybenzophenone

3. What are the following reagents? Give one use for each in organic chemistry.

   (a) Sodium borohydride

   (b) Lithium aluminium hydride

   (c) Zinc amalgam

   (d) Aluminium isopropoxide

   (e) Tollen's reagent

   (f) Fehling's solution

   (g) Borsche's reagent

   (h) Sodium hypobromite

SECTION II

1. Give the mechanisms of the following reactions.

   (a) Acetone upon heating with HCl gives phorone.

   (b) Glyoxal, CHO—CHO, on treatment with alkali gives $HO-CH_2-COOH$ in an intramolecular Cannizzaro reaction.

   (c) Hydrolysis of ketals by aqueous acid,
   $R_2C(OCH_3)_2 + H_2O, (HCl)$ → $R_2C{=}O + 2CH_3OH$

   (d) Formation of the ketal, 2,2-dimethoxypropane, from the acetone and methanol in the presence of mineral acids.

   (e) Condensation of hydroxylamine with an aldehyde to form the oxime.

   (f) Condensation of semicarbazide with a ketone to form the semicarbazone.

2. Account for the following observations.

   (a) In the Knovenagel reaction between $CH_3-CH_2-CH_2-CHO$ and diethyl malonate, an intermediate hydroxyl compound, (I), is formed. It then undergoes dehydration to give the α, β-unsaturated diester (II). The β, γ-unsaturated diester (III) is not formed.

$$CH_3-CH_2-CH_2-CHO + CH_2(COOR)_2$$
$$\rightarrow CH_3-CH_2-CH_2-CH(OH)-CH(COOR)_2 \quad (I)$$
$$\rightarrow CH_3-CH_2-CH_2-CH=C(COOR)_2 \quad (II)$$
$$(not, \ CH_3-CH_2-CH=CH-CH(COOR)_2) \quad (III)$$

(b) Crossed Cannizzaro reaction between benzaldehyde and formaldehyde gives benzyl alcohol and formic acid, not benzoic acid and methanol.

(c) $C_6H_5-CO-CH_2-COOCH_3$ dissolves in aqueous alkali in the cold. $C_6H_5-CH_2-COOCH_3$ does not dissolve.

(d) In the reaction of a carbonyl compound with semicarbazide, of the two NH_2 groups available in the reagent, only one reacts (see the structure of semicarbazone in Table 16.3 and also see Question 1(f), above).

(e) In the Claisen reaction between benzaldehyde and acetaldehyde, cinnamalehyde is the main product. Self condensation of acetaldehyde to give aldol and crotonaldehyde takes place only to a minor degree.

3. Identify the unknowns.

   (a) $C_5H_{10}O$, does not decolourise Baeyer's reagent, does not answer Tollen's or Fehling's test and answers positive for the iodoform test. It shows a strong band in the IR spectrum at $1720\,cm^{-1}$.

   NMR spectrum: 1.1, doublet, 6 H

   2.1, singlet, 3 H

   2.5, septuplet, 1 H

   (b) C_8H_8O gives a positive test with Tollen's reagent and Fehling's solution. It shows characteristic bands in the IR spectrum at $1720\,cm^{-1}$ and $2720\,cm^{-1}$.

   NMR spectrum: 3.5, doublet, 2 H

   7.1, singlet, 5 H

   9.2, triplet, 1 H

CHALLENGING QUESTIONS

1. Cyclopentene can be converted to cyclopentane carboxaldehyde, *cyclo*-C_5H_9-CHO by the following route. For each stage, show the type of reaction involved, the reaction conditions and other reagents required, and structures.
   Cyclopentene $\rightarrow$ cyclopentanone $\rightarrow$ methylenecyclopentane $\rightarrow$ cyclopentylcarbinol $\rightarrow$ cyclopentane carboxaldehyde.

2. Reaction of acetaldehyde with excess formaldehyde under alkaline conditions gives pentaerythritol, $C(CH_2OH)_4$, which is used for the manufacture of the explosive, pentaerythritol tetranitrate (PETN). Give the mechanism of the formation of pentaerythritol. (Hint: It involves three successive aldol condensation reactions followed by a crossed

Cannizzaro reaction.) Give the structure of PETN. Which other polyhydric alcohol is used for the manufacture of a similar explosive?

3.   Refer to Eq. 16.25. Can you visualise another mechanism for the elimination of acetic acid from III to IV? (Hint: Ester pyrolysis involving a cyclic transition state.)

4.   Refer to Eq. 16.27. The rearrangement of I → II involves the heterolysis of the C—H bond and the formation of the O—H bond. Answer the following questions with supporting arguments.

(a)   Is the migration of H from C to O likely to be intramolecular?

(b)   If not, which step—heterolysis of the C—H bond or protonation of O^-, is likely to take place first?

17 Carboxylic Acids

OBJECTIVES In this chapter, you will learn about,

- the factors affecting the acidity of carboxylic acids
- the methods of synthesis of carboxylic acids
- the conversion of acids to esters, acid halides, anhydrides and amides
- the mechanism of esterification and hydrolysis
- the synthesis of dicarboxylic acids
- the applications of malonic ester and cyanoacetic ester for synthesis
- the preparation of acetoacetic ester
- tautomerism
- the applications of acetoacetic ester in synthesis

17.1 ACIDITY OF CARBOXYLIC ACIDS

Aliphatic carboxylic acids are also called *fatty acids* because the higher members of the series were originally isolated by the hydrolysis of fats, which are triesters of long chain carboxylic acids with glycerol (Chapter 14, Section 14.2.3). Carboxylic acids are stronger than phenols but weaker than most mineral acids. Their acidity is due to the stabilisation of the carboxylate ion due to resonance (Eq. 17.1).

$$
R-C\overset{O}{\underset{O-H}{\big\backslash}} + H_2O \rightleftharpoons H_3O^+ + \left[R-C\overset{O}{\underset{O^-}{\big\backslash}} \longleftrightarrow R-C\overset{O^-}{\underset{O}{\big\backslash}} \right] \tag{17.1}
$$

Electronic effects of the R group influence the acidity. Electron-withdrawing effects – whether due to resonance or due to inductive effect – will increase the acidity by helping to further delocalise and stabilise the anion. Conversely, electron-releasing effects lower the acidity. A list of carboxylic acids, their boiling points and pK_a values is given in Table 17.1.

Table 17. 1 Carboxylic acids

Name (trivial name), structure	m.p. (°C)	b.p. (°C)	pK$_a$
Methanoic acid (formic acid) HCOOH	8.4	100.5	3.77
Ethanoic acid (acetic acid) CH_3-COOH	16.6	118	4.76
Propanoic acid (propionic acid) CH_3-CH_2-COOH	−22	141	4.88
Chloroethanoic acid (chloroacetic acid) $CH_2(Cl)-COOH$	63	189.5	2.81
Dichloroethanoic acid (dichloroacetic acid) $CH(Cl_2)-COOH$	10	193.5	1.29
Trichloroethanoic acid (trichloroacetic acid) $C(Cl_3)-COOH$	58	196	0.08
2-Chloropropanoic acid (α-chloropropionic acid) $CH_3-CH(Cl)-COOH$		186	2.8
3-Chloropropanoic acid (β-chloropropionic acid) $CH_2(Cl)-CH_2-COOH$		204	4.1
Fluoroethanoic acid (fluoroacetic acid) $CH_2(F)-COOH$	33	165	2.66
Benzoic acid C_6H_5-COOH	121		4.17
p-Methoxybenzoic acid (anisic acid) p-$CH_3O-C_6H_4-COOH$	184		4.49
p-Chlorobenzoic acid p-$Cl-C_6H_4-COOH$	243		4.03
p-Nitrobenzoic acid p-$NO_2-C_6H_4-COOH$	240		3.40
Ethanedioic acid (oxalic acid) $HOOC-COOH$	189		1.46, 4.40*
Propanedioic acid (malonic acid) $HOOC-CH_2-COOH$	135		2.80, 5.85*
Butanedioic acid (succinic acid) $HOOC-CH_2-CH_2-COOH$	185		4.17, 5.64*
Pentanedioic acid (glutaric acid) $HOOC--CH_2-CH_2-CH_2-COOH$	97.5		4.33, 5.57*
Hexanedioic acid (adipic acid) $HOOC-CH_2-CH_2-CH_2-CH_2-COOH$	151		4.43, 5.52*

The two values for the dicarboxylic acids are pK$_a^1$ and pK$_a^2$, for the first ionisation constant and the second ionisation constant.

The data in Table 17.1 bring out several aspects of the influence of structure on acidity. Recall that $pK_a = -\log K_a$, and stronger acids have lower pK_a values. In the series, formic acid, acetic acid and propionic acid, the influence of the electron-releasing alkyl group is observed. Acetic and propionic acids (and the higher acids) are weaker than formic acid. The presence of an electron-withdrawing group like chlorine markedly increases the acidity, as seen by comparing acetic acid with the chloroacetic acids. Chlorine has an electron-withdrawing inductive effect. The more the number of chlorines, the stronger the acids as observed in the series—monochloro, dichloro and trichloroacetic acids. As expected, fluoroacetic acid is stronger than chloroacetic acid. Since inductive effect decreases with distance, the weaker acidity of 3-chloropropanoic acid compared to 2-chloropropanoic acid is not surprising.

Comparing the aliphatic acids with the aromatic acids, benzoic acid is stronger than acetic acid. This is due to the electron-withdrawing inductive effect of the benzene ring. Electron-releasing substituents like the methoxy group decrease acidity. In the case of *p*-methoxybenzoic acid, resonance increases the negative charge on the carboxylate ion, leading to destabilisation (Eq. 17.2a). Electron-withdrawing groups like chloro and nitro increase the acidity. The resonance effect due to the nitro group is shown in Eq. 17.2b. The positive charge on the carbon to which the carboxy group is attached, stabilises the negative charge.

$$(17.2)$$

Dicarboxylic acids are dibasic acids and have two ionisation constants. After the first ionisation, the presence of the negative charge affects the second ionisation, making it weaker. The low pK_a^1 of oxalic acid and malonic acid show that they are much stronger than acetic acid. This is due to the electron-withdrawing effect of the other unionised carboxy group. The effect decreases as the number of CH_2 groups increases.

17.2 GENERAL METHODS OF PREPARATION

17.2.1 OXIDATION OF PRIMARY ALCOHOLS AND ALDEHYDES

Carboxylic acids can be prepared by the oxidation of aldehydes or primary alcohols using permanganate or dichromate. Side chains of alkylbenzenes can also be oxidised by permanganate to the corresponding benzoic acids. Regardless of the number of carbon atoms in the side chain, and regardless of the presence of functional groups, the product is benzoic acid (Eq. 17.3).

$$\underset{\text{R = any alkyl or substituted alkyl}}{\overset{R}{\bigcirc}\quad\xrightarrow[\Delta]{\text{KMnO}_4,\ \text{alkali}}\quad\overset{COOH}{\bigcirc}} \tag{17.3}$$

R = any alkyl or substituted alkyl

17.2.2 FROM ALKYL HALIDES

There are two reactions by which the conversion, $RX \rightarrow RCOOH$, can be brought about. In one, the extra carbon is introduced in the form of carbon dioxide. In the other, it comes in the form of the cyanide ion.

17.2.2.1 Carboxylation of Grignard Reagents

Upon bubbling carbon dioxide into a solution of a Grignard reagent, a carboxylic acid is obtained (Eq. 17.4).

$$R\!-\!\underset{X}{\overset{|}{Mg}} + \overset{O\uparrow}{\underset{O}{\overset{\|}{C}}} \longrightarrow R\!-\!\underset{\underset{O}{\|}}{\overset{OMgX}{\overset{|}{C}}} \xrightarrow{H_3O^+} R\!-\!\overset{OH}{\overset{|}{C}}\!=\!O \tag{17.4}$$

This reaction is widely applicable and is useful for both aliphatic and aromatic acids. Here, carbon dioxide acts as a carbonyl compound and as an electrophile. The product of the reaction is a carboxylate anion and no further reaction takes place. Introduction of a carboxyl group is called *carboxylation* and is the reverse of decarboxylation (Chapter 16, Section 16.3.2.2). In this reaction, carbon dioxide is a *synthetic equivalent* of COOH. Recall the Kolbe-Schmitt synthesis of salicylic acid, where again, carbon dioxide functions as an electrophile for the introduction of the COOH group into the electron-rich phenol molecule.

17.2.2.2 Carboxylic Acids by the Hydrolysis of Nitriles

In contrast to carbon dioxide, the cyanide ion is a nucleophile. The cyanide ion is also a synthetic equivalent of carboxylate. Nucleophilic displacement of a halide by cyanide followed by hydrolysis of the nitrile is a convenient method for synthesising carboxylic acids, other than aromatic acids (Eq. 17.5).

$$R\!-\!Cl + CN^- \longrightarrow R\!-\!CN \xrightarrow[\Delta]{H_3O^+} R\!-\!COOH \tag{17.5}$$

R = primary or secondary alkyl

For aromatic acids, the Grignard route is applicable. Acetylation followed by oxidation of $COCH_3$ to COOH using $KMnO_4$ is also a good method for aromatic acids. Another method to oxidise the $COCH_3$ group is the haloform reaction.

Primary and secondary alkyl halides react with cyanide to give the nitrile. Tertiary alkyl halides undergo elimination rather than substitution (Chapter 11, Section 11.4).

17.3 OCCURRENCE, PROPERTIES AND USES

The higher fatty acids occur in nature as triglycerides in oils and fats. Salts of fatty acids are soaps. The methyl and ethyl esters obtained from oils and fats find use as biofuel. Fatty alcohols obtained by reduction are used in the manufacture of detergents.

Formic acid is manufactured by heating carbon monoxide with sodium hydroxide under pressure (Eq. 17.6).

$$CO + NaOH \xrightarrow{\Delta,\ pressure} H-\overset{\overset{\textstyle O}{\|}}{C}-ONa \tag{17.6}$$

Formic acid is dehydrated to carbon monoxide when heated with con. sulphuric acid (Eq. 17.7).

$$HCOOH \xrightarrow{H_2SO_4,\ heat} CO + H_2O \tag{17.7}$$

Formic acid is a reducing agent. It reduces ammoniacal silver nitrate to liberate metallic silver (Tollen's test).

Acetic acid is obtained by the oxidation of acetaldehyde or ethanol. Vinegar is an aqueous solution of acetic acid obtained by the bacterial oxidation of ethanol. Pure anhydrous acetic acid is known as glacial acetic acid. It freezes to a glassy solid when cooled to 15°C. Acetic acid and its esters are important industrial chemicals.

17.4 REACTIONS

Carboxylic acids form salts called carboxylates. They can be differentiated from phenols by their ability to dissolve in aqueous alkali as well as in aqueous sodium bicarbonate. The OH group can be replaced by OR (esterification), by halogens (formation of acid halides or acyl halides), by NH_2 or substituted amino groups (amide formation) or by carboxylate group (anhydride formation).

17.4.1 ESTERIFICATION

Carboxylic acids react with alcohols in the presence of mineral acid catalysts, to form esters (Eq. 17.8).

$$R-\overset{\overset{\textstyle O}{\|}}{C}-OH + R'-OH \underset{\xrightarrow{H^+}}{\overset{\textstyle H^+}{\rightleftharpoons}} R-\overset{\overset{\textstyle O}{\|}}{C}-OR' + H_2O \tag{17.8}$$

This is an equilibrium reaction. The reaction can be driven in the forward direction by using one of the reactants – acid or alcohol – in excess or by removing the product, water, from the reaction medium by distillation. A common practice is to carry out the esterification in the presence of, benzene and a catalytic quantity of a mineral acid like sulphuric acid in a *Dean–Stark apparatus* and distilling out water as an azeotrope. In the *Fischer–Speier method* of esterifiction, hydrogen chloride gas is passed into the mixture of acid and alcohol and refluxed. The mechanism of esterification can differ depending on the structure of the reactants and reaction conditions. The most common mechanism is outlined in Eq. 17.9.

$$
R-\overset{\overset{O}{\parallel}}{C}-OH + H^+ \quad \underset{}{\overset{(1)}{\rightleftarrows}} \quad \left[R-\overset{\overset{+}{O}-H}{\underset{O-H}{C}} \quad \longleftrightarrow \quad R-\overset{O-H}{\underset{\overset{+}{O}-H}{C}} \right]
$$

$$
\begin{array}{c}
R-\overset{\overset{+}{O}-H}{\underset{\underset{H-\ddot{O}-R'}{O-H}}{C}} \quad \overset{(2)}{\rightleftarrows} \quad R-\overset{\overset{OH}{|}}{\underset{\underset{+}{H-O-R'}}{C}}-OH \quad \overset{(3)}{\rightleftarrows} \quad R-\overset{\overset{\ddot{O}H}{|}}{\underset{\underset{O-R'}{H}}{C}}\overset{+}{-}\overset{}{O}-H \\
\\
\qquad\qquad\qquad\qquad\qquad (I) \qquad\qquad\qquad\qquad\qquad (II)
\end{array}
\qquad (17.9)
$$

$$
\overset{(4)}{\rightleftarrows} \quad R-\overset{\overset{\overset{+}{O}-H}{\parallel}}{\underset{O-R'}{C}} + H_2O \quad \overset{(5)}{\rightleftarrows} \quad R-\overset{\overset{O}{\parallel}}{\underset{O-R'}{C}} + H_3O^+
$$

The protonated acid formed in step (1) reacts with a molecule of alcohol, step (2), to give a tetrahedral intermediate, (I) or (II). This is called a tetrahedral intermediate because the hybridisation of the central carbon atom has changed from sp^2 (trigonal) in the starting carboxylic acid to sp^3 in the intermediate. This intermediate loses a molecule of water and then deprotonates to give the ester, steps (4) and (5). Each step in this mechanism is reversible. The reverse reaction is acid catalysed ester hydrolysis. Ester hydrolysis can also be brought about by alkali (Eq. 17.10). Esterification cannot be carried out under alkaline conditions because the carboxylate anion formed in the alkaline medium cannot undergo nucleophilic attack. The delocalised structure of the carboxylate ion (Eq. 17.1) has little residual positive charge at the carbonyl carbon.

$$
R-\overset{\overset{O}{\parallel}}{\underset{\overset{|}{OH}}{C}}-OR' \quad \rightleftarrows \quad R-\overset{\overset{O^-}{|}}{\underset{\overset{|}{OH}}{C}}-OR' \quad \rightleftarrows \quad R-\overset{\overset{O}{\parallel}}{\underset{\overset{|}{OH}}{C}} + R'O^- \quad \longrightarrow \quad R-\overset{\overset{O}{\parallel}}{C}-O^- + R'OH
$$

$$
(17.10)
$$

The first two steps in the alkaline hydrolysis of ester are reversible (Eq. 17.10), but the last step is irreversible. Thus overall, alkaline ester hydrolysis (also called saponification because this is the reaction involved in soap making) is irreversible. This reaction can proceed to completion, unlike the acid catalysed reaction.

One aspect of ester hydrolysis and esterification, namely, which bond in the ester or acid is cleaved in the reaction, deserves attention. There are two possibilities (Eqs. 17.11a and 17.11b).

In 17.11a, the bond between the RCO (acyl group of the acid) and the oxygen is cleaved and is known as acyl-oxygen cleavage. In 17.11b, the bond between the R′ (the alkyl group of the alcohol) and oxygen is cleaved and is known as alkyl-oxygen cleavage. For most simple esters, the former cleavage occurs and is shown in the mechanisms (Eqs. 17.9 and 17.10). This has been proved by carrying out the hydrolysis using H_2O in which most of the oxygen is the heavier isotope, ^{18}O. Such a water sample is called isotopically labelled water. After the hydrolysis, the acid and alcohol are analysed for their ^{18}O content. It is found that the carboxylic acid contains the labelled oxygen and not the alcohol (Eq. 17.12). This is also valid for esterification. In the case of simple acids and alcohols, both esterification and ester hydrolysis take place by acyl-oxygen cleavage in the acid part. In other words, esterification is acylation of the alcohol. (Recall acylation of phenols Chapter 15, Section 15.4.7.)

$$
\begin{array}{cc}
\underset{\substack{\\ H-O \overset{\displaystyle }{} H}}{R-C \overset{\displaystyle O}{\overset{\|}{}} O-R'} & \rightleftharpoons \quad \underset{\substack{\\ H-O \quad H}}{R-C \overset{\displaystyle O}{\overset{\|}{}} + \; O-R'} \qquad (a)
\end{array}
$$

$$
\begin{array}{cc}
\underset{\substack{\\ H \; O-H}}{R-C \overset{\displaystyle O}{\overset{\|}{}} -O \; R'} & \rightleftharpoons \quad \underset{\substack{\\ H \quad O-H}}{R-C \overset{\displaystyle O}{\overset{\|}{}} -O \; + \; R'} \qquad (b)
\end{array}
$$

(17.11)

$$R-CO-OR' + H_2^{18}O \rightarrow R-CO-^{18}OH + R'OH \qquad (17.12)$$

17.4.2 CONVERSION TO ACYL HALIDES, ANHYDRIDES AND AMIDES

Acids can be converted to acyl halides or acid halides, RCO–X using the same reagents which are used for converting alcohols to alkyl halides. Thus, benzoic acid can be converted to benzoyl chloride by reaction with phosphorous pentachloride, phosphorous trichloride or thionyl chloride (Eq. 17.13).

$$
\text{C}_6\text{H}_5\text{—C(O)—OH} + \begin{array}{l} \text{PCl}_5 \text{ (or)} \\ \text{PCl}_3 \text{ (or)} \\ \text{SOCl}_2 \end{array} \longrightarrow \text{C}_6\text{H}_5\text{—C(O)—Cl} \qquad (17.13)
$$

Acyl halides react with carboxylate salts, in a Williamson-type reaction to give anhydrides (Eq. 17.14a). Anhydrides, which are formally like ethers and correspond to the intermolecular dehydration of two molecules of the acid, are rarely prepared by direct dehydration.

Both acyl halides and acid anhydrides can be used for acylation of alcohols (esterification, Eqs. 17.14b, 17.14c), of arenes (Friedel–Crafts acylation, Eq. 17.14d), and for acylation of amines to form amides (Eq. 17.14e). The general mechanism of acylation of amines is shown in Eq. 17.14f, where Z = halogen or carboxylate.

$$R - \overset{\overset{O}{\parallel}}{C} - Cl$$
$$+ R' - \overset{\overset{\parallel}{O}}{C} - O^- \quad Na^+ \longrightarrow R - \overset{O^-}{\underset{R' - C - O}{\underset{\parallel}{\overset{\mid}{C}}}} Cl \longrightarrow R - C \overset{O}{\underset{R' - C}{\underset{\parallel}{O}}} + Cl^- \qquad (a)$$

$$R - \overset{\overset{O}{\parallel}}{C} - Cl + R' - OH \longrightarrow R - \overset{\overset{O}{\parallel}}{C} - OR' + HCl \qquad (b)$$

$$R - \overset{\overset{O}{\parallel}}{C} - O - \overset{\overset{O}{\parallel}}{C} - R + R' - OH \longrightarrow R - \overset{\overset{O}{\parallel}}{C} - OR' + R - COOH \qquad (c)$$

$$R - \overset{\overset{O}{\parallel}}{C} - Cl + \bigcirc \xrightarrow{AlCl_3} R - \overset{\overset{O}{\parallel}}{C} \bigcirc + HCl \qquad (d)$$

$$R - \overset{\overset{O}{\parallel}}{C} - Cl + H_2N - R' \xrightarrow{OH^-} R - \overset{\overset{O}{\parallel}}{C} - NHR' + (HCl) \qquad (e)$$

$$R - \overset{\overset{O}{\parallel}}{C} - Z$$
$$+ R' - \ddot{N}H_2 \longrightarrow R - \overset{O^-}{\underset{R' - NH_2}{\underset{+}{\overset{\mid}{C}} Z}} \longrightarrow R - \overset{O}{\underset{R' - NH_2}{\underset{+}{\overset{\parallel}{C}}}} + Z^- \longrightarrow R - \overset{\overset{O}{\parallel}}{C} - NHR' + ZH \qquad (f)$$

$$(17.14)$$

When Z = OH, this mechanism corresponds to the use of the free acid as the alkylating agent. When Z = halogen or carboxyate, the reactions proceed better since these are better leaving groups.

Acyl halides, anhydrides and amides can be hydrolysed to the acid by using aqueous acid or alkali. The mechanism is analogous to the mechanism for the hydrolysis of esters. These reactions are nucleophilic displacement reactions at the carbonyl carbon. They are mechanistically different from the nucleophilic displacement reactions at saturated carbon atom discussed in Chapter 11. These displacements are not concerted as in S_N2 reactions, nor do they proceed through carbocations as in the case of S_N1 reactions. They involve addition of the nucleophile to the carbonyl carbon to form a tetrahedral intermediate, followed by elimination of a leaving group (Eqs. 17.9, 17.10, 17.14a and 17.14f). It is an addition–elimination mechanism and in that sense, it is similar to nucleophilic substitution reactions on the aromatic ring.

17.4.3 LACTONES AND LACTAMS

Cyclic esters and cyclic amides are called lactones and lactams, respectively. They are formed from the corresponding hydroxy acids and amino acids (Eq. 17.15).

$$\text{(17.15)}$$

γ-Hydroxybutyric acid (4-hydroxybutanoic acid) upon heating loses a molecule of water to give γ-butyrolactone. δ-Aminovaleric acid (5-aminopentanoic acid) on heating similarly undergoes intramolecular dehydration to give δ-valerolactam.

17.5 DICRBOXYLIC ACIDS

Dicarboxylic acids are dioic acids, containing two COOH groups. The first few members of the homologous series are listed in Table 17.1.

Oxalic acid occurs in nature as the sour component in certain fruits. It is manufactured by heating sodium formate (Eq. 17.16). Oxalic acid is a reducing agent and finds use in analytical chemistry.

$$2 \ NaO-\overset{\overset{\textstyle O}{\|}}{C}-H \xrightarrow{350^{\circ}C} \begin{array}{c} NaO-\overset{\overset{\textstyle O}{\|}}{C} \\ | \\ NaO-\underset{\underset{\textstyle O}{\|}}{C} \end{array} + H_2 \qquad (17.16)$$

Malonic acid can be prepared from chloroacetic acid (Eq. 17.17).

$$Cl-CH_2 \, COO^-K^+ \xrightarrow{KCN} N\equiv C-CH_2-COO^- \xrightarrow[\Delta]{H_2O, \, OH^-} CH_2 \overset{\diagup COO^-}{\underset{\diagdown COO^-}{}}$$

$$\xrightarrow{H_3O^+} CH_2 \overset{\diagup COOH}{\underset{\diagdown COOH}{}}$$

$$(17.17)$$

Malonic acid and other dicarboxylic acids containing two COOH groups attached to the same carbon atom undergo decarboxylation to give the monocarboxylic acid. This reaction proceeds through a cyclic transition state (Eq. 17.18).

$$(17.18)$$

Diethyl ester of malonic acid is an important laboratory reagent for the synthesis of carboxylic acids by the *malonic ester synthesis (vide infra)*. Succinic acid may be prepared from the corresponding nitrile (Eq. 17.19).

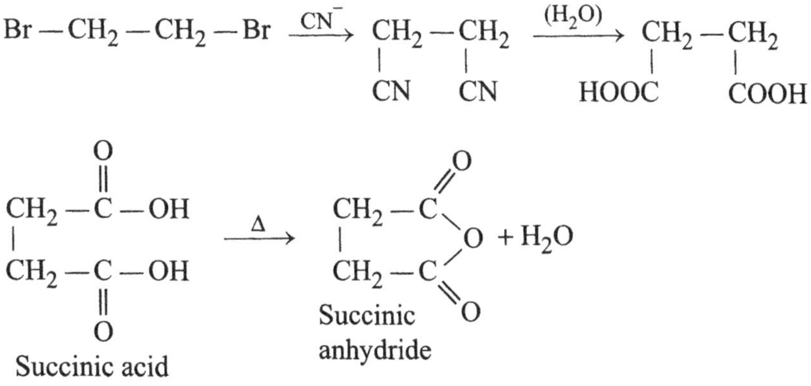

$$\text{Br}-\text{CH}_2-\text{CH}_2-\text{Br} \xrightarrow{\text{CN}^-} \underset{\underset{\text{CN}}{|}}{\text{CH}_2}-\underset{\underset{\text{CN}}{|}}{\text{CH}_2} \xrightarrow{(\text{H}_2\text{O})} \underset{\underset{\text{HOOC}}{|}}{\text{CH}_2}-\underset{\underset{\text{COOH}}{|}}{\text{CH}_2} \qquad \text{(a)}$$

(17.19)

Succinic acid $\xrightarrow{\Delta}$ Succinic anhydride $+ \text{H}_2\text{O}$ (b)

Upon heating, succinic acid undergoes intramolecular dehydration to give succinic anhydride which is a cyclic molecule (Eq. 17.19b). Glutaric acid can be prepared by a route similar to that for succinic acid. It also forms a cyclic anhydride upon heating (Eq. 17.20).

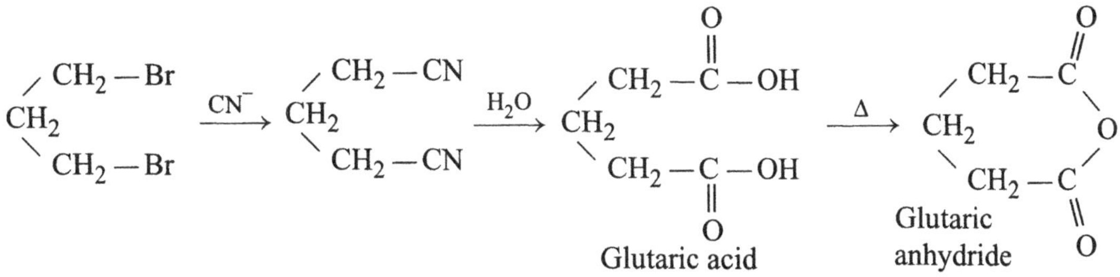

(17.20)

Adipic acid can be prepared by the oxidation of cyclohexene, cyclohexanone or cyclohexanol (Eq. 17.21).

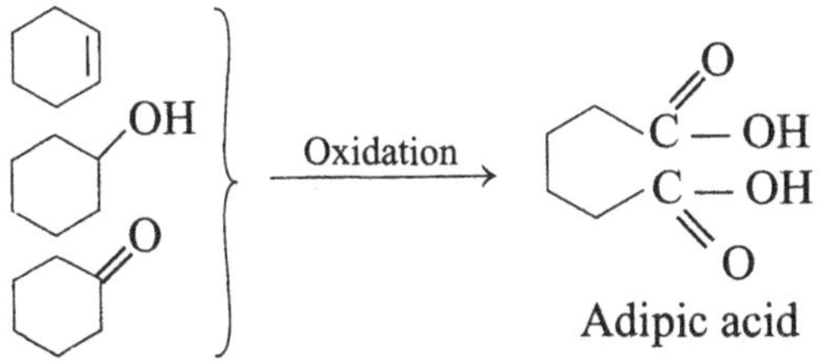

(17.21)

The main use of adipic acid is in the manufacture of polyamides like *Nylon 66*.

17.5.1 MALONIC ESTER SYNTHESIS

The methylene group (CH_2) in malonic ester is *active* in the sense that the two hydrogens are relatively acidic. One of them can be abstracted by bases to give the carbanion, which is a resonance stablised enolate ion (Chapter 16, Section 16.1.2). We have seen how this can add to the carbonyl group of aldehydes in the Knoevenagel reaction (Section 16.3.2.2). The enolate ion can function as a nucleophile and react with alkyl halides in a nucleophilic displacement reaction (Eq. 17.22).

$$\underset{\substack{\diagup COOC_2H_5 \\ CH_2 \\ \diagdown COOC_2H_5}}{} \xrightarrow{NaOC_2H_5} \underset{\substack{\diagup COOC_2H_5 \\ \overset{-}{CH} \\ \diagdown COOC_2H_5}}{} \xrightarrow{RX} \underset{\substack{\diagup COOC_2H_5 \\ R-CH \\ \diagdown COOC_2H_5}}{}$$

(17.22)

$$\xrightarrow{OH^-, H_2O} \underset{\substack{\diagup COO^- \\ R-CH \\ \diagdown COO^-}}{} \xrightarrow{H^+} \underset{\substack{\diagup COOH \\ R-CH \\ \diagdown COOH}}{} \xrightarrow{\Delta} R-CH_2-COOH + CO_2$$

This reaction can be described as alkylation of the active methylene position of malonic ester. Since there are two acidic hydrogens, two alkyl groups (either same or different) can be introduced one after the other. The resultant mono- or di-alkylated malonic ester upon hydrolysis and decarboxylation by heating, gives a substituted acetic acid (Eq. 17.23).

$$\underset{\substack{\diagup COOC_2H_5 \\ CH_2 \\ \diagdown COOC_2H_5}}{} \xrightarrow{NaOC_2H_5, RX} \underset{\substack{\diagup COOC_2H_5 \\ R-CH \\ \diagdown COOC_2H_5}}{} \xrightarrow{NaOC_2H_5, R'X} \underset{\substack{R \diagdown \diagup COOC_2H_5 \\ C \\ R' \diagup \diagdown COOC_2H_5}}{}$$

$$\xrightarrow[\text{(ii) } H_3O^+, \Delta]{\text{(i) } OH^-, H_2O, \Delta} \underset{\substack{R \diagdown \\ CH-COOH \\ R' \diagup}}{}$$

(17.23)

This is the *malonic ester synthesis of carboxylic acids*. By choosing appropriate alkyl halides, any desired mono- or disubstituted acetic acid can be synthesised.

Cyanoacetic ester (prepared as in Eq. 17.24a), is another active methylene compound which can be used instead of malonic ester, for the same kind of synthesis.

$$KCN + Cl-CH_2-COOC_2H_5 \rightarrow N\equiv C-CH_2-COOC_2H_5 + KCl \qquad \text{(a)}$$

$$N\equiv C-CH_2-COOC_2H_5 \xrightarrow{NaOC_2H_5, RX} \underset{\substack{\\ R}}{N\equiv C-\overset{|}{\underset{|}{CH}}-COOC_2H_5}$$

$$\xrightarrow{NaOC_2H_5, R'X} \underset{\substack{R' \\ | \\ N\equiv C-\overset{|}{C}-COOC_2H_5 \\ | \\ R}}{} \xrightarrow{OH^-, \Delta} \underset{\substack{R' \\ | \\ R-\overset{|}{C}-COO^- \\ | \\ COO^-}}{}$$

$$\xrightarrow{H^+, \Delta} \underset{\substack{R' \diagdown \\ CH-COOH + CO_2 \\ R \diagup}}{} \qquad \text{(b)}$$

(17.24)

17.6 ACETOACETIC ESTER

17.6.1 SYNTHESIS OF ACETOACETIC ESTER: CLAISEN CONDENSATION

Acetoacetic acid (3-oxobutanoic acid) is a β-*ketoacid*. It has an active methylene group. Ethyl acetoacetate is used in synthesis in reactions similar to those with malonic ester. Acetoacetic ester is prepared by *Claisen condensation* of ethyl acetate (Eq. 17.25).

$$2\ CH_3\ COOC_2H_5 \xrightarrow{\ NaOC_2H_5\ } CH_3-CO-CH_2-COOC_2H_5 \qquad\text{(a)}$$

Acetoacetic ester
(ethyl acetoacetate)

$$CH_3-\overset{\overset{\displaystyle O}{\|}}{C}-OC_2H_5 \ \underset{\longleftarrow}{\overset{C_2H_5O^-}{\rightleftharpoons}}\ \left[\bar{C}H_2-\overset{\overset{\displaystyle O}{\|}}{C}-OC_2H_5 \longleftrightarrow CH_2=\overset{\overset{\displaystyle O^-}{|}}{C}-OC_2H_5 \right]$$

(I)

$$\begin{array}{l} CH_3-\overset{\overset{\displaystyle O}{\|}}{C}-OC_2H_5 \\ \quad\bar{C}H_2-\underset{\underset{\displaystyle O}{\|}}{C}-OC_2H_5 \end{array} \rightleftharpoons CH_3-\overset{\overset{\displaystyle O^-}{|}}{C}-OC_2H_5 \\ \qquad\qquad CH_2-\underset{\underset{\displaystyle O}{\|}}{C}-OC_2H_5$$

(II)

$$\rightleftharpoons CH_3-\overset{\overset{\displaystyle O}{\|}}{C}-CH_2-\overset{\overset{\displaystyle O}{\|}}{C}-OC_2H_5 \longrightarrow \left[\begin{array}{c} CH_3-\overset{\overset{\displaystyle O}{\|}}{C}-\bar{C}H-\overset{\overset{\displaystyle O}{\|}}{C}-OC_2H_5 \\ \updownarrow \\ CH_3-\overset{\overset{\displaystyle O^-}{|}}{C}=CH-\overset{\overset{\displaystyle O}{\|}}{C}-OC_2H_5 \end{array} \right] \xrightarrow{\ H^+\ } \text{(III)}$$

$$+\ C_2H_5O^- \qquad\text{(III)} \qquad\qquad\qquad\qquad\qquad\qquad\qquad\qquad\qquad \text{(IV)} \qquad\qquad\qquad\text{(b)}$$

$$2\ CH_3-\overset{\overset{\displaystyle H}{|}}{\underset{\underset{\displaystyle CH_3}{|}}{C}}-COOC_2H_5 \underset{\longleftarrow}{\overset{NaOC_2H_5}{\rightleftharpoons}} CH_3-\overset{\overset{}{}}{\underset{\underset{\displaystyle CH_3}{|}}{CH}}-\overset{\overset{\displaystyle O}{\|}}{C}-\overset{\overset{\displaystyle CH_3}{|}}{\underset{\underset{\displaystyle CH_3}{|}}{C}}-\overset{\overset{\displaystyle O}{\|}}{C}-OC_2H_5 \qquad\text{(c)}$$

(V) (VI)

(17.25)

Recall Claisen–Schmidt reaction also called Claisen reaction (Chapter 16, Section 16.3.2.1), which is not the same as Claisen condensation. The mechanism of Claisen condensation leading to acetoacetic ester is outlined in Eq. 17.25b. The enolate anion (I), from ethyl acetate reacts with another molecule of ethyl acetate in a nucleophilic addition, very similar to the corresponding step in aldol condensation. The fate of the tetrahedral intermediate (II), formed by this addition is the same as in ester hydrolysis; ejection of of the alkoxide, $C_2H_5O^-$. The product is ethyl acetoacetate (III). All the steps upto this stage are reversible and the equilibrium is in favour of the reverse reactions, so much so that if the reaction were to end here, the yield of ethyl acetoacetate would have been very low. However, this is not the end of the reaction sequence. The final step is the conversion of ethyl acetoacetate (III), to the enolate anion (IV), by the action of the base. This step is irreversible since it is a reaction of a relatively strong acid with a strong base. It is this that drives the reaction to completion. Esters like ethyl isobutyrate (V), which should give the ketoester (VI) by Claisen condensation, does not undergo the reaction under these conditions because the ketoester (VI) which does not have even one active hydrogen, cannot form the enolate anion to drive the reaction in the forward direction (Eq. 17.25c).

17.6.2 KETO–ENOL TAUTOMERISM

The enolate anion (IV) (Eq. 17.25b) of ethyl acetoacetate is resonance stabilised. The ketoester (I) exists as an equilibrium mixture of the keto form and the enol form (Eq. 17.26a).

(a)

Keto form   Enol form

(17.26)

(b)

(c)

The two structures are not canonical structures; they are isomers, called *tautomers*. This kind of isomerism is called tautomerism. In this case it is specifically called keto–enol tautomerism. The acidic hydrogen of the active methylene group is the labile hydrogen, which apparently moves back and forth from the carbon to the oxygen. Such a shift of the hydrogen from one atom to another is called *prototropy*. Even simple ketones like acetone can exhibit tautomerism (Eq. 16.26c). However, in the case of simple ketones, the equilibrium is very much in favour of the ketone and the enol form exists only in traces at equilibrium.

In the case of ethyl acetoacetate, the enol form is relatively stable and any sample of the ester is an equilibrium mixture of the two tautomers. Their presence can be detected by physical measurements. They can even be separated by careful fractionation. However they are very labile, meaning that they isomerise from one form to the other very easily. A sample of the pure keto form or the pure enol form readily isomerises to give the equilibrium mixture.

The relative stability of the enol form of ethyl acetoaetate compared to that of simple ketones, is due to the stabilisation of the former by intramolecular hydrogen bonding (Eq. 17.26b). In the case of the enol form of acetone (Eq. 17.26c), such intramolecular hydrogen bonding is not possible.

Thus, the conditions for the existence of a relatively stable enol form that is in equilibrium with the ketone are: (i) the labile hydrogen is relatively acidic and (ii) the enol form can be stabilised by intramolecular hydrogen bonding. At room temperature, a sample of ethyl acetoacetate is an equilibrium mixture of about 93% ketone and 7% enol. In the case of 2,4-pentanedione, it is 24% ketone and 76% enol. In molecules like acetone and acetaldehyde, the enol form exists only in traces. A method of estimating the concentration of enol in a sample is by bromination. The enol form reacts almost instantaneously with bromine (Eq. 17.27). When the enol form is consumed, more of the ketone will be converted to the enol to restore equilibrium. Eventually all the molecules will be brominated, via the enol form. However, the isomerisation of the ketone to the enol is slower than the reaction of the enol with bromine.

$$
\begin{array}{cc}
\underset{\underset{\displaystyle \text{Br} - \text{Br}}{\Big\downarrow}}{\underset{\displaystyle R - C = CH - \overset{\displaystyle \text{O}}{\overset{\displaystyle \|}{C}} - R'}{H - \overset{\cdot\cdot}{O}\text{:}}} & \longrightarrow \quad R - \overset{\overset{\displaystyle +}{\overset{\displaystyle H - O}{\|}}}{C} - \underset{\underset{\displaystyle \text{Br}}{\displaystyle |}}{CH} - \overset{\displaystyle O}{\overset{\displaystyle \|}{C}} - R' \\
\end{array}
\qquad (17.27)
$$

Hence, by adding bromine to a sample and measuring the amount of bromine consumed within a few seconds, the enol content can be caculated. Conversion of the keto form to the enol or the reverse can be catalysed by both acids and by bases (Eqs. 17.28a and 17.28b).

Acid catalysis

$$
R - \overset{\displaystyle O}{\overset{\displaystyle \|}{C}} - CH_2 - \overset{\displaystyle O}{\overset{\displaystyle \|}{C}} - R' + H^+ \rightleftharpoons R - \overset{\overset{\displaystyle H - \overset{+}{O}}{\|}}{C} - \underset{\underset{\displaystyle H}{\displaystyle |}}{C} - \overset{\displaystyle O}{\overset{\displaystyle \|}{C}} - R' \rightleftharpoons R - \overset{\overset{\displaystyle H - O}{\|}}{C} = \overset{\overset{\displaystyle H}{\displaystyle |}}{C} - \overset{\displaystyle O}{\overset{\displaystyle \|}{C}} - R' \qquad (a)
$$

Base catalysis

$$
R - \overset{\overset{\displaystyle \text{:B}^-}{\Big\downarrow}}{\underset{\underset{\displaystyle H}{\displaystyle |}}{\overset{\overset{\displaystyle O}{\|}}{C}}} - \overset{\overset{\displaystyle H}{\displaystyle |}}{C} - \overset{\displaystyle O}{\overset{\displaystyle \|}{C}} - R' \rightleftharpoons R - \overset{\overset{\displaystyle H - B}{}}{\underset{\displaystyle O^-}{C}} = \overset{\overset{\displaystyle H}{\displaystyle |}}{C} - \overset{\displaystyle O}{\overset{\displaystyle \|}{C}} - R' \rightleftharpoons R - \overset{\overset{\displaystyle H \quad \text{:B}^-}{}}{\underset{\displaystyle O}{C}} = \overset{\overset{\displaystyle H}{\displaystyle |}}{C} - \overset{\displaystyle O}{\overset{\displaystyle \|}{C}} - R' \qquad (b)
$$

$$(17.28)$$

Tautomerism due to prototropy is also present in other classes of compounds like amides and aliphatic nitro compounds. The tautomer of an amide is called the imidol and that of a nitro compound, the acinitro compound.

Amido–imidol tautomerism: $R-NH-CO-R' \rightleftarrows R-N=C(OH)-R'$ (Imidol)

Nitro–acinitro tautomerism: $R_2CH-NO_2 \rightleftarrows R_2C=NO(OH)$ (Acinitro compound)

17.6.3 ACETOACETIC ESTER SYNTHESIS

Like malonic ester, acetoacetic ester is a useful reagent in organic synthesis. Acetoacetic ester synthesis can be adapted to synthesise methyl ketones and carboxylic acids. Both syntheses make use of the presence of the active methylene group where one or two alkyl groups can be introduced by alkylation (Eqs. 17.29 and 17.30).

(17.29)

The first part is common for the synthesis of both ketones and carboxylic acids (Eq. 17.29a). Treatment of ethyl acetoacetate with sodium ethoxide in ethanol gives the anion which is alkylated by RX to produce the monoalkyl derivative. If two alkyl groups, R and R′ are to be introduced, the process is repeated. The alkylated acetoacetic ester is hydrolysed by dilute sodium hydroxide to the sodium salt of the carboxylic acid, from which the free acid is obtained by acidification. This β-ketoacid upon heating undergoes decarboxylation to give the desired ketone. The decarboxylation of β-ketoacids is similar to the decarboxylation of malonic acid and proceeds through a cyclic transition state (Eq. 19.29b). By appropriate choice of R and R′, ketones of the type $CH_3 — CO — CH(R)R′$ can be synthesised.

Synthesis of carboxylic acids by the acetoacetic ester route is shown in Eq. 17.30. The mono- or dialkylacetoacetic ester is prepared (Eq. 17.29a). This is heated with concentrated alkali. Under these conditions, a reversal of Claisen condensation takes place [steps (2) and (3) of Eq. 17.30]. The products are one molecule of acetic acid and one molecule of the desired substituted acetic acid (I). Using this procedure, disubstituted acetic acids can be prepared. Notice that these can also be synthesised using malonic ester. Malonic ester sythesis is better suited for this purpose.

$$(17.30)$$

These alkylations of the enolate anion are true S_N2 reactions, the enolate being the nucleophile and the alkyl halide being the substrate.

17.7 SPECTROSCOPY OF CARBOXYLIC ACIDS AND DERIVATIVES

The relevant IR bands for the carboxylic acids and their derivatives are listed in Table 17.2.

The proton of the COOH group is highly deshielded and appears in the NMR spectrum way downfield, in the region δ 10.5–12.

Table 17. 2   IR frequencies of carboxylic acids and derivatives

Molecule	Bond	Frequency (cm^{-1})
RCOOH	O—H	2500–3000
	C—O	1250
	C=O	1700–1725
Ar—COOH C=C—COOH	C=O	1680–1700
RCOOR	C=O	1740
ArCOOR C=C—COOR	C=O	1715–1730
RCOOAr	C=O	1770
RCONH$_2$	C=O	1650–1690
RCOCl	C=O	1750–1810
δ-Lactones	C=O	1740
γ-Lactones	C=O	1770

KEY POINTS

- Acidity of carboxylic acids is due to the resonance in the carboxylate ion which delocalises the negative charge.
- Formic acid is the strongest of the aliphatic acids. Higher aliphatic acids are weaker due to the electron releasing effect of the alkyl group. Presence of electron-withdrawing groups like the halogens increases the acidity.
- Benzoic acid is stronger than acetic acid because of the electron-withdrawing effect of the phenyl group. Electron-withdrawing groups like halogen or nitro on the ring increase the acidity. Electron-releasing groups like methoxy decrease the acidity.
- Acids can be prepared by oxidation of primary alcohols or aldehydes. Aromatic acids can be prepared by the oxidation of aromatic compounds with alkyl or substituted alkyl side chains.
- Grignard reagents react with carbon dioxide to give carboxylic acids. Carbon dioxide functions as an electrophile, a synthetic equivalent of the carboxylate.
- Primary and secondary alkyl halides undergo nucleophilic displacement by the cyanide ion to give nitriles, which upon hydrolysis yield carboxylic acids. CN^-, a nucleophile, is also a synthetic equivalent of COOH.

- Carboxylic acids are converted to esters by reaction with alcohols catalysed by mineral acids in a reversible manner. Esters can be hydrolysed back to the acid and alcohol, under acidic or alkaline conditions. These reactions involve acyl-oxygen cleavage.
- Acid halides are prepared from acids by using PCl_5, PCl_3 or $SOCl_2$. Condensation of acid chloride with a carboxylate ion gives the acid anhydride. Either acid chloride or acid anhydride can react with amines to give amides, and with alcohols and phenols to give esters.
- Succinic acid and its higher homologues can be prepared by the hydrolysis of dinitriles. Succinic and glutaric acids give cyclic anhydrides on heating. Adipic acid is prepared by the oxidation of cyclohexene, cyclohexanol or cyclohexanone.
- Malonic ester contains an active methylene group. Alkylation of the methylene position followed by hydrolysis of the ester followed by decarboxylation, is a useful method for the synthesis of carboxylic acids. Cyanoacetic ester can also be used for the same kind of syntheses.
- Claisen condensation of ethyl acetate gives acetoacetic ester. This is also an active methylene compound. Alkylation followed by hydrolysis and decarboxylation produces ketones.
- Malonic acid, substituted malonic acids and β-ketoacids undergo decarboxylation upon heating via cyclic transition states.
- Carboxylic acids and their derivatives have a characteristic $C=O$ stretching frequency. The NMR spectrum of carboxylic acids will show a singlet signal for the acidic proton in the δ 9.5–12 region.

EXERCISES

SECTION I

1. How can the following be converted to benzoic acid?

   (a) Ethylbenzene   (b) Benzyl alcohol

   (c) Bromobenzene   (d) Acetophenone

2. How can propanoic acid be converted to the following?

   (a) Propanol
   (b) Propanoyl chloride
   (c) Propanoic anhydride
   (d) The mixed anhydride of propanoic acid and ethanoic acid, $CH_3-CH_2-CO-O-CO-CH_3$
   (e) Propanamide
   (f) Methyl propanoate
   (g) $C_6H_5-CO-CH_2-CH_3$

3. Arrange the following in the increasing order of acidity.

    (a) $ClCH_2$—$COOH$, $BrCH_2$—$COOH$, FCH_2—$COOH$, ICH_2—$COOH$
    (b) CH_3—$CH(Br)$—CH_2—$COOH$, $CH_2(Br)$—CH_2—CH_2—$COOH$,
         CH_3—CH_2—$CH(Br)$—$COOH$
    (c) Benzoic acid, *p*-methoxybenzoic acid, *p*-methylbenzoic acid, *m*-chlorobenzoic acid
    (d) Benzoic acid, 2,4-dinitrobenzoic acid, *p*-nitrobezoic acid.

4. How can the following conversions be brought about in one or two steps?

    (a) Benzyl chloride to phenylacetic acid
    (b) 3-Methyl-2-butanone to 2-methylpropanoic acid
    (c) Benzoic acid to benzamide, C_6H_5—CO—NH_2
    (d) Benzoic acid to benzyl acetate
    (e) Diethyl malonate to C_6H_5—CH=$C(COOC_2H_5)_2$
    (f) Succinic acid to succinic anhydride
    (g) Benzoic acid to benzoic anhydride
    (h) Benzaldehyde to cinnamic acid
    (i) Oxalic acid to dimethyl oxalate

5. Indicate the reagent(s) and/or reaction conditions which can bring about the following conversions.

    (a) Benzoic acid to benzoyl chloride
    (b) CH_3MgBr to CH_3COOH
    (c) Benzonitrile to benzoic acid
    (d) *p*-nitroacetanilide to *p*-nitroaniline
    (e) Malonic acid to acetic acid
    (f) Ethyl acetoacetate to acetone

SECTION II

1. Show how the following can be synthesised using malonic ester.

    (a) 4-Methylpentanoic acid
    (b) 2-Methylbutanoic acid
    (c) Dibenzylacetic acid, $(C_6H_5CH_2)_2CH$—$COOH$
    (d) 4-Phenylbut-2-enoic acid, C_6H_5—CH_2—CH=CH—$COOH$

2. Show how ethyl cyanoacetate can be used to synthesise the following.

    (a) 2-Benzylpropanoic acid   (b) Isobutyric acid

3. Show how the following ketones can be synthesised by acetoacetic ester synthesis.

    (a) 4-Methyl-2-pentanone
    (b) 3-Methyl-2-pentanone
    (c) 3-Benzyl-4-methyl-2-hexanone

4. Outline the mechanism of the conversion of R—COOH to R—COCl using the following. (a) PCl$_5$ (b) SOCl$_2$. Take hints from the conversion of alcohol to alkyl chloride.
5. Esters undergo trans-esterification (ester exchange) as below, catalysed by alkoxide. Outline the mechanism.

$$RCOOR + R'OH/R'ONa \rightarrow RCOOR' + ROH$$

6. Identify the unknown compound.

    Molecular formula: C$_9$H$_{10}$O$_2$
    IR band at 1720 cm^{-1}
    NMR δ 3.4, singlet, 2H; 3.5, singlet, 3H; 7.4, singlet, 5H

7. Two isomeric molecules, C$_4$H$_8$O$_2$, (A) and (B) are suspected to be ethyl acetate and methyl propanoate; both have a strong IR band at 1740 cm^{-1}. The NMR spectra of A and B are as follows.

    A : δ 1.2, triplet, 3H; 2.0, singlet, 3H; 4.1, quartet, 4H

    B : 1.0, triplet, 3H; 2.3, quartet, 2H; 3.7, singlet, 3H

    Identify which is which.
8. Phenol is the enol form of cyclohexadienone. Visualise this tautomerism and explain why the keto form does not exist, for all practical purposes, at equilibrium.
9. Outline the mechanism of acid catalysed ester hydrolysis. How can the reaction be driven to completion?
10. Draw the structures of the enol forms with intramolecular hydrogen bonds, of 2,4-pentanedione and butanal-3-one, CH$_3$—CO—CH$_2$—CHO.

CHALLENGING QUESTIONS

1. Benzene reacts with γ-butyrolactone in the presence of anhydrous aluminium chloride to give 1-tetralone. Suggest the mechanism.
2. Suggest the mechanism for the conversion of 4-hydroxybutanoic acid to the lactone upon heating. Speculate why the reaction proceeds to completion.
3. A process for the manufacture of 'biodiesel', a mixture of ethyl esters of fatty acids present in oils and fats, involves heating the oil with ethanol and catalytic quantity of solid potassium hydroxide. The ethyl ester separates out as the top layer, glycerol and other components forming the lower layer. Suggest the mechanism for the reaction.
4. Consider the reaction, R—CO—Z + C$_2$H$_5$O$^-$ → R—COOC$_2$H$_5$ + Z$^-$
   Compare the feasibility of the reaction when Z=OR, OH, OCOR, CH$_2$—COOR, CH$_2$—CO—R, CH$_2$—R, NH—R.

5. Design synthesis for the following molecules using simpler starting materials.

 (a) *p*-isopropylacetophenone
 (b) $H_2N - CH_2 - CH(CH_2OH) - CH(CH_3)_2$
  (Hint: use ethyl cyanoacetate)
 (c) 4-(*p*-nitrophenyl)-2-butanone
 (d) *p*-$CH_3CONH - C_6H_4 - CH_2 - COOH$, (*p*-acetamidophenyl)acetic acid
 (e) $(CH_3)_3C - COOH$

6. Find out what the Dean–Stark apparatus is and how it is useful for reactions such as esterification and dehydration of alcohols, by azeotropic removal of water (Section 17.4.1).

7. Friedel–Crafts acetylation of benzene can be classified as either an electrophilic substitution reaction or as a nucleophilic substitution reaction. Elaborate.

8. Even though the conditions in Section 17.6.2 are satisfied, the enol content of malonic ester is negligible. Give likely reasons.

18 Nitrogen Compounds

OBJECTIVES In this chapter, you will learn about,

- preparation of nitroalkanes by the reaction of silver nitrite with alkyl halides
- ambident nucleophiles
- the nature of nitroarenes
- reduction of nitrobenzene under different conditions
- classification and methods of preparation of amines
- basicity of amines: comparison of methyl, dimethyl and trimethylamines
- basicity of aromatic amines: effect of substituents
- reaction of amines with nitrous acid: diazotisation
- applications of diazonium salts
- preparation and properties of specific compounds—diazomethane, diazoacetic ester, nitroanilines and phenylene diamines, sulphanilic acid, sulphanilamide, saccharin, chloramine-T
- synthetic applications of nitration and diazotisation reactions

18.1 NITRO COMPOUNDS

Nitro compounds contain NO_2 bound to carbon through the nitrogen. They are isomeric with nitrites, where the bonding is between the carbon and oxygen. The inorganic nitrite ion can act as a nucleophile, either through the nitrogen or through the oxygen and can give rise to either to a nitroalkane or to an alkyl nitrite by nucleophilic displacemnt (Eq. 18.1). Such a nucleophile which can react at two sites, is called an *ambident nucleophile*.

$$\text{Na}^+ \; O = \ddot{\text{N}} - \ddot{\underset{..}{\text{O}}}: \quad R - X \longrightarrow R - O - N = O + \text{NaX}$$
Alkyl nitrite

(18.1)

$$\text{Ag}^+ \; O = \ddot{\text{N}} - \ddot{\underset{..}{\text{O}}}: \quad R - X \longrightarrow R - \overset{+}{\text{N}} \overset{\displaystyle O}{\underset{\displaystyle O^-}{\diagup}} + \text{AgX}$$
Nitroalkane

If sodium nitrite is used in this reaction, the product is mainly the nitrite. The nitroalkane is formed as the major product if silver nitrite is used.

Another familiar example of an ambident nucleophile is the cyanide ion, which is the nucleophile used to prepare nitriles from alkyl halides (Chapter 11, Section 11.4). The cyanide can also react as nucleophile to give an isocyanide (if silver cyanide is used), where the bonding is through the N.

$$\text{Na}^+ \; :\bar{\text{C}} \equiv \text{N}: \quad R - X \longrightarrow R - C \equiv N + \text{NaX}$$

(18.2)

$$\text{Ag}^+ \; :\bar{\text{C}} \equiv \text{N}: \quad R - X \longrightarrow R - \overset{+}{\text{N}} \equiv \bar{\text{C}} + \text{AgX}$$

Aliphatic nitro compounds containing α-hydrogens are acidic. The α-hydrogen can be abstracted by a base, like alkoxide, to give a carbanion which is resonance stabilised (Eq. 18.3a). In the presence of alkali, nitroalkanes can condense with aldehydes in a reaction similar to aldol condensation (Eq. 18.3b).

$$R - CH_2 - \overset{+}{\text{N}} \overset{\displaystyle O}{\underset{\displaystyle O^-}{\diagup}} + RO^- \longrightarrow \left[R - \overset{..}{\underset{}{\bar{\text{C}}\text{H}}} - \overset{+}{\text{N}} \overset{\displaystyle O}{\underset{\displaystyle O^-}{\diagup}} \longleftrightarrow R - CH = \overset{+}{\text{N}} \overset{\displaystyle O^-}{\underset{\displaystyle O^-}{\diagup}} \right] \quad \text{(a)}$$

(18.3)

$$C_6H_5 - CHO + CH_3NO_2 \xrightarrow{\;OH^-\;} C_6H_5 - CH = CH - NO_2 \quad \text{(b)}$$

Aromatic nitro compounds are prepared by the nitration of arenes using a mixture of nitric acid and sulphuric acids (Chapter 9).

18.1.1   NITROARENES

The nitro group is a deactivating group that is *meta* orienting in aromatic electrophilic substitution. Benzene can be nitrated to nitrobenzene under mild conditions. Further nitration under more drastic conditions gives mainly *m*-dinitrobenzene (Chapter 9, Section 9.5.1.2). Dinitrobenzene is so deactivated that the introduction of a third nitro group by direct nitration is virtually impossible. Toluene is more reactive than benzene. It can be nitrated at 30°C to give *o*- and *p*-nitrotoluenes along with a small amount of *m*-nitrotoluene. Nitration of toluene at 100°C gives 2,4-dinitrotoluene and under more drastic conditions 2,4,6-trinitrotoluene (TNT), an explosive, can be prepared. Oxidation of trinitrotoluene to 2,4,6-trinitrobenzoic acid can be done using dichromate or permanganate. This, on mild heating undergoes decarboxylation to give 1,3,5-trinitrobenzene which is also an explosive.

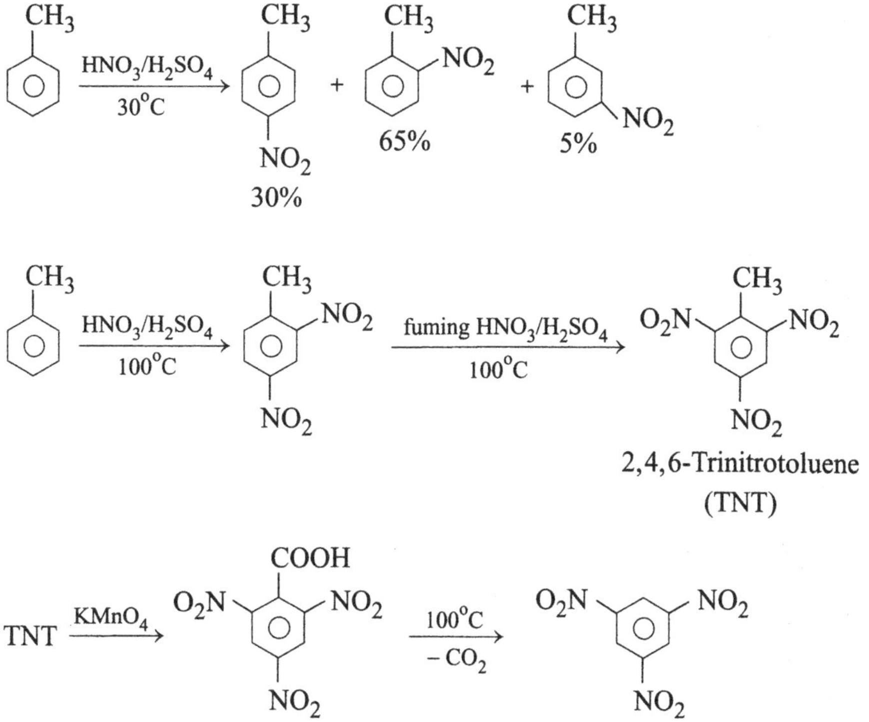

(18.4)

2,4,6-Trinitrotoluene
(TNT)

18.1.2 REDUCTION OF THE NITRO GROUP

Introduction of a nitro group is an entry point for the synthesis of many substituted benzene derivatives. This is because the nitro group can be reduced to the amino group, which can be converted to several functional groups via diazotisation. Nitrobenzene can be reduced to different products, depending upon reaction conditions (Fig. 18.1).

The sequence of formation of the products is outlined in Eq. 18.5. The first step in the reduction is the formation of nitrosobenzene which cannot be isolated. It is reduced in successive steps to phenylhydroxylamine and finally to aniline. A condensation product of nitrosobenzene and phenylhydroxylamine, azoxybenzene, can be isolated which then undergoes reduction to azobenzene and further to hydrazobenzene (Eq. 18.5). By controlling the reaction conditions (Fig. 18.1), all these products can be isolated. The condensation of nitrosobenzene with phenylhydroxylamine to form azoxybenzene is depicted in Eq. 18.6.

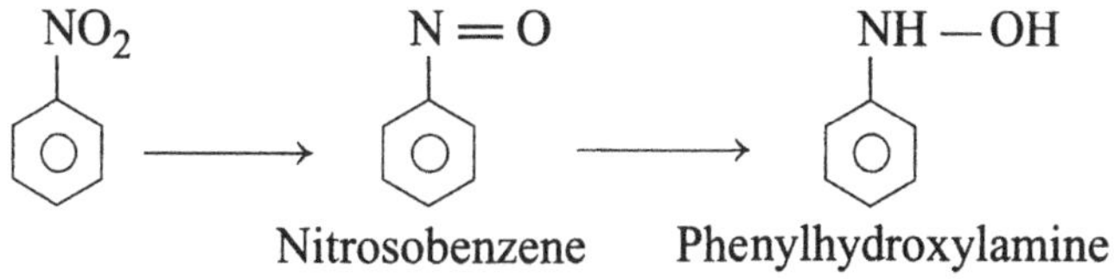

Nitrosobenzene   Phenylhydroxylamine

N=O   NH—OH   N = $\overset{+}{N}$—$\overset{-}{O}$

⬡ + ⬡ ⟶ ⬡ ⬡

Azoxybenzene

N = N   NH—NH   NH₂

⟶ ⬡ ⬡ ⟶ ⬡ ⬡ ⟶ ⬡

Azobenzene   Hydrazobenzene   Aniline

(18.5)

O   O⁻   O⁻   O⁻

‖   │   │   │

N :NH—OH   N—$\overset{+}{N}$H—OH   :N—N—$\overset{+}{O}$H₂ $\overset{+}{N}$=N

⬡ ⬡ ⟶ ⬡⬡ ⟶ ⬡⬡ ⟶ ⬡ ⬡   (18.6)

The reaction conditions required for the isolation of different products are summarised in Fig. 18.1. (1) The reduction in neutral medium using zinc dust/ammonium chloride/water enables us to isolate phenylhydroxylamine. Ammoium chloride is used to control the pH.

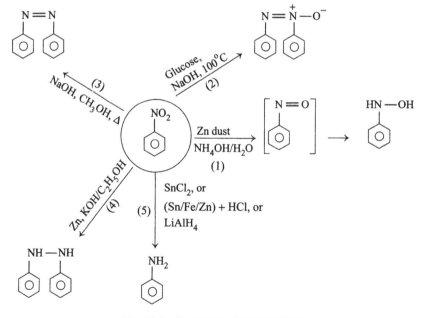

Fig. 18.1 Reduction of nitrobenzene

Phenylhydroxylamine is a reducing agent which can reduce the Ag^+ ion in Tollen's reagent just like the aldehydes. This reaction is used as a test for nitro compounds in qualitative analysis (Mulliken–Baker test). (2) In the reduction to azoxybenzene in mild alkali, glucose (an aldehyde) is used as the reducing agent. (3) In the reduction to azobenzene, where the medium is more alkaline, methanol (sodium methoxide) is the reducing agent which gets itself oxidised to formic acid in the process. (4) In even stronger alkali, using a metal like zinc as the reducing agent (electron donor), further reduction to hydrazobenzene takes place. (5) In acidic medium, a variety of metals like Sn, Zn or Fe, or strong reducing agents like $SnCl_2$, yield aniline as the product. Nitrobenzene can also be reduced to aniline using lithium aluminum hydride, but not sodium borohydride.

18.2 AMINES

18.2.1 CLASSIFICATION

Amines are organic derivatives of ammonia. They are obtained when one or more of the hydrogens of ammonia are replaced by alkyl or aryl groups. When the groups are all alkyl, like methylamine, triethylamine and methylethylamine, they are called aliphatic amines. When at least one group is aryl, it is an aromatic amine, like aniline, N-methylaniline and N,N-diethylaniline. Amines are classified as primary, secondary and tertiary. This classification is different from the classification of alcohols as primary, secondary and tertiary. Primary amines are those in which one H of ammonia has been replaced by an alkyl or aryl group. They contain the NH_2 group which is called a primary amino group. Methylamine and aniline are examples. When two hydrogens have been replaced as in dimethylamine and N-methylaniline, it is a secondary amine. When all three hydrogens have been replaced as in trimethylamine, it is tertiary.

18.2.2 PREPARATION OF AMINES

As already mentioned, aromatic primary amines are prepared by the reduction of nitro compounds. Nitroalkanes are not as readily available as nitroarenes. There are a number of other methods for the preparation of aliphatic amines. The more important ones are listed below.

18.2.2.1 Alkylation of Ammonia and Amines

Alkylation of ammonia is a nucleophilic displacement on a substrate like an alkyl halide by ammonia or an amine (see Chapter 11, Eq. 11.19d). It is difficult to stop the reaction after the introduction of only one alkyl group, that is, it is difficult to obtain the primary amine selectively. Secondary and tertiary amines may be formed to some extent, even if the two reagents are taken in an equimolar ratio. Tertiary amines can be further alkylated to the quaternary ammonium salt. The reaction of a primary amine RNH_2, with excess methyl iodide gives the quaternary salt, $R-N(CH_3)_3^+I^-$. This reaction is called exhaustive methylation. Such quaternary salts can be subjected to Hofmann elimination (Chapter 12, Section 12.3.2). Alkylation can be applied to aromatic primary amines like aniline to prepare N-alkyl and N,N-dialkyl anilines.

18.2.2.2   Gabriel Synthesis of Primary Amines

The drawback of the direct alkylation of ammonia for the synthesis of primary amines, namely contamination with secondary and tertiary amines, is overcome by employing the Gabriel synthesis. In the Gabriel synthesis, phthalimide is alkylated to N-alkylphthalimide which on hydrolysis gives the alkylamine selectively (Eq. 18.7).

Phthalimide (I)   (II)   (III)

(18.7)

Phthalimide (I) is moderately acidic and reacts with alkali to form a salt (II). The acidity is due to the presence of the two carbonyl groups. Recall the discussion of the acidity of dicarbonyl compounds like 2,4-pentanedione, acetoacetic ester and diethyl malonate (Chaper 16, Table 16.1). The anion of phthalimide (II) is alkylated by the alkyl halide RX to give the N-alkylphthalimide (III), which upon hydrolysis yields the primary amine RNH_2.

18.2.2.3   Reduction of Azides

Alkyl halides react with inorganic azides to give alkyl azides which can be reduced to primary amines (Chapter 11, Eq. 11.19c).

18.2.2.4   Reduction of Nitriles

Primary and secondary alkyl halides react with the cyanide ion to give the corresponding nitrile which can be reduced either catalyically or by lithium aluminium hydride to obtain a primary amine with one carbon more than the alkyl halide (Chapter 11, Eq. 11.19b).

18.2.2.5   Reductive Amination of Carbonyl Compounds

Reductive amination involves the conversion of an aldehyde or ketone to the imine by reaction with reagents containing the NH_2 group (primary amine, hydroxylamine) followed by hydrogenation of the carbon–nitrogen double bond (Chapter 16, Section 16.3.1.4). By this reaction, an amine with the same number of carbons as the aldehyde or ketone can be synthesised.

18.2.2.6 Hofmann Rearrangement and Curtius Rearrangement

Conversion of carboxylic acid to the amine with one carbon less, through the Hofmann or the Curtius rearrangement is discussed in Chapter 19, Section 19.3.2 and Section 19.3.3.

18.2.2.7 Beckmann Rearrangement

In the Beckmann rearrangement (Chapter 19, Section 19.3.1), the oxime of a ketone rearranges to the amide which can be hydrolysed to the amine, with fewer number of carbon atoms. Thus, acetophenone oxime gives acetanilide by Beckmann rearrangement. Acetanilide can be hydrolysed to aniline and acetic acid.

18.3 PROPERTIES OF AMINES

18.3.1 BASICITY OF AMINES

Amines are bases. The lone pair of electrons on the nitrogen is responsible for basicity (Eq. 18.8).

$$R - \overset{..}{N}H_2 + H_3O^+ \rightleftharpoons R - \overset{+}{N}H_3 + H_2O$$

$$R - \overset{..}{N}H_2 + H_2O \rightleftharpoons R - \overset{+}{N}H_3 + \overset{-}{O}H$$

$$K_b = \frac{\left[R - \overset{+}{N}H_3\right]\left[OH^-\right]}{\left[RNH_2\right]} \tag{18.8}$$

$$pK_b = - \log K_b$$

$$K_a \text{ (of } RNH_3^+) = \frac{\left[RNH_2\right]\left[H_3O^+\right]}{\left[RNH_3^+\right]}$$

$$pK_a = - \log K_a$$

The basicity of some amines is listed in Table 18.1. The pK_b value, as a measure of base strength, is defined in Eq. 18.8. Stronger bases have lower pK_b values. The basicity of amines can also be expressed in terms of the pK_a values of the corresponding conjugate acids (Table 18.1). When applied to bases, higher pK_a values of the conjugate acids correspond to stronger bases.

$$pK_a \approx 14 - pK_b$$

The factors which increase the stability of the cation (the ammonium ion) increase basicity. In general, the electron-releasing effect of R will increase basicity and electron-withdrawing groups will decrease basicity. Dimetylamine is a stronger base than methylamine due to the presence of two electron-releasing methyl groups. However, this is not the only effect operating here. This is evident from the pK_b value of trimethylamine, which is higher than that of dimethylamine. If

Table 18. 1 Basicity of amines

Amine	Structure	pK$_b$ value	pK$_a$ (of the conjugate acid)
Ammonia	NH_3	4.7	9.3
Methylamine	$CH_3 - NH_2$	3.4	10.6
Dimethylamine	$(CH_3)_2NH$	3.2	10.8
Trimethylamine	$(CH_3)_3N$	4.2	9.8
Aniline	$C_6H_5 - NH_2$	9.4	4.6
Methylaniline	$C_6H_5 - NHCH_3$	9.2	4.8
Dimethylaniline	$C_6H_5 - N(CH_3)_2$	9.0	5.1
p-Toluidine	$p\text{-}CH_3 - C_6H_4 - NH_2$	8.9	5.3
p-Anisidine	$p\text{-}CH_3O - C_6H_4 - NH_2$	8.8	5.0
p-Nitroaniline	$p\text{-}NO_2 - C_6H_4 - NH_2$	13.0	1.0
o-Nitroaniline	$o\text{-}NO_2 - C_6H_4 - NH_2$	14.3	-0.13
p-Chloroaniline	$p\text{-}Cl - C_6H_4 - NH_2$	10.0	4.0

the electron-releasing inductive effect of the methyl groups was the only factor to be considered, the prediction would be that trimethylamine would be stronger than dimethylamine. The stability of an ion is decided not only by effects present within the ion (internal factors), but also by the delocalisation of the charge by effects such as solvation (external factors). In the case of solvation of an ammonium ion in water, stabilisation is afforded mainly by hydrogen bonding with the water molecules. It is mainly through the N—H bonds (Eq. 18.9).

$$(18.9)$$

Due to the presence of more number of alkyl groups, solvation is less effective in the trimethylammonium cation in comparison with the dimethylamonium cation. This overrides the electron-releasing effect of the methyl groups, making trimethylamine weaker than dimethylamine. As expected, aliphatic amines are stronger bases than aromatic amines. In the case of the latter, it is not the electron-withdrawing effect of the benzene ring alone that is responsible. In the free base (unprotonated aniline), the lone pair on the nitrogen is delocalised and contributes to resonance stabilisation. This stabilisation is absent in the ammonium ion. Hence, the conversion of the free base to the protonated form requires more energy than that for an aliphatic amine. Another way of expressing this idea is that the lone pair of electrons on the nitrogen in aniline is less available for bonding with a proton than in aliphatic amines. The behaviour of substituted anilines is along predictable lines. Those with electron releasing groups like CH_3 and OCH_3 are stronger bases; those with electron withdrawing groups like Cl and NO_2 are weaker bases.

Amides, $RCONH_2$, in which the lone pair of electrons on the nitrogen is effectively delocalised are not basic (Eq. 18.10).

$$R-C\underset{NH_2}{\overset{O}{\big\|}} \longleftrightarrow R-C\underset{\overset{+}{N}H_2}{\overset{\ddot{O}^-}{\diagup}} \tag{18.10}$$

Amines dissolve in dilute acids in the cold while amides do not.

18.3.2 REACTION OF AMINES WITH NITROUS ACID: DIAZOTISATION

Nitrous aid, HNO_2, formed in solution by the action of mineral acids on nitrites, is a source of the nitrosyl cation $(NO)^+$, and is a means of introducing the nitroso group, $N=O$, in electrophilic reactions. This is called nitrosation. The reaction of nitrous acid with a primary amine is elaborated in Eq. 18.11.

$$R-\overset{H}{\underset{H}{N:}} + \overset{+}{NO} \xrightarrow{(1)} R-\overset{H}{\underset{\underset{(I)}{H}}{\overset{+}{N}}}-N=O \overset{(2)}{\rightleftharpoons} \left[\begin{array}{c} R-\overset{H}{\underset{H}{N}}-N=\overset{+}{OH} \\ \updownarrow \quad (II) \\ R-\overset{+}{\underset{H}{N}}=N-OH \end{array}\right]$$

$$R-\overset{+}{\underset{\underset{(II)}{H}}{N}}=N-OH \xrightarrow{(3)} R-\overset{..}{N}=N-\overset{+}{OH_2} \xrightarrow{(4)} R-\overset{+}{N}\equiv\overset{..}{N} + H_2O$$
$$\qquad\qquad\qquad\qquad\qquad\qquad\qquad\qquad (III) \tag{18.11}$$

$$R-\overset{+}{N}\equiv N \xrightarrow{(5)} R^+ + N_2 \xrightarrow{H_2O} ROH + H^+$$

$$\langle O \rangle -NH_2 \xrightarrow{HCl, NaNO_2, 0^\circ C} \langle O \rangle -\overset{+}{N}\equiv N: \ Cl^-$$
Benzenediazonium chloride

$$\left[\langle O \rangle -\overset{+}{N}\equiv N: \longleftrightarrow \langle O \rangle -\overset{..}{N}=\overset{+}{N}: \longleftrightarrow \langle O \rangle =\overset{+}{N}=\overset{..}{N}: \cdots\right]$$

The initial product is an N-nitroso compound (I), which tautomerises to a resonance stabilised hydroxy compound (II). This after prototropy, step (3), and loss of water, step (4), gives the cation (III). This is called a diazonium cation. The most likely fate of this cation is the loss of a molecule of nitrogen, step (5), to give a carbocation. This is what happens when R is alkyl. When a primary alkyl amine dissolved in dilute HCl is treated with sodium nitrite and dilute hydrochloric acid at 0°C, a brisk effervescence due to the evolution of nitrogen takes place. The alkyl group usually ends up as the alcohol (ROH) or as an alkene. Low temperature should be maintained because HNO$_2$ is thermally unstable and decomposes and brown fumes of NO$_2$ can be observed if the reaction is attempted at room temperature. When the amine is aromatic, the diazonium cation is resonance stabilised and the diazonium salt is fairly stable at 0°C. Aniline gives a solution of benzenediazonium chloride when the reaction is carried out in aqueous hydrochloric acid. It does not decompose if mainained at 0°C and can be made to undergo a variety of useful transformations. This reaction is called the diazotisation of aromatic primary amines.

18.3.3 SYNTHETIC APPLICATIONS OF DIAZONIUM SALTS

The diazotisation reaction is useful for the preparation of various benzene derivatives. These conversions are summarised in Fig. 18.2.

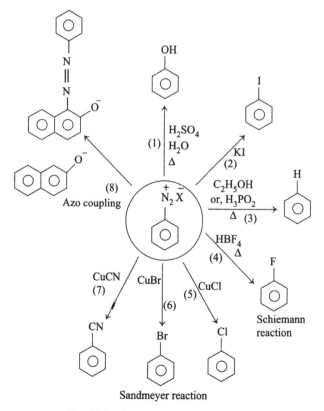

Fig. 18.2 Reactions of diazonium salt

18.3.3.1   Hydrolysis to Phenols

After diazotisation at 0°C, if the solution is heated, loss of nitrogen will take place [step (5) of Eq. 18.11]. The phenyl cation then reacts with the nucleophile, water, to form phenol (Reaction 1, Fig. 18.2, Eq. 18.12).

$$\underset{\text{0°C}}{\xrightarrow{\text{NaNO}_2,\ \text{H}_2\text{SO}_4,\ \text{H}_2\text{O}}} \qquad \underset{\Delta}{\xrightarrow{\text{H}_2\text{O}}} \qquad + \text{N}_2 + \text{H}_2\text{SO}_4 \qquad (18.12)$$

For this reaction, dilute sulphuric acid is preferred over hydrochloric acid. If the chloride ion is present in the medium, part of the phenyl carbocation could react with chloride to form chlorobenzene. This kind of side reaction is less when the less nucleophilic sulphate is the counter ion in the diazoium salt. When benzenediazonium chloride is heated, though some cholrobenzene is formed along with phenol, it is not a useful method of preparation of chlorobenzene. Sandmeyer reaction, discussed below, is the best method to obtain chloro and bromobenzene. On the other hand, iodobenzene can be prepared by merely warming the diazonium salt with potassium iodide (Reaction 2, Fig. 18.2). The greater nucleophilicity of the iodide ion makes this possible.

18.3.3.2   Reduction of the Diazonium Salt to the Hydrocarbon: Deamination

The phenyl cation formed by the loss of nitrogen, can be reduced by a variety of reducing agents to the hydrocarbon, effectively replacing the amino group in the original molecule by hydrogen. This is called deamination. A convenient reducing agent for this purpose is ethanol, which functions as a hydride donor. After diazotisation at 0°C, ethanol is added to the solution and warmed (Reaction 3, Fig. 18.2). Hypophosphorous acid can also be used as the reducing agent.

18.3.3.3   Synthesis of Aryl Fluorides: Schiemann Reaction

After diazotisation, fluoroboric acid HBF_4, or a fluoroborate salt is added, and then warmed. Schiemann reaction is the most convenient route for aryl fluorides (Reaction 4, Fig. 18.2).

18.3.3.4   Sandmeyer Reaction: Synthesis of Chloro-, Bromo- and Cyanoarenes

In these reactions, the diazonium salt is treated with the cuprous salt (cuprous chloride, bromide or cyanide) and heated. These reactions are believed to involve a free radical mechanism. The diazonium cation is reduced by the cuprous ion to generate a phenyl radical (Eq. 18.13). The radical then reacts with cupric chloride to form the chloroarene and regenerates the cuprous ion (Reactions 5, 6 and 7, Fig. 18.2).

$$\text{Ar}\,\overset{+}{\text{N}}_2 + \overset{+}{\text{Cu}} \longrightarrow \text{Ar}^{\bullet} + \overset{++}{\text{Cu}} + \text{N}_2$$

$$\text{Ar}^{\bullet} + \text{CuCl}_2 \longrightarrow \text{ArCl} + \overset{+}{\text{Cu}} \qquad (18.13)$$

18.3.3.5   Azo Coupling: Formation of Azo Dye

The diazonium cation can enter into aromatic electrophilic substitution reactions with aromatic compounds which are highly electron rich; for example, with the phenolate anion. Thus, when the solution of benzene diazonium chloride is added to a cold solution of 1-naphthol in alkali, an intensely coloured azo dye is formed (Reaction 8, Fig. 18.2, also see Chapter 15, Section 15.4.6, Eq. 15.14).

18.4   DIAZOMETHANE AND DIAZOACETIC ESTER

Diazo compounds are derivatives of diazomethane, CH_2N_2, whose structure was discussed in Chapter 1, Section 1.3, as an exercise in drawing resonance structures. It is an unstable, highly reactive gaseous molecule which is toxic and explosive in the pure form. Its solution in ether is relatively stable. Such a solution is usually prepared as and when required from a precursor reagent. A commercially available precursor is N-methyl-N-nitroso-*p*-toluenesulphonamide, which when treated with alkali liberates diazomethane (Eq. 18.14).

$$CH_3 - \!\!\bigcirc\!\!- SO_2 - \overset{|}{\underset{NO}{N}} - CH_3 \xrightarrow{OH^-} \bar{C}H_2 - \overset{+}{N} \equiv N \tag{18.14}$$

An important application of diazomethane in organic synthesis is in the preparation of methyl esters of carboxylic acids. Treatment of a carboxylic acid with the ether solution of diazomethane gives the methyl ester. This is done under very mild conditions and is applicable to all carboxylic acids. Sterically hindered acids like 2,2-dimethylpropanoic acid cannot be converted to the methyl ester by conventional acid catalysed esterification, but can readily be converted to the methyl ester using diazomethane (Eq. 18.15).

$$R - \overset{O}{\overset{||}{C}} - OH + \bar{C}H_2 - \overset{+}{N} \equiv N \longrightarrow R - \overset{O}{\overset{||}{C}} - O^- + CH_3 - \overset{+}{N} \equiv N \tag{18.15}$$
$$\longrightarrow R - \overset{O}{\overset{||}{C}} - OCH_3 + N_2$$

Another synthetic application of diazomethane is in lengthening the chain of a carboxylic acid. That is, converting it to the higher homologue as in $R-COOH \rightarrow R-CH_2-COOH$, by *Arndt–Eistert reaction* (Eq. 18.16).

$$R - \overset{O}{\overset{||}{C}} - Cl + CH_2N_2 \longrightarrow R - \overset{O}{\overset{||}{C}} - CHN_2 + HCl$$
$$\text{Diazoketone}$$

$$R-\overset{\overset{\displaystyle O}{\|}}{C}-CHN_2 + H_2O \xrightarrow{\text{colloidal Ag}} R-CH_2-\overset{\overset{\displaystyle O}{\|}}{C}-OH + N_2 \qquad (18.16)$$

In the first step, an acyl chloride is treated with two equivalents of diazomethane to form a diazoketone. Two equivalents are required because the HCl formed in the reaction destroys one equivalent of the reagent. The diazoketone on warming with water and colloidal silver produces the desired carboxylic acid. The two steps are together known as the Arndt–Eistert synthesis. The second step, which involves a molecular rearrangement, is the *Wolff rearrangement* (Chapter 19).

Ethyl diazoacetate (diazoacetic ester) is preapared by the action of nitrous acid on ethyl α-aminoacetate, also known as ethyl glycinate (Eq. 18.17).

$$H_2N-CH_2\,COOC_2H_5 \xrightarrow{\text{HCl, NaNO}_2} N\equiv\overset{+}{N}-\overset{-}{C}H-COOC_2H_5$$

Ethyl α-aminoacetate   Ethyl diazoacetate   (18.17)
(ethyl glycinate)

It also reacts with carboxylic acids to form esters (Eq. 18.18).

$$R-COOH+N_2CH-COOC_2H_5 \longrightarrow R-\overset{\overset{\displaystyle O}{\|}}{CO}-CH_2-\overset{\overset{\displaystyle O}{\|}}{C}-OC_2H_5 \qquad (18.18)$$

18.5 MORE AROMATIC COMPOUNDS CONTAINING NITROGEN

18.5.1 DINITROBENZENES AND NITROANILINES

m-Dinitrbenzene is directly obtained by the nitration of nitrobenzene (Chapter 9, Section 9.5.1). The *o*- and *p*-isomers are also formed in small quantities and can be separated. *m*-Dinitrobenzene can be reduced to *m*-nitroaniline by ammonium polysulphide, obtained by bubbling hydrogen sulphide into aqueous ammonia (Eq. 18.19a). Direct nitration of aniline to prepare *o*- and *p*-nitroaniline is unsatisfactory because of the high reactivity of aniline and the side reactions due to the oxidation of aniline by nitric acid. The same limitation is experienced in the direct halogenation of aniline. This is overcome by protecting the amino group by acetylation, performing the nitration or halogenation, and then deprotecting the amino group by hydrolysis of the acetylamino group to release the free amine (Eq. 18.19b). This protection–deprotection methodology has already been discussed in connection with mononitration and monobromination of phenol (Chapter 15, Section 15.5.2).

o-Nitroaniline is best prepared by protecting both the amino group and the para position of aniline by acetylation and sulphonation, followed by nitration and removal of the acetyl and sulphonic acid groups by hydrolysis (Eq. 18.19c).

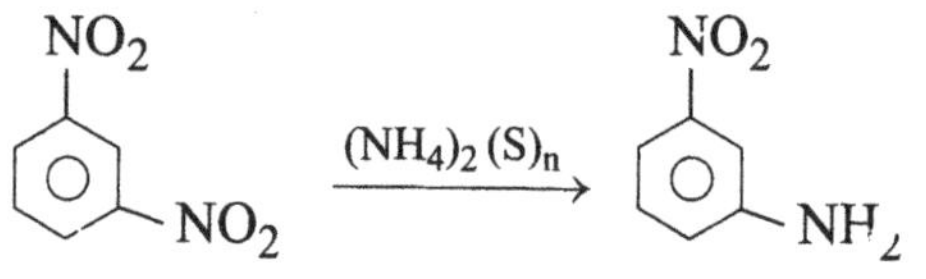

(a)

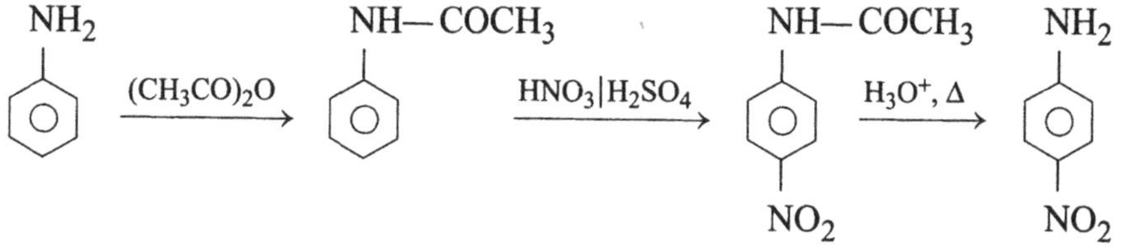

(b)

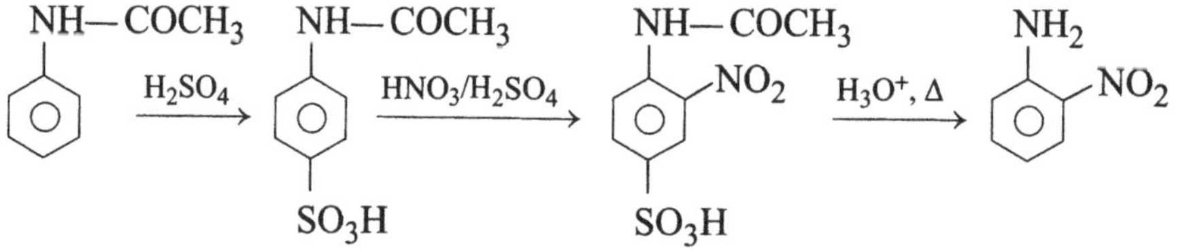

(c)

(18.19)

18.5.2 PHENYLENE DIAMINES

There are three isomeric diaminobenzenes, also called phenylenediamines. All the three are synthesised from the corresponding dinitrobenzenes or nitroanilines by reduction using metal (Sn, Zn or Fe) and hydrochloric acid. The *ortho* isomer finds use in the synthesis of heterocyclic compounds. The *meta* and *para* isomers are used in the preparation of azo dyes.

18.5.3 SULPHANILIC ACID, SULPHANILAMIDE AND SULPHONAMIDES

When aniline is sulphonated with oleum, *p*-aminobenzenesulphonic acid or sulphanilic acid is obtained. This can also be obtained by 'baking' the sulphate of aniline at 200°C (Eq. 18.20a). *p*-Acetamidobenzenesulphonyl chloride is obtained by heating acetanilide with chlorosulphonic acid. This is the general method for the preparation of suphonyl chlorides (Eq. 18.20b). Sulphonyl chloride on treatment with ammonia or amines gives sulphonamide (Eq. 18.20c). *p*-Aminobenzenesulphonamide is called sulphanilamide. *Sulpha drugs* are derivatives of sulphanilamide.

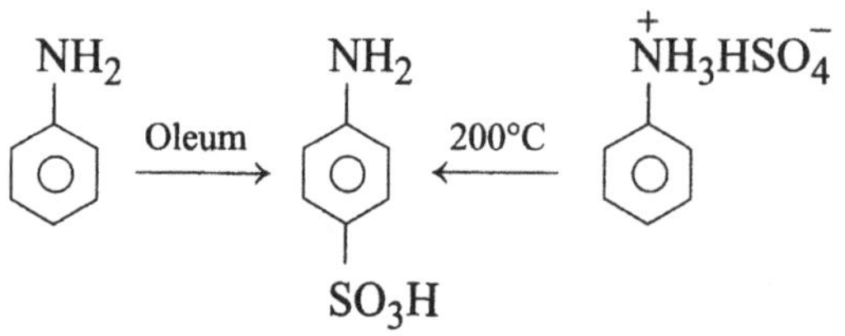

(a)

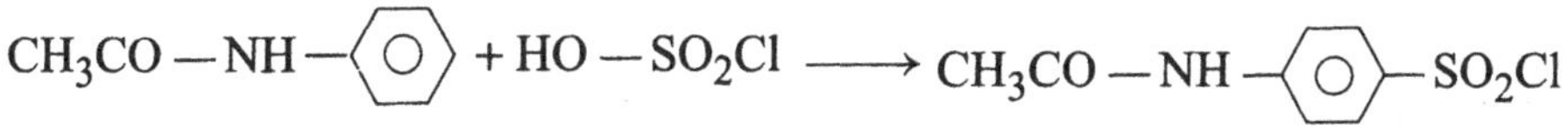

$$CH_3CO-NH-\langle O \rangle + HO-SO_2Cl \longrightarrow CH_3CO-NH-\langle O \rangle-SO_2Cl \quad (b)$$

Chlorosulphonic
acid

p-Acetamidobenzene-
sulphonyl chloride

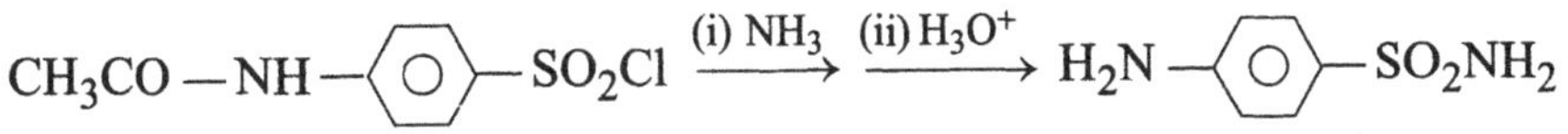

$$CH_3CO-NH-\langle O \rangle-SO_2Cl \xrightarrow{\text{(i) }NH_3} \xrightarrow{\text{(ii) }H_3O^+} H_2N-\langle O \rangle-SO_2NH_2 \quad (c)$$

Sulphanilamide

(18.20)

18.5.4 SACCHARIN

Saccharin is a synthetic sweetening agent. There are many methods for the manufacture of saccharin. A laboratory method is outlined in Eq. 18.21. Chlorosulphonation of toluene gives a mixture of *p*- and *o*-toluenesulphonyl chloride. The major product, the *para* isomer, is a solid which is removed by crystallisation. The *ortho* isomer on treatment with ammonia gives the sulphonamide which upon oxidation with permanganate gives benzoic acid with the sulphonamide group in the *ortho* position. This spontaneously undergoes dehydration to give saccharin. The hydrogen on the nitrogen is acidic because of the electron-withdrawing groups on either side. Treatment with KOH gives the potassium salt which is soluble in water.

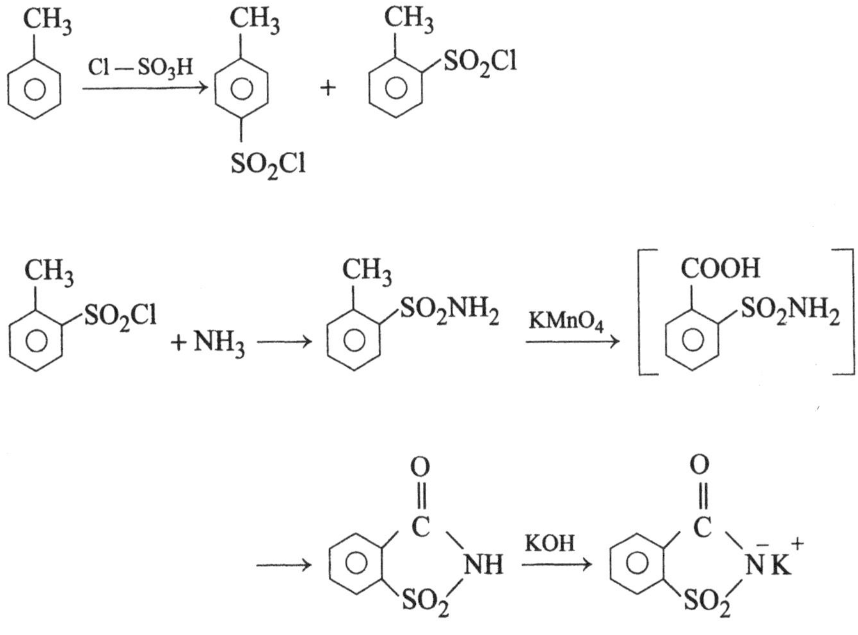

Saccharin   Soluble saccharin   (18.21)

18.5.5 CHLORAMINE-T

p-Toluenesulphonyl chloride called tosyl chloride, is a laboratory reagent for the conversion of the OH group of alcohols to tosylate, in order to make it a good leaving group (Chapter 11, Section 11.4.1). The amide prepared from tosyl chloride, *p*-toluenesulphonamide, upon chlorination with sodium hypochlorite gives the sodium salt of the chloro derivative, which is known as chloramine-T. This is used as an antiseptic and also as an oxidising agent in the laboratory. The oxidising property is due to the N—Cl bond whose polarity is such that it is a source of Cl^+ or of hypochlorous acid. With excess of sodium hypochlorite, dichloramine-T is obtained (Eq. 18.22).

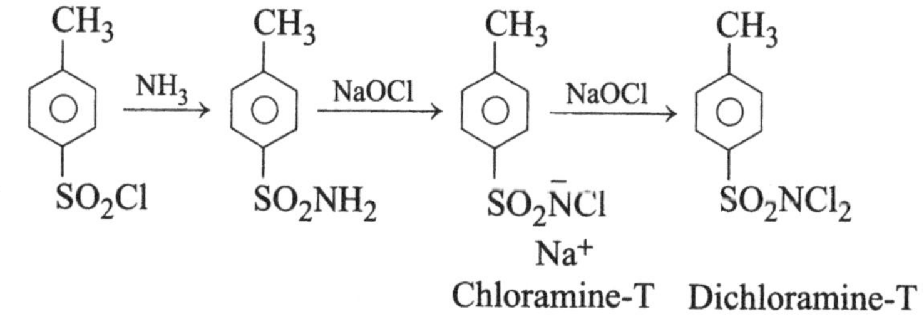

$$(18.22)$$

Chloramine-T   Dichloramine-T

18.6 SYNTHETIC APPLICATIONS OF NITRATION AND DIAZOTISATION REACTIONS

Some of the applications of the reactions studied so far in synthesis, are summarised below (Eq. 18.23).

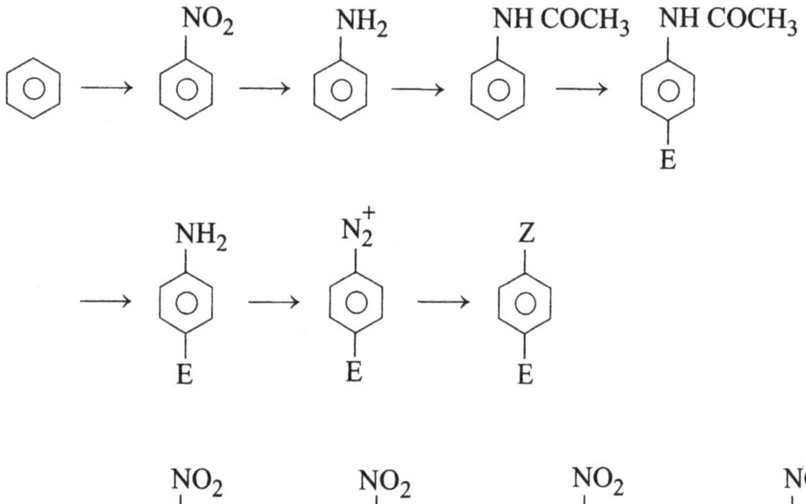

(a)

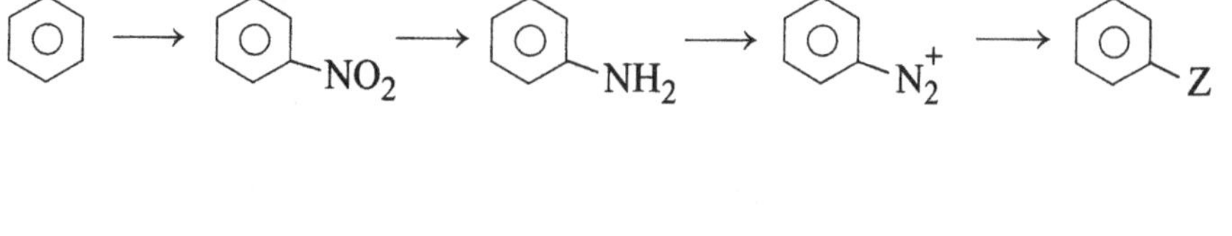

(b)

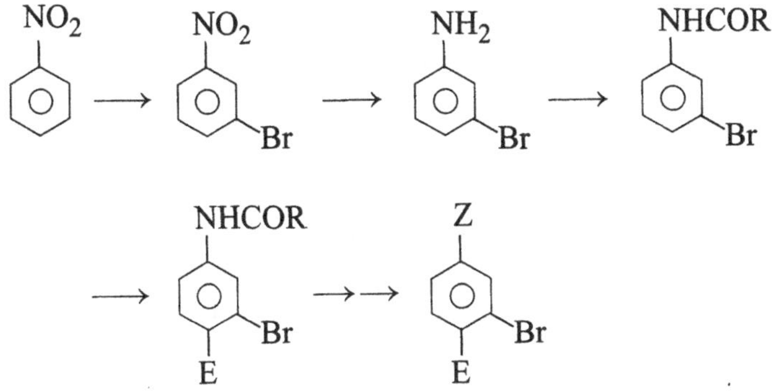

(c)

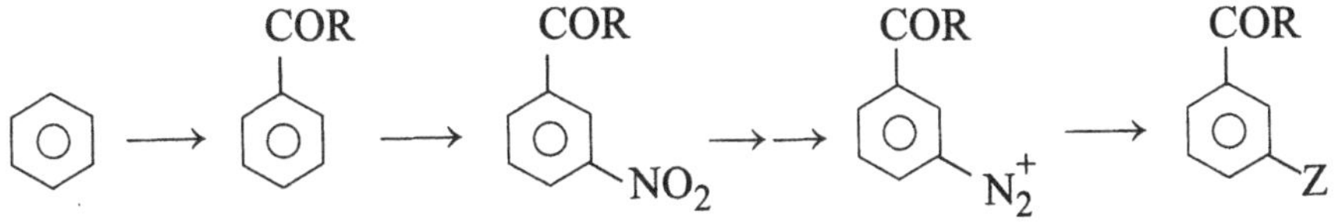

(d)

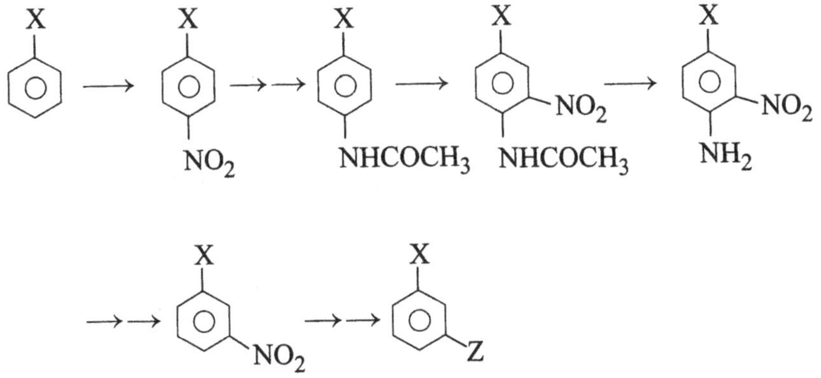

(e)

E = electrophile like NO$_2$, halogen
Z = H, halogen, CN
X = alkyl, halogen or other *o/p*-orienting group

(18.23)

In Eq. 18.23a, the route nitrobenzene → aniline → *p*- and *o*-substituted anilines is outlined. As a rule, it is safer to acetylate the amino group before electrophilic substitutions are carried out. The usual electrophiles introduced (E), are halogen or nitro. The amino group is liberated by the hydrolysis of the acetyl group and then converted to other groups like OH, halogen, cyano or H (deamination) via diazotisation. The cyano group can be further converted to COOH or CH$_2$NH$_2$. The *ortho* isomer formed in the electrophilic substitution reaction can be separated and subjected to the same kind of reactions. Applications of *m*-dinitrobenzene are given in Eq. 18.23b.

Halogenation of nitrobenzene, reduction of *m*-halonitrobenzene to *m*-haloaniline opens a route to compounds of the type shown in Eq. 18.23c. Friedel–Crafts acylation followed by nitration is also a route for the meta series (Eq. 18.23d). The reverse approach, acylation of nitrobenzene is not applicable because nitrobenzene cannot be readily acylated. Nitration of halobenzene or alkylbenzene gives *p*-nitrohalobenzene or *p*-nitroalkylbenzene, along with some ortho which can be separated (Eq. 18.23e). After acetylation, the *p*-amino compound can be subjected to reactions such as halogenation and nitration. The reaction will take place at the position *ortho* to the amino group rather than at the position *ortho* to the halogen or alkyl. This is due to the greater activating effect of the amino group. If necessary, the acetyl amino group can be altogether removed by deacetylation and deamination. This is the route for *m*-aminoalkylbenzenes and *m*-dihalobenzenes, where the two halogens may be the same or different.

KEY POINTS

- Reaction of an alkyl halide with inorganic nitrite can give either the nitroalkane or the alkyl nirite. Siver nitrite gives mainly the former. Sodium nitrite gives mainly the alkyl nitrite. Nitrite is an example of an ambident nucleophile. Cyanide is another example.
- The α-hydrogen in nitroalkanes is acidic.
- Aromatic nitro compounds are prepared by aromatic electrophilic substitution. The nitro group is a meta directing and deactivating group.
- There are several methods which are specifically applicable for the synthesis of aliphatic amines. These include alklylation of ammonia or amines, Gabriel synthesis involving alkylation of phthalimide, reductive alkylation of carbonyl compounds, reduction of azides and nitriles, Hofmann and Curtius rearrangements and Beckmann rearrangement.
- Reduction of nitrobenzene goes through many steps, eventually giving aniline. The intermediates are nitrosobenzene, phenylhydroxylamine, azoxybenzene, azobenzene and hydrazobenzene. Reduction in neutral medium, using zinc dust and ammonium chloride in water gives phenylhydroxylamine. Reduction in alkaline medium gives azoxybenzene, azobenzene and hydrazobenzene, each of which can be isolated by adjusting the reaction conditions. Reduction in acidic medium gives aniline.
- Reaction of primary amines with nitrous acid (sodium nitrite and dilute hydrochloric acid at $0°C$) gives N-nitrosoamine, which transforms itself to the diazonium cation, $R-N_2^+$. In the case of aliphatic primary amines, this cation is unstable even at $0°C$; it decomposes giving out nitrogen and gets converted to the alkyl cation. In the case of aromatic primary amines, the diazonium cation is stable at $0°C$. Benzenediazonium chloride can be transformed to a variety of benzene derivatives.

- These transformations include (i) hydrolysis to phenol, conversion to (ii) iodobenzene with potassium iodide, (iii) fluorobenzene when treated with fluoroboric acid (Schiemann reaction), (iv) chloro or bromobenzene by treatment with cuprous chloride or bromide (Sandmeyer reaction), (v) benzonitrile by treatment with cuprous cyanide and (vi) deamination to benzene by reduction with ethanol or other reducing agents.
- Diazonium salts couple with reactive molecules like phenols to give azo dyes.
- Diazomethane, a toxic gas, is prepared by the action of alkali on a precursor molecule like nitrosomethyl-*p*-toluenesulphonamide. It is useful for converting carboxylic acids to the methyl esters. Diazoacetic ester is prepared by the action of nitrous acid on aminoacetic ester.
- Saccharin is the cyclic imide of *o*-carboxybenzenesulphonic acid. Sulphanilamide is *p*-aminobenzenesulphonamide. It is the basic structure present is sulpha drugs.
- Different substituted benzene derivatives can be synthesised by appropriate choice of reactions such as nitration, halogenation, acyation, reduction of nitro group and diazotisation reactions in the right sequence.

EXERCISES

SECTION I

1. Rearrange the following in the order of increasing base strength.

   (a) *p*-Chloroaniline, *m*-chloroaniline, 2,4-dichloroaniline
   (b) Cyclohexylamine, aniline, *p*-anisidine
   (c) *o*-Chloroaniline, *m*-chloroaniline, *p*-chloroaniline

2. Give the products of the following reactions

   (a) Bromobenzene $\xrightarrow{\text{HNO}_3/\text{H}_2\text{SO}_4}$

   (b) Methyl benzoate $\xrightarrow{\text{HNO}_3/\text{H}_2\text{SO}_4}$

   (c) Acetanilide $\xrightarrow{\text{HNO}_3/\text{H}_2\text{SO}_4}$

   (d) Nitrobenzene $\xrightarrow{\text{Cl}_2/\text{Fe}}$

   (e) *p*-Nitrotoluene $\xrightarrow{\text{Zn dust}/\text{H}_2\text{O}/\text{NH}_4\text{Cl}}$

   (f) *o*-Anisidine $\xrightarrow{\text{NaNO}_2/\text{HCl}/0°\text{C}}$

   (g) *p*-Nirobenzoic acid $\xrightarrow{\text{diazomethane}/\text{ether}}$

SECTION II

1. Show how the following can be synthesised from benzene

    (a) *p*-Chloroaniline

    (b) *m*-Bromochlorobenzene

    (c) *m*-Aminophenol

    (d) *p*-Ethylbenzoic acid

    (e) *m*-Chlorobenzoic acid

    (f) 2, 4-Dinitrophenol

    (g) *o*-Bromofluorobenzene

    (h) *m*-Ethylaniline

2. Draw resonance structures and show how resonance will affect the stability of the protonated form of *p*-anisidine (I), and *p*-nitrotoluene (II), as compared to the anilinium ion (III).

    $$CH_3O - C_6H_4 - NH_3^+ \quad (I)$$
    $$NO_2 - C_6H_4 - NH_3^+ \quad (II)$$
    $$C_6H_5 - NH_3^+ \quad (III)$$

3. HNO_2 is a source of the nitrosyl cation, NO^+, which reacts as an electrophile and brings about nitrosation. Its reaction with primary amines has been discussed in Section 18.2.2. Suggest mechanisms for the following nitrosations.

    (a) N,N-dimethylaniline + HNO_2 → *p*-nitroso-N,N-dimethylaniline
    (b) N-methylaniline + HNO_2 → N-nitroso-N-methylaniline

4. Outline the mechanism for the coupling reaction of benzenediazonium chloride with phenol.

CHALLENGING QUESTIONS

1. Guanidine, $(NH_2)_2C = NH$, is a strong base. Draw the structure of the protonated form of guanidine and account for its base strength.
2. Nitriles, $R - CN$, are not basic in spite of the presence of a lone pair of electrons on the nitrogen. Account for this. Draw the structure of the protonated form (conjugate acid) and compare it with a terminal alkyne.
3. Suggest the mechanism for the liberation of diazomethane by the action of alkali on N-nitroso-N-methyl-*p*-toluenesulphonamide.
4. Compare the base strength of *o*-nitroaniline and *p*-nitroaniline (Table 18.1). Does the existence of intramolecular hydrogen bonding in the *ortho* isomer have any effect on base strength?

PROJECT

Prepare a report on (a) artificial sweetening agents, (b) sulpha drugs.

19 Molecular Rearrangements

OBJECTIVES   In this chapter, you will learn about,

- ■   classification and definitions pertaining to molecular rearrangements
- ■   rearrangements involving migration to electron deficient atoms
- ■   Wagner–Meerwein and pinacol–pinacolone rearrangements
- ■   migratory aptitudes
- ■   benzilic acid rearrangement
- ■   Beckmann rearrangement: anti migration
- ■   migration to carbene: Wolff rearrangement
- ■   migration to nitrene: Hofmann and Curtius rearrangements
- ■   sigmatropic rearrangements: Cope, oxy-Cope and Claisen rearrangements
- ■   evidence for intramolecular nature, cyclic transition state and allylic shift
- ■   electrophilic rearrangement: Fries rearrangement

19.1   INTRODUCTION AND CLASSIFICATION

Molecular rearrangement is the name given to a transformation which occurs when a group or atom gets detached from its point of attachment and becomes reattached elsewhere in the same molecule. The new structure is an isomer of the parent molecule. The new structure may have a different carbon skeleton or the same carbon skeleton with certain atoms or groups having changed positions. Rearrangements of carbocations – many examples of which have been discussed already – come under molecular rearrangements. Allylic rearrangement (Eq. 19.1a) is an example. We shall use this example to define certain terms. The group Z *migrates* from position 1 to position 3, it is therefore called the *migrating group*. Position 1 is the *migration origin* and position 3 the *migration terminus*.

$$R-\overset{3}{CH}=\overset{2}{CH}-\overset{\overset{\displaystyle Z}{\displaystyle |}}{\overset{1}{CH}}-R' \longrightarrow R-\overset{\overset{\displaystyle Z}{\displaystyle |}}{CH}-CH=CH-R' \qquad (a)$$

$$R-CH\underset{\diagdown CH}{\overset{Z\diagdown}{\diagup}}CH-R' \longrightarrow R-CH\underset{\diagdown CH}{\overset{Z\diagdown}{\diagdown}}CH-R' \longrightarrow R-CH\underset{\diagdown CH}{\overset{\diagup Z}{\diagup}}CH-R' \qquad (b)$$

Step 1: $R-CH=CH-\overset{\overset{\displaystyle Z}{\displaystyle |}}{CH}-R' \longrightarrow R-CH=\overset{\cdot}{CH}-\overset{*}{CH}-R' + \overset{*}{Z}$

$\qquad\qquad\qquad\qquad\qquad\qquad\qquad\qquad\quad (I) \qquad\qquad (II)$

Step 2: $R-CH=CH-\overset{*}{CH}-R' + Z^{*} \longrightarrow R-\overset{\overset{\displaystyle Z}{\displaystyle |}}{CH}-CH=CH-R' \qquad (c)$

$$(19.1)$$

19.1.1 INTERMOLECULAR AND INTRAMOLECULAR REARRANGEMENTS

The rearrangement or migration of Z can be *intramolecular* or *intermolecular* (Eq. 19.1b and 19.1c). In intramolecular rearrangement, the migrating group remains a part of the same molecule throughout the whole process. It involves only one step. An intermolecular rearrangement involves at least two steps (Eq. 19.1c). In the first step, the migrating group dissociates from the parent molecule either homolytically or heterolytically to form free radical or ionic fragments. In a second step, they recombine to give an isomeric structure. A significant aspect of this mechanism is that the fragments, (I) and (II), which recombine need not have come from the same parent molecule.

19.1.2 NUCLEOPHILIC, ELECTROPHILIC AND SIGMATROPIC REARRANGEMENTS

Another type of classification is based on the nature of the breaking bond and the bond that is formed during the rearrangement.

19.1.2.1 Nucleophilc Rearrangement: Migration to an Electron Deficient Atom, Anionotropy

These terms refer to the same type of rearrangement. Many of the familiar rearrangements in organic chemistry fall under this category. The rearrangement of a secondary carbocation to a more stable tertiarty carbocation is illustrated in Eq. 19.2.

$$CH_3 - \underset{\underset{CH_3}{|}}{\overset{\overset{CH_3}{|}}{C}} - \overset{\overset{OH}{|}}{CH} - CH_3 \xrightarrow{H^+, -H_2O} CH_3 - \underset{\underset{CH_3}{|}}{\overset{\overset{CH_3}{|}}{C}} - \overset{+}{CH} - CH_3$$

(I)

$$\longrightarrow CH_3 - \overset{+}{\underset{\underset{CH_3}{|}}{C}} - \underset{\underset{CH_3}{|}}{CH} - CH_3 \xrightarrow{-H^+} CH_3 - \underset{\underset{CH_3}{|}}{C} = \underset{\underset{CH_3}{|}}{C} - CH_3$$

(II)

(19.2)

When 3,3-dimethyl-2-butanol is heated with suphuric acid to bring about dehydration, the product obtained is mostly 2,3-dimethyl-2-butene, a product of rearrangement. The mechanism of this reaction has been discussed in Chapter 4. 1,2,2-trimethylpropyl, I, the carbocation formed initially, rearranges to the more stable 1,1,2-trimethylpropyl carbocation II, which then undergoes deprotonation to give the observed products. In the rearrangement (I) to (II), the migration terminus is a cationic centre, that is, an electron deficient atom. The migrating group, CH_3, has to necessarily disscociate heterolytically from one carbon and move to another, carrying a pair of electrons with it as a nucleophile. The methyl group *formally* migrates as an anion, hence the term anionotropy is used. The terms, nucleophilic rearrangement, and migration to an electron deficient atom are more widely used.

19.1.2.2 Electrophilic Rearrangement: Migration to an Electron-Rich Atom, Cationotropy

A typical and familiar example of electrophilic rearrangement is keto–enol tautomerism.

$$R - \underset{\overset{|}{\underset{}{}}}{\overset{\overset{H}{|}}{CH}} - \overset{\overset{O}{||}}{C} - R \longrightarrow R - CH = \overset{\overset{H-O}{|}}{C} - R$$

(19.3)

The migrating atom here is the proton. It migrates as an electrophile to the electron-rich oxygen atom. This rearrangement is also known as prototropy.

19.1.2.3 Sigmatropic Rearrangement

The two categories discussed above – nucleophilic and electrophilic rearrangements – can be intramolecular or intermolecular. The third class of rearrangements – sigmatropic rearrangements – are truly intramolecular. They fall under the group of reactions known as *pericyclic reactions*. (We came across this term under Diels–Alder reactions in Chapter 6). The bond dissociations in sigmatropic rearrangements cannot be classified as homolytic or heterolytic. The changes take place as a result of a reorganisation of molecular orbitals through a cyclic transition state. The name is derived from the fact that in effect, a sigma bond shifts. Claisen and Cope rearrangements come under this category.

19.2   REARRANGEMENTS TO ELECTRON DEFICIENT CARBON

19.2.1   WAGNER–MEERWEIN AND PINACOL–PINACOLONE REARRANGEMENT

The rearrangement of carbocations – such as the ones occurring during the dehydration of alcohols – is called Wagner–Meerwein rearrangement (Eq. 19.2). The driving force for the rearrangement is carbocation stability. A related rearrangement is the pinacol–pinacolone rearrangement (Eq. 19.4).

$$
\begin{array}{ccc}
\underset{\substack{|\\ \text{CH}_3}}{\overset{\substack{\text{OH OH}\\|\ \ |}}{\text{CH}_3-\text{C}-\text{C}-\text{CH}_3}} & \xrightarrow{\text{H}^+} & \underset{\substack{|\\ \text{CH}_3 \text{ CH}_3}}{\text{CH}_3-\text{C}-\text{C}-\text{CH}_3} \longrightarrow \text{CH}_3-\overset{+}{\text{C}}-\text{C}-\text{CH}_3 \\
\text{Pinacol} & & (\text{I})
\end{array}
$$

$$
\longrightarrow \left[\begin{array}{ccc}
\underset{\substack{|\\ \text{CH}_3}}{\overset{\substack{\text{CH}_3 \text{ O}-\text{H}\\|\ \ |}}{\text{CH}_3-\text{C}-\overset{+}{\text{C}}-\text{CH}_3}} & \longleftrightarrow & \text{CH}_3-\text{C}-\text{C}-\text{CH}_3 \\
(\text{II}) & &
\end{array} \right] \xrightarrow{-\text{H}^+} \underset{\substack{|\\ \text{CH}_3}}{\overset{\substack{\text{CH}_3 \text{ O}\\|\ \ \|}}{\text{CH}_3-\text{C}-\text{C}-\text{CH}_3}}
$$

3, 3-Dimethyl-2-butanone
(Pinacolone)

(19.4)

Pinacol is the trivial name of 2,3-dimethyl-2,3-butanediol which is obtained by the bimolecular reduction of acetone using magnesium (Chapter 16, Section 16.3.5). When pinacol is heated with sulphuric acid, dehydration takes place but the product is not an alkene. It is a ketone, 3,3-dimethyl-2-butanone, also known as pinacolone, which is the result of a rearrangement (Eq. 19.4). Protonation of one of the OH groups of pinacol followed by the elimination of a molecule of water gives the tertiary carbocation (I). This rearranges by the migration of a CH_3 group to the adjacent carbon to give the resonance stabilised cation (II). Note that (I) is already a tertiary cation and is stabilised by the alkyl groups, but (II) is more stable because of resonance involving the oxygen. This is the driving force for the rearrangement. Deprotonation of this cation gives the final product, 3,3-dimethyl-2-butanone (pinacolone). This kind of rearrangement is a general reaction of *vic*-glycols. The migrating group can be alkyl as in this case, or aryl or hydrogen.

This is a nucleophilic rearrangement. The migration takes place to an electron deficient carbon, namely a carbocation. The migrating group takes the bonding electron pair with it and moves formally as a methyl carbanion. Thus, the reaction is categorised as anionotropy. However, the methyl group never dissociates from the parent molecule during the course of the reaction to form an independent carbanion. A transition state can be visualised where the migrating group is partially bound to the migration origin and the migration terminus (Eq. 19.5).

$$
-\overset{\underset{\displaystyle |}{CH_3}}{\underset{\underset{\displaystyle O-H}{|}}{C}}-\overset{|}{\underset{|}{C}}- \longrightarrow
\left[-\overset{CH_3}{\overset{.\,.}{\underset{\underset{OH}{|}}{C}}}\overset{+}{-}\overset{|}{C}- \right] \rightleftharpoons
\left[-\overset{\underset{|}{CH_3}}{\overset{|}{C}}-\overset{+}{\underset{\underset{O-H}{|}}{C}}- \longleftrightarrow
-\overset{\underset{|}{CH_3}}{\overset{|}{C}}-\underset{\underset{\underset{+}{OH}}{\|}}{\overset{\overset{H}{|}}{C}}- \right] \quad (a)
$$

$$
(19.5)
$$

(b)

The transition state can revert to the starting cation or proceed to the rearranged cation. The equilibrium is controlled by the stability of the two cations. The rearrangement proceeds to give more of the more stable cation. In the pinacol–pinacolone rearrangement, the more stable cation is the one which has an oxygen attached to the positive carbon.

The fact that the rearrangement expected of carbocations occurs, points to the intermediacy of carbocations. Migratory aptitudes, discussed below, also show that the intermediates in the rearrangement are electron deficient. Some other aspects of the rearrangement are brought out by studying the examples (Eq. 19.6).

$$
C_6H_5-\overset{\overset{\displaystyle OH}{|}}{\underset{\underset{\displaystyle CH_3}{|}}{C}}-\overset{\overset{\displaystyle OH}{|}}{\underset{\underset{\displaystyle CH_3}{|}}{C}}-C_6H_5 \longrightarrow
C_6H_5-\overset{\overset{\displaystyle {}^+OH_2}{|}}{\underset{\underset{\displaystyle CH_3}{|}}{C}}-\overset{\overset{\displaystyle OH}{|}}{\underset{\underset{\displaystyle CH_3}{|}}{C}}-C_6H_5 \longrightarrow
C_6H_5-\overset{+}{\underset{\underset{\displaystyle CH_3}{|}}{C}}-\overset{\overset{\displaystyle OH}{|}}{\underset{\underset{\displaystyle CH_3}{|}}{C}}-C_6H_5
$$

(I)   (II)

$$
\longrightarrow C_6H_5-\overset{\overset{\displaystyle H_5C_6}{|}}{\underset{\underset{\displaystyle CH_3}{|}}{C}}-\overset{\overset{\displaystyle O-H}{|}}{\underset{+}{C}}-CH_3
\qquad
\left[\text{not } \; C_6H_5-\overset{\overset{\displaystyle CH_3}{|}}{\underset{\underset{\displaystyle CH_3}{|}}{C}}-\overset{\overset{\displaystyle OH}{|}}{\underset{+}{C}}-C_6H_5 \right]
$$

(III)   (IV)

$$
\longrightarrow C_6H_5-\overset{\overset{\displaystyle H_5C_6}{|}}{\underset{\underset{\displaystyle CH_3}{|}}{C}}-\overset{\overset{\displaystyle O}{\|}}{C}-CH_3
\qquad
\left[\text{not } \; C_6H_5-\overset{\overset{\displaystyle CH_3}{|}}{\underset{\underset{\displaystyle CH_3}{|}}{C}}-\overset{\overset{\displaystyle O}{\|}}{C}-C_6H_5 \right] \quad (a)
$$

(V)   (VI)

$$
\underset{(VII)}{\text{C}_6\text{H}_5 - \overset{\overset{\text{OH}}{|}}{\underset{\underset{\text{H}}{|}}{\text{C}}} - \overset{\overset{\text{OH}}{|}}{\underset{\underset{\text{H}}{|}}{\text{C}}} - \text{CH}_3} \longrightarrow \underset{(VIII)}{\text{C}_6\text{H}_5 - \overset{+}{\underset{\underset{\text{H}}{|}}{\text{C}}} \overset{\overset{\text{OH}}{|}}{\underset{(\text{H})}{\text{C}}} - \text{CH}_3} \longrightarrow \underset{(IX)}{\text{C}_6\text{H}_5 - \overset{\overset{\text{H}}{|}}{\underset{\underset{\text{H}}{|}}{\text{C}}} - \overset{\overset{\text{O}}{\|}}{\text{C}} - \text{CH}_3}
$$

$$
\left[\text{not } \text{C}_6\text{H}_5 - \overset{\overset{\text{H}}{|}}{\underset{\underset{\text{CH}_3}{|}}{\text{C}}} - \overset{\overset{\text{O}}{\|}}{\text{C}} - \text{H}\right] \qquad \text{(b)}
$$

(X)   (19.6)

$$
\longrightarrow \underset{(XI)}{\text{C}_6\text{H}_5 - \overset{\overset{\text{OH}}{|}}{\underset{\underset{\text{H}}{|}}{\text{C}}} - \overset{+}{\underset{\underset{\text{H}}{|}}{\text{C}}} - \text{CH}_3} \rightarrow\rightarrow \underset{(XII)}{\text{C}_6\text{H}_5 - \overset{\overset{\text{O}}{\|}}{\text{C}} - \overset{\overset{\text{H}}{|}}{\underset{\underset{\text{H}}{|}}{\text{C}}} - \text{CH}_3}
$$

$$
\underset{(XIII)}{\text{H} - \overset{\overset{\text{O}}{\|}}{\text{C}} - \overset{\overset{\text{C}_6\text{H}_5}{|}}{\underset{\underset{\text{H}}{|}}{\text{C}}} - \text{CH}_3}
$$

not formed   (c)

Glycol (I) gives the carbocation (II). From this, the migration of phenyl or methyl groups can give the cations (III) or (IV), and the final products (V) or (VI), respectively. In actual experiments, it is found that only (V) corresponding to phenyl migration, is obtained. Similarly, from (VII), the only product obtained is (IX). The other expected products (X), (XII) and (XIII) corresponding to the alternate routes indicated do not get formed. Based on studies using this and other similar molecules, various groups have been classified on the basis of their *relative migratory aptitudes*. The order of migratory aptitude is: Aryl > H > alkyl.

The greater migratory aptitude of the aryl group can be understood on the basis of its ability to stabilise the transition state better. This can be seen in the structures (Eq. 19.5b) where the delocalisation of the positive charge by the phenyl group is picturised. This stabilisation is better when electron-releasing groups are present. The following listing of migratory aptitudes of different aryl groups, illustrate this order:

p-methoxyphenyl > *p*-methylphenyl > phenyl > *p*-chlorophenyl

Consider molecule (VII) in Eq. 19.6b. This has two non-equivalent OH groups. Protonation of these can give rise to different cations, (VIII) or (XI), respectively. No product corresponding to (XI) is obtained. The other cation, (VIII) is the more stable one. It is benzylic and is stabilised by resonance involving the phenyl group. On the basis of studies with similar molecules, it has been generalised that in unsymmetrical pinacols, where the two hydroxyl groups are non-equivalent, protonation will take place in such a way that the more stable cation is formed and rearranged products arising from that cation alone will be obtained.

Rearrangement can take place if a carbocation is generated next to a carbon bearing a hydroxyl group by even some other reaction. The reaction of a vicinal aminoalcohol (Eq. 19.7a) with nitrous acid is one such situation.

This is called pinacolic deamination or Demjanov (Demyanov) rearrangement. The conversion of cyclohexanone to cycloheptanone (ring expansion) is an application of this reaction (Eq. 19.7b).

19.2.2 WOLFF REARRANGEMENT

This is a rearrangement in which the migration terminus is an electron deficient carbon that is not a carbocation but a carbene. Wolff rearrangement is a component of Arndt–Eistert synthesis (Chapter 18, Section 18.3). In this reaction, a diazoketone which is formed from an acyl chloride and diazomethane, is the starting material.

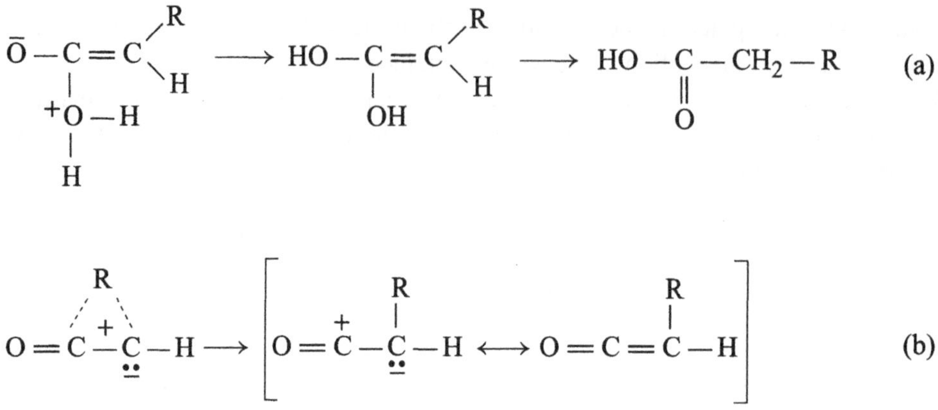

$$\bar{O}-C=C\underset{H}{\overset{R}{\diagup}} \longrightarrow HO-C=C\underset{H}{\overset{R}{\diagup}} \longrightarrow HO-\underset{O}{\overset{||}{C}}-CH_2-R \quad (a)$$

with $\overset{+}{O}-H$ / H below first structure, and OH below second structure.

$$O=\overset{R}{\overset{\diagup}{C}}\overset{+}{\underset{\bullet\bullet}{-C}}-H \longrightarrow \left[O=\overset{+}{\overset{|}{C}}-\overset{R}{\underset{\bullet\bullet}{\overset{|}{C}}}-H \longleftrightarrow O=C=\overset{R}{\overset{|}{C}}-H \right] \quad (b)$$

$$(19.8)$$

The diazoketone loses a molecule of nitrogen in a reaction catalysed by colloidal silver to form a carbene (carbon with only 6 electrons around it) which is an electron deficient species. The migration of a group from the adjacent carbon to the electron deficient centre gives a ketene, which in the aqueous medium, adds on a molecule of water to give the final product, the carboxylic acid.

19.2.3 BENZILIC ACID REARRANGEMENT

When 1,2-diketones without α-hydrogens are treated with alkali, a rearrangement takes place to give an α-hydroxycarboxylic acid. The representative example of this is the conversion of benzil to benzilic acid (Eq. 19.9).

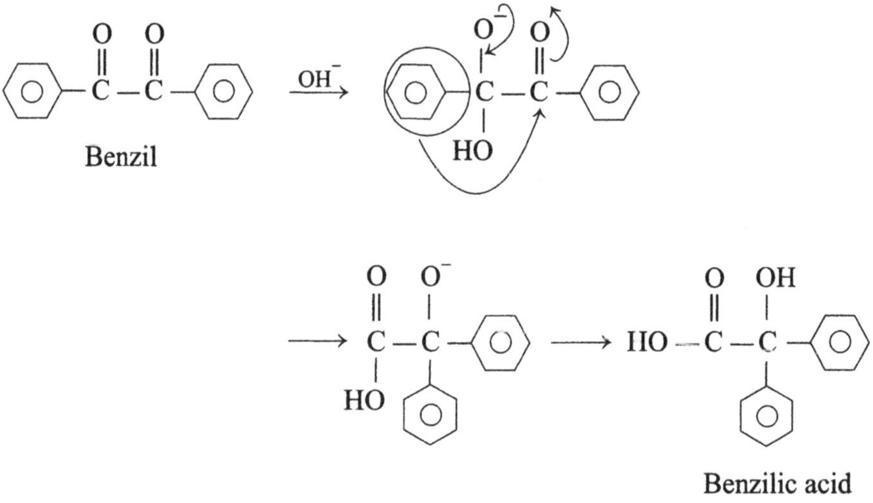

Benzil

Benzilic acid   (19.9)

Benzil is the diketone obtained by the oxidation of benzoin which itself can be synthesised from benzaldehyde by the benzoin condensation. The phenyl group with its bonding electron pair is the migrating group and migrates to the electron deficient carbonyl carbon. The internal oxidation–reduction reaction of glyoxal (Eq. 19.10a) can be considered as either an intramolecular Cannizzaro reaction or a benzilic acid rearrangement.

$$
\underset{\text{Glyoxal}}{\text{H}-\overset{\overset{\text{O}}{\|}}{\text{C}}-\overset{\overset{\text{O}}{\|}}{\text{C}}-\text{H}} \xrightarrow{\text{OH}^-} \left(\text{H}-\overset{\overset{\text{O}^-}{|}}{\underset{\underset{\text{OH}}{|}}{\text{C}}}-\overset{\overset{\text{O}}{\|}}{\text{C}}-\text{H} \right) \longrightarrow \text{HO}-\overset{\overset{\text{O}}{\|}}{\text{C}}-\text{CH}_2-\text{O}^-
$$

$$
\longrightarrow \underset{\text{Glycolic acid}}{\text{HO}-\overset{\overset{\text{O}}{\|}}{\text{C}}-\text{CH}_2\text{OH}} \qquad \text{(a)}
$$

$$
\underset{\text{Phenylglyoxal}}{\text{C}_6\text{H}_5-\overset{\overset{\text{O}}{\|}}{\text{C}}-\overset{\overset{\text{O}}{\|}}{\text{C}}-\text{H}} \xrightarrow{\text{OH}^-} \text{C}_6\text{H}_5-\overset{\overset{\text{O}}{\|}}{\text{C}}-\overset{\overset{\text{O}^-}{|}}{\underset{\underset{\text{HO}}{|}}{\text{C}}}-\text{H} \longrightarrow \text{C}_6\text{H}_5-\overset{\overset{\text{O}^-}{|}}{\underset{\underset{\text{H}}{|}}{\text{C}}}-\overset{\overset{\text{O}}{\|}}{\text{C}}-\text{OH}
$$

$$
\longrightarrow \underset{\text{Mandelic acid}}{\text{C}_6\text{H}_5-\overset{\overset{\text{OH}}{|}}{\underset{\underset{\text{H}}{|}}{\text{C}}}-\overset{\overset{\text{O}}{\|}}{\text{C}}-\text{OH}} \qquad \text{(b)} \qquad\qquad \text{(19.10)}
$$

The rearrangement of phenylglyoxal to mandelic acid by the action of alkali (Eq. 19.10b) can also be considered as either a benzilic acid rearrangement or an intramolecular Cannizzaro reaction.

19.3 REARRANGEMENTS INVOLVING MIGRATION TO ELECTRON DEFICIENT NITROGEN

19.3.1 BECKMANN REARRANGEMENT

An oxime obtained from a ketone and hydroxylamine is called a ketoxime (Chapter 16, Section 16.3.1.4). Upon reaction with acidic reagents like H_2SO_4, H_3PO_4, $SOCl_2$, PCl_5 and BF_3, the ketoxime undergoes a rearrangement to give an amide (Eq. 19.11). For example, the oxime of acetophenone undergoes the Beckmann rearrangement (Eq. 19.11) to give acetanilide.

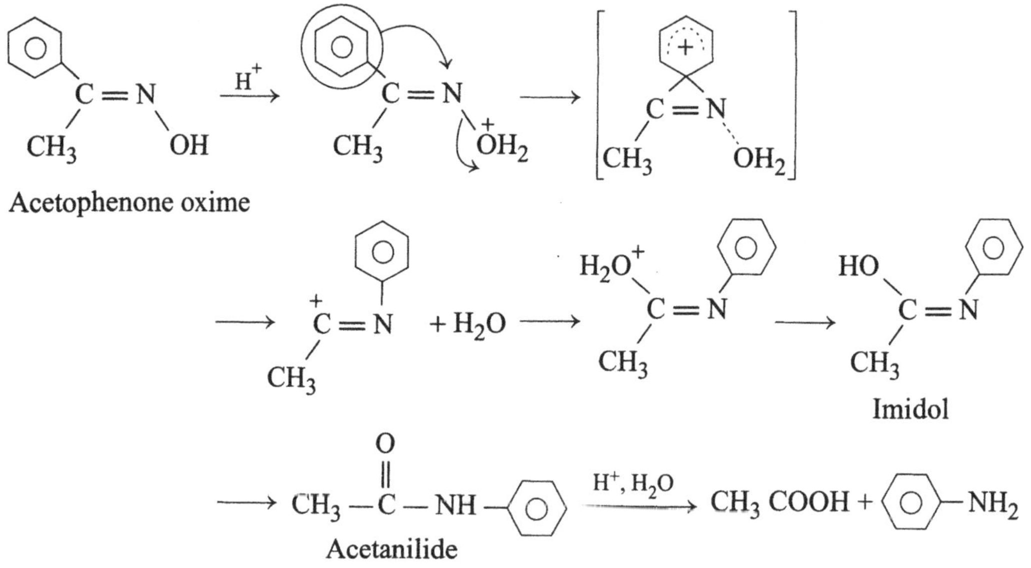

Acetophenone oxime

Imidol

Acetanilide

$$\text{(19.11)}$$

Oximes exhibit geometrical isomerism when the two groups of the ketone are different. The group on the same side as the OH is called the *syn* group and the one on the opposite side, the *anti* group. In this case, the phenyl group is *anti* and the methyl is *syn*. It has been established that the group that is *anti* to the OH group migrates, not the *syn* group (Eq. 19.11). The rearrangement is intramolecular. In effect, the migrating group performs an intramolecular nucleophilic displacement of the hydroxyl group. The role of reagents such as sulphuric acid or phosphorus pentachloride is to make the OH a good leaving group. The carbocation formed as a result of rearrangement is quenched by water to give the imidol which tautomerises to the amide. The amide can be hydrolysed to the carboxylic acid and the amine. Acetophenone, in the given example, can be converted to acetic acid and aniline through the oxime by Beckmann rearrangement followed by hydrolysis. This reaction can also be used to assign the configuration of the oxime since the group that ends up as the amine is the one that has the *anti* configuration (see Chapter 22, Section 22.1.2.5 for further discussions on the stereochemistry of the Beckmann rearrangement).

Beckmann rearrangement is used for the manufacture of carprolactam which is the monomer for Nylon-6 (Eq. 19.12).

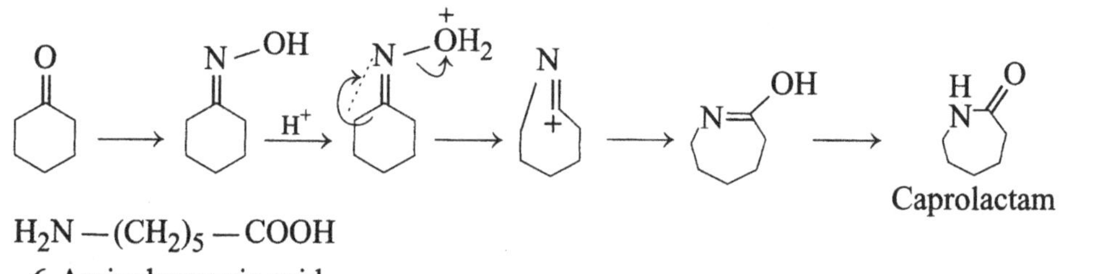

Caprolactam

$H_2N-(CH_2)_5-COOH$
6-Aminohexanoic acid
(ε-Aminocaproic acid)

$$\text{(19.12)}$$

Cyclohexanone – the starting material for the manufacture of caprolactam by this process – is obtained as a petrochemical from benzene. Beckmann rearrangement of the oxime of cyclohexanone results in ring expansion to give the cyclic amide (lactam) of 6-aminohexanoic acid, also known as ε-aminocaproic acid.

19.3.2 HOFMANN REARRANGEMENT

The rearrangement of amides by the action of sodium hypobromite in alkaline medium to give an amine with one carbon less is called the Hofmann rearrangement or the Hofmann degradation.

$$R-\overset{\overset{\displaystyle O}{\|}}{C}-\underset{\cdot\cdot}{N}H_2 + NaOBr \longrightarrow R-\overset{\overset{\displaystyle O}{\|}}{C}-\underset{\cdot\cdot}{N}HBr + NaOH$$

Bromoamide

$$R-\overset{\overset{\displaystyle O}{\|}}{C}-\underset{\cdot\cdot}{N}HBr + NaOH \longrightarrow R-\overset{\overset{\displaystyle O}{\|}}{C}-\underset{\cdot\cdot}{\overset{\_}{N}}-Br\ Na^+ \xrightarrow{-Br^-} \left[R-\overset{\overset{\displaystyle O}{\|}}{C}-N\colon\right]$$

Nitrene

$$R-N=C=O \longrightarrow \text{Isocyanate}$$

$$R-N=C=O \longrightarrow R-\overset{}{N}=C-O^- \longrightarrow R-NH-C=O$$
$$HO^- \qquad\qquad H^+\ OH \qquad\qquad OH$$

Carbamic acid

(19.13)

$$R-NH_2 + CO_2$$

The first step is the bromination of the amide at the nitrogen to give an N-bromoamide. The hydrogen on the nitrogen, in the bromoamide, is acidic due to the presence of the electron-withdrawing halogen and carbonyl groups. The alkali removes this proton to create a nitrogen anion. Migration of the R group to the nitrogen with simultaneous loss of bromide takes place to give an isocyanate, $R-N=C=O$ (Eq. 19.13). Isocyanates are readily hydrolysed to give an unstable carbamic acid, $R-NH-COOH$, which spontaneously decarboxylates to form the amine. The intermediates—bromoamide, its anion and isocyanate, have been detected under carefully controlled conditions. The anion of the bromoamide is structurally similar to the diazoketone in

Wolff rearrangement (Eq. 19.8). Loss of Br⁻ from this anion can produce a nitrene, just as loss of N_2 from the diazoketone gives a carbene. Both carbenes (derived from $\ddot{:}CH_2$) and nitrenes (derived from $\ddot{:}NH$) are neutral, electron deficient species with only 6 electrons around the C or N. There is no evidence for the presence of a free nitrene in the Hofmann rearrangement and the rearrangement is believed to be a concerted reaction, that is, the migration of the R group and the departure of bromide ion take place simultaneously. The similarity with Wolff rearrangement is pointed out to emphasise the fact that Hofmann rearrangement also involves migration to an electron deficient atom, a potential nitrene.

Hofmann rearrangement is a useful reaction to prepare primary amines from carboxylc acids (*decarboxylative deamination*). Hofmann rearrangement is not to be confused with Hofmann elimination, which is the elimination of a tertiary amine from a quaternary ammonium hydroxide (Chapter 4).

19.3.3 CURTIUS REARRANGEMENT

A related reaction is the Curtius rearrangement, which is another method for converting a carboxylic acid to the amine, with one less carbon atom.

$$R-\overset{\overset{O}{\|}}{C}-Cl + H_2N-NH_2 \longrightarrow R-\overset{\overset{O}{\|}}{C}-NH-NH_2$$

Acid hydrazide

$$\xrightarrow{NaNO_2,\ HCl}\ R-\overset{\overset{O}{\|}}{C}-\overset{-}{N}-\overset{+}{N}\equiv N \longrightarrow O=C=N-R+N_2$$

Acyl azide   Isocyanate

$$\xrightarrow{H_2O}\ R-NH-\overset{\overset{O}{\|}}{C}-OH \longrightarrow R-NH_2+CO_2$$

(19.14)

The starting material is the hydrazide of the carboxylic acid, which can be prepared by the reaction of the acid chloride and hydrazine. The hydrazide upon reaction with nitrous acid is converted to the acyl azide. This is similar to the diazoketone (Eq. 19.8) and the bromoamide anion (Eq. 19.13). Loss of nitrogen with simultaneous migration of the R group gives the isocyanate as in the Hofmann rearrangement. Isocyanate gets hydrolysed to give the amine.

Incidentally, a carboxylic acid can be converted to the amine with the same number of carbon atoms, $RCOOH \rightarrow RCH_2-NH_2$, by first converting it to the amide, $R-CO-NH_2$, and then reducing the amide using lithium aluminium hydride.

19.4 SIGMATROPIC REARRANGEMENTS

Sigmatropic rearrangements belong to a class of reactions known as pericyclic reactions. These are reactions that take place through cyclic transition states, without the involvement of ionic or free radical intermediates. Cycloaddition, specifically the Diels–Alder reaction, which is an example of pericyclic reactions, has been discussed in Chapter 6, Section 6.4.3.

19.4.1 COPE REARRANGEMENT

Some of the general aspects, terminology and definitions related to sigmatropic rearrangements are illustrated using Cope rearrangement as an example.

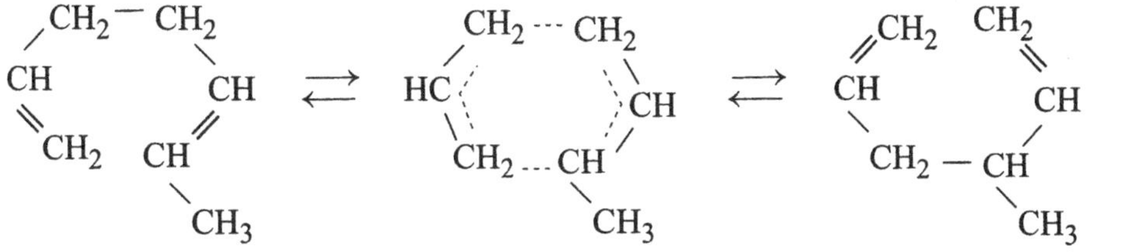

1,5-Heptadiene (I)   Cyclic tansition state (II)   3-Methyl-1,5-hexadiene (III)

(a)

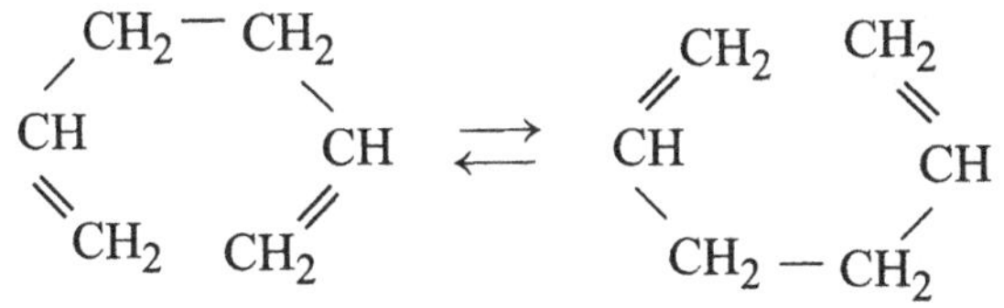

(b)

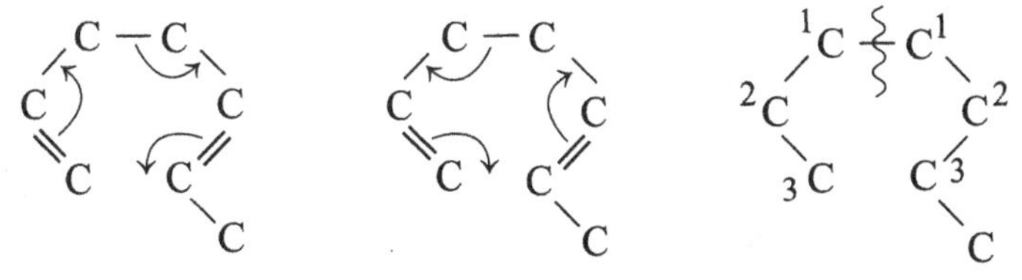

(IV)   (V)   (VI)

(c)

$$(19.15)$$

1,5-Heptadiene (I) and 3-methyl-1,5-hexadiene (III), are mutually interconvertable upon heating (Eq. 19.15a). This is a general reaction of 1,5-dienes. A cyclic transition state (II) is involved. Even simple 1,5-hexadiene will undergo the rearrangement upon heating (Eq. 19.15b). In this case, the starting material and product are the same, therefore, the reaction will not be visible. It is an example of degenerate Cope rearrangement. For purposes of arriving at the structure of the product, electron shifts as in (IV) or (V) can be visualised. However, there is no evidence to show that electrons are moving clockwise as in (IV) or counter-clockwise as in (V). What is happening is, a reorganisation of bonds in a cyclic transition state, (II). The transition state has no charge separation, that is, it is not polar. Free radicals are also not involved. The reorganisation will take place only if the symmetry properties of the involved molecular orbitals allow it.

In this rearrangement (Eq. 19.15), the σ-bond between C3 and C4 is broken and a new σ-bond is established between C1 and C6. Since a σ-bond has shifted position, it is called a *sigmatropic rearrangement*. One can imagine two fragments as a result of the cleavage of the C3 — C4 bond [Eq. 19.15c, (VI)]. The atoms of the two fragments are numbered as in (VI). Upon completion of the rearrangement, the new bond has been established between C3 of one fragment and C3 of the other fragment. Hence it is called *a 3,3-sigmatropic rearrangement*. A simple allylic rearrangement is a 1,3-sigmatropic rearrangement (Eq. 19.16).

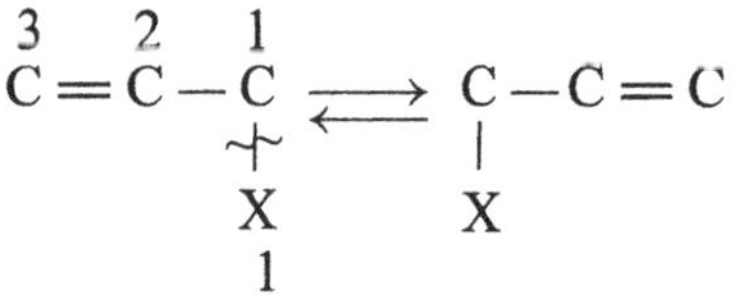

$$\overset{3}{C}=\overset{2}{C}-\overset{1}{\underset{\underset{\underset{1}{X}}{\overset{|}{X}}}{C}} \rightleftharpoons C-C=C \qquad (19.16)$$

19.4.2 OXY-COPE REARRANGEMENT

1,5-Dienes undergo rearrangement on heating, even when atoms other than carbon are present in the chain. Allyl vinyl ethers belong to this category. For example, the reaction that takes place when the allyl ether of the enol of acetone is heated (Eq. 19.17).

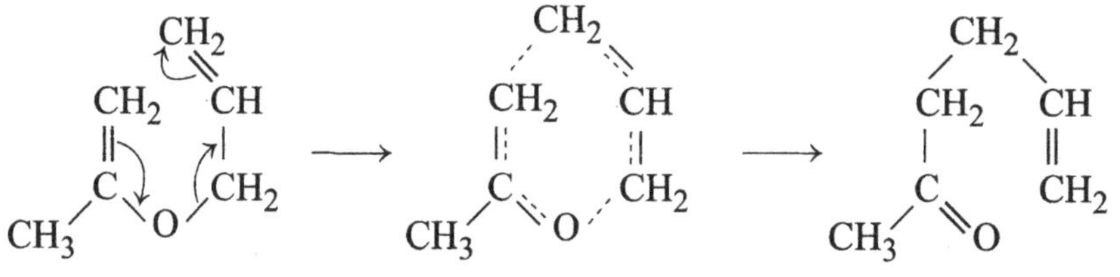

$$(19.17)$$

It is emphasised that the curved arrows shown here do not have the same significance as in the representations of other reaction mechanisms. They do not imply that there is a shift of a pair of electrons in the direction indicated. The curved arrows are only meant to keep track of overall bond alterations. The thermal isomerisation of the allyl vinyl ethers goes by the name oxy-Cope rearrangement (Eq. 19.17).

19.4.3 CLAISEN REARRANGEMENT

The best known and the most extensively studied 3,3-sigmatropic rearrangement is Claisen rearrangement. This is really the aromatic analogue of oxy-Cope rearrangement. Basic Claisen rearrangement is the thermal isomerisation (isomerisation by the action of heat) of allyl phenyl ether to *o*-allylphenol.

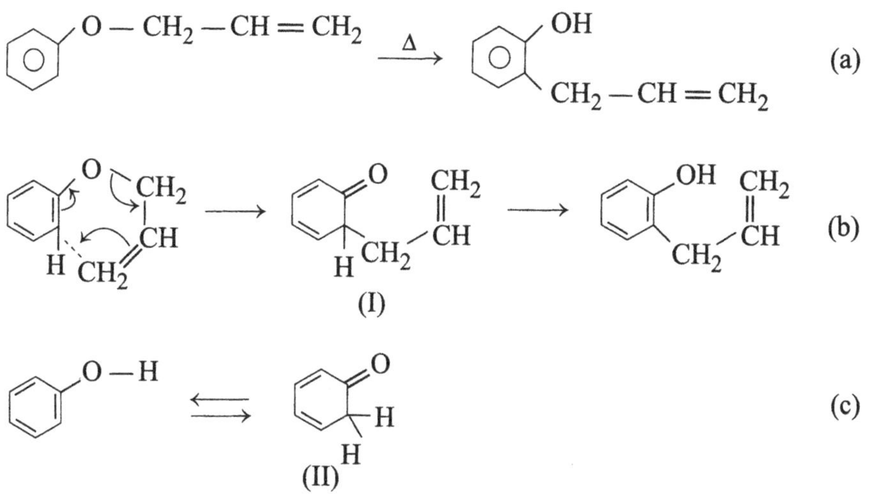

(a)

(b)

(19.18)

(I)

(c)

(II)

Recall that phenol is the enol form of cyclohexadienone (II) (Eq. 19.18c). Allyl ether of phenol is readily prepared by the reaction of sodium phenoxide with allyl chloride. The mechanism shown in Eq. 19.18b (with the disclaimer about the use of curved arrows), which is similar to the oxy-Cope rearrangement, leads to the formation of allylcyclohexadienone (I), which tautomerises irreversibly to *o*-allylphenol.

Two questions about this reaction have been convincingly answered as a result of carefully designed experiments. (i) Is the reaction intermolecular or intramolecular? (ii) Is there an allylic shift in the migrating group?

One of the methods usually employed to establish whether a rearrangement is intra or intermolecular is to conduct crossover studies. In this kind of study two similar molecules, both of which undergo rearrangement, are made to react in the same flask. After the reaction, the products are analysed to see if the migrating group from one molecule has ended up at the migration terminus of the other molecule, that is, whether it has crossed over to the other molecule. To study the Claisen rearrangement, a suitable pair of reactants will be (I) and (II) (Eq. 19.19).

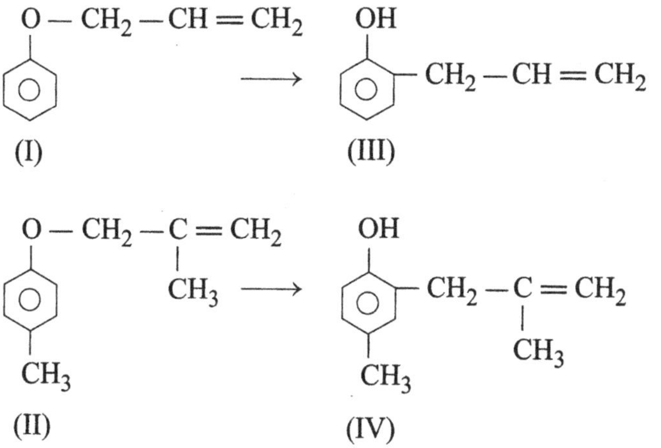

(I)

(III)

(II)

(IV)

$$
\underset{(V)}{\overset{\displaystyle OH}{\bigodot}} CH_2-\underset{\underset{CH_3}{|}}{C}=CH_2 \qquad \underset{(IV)}{\overset{\displaystyle OH}{\underset{\underset{CH_3}{|}}{\bigodot}}} CH_2-CH=CH_2 \qquad (19.19)
$$

When a mixture of these two compounds is heated, the expected product from (I) and (II), namely (III) and (IV) will be formed. Products (V) and (VI) are crossover products. If only (III) and (IV) are formed, it can be confirmed that the reaction is intramolecular. If the crossover products are also formed, the rearrangement is surmised to be intermolecular. Such studies have shown that Claisen rearrangement is truly intramolecular. With the help of kinetic studies, the entropy of activation, ΔS^*, of the reaction has been determined and has been found to have a large negative value. This implies that the transition state is more ordered than the starting material and for unimolecular reactions, this is taken as evidence for a cyclic transition state.

Allylic shift has been confirmed by using substrates in which the 1 or 3 position of the allyl group is 'labelled' either by placing a carbon isotope, usually ^{14}C, at that position (isotopic labelling) or by placing a substituent at that position. These are illustrated in Eq. 19.20a and 19.20b, respectively. In Eq. 19.20a, the carbon bearing the asterisk in structure I, is rich in ^{14}C. The product is analysed to locate the position of the 'label' in the side chain, by appropriate chemical degradation. The product is found to be exclusively II, not III. In the other example (Eq. 19.20b) the substrate is 3-phenoxy-l-butene, IV. The product of rearrangement is V, not VI. Such studies have established that the Claisen rearrangement is a 3, 3-sigmatropic rearrangement. This is what is implied in the statements that the allyl group undergoes an allyic shift during the rearrangement or that the allyl group undergoes inversion.

$$
\text{(I)} \qquad \longrightarrow \qquad \text{(II)} \qquad \qquad \text{(III)} \qquad \qquad \text{(a)}
$$

$$
\text{(IV)} \qquad \longrightarrow \qquad \text{(V)} \qquad \qquad \text{(VI)} \qquad \qquad \text{(b)}
$$

$$
(19.20)
$$

19.4.3.1 *p*-Claisen Rearrangement

When the two *ortho* positions of the ether are both occupied by alkyl groups, the allyl group migrates to the *para* position. This is the result of a Claisen rearrangement followed by a Cope rearrangement (Eq. 19.21a).

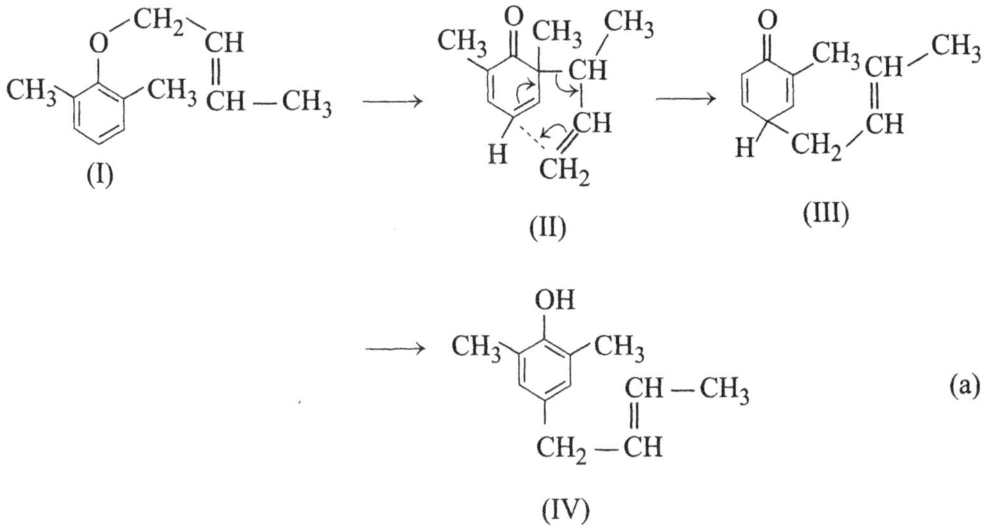

(a)

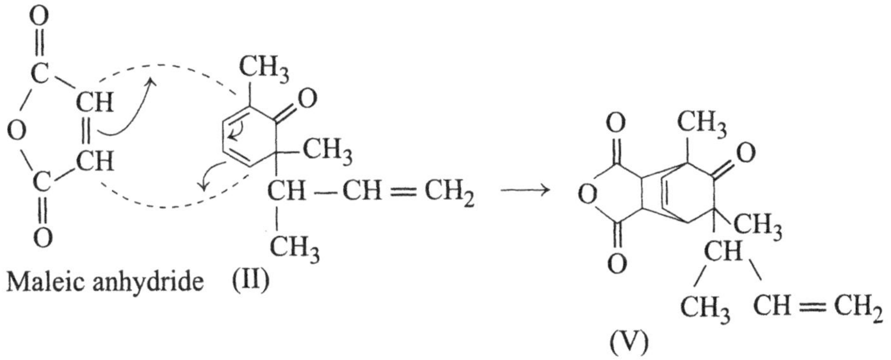

(b)

(19.21)

The starting material, (I), upon heating gives (IV). The first step is a normal Claisen rearrangement to give the dienone (II), except that this dienone cannot tautomerise to the phenol. Examination of this dienone shows that it is a 1,5-diene, a candidate for Cope rearrangement, (II) → (III). The product of this second rearrangement (III) then tautomerises to the final product, (IV). Notice that there is inversion of the allyl group at both stages, so that the structure of the allyl group in the final product is the same as that in the starting ether, The intermediate, (II) can be trapped as a Diels–Alder adduct (V) (Chapter 6, Section 6.4.3), if a dienophile like maleic anhydride is present in the medium (Eq. 19.21b).

19.5 FRIES REARRANGEMENT

Fries rearrangement (Chapter 15, Section 15.5.6) is an example of an electrophilic or cationotropic rearrangement. When a phenol ester like phenyl acetate is heated with anhydrous aluminium chloride, the acyl group migrates to the *o*- or *p*-position to give *o*-acyl or *p*-acylphenol.

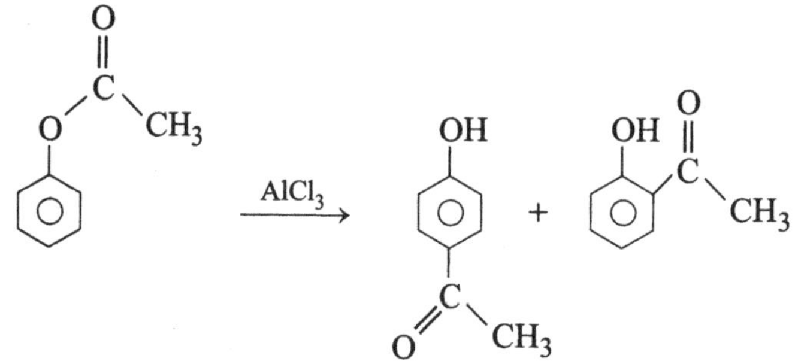

(a)

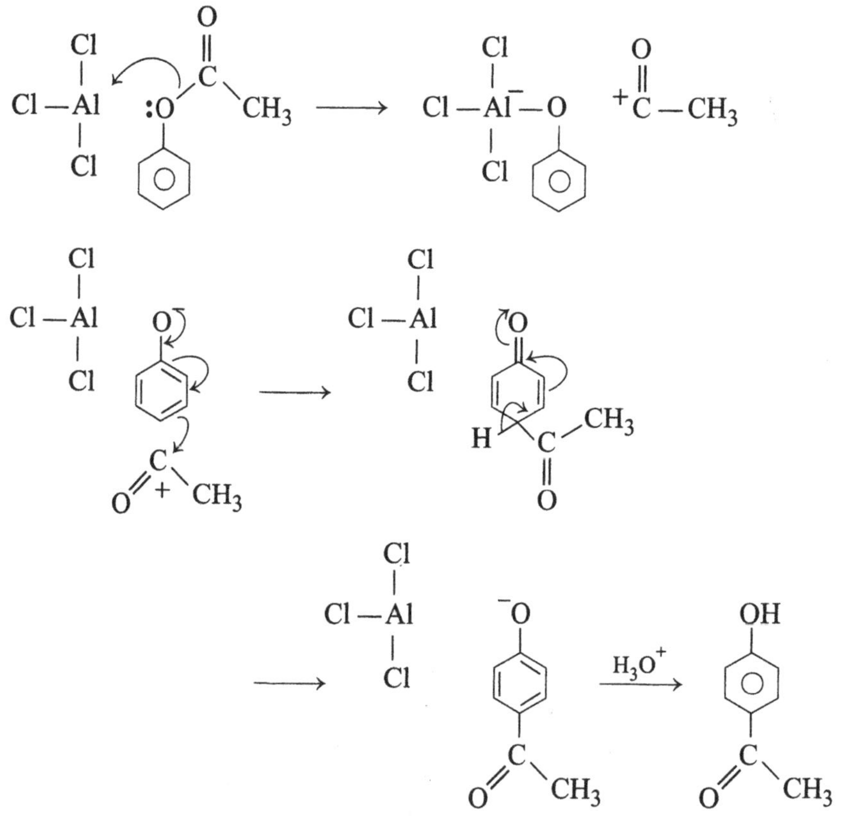

(b)

(19.22)

The *ortho/para* ratio is variable and is decided by the temperature and other conditions. Crossover experiments have been carried out to decide whether the rearrangement is inter or intramolecular.

The bulk of the evidence is in favour of the intermolecular mechanism (Eq. 19.22b). Under certain conditions evidence suggests that it can also be intramolecular. In the latter case, the acyl cation may not be fully separated from the parent molecule at any time during the rearrangement. These situations are exceptional. Fries rearrangement is generally classified as an intermolecular rearrangement. As shown, it is an electrophilic substitution reaction by the acyl cation. The cation can acylate the parent molecule or another molecule.

KEY POINTS

- Rearrangements can involve nucleophilic migration or electrophilic migration. In the former case, the migrating group shifts with its bonding electron pair, in effect as an anion. Therefore, it is also called anionotropy. The latter class involves migration of a cation and is thus called electrophilic rearrangement or cationotropy.
- Sigmatropic rearrangements are a subclass of pericyclic reactions, which proceed through cyclic transition states.
- Wagner–Meerwein and pinacol–pinacolone rearrangement are rearrangements of carbocations dictated by the relative stabilities of carbocations. They are intramolecular rearrangements. Depending upon their propensity for migration, groups are graded according to their migratory aptitudes.
- Benzilic acid rearrangement is the rearrangement of vicinal diketones, where the migration terminus is the positive carbon of a carbonyl group. There is similarity between benzilic acid rearrangement and the internal Cannizzaro reaction of glyoxal.
- Beckmann rearrangement, Hofmann rearrangement, Curtius rearrangement and Wolff rearrangement are examples of nucleophilic rearrangements where the migration terminus is an electron-deficient nitrogen.
- Cope, oxy-Cope and Claisen rearrangements are 3,3-sigmatropic rearrangements. The mechanism of Claisen rearrangement has been well studied. There is evidence for its intramolecular nature, (crossover experiments) and cyclic transition state as suggested by the negative entropy of activation. With the help of isotope labelling and by the use of substituted allyl groups, it has been shown that there is allylic shift in the migrating allyl group. Allyl phenyl ethers in which the *ortho* positions are occupied, undergo *p*-Claisen rearrangement which involves two successive 3,3-sigmatropic rearrangements.
- Fries rearrangement is the rearrangement of phenyl esters to *p*- and *o*-acylphenols, in the presence of $AlCl_3$. It is an electrophilic rearrangement. The migrating group is an acyl cation. It is generally intermolecular. Under certain conditions, it can also be intramolecular.

EXERCISES

SECTION I

1. How can the following conversions be brought about?

    (a) Phenyl *p*-methylbenzoate to *p*-(*p*-methylbenzoyl) phenol

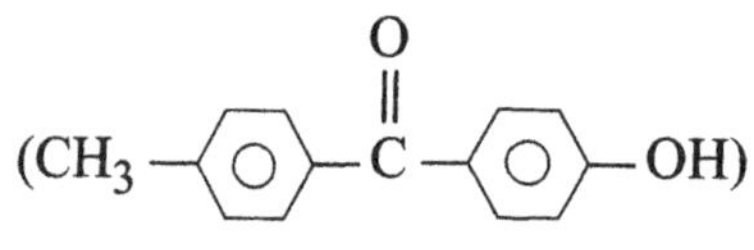

$$(CH_3 \!-\!\!\langle O \rangle\!\!-\! \overset{\overset{\displaystyle O}{\displaystyle \|}}{C} \!-\!\!\langle O \rangle\!\!-\! OH)$$

    (b) Phenylacetamide, $C_6H_5-CH_2-CONH_2$, to benzylamine
    (c) Phenylacetyl hydrazide, $C_6H_5-CH_2-CONHNH_2$ to benzylamine
    (d) *p*-Cresol to 2-allyl-4-methylphenol
    (e) $CH_3-CH=CH-O-CH(CH_3)-CH=CH_2$ to
        $CH_3-CH(CHO)-CH_2-CH=CH-CH_3$
    (f) 1,2-Diphenyl-1,2-cyclohexanediol to 2,2-diphenylcyclohexanone

2. Give the products of the following reactions.

    (a) 2-Amino-1-phenylethanol + HNO_2
    (b) Cyclopentanone oxime + PCl_5
    (c) Oxime of 3-pentanone + $SOCl_2$
    (d) *p*-Methoxybenzoic hydrazide + HNO_2
    (e) $(CH_3)_2CH-CH_2-CO-NH_2 + NaOBr$

SECTION II

1. N-substituted amides, $R-CO-NHR'$, do not undergo Hofmann rearrangement. Why?
2. When the Wolff rearrangement of $R-CO-CHN_2$, is carried out in the presence of an amine $R'NH_2$, or an alcohol $R'OH$, the products are $RCH_2-CO-NHR'$ and $RCH_2-CO-OR'$ respectively. Explain how these are formed from the ketene intermediate.
3. Design crossover experiments to decide whether the following reactions are inter or intramolecular. You should have two molecules undergoing the reaction and the products should be such that intermolecular reaction, if any, can be confirmed. If the reaction is not suitable for crossover study, explain why.

    (a) Pinacol–pinacolone rearrangement
    (b) Hofmann rearrangement
    (c) Fries rearrangement

4. A molecule, $C_6H_{14}O_2$, has a strong IR band at about $3300\,\text{cm}^{-1}$, and no band between 1650 and $1780\,\text{cm}^{-1}$. Its NMR spectrum has a singlet signal at near δ 1.0 and a smaller, broad signal at around δ 4. Upon warming with con. sulphuric acid, it gave a new compound $C_6H_{12}O$,

which has a strong IR band at $1700 \, cm^{-1}$, none at $3300 \, cm^{-1}$. Its NMR spectrum has two singlets in the area ratio 3:1 at δ 1.0 and 2.5. Assign structures to these compounds and explain the reaction.

CHALLENGING QUESTION

1. Bullvalene (I), is called a fluxional molecule. It can undergo a series of degenerate Cope rearrangements. The number of equivalent structures possible is more than 1.2 million. Draw two of them. Find out from the literature about the interesting NMR spectrum of the molecule.

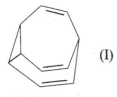   (I)

Bullvalene

20 Heterocyclic Compounds

OBJECTIVES   In this chapter, you will learn about,

- structures and numbering of simple heterocyclic compounds
- aromaticity of pyrrole, furan and thiophene
- general methods of synthesis
- the reason for pyrrole not being basic
- examples of electrophilic reactions of the three 5-membered heterocyclics
- Diels–Alder reaction of furan
- hydrogenation of pyrrole and furan
- synthesis of the pyridine ring system
- structure and basicity of pyridine
- electrophilic and nucleophilic substitution reactions of pyridine
- acidity of the picolines
- hydrogenation of pyridine
- properties of piperidine
- synthesis of indole: Fischer indole synthesis
- chemistry of indole
- synthesis of quinoline and isoquinolines: Skraup and Bischler–Napieralski syntheses
- chemistry of quinoline and isoquinoline

20.1 INTRODUCTION

Cyclic compounds containing at least one heteroatom in the ring – that is, atoms other than carbon – are called heterocyclic compounds or heterocyclics. The most common hetero atoms are nitrogen, oxygen and sulphur. They may be aromatic or non-aromatic. Many natural products of plant and animal origin, and biomolecules contain heterocyclic structures. Some of the common heterocyclic compounds are listed in Table 20.1.

The non-aromatic molecules behave as cyclic ethers or amines, and their chemistry is similar to that of their open chain analogues. In this chapter, we will study mainly the aromatic heterocyclic compounds.

Table 20. 1 Common heterocyclic compounds

Name	Structure	Name	Structure
Aromatic			
Pyrrole		Pyridine	
Furan		Indole	
Thiophene		Quinoline	
Pyrazole		Isoquinoline	
Imidazole		Pyrimidine	
Thiazole		Purine	

Table 20. 1   Continued

Name	Structure	Name	Structure
Nonaromatic			
Oxirane	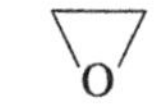	Pyrrolidine	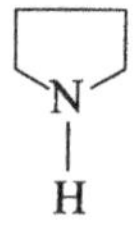
Oxetane	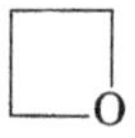	Piperidine	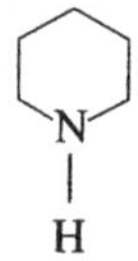
Tetrahydrofuran	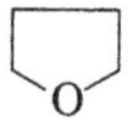	Pyran	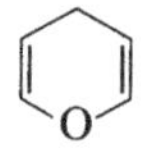
		1,4-Dioxan	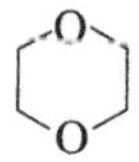

20.2   5-MEMBERED HETEROCYCLICS: PYRROLE, FURAN AND THIOPHENE

20.2.1   AROMATICITY

The two double bonds account for 4π-electrons. The aromatic sextet is completed by one lone pair of electrons on the heteroatom which is sp^2 hybridised. The electrons in the unhybridised *p*-orbitals of the five atoms are involved in the delocalised molecular orbitals (Eq. 20.1).

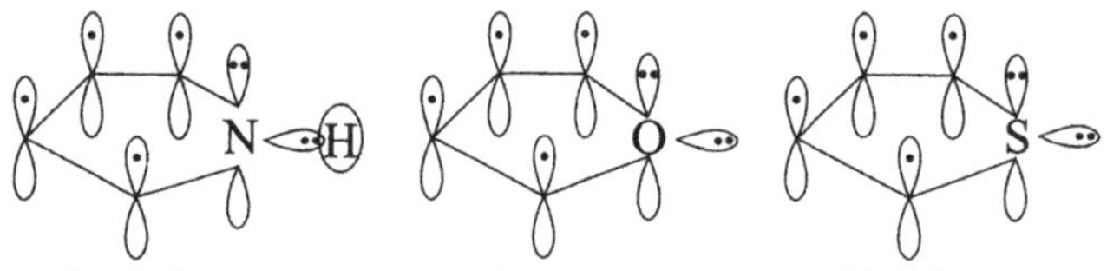

Pyrrole   Furan   Thiophene   (a)

(20.1)

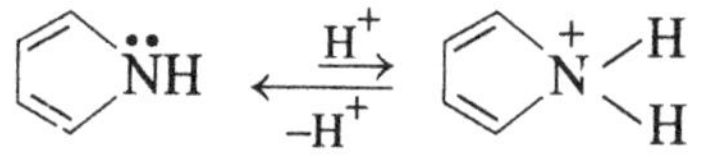

(b)

The remaining pair of electrons which are in sp^2 hybrid orbitals on O and S as well as the pair of electrons of the NH bond in pyrrole, are not part of the aromatic sextet. One of the obvious consequences of the aromaticity of pyrrole is that, though it appears to be a secondary amine, it is not basic. Its pK_b value is around 14. This is because protonation of pyrrole utilising the lone pair of electrons of nitrogen will destroy aromaticity (Eq. 20.1b). In fact, both pyrrole and furan are unstable and tend to polymerise in the presence of strong mineral acids. The resonance energies (listed below) are less than that of benzene.

Furan   : 67 kJ mol^{-1}
Pyrrole   : 92 kJ mol^{-1}
Thiophene : 121 kJ mol^{-1}
Benzene   : 152 kJ mol^{-1} (36 kcal mol^{-1})

20.2.2 GENERAL METHODS OF SYNTHESIS

All the three compounds can be considered to be derivatives of the enol form of 1,4-dicarbonyl compounds. Equation 20.2 is a schematic representation of this relationship.

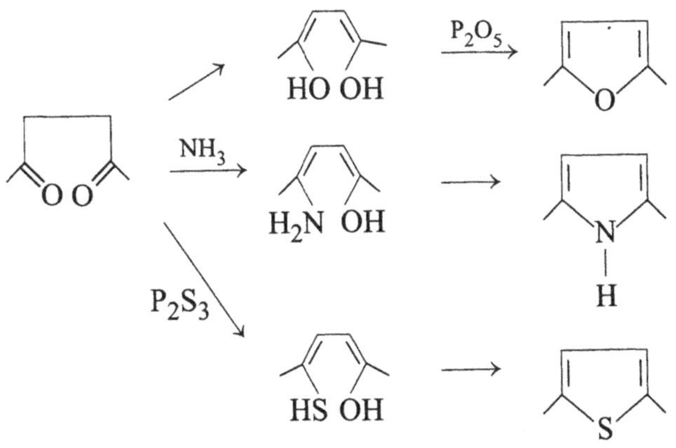

$$(20.2)$$

A general method to obtain 1,4-diketones is the alkylation of an acetoacetic ester by a chloroketone (Eq. 20.3). The carboxylate function can be removed if necessary by hydrolysis and decarboxylation since it is a β-ketoester.

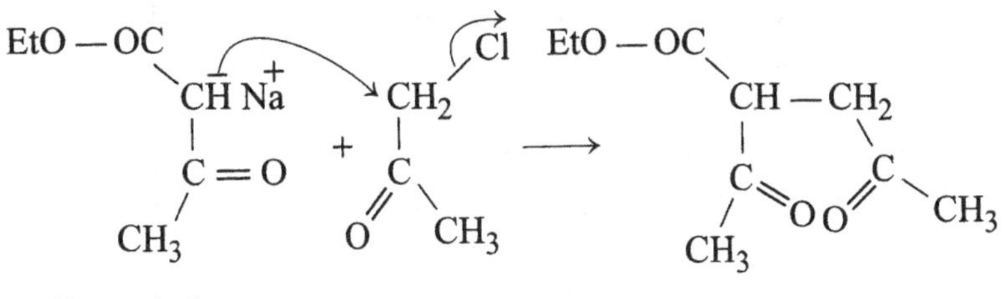

$$(20.3)$$

Et = ethyl

This methodology is illustrated in the *Hantzsch synthesis of pyrrole* (Eq. 20.4a). The mechanism of this reaction is outlined in Eq. 20.4b.

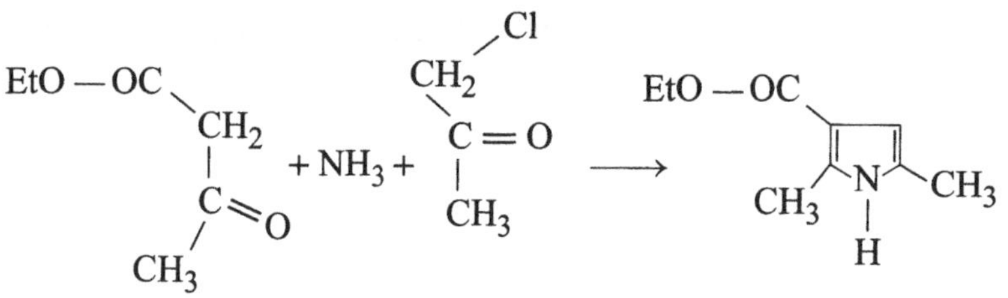

(a)

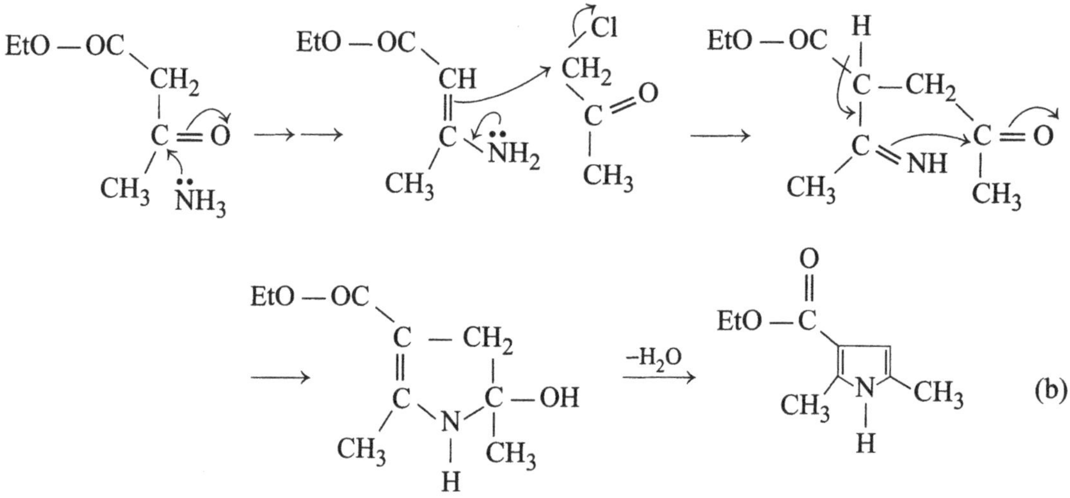

(b)

(20.4)

20.2.3 REACTIONS

20.2.3.1 Electrophilic Substitution

The lone pair on the heteroatom increases the electron density of the ring, as seen in the resonance structures (Eq. 20.5a) for pyrrole. Even though both the 2 and 3 positions are electron rich, substitution takes place preferentially at the 2-position. The σ-complex for 2-substitution is more stable than that for 3-substitution, as seen from the comparison of their resonance structures (Eq. 20.5b, c). They are more reactive than benzene in electrophilic substitution. This reactivity has limited application in the case of pyrrole and furan because they are not stable in acidic medium. However, electrophilic substitutions can be carried out under conditions where strong acids are avoided. Pyrrole is similar to phenol in reactivity. In fact, the hydrogen on the nitrogen is acidic and an anion can be formed by the action of base (Eq. 20.5d). Reactions characteristic of phenols such as azo coupling (Chapter 15, Section 15.4.6) can be carried out on pyrrole (Eq. 20.5e). Sulphonation of furan can be carried out using pyridine–sulphur trioxide complex (Eq. 20.5f). Acylation can be done with acid anhydride using the mild Lewis acid BF$_3$–diethyl ether complex (Eq. 20.5g). Thiophene, which is stable in acid medium can be nitrated, suphonated and halogenated to produce the 2-substituted thiophene. Some of these reactions are listed below (Eq. 20.5).

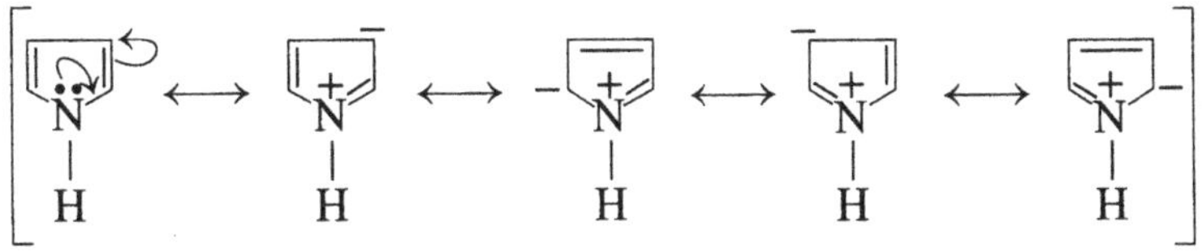

(a)

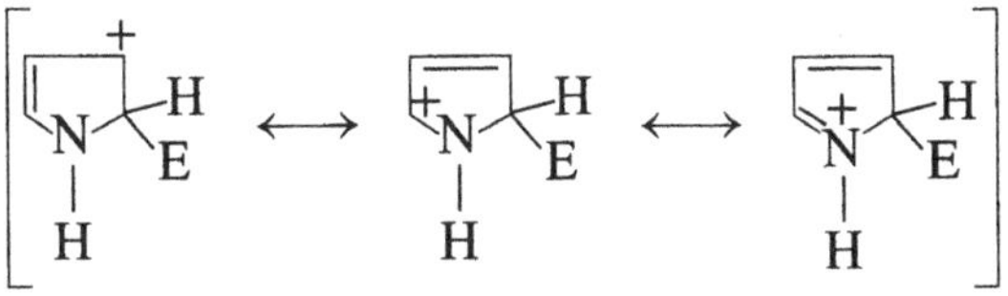

(b)

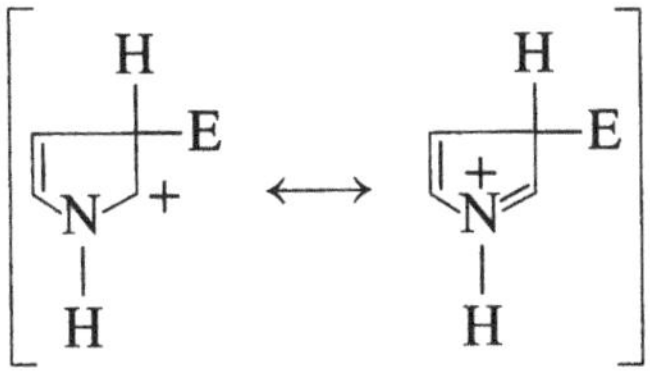

(c)

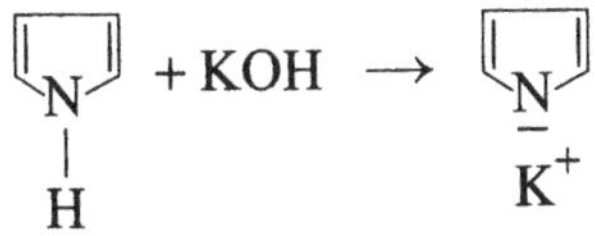

(d)

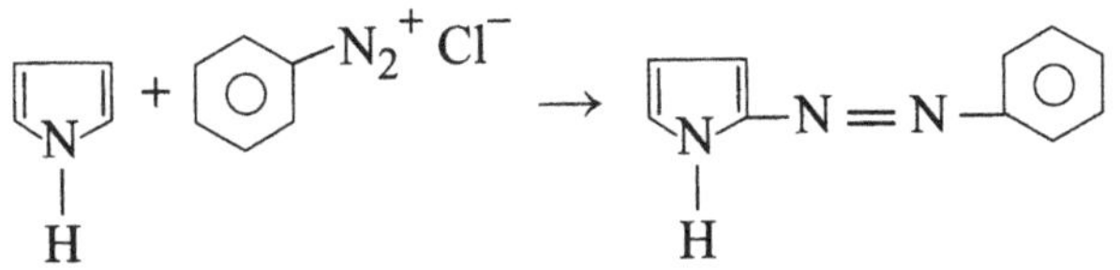

(e)

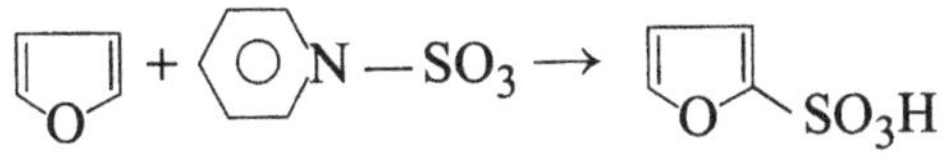

(f)

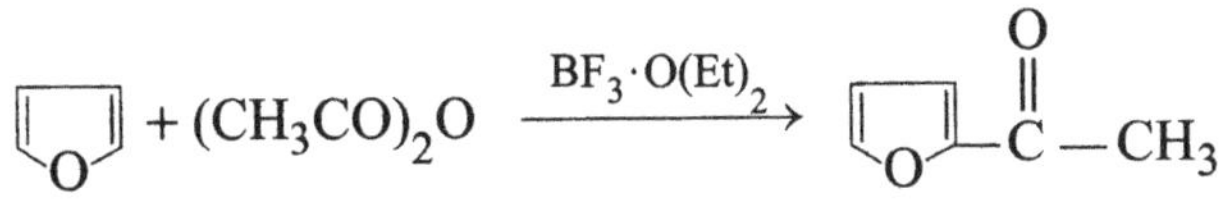

(g)

(20.5)

20.2.3.2 Hydrogenation

Pyrrole and furan can be hydrogenated catalytically using palladium to give the tetrahydro derivatives, pyrrolidine and tetrahydrofuran respectively.

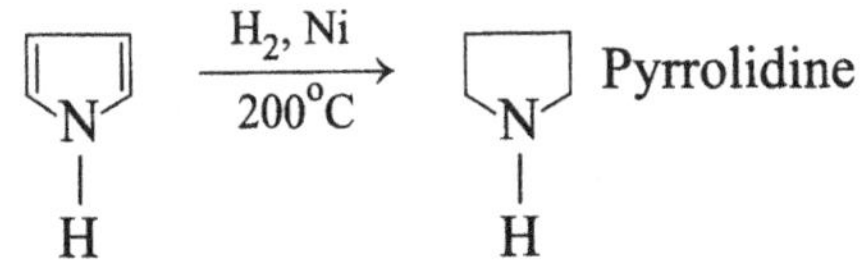

$$\text{(pyrrole)} \xrightarrow{\text{Zn/CH}_3\text{COOH}} \text{(pyrroline)} \quad \text{Pyrroline}$$

$$\text{(20.6)}$$

$$\text{(furan)} \xrightarrow{\text{H}_2,\text{ Pd}} \text{(THF)} \quad \text{Tetrahydrofuran}$$

Pyrrolidine is a secondary amine and is stable in acid medium. Dihydropyrrole is called pyrroline and can be prepared by the reduction of pyrrole with zinc and acetic acid. The pyrrole ring is present in the structure of porphyrins which are present in the biological pigments, hemin (blood) and chlorophyll (plants). The pyrrolidine ring is present in many alkaloids.

Tetrahydrofuran in a cyclic ether and is used as a solvent. An important derivative of furan is furfural, which is furan-2-carboxaldehyde.

Furfural

It is manufactured as an agricultural chemical from the husk of grains such as rice, oats and maize. The husk contains pentosans, which are pentose polymers. Digestion with sulphuric acid followed by steam distillation gives furfural. Furfural is similar to benzaldehyde in chemical properties and undergoes reactions such as benzoin condensation and Cannizzaro reaction. Fufural is used in the manufacture of dyes and resins and as an industrial solvent.

Furan, though aromatic, shows many chemical reactions of dienes. It behaves as a diene in the Diels–Alder reaction (Eq. 20.7, see also Chapter 6, Section 6.4.3).

$$\text{(20.7)}$$

Maleic anhydride   Adduct

20.3 6-MEMBERED HETEROCYCLICS: PYRIDINE

Pyridine is present in coal tar. Methylpyridines are called picolines. The three isomers are designated by the prefixes, α, β and γ. All the three can be oxidised to the corresponding carboxylic acids by permanganate. Pyridine-3-carboxylic acid is also called nicotinic acid because it can be obtained by oxidation of the alkaloid nicotine using $KMnO_4$. This acid, under the name niacin, is one of the vitamins of the B group. The hydrazide of pyridine-4-carboxylic acid, called isonicotinic acid hydrazide or isoniazid, is an antitubercular drug. Pyridoxine – another vitamin – is a tetrasubstituted pyridine (discussed in Chapter 23).

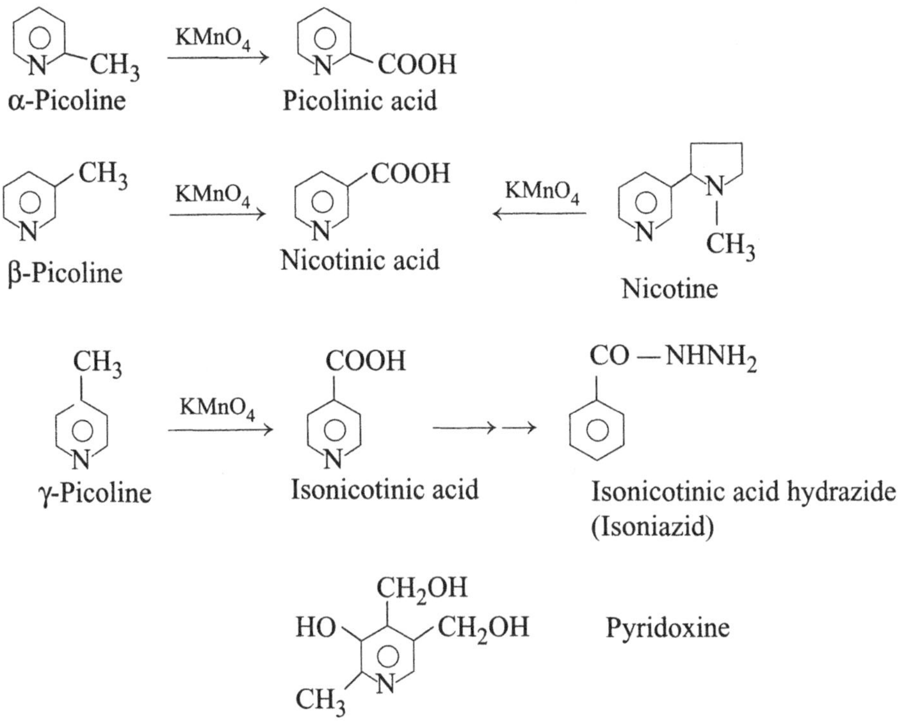

α-Picoline → Picolinic acid (KMnO₄)

β-Picoline → Nicotinic acid (KMnO₄) ← Nicotine (KMnO₄)

γ-Picoline → Isonicotinic acid (KMnO₄) → → Isonicotinic acid hydrazide (Isoniazid)

Pyridoxine

(20.8)

20.3.1 STRUCTURE AND BASICITY

The nitrogen in pyridine is sp^2 hybridised. The lone pair of electrons on the nitrogen is in an sp^2 hybrid orbital, which lies in the plane of the carbon framework, and is not a part of the aromatic sextet (Eq. 20.9b).

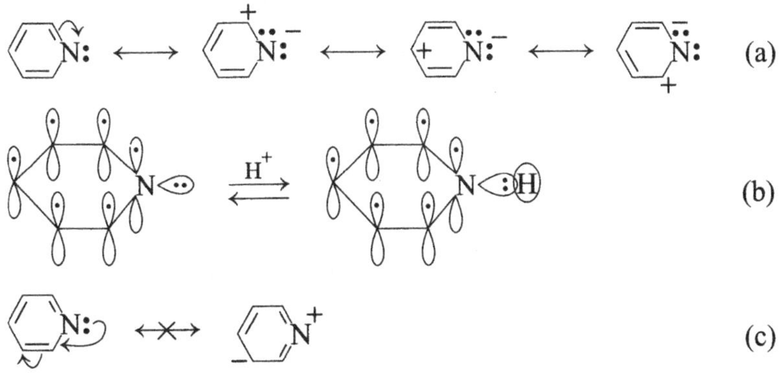

(a)

(b)   (20.9)

(c)

Protonation of the nitrogen does not disturb aromaticity but does reduce the electron density of the ring. Pyridine is basic with a pK_b value of 8.8. It is more basic than aniline ($pK_b = 9.4$) but less basic than dimethylamine ($pK_b = 3.2$). It is often used as an organic base for reactions such as the Knoevenagel reaction and where non-aqueous conditions are required. Pyridine is soluble in water and can be removed from the reaction mixture by washing with water.

Resonance in pyridine involves structures that make the ring electron deficient (Eq. 20.9a). The lone pair on the nitrogen cannot be delocalised into the benzene ring. Structures like 20.9c are untenable. Pyridine is less reactive than benzene in electrophilic substitution. The reactivity of pyridine is similar to that of nitrobenzene.

20.3.2 SYNTHESIS

A general method of synthesis of pyridine derivatives is illustrated in Eq. 20.10.

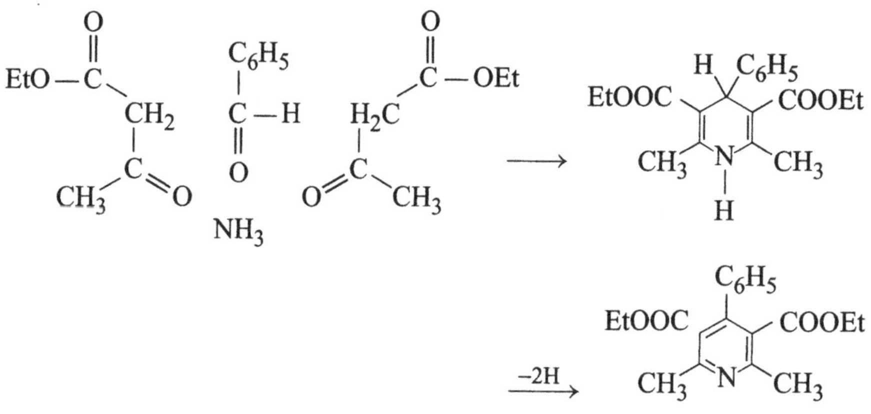

(a)

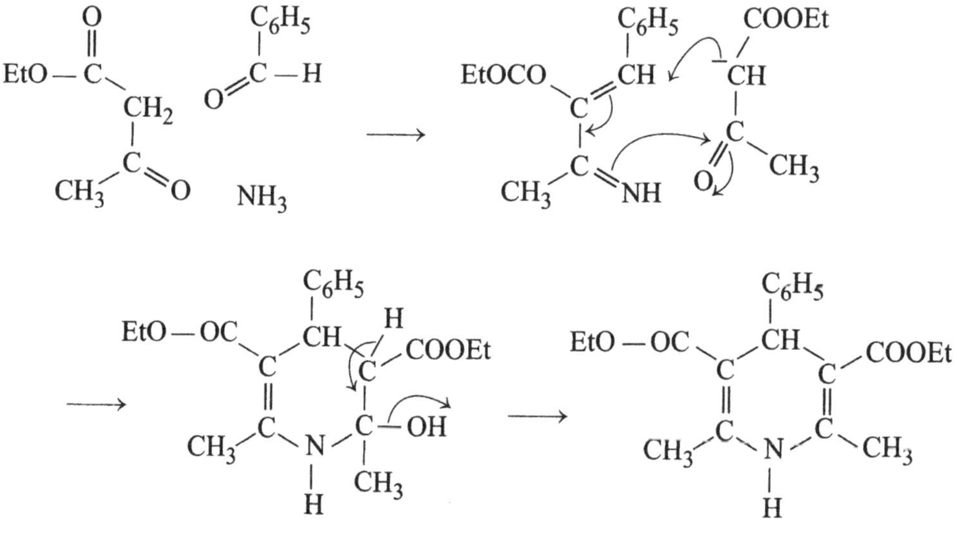

(b)

(I)

(20.10)

The reactants are two equivalents of acetoacetic ester and one each of benzaldehyde and ammonia. The reaction involves a Knovenagel reaction and a Michael addition (Eq. 20.10b). The product (I), is a dihydropyridine derivative. This can be oxidised to the pyridine derivative using nitric acid. This is the *Hantzsch synthesis* of pyridine.

20.3.3 REACTIONS

Pyridine is electron deficient (Eq. 20.9a) and is deactivated with respect to electrophilic substitution reactions. This is further aggravated in acidic medium due to the protonation of the nitrogen (Eq. 20.9b). Pyridine can be nitrated only by heating with fuming nitric acid and sulphuric acid to yield 3-nitropyridine.

This selectivity for 3-nitration can be understood by examining the resonance structures for the σ-complexes for 3-nitration (b), and for 4-nitration (a), (Eq. 20.11) which show how the former is more stable than the latter. Even though three canonical structures can be drawn for both, the structure marked (I) is relatively unstable because (i) the nitrogen has only a sextet of electrons and (ii) there is a positive charge on the electronegative nitrogen. Therefore, 3-nitration takes place at a faster rate.

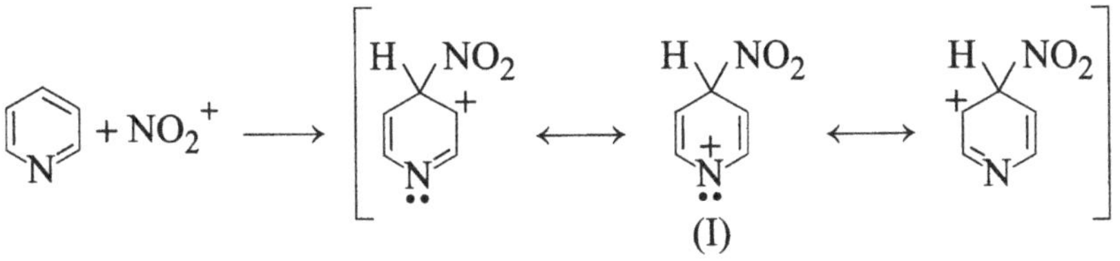

(a)

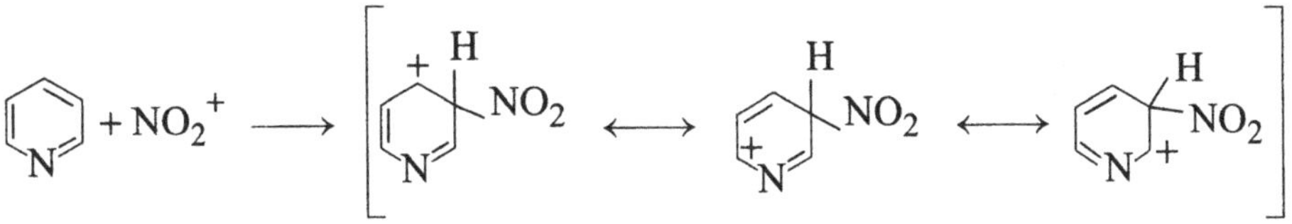

(b)

(20.11)

Sulphonation and halogenation also take place at high temperatures and result in 3-substitution. Friedel–Crafts acylation does not take place (cf. nitrobenzene).

Nucleophilic substitution reactions on the pyridine ring do take place. Notable among these is the reaction with sodium amide to obtain 2-aminopyridine (Eq. 20.12a). This is known as the *Chichibabin reaction.*

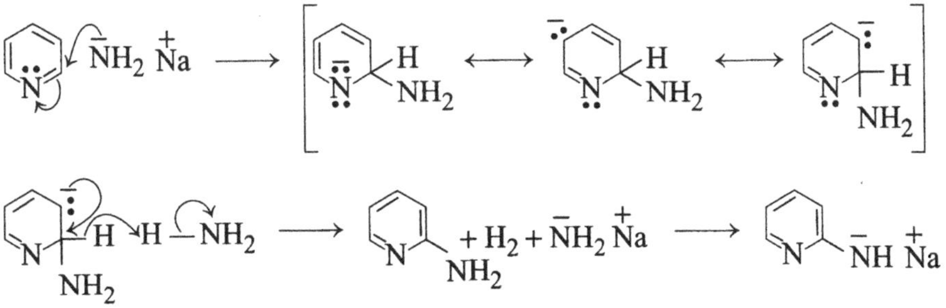

(a)

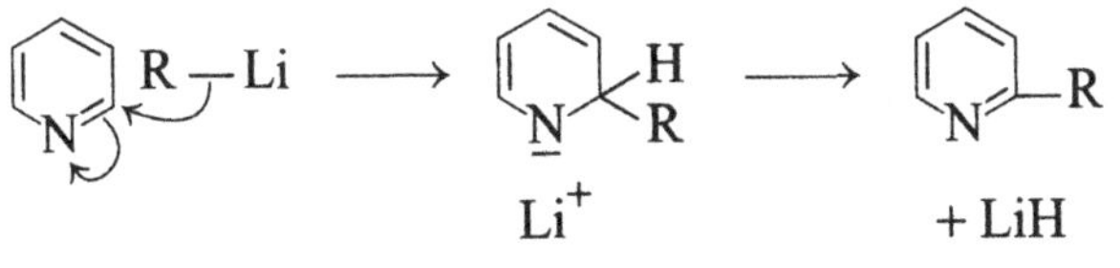

(b)

(20.12)

A similar reaction with alkyllithium or aryllithium gives 2-alkylpyridine or 2-arylpyridine. This reaction is known as *Ziegler alkylation* or arylation (Eq. 20.12b).

Pyridine can be hydrogenated under milder conditions than are required for the hydrogenation of benzene. Since reduction involves the addition of electrons, it is not surprising that the electron deficient pyridine ring gets readily reduced. The reaction takes place better under acidic conditions. Protonated pyridine – the pyridinium ion – is more electrophilic than pyridine itself. The product of reduction, hexahydropyridine, is *piperidine*. Pyridine can be reduced to piperidine also by reagents such as sodium and ethanol.

$$\text{pyridine} \xrightarrow{\text{H}_2, \text{Pt}, \text{H}^+, 25^\circ\text{C}} \text{piperidine} \xleftarrow{\text{Na}, \text{C}_2\text{H}_5\text{OH}} \text{pyridine} \tag{20.13}$$

Piperidine is a typical secondary amine and is a stronger base ($pK_b = 2.8$) than pyridine or even open chain secondary amines. Piperidine finds use in base catalysed reactions. It can be alkylated and acylated at the nitrogen.

$$\xrightarrow{\text{CH}_3\text{I}} \xrightarrow{\text{CH}_3\text{I}} \quad \text{(a)}$$

$$\xrightarrow{\text{CH}_3\text{COCl}} \quad \text{(b)} \tag{20.14}$$

The hydrogens of the methyl group of α-picoline and of γ-picoline are more acidic than those of toluene (see resonance structures, Eq. 20.15a). In the alkaline medium, they condense with aldehydes (Eq. 20.15b). β-Picoline does not react in this manner.

$$\xrightarrow{\text{OH}^-} \quad \text{(a)}$$

$$+ \text{C}_6\text{H}_5 \cdot \text{CHO} \xrightarrow{\text{OH}^-} \quad \text{(b)} \tag{20.15}$$

20.4 FUSED RING HETEROCYCLIC COMPOUNDS

20.4.1 INDOLE

Indole is present in coal tar. It has a pyrrole ring fused to a benzene ring. Indole-3-acetic acid is a plant growth hormone. Indole is the precursor of the dye, indigo.The best known method of synthesis of indole is the *Fischer indole synthesis*. The phenylhydrazone of a ketone or aldehyde is treated with strong acids or Lewis acids like zinc chloride. The reaction of the phenylhydrazone of pyruvic acid is shown below (Eq. 29.16).

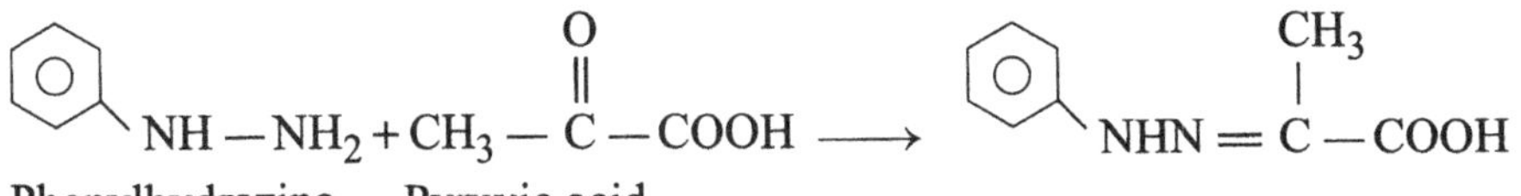

Phenylhydrazine   Pyruvic acid

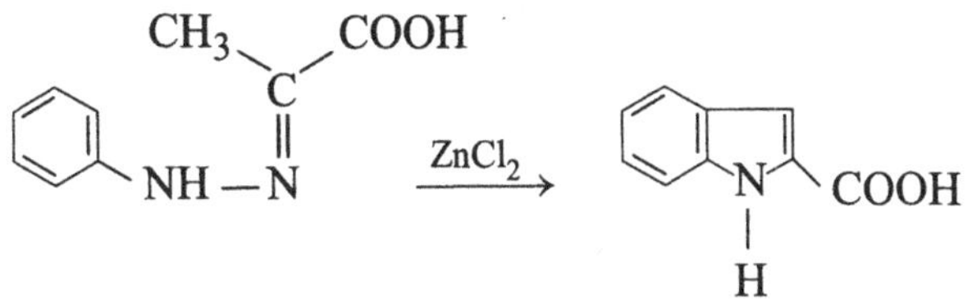

Indole-2-carboxylic acid   (20.16)

The end product is indole-2-carboxylic acid. This may be deacrboxylated by heat to obtain indole. The reaction involves a sequence of steps, in the course of which one of the nitrogens is eliminated as ammonia. Substituents on the benzene ring of phenylhydrazine end up on the benzene ring of indole; those on the carbonyl compound end up on the pyrrole ring. Details of the mechanism are outlined for the reaction of the phenylhydrazone of 3-pentanone (Eq. 20.17).

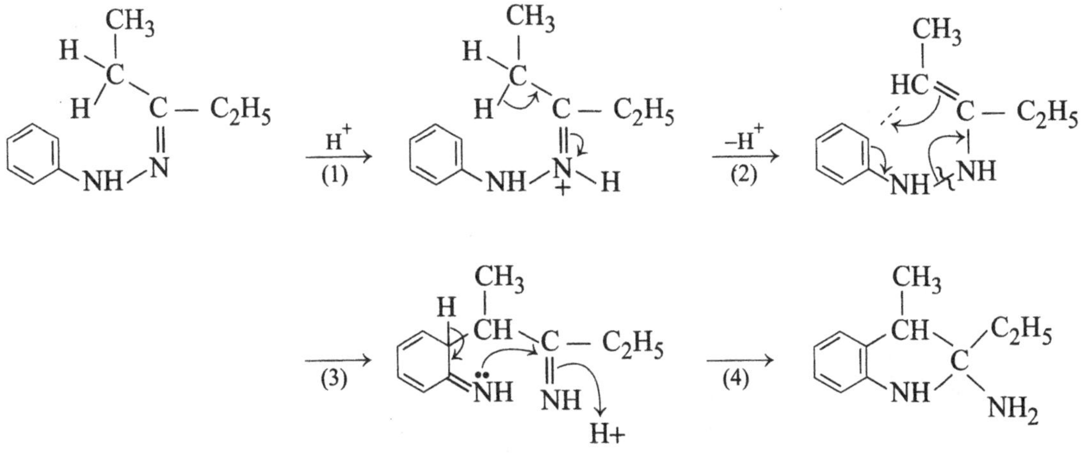

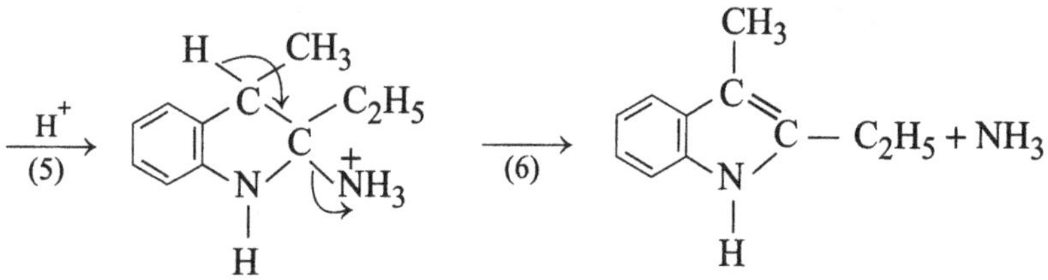

(20.17)

Steps (1) and (2) represent the acid catalysed shift of the double bond. The product of step (2) is a 1,5-diene which undergoes a Cope rearrangement in step (3) (Chapter 19, Section 19.4.1 for a discussion of Cope rearrangement). This is the crucial step in the reaction. The product of this rearrangement undergoes a cyclisation, step (4), followed by the loss of a molecule of ammonia, steps (5) and (6), to give the final product. The driving force for the last step is the attainment of aromaticity.

The pyrrole ring of indole is richer in electrons than the benzene ring. Indole undergoes reactions readily at the 3-position in contrast to pyrrole, which reacts at the 2-position.

20.4.2 Quinoline

Quinoline has a pyridine ring fused to a benzene ring. Quinoline along with isoquinoline is present in coal tar. It is basic but slightly weaker than pyridine. Electrophilic substitutions take place on the benzene ring. The most important method of synthesis of quinoline is the *Skraup synthesis*. When applied to unsubstituted quinoline, the synthesis involves heating aniline, glycerol and nitrobenzene together in the presence of sulphuric acid and ferrous sulphate (Eq. 20.18).

$$\underset{\text{OH}}{\text{CH}_2}-\underset{\text{OH}}{\text{CH}}-\underset{\text{OH}}{\text{CH}_2} \xrightarrow{H^+} \text{H}-\underset{\text{Acrolein}}{\overset{\text{O}}{\overset{\|}{\text{C}}}-\text{CH}=\text{CH}_2} + 2\text{H}_2\text{O} \quad \text{(a)}$$

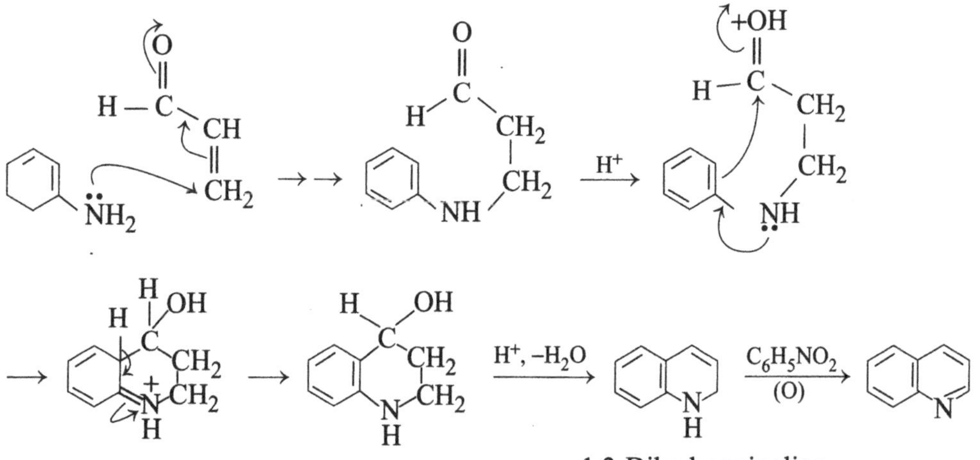

1,2-Dihydroquinoline

(b)

(20.18)

The function of glycerol is to provide propenal (acrolein) in situ (Eq. 20.18a).This method is more effective than using free acrolein. Aniline adds to acrolein in a reaction similar to Michael addition to produce 3-(phenylamino)propanal which undergoes acid catalysed cyclisation to give dihydroquinoline. This gets oxidised by nitrobenzene to quinoline. It is an *oxidative dehydrogenation*. The driving force is the conversion of the partially aromatic dihydroquinoline to the fully aromatic quinoline. Nitrobenzene is an unconventional oxidising agent, but is compatible with the reactants used in this synthesis. Ferrous sulphate exercises a moderating influence on the reaction. Quinoline with substituents on the pyridine ring can be prepared by using suitable α, β-unsaturated carbonyl compounds in place of acrolein. *o*- and *p*-toluidines will give 8-methylquinoline and 6-methylquinoline, respectively. *m*-Toluidine will give a mixture of 5-methyl and 7-methylquinoline. Other substituents such as methoxy, halogen and nitro can be present (Eq. 20.19).

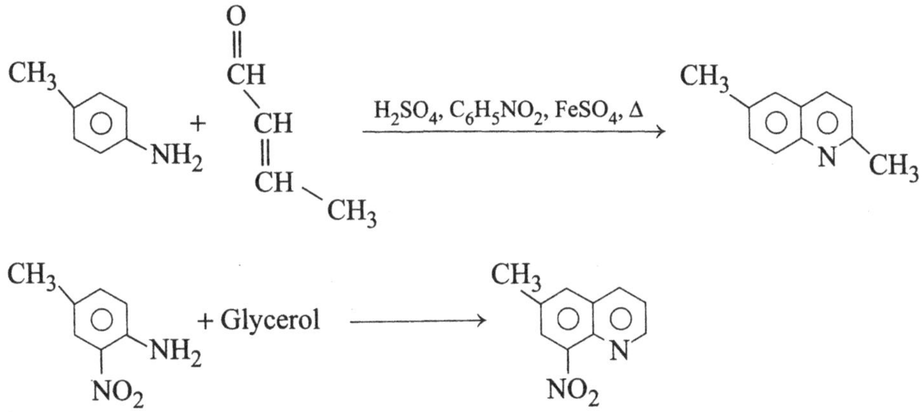

(20.19)

Electrophilic substitutions like nitration and sulponation take place at the 5- and 8-positions of quinoline. Chichibabin reaction can be carried out with quinoline to prepare 2-aminoquinoline. The pyridine ring is more resistant to oxidation than the benzene ring. Upon oxidation with permanganate, quinoline gives pyridine-2,3-dicarboxylic acid, also known as quinolinic acid.

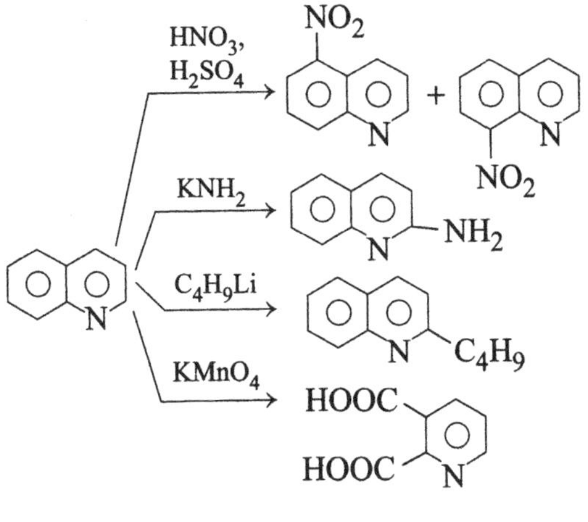

(20.20)

20.4.3 ISOQUINOLINE

The isoquinoline structure is present in many alkaloids. Many methods have been developed for the synthesis of isoquinolines. One of the more important methods is the Bischler–Napieralski synthesis. This has been adapted for the total synthesis of some alkaloids. The basic reaction is illustrated in Eq. 20.21.

$$(20.21)$$

The starting material is the N-acyl derivative of 2-phenylethylamine. This gives dihydroisoquinoline upon acid-catalysed cyclisation. The dihydro compound is aromatised by heating with a dehydrogenating agent.

Electrophilic substitution on isoquinoline takes place at position 5. Chichibabin reaction gives 1-aminoisoquinoline. Reaction with phenyllithium gives 1-phenylisoquinoline. Oxidation using permanganate gives pyridine-3,4-dicarboxylic acid (cinchomeronic acid) along with some phthalic acid.

$$(20.22)$$

Cinchomeronic acid

KEY POINTS

- Furan, pyrrole and thiophene are aromatic.

- The aromatic sextet is completed by the lone pair of electrons on the hetero atom. Protonation will destroy the aromaticity. Hence, pyrrole is not basic.

- All the three ring systems can be synthesised from 1,4-dicarbonyl compounds.

- The lone pair of electrons on the heteroatom is delocalised into the ring, making the ring electron rich. Electrophilic reactions take place readily under non-acidic conditions. In acidic media, they are unstable and get polymerised. Substitution takes place at position 2.

- Pyrrole is similar to phenol in reactivity. It forms a salt with alkali and undergoes reactions such as diazo coupling.

- Fully hydrogenated pyrrole is called pyrrolidine which behaves as a typical secondary amine.

- Hydrogenation of furan gives tetrahydrofuran which is an important solvent.

- Furfural, furan-2-carboxaldehyde, is similar to benzaldehyde in reactions. It undergoes benzoin condensation and Cannizzaro reaction.

- Indole has a pyrrole ring fused to a benzene ring. The most important method of synthesis of indole is the Fischer indole synthesis. In this, the phenylhydrazone of an aldehyde or ketone is treated with a strong acid. The key step in this reaction is a Cope rearrangement.

- Indole is reactive like pyrrole, but the preferred position of electrophilic substitution is the 3-position.

- Pyridine has an electron deficient aromatic ring. The lone pair of electrons on the nitrogen in pyridine is located in an sp^2 hybrid orbital and is not part of the aromatic sextet, unlike in pyrrole.

- Electrophilic substitution on pyridine requires drastic conditions. Substitution takes place at position 3.

- Nucleophilic replacement of hydrogen at the 2-position is possible. Reaction with sodium amide gives 2-aminopyridine (Chichibabin reaction). Reaction with alkyl or aryllithium gives 2-alkyl or 2-arylpyridine (Ziegler alkylation or arylation).

- The methyl groups in 2-picoline and 4-picoline are acidic and can be condensed with aldehydes.

- Hydrogenation of pyridine gives piperidine which is a stronger base than pyridine.

- Quinoline and isoquinoline have a pyridine ring fused to a benzene ring, with the nitrogen located at the 1-position and the 2-position, respectively.

- The best known methods of synthesis are the Skraup synthesis for quinoline and the Bischler–Napieralski synthesis for isoquinoline.

- In these molecules, electrophilic substitution takes place preferentially on the benzene ring. Nucleophilic substitutions, as with pyridine, take place on the pyridine ring.

EXERCISES

SECTION I

1. Which of the following are aromatic?

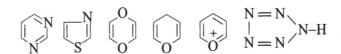

2. Give the products of the following reactions.

    (a) Furfural + diethyl malonate (in the presence of pyridine)
    (b) Furfural + KOH
    (c) Furfural + KCN
    (d) 4-Methylquinoline + benzaldehyde (in the presence of alkali)
    (e) Isoquinoline + butyllithium
    (f) Pyrrolidine + benzoyl chloride
    (g) Indole + benzenediazonium chloride

3. How can the following compounds by synthesised?

    (a) 2-Phenylquinoline   (b) 1-Benzylisoquinoline
    (c) 8-Chloroquinoline   (d) 2-Phenylindole
    (e) 5-Methylindole-2-carboxylic acid

SECTION II

1. How can β- and γ-picolines be converted to β- and γ-aminopyridines respectively?
2. Rationalise the observation that imidazole (see Table 20.1 for structure) is a fairly strong base unlike pyrrole.
3. Account for the observation that the reaction of isoquinoline with sodium amide gives 1-aminoisoquinoline and not 3-aminoisoquinoline.
4. α-Picoline condenses with furfural in the presence of alkali. Give the structure of the product and the mechanism of the reaction.
5. Give all the steps in detail, of the mechanisms outlined in Eqs. 20.10, 20.18, and 20.21.
6. Draw potential energy diagrams and show how nitration of pyridine is favoured at the 3-position and not at the 4-position.
7. Give the mechanism of the conversion of 2,5-dimethylfuran to an open chain diketone by hydrolysis in aqueous acid.
8. With the help of resonance structures of the sigma complexes, show why nitration of quinoline takes place at the 5-position in preference to the 6-position.
9. Explain why piperidine is a stronger base than pyridine.

CHALLENGING QUESTIONS

1. Treatment of furan with bromine in methanol gives 2,5-dimethoxy-2,5-dihydrofuran (I). The reaction probably involves the following steps: (i) electrophilic attack by Br^+ (ii) quenching of the carbocation by the solvent (methanol) at the allylic position and (iii) solvolysis of the allyl bromide. Write down all the steps.

$$(I) \quad CH_3O \diagup \!\!\! \diagdown O \diagdown OCH_3$$

2. Structure (I), in question 1 is an acetal of which aldehyde? Give the mechanism of acid catalysed hydrolysis of (I).

3. (a) Show how furfural may be converted to tetrahydrofurfuryl alcohol (II).

   (b) When (II) is dehydrated under acidic conditions, a ring expansion to dihydropyran (III), takes place. Give the mechanism of this rearrangement. It is similar to pinacolic rearrangement.

$$\underset{(I)}{\diagup \!\!\! \diagdown O \diagdown CHO} \longrightarrow \underset{(II)}{\diagup \!\!\! \diagdown O \diagdown CH_2OH} \xrightarrow{H^+} \underset{(III)}{\diagup \!\!\! \diagdown O}$$

21 Stereochemistry – I

OBJECTIVES In this chapter, you will learn about,

- types of stereoisomerism, optical and geometrical isomerism and their definitions
- optical activity, specific rotation and molecular rotation
- conditions for optical activity
- representation of optical isomers, D,L-notation, Fischer projection
- R,S-notation, sequence rules
- optical isomerism in molecules containing more than one asymmetric carbon
- flying-wedge, saw-horse and Newman projections
- threo, erythro, meso nomenclature
- methods of resolution
- asymmetric synthesis, Cram's rule, definition of enantiomeric excess
- stereoselectivity and stereospecificity
- racemisation, Walden inversion
- atropisomerism: optical isomerism of biphenyls, allenes and spirans

21.1 INTRODUCTION

Stereochemistry is chemistry in relation to the three-dimensional structure and reactivity of molecules. *Steroisomerism* may be defined as isomerism in molecules that have the same constitution, but differ in the arrangement of atoms in space. Van't Hoff and Le Bel proposed, independently, in 1874 that the four bonds of carbon are directed towards the four corners of a tetrahedron. This fact lead to the understanding of the three dimensional nature of molecules and the study of stereochemistry.

Stereoisomerism is classified into two types, (i) optical isomerism and (ii) geometrical isomerism. In this chapter, we shall deal mainly with optical isomerism. Geometrical isomerism, which arises as a result of restricted rotation around the double bond and in cyclic molecules, is also called *cis–trans* isomerism (discussed in Chapter 22).

Many objects in nature, while possessing the same structure, cannot occupy the same three dimensional space. A typical example is left and right feet. The left foot will not fit into the shoe meant for the right foot and vice versa. Left and right feet are mirror images of each other, so are the left and right shoes and the left and right hands. Two objects which are structurally identical but which are mirror images of each other and cannot occupy the same three dimensional space are called *chiral objects* and are said to possess the property of *chirality*. Chirality is 'handedness' and refers to the left hand–right hand relationship. Molecules also can have this property. Many of the organic molecules of plant and animal origin are chiral.

The molecule dibromochloromethane can be projected in two dimensions in different ways (structures (I) to (IV), Fig. 21.1). In structure (I), the groups are shown at the four corners of a tetrahedron. One edge of the tetrahedron, shown as a horizontal thick line, is in front of the plane of the paper. Another edge, vertical dotted line, is behind the plane of the paper. The same structure is shown in more conventional ways in (II) and (III). In (III), the broken lines are pointing away from us, behind the plane of the paper, and the wedge-shaped lines are pointing towards us and are in front of the paper surface. This has been further simplified in structure (IV) where the broken lines and wedge-shaped lines have been omitted and replaced by simple straight lines. The central carbon atom is also not shown. This simplification is meaningful only because of the introduction of some conventions. These are:

- the structure will be drawn as a cross as shown,

- the horizontal lines represent bonds in front of the plane of the paper, and

- the vertical lines represent bonds behind the plane of the paper.

This method of drawing the structure while following the conventions leads to *Fischer projections*.

Jacobus van't Hoff (1852–1911), a Dutch chemist, had made significant contributions not only in stereochemistry, but also in other areas of chemistry. The first Nobel Prize in chemistry was awarded to van't Hoff in 1901 for his contributions in chemical dynamics and osmotic pressure.

Emil Fischer (1852–1919), a German chemist, had made outstanding contributions in many areas of organic chemistry. He designed this representation of configurations in connection with his studies in sugar chemistry. His studies using phenylhydrazine lead to the indole synthesis named after him. It is believed that long-term exposure to the toxic chemical, phenylhydrazine, affected his health and contributed to his death. He was awarded the Nobel Prize in chemistry in 1902.

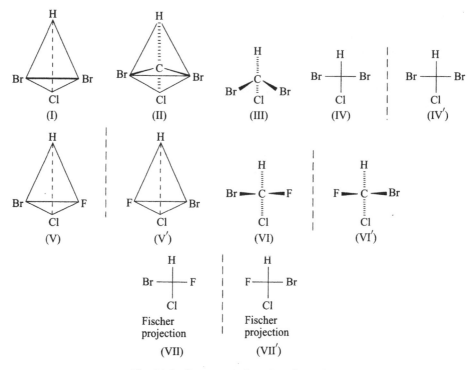

Fig. 21.1 Representation of configurations

The Fisher projection structure should not be lifted from the plane of the paper. It can be rotated in the plane of the paper by $180°$, or either the lower three groups or the upper three groups can be rotated clockwise or counter-clockwise keeping the fourth group stationary.

(IV) and (IV′) are mirror images. In these drawings, the broken line between the structures represents a mirror plane. The structures on either side of the line are mirror images of each other. Examination of these structures (IV) and (IV′), shows that they can be placed one on top of the other in such a way that the same bonds and atoms occupy the same space. They are *superimposable*. The plane passing through the H, central C and Cl is a plane of symmetry in the molecule. There is no chirality and therefore the molecule is said to be *achiral*.

Now, let us examine structures (V), (V′), (VI), (VI′), (VII) and (VII′) which represent bromochlorofluoromethane. Structures (V) to (VII) represent the same molecule in the same way as structures (I)–(IV) represent the same molecule. Similarly, structures (V′)–(VII′) are the same. Structures (VII) and (VII′) are Fischer projections. These pairs of structures are not identical. They are not superimposable and are mirror images of each other. The plane passing through H, C and Cl is not a symmetry plane for the molecule. In fact, the molecule possesses no plane of symmetry. These structures have the property of 'handedness' or chirality. If one is left handed, the other is right handed since they are mirror images of each other. (VII) and (VII′) are *optical isomers*, specifically they are called *enantiomers*.

They are called optical isomers because they can be differentiated by an optical measurement using the instrument called polarimeter. The plane of a monochromatic, plane polarised beam of light gets rotated when passed through a medium containing either one of the enantiomers. The rotation can be to the left or to the right, depending on which enantiomer is used. The enantiomers are said to be optically active – measured in degrees with a positive or negative sign – for rotation to the right and to the left, respectively. The magnitude of rotation will also depend upon factors like wave length of the light, concentration of the solution, path length that the beam travels, temperature and solvent used. All these have to be specified while reporting optical activity.

Optical activity is reported as specific rotation $[\alpha]$, or molecular rotation $[M]$ or $[\Phi]$. These are defined by Eq. 21.1.

$$\text{Specific rotation,} \quad [\alpha] = \frac{\alpha}{l(dm) \cdot c(g/100\,ml)} \qquad \text{(a)}$$

$$\text{Molecular rotation,} \quad [M] = [\phi] = \frac{[\alpha] \cdot M}{100} = \frac{\alpha}{l(dm) \cdot C(moles/100\,ml)} \qquad \text{(b)}$$

$$[\alpha]_D^{25}(\text{ethanol}) = -17.5° \qquad \text{(c)} \quad (21.1)$$

where, α = measured rotation in degrees with the sign $-$ or $+$, l = length of the polarimeter cell in dm, c = concentration in g per 100 ml, C = moles per 100 ml and M = molecular weight. In Eq. 21.1c, the specific rotation of a particular compound is reported. From the equation, a lot of information can be gathered—the solvent is ethanol, D specifies that the sodium-D line with a wave length of 589 nm was used (if another wave length is used, it should be specified), 25 is the temperature, at which the measurement was made in °C and the rotation is anticlockwise (denoted by the $-$ve sign).

21.1.1 CONDITIONS FOR CHIRALITY, CONCEPT OF CHIRALITY CENTRE

A molecule which is asymmetric is usually chiral. The chiral molecules are those which lack a *plane of symmetry* or a *centre of symmetry*. The most common situation for this to exist is when one carbon contains four different groups. The central carbon of (VII) in Fig. 21.1 is asymmetric since it carries a H, a Cl, a Br and a F. It is called the *chirality centre* or the *asymmetric centre*. Chirality can exist even without the presence of an asymmetric carbon atom. This will be discussed in a later section. Some molecules which contain more than one chiral centre could be optically inactive, because the chirality is internally compensated. Such situations will also be discussed in another section.

21.1.2 NOMENCLATURE

A familiar example of an asymmetric molecule is 2-butanol (Fig. 21.2). I and II are enantiomers. Commercially obtained 2-butanol – the chemical that is available from the laboratory shelf – is a 1:1 mixture of the two enatiomers. This is because conventional methods of manufacture such as hydration of 2-butene or reduction of 2-butanone, result in the formation of equal amounts of the two enantiomers. These reactions are not *enantioselective*. In other words, they are not *stereoselective*. Such a sample of 2-butanol which is a mixture of the two enantiomers in equal

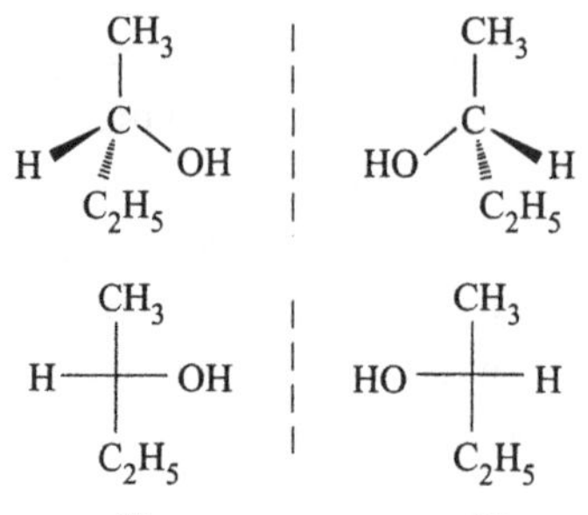

Fig. 21.2 Configurations of 2-butanol

amounts, is called a *racemic mixture*. Each individual molecule in the sample is chiral. But when the bulk sample is placed in the polarimeter, it will register zero optical activity. The chirality of the molecules has been externally compensated. The pure enantiomers can be separated from the racemic mixture by several methods. The separation of enantiomers from a racemic mixture is called *resolution*.

Enantiomers have identical physical properties like melting and boiling points, density and refractive index. Their chemical properties are also the same. The only way that they can be differentiated is by a polarimetric measurement of optical activity. They show identical rotation either to the right or to the left, designated as positive rotation (+) or negative rotation (−). After such a measurement, the two samples can be labelled (+)-2-butanol and (−)-2-butanol. However, we do not have the answer to the question, 'which sample, (+) or (−), has structure (I) and which has structure (II)?' Structures (I) and (II) are two isomers which represent the actual arrangement of the atoms in space, and is referred to as their *configuration*. The magnitude or sign of the measured optical rotation does not give information about the configuration. Absolute configurations of some molecules have been determined by specialised X-ray diffraction studies and other techniques. Configurations of most other molecules have been assigned with the help of chemical reactions which relate them to one or another of the molecules whose absolute configuration is known. This method is called, assigning relative configuration. Today, the configurations of almost all chiral molecules are known.

21.1.2.1 Determining Relative Configurations

A simple illustration of the steps involved in assigning the relative configuration is given in Fig. 21.3.

(−)-Mandelic acid (I) is converted to ethyl 1-phenylethyl ether (III) in several steps, none of which affects the chiral centre. (III) could also be prepared from (−)-1-phenylethanol, (IV) by reaction with ethyl iodide which also does not affect the chiral centre. Samples of ethyl 1-phenylethyl ether (III) obtained by both routes were found to be identical as characterised by optical acivity measurements, showing that the two samples have the same configuration at the asymmetric centre. Hence (−)-mandelic acid and (−)-1-phenylethanol have the same configuration. If, in such conversions, a bond between the asymmetric carbon and another atom is broken, uncertainty will arise regarding whether the configuration has remained the same

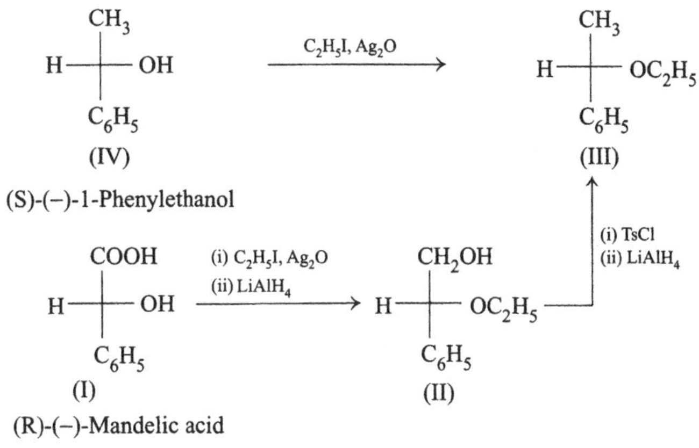

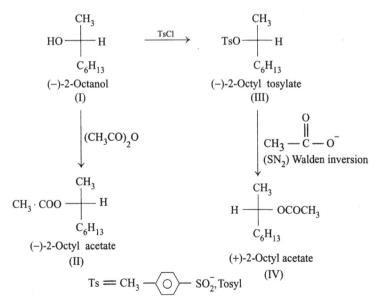

Fig. 21.3 Relative configurations

(retention), has become the opposite configuration (inversion), or has racemised. Fortunately, from the understanding of the mechanisms of most reactions, the stereochemical outcome of such changes can generally be predicted. Thus, if the reaction involves S_N2 displacement, inversion of configuration can be expected. In fact, such studies form the experimental basis for the conclusion that S_N2 reactions involve inversion of configuration (called Walden inversion, Chapter 11, Section 11.2.1.3).

21.1.2.2 Walden Inversion

An illustration of a study involving Walden inversion is given in Fig. 21.4. In this example, $(-)$-2-octanol (I) is converted by two routes to 2-octyl acetate. In route (I) → (II) which does not

Fig. 21.4 Walden inversion

involve any bond cleavage at the chiral centre, the product is (−)-2-octyl acetate (II) which must have the same configuration as the starting alcohol. In the other route, (I) → (III) → (IV), the product is (+)-2-octyl acetate. Since the sign of rotation has changed, the configuration must have undergone inversion. The step where it could have happened is, (III) → (IV), an S_N2 reaction. From this and similar studies, it has been generalised that S_N2 reactions involve inversion of configuration.

The earliest convention for differentiating enantiomers is by attaching the prefixes (+) or (−) as illustrated in these examples. Positive rotation is also called *dextrorotation* and negative as *levorotation*. These words stand for right and left and are sometimes abbreviated as *d* and *l*. More commonly, only the signs (+) or (−) are used. The racemic mixture is designated by the prefix (±) or *dl-*. Once the absolute configurations are known, it becomes necessary to have systematic nomenclature to designate the configurations unambiguously.

21.1.3   D,L-NOTATION

Much of the early efforts were oriented towards naming sugars and amino acids which are naturally occurring chiral molecules. All sugars can be considered to be related to glyceraldehyde which is a chiral molecule. Fischer projections of the two enantiomers of glyceraldehyde are shown in Fig. 21.5a. We have already learnt the conventions for drawing Fischer projections.

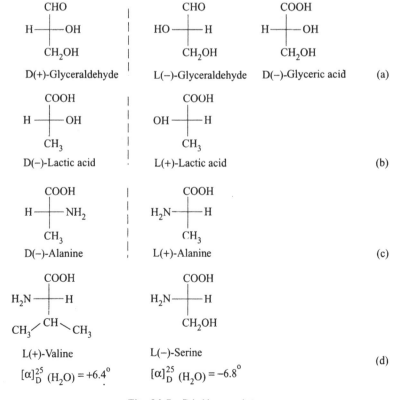

Fig. 21.5   D,L-Nomenclature

For applying the D,L-nomenclature, an additional condition is imposed. The carbon skeleton is drawn vertically with the most oxidised group (the aldehyde group in this case) at the top. Drawn this way, the configuration with the OH group to the right is designated D and the other one, L. It has been established that the dextrorotatory enantiomer of glyceraldehyde has the D configuration. However, it should not be assumed that dextro or (+) which refers to the sign of optical rotation, and D which denotes absolute configuration, are the same. D(+)-glyceraldehyde upon oxidation yields D(−)-glyceric acid (Fig. 21.5a). They have the same configuration but the sign of optical rotation is reversed. In the amino acid series, the convention is illustrated in the nomenclature of alanine, valine and serine (Fig. 21.5c, 21.5d). For using the D/L notation, the Fischer projection formula is drawn with the COOH group at the top and the carbon skeleton vertical. The configuration of all the natural amino acids – isolated from proteins – is such that they have the amino group to the left and are called L-amino acids. They can be dextrorotatory or levorotatory, as the examples illustrate. Further extension of this notation is possible for structures which can be directly correlated to glyceraldehydes like lactic acid (Fig. 21.5b) or to the amino acids.

21.1.4 R,S-NOTATION: CAHN–INGOLD–PRELOG NOMENCLATURE

D,L-nomenclature will fail when applied to more complex molecules. Today, configurations are designated using the R,S-notation based on the Cahn–Ingold–Prelog rules or sequence rules. According to this, the four groups on the chiral centre are arranged in sequence decided by the sequence rules. Let us consider the four groups in 2-butanol (Fig. 21.6). They are H, CH_3, C_2H_5 and OH.

(i) The groups are listed in the order of decreasing atomic number of the atom directly attached to the chiral centre.

$$OH-, C_2H_5-, CH_3-, H$$

(ii) When two or more of the atoms are the same like the C in CH_3 and C_2H_5, the next atom(s) attached to that are considered. C(H,H,H) and C(H,H,C). The second one (C_2H_5) outranks the first (CH_3) since the former has one C and two Hs while the latter has three Hs. For 2-butanol, sequence of decreasing rank is therefore as in (i) above.

(iii) If an atom is bound by a multiple bond, it is considered to be replicated. CHO is considered to be equivalent to C(H,O,O) and it outranks CH_2OH. For glyceraldehydes, the sequence of rank (highest rank listed first) is OH, CHO, CH_2OH, H.

A list of commonly encountered groups arranged according to decreasing rank is given below.
I−, Br−, Cl−, F−, HO−, NH_2−, HOC(=O)−, O=C(CH_3)−, O=CH−, $HOCH_2$−, C_6H_5−, $(CH_3)_3C$−, CH_2=CH−, $(CH_3)_2CH$−, $(CH_3)_2CHCH_2$−, CH_3CH_2−, CH_3−, H−

After listing the groups in the correct sequence, the molecule is projected on paper in such a way that the group of lowest rank is behind the plane of the paper, and the other three groups are projecting in front of the paper, and you are viewing the molecule from the front. You see the asymmetric carbon and the groups of ranks 1,2 and 3. The group of rank 4 is invisible, hidden behind the carbon. One way of doing this is illustrated in Fig. 21.6.

Structures (I) and (II) represent 2-butanol in the tetrahedral format and the corresponding Fischer projection. For applying the sequence rules, H should be pointing away from us. This has been done in structures (I) → (III) → (V) in two operations. (i) The tetrahedron is held at the apex (CH_3), and the bottom triangle rotated so that the H is behind as in (III). (ii) Holding on to the apex, the tetrahedron is tilted forward so that the H goes all the way to the rear, as in (V). In (V) we are looking at the molecule from one trianglular face of the tetrahedron with the

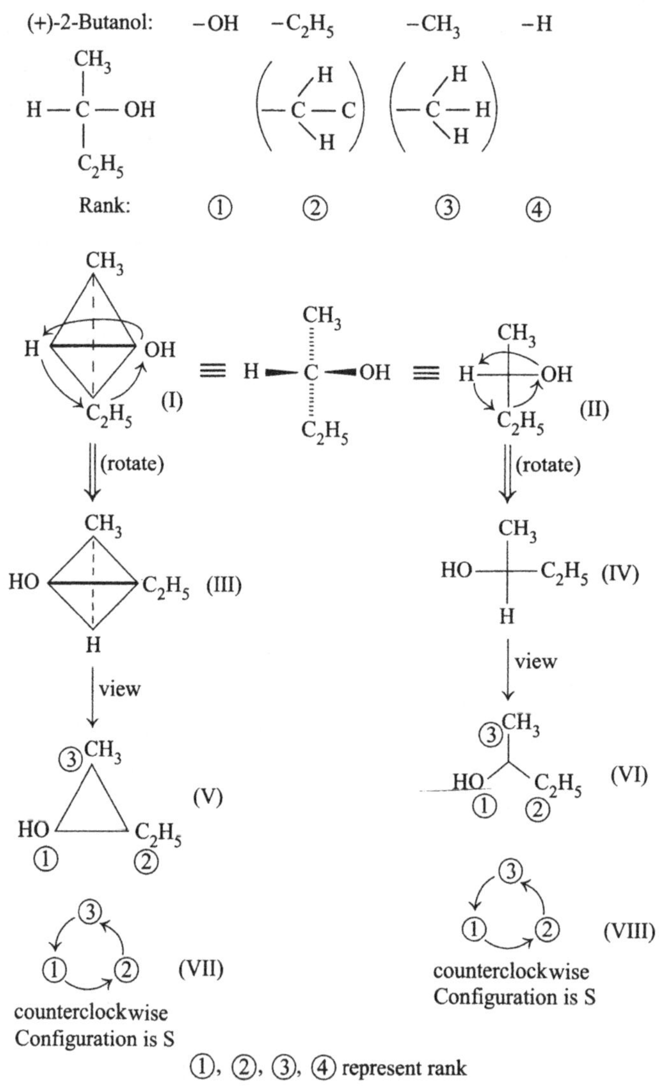

Fig. 21.6 (Continued)

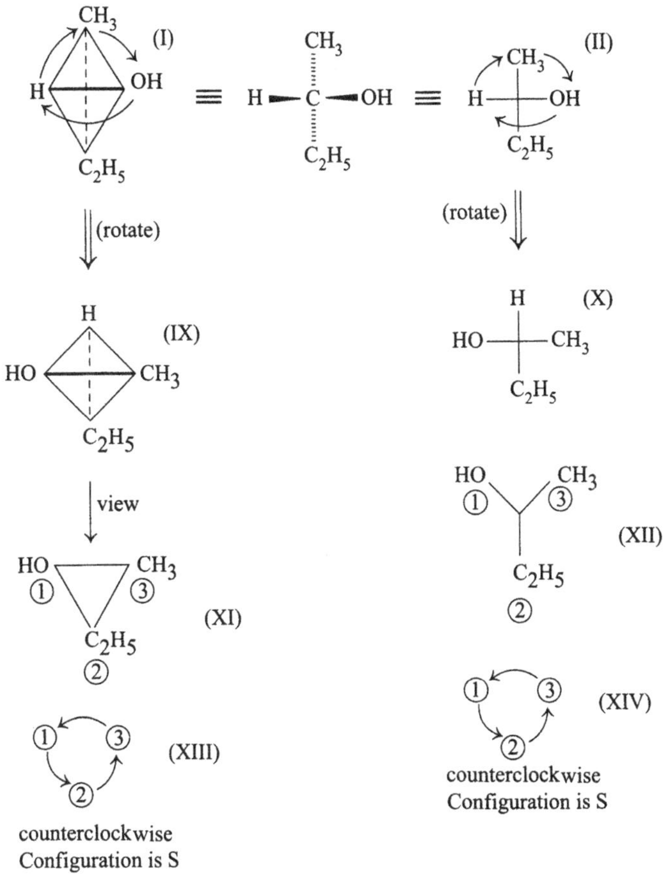

Fig. 21.6 Configuration of (+)-2-butanol

three top ranking groups at the three corners. The fourth – the lowest ranking group – is behind. Groups 1, 2 and 3 are now counterclockwise as in (VII). The molecule has S-configuration. The same operations have been done in (II) → (IV) → (VI) with the Fischer projection formula to arrive at (VIII), S configuration. The molecule is (S)-2-butanol. This is the enantiomer with an optical rotation, $[\alpha]_D^{25} = +13.5°$ and is named (S)-(+)-2-butanol. Its enantiomer is (R)-(−)-2-butanol and its optical rotation $[\alpha]_D^{25} = -13.5°$.

In the operation, (I) → (III) or (II) → (IV), we have made the least-priority group H, point downwards. Alternately, we can make the H point upwards, by the operations (I) → IX or (II) → (X) leading to the trigonal projections, (XI) and (XII). These also give S configuration. Either approach can be followed. The aim is to draw a Fischer projection in which the least-priority group is pointing either down or up. Once that is done, the group with the least priority can be ignored and the remaining groups, in the order of rank 1, 2 and 3 can be identified to be clockwise (R configuration) or counterclockwise (S configuration). R and S stand for *rectus* and *sinister* meaning, right and left.

This exercise of assigning the R,S-prefix has been done for several molecules in Fig. 21.7.

In Fig. 21.3, it was established that (−)-mandelic acid and (−)-1-phenylethanol have the same configuration. By the application of sequence rules, it can be seen that the former is (R)-(−)-mandelic acid and the latter is (S)-(−)-1-phenylethanol. Neither the sign of rotation, nor

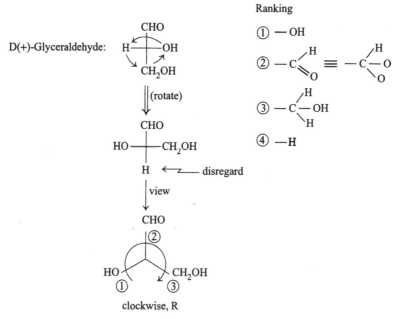

Fig. 21.7   (Continued)

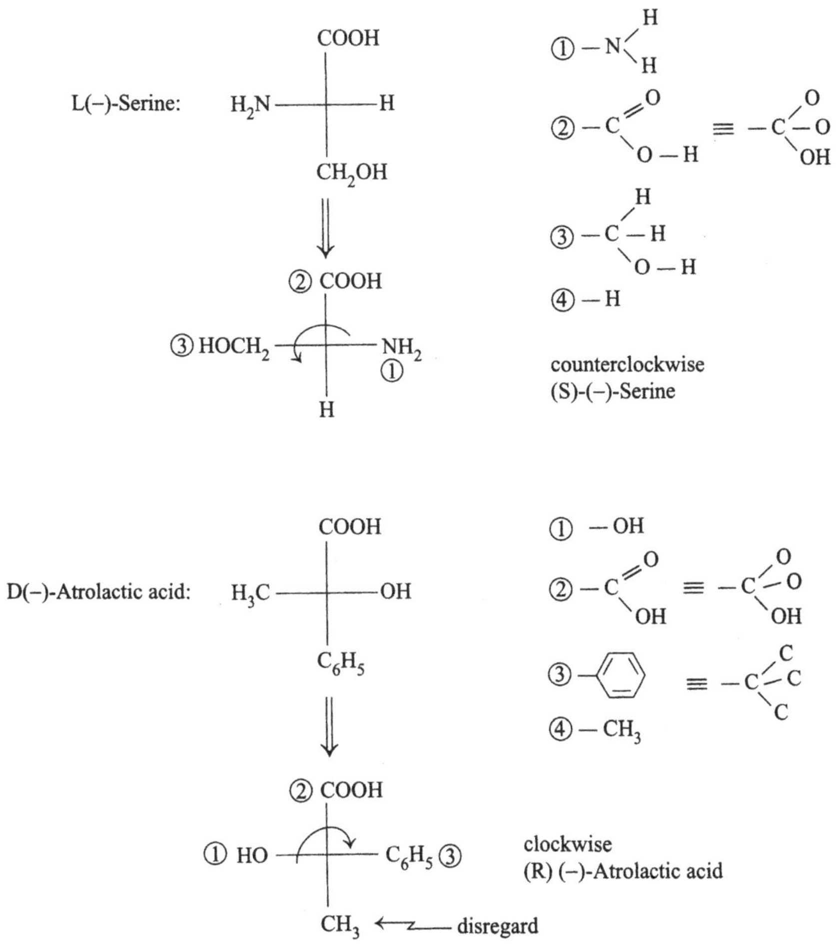

L(−)-Serine:

COOH

H_2N——H

CH_2OH

② COOH

③ $HOCH_2$ —— NH_2 ①

H

① $-N\langle^H_H$

② $-C{\Large\langle}^O_{O-H} \equiv -C{\Large\langle}^O_{OH}$

③ $-C{\Large\langle}^H_{O-H}$, H

④ $-H$

counterclockwise
(S)-(−)-Serine

D(−)-Atrolactic acid:

COOH

H_3C——OH

C_6H_5

② COOH

① HO —— C_6H_5 ③

CH_3 ←——— disregard

① $-OH$

② $-C{\Large\langle}^O_{OH} \equiv -C{\Large\langle}^O_{OH}$

③ ⬡ $\equiv -C{\Large\langle}^C_C C$

④ $-CH_3$

clockwise
(R) (−)-Atrolactic acid

Fig. 21.7 Assigning configurations for selected molecules

the R,S designation can be used to decide whether two chiral molecules have the same configuration or not.

21.2 MOLECULES WITH MORE THAN ONE ASYMMETRIC CENTRE

21.2.1 THE TETROSES

If there are n asymmetric carbons, the number of possible stereoisomers is 2^n. An example of a molecule containing two asymmetric carbons is 2,3,4-trihydroxybutanal. This is classified as a 4-carbon sugar, a tetrose. There are four isomeric tetroses. They are two pairs of enantiomers shown using the Fischer projection and D,L-notation (Fig. 21.8). D(−)-threose and L(+)-threose are enantiomers; D(−)-erythrose and L(+)-erythrose are enantiomers. All four are stereoisomers, but (I) and (III), and (I) and (IV) are not enantiomers. Similarly (II) is not the enantiomer of (III)

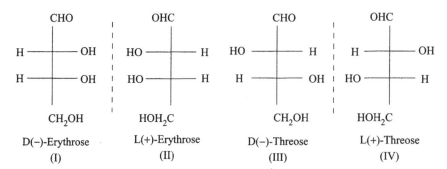

Fig. 21.8 Configurations of the tetroses

or (IV). They are called *diastereomers* of each other. Enantiomers have opposite configurations at each chiral centre. Diastereomers have the same configuration at one chiral centre and opposite configurations at the other chiral centre. While enantiomers have identical properties, diasteromers may differ in properties.

In drawing the Fischer projection formulae of these molecules containing two asymmetric carbons, we have followed the same conventions as applied to glyceraldehde except that the vertical bond connecting the two chiral carbons C2 and C3, is assumed to be in the plane of the paper. The two other vertical bonds (C2 $-$CHO and C3$-$CH$_2$OH) are behind the plane of the paper. The horizontal bonds are in front of the plane of the paper.

21.2.2 FLYING-WEDGE, SAW-HORSE AND NEWMAN PROJECTION FORMULAE

When used for representing configurations, Fischer projections have certain limitations, the most important of which is that it can represent only one particular conformation or rotational form, namely the eclipsed conformation. There are other ways of projecting a molecule like a tetrose (Fig. 21.9).

D($-$)-erythrose, structure (I), is the tetrahedral representation which leads to the Fischer projection (III) via (II). In these, the carbon skeleton of the molecule is held vertically, with both OH groups to the right and in front, and both Hs to the left and in front. The CHO and CH$_2$OH are arranged up and down respectively, pointing to the rear. These are the rigid requirements of the Fischer projection convention. (IV) is a projection where the same molecule is held nearly horizontally. The wedges and the broken lines have the same significance as defined earlier. Structure (V) is obtained by rotating C2 through 180° with reference to C3, and represents a different rotational form or conformation. (IV) and (V) represent the same configuration but different *conformations*. Structures (IV) and (V) are sometimes called *flying-wedge* representations. At this stage, let us recall the meanings of the terms, conformation and configuration. Conformations are the rotational forms of a molecule and are also called rotamers or rotational isomers though they are not isomers in the usual sense. Conformations are mutually interconvertible by 'free rotation' around single bonds. Configurations are not mutually interconvertible by rotation around bonds. One configuration can be converted to another only by breaking and re-forming a bond.

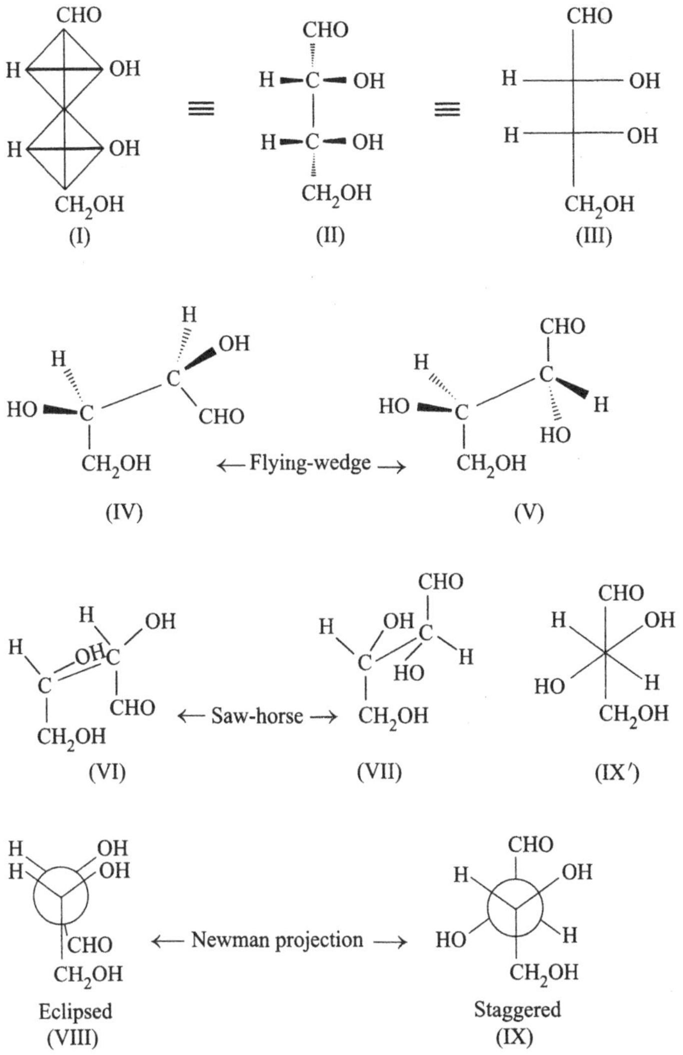

Fig. 21.9 D-Erythrose, various projections

Structures (VI) and (VII) are equivalent to (IV) and (V) respectively, without using wedges and broken lines. (VI) and (VII) are *perspective drawings*, sometimes referred to as *saw-horse* representations. Both flying-wedge and saw-horse projections can represent any desired conformation. (VI) and (VII) are eclipsed and staggered (anti) conformations respectively. These terms will be clearer when we look at Newman projections. Let us view structure (VII), keeping our eye on the left in a straight line with the C3—C2 bond. What we see is something like (IX'). C2 is hidden by C3 and is not seen. Three of the bonds radiate from C3 and three from C2. In the Newman projection, these two sets of bonds are separated from each other by an imaginary disc or circle placed between C2 and C3 as in (IX). The Newman projection corresponding to (VI) is (VIII). The three bonds on the front carbon C3, eclipse the three bonds on the rear carbon C2, in structure (VIII). Hence this is called an *eclipsed conformation*. (IV) and (VI) are also eclipsed conformations. In (IX), the bonds on the front and rear carbons are staggered. Hence this and also (V) and (VII) are called *staggered conformations*. The groups CHO and CH$_2$ OH are oriented *anti* to each other.

An exercise in converting a structure in Fischer projection to saw-horse or Newman projection and vice versa is illustrated in Fig. 21.10 using L-threose as example.

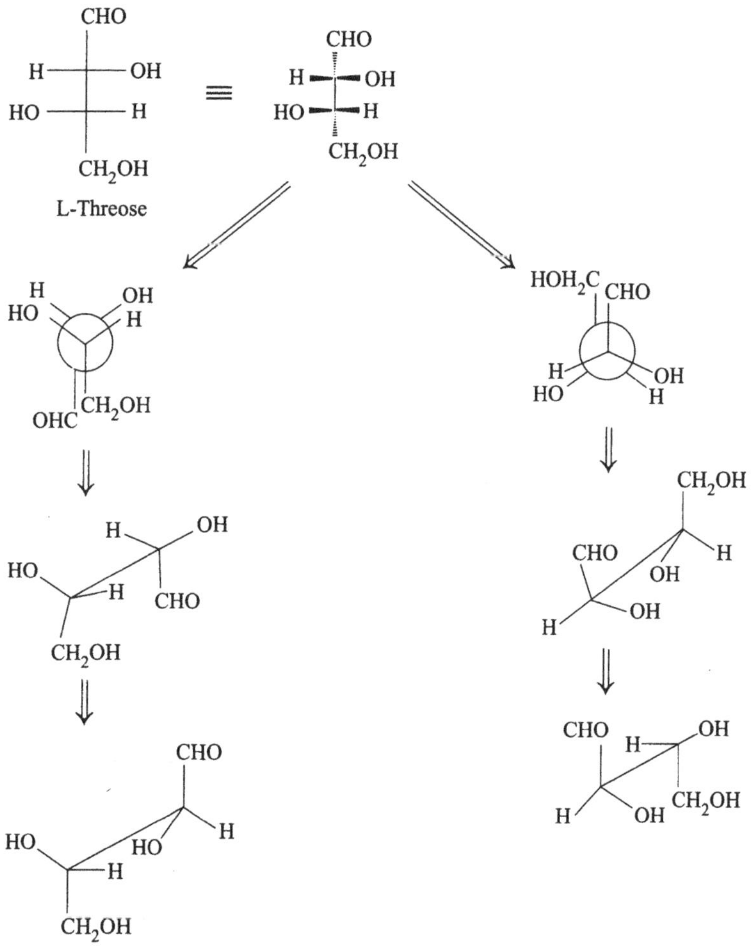

Fig. 21.10   Interconversion of different projection formulae

21.2.3 APPLYING CAHN–INGOLD–PRELOG NOMENCLATURE TO MOLECULES CONTAINNG TWO ASYMMETRIC CARBONS

R,S-designation for the two asymmetric centres of D(−)-erythrose are carried out as in Fig. 21.11.

The two asymmetric centres are taken individually. The groups present on C2 are listed, their priorities assigned and the configuration identified as R. Similarly the configuration of C3 is

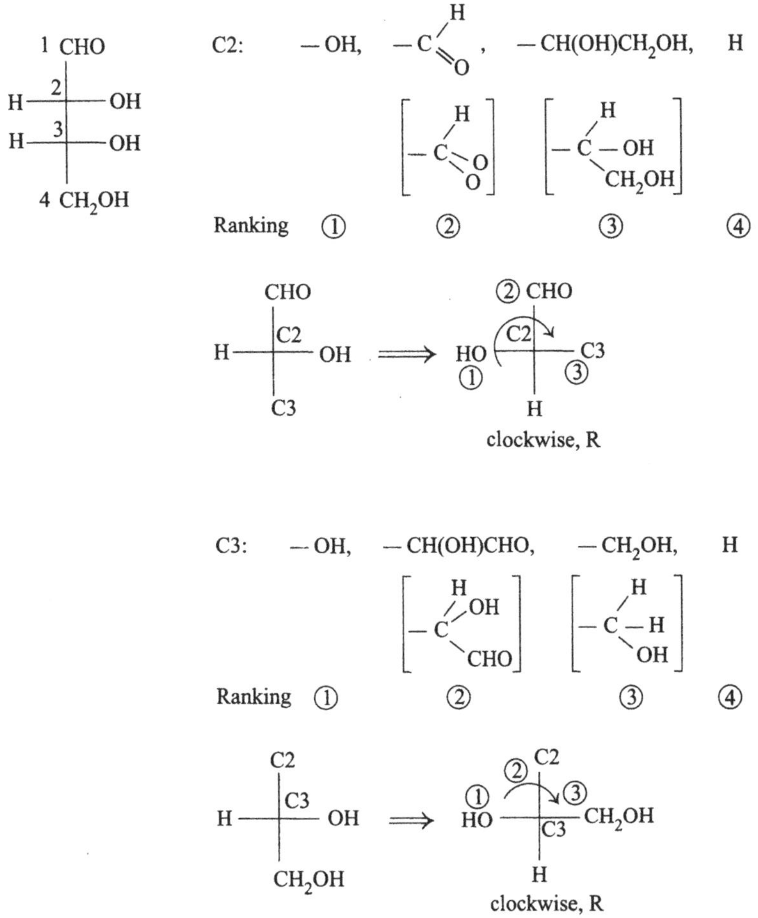

Fig. 21.11 R,S-nomenclature of D-erythrose

identified to be R. Hence, the molecule is 2(R),3(R)-2,3,4-trihydroxybutanal. Similarly the other tetroses are identified to be:

> L(+)-Erythrose is 2(S),3(S)-2,3,4-trihydroxybutanal
>
> D(−)-Threose is 2(S),3(R)-2,3,4-trihydroxybutanal
>
> L(−)-Threose is 2(R),3(S)-2,3,4-trihydroxybutanal.

21.2.4 THE TARTARIC ACIDS

Tartaric acid, 2,3-dihydroxybutanedioic acid, has two chiral centres, each with identical groups (Fig. 21.12). Structures (I) and (I′) corresponding to the enantiomeric erythroses are actually identical. Even though this structure has two chiral centres, the molecule is not chiral. The asymmetry is internally compensated. If one of the carbons is S, the other is R. It has a plane of symmetry which bisects the C2—C3 bond. This diastereomer is called *meso*-tartaric acid. (II) and

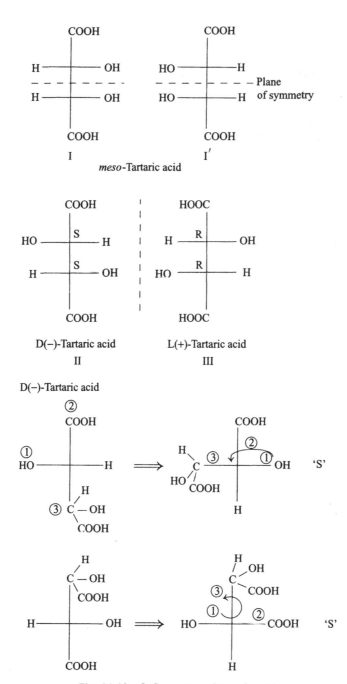

Fig. 21.12 Cofigurations of tartaric acid

(III) are enantiomers, D-tartaric acid and L-tartaric acid. As shown in Fig. 21.12, D(−)-tartaric acid is 2(S),3(S)-tartaric acid or simply, (S)-tartaric acid and L(+)-tartaric acid is 2(R),3(R)-tartaric acid or simply, (R)-tartaric acid.

Molecules such as tartaric acid, in which the two asymmetric centres are identically substituted, have only three stereoisomers comprising of two enantiomers and one *meso* form. Some other examples are 3,4-dibromopentane and 2,3-butanediol.

21.2.4.1 Nomenclature—Threo, Erytho and Meso

As seen above, in molecules like tartaric acid, one diasteromer is achiral due to internal compensation. It is named with the prefix '*meso-*'. For molecules with two asymmetric centres and dissimilar substituents, unambiguous nomenclature is based on the R,S notation. There is a trivial system of nomenclature based on the names, threose and erythrose (Fig. 21.13). Threose-like molecules are named with the prefix '*threo-*', and erythrose-like molecules, with the prefix '*erythro*'. Erythrose-like molecules are also *meso*-like. If one pair of groups on each of the asymmetric carbons is the same or similar, the diastereomer in which the same groups are eclipsed is *erythro-* and the other is *threo-*.

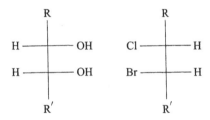

erythro-(Erythrose-like; identical or similar groups are eclipsed)

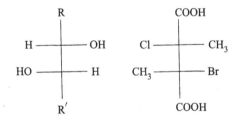

threo-(Threose-like; identical or similar groups do not eclipse)

Fig. 21.13 *erythro-* and *threo*-nomenclature

21.3 OBTAINING ENANTIOMERICALLY PURE MOLECULES

Many naturally occurring compounds are chiral. These include sugars and carbohydrates, amino acids and proteins including enzymes, alkaloids, terpenes, and many biological molecules. Many synthetic drugs are chiral molecules. Generally, only one of the enantiomers is effective as the drug. If the drug is taken as the racemic mixture, only the active enantiomer is useful. Drug manufacturers try to ensure that, if the racemic mixture is marketed, the unwanted enantiomer is harmless. The desired practice is to market only the active enantiomer. The tragic story of the drug thalidomide is often quoted.

Thalidomide

One of the enantiomers of this drug was proved to be effective against nausea. The other enantiomer was believed to be harmless. In the 1960s, the racemic drug was administered to pregnant women. Babies were born with genetic defects which were later recognised as due to the 'inert' enantiomer.

There are primarily two approaches to obtaining enantiomerically pure molecules. One is *resolution*, which is the separation of the enantiomers from the racemic mixture. The other is *asymmetric synthesis*, which is the selective formation of one enantiomer at the synthesis stage itself.

21.3.1 RESOLUTION

21.3.1.1 Resolution by Conversion to Diastereomers

Conventional methods of separation of the components of a mixture like fractional distillation or fractional crystallisation are ineffective because the enantiomers have identical properties. The most important method of resolution of racemic mixtures exploits the fact that, while enantiomers have identical properties such as boiling point and solubility, diastereomers differ in their physical and chemical propterties. This is illustrated by the following example (Fig. 21.14).

A racemic mixture of an acid A, composed of equal amounts of (+)-A and (−)-A is treated with an enantiomerically pure amine, say (+)-X, in excess. A mixture of salts, usually crystalline solids (I) and (II), containing both in equal amounts is formed. These are not enantiomers, but are diastereomers and usually differ in their solubility in suitable solvents. They can be separated by fractional crystallisation to obtain pure (I) and (II). They are taken to be pure if the optical activity does not change after repeated crystallisations. The pure acids, (+)-A and (−)-A are obtained by acidification of the respective salts using mineral acid. This example is among the simplest and uses a racemic mixture which has the carboxylic acid functional group and a chiral amine as resolving agent because,

(i) salt formation takes place under mild conditions by just mixing the reagents,

(ii) salts are usually crystalline solids which can be separated by fractional crystallisation, and

(iii) after the separation, the enantiomerically pure materials can be isolated by the simple process of acidification.

Also, the amine which is the resolving agent, is generally a costly chemical and can be recovered and reused. Alkaloids like brucine and strychnine are naturally occurring chiral amines

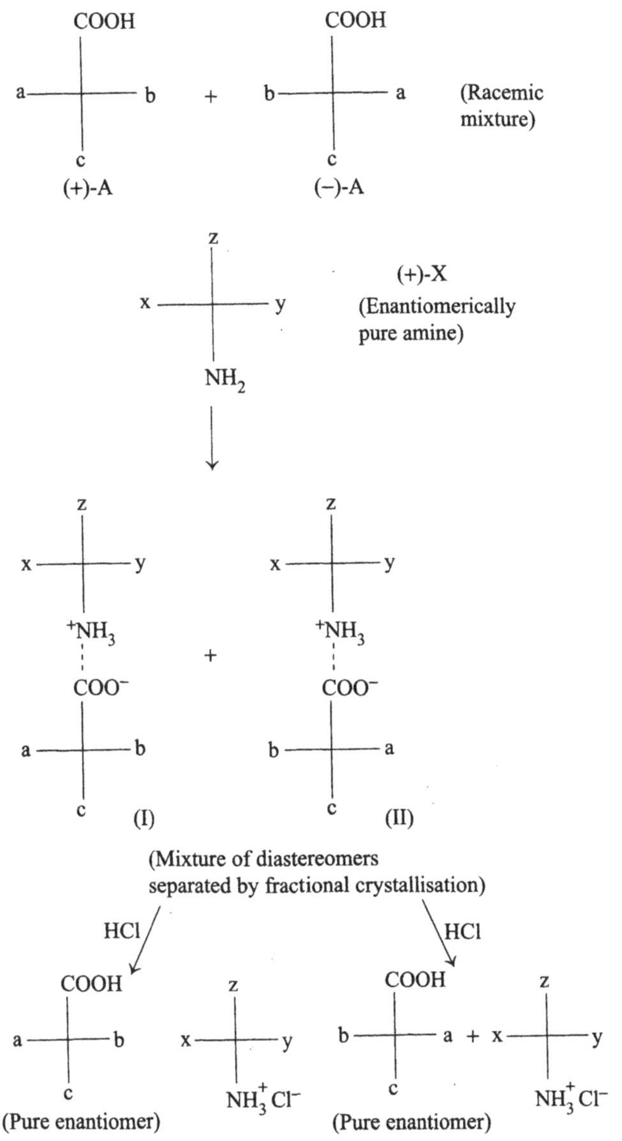

$$
\begin{array}{ccc}
\underset{\substack{| \\ c \\ (+)\text{-}A}}{a\!-\!\!\!-\!\!\!\overset{\displaystyle \text{COOH}}{\underset{}{|}}\!\!\!-\!\!\!-b} & + & \underset{\substack{| \\ c \\ (-)\text{-}A}}{b\!-\!\!\!-\!\!\!\overset{\displaystyle \text{COOH}}{\underset{}{|}}\!\!\!-\!\!\!-a}
\end{array}
\qquad \text{(Racemic mixture)}
$$

$$
\underset{\substack{| \\ NH_2}}{x\!-\!\!\!-\!\!\!\overset{\displaystyle z}{\underset{}{|}}\!\!\!-\!\!\!-y}
\qquad
\begin{array}{l}
(+)\text{-}X \\
\text{(Enantiomerically} \\
\text{pure amine)}
\end{array}
$$

$$
\begin{array}{ccc}
\begin{array}{c}
x\!-\!\!\!-\!\!\overset{z}{\underset{}{|}}\!\!\!-\!\!\!-y \\
{}^{+}NH_3 \\
\vdots \\
COO^{-} \\
a\!-\!\!\!-\!\!\overset{}{\underset{}{|}}\!\!\!-\!\!\!-b \\
c \\
\text{(I)}
\end{array}
& + &
\begin{array}{c}
x\!-\!\!\!-\!\!\overset{z}{\underset{}{|}}\!\!\!-\!\!\!-y \\
{}^{+}NH_3 \\
\vdots \\
COO^{-} \\
b\!-\!\!\!-\!\!\overset{}{\underset{}{|}}\!\!\!-\!\!\!-a \\
c \\
\text{(II)}
\end{array}
\end{array}
$$

(Mixture of diastereomers
separated by fractional crystallisation)

HCl   HCl

$$
\underset{\substack{| \\ c \\ \text{(Pure enantiomer)}}}{a\!-\!\!\!-\!\!\overset{\displaystyle \text{COOH}}{\underset{}{|}}\!\!\!-\!\!\!-b}
\quad
\underset{\substack{| \\ NH_3^{+}Cl^{-}}}{x\!-\!\!\!-\!\!\overset{z}{\underset{}{|}}\!\!\!-\!\!\!-y}
\qquad
\underset{\substack{| \\ c \\ \text{(Pure enantiomer)}}}{b\!-\!\!\!-\!\!\overset{\displaystyle \text{COOH}}{\underset{}{|}}\!\!\!-\!\!\!-a}
\; + \;
\underset{\substack{| \\ NH_3^{+}Cl^{-}}}{x\!-\!\!\!-\!\!\overset{z}{\underset{}{|}}\!\!\!-\!\!\!-y}
$$

Fig. 21.14 Resolution via diastereomers

which are available in enantiomerically pure form and are often used as resolving agents. If the compound that has to be resolved is an amine, then an acid resolving agent like camphor-10 sulphonic acid or tartaric acid is used. When the compound to be resolved does not have acidic or basic functional groups, indirect methods are adopted. The resolution of 2-butanol after its conversion to a derivative with a COOH group is illustrated (Fig. 21.15). The method adopted to introduce the required COOH group involves reaction with phthalic anhydride. In the end, the alkaloid and phthalic acid can be removed by hydrolysis or by reduction with lithium aluminium hydride.

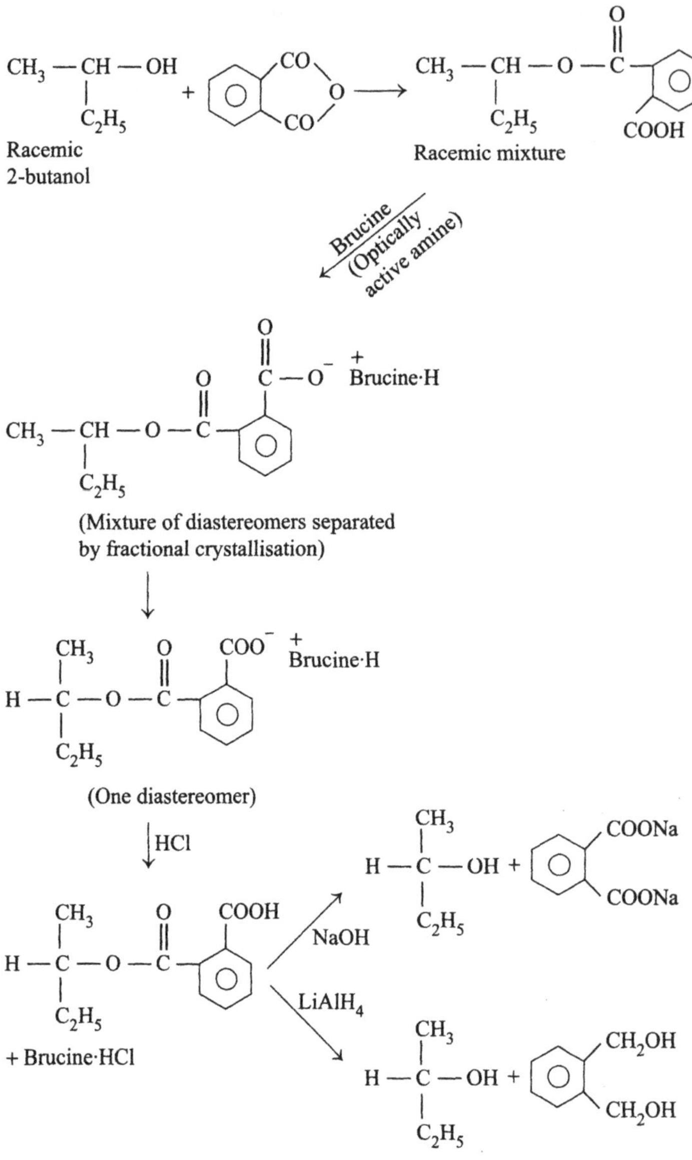

Fig. 21.15   Resolution of 2-butanol

21.3.1.2   Miscellaneous Methods of Resolution

Sometimes the two enantiomers crystallise out from solution as separate enantiomeric crystals. Pasteur was successful in resolving tartaric acid by crystallising out sodium ammonium tartarate from its aqueous solution. The enantiomers crystallised out in crystals which were visually different. He could pick out the two kinds of crystals. This is called *resolution by mechanical separation of the crystals*. Naturally, this is applicable only to a very limited number of cases and on a small scale.

A more successful method is *inoculation* or *seeding*. To a saturated solution of a racemic mixture, if a crystal of one enantiomer is added, it will act as a seed around which that enantiomer will crystallise. This method is especially applicable for the final purification of enantiomers, after achieving partial resolution by other methods.

Chromatography over a stationary phase which is chiral, is often successful for the separation of enantiomers. This is based on the principle that the two enantiomers are not adsorbed or bound with equal strength to an asymmetric surface or substrate.

In the *biochemical method* of resolution, a selected bacterium is allowed to grow in a solution containing the racemic mixture. Bacteria are selective towards enantiomers and preferentially destroy one enantiomer, leaving the other intact. This method is effective if the appropriate bacteria can be identified. The drawback is that one of the enantiomers is lost.

21.3.2 ASYMMETRIC SYNTHESIS

One of the rules of asymmetric synthesis is that a chiral molecule can be obtained only with the help of another chiral reagent or chiral physical agent like circularly polarised light. This is true for resolution also. Asymmetric synthesis is the selective transformation of an achiral molecule into a chiral product. An achiral molecule which can give a chiral product upon reaction is called a *prochiral molecule*. 2-Butanone and both the geometrical isomers of 2-butene are prochiral molecules (Eq. 21.2).

$$CH_3 - \overset{\overset{\displaystyle O}{\|}}{C} - CH_2 - CH_3 + H_2 \longrightarrow CH_3 - \overset{\overset{\displaystyle OH}{|}}{\underset{\underset{\displaystyle H}{|}}{\overset{*}{C}}} - CH_2 - CH_3$$

$$(21.2)$$

$$\underset{CH_3}{\overset{H}{\diagdown}} C = C \underset{H}{\overset{CH_3}{\diagup}} + HBr \longrightarrow H_3C - CH_2 - \overset{\overset{\displaystyle H}{|}}{\underset{\underset{\displaystyle Br}{|}}{\overset{*}{C}}} - CH_3$$

In the two examples given, the starting materials have no chiral centre while the products do. The objective of asymmetric synthesis is to obtain exclusively, or at least, more of one of the enantiomers.

21.3.2.1 Absolute Asymmetric Synthesis

Early attempts to bring about the selective formation of one enantiomer from a prochiral molecule in a photochemical reaction using circularly polarised light, belong to this category. An example of such a synthesis is the addition of bromine to a prochiral stilbene molecule in the presence of right-circularly polarised light, to obtain the dextro rotatory dibromide. These studies are only of theoretical interest. All practical asymmetric synthesis involve the use of another chiral reagent. Such syntheses are called *partial asymmetric syntheses*. There are two approaches to this, (i) using a chiral reagent and (ii) using a chiral auxiliary.

21.3.2.2 Partial Asymmetric Synthesis Using Chiral Reagents

This is best illustrated with the help of a specific example—addition of water to *cis*-2-butene to synthesise 2-butanol (Fig. 21.16).

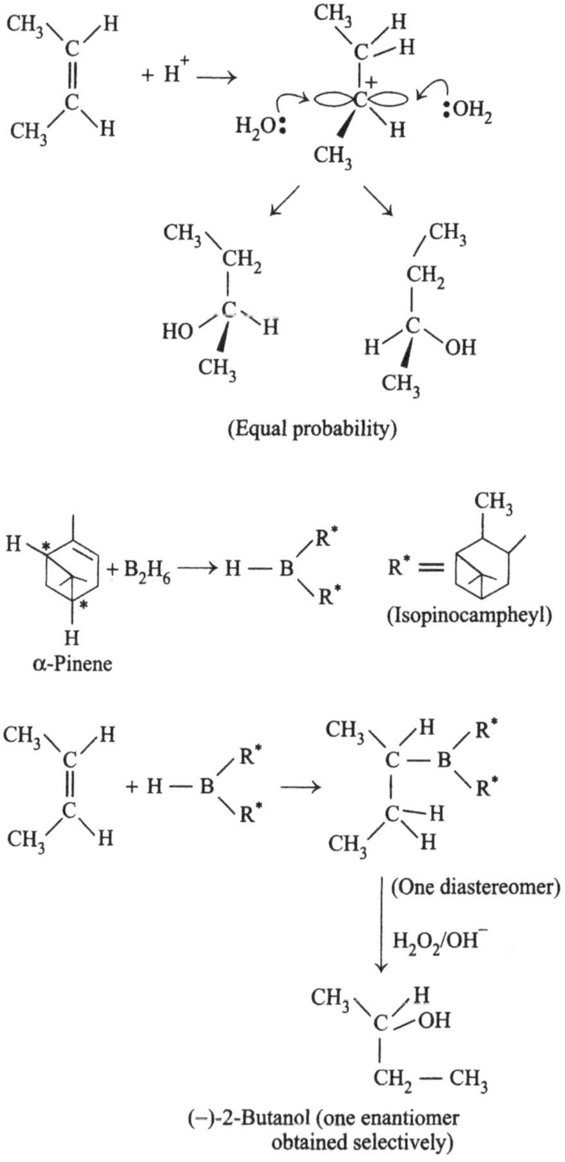

(Equal probability)

Fig. 21.16 Hydration of *cis*-2-butene

Acid catalysed hydration involves the formation of the 2-butyl carbocation which is a prochiral intermediate. The two faces of the carbocation are *prochiral faces* or *enantiotopic faces*. Addition of a nucleophile can take place on either face to give one or the other of the two enantiomers. The carbocation is symmetrical, the two faces are identical, and the approach of a water molecule can take place with equal facility from either face. The two enantiomers will be formed in equal amounts. The 2-butanol obtained will be racemic. This is true whatever be the finer details of the hydration such as, acid catalysed hydration, hydration via oxymercuration or hydroboration using diborane because in all such reactions, the reagent which combines with the prochiral carbon is achiral. The situation will be different if the reagent is chiral. The approach of a chiral reagent to a prochiral face will not be of equal facility from either side. Hydroboration has been adapted for chiral synthesis (Chapter 4, Section 4.4.2 and 4.4.2.1, for discussion of hydroboration). H C Brown used a dialkylborane prepared by the addition of B_2H_6 to a bulky chiral alkene, α-pinene. The alkyl group corresponding to α-pinene is called isopinocampheyl. Because of the bulk of the alkene, only a dialkylborane, diisopinocampheylborane is formed. It can still react with the smaller alkene, *cis*-2-butene to give 2-butyldiisopinocampheylborane. Since the dialkylborane is chiral due to the presence of the two α-pinene moieties, it adds selectively to one face of the butene and after oxidation by H_2O_2, the 2-butanol obtained is optically active with more than 90% optical purity. From *cis*-2-butene, using the naturally occurring $(-)$-α-pinene, the product is R-2-butanol. Such a synthesis is called a *stereospecific* synthesis. *Stereospecificity* refers to the situation where one particular stereoisomeric starting material gives one particular stereoisomer of the product. In the above example, the *cis*-alkene gives R-2-butanol; *trans*-2-butene yields S-2-butanol. Another example is the formation of *meso*-2,3-dibromobutane from *trans*-2-butene and *dl*-2,3-dibromobutane from *cis*-butene, by bromine addition. *Stereoselectivity* is a more general term, used to denote the preferential formation of one stereoisomeric product over the other, from a given starting material which may be chiral, racemic or achiral. Formation of *trans*-2-butene in preference to *cis*-2-butene, in the dehydrobromination of 2-bromobutane, is an example of stereoselectivity. All stereospecific reactions are stereoselective. The reverse may not be true.

Another well known asymmetric synthesis is asymmetric epoxidation of alkenes, discovered by Sharpless. In Sharpless asymmetric epoxidation, the chiral reagent is a titanium peroxy complex containing diethyl tartarate.

There are many asymmetric syntheses using chiral catalysts, including enzyme catalysis. Microbiological transformations for the synthesis of chiral molecules also belong to this category. Products like lactic acid and ephedrine are manufactured by microbiological processes. Several homogeneous chiral hydrogenation catalysts, which are metal complexes containing chiral ligands, are available which bring about asymmetric hydrogenation of prochiral $C = C$ and $C = N$ double bonds with high optical purity.

Optical purity or enantiomeric excess (e.e), is expressed as the ratio of the specific rotation of the product of synthesis with that of the pure enantiomer and expressed as a percentage. Today, e.es close to 100% are expected in a process of asymmetric synthesis to be considered useful for practical application.

21.3.2.3 Use of Chiral Auxiliary

If a reaction is carried out at a prochiral centre in a molecule which already contains another chiral centre, the new chiral centre is likely to be generated with stereospecificity. The pre-existing centre of chirality is called a chiral auxiliary and it is said to exert asymmetric induction at the new asymmetric centre.

This principle is best illustrated by the examples given in Fig. 21.17. What occurs when methylmagnesium iodide reacts with an achiral ketone (I), is shown in Fig. 21.17a. The reagent can approach the two faces of the ketone with equal facility and a racemic mixture is formed as product. In 21.17b, the ketone (II) is chiral. From the figure, it is obvious that the approach of the reagent to the two faces meet with different degrees of steric hindrance. The approach from the left

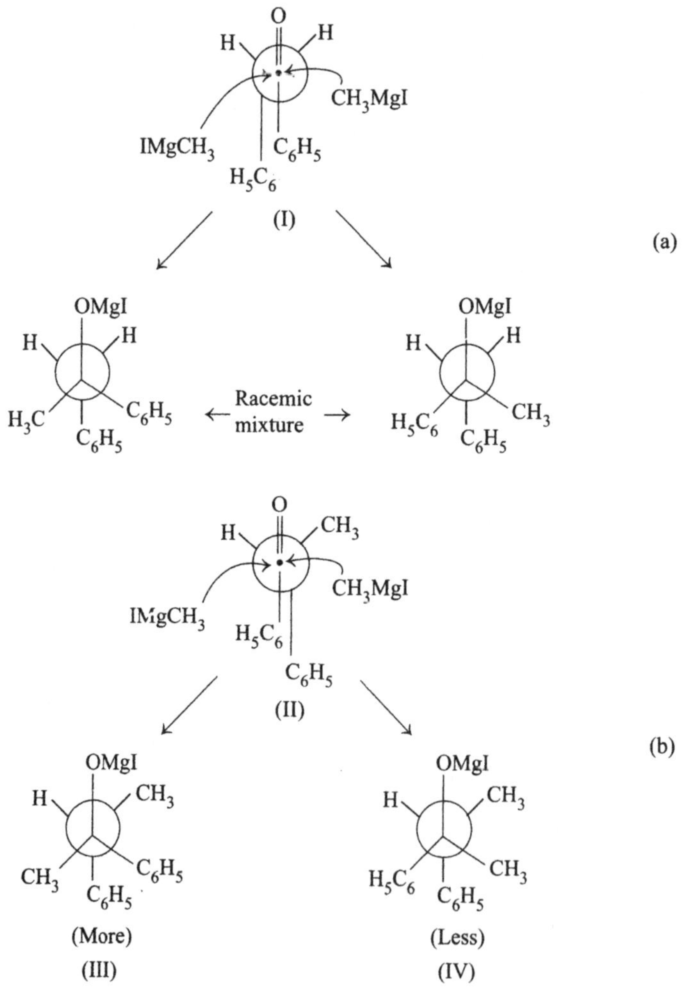

Fig. 21.17 (Continued)

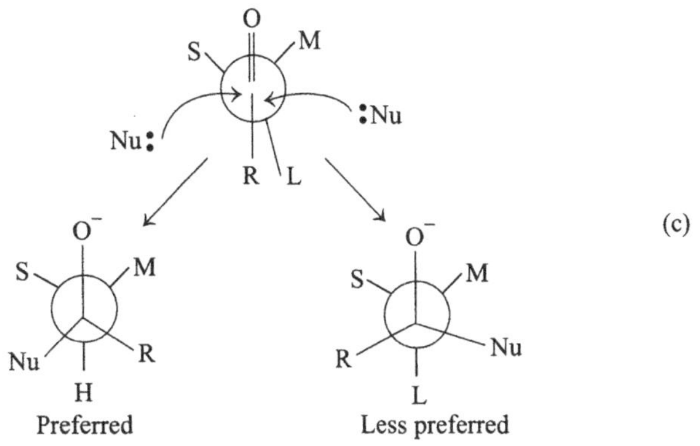

Fig. 21.17   Assymmetric induction: Cram's rule

side where the smaller group H is present, is less hindered than the approach from the right, where the larger group CH_3 is present. Diastereomer (III) is formed in preference to the diastereomer (IV). The diagram may give the impression that the reaction takes place only when the ketone is in the specific conformation shown. This is not true. The ketone can react in any conformation. But because of the asymmetric nature of the molecule, some conformations are preferred over others, and the net result is that more of one diastereomer is formed. The result with the conformation shown where the largest group on the asymmetric carbon – phenyl in this case – is close to the alkyl or aryl group on the carbonyl carbon, has been found to predict the preferred product in such reactions accurately. This has been generalised as *Cram's rule* (Fig. 21.17c). In this illustration, S, M and L represent the smallest, medium sized and largest group on the asymmetric carbon. For the application of Cram's rule, the largest group L is just behind, eclipsing the group on the carbonyl carbon. The preferred product corresponds to the approach of the nucleophile from the side of the smallest group. The nucleophilc addition to the carbonyl compound can be a Girgnard reaction as in the illustrated example or any other reaction like hydride reduction.

The asymmetric carbon which has induced chirality in the reaction is called a chiral auxiliary. Ideally, we should be able to attach a chiral auxiliary to the prochiral molecule, carry out the stereoselective reaction and then remove the chiral auxiliary. For example, reaction of the ester of phenylglyoxylic acid with methylmagnesium iodide (Fig. 21.18). To make the ester chiral, the esterification is carried out with an optically active alcohol like naturally occurring menthol. The attack of the Grignard reagent takes place selectively from one face of the keto group. The secondary alcohol obtained corresponds to one enantiomer. The chiral auxiliary, menthol, can be removed by hydrolysis after the reaction. Correlations arising from reactions of the type outlined in Fig. 21.18 have been generalised as *Prelog's rule*.

21.4   RACEMISATION

An enantiomerically pure material can lose its optical activity in many ways. Racemisation involves conversion of one enantiomer to another in a reversible reaction, so that in the course of time, a 1:1 mixture of the two enantiomers is obtained. All such reactions involve breaking of at least one

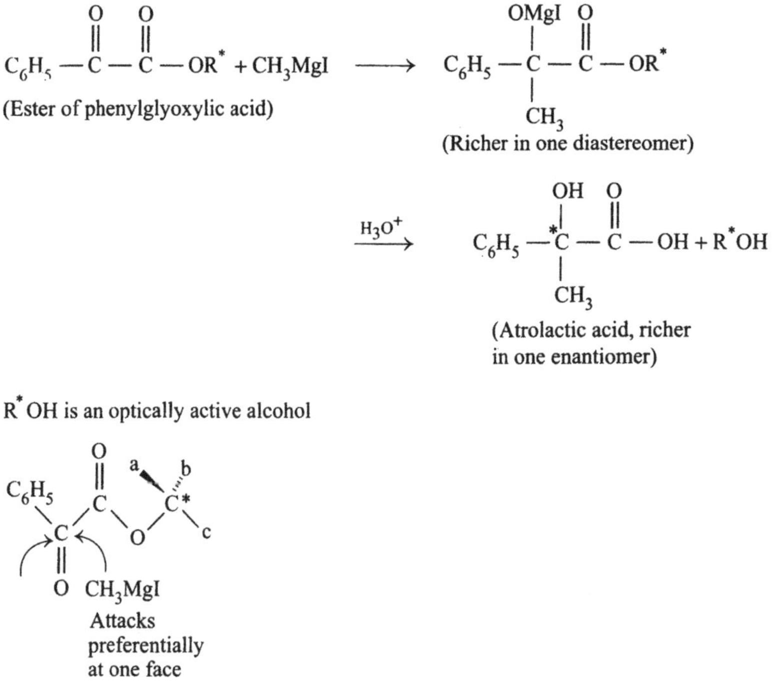

$C_6H_5 - \overset{\overset{O}{\|}}{C} - \overset{\overset{O}{\|}}{C} - OR^* + CH_3MgI \longrightarrow$ $C_6H_5 - \overset{\overset{OMgI}{|}}{\underset{\underset{CH_3}{|}}{C}} - \overset{\overset{O}{\|}}{C} - OR^*$

(Ester of phenylglyoxylic acid)　　　　　　　　(Richer in one diastereomer)

$\xrightarrow{H_3O^+}$ $C_6H_5 - \overset{\overset{OH}{|}}{\underset{\underset{CH_3}{|}}{\overset{*}{C}}} - \overset{\overset{O}{\|}}{C} - OH + R^*OH$

(Atrolactic acid, richer in one enantiomer)

R^*OH is an optically active alcohol

CH$_3$MgI
Attacks
preferentially
at one face

Fig. 21.18 Asymmetric synthesis induced by a chiral auxiliary

bond at the chiral centre and formation of a symmetrical intermediate or transition state. Three of the most common mechanisms of racemisation are illustrated in Eq. 21.3.

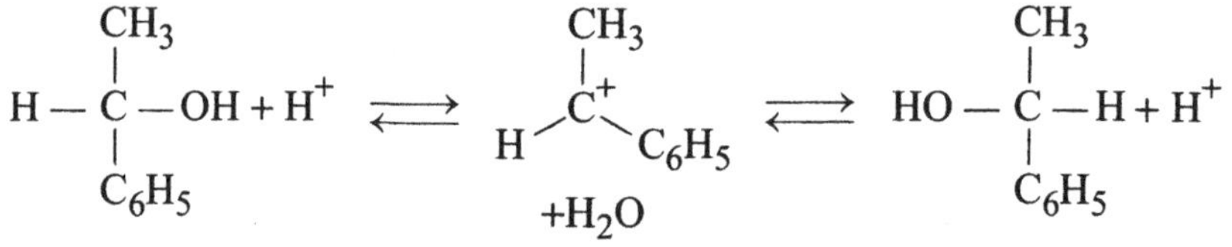

$H - \overset{\overset{CH_3}{|}}{\underset{\underset{C_6H_5}{|}}{C}} - OH + H^+ \rightleftharpoons \underset{+H_2O}{H - \overset{\overset{CH_3}{|}}{\underset{C_6H_5}{\overset{+}{C}}}} \rightleftharpoons HO - \overset{\overset{CH_3}{|}}{\underset{\underset{C_6H_5}{|}}{C}} - H + H^+$ (a)

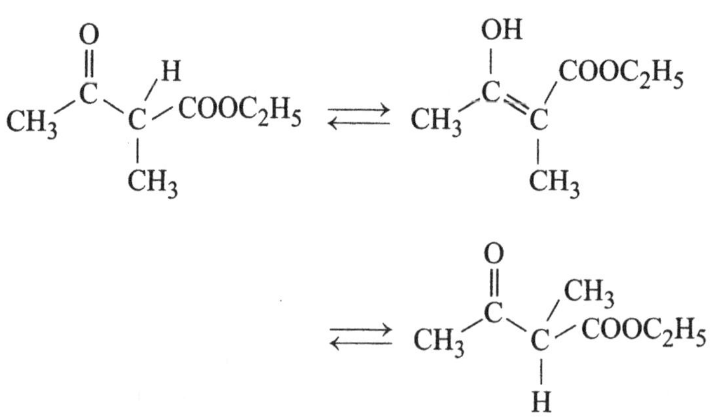

$CH_3 - \overset{\overset{O}{\|}}{C} - \overset{\overset{H}{\diagup}}{\underset{\underset{CH_3}{|}}{C}} - COOC_2H_5 \rightleftharpoons CH_3 - \overset{\overset{OH}{|}}{C} = \overset{\diagup COOC_2H_5}{\underset{\underset{CH_3}{|}}{C}}$

$\rightleftharpoons CH_3 - \overset{\overset{O}{\|}}{C} - \overset{\overset{CH_3}{\diagup}}{\underset{\underset{H}{|}}{C}} - COOC_2H_5$ (b)

$$\text{I}^- + \begin{matrix} \text{H} \\ \diagdown \\ \text{C} - \text{I} \\ \diagup \\ \text{C}_6\text{H}_{13} \end{matrix} \overset{\text{CH}_3}{} \rightleftharpoons \left[\begin{matrix} \text{H} \diagdown \quad \diagup \text{CH}_3 \\ \text{I} \cdots \text{C} \cdots \text{I} \\ \mid \\ \text{C}_6\text{H}_{13} \end{matrix} \right]^- \rightleftharpoons \begin{matrix} \text{CH}_3 \\ \diagup \text{H} \\ \text{I} - \text{C} \\ \diagdown \\ \text{C}_6\text{H}_{13} \end{matrix} + \text{I}^- \qquad (c)$$

$$\text{C}_6\text{H}_{13}\text{CH}(\text{CH}_3)\text{I} + \overset{+}{\text{Na}}\overset{-}{\text{I}}^* \rightleftharpoons \text{C}_6\text{H}_{13}\text{CH}(\text{CH}_3)\text{I}^* + \overset{+}{\text{Na}}\overset{-}{\text{I}} \qquad (d)$$

$$(21.3)$$

Equation 21.3a illustrates racemisation through a carbocation by an S_N1 reaction. The symmetrical intermediate here is the cation. In 21.3b, racemisation via tautomerism is illustrated via the enol which is achiral. Even simple aldehydes, ketones and carboxylic acid derivatives with asymmetry at the α-carbon and with a hydrogen at the asymmetric carbon will racemise slowly by the reversible formation of the enol or the enolate anion. Racemisation will take place faster if an acid or base is present, because these catalyse tautomerism (Chapter 17, Section 17.6.2). Equation 21.3c illustrates how racemisation can take place even in S_N2 reactions. A solution of optically active 2-iodooctane in acetone to which a small quantity of sodium iodide is added, slowly loses its optical activity on standing. The reaction shown is a nucleophilic displacement of iodide ion by iodide ion. Recall that iodide is a good nucleophile and a good leaving group (Chapter 11, Section 11.2.1.4). That this reaction really occurs has been established by using radioactive iodide, in which case the labelled iodine enters the molecule (Eq. 21.3d). As can be seen, the transition state has a plane of symmetry. Any reaction proceeding through a symmetrical intermediate or transition state will result in racemisation.

21.5 OPTICAL ACTIVITY IN COMPOUNDS NOT CONTAINING ASYMMETRIC CARBONS

21.5.1 BIPHENYLS: ATROPISOMERISM

We have seen earlier, in the discussion on conformations and configurations, that the former are rotational forms resulting from free rotation around single bonds. Rotation around single bonds is free in the sense that the molecule possesses enough energy at room temperature to make this happen without the supply of external energy. However, there are barriers to rotation around single bonds, as we shall see in Chapter 22. There are molecules where the barrier to rotation is high enough that interconversion of conformations does not take spontaneously at room temperature. Conformations of such molecules can be isolated, and if they are devoid of symmetry they behave as stereoisomers. The most widely studied class of such molecules are the *ortho*-substituted biphenyls. Biphenyl is capable of free rotation around the $C-C$ bond connecting the phenyl rings. Of all the possible conformations, the one where the two rings are coplanar is the most stable because this allows for extended conjugation as in structure (I), Fig. 21.19.

When there are bulky groups at the *ortho* positions of the two rings (2,2',6,6'-positions) as in structure (II), planarity is not possible and the two rings assume a mutually perpendicular conformation. If the substituents at the two *ortho* positions are not the same in each ring as in (II),

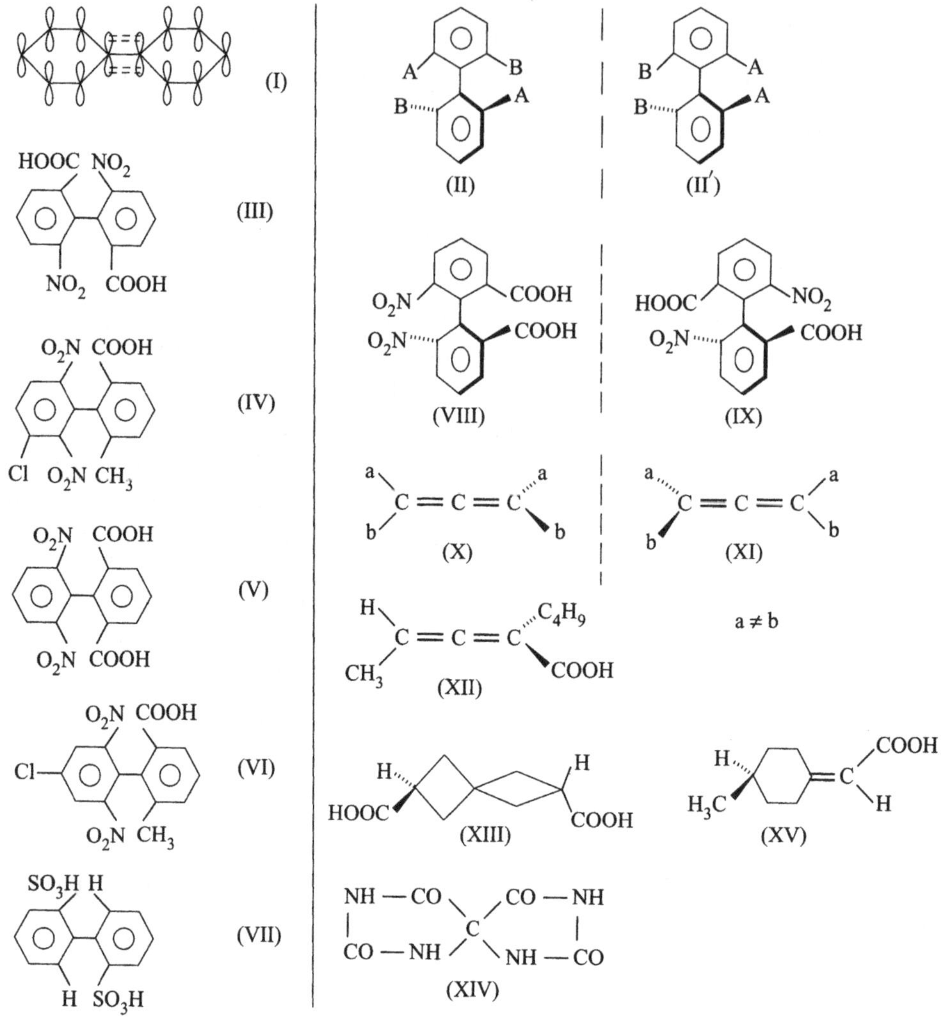

Fig. 21.19

A $\neq$ B, the nonplanar structure is chiral and can be resolved into the enantiomers. The necessary conditions are:

- the substituents at the 2,2',6,6'-positions should be sufficiently bulky so that the rings are held in the nonplanar conformation and

- neither ring should have a vertical plane of symmetry.

Structures (III) and (IV) can be resolved, not structures (V) and (VI). The last two have a vertical plane of symmetry in one of the rings. This kind of optical isomerism is called *atropisomerism.*

The two enantiomers of 6,6'-dinitro-2,2'-dicarboxybiphenyl (6,6'dinitrodiphenic acid) are shown in structures (VIII) and (IX) in Fig. 21.19. This compound could be resolved via the brucine salt. Most of the studies of optically active biphenyls have been with molecules having at least one carboxyl group to enable resolution by salt formation. It should be kept in mind that these enantiomers are really conformations which are prevented from being interconverted because of restricted rotation. The barrier to rotation involves more energy than what is available in the molecule as thermal energy, at room temperature. The optically active forms are racemised by heat. Racemisation takes place by rotation around the $C-C$ bond after overcoming the barrier. If the *ortho* substituents are large in size and this barrier is more than $\sim 20\,kcal\,mol^{-1}$, then resolution is possible. The ease of racemisation is expressed as half life at a particular temperature. Even some disubstituted biphenyls like (VII) have been resolved. They are readily racemised by heat and have very short half lives.

21.5.2 ALLENES AND SPIROCOMPOUNDS

Allenes of the type (X) to (XII) (Fig. 21.19) lack a plane of symmetry and are resolvable. The necessary condition is that neither of the sp^2 hybridised carbons should have a vertical plane of symmetry. Unlike biphenyls, these cannot be racemised by rotation around a bond. Spiro compounds are bicyclic molecules where the two rings share one atom. Substituted spiro compounds like (XIII) and (XIV), and alkylidenecyclohexanes like (XV) (Fig. 21.19) which lack a plane of symmetry, are resolvable.

KEY POINTS

- When two molecules have the same constitution but differ in the arrangement of atoms in space, they are called stereoisomers. Stereoisomers are classified into optical and geometrical isomers. Optical isomerism arises when the molecule and its mirror image are not superimposable. The property is called chirality, which stands for 'handedness'. These isomers are mirror images in the same way as the left and the right hand are mirror images of each other. They are called enantiomers. They do not possess symmetry; more particularly, they do not have a plane of symmetry. The most common examples are of molecules containing four different groups on one carbon atom. Chiral molecules are asymmetric. Symmetric molecules are achiral. Geometrical isomerism or *cis–trans* isomerism arises due to restricted rotation around the double bond and in rings. Optical isomers are characterised by optical rotation measured in a polarimeter.

- The three dimensional structure of stereoisomers can be represented on paper by different projection conventions. Fischer projections, flying-wedge and saw-horse projections and Newman projections are commonly used.

- Configurations of optical isomers have to be determined experimentally. They are not related in a simple manner to the sign of optical rotation. Absolute configurations of some molecules have been determined by X-ray diffraction using circularly polarised X-rays. For most chiral molecules, the configuration has been assigned by correlation with molecules of known configuration.

- D,L-notation is an early, arbitrary method of designating the absolute configuration. The method based on sequence rules put forth by Cahn, Ingold and Prelog is today used to unambiguously designate configuration. In this method, configurations are denoted by the prefixes R and S.

- Molecules containing n asymmetric atoms have 2^n optical isomers. Molecules containing two asymmetric carbons have four optical isomers, which are two pairs of enantiomers. Either structure of one pair of enantiomers, is not the mirror image of either structure of the other pair. Such stereoisomers which are not enantiomers are called diastereomers. Enantiomers have identical properties, except the sign of optical rotation. Diasteromers differ in properties such as melting point, boiling point and solubility. The tetroses which are stereoisomers of 2,3,4-trihydroxybutanal are four isomeric molecules, $(+)$ and $(-)$-threose and $(+)$ and $(-)$-erythrose. The prefixes threo- and erythro- have been used on other molecules with two asymmetric carbons which are considered structurally similar to threose and erythrose. Molecules like tartaric acid, 2,3-dihydroxybutanedioic acid, have only three stereoisomers and not four—one pair of enantiomers and one achiral isomer. The latter arises because the two asymmetric centres have opposite configurations and they cancel each other. These molecules possess a plane of symmetry. They are designated by the prefix, meso-.

- An achiral molecule which upon reaction, can generate a chiral centre is called a prochiral molecule. 2-Butanone is a prochiral molecule. Upon reduction it gives 2-butanol which is chiral. Conventional methods of reduction of 2-butanone using achiral reducing agents give equal quantities of the two enantiomeric forms of 2-butanol. Such an optically inactive mixture is called a racemic mixture.

- There is often a need to synthesise enantiomerically pure compounds, especially for pharmaceutical applications. This can be done either by separating the enantiomers from a racemic mixture by a procedure called resolution, or the synthesis can be designed in such a way that only one enantiomer is formed. The latter is called asymmetric synthesis.

- *Optical purity or enantiomeric excess (e.e)* is the difference between the percentage of the more abundant enantiomer and that of the less abundant enantiomer in a mixture of the two. It is expressed as a percentage, e.e = (measured optical activity of the mixture/optical activity of the pure enantiomer) × 100.

- In the reaction of a starting material which can form two or more stereoisomeric products, if one of the stereoisomers is formed in greater amounts than the others, the reaction is said to be *stereoselective*. When a particular stereoisomer of a starting material reacts to form one specific stereoisomer of the product, the reaction is said to be *stereospecific*.

- There are several methods for resolution. The most important of these is to react the racemic mixture with an enantiomerically pure chiral reagent to make a derivative containing two asymmetric centres. This derivative is a mixture of two diastereomers. Since they differ in physical properties, they can be separated by physical methods like fractional crystallisation. After separation, the pure diastereomers are liberated by suitable reactions.

- Asymmetric synthesis makes use of one of two methodologies. (i) Use of a chiral reagent to bring about the transformation of a prochiral molecule. For example, the conversion of *cis*-2-butene to optically active 2-butanol by the hydroboration route using a chiral dialkylborane. (ii) Attach a chiral portion to the prochiral substrate and then bring about the desired reaction. The chiral portion, called a chiral auxiliary, will induce chirality at the newly formed centre. After the reaction, the chiral auxiliary can be removed by chemical reaction. An example is, the reaction at the keto group of an α-ketoacid, after esterifying it with a chiral alcohol. This is useful even if the chiral auxillary is not removed from the final product, as in the hydride reduction or Grignard reaction of chiral ketones and aldehydes. The outcome of such reactions is predicted by Cram's rule.

- Optically active molecules can lose optical activity by racemisation. A reversible reaction of the chiral molecule, which proceeds through a symmetric intermediate or transition state, results in racemisation. Examples of such reactions are, S_N1 reactions proceeding through the planar carbocation, keto–enol tautomerism through the planar enol form and S_N2 reactions involving the same nucleophile as the leaving group, which proceed through a symmetric transition state.

- Optical isomerism can exist even without an asymmetric carbon. Substituted biphenyl, where each ring lacks a plane of symmetry and where the *ortho* positions carry bulky substituents, exist in conformations in which the two phenyl rings are not coplanar. Two such conformations are possible, which are not mutally interconvertible because of steric hindrance to rotation by the bulky groups in the *ortho* positions. Such molecules are resolvable into their enantiomers. 6,6'-dinitrodiphenic acid is an example. This kind of isomerism is called atropisomerism and is exhibited due to resticted rotation around a single bond.

- Substituted allenes and substituted spiro compounds are among other classes of compounds which show optical activity without possessing any asymmetric carbon atoms.

EXERCISES

SECTION I

1. Draw the Fischer projection formulae for the following.

   (a) D-Mandelic acid, $C_6H_5 — CH(OH) — COOH$

   (b) L-Cysteine, $HSCH_2 — CH(NH_2) — COOH$

   (c) L-Glutamic acid, $HOOC — CH_2 — CH(NH_2) — COOH$

   (d) D-Isoserine, $H_2N — CH_2 — CH(OH) — COOH$

   (e) (R)-Malic acid, $HOOC — CH_2 — CH(OH) — COOH$

   (f) (S)-Bromolactic acid, $BrCH_2 — CH(OH) — COOH$

   (g) (S)-3-Methyl-1-penten-3-ol

   (h) (R)-3-Methyl-2-phenylbutane

2. Draw Fischer, saw-horse and Newman projection formulae of any one enantiomer (except for d) for the following.

    (a) threo-1,2-Diphenyl-1,2-propanediol
    (b) erythro-2,3-Dibromopentane
    (c) threo-4-Amino-3-hexanol
    (d) 2R,3R-2,3-Dihydroxybutanoic acid (make sure that the configuration is the same in all the projections)

3. Assign R,S-configuration for all the asymmetric centres in the following molecules.

4. Decide whether the following compounds are chiral.

5. Locate the asymmetric centres, if any, in the molecules given in Question 4, above.

SECTION II

1. Explain how the following optically active molecules get racemised.

    (a) 3-Phenyl-2-butanone kept in contact with aqueous acid.
    (b) Ethyl 2-phenylpropanoate kept in contact with aqueous alkali.
    (c) 2,2'-Dichloro-6,6'-dimethylbiphenyl on heating to 100°C.
    (d) 2-Phenyl-2-butanol kept in contact with aqueous acid.

2. An unknown compound, $C_4H_6O_3$, is soluble in dilute alkali and reduces Tollen's reagent. It is optically active. Upon mild oxidation, a compound $C_4H_6O_4$ is obtained which is still soluble in dilute alkali, but does not reduce Tollen's reagent and is optically inactive. Identify the structures.

3. Explain the process by which one enantiomer of 2,2'-dinitro-6,6'-dimethylbiphenyl is converted to the other enantiomer upon heating. Explain the nature of the hindrance to rotation. Draw a potential energy diagram for the rotation.

CHALLENGING QUESTIONS

1. Draw stereochemical structures to show the steps in the following reactions and their mechanisms. Look up the reactions in the relevant, earlier chapters. Draw appropriate projection formulae like Fischer, saw-horse or Newman, wherever necessary. Give suitable names for the stereoisomers.

    (a) *cis*-Stilbene + Br_2 → dibromide → debromination to one of the stilbenes

    (b) *trans*-2-Butene + OsO_4 → diol

    (c) *cis*-2-Butene + perbenzoic acid → epoxide → sodium ethoxide → monoether of a glycol

    (d) R-2-Butanol + $SOCl_2$ → 2-chlorobutane $\xrightarrow{CH_3SNa}$ thioether

    (e) S-2-Butanol + *p*-toluenesulphonyl chloride

         → tosylate $\xrightarrow{\text{sodium acetate}}$ 2-butyl acetate

    (f) erythro-1,2-Diphenyl-1-bromopropane + ethanolic KOH $\xrightarrow{\Delta}$
         (dehydrobromination, trans elimination) (Assume E2 mechanism)

    (g) threo-1,2-Diphenyl-1-propanol → S-methyl xanthate
         $\xrightarrow{\Delta}$ (Chugayev reaction)

2. Dehydrochlorination of 2-chlorobutane using ethanolic KOH, under E2 conditions gives a mixture of 1-butene, *trans*-2-butene and *cis*-2-butene (more of *trans*- than *cis*-2-butene). Using Newman projection formulae, draw transition states for the formation of all three products. Show why *trans*-2-butene is preferred over *cis*-2-butene.

3. Show what will be the stereochemical outcome if, in the reaction outlined in Fig. 21.17b, the racemic mixture corresponding to II is made to react with CH_3MgI.

PROJECT

It is accepted that asymmetric molecules cannot be created without the help of other asymmetric agents. A puzzling question has been how asymmetric molecules came into existence in the very beginning of the universe. Most biomoleules are chiral. There have been many theories about this. Prepare a project report on the 'Origin of asymmetric molecules on earth'.

22 Sterochemistry – II

OBJECTIVES   In this chapter, you will learn about,

- geometrical isomerism
- nomenclature of geometrical isomers
- methods of distinguishing between geometrical isomers including methods based on physical constants, dipole moments, heats of hydrogenation, chemical methods such as cyclisation and Beckmann rearrangement
- conformations of open chain molecules, ethane, butane, 1,2-dichloroethane and ethylene glycol, projection formulae, nomenclature, potential energy diagrams
- conformations of cyclohexane—chair, boat, twist-boat forms, potential energy diagram, axial and equatorial bonds, chair–chair flipping
- monosubstituted cyclohexanes, methylcyclohexane, energy difference between axial and equatorial conformations, potential energy diagram
- conformational analysis of dimethylcyclohexanes

22.1 GEOMETRICAL ISOMERISM

Geometrical isomerism arises due to restricted rotation around double bonds. In earlier chapters we have had several occasions to deal with geometrical isomers or *cis–trans* isomers of alkenes. Alkenes bearing non-identical groups on each sp^2 hybridised carbon can exist in two stereoisomeric forms (Figs. 22.1a and 22.1b). They are not enantiomers and need not be optically active. They are diastereomers.

Restricted rotation is the direct result of the requirement that the unhybridised p orbitals should be parallel to each other in order to obtain maximum overlap. Overlap is minimum in the

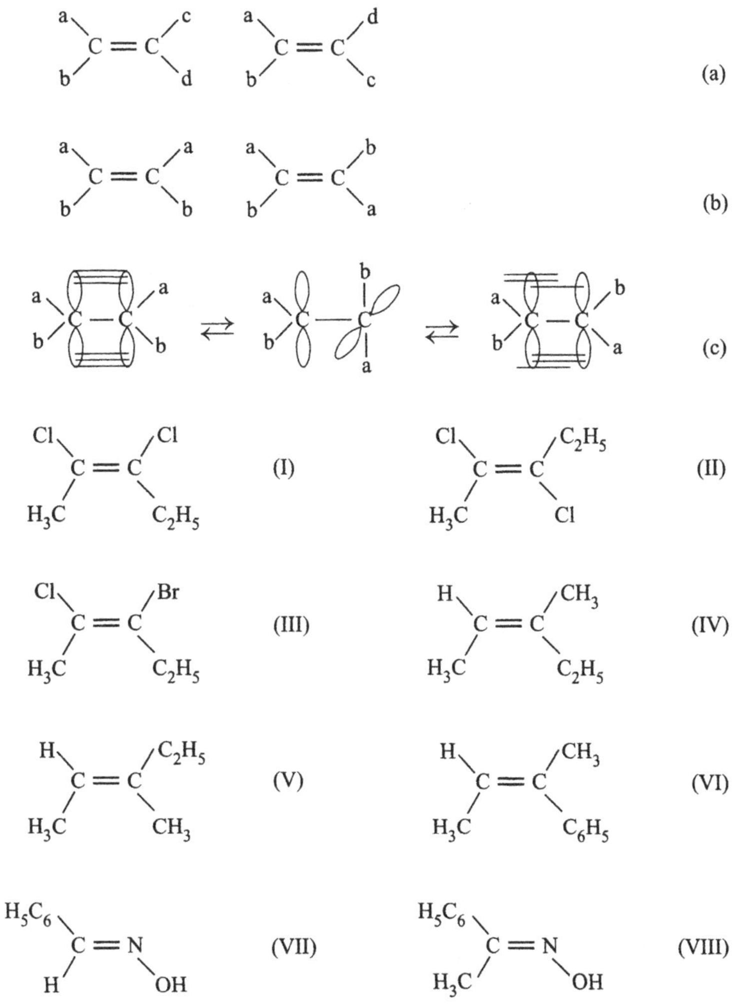

(a)

(b)

(c)

(I)

(II)

(III)

(IV)

(V)

(VI)

(VII)

(VIII)

Fig. 22.1 Geometrical isomers

conformation where the planes of the sp^2 carbons are orthogonal (Fig. 22.1c). This arrangement can be obtained only if the π-bond of the double bond is broken. Interconversion can occur only if the molecule passes through this conformation and this can happen only as a result of chemical reaction involving bond cleavage (also see discussion in Chapter 4, Section 4.3).

22.1.1 NOMENCLATURE

The *cis–trans* system of nomenclature is very convenient and widely used in molecules of the type, $C(a, b) = C(a', c)$, when a and a′ are the same or are unambiguously similar. The structures in Fig. 22.1 can be named 2,3-dichloro-*cis*-2-pentene (I), 2,3-dichloro-*trans*-2-pentene (II) and 3-bromo-2-chloro-*cis*-2-pentene (III). (IV) and (V) are respectively, 3-methyl-*cis*-2-pentene and 3-methyl-*trans*-2-pentene. These last two names can lead to some confusion. Even though CH_3 is present on both carbons, the nomenclature is not based on their being on the same side or opposite sides. Also, when the groups are entirely different, the *cis–trans* nomenclature will fail.

Today, the E — Z notation based on the Cahn–Ingold–Prelog sequence rules is the accepted system of nomenclature. In this system, the groups on each carbon are ranked as 1 and 2, according to the sequence rules. The isomer with the groups ranked 1 and 1 on the same side is named with the prefix Z (for zusammen meaning 'together' in German). When 1 and 1 are on opposite sides, the prefix, E (for entgegen, meaning 'opposite') is used (See Chapter 21, Section 21.1.4 for sequence rules). Consider Fig. 22.1, structure (IV). The groups on C2 are CH3 and H, with priority 1 and 2 respectively. On C3, the groups, C_2H_5 and CH3 have priorities 1 and 2 respectively. Thus, it is named (Z)-3-methyl-2-pentene. Structure (V) is (E)-3-methyl-2-pentene. Structure (VI) is (Z)-2-phenyl-2-butene. Z and E are not synonymous with *cis* and *trans*, even though it may appear so from these examples.

C=N and N=N double bonds also exhibit geometrical isomerism. The prefixes *syn* and *anti* rather than *cis* and *trans* are used to refer to the geometrical isomers of oximes. In the case of aldoximes, the isomer where the H and OH are on the same side is called *syn*-aldoxime. Structure (VII) is *syn*-benzaldoxime. In the case of the geometrical isomers of unsymmetrical ketoximes, the group that is *syn* or *anti* should be specified. Structure (VIII) is named *syn*-methyl phenyl ketoxime or *anti*-phenyl methyl ketoxime.

22.1.2 METHODS OF DISTINGUISHING GEOMETRICAL ISOMERS

Several methods have been used to assign configurations to geometrical isomers. Methods based on chemical reactions such as cyclisation are generally reliable. Methods based on physical measurements such as dipole moment, melting point and boiling point are not always dependable. It is advisable to use results based on more than one type of property. Spectroscopic methods are useful in many cases.

22.1.2.1 Dehydration, Cyclisation

When two groups in the *cis* position can interact and cyclise, it provides a reliable method of assigning configuration. Maleic and fumaric acids are the two geometrical isomers of butenedioic acid. The lower melting isomer maleic acid, readily loses water upon mild heating and cyclises to an anhydride (Eq. 22.1a). It is assigned the *cis* configuration. The higher melting isomer, fumaric acid, also cyclises but only on prolonged heating at a higher temperature. It first isomerises to maleic acid and then cyclises to the same anhydride (Eq. 22.1b).

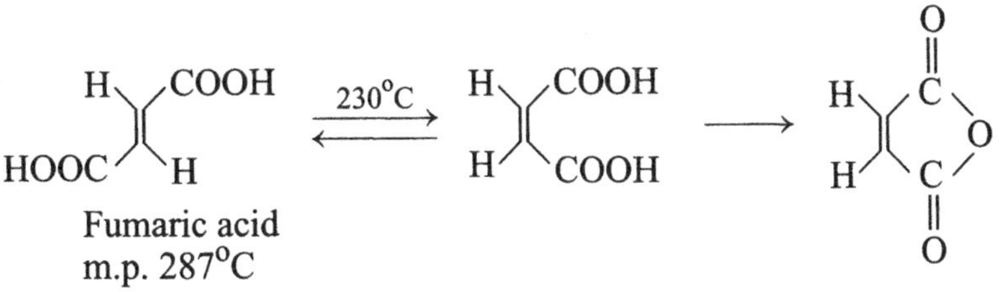

(b)

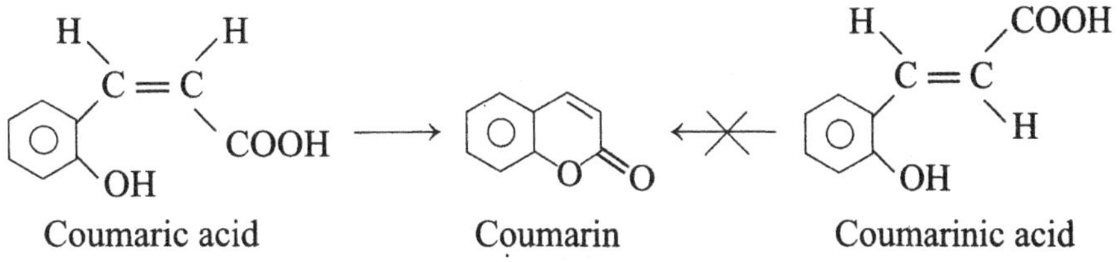

(c)

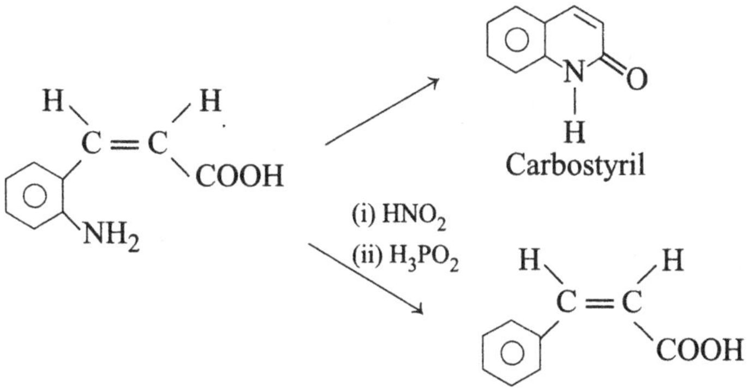

(d)

(e)

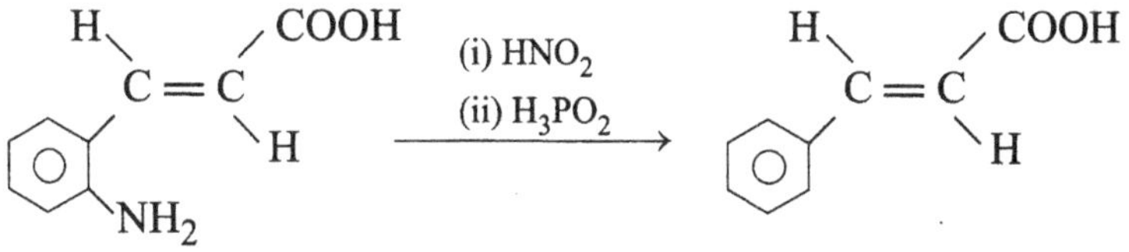

(f)

(22.1)

One of the geometrical isomers of *o*-hydroxycinnamic acid – called coumaric acid – cyclises upon heating to give a lactone called coumarin. This isomer is assigned the *cis* configuration (Eq. 22.1c). The other geometrical isomer, coumarinic acid, does not cyclise and is assigned the *trans* configuration. Both isomers are obtained by the Perkin reaction between salicylaldehyde and acetic anhydride (Chapter 16, Section 16.3.2.3).

Nitration of cinnamic acid gives a mixture of *p*- and *o*-nitrocinnamic acids. *o*-Nitrocinnamic acid upon reduction gives *o*-aminocinnamic acid, for which two geometrical isomers are known. The *cis*-isomer cyclises to a lactam, carbostyril (Eq. 22.1d). The other isomer does not give a lactam. Both isomers upon deamination via the diazonium salt give the corresponding cinnamic acids (Eq. 22.1e, f). The familiar, commercially available cinnamic acid is the *trans* isomer. The configurations of the two cinnamic acids have been assigned based on their relationship to the *o*-aminocinnamic acids, whose configurations are known from the cyclisation reaction.

22.1.2.2 Dipole Moments

Trans-1,2-dichloroethene has zero dipole moment. Eventhough the $C—Cl$ bonds are polar, they point in opposite directions and cancel each other [Fig. 22.2, (I)]. The symbol, $\leftrightarrow$, is used to indicate the dipole with the arrowhead pointing to the negative pole. In the *cis* isomer (II), the vectors do not cancel each other and *cis*-1,2-dichloroethene has a dipole moment, $\mu = 1.85$ D. In molecules of the type $ACH=CHB$, where the dipoles, $A—C$ and $B—C$ are in the same direction, that is, towards C in both cases (as in $CH_3—C$) or away from C in both cases (as in $Cl—C$), the *cis* isomer will possess a higher dipole moment than the *trans* isomer. When the dipoles $A—C$ and $B—C$ are in opposite directions, that is, towards C in one bond (as in $CH_3—C$) and away from C in the other bond (as in $Cl—C$), the *cis* will have a lower dipole moment than the *trans* isomer. An example of the latter type is 1-chloropropene. CH_3 is an electron donating group and Cl is electron withdrawing. The dipoles of the $CH_3—C$ and $Cl—C$ bonds are as shown [Fig. 22.2, (III) and (IV)]. In the *trans* isomer, the dipoles are additive and it will have a higher dipole moment. For more complicated molecules, the dipole moments cannot be predicted so easily.

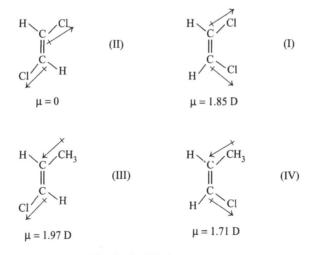

Fig. 22.2 Dipole moments

22.1.2.3 Physical Constants: Melting Point, Boiling Point

In general, the *trans* isomers have higher melting points as seen from the examples in Table 22.1. The *trans* isomer has greater symmetry and packs better in the crystal lattice accounting for the higher melting point. Correlation with the boiling points is not so reliable. For structures of the type (I) and (II) of Fig. 22.2, the *cis*-isomer has the higher boiling point. For structures of the type

(III) and (IV) of Fig. 22.2, the reverse is true. Rather than using the *cis–trans* label, it is safer to say that the isomer with the higher dipole moment has the higher boiling point. Intermolecular forces are stronger in molecules with the higher dipole moment (Table 22.1).

Table 22. 1 Physical constants of geometrical isomers

Compound	Melting point (°C)	Boiling point (°C)
cis-Cinnamic acid	68	
trans-Cinnamic acid	133	
Maleic acid	130	
Fumaric acid	287	
cis-Crotonic acid	15.5	
trans-Crotonic acid	72	
cis-1,2-Dichloroethene		60.3
trans-I,2-Dichloroethene		48.4
cis-2-Butene		3.7
trans-2-Butene		0.9
1-chloropropene, $CH_3-CH=CHCl$ (*trans*, $\mu = 1.97$ D)		37.4
1-chloropropene, $CH_3-CH=CHCl$ (*cis*, $\mu = 1.71$ D)		32.8

The *Auwers-Skita rule* is a generalisation according to which, in a pair of geometrical isomers, the *cis* isomer has higher boiling point, refactive index and density. This rule is of limited utility and has been applied mainly to terpenoids.

22.1.2.4 Stability and Heats of Hydrogenation

The heats of hydrogenation of *cis*- and *trans*-2-butene are about 119 and 115 kJ mol^{-1} (28.6 and 27.6 kcal mol^{-1}) respectively (Chapter 4, Section 4.4.4.1). Since both give butane upon hydrogenation, the difference of about 4 kJ mol^{-1} is taken as a measure of the greater thermodynamic stability of the *trans* isomer. The lower stability of the *cis* isomer is due to the existence of non-bonded interactions between the methyl groups. The larger the groups, the greater the non-bonded interaction and greater the difference in their heats of hydrogenation. For 4,4-dimethyl-2-pentene, the difference is 18.6 kJ mol^{-1} (4.3 kcal mol^{-1}) and for 2,2,5,5-tetramethyl-3-hexene it is 40 kJ mol^{-1} (9.4 kcal mol^{-1}), respectively. The heat of hydrogenation of *cis*-stilbene is higher than that of *trans*-stilbene by about 24 kJ mol^{-1} (5.7 kcal mol^{-1}). We have already seen how the phenyl rings of *cis*-stilbene are not coplanar which contributes to its greater instability due to steric inhibition of resonance (Chapter 13, Section 13.3). The stilbenes also illustrate the application of UV spectroscopy to assign the configuration.

22.1.2.5 Configuration of Ketoximes

The Beckmann rearrangement is a ready method for deciding the configuration of unsymmetrical ketoximes. This reaction has been discussed in Chapter 19, Section 19.3.1. The group that migrates and finally ends up on the nitrogen of the product amide is the *anti* group in the oxime. Benzil forms two isomeric monoximes, one with the phenyl *anti* and the other with the benzoyl group *anti*, designated as β-(II) and α-(III) respectively (Eq. 22.2). The configuration of the β-isomer was independently established by studying its formation by ozonolysis from 3,4-diphenylisoxazole-5-carboxylic acid (I). Both undergo Beckmann rearrangement and the products correspond to the migration of the *anti* group (Eq. 22.2). Based on several such observations, the generalisation that the anti group migrates in Beckmann rearrangement is generally accepted as valid. Conversely, the Beckmann rearrangement can be used as a tool to assign the configuration of unsymmetrical ketoximes.

22.1.3 GEOMETRICAL ISOMERISM IN NATURALLY OCCURRING ALKENES

Fatty acids present in oils and fats may be saturated, monounsaturated (one double bond) or polyunsaturated (more than one double bond). Oleic acid [*cis*-9-octadecenoic acid, $CH_3-(CH_2)_7-CH=CH-(CH_2)_7-COOH$] is one of the most widely occurring fatty acids. The double bond in this molecule is *cis*. The double bonds in most other natural fatty acids are also *cis*. During catalytic hydrogenation of oils to obtain hydrogenated fats, some of the double bonds get isomerised to the *trans* isomers which are referred to as *trans* fats. These are considered to be harmful from the dietary standpoint. Natural rubber (Chapter 6, Section 6.5.2) is a polyene with all the double bonds being *cis*. The corresponding all-*trans* polymer called gutta percha, does not have the elastic properties of rubber. Vitamin A and its precursor, β-carotene, are polyenes with an all-*trans* configuration. The key molecule involved in the chemistry of vision is retinal – a transformation product of vitamin A – with one double bond in the *cis*-configuration [Fig. 22.3, (I)]. This molecule when bound to a protein, undergoes photochemical *cis–trans* isomerisation. This is the basic chemical reaction [Fig. 22.3, (II) → (III)] which generates the nerve impulses that are transmitted to the brain and result in vision.

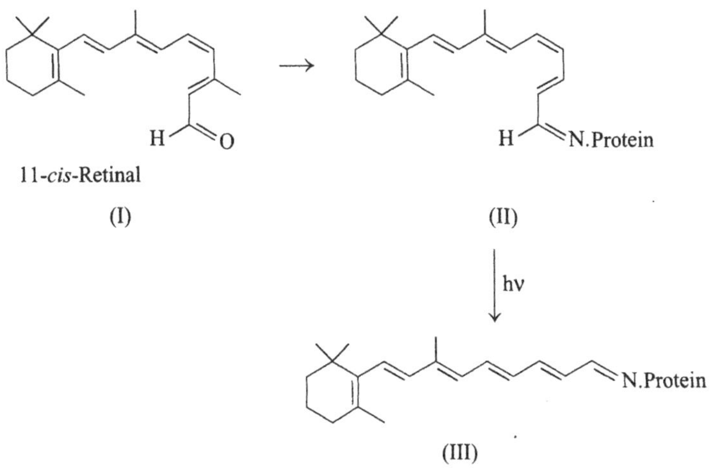

Fig. 22.3 *Cis–trans* isomerisation of retinal

22.2 CONFORMATIONAL ANALYSIS OF OPEN CHAIN MOLECULES

This topic has been introduced and the terms defined in Chapter 8, Section 8.2.3. The terms, conformation and configuration, have been discussed and defined in Chapter 21, Section 21.2.2. The name conformation or conformer is given to any one of the various orientations in space that atoms in a molecule can possess as a result of rotation around bonds. Rotation around a single bond is described as free rotation. However, certain energy barriers have to be crossed in going from one conformation to another. When these barriers are small and can be overcome by the thermal energy present in the molecule at room temperature, the rotation takes place readily and is described as free rotation. When the barrier to rotation is large, as in the case of *ortho*-substituted biphenyls, conformations cannot interconvert at room temperature and can be isolated (Chapter 21, Section 21.5.1).

22.2.1 CONFORMATIONS OF ETHANE

In molecules like ethane, the two carbons can be rotated relative to each other readily and an infinite number of conformations are possible. Potential energy of the molecule can be plotted as a function of the angle of rotation, also referred to as angle of torsion or dihedral angle (Fig. 22.4).

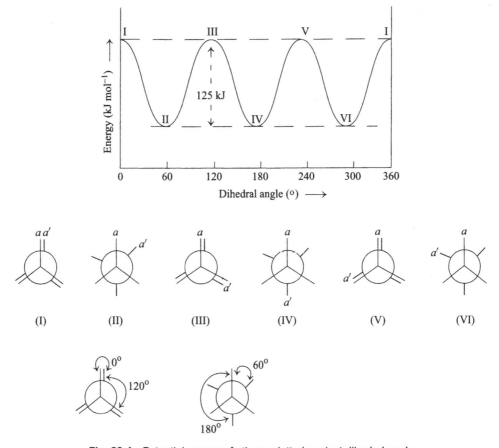

Fig. 22.4 Potential energy of ethane plotted against dihedral angle

These conformations are best represented by Newman projections. In the drawings, one pair of hydrogens has been labelled *a* and *a'*. They are of course identical. Conformations (I), (III) and (V), which are identical, belong to one type, and (II), (IV) and (VI), belong to another group. The angle between the $H—C—C$ plane on one carbon and the $C—C—H$ plane on the other carbon is called the *dihedral angle* or the *torsional angle*. In (I), the dihedral angle of $H_a—C—C—H_{a'}$ is zero. The bonds in the projection are directly behind one another, that is, eclipsed. This group of conformations are designated as *eclipsed* conformations. At the other extreme are conformations (II), (IV) and (VI), where the bonds are as far separated as possible. The dihedral angle of $H_a—C$ and $H_{a'}—C$ is 60°. This relationship is referred to as *gauche* or *skew*. In (IV), the dihedral angle of $H_a—C$ and $H_{a'}—C$ is 180°. In this conformation, H_a and $H_{a'}$ are *anti* to each other. The conformations (II), (IV) and (VI) are termed *staggered* conformations. An infinite number of intermediate conformations are possible. The different conformations are in equilibrium, with the greatest population existing in conformations corresponding to the energy minima—the staggered conformations. For ethane, the three staggered conformations are identical.

Based on thermodynamic calculations, the energy difference between these two extremes has been estimated to be about $12.5\,kJ\,mol^{-1}$ ($2.9\,kcal\,mol^{-1}$). This is the energy barrier to rotation in ethane. The barrier is not due to steric hindrance as in the case of the biphenyls since the hydrogen atom is very small. The eclipsed conformation is said to possess *torsional strain*. The source of this strain is believed to be two fold—destabilisation of the eclipsed form due to repulsion between the bonding electron pairs of the $C-H$ bonds and stabilisation of the staggered conformation due to more effective delocalisation of electrons. The barrier, $12.5\,kJ\,mol^{-1}$ is the sum of the torsional strain of the three pairs of eclipsed $C-H$ bonds. Each pair of eclipsed $C-H$ bonds is assigned a torsional strain of about $4.2\,kJ$ or $1\,kcal\,mol^{-1}$.

22.2.2 CONFORMATIONS OF BUTANE

The potential energy diagram for the rotation of butane around the $C2-C3$ bond is given in Fig. 22.5 (see also Chapter 8, Fig. 8.4).

There are two energy minima, corresponding to conformations (II)/(VI) and (IV) and two maxima corresponding to (I) and (III)/(V). The maxima are due to the eclipsed conformations. In (I), there are one set of CH_3, CH_3 and two sets of H, H eclipsing each other. In (III) and (V) there are two H, CH_3 and one H, H eclipsing each other. The energy minima correspond to the staggered conformations, (II)/(VI) and (IV). In (II) and (VI), the two CH_3 groups are *gauche* or *skew* and in (IV) they are *anti*. The energy of structure (IV) is arbitrarily taken as zero. The other energy levels are as shown in Fig. 22.5. The strain is maximum in conformation (I) where the two large groups,

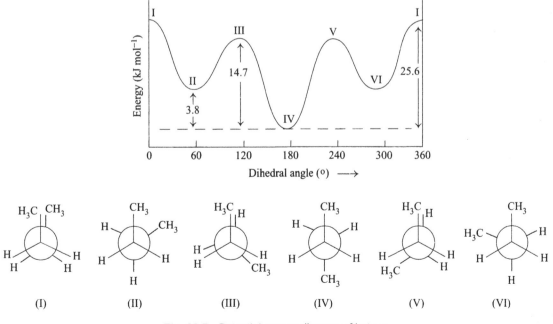

Fig. 22.5 Potential energy diagram of butane

CH_3, are eclipsed. This is not only torsional strain but also includes *steric strain*, otherwise called *van der Waals' strain*. At any given temperature, butane molecules will exist mostly in the staggered conformations, the population being more in the more stable conformation (IV), with the CH_3 groups *anti* to each other. The strain in the gauche conformation of butane is described as due to the *gauche interaction* between two CH_3 groups. Its magnitude is about $3.8 \, kJ \, mol^{-1}$ ($0.9 \, kcal \, mol^{-1}$). This interaction is called non-bonded interaction and is a repulsive interaction. It is essentially steric repulsion. Larger the size of the group, greater will be the strain due to this interaction.

In (III) and (V) the strain is mainly due to the eclipsing interaction of H and CH_3; two such interactions exist in each structure with a magnitude of about $7.4 \, kJ$ or $1.7 \, kcal \, mol^{-1}$ each.

22.2.3 CONFORMATIONS OF 1,2-DICHLOROETHANE AND ETHYLENE GLYCOL

1,2-Dichloroethane is similar to butane in its conformations. The nature of the non-bonded interaction between the two Cl atoms is not only steric. There is also electrostatic repulsion due to the polarity of the C—Cl bond which places a partial negative charge on the chlorine atom. Evidence in support of the *anti* and the *gauche* conformations is obtained from dipole moment measurement. The pure *anti* conformation is expected to have nearly zero dipole moment. In practice, 1,2-dichloroethane has a finite dipole moment of about 1 D due to the presence of the gauche conformer also, which has a non-zero dipole moment. The dipole moment increases with temperature because the equilibrium shifts and the population of the gauche conformation increases.

In 1,2-disubstituted ethanes, it is generally safe to assume that the *anti* conformation is more stable than the *gauche*. This situation may change if the *gauche* form is stabilised by intramolecular hydrogen bonding. The IR spectrum shows that ethylene glycol exists in the *gauche* conformation to a large extent (Fig. 22.7), with intramolecular hydrogen bonding.

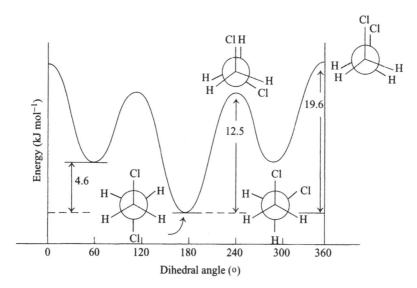

Fig. 22.6 Potential energy diagram of 1,2-dichloroethane

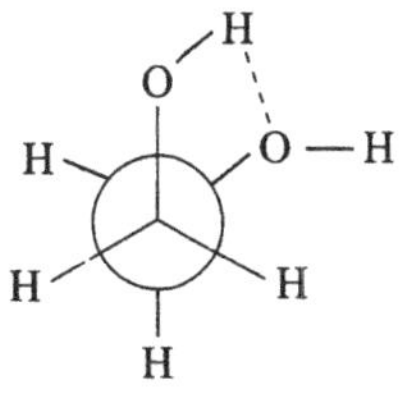

Fig. 22.7 Gauche conformation of ethyleneglycol

22.3 CONFORMATIONS OF CYCLOHEXANE

As seen in Chapter 8, cycloalkanes other than cyclopropane have non-planar conformations. Cyclohexane and larger rings can exist in non-planar conformations without much deviation from the tetrahedral bond angle and are therefore essentially free of angle strain. The stable conformations are those with minimum torsional strain, steric strain and non-bonded repulsive interactions. Cyclohexane has two conformations where there are no eclipsing interactions and all contiguous $C-C-C-C$ segments (butane units) are in staggered conformation with their $C-C$ bonds being *gauche*. These are called the *chair conformations* of cyclohexane. There are the two interconvertible chair forms, (I) and (II) (Fig. 22.8).

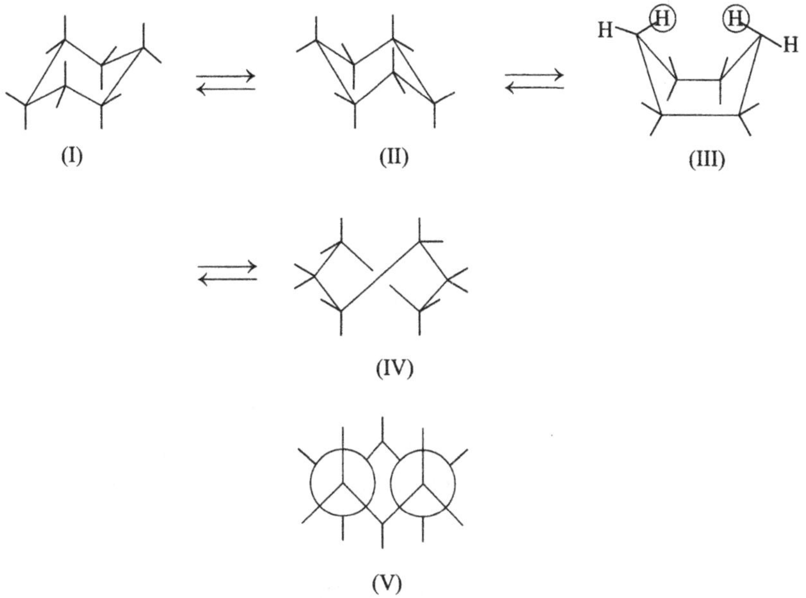

Fig. 22.8 Conformations of cyclohexane

Each 4-carbon segment of the ring is of the type $-CH_2-CH_2-CH_2-CH_2-$, a butane chain. There are six such butane units in the ring. Each of these is in the staggered conformation with the CH_2 groups *gauche*. Angle strain is absent. There are no eclipsing interactions (Fig. 22.9). The chair form can also be represented using the Newman projection [Fig. 22.8, structure (V)].

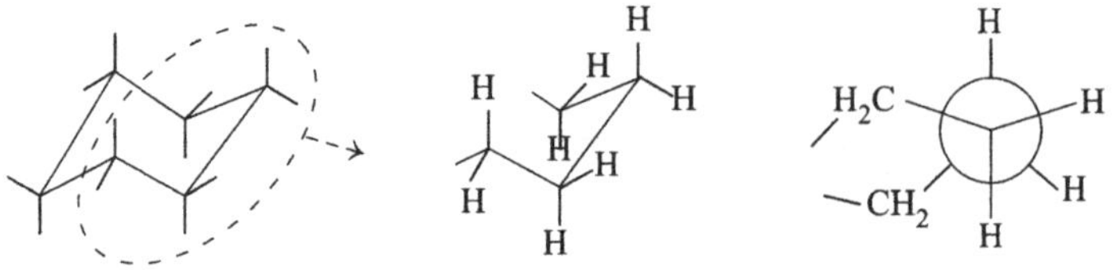

Fig. 22.9   Butane unit in cyclohexane in the chair form

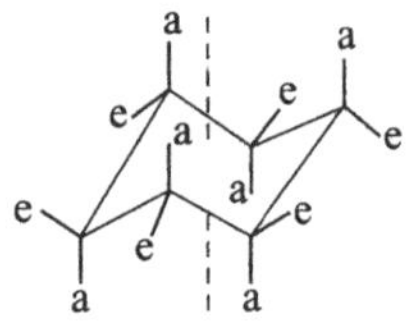

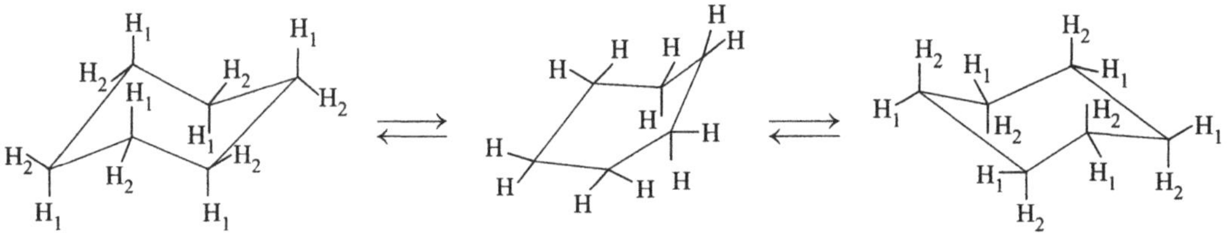

Fig. 22.10   Flipping of cyclohexane chair form

The chair has a six-fold alternating axis of symmetry with the axis passing through the centre of the molecule [Fig. 22.10, (I)]. Examination of the of the chair structure (I) shows that there are two types of C—H bonds. Six of the bonds are parallel to the axis, directed straight up and straight down, on alternate carbon atoms. These are called axial bonds and are marked 'a' in structure (I), Fig. 22.10. The other six are directed away from the ring and are called equatorial bonds, labelled 'e' in Fig. 22.10, (I) (the word 'equatorial' is related to the word, equator). The chair can be converted to another chair, by some degree of twisting of the molecule. This is called *flipping*. (II) and (IV), Fig. 22.10, represent the two chair forms. The axial and equatorial hydrogens in (II) are labelled H_1 and H_2 respectively for identification. Strucure (IV) is formed as a result of flipping. The axial bonds of (II) become equatorial bonds in (IV) and vice versa. Flipping takes place through one or more intermediate conformations, in which the strain is much more than in the chair forms. Structure (III) is an attempt to depict one such intermediate structure. Cyclohexane has other conformations which are also free of angle strain. One such is the '*boat form*' [Fig. 22.8, (III)]. More details of the boat form are shown in Fig. 22.11.

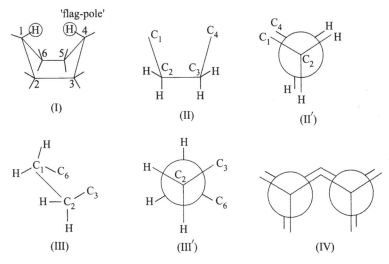

Fig. 22.11 Details of the boat conformation

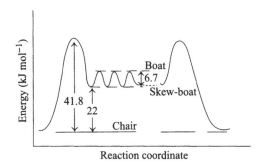

Fig. 22.12 Potential energy diagram of cyclohexane conformations

The butane unit, (II) and (II′) comprising of the carbons C2—C3 is in the eclipsed conformation. So is C5—C6. The other butane units such as those involving C1—C2, (III) and (III′), are in nearly staggered conformation. However, two sets of butane eclipsing interactions make the potential energy of the boat form quite high. Besides, the two hydrogens on C1 and C4 – called flag-pole hydrogens – are very close to each other, closer than the sum of their van der Waals' radii. This contributes to additional strain. All these interactions make the boat a very unstable conformation and it occupies an energy maximum in the potential energy curve. A small distortion of the boat gives conformations in which these unfavourable interactions are very much reduced. This distorted conformation, called a *twist-boat* or *skew-boat*, is depicted in structure (IV) (Fig. 22.8). The twist-boat is a highly flexible conformation capable of interconversion among a large number of such equivalent forms. The chair form on the other hand is relatively rigid even though it can also be converted to other conformations, but only at the expense of more energy. The energies involved are shown in the potential energy diagram (Fig. 22.12).

The chair form and the twist-boat represent energy minima, the boat form is at an energy maximum.

The energy values given in these discussions are enthalpy differences. The high degree of flexibility of the skew-boat confers it with a favourable entropy factor. The enthalpy difference, $-\Delta H$ of about 22 kJ mol^{-1}, between the chair and the twist-boat corresponds to a free energy difference of about 17.5 kJ mol^{-1}. From the free energy values (ΔG values), one can calculate the equilibrium composition of any two conformers using the equation,

$$\Delta G = -2.3\, RT \log K$$

Such values are useful in determining equilibrium composition of conformational mixtures. Some values relating ΔG and composition are listed in Table 22.2. From the data in this table – keeping in mind that the free energy difference between the chair and the twist boat is about 17.5 kJ mol^{-1} – it can be seen that at 25°C, only about 0.1% of the cyclohexane molecules will be in the twist-boat conformation.

Table 22. 2 ΔG and conformation equilibrium

$-\Delta G$ (kJ mol^{-1})	1.7	3.4	5.7	11.4	17.0	23.0
K (298°C)	2	4	10	100	1000	10000
Concentration of the more stable form at equilibrium (%)	67	80	91	99	99.9	99.99

22.3.1 MONOSUBSTITUTED CYCLOHEXANES

Henceforth, in our discussions we shall consider only .the chair form of cyclohexane. Monosubstituted cyclohexanes can exist in two conformations—with the substituent in the equatorial conformation or in the axial conformation (Fig. 22.13).

The two forms are interconvertible and are not of equal energy. A discussion of how a rough estimate of the energy difference between the two can be made is given below. Let us consider the axial form (I). The segment comprising of $C5 - C6 - C1 - CH_3$ of structure (I), is taken as a butane unit, (III), and the segment $C3 - C2 - C1 - CH_3$ is another butane unit (IV). Both are in staggered conformations. The orientation of the CH_3 is *gauche* with respect to C5 in (III). It is also *gauche* with respect to C3 in (IV). The cost of placing a CH_3 in a *gauche* conformation in butane in terms of energy is about 3.8 kJ mol^{-1} (Fig. 22.5). In axial-methylcyclohexane there are two such butane-like *gauche* interactions. The energy involved is 7.6 kJ mol^{-1}. One of the *gauche* interactions can be seen in the Newman projection in structure (V). The molecule should be projected from another angle to see the other *gauche* interaction. Considering the equatorial conformation (II), we can see from structures (VI), (VII) and (VIII), that the CH_3 group is *anti* to both C5 and C3. There are no *gauche* interactions. Arbitrarily assigning the equatorial-methylcyclohexane an energy level of zero, the potential energy diagram can be drawn (Fig. 22.14). The small energy minimum between

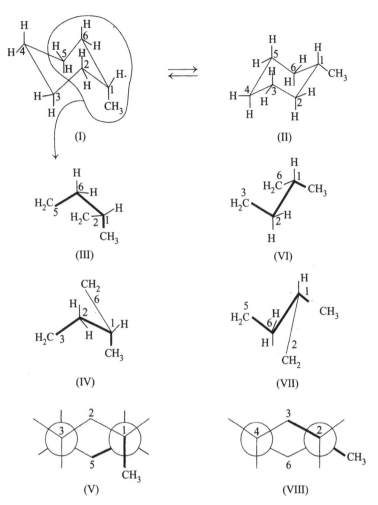

Fig. 22.13 Interactions in methylcyclohexane

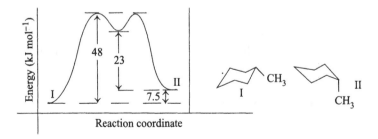

Fig. 22.14 Potential energy diagram for the interconversion of equatorial and axial methylcyclohexane

the two structures may represent skew-boat forms which may or may not be intermediates in flipping.

Taking the free energy difference between the two forms to be about $7.5\,kJ\,mol^{-1}$, it can be seen from Table 22.2, that at room temperature, a sample of methyl cyclohexane contains about 95% of the equatorial and 5% axial conformers. In monosubstituted cyclohexanes, the conformation in which the substituent is equatorial is more stable than the one where it is axial.

Table 22. 3 Conformational free energies of groups

Group	$-\Delta G\,(kJ\,mol^{-1})$
OH	2.9
Cl	1.7
Br	1.7
CH_3	7.5
C_2H_5	8.8
C_6H_5	10.9
COOH	7.1
tert-Butyl	> 21

In the case of large groups – as in *tert*-butylcyclohexane – the molecule is almost exclusively in the equatorial conformation. Such a molecule is said to be 'locked' in one conformation and the *tert*-butyl group is called the *anchoring group*.

The free energy difference between the equatorial and axial orientations of a group depends mainly on the size of the group. This is called conformational free energy. Some values are listed in Table 22.3. These are approximate and are useful for qualitative comparison only.

22.3.2 DISUBSTITUTED CYCLOHEXANES

There are four structurally isomeric dimethylcyclohexanes, 1,1-, 1,2-, 1,3- and 1,4-dimethylcyclohexane. 1,1-dimethylcyclohexane is not interesting from a stereochemical point of view. Each of the other three isomers possess two geometrical isomers, *cis* and *trans*. Two of the geometrical isomers are racemic ($\pm$). The other four are achiral. Let us carry out a partial conformational analysis of these isomers, estimate the energy differences between conformations and evaluate their relative stabilities.

22.3.2.1 1,2-Dimethylcyclohexane

There are two geometrically isomeric 1,2-dimethylcyclohexanes, *trans* and *cis*, and each of them has two chair conformations.

If we consider the six membered ring to be an uneven disc held perpendicular to the plane of the paper, the axial bonds are all directed straight up or straight down. The equatorial bonds are also directed up or down, but at an angle. In structure (II), the methyl groups are both axial, pointing in opposite directions and obviously *trans*. In the flipped conformation (I), the two methyl

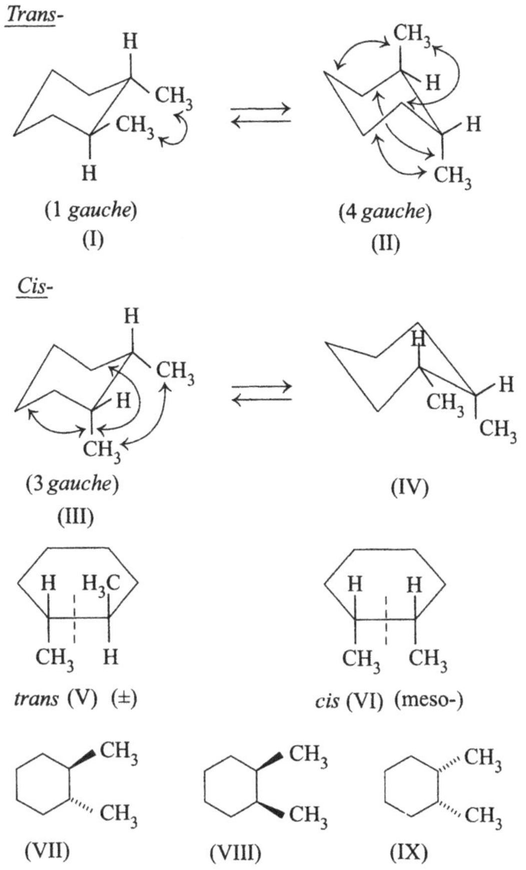

Fig. 22.15 1,2-Dimethylcyclohexanes

groups are equatorial, pointing up and down at an angle, and this is also classified as *trans*. The hydrogens on C1 and C2 in (I) are both axial and obviously *trans*. In 1,2-disubstituted molecules, the diaxial conformation and the diequatorial conformation are *trans*.

Structures (III) and (IV) are identical. The CH_3 groups are axial–equatorial. This is *cis*.

Let us look at the two conformations of the *trans* isomer. In conformation (I), each methyl group is equatorial. An equatorial CH_3 has no butane-type *gauche* interaction with the ring carbons (as observed in our analysis of methylcyclohexane). However, the two CH_3 groups are themselves in *gauche* orientation and account for one *gauche* interaction, indicated by a curved double-headed arrow in the drawing. This amounts to $3.8 \, \text{kJ mol}^{-1}$.

In the diaxial conformer (II), there are two axial CH_3 groups each with two butane-type *gauche* interactions, accounting for a total $4 \times 3.8 = 15.2 \, \text{kJ mol}^{-1}$ of energy. The methyl groups themselves are oriented *anti* to each other and do not interact with each other. The difference in energy between the diequatorial conformation (I), and the diaxial conformation (II), is about $11.4 \, \text{kJ mol}^{-1}$. From Table 22.2, we can estimate that at room temperature, 99% of the molecules will be in the diequatorial conformation.

Coming to the *cis*-isomer, the conformations (III) and (IV) are identical. The butane-type gauche interactions are:

for 1 axial CH_3 group: $2 \times 3.8 = 7.6 \, kJ \, mol^{-1}$

1 gauche interaction between the CH_3 groups: $1 \times 3.8 = 3.8 \, kJ \, mol^{-1}$

Total: $11.4 \, kJ \, mol^{-1}$.

The more stable conformer of the *trans* isomer (I), is thermodynamically more stable than the *cis* isomer. Needless to say, each isomer is a stable molecule, and the *cis*-isomer notwithstanding its lower thermodynamic stability, does not get converted to or equilibrate with the other isomer, spontaneously.

The *trans* isomer is chiral and exists as a a racemic or (±) mixture. The *cis*-isomer is meso. The planar representations of the two isomers (V) and (VI) help in arriving at this conclusion. (VI) has a plane of symmetry (dotted line); (V) does not have a plane of symmetry. This treatment is valid because even though the stable conformations are nonplanar, at some stage during interconversion, the molecules do assume planar shapes. Geometrical isomers are also represented as in (VII) for *trans* and (VIII) or (IX) for *cis*.

22.3.2.2 1,3-Dimethylcyclohexane

The conformations of *cis* and *trans*-1,3-dimethylcyclohexane are given Fig. 22.16. (I) and (II) represent the *cis* isomer, while (III) and (IV) represent the *trans*. Non-bonded interactions in each conformer are listed below.

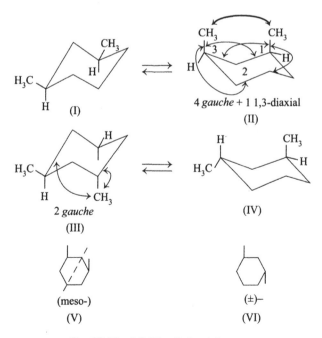

Fig. 22.16 1,3-Dimethylcyclohexane

cis-1,3-Diequatorial (I):

  gauche interactions: 0

cis-1,3-Diaxial (II):

  gauche interactions for 2 axial groups: $4 \times 3.8 = 15.2 \, \text{kJ mol}^{-1}$

  $+1,3\text{-diaxial interaction}$ between CH_3 groups: $\sim 7.6 \, \text{kJ mol}^{-1}$

  total: $\sim 22.8 \, \text{kJ mol}^{-1}$

The two axial CH_3 groups in (II) are very close to each other and offer a high degree of steric strain. This is what is referred to as 1,3-diaxial interaction in the listing above. Its magnitude is estimated to be about $7.6 \, \text{kJ mol}^{-1}$. The *cis* isomer exists almost entirely in the diequatorial conformation.

The two conformations (III) and (IV) of the trans isomer are identical. It has *gauche* interactions for one axial CH_3: $2 \times 3.8 = 7.6 \, \text{kJ mol}^{-1}$.

cis-1,3-Dimethylcyclohexane is thermodynamically more stable than the *trans* isomer. The *cis* isomer (V), is meso and the *trans*-isomer (VI) is chiral.

1,3-Diaxial interaction can be a stabilising factor if intramolecular hydrogen bonding exists. This is the case with *cis*-1,3-cyclohexanediol (Fig. 22.17). In this molecule, the diaxial form is more stable than the diequatorial form.

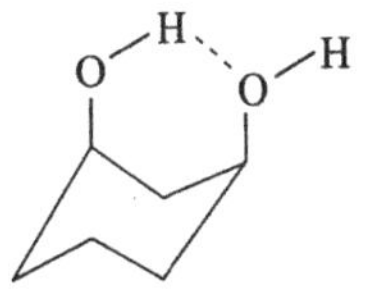

Fig. 22.17 Hydrogen bond in *cis*-1,3-cyclohexanediol

22.3.2.3 1,4-Dimethylcyclohexane

The trans isomer has two conformations, diequatorial (I) and diaxial (II) (Fig. 22.18).

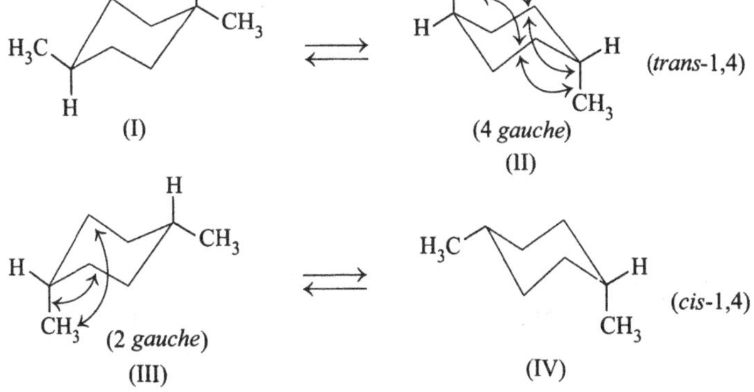

Fig. 22.18 1,4-Dimethylcyclohexane

trans-1,4-Diequatorial (I) has no *gauche* interactions.

trans-1,4-Diaxial (II) has 4 *gauche* interactions: $4 \times 3.8 = 15.2 \, \text{kJ mol}^{-1}$.

The diequatorial conformation is more stable.

cis-Equatorial–axial conformations (III) and (IV) are identical. Each has one axial group, accounting for 2 *gauche* interactions: $7.6 \, \text{kJ mol}^{-1}$.

The *trans*-isomer is thermodynamically more stable than the *cis* isomer. Both isomers are achiral. The plane passing through C1 and C4 is a plane of symmetry in each case. The stereochemical properties of the dimethylcyclohexanes are listed in Table 22.4.

Table 22. 4 Sterochemical properties of dimethylcyclohexanes

Name*	Disposition of CH$_3$ groups a = axial, e = equatorial	No. of *gauche* interactions**	Energy (kJ mol^{-1})	Chirality
cis-1,2-	a,e	3	11.3	meso
trans-1,2-	a,a	4	15.1	chiral
	e,e +	1	13.8	
trans-1,3-	a,e	1	13.8	chiral
cis-1,3-	a,a	4+ 1,3-diaxial interaction	22.6	meso
	e,e +	nil	0	
cis-1,4-	a,e	2	7.5	achiral
trans-1,4-	a,a	4	15.1	achiral
	e,e +	nil	0	

Less stable isomer listed first

+ More stable conformation

*** Includes 2 gauche interactions for each axial CH$_3$ and 1 for a pair of a,e and e,e in the case of 1,2-dimethylcyclohexane*

22.3.3 OTHER SUBSTITUTED CYCLOHEXANES

In general, a substituent prefers to be in the equatorial conformation. If more than one substituent is present, if there is a choice, the conformation with the bulkier substituent in the equatorial position is more stable. The more stable conformation of *cis*-2-methylcyclohexanol is the one in which the CH$_3$ group is equatorial [Fig. 22.19, (I)].

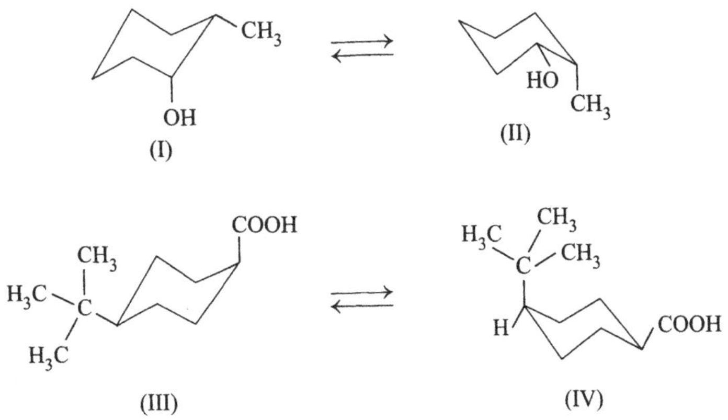

Fig. 22.19 2-methylcyclohexanol and 4-*tert*-butylcyclohexanecarboxylic acid

The more stable conformation of *cis*-4-*tert*-butylcyclohexanecarboxylic acid is (III), where the *tert*-butyl group is equatorial, predominantly so because of the large size of the *tert*-butyl group.

KEY POINTS

- Geometrical isomerism arises due to restricted rotation around double bonds. Alkenes of the type $(a, b)C-C(a, b)$ and oximes of the type $(a, b)C=NOH$, exhibit geometrical isomerism. In simple cases, the isomers can be designated by the prefixes *cis* and *trans* for alkenes and *syn* and *anti* for oximes. An unambiguous nomenclature is by the application of the sequence rules of Cahn–Ingold–Prelog which uses the prefixes Z and E.

- Generally, *trans* isomers have higher melting points, because the they pack better in the crystal lattice. In an alkene $HCX=CYH$, if the dipoles of the $C-X$ and $C-Y$ bonds are in the same direction with reference to C, they partially cancel each other in the *trans*-isomer. In the *cis*-isomer, the vectors of the two dipole moments do not cancel out. In such alkenes, the *cis*-isomer has a higher dipole moment. An example is 1,2-dichloroethene. If the dipoles of $C-X$ and $C-Y$ are in opposite directions with reference to C, the reverse is true. *trans*-1-Chloropropene has a higher dipole moment than the *cis*-isomer. The isomer with the higher dipole moment usually has the higher boiling point.

- Maleic and fumaric acids are assigned the *cis* and *trans* configurations based on the easy intramolecular dehydration of the former to form a cyclic anhydride. Similar chemical reactions often come in handy to assign the configuration of other geometrical isomers.

- Beckmann rearrangement is useful to assign the configuration of ketoximes, because the group that migrates from C to N is the one *anti* to the OH.

- *Cis* isomers usually have higher heats of hydrogenation. The heat of hydrogenation of *cis*-2-butene is about $4\,kJ\,mol^{-1}$ more than that of *trans*-2-butene.

- A molecule like ethane has an infinite number of conformations as a result of free rotation around the $C—C$ single bond. Although rotation around the single bond takes place readily at room temperature, it is not strictly 'free'. There is a barrier to such rotation. A plot of potential energy against dihedral or torsional angle has maxima and minima. Dihedral angle is best seen when Newman projections are used. The potential energy graph has maxima corresponding to dihedral angle of $0°$, and minima corresponding to a dihedral angle of $60°$. The former conformation is referred to as eclipsed and the latter, the staggered conformation. The energy barrier in ethane is about $12.5\,kJ\,mol^{-1}$ (corresponding to about $4.2\,kJ\,mol^{-1}$ for each pair of eclipsed $C—H$ bonds).

- In butane, rotation around the C2—C3 bond gives rise to two energy minima, one in which the two CH_3 groups are at $180°$ (*anti*) and the other where the CH_3 groups are at $60°$ (*gauche* or *skew*). Both are staggered conformations. The graph has two energy maxima, one in which the CH_3 groups are at a dihedral angle of $0°$ and the other where two pairs of H, CH_3 are at $0°$. Both are eclipsed conformations. The energy difference between the energy minima corresponding to the *skew* and *anti* conformations is about $3.8\,kJ\,mol^{-1}$. This is the magnitude of one $CH_3—CH_3$ *skew* or *gauche* interaction.

- The pattern is the same in other disubstituted ethanes as well. The magnitude of the *gauche* interaction increases with the size of the groups concerned. In 1,2-dichloroethane, the chlorine atoms repel each other due to electrostatic repulsion as well. The dipole moment of 1,2-dichloroethane is low due to the fact that most of the molecules are in the *anti* conformation (expected dipole moment $\mu = 0$), but is non-zero due to the fact that the *gauche* conformation is also present. The dipole moment is more at higher temperatures, because at the higher temperature the population of the *gauche* conformation increases.

- Cyclohexane is a non-planar molecule. Several strain-free conformations are possible. Of these, one pair of identical conformations – called the chair conformations – has the lowest energy. Another conformation, called the twist-boat, is much less stable but is more flexible than the chair. The energy difference between the chair and the twist-boat is quite high so that at room temperature, in a sample of cyclohexane almost all the molecules will exist in the chair conformation. Yet another conformation, called the boat conformation, occupies a maximum in the potential energy graph.

- In discussions of the conformations of cyclohexane, only the chair conformation is taken into account. In this conformation, six $C—H$ bonds are parallel to the axis of the ring and are called axial bonds. The other six are called equatorial bonds. One chair form of cyclohexane can be 'flipped' to another equivalent chair form. In this process, the bonds which were axial in the original structure, become equatorial and vice versa.

- Monosubstituted cyclohexane, as illustrated by methylcyclohexane, can have two chair conformations—one in which the CH_3 is equatorial and the other in which the CH_3 is axial. An axial methyl group has two butane-type gauche interactions while the equatorial CH_3 has none. Axial methylcyclohexane is about $2 \times 3.8 \, kJ \, mol^{-1}$ higher energy than the equatorial. In a sample of methylcyclohexane, most of the molecules will be in the equatorial methyl conformation.

- There are four dimethylcyclohexanes, 1,1-dimethyl, 1,2-dimethyl, 1,3-dimethyl and 1,4-dimethylcuclohexane. Each of the last three has two geometrical isomers, *trans* and *cis*. Each geometrical isomer can be flipped to another conformation where the axial and equatorial nature of the groups gets reversed. Relative stabilities of the conformations can be compared by counting butane-type *gauche* interactions.

EXERCISES

SECTION I

1. Designate the geometrical isomers of the following molecules except (b), using the E, Z notation.

    (a) 3-Chloro-3-phenylpropenoic acid
    (b) The oximes of isopropyl phenyl ketone (syn/anti notation)
    (c) Oleic acid
    (d) Crotonic acid
    (e) α-Chlorocinnamic acid
    (f) 3-Methyl-2-phenyl-2-pentene

2. Draw the most stable conformation of each of the following.

    (a) Chlorocyclohexane
    (b) *cis*-3-Methylcyclohexanol
    (c) 1,2-Diphenylethane
    (d) Dimethyl succinate
    (e) *cis*-4-*tert*-Butylcyclohexyl chloride

SECTION II

1. Identify the structure and stereochemistry of the oxime which gives acetanilide upon Beckmann rearrangement.
2. Draw the structures of the geometrical isomers of cyclohexane-1,2-dicarboxylic acid. Both form cyclic anhydrides upon heating. Draw the structures of the anhydride. How can the isomers be differentiated? (Hint: Can they be resolved?)
3. Which isomer of cyclohexane-1,3-dicarboxylic acid will form a cyclic anhydride upon heating? Draw the structure of the anhydride.

4. Draw all the geometrical isomers of $CH_3-CH=CH-CH=CH-COOH$. Name them using the E, Z notation.

CHALLENGING QUESTIONS

1. Do a conformational analysis of methylcyclohexanecarboxylic acids (I), (II) and (III), the same way as was done for dimethylcyclohexanes. Assume that the COOH group is about the same size as CH_3, for the purpose of calculating gauche interaction energies.

2. One of the geometrical isomers of 3-hydroxycyclohexanecarboxylic acid forms a lactone readily by an acid catalysed intramolecular esterification. The same lactone can be obtained from the potassium salt of one of the geometrical isomers of 3-chlorocyclohexanecarboxylic acid. Identify the reactive isomer in each reaction. Give the mechanisms of the reactions. Draw stereochemical structures. (Hint: The second reaction is a nucleophilic displacement of the chloride ion by the carboxylate.)

23 Bioorganic Chemistry – Carbohydrates and Vitamins

(Natural Products – I)

OBJECTIVES　In this chapter, you will learn about,

- how carbohydrates are classified
- names and structures of aldoses upto 6 carbons
- cyclic structures
- representation of structures by Fischer projection, Haworth formulae and in the chair conformations
- mutarotation of glucose, mechanism
- constitution and configuration of glucose
- structure of fructose
- oxidation, reduction, osazone formation, epimerisation, chain lengthening, chain shortening, glycoside formation, ether formation and acetylation of glucose
- disaccharides: reactions and structure of sucrose
- polysaccharides: reactions and structure of starch and cellulose
- structure of ascorbic acid, synthesis
- structure of pyridoxine

23.1 CARBOHYDRATES: MONOSACCHARIDES

Carbohydrate is the name given to a group of natural products comprising of simple sugars like glucose and polymeric substances like starch and cellulose which are derived from simple sugars. The molecular formula of glucose and other isomeric sugars is $C_6H_{12}O_6$, which can be written as $C_n(H_2O)_n$, formally hydrates of carbon—hence the name carbohydrates. The sugars are polyhydroxy aldehydes or polyhydroxy ketones with the root name saccharide. Glucose which has one sugar unit is a *monosaccharide*. Sucrose and lactose which are made up of two sugar units each, are *disaccharides*. A molecule made up of a few, finite number of sugar units is called an *oligosaccharide*. Molecules made up of a large number of sugar units, like starch and cellulose, are called *polysaccharides*.

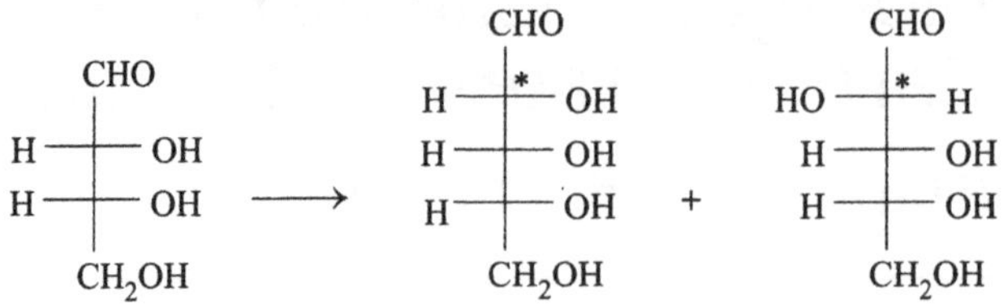

$$
\begin{array}{c}
\text{CHO} \\
\text{H} \longmapsto \text{OH} \\
\text{H} \longmapsto \text{OH} \\
\text{CH}_2\text{OH}
\end{array}
\longrightarrow
\begin{array}{c}
\text{CHO} \\
\text{H} \overset{*}{\longmapsto} \text{OH} \\
\text{H} \longmapsto \text{OH} \\
\text{H} \longmapsto \text{OH} \\
\text{CH}_2\text{OH}
\end{array}
\;+\;
\begin{array}{c}
\text{CHO} \\
\text{HO} \overset{*}{\longmapsto} \text{H} \\
\text{H} \longmapsto \text{OH} \\
\text{H} \longmapsto \text{OH} \\
\text{CH}_2\text{OH}
\end{array}
$$

(I)   (II)   (III)

Fig. 23.1 Creation of a new asymmetric centre

Glucose is 2,3,4,5,6-pentahydroxyhexanal. Fructose is 1,3,4,5,6-pentahydroxy-2-hexanone. The former is an *aldose*, specifically an *aldohexose*. The latter is a *ketose*, specifically a *ketohexose*. The simplest aldoses are the tetroses which have two pairs of enantiomers—threose and erythrose (Chapter 21, Section 21.2.1). The tetroses have two asymmetric centers and have $2^n = 4$ stereoisomers (n = the number of asymmetric carbons). The pentoses with three asymmetric centres have $2^3 = 8$ stereoisomers composed of 4 pairs of enantiomers. The hexoses have 16 stereoisomers, that is, 8 pairs of enantiomers.

By inserting a H—C—OH unit between C1 and C2 of D-erythrose (I), a third asymmetric centre, (∗), is created (Fig. 23.1). The new pentose can be either one of the two isomers, (II) or (III). Similarly, two isomers from L-erythrose and two each from D- and L-threose are possible.

Building of the D-aldoses from D-glyceraldehyde upto the eight D-aldohexoses, is shown in Fig. 23.2. The Fischer convention for the D-configuration, with the OH present on the right on the bottom-most asymmetric carbon, is followed. This also happens to be the absolute configuration.

23.1.1 CONFIGURATION AND NAMES OF THE ALDOHEXOSES

Configurations and names of the aldohexoses are as shown in Fig. 23.2. A simple procedure to draw these structures and to name them is described below. Eight carbon skeletons are drawn as in (I), Fig. 23.3. A hydroxyl group is placed on the right (R) on C5 in all the structures (R, R, R, R, R, R, R, R). Proceed to C4, place 4 OH on the right and 4 on the left, R, R, R, R, L, L, L, L. Then on C3, R, R, L, L, R, R, L, L, and on C2, R, L, R, L, R, L, R, L. Constructed this way, the third structure will look like (II). After drawing all the eight structures in this sequence, as in Fig. 23.2, names are given, following the mnemonic,

All	**Al**truists	**Gl**adly	**Ma**ke	**Gu**m	**I**n	**Gal**lon	**Ta**nks
Allose	**Al**trose	**Gl**ucose	**Ma**nnose	**Gu**lose	**I**dose	**Gal**actose	**Ta**lose

23.1.2 RING STRUCTURE

The polyhydroxyaldehyde structures in Fig. 23.2 do not truly represent the structures of aldohexoses. Consider D-glucose, the equilibrium represented in Fig. 23.4, (I) ⇌ (II) ⇌ (III), exists in solution. This is a type of tautomerism and is called *ring-chain tautomerism*.

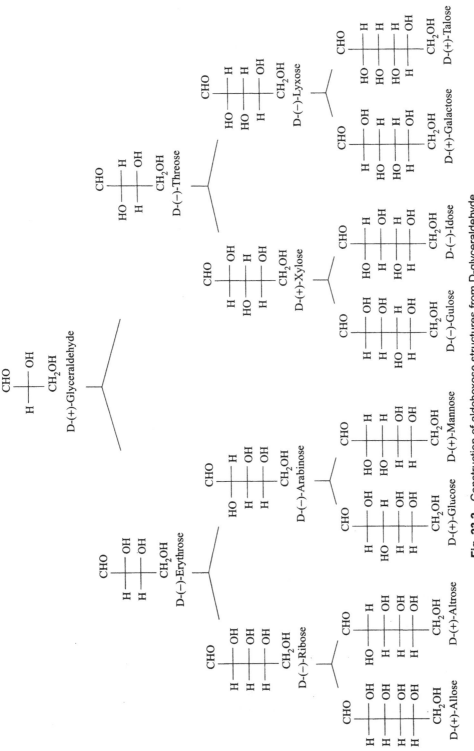

Fig. 23.2 Construction of aldohexose structures from D-glyceraldehyde

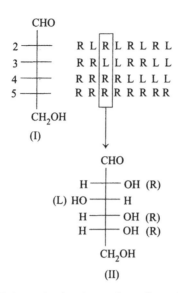

Fig. 23.3 Scheme for drawing conformations of D-hexoses

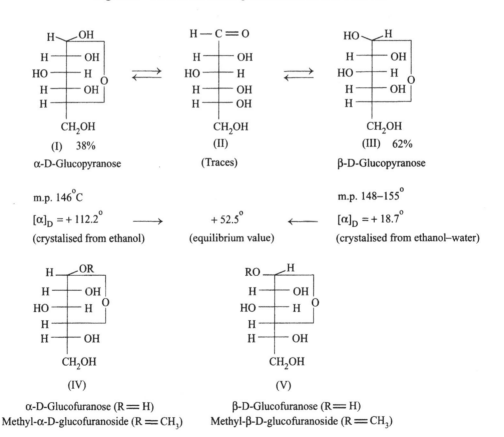

Fig. 23.4 Mutarotation of glucose

When D-(+)-glucose is crystallised from ethanol or from a cold saturated aqueous solution, crystals are obtained with a melting point of 146°C. This is ordinary glucose and is called α-D-glucose. When crystallised from a water–ethanol mixture or from hot aqueous solution, crystals with a melting point of 148–155°C are obtained which are designated β-D-glucose. When an aqueous solution of each is freshly prepared, the α-form has a specific rotation of +112.2° and the β-form, +18.7°. On standing, the rotation of both solutions change and a final value of +52.5° is reached. This change in rotation is called *mutarotation.*

It is unequivocally known by X-ray diffraction studies that the two forms are hemiacetals and are stereoisomers called *anomers*, with structures (I) and (III). In solution, the two isomers interconvert through the aldehyde form (II), to reach an equilibrium mixture. From the specific rotation of the equilibrium mixture, the composition can be calculated to be 38% and 62% respectively. Mutarotation takes place faster if the solution is heated or if it is made mildly acidic or alkaline. The mechanism of hemiacetal formation is given in Fig. 23.5.

Structures (I) and (II) are hemiacetals formed by the addition of the C5-hydroxyl group to the carbonyl group (Chapter 16, Section 16.3.1.1). Usually, hemiacetals are unstable molecules and revert to the parent aldehyde and alcohol. However, when the reaction is intramolecular and the hemiacetal is cyclic, it is stable and the equilibrium is in favour of the hemiacetal [Fig. 23.5, structures (I) and (II)]. In the case of glucose, at the equilibrium of mutarotation, the aldehyde form is found only in traces. When the hemiacetal is formed, a new asymmetric centre is created with two possible configurations, which are the two anomers.

Addition of the C5-hydroxyl to the aldehyde gives the 6-membered cyclic hemiacetal. Because of its relationship with pyran – the 6-membered heterocycle with one oxygen – it is described as a pyranose structure. The names of (I) and (III) in Fig. 23.4 are, α-D-glucopyranose and β-D-glucopyranose respectively. Two anomeric 5-membered hemiacetals are also possible. They are called α-D-glucofuranose and β-D-glucofuranose respectively [(IV) and (V), R = H, Fig. 23.4]. These are not included as components of the mutarotation equilibrium, but have been detected in very small concentration by NMR spectroscopy. The furanose form of glucose has been isolated only in the form of the methyl glycosides, methyl glucofuranosides (IV) and (V); R = CH₃ (Fig. 23.4).

23.1.2.1 Conformation of the Pyranose Ring

In the Haworth representation of the pyranose ring, structures (I) and (II) in Fig. 23.5, the 6-membered ring is drawn as a planar structure, perpendicular to the plane of the paper. The conformations in chair form are shown in (III) and (IV). In the Haworth structure and in the chair form, the ring oxygen is shown at the top right by convention. The lower portion of the ring is assumed to be in front of the plane of the paper, that is, the plane passing through C1 and C4.

23.1.2.2 Oxidation of Glucose

Even though the concentration of the open chain aldehyde form is very small, glucose and the other monosaccharides are *reducing sugars* in the sense that they reduce reagents like Tollen's reagent

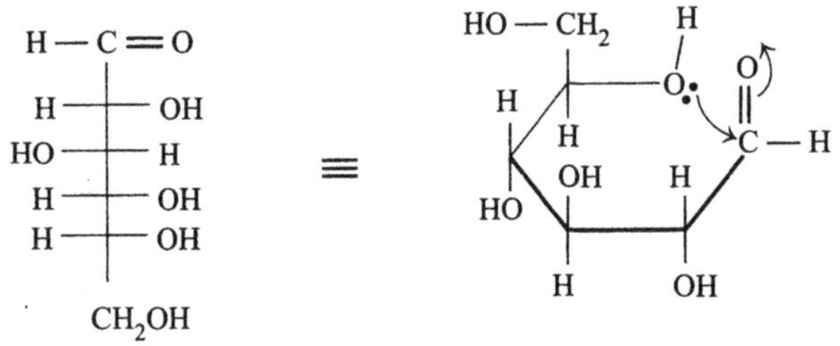

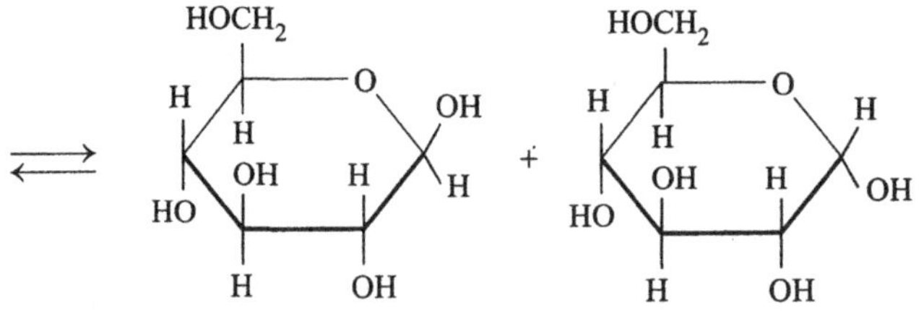

β-D-Glucopyranose (Haworth formula)

(I)

α-D-Glucopyranose

(II)

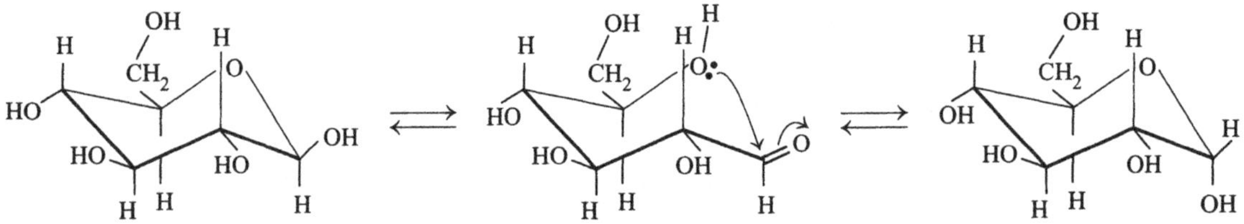

β-D-Glucopyranose (chair conformation)

(III)

Aldehyde form

α-D-Glucopyranose

(IV)

Fig. 23.5 Hemiacetal formation: Haworth formulae and chair conformations

and Fehling's solution, which is behaviour characteristic of aldehydes. The product of oxidation by Tollen's reagent is D-gluconic acid or its salt, gluconate. A better reagent for oxidation is bromine in buffered solution (Fig. 23.6). Gluconate upon acidification forms the five membered γ-latone preferentially, and also the six membered δ-lactone.

In Fig. 23.6, the structure of glucose (I), is drawn using the Haworth projection formula. The wavy lines for the anomeric bonds imply that the stereochemistry is not specified and the OH can be either α or β. Oxidation of glucose or gluconic acid by nitric acid will yield the dicarboxylic acid, glucaric acid. The general name of such dicarboxylic acids is *aldaric acid.* The general name for the monocarboxylic acids like gluconic acid is *aldonic acid.*

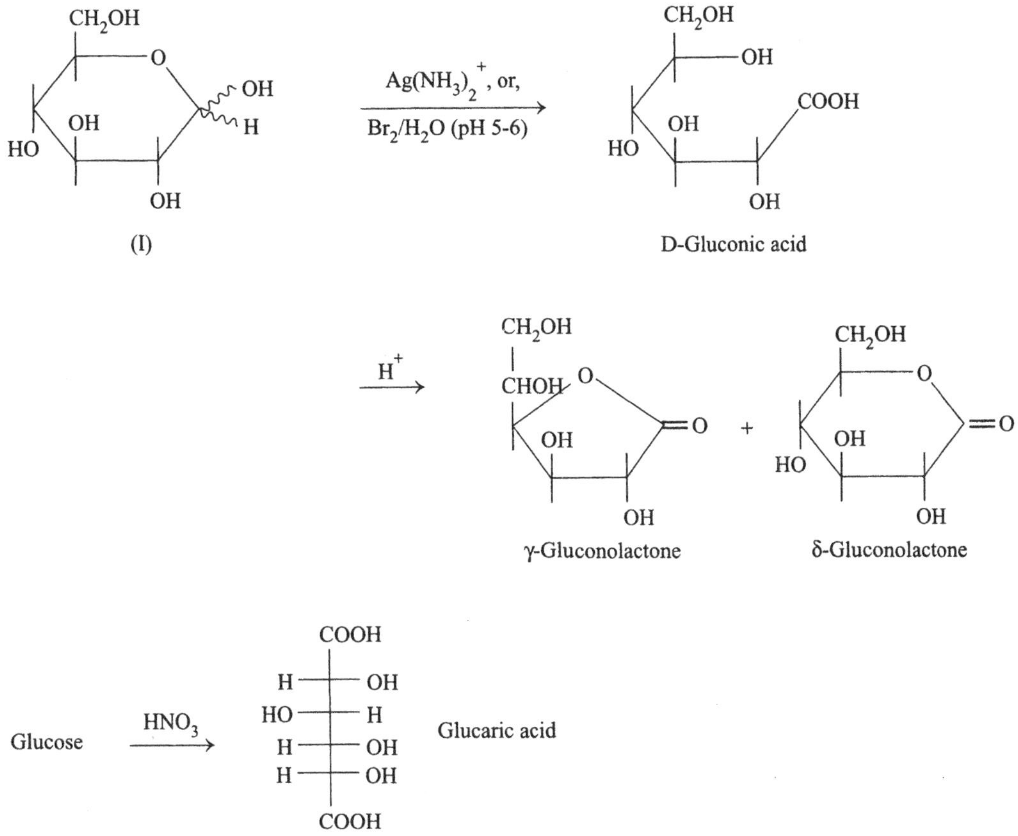

CH$_2$OH ─── O OH H

Ag(NH$_3$)$_2$$^+$, or,
Br$_2$/H$_2$O (pH 5-6)

(I)

D-Gluconic acid

H$^+$

γ-Gluconolactone   δ-Gluconolactone

Glucose   HNO$_3$

COOH
H ──── OH
HO ──── H
H ──── OH
H ──── OH
COOH

Glucaric acid

Fig. 23.6 Oxidation of glucose

23.1.2.3 Constitution and Configuration

The open chain aldohexose structure of glucose was already known in 1888, when Emil Fischer started the studies to establish its configuration. Some of the important reactions which lead to the structure and configuration are outlined below.

Lengthening of the chain, Kiliani–Fischer synthesis: D-(−)-Arabinose, an aldopentose, is converted to two isomeric aldohexoses (which differ in the configuration at C2) by the sequence of reactions outlined in Fig. 23.7. The two hexoses obtained from D-arabinose are D-glucose and D-mannose. This is known as the Kiliani–Fischer synthesis for lengthening of the aldose chain. The cofigurations at C3, C4 and C5 of D-glucose and D-mannose are the same as in D-arabinose.

Arabinose itself can be obtained in two successive Kiliani–Fischer syntheses from D-glyceraldehyde.

Glyceraldehyde → erythrose + threose
Erythrose → arabinose + ribose

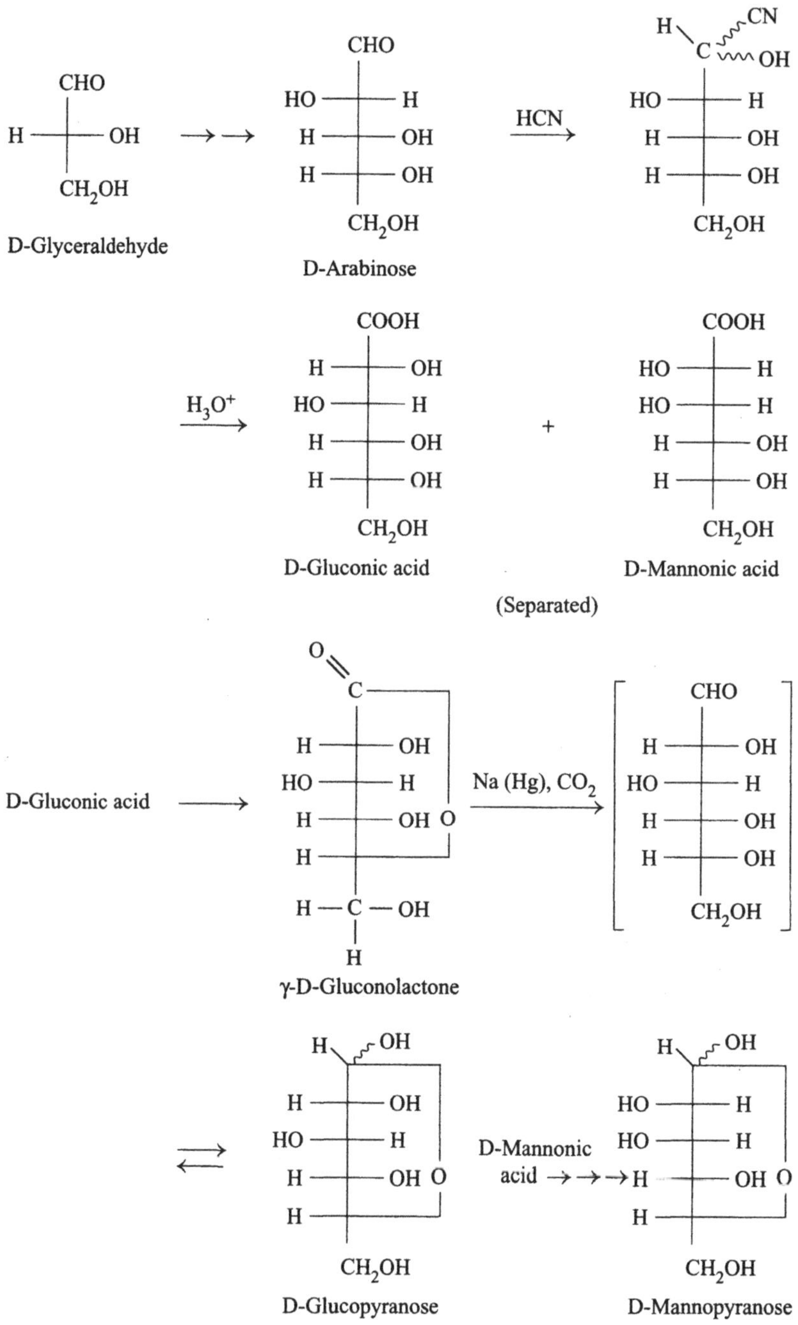

Fig. 23.7 · Kiliani–Fischer synthesis

Configuration of glucose and mannose: Oxidation to aldaric acids was used by Fischer to establish the configuration of glucose (Fig. 23.8).

This scheme is not exactly same as the method employed by Fischer. The scheme given here is only to illustrate the methodology. The structure and configuration of erythrose (I), is known from its relationship to *meso*-tartaric acid. Of the two pentoses obtained from erythrose, (III) and

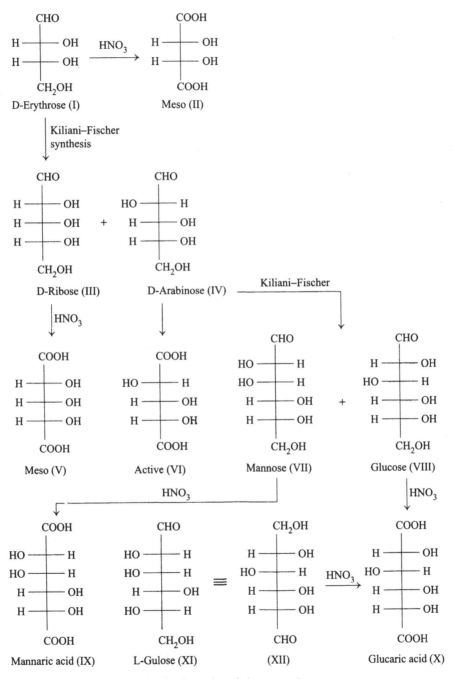

Fig. 23.8 Configuration of glucose and mannose

(IV), the one which gives a *meso*-aldaric acid (V), is ribose. The one which gives an optically active aldaric acid (VI), is arabinose. Arabinose upon undergoing Kiliani–Fischer synthesis, gives mannose (VII) and glucose (VIII), which further yield the corresponding aldaric acids, (IX) and (X). Mannose and glucose differ only in the configuration at C2. Such stereoisomers are called *epimers*. Differentiation of the configuration at the epimeric centres of D-glucose and D-mannose was carried out by Fischer in an ingenious way. A sugar of the L-series, L-gulose (XI = XII), also gives the same aldaric acid (X) as glucose. On the other hand, the aldaric acid (IX), obtained from D-mannose is unique. No other sugar can give this aldaric acid. Hence, D-glucose is assigned the configuration (VIII) and mannose, (VII).

23.1.2.4 Fructose

Fructose and glucose are the two monosaccharides which combine to form the disaccharide, sucrose. Fructose is a ketohexose—1,3,4,5,6-pentahydroxy-2-hexanone (I) (Fig. 23.9). Even though it is a ketone, fructose is a reducing sugar. This is because in the alkaline solution in which reactions such as Tollen's test and Fehling's test are carried out, isomerisation to the aldehyde takes place (Fig. 23.9).

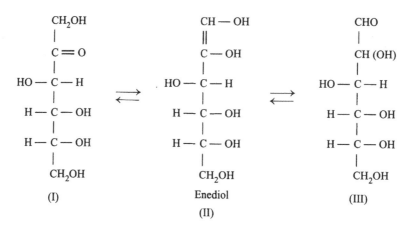

Fig. 23.9 Aldose–ketose interconversion

The keto form tautomerises to the enol, which in this case is an *enediol* (II). This can tautomerise either to the original ketone or to the aldehyde form (III). Like glucose, fructose also exists in the cyclic hemiacetal (hemiketal) form [Fig. 23.10, structure (II)].

The position of the keto-group in the chain is established by its conversion to 2-methylhexanoic acid, by the sequence of reactions, (I) → (III) → (IV) → (V). As with glucose, both the 6-membered pyranose ring and the 5-membered furanaose ring are possible and have been prepared as the respective methyl fructosides, (VI) and (VII) (Fig. 23.10).

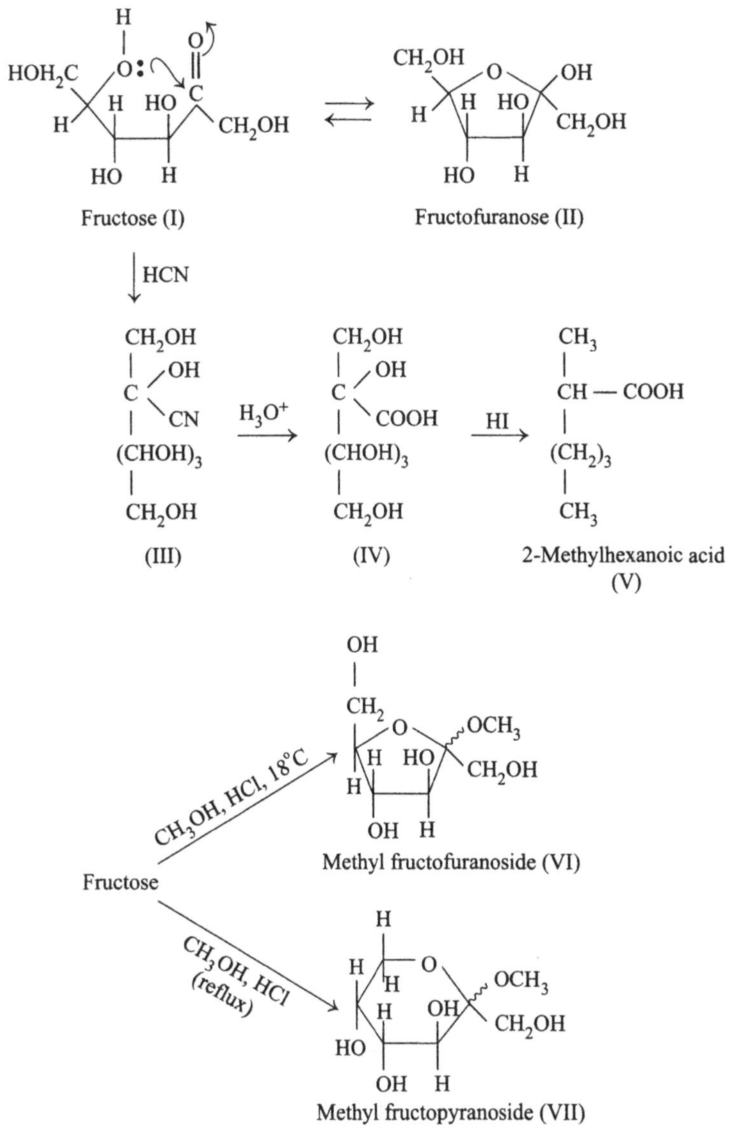

Fructose (I)
Fructofuranose (II)
HCN
H_3O^+
HI
(III)
(IV)
2-Methylhexanoic acid
(V)
Fructose
CH_3OH, HCl, 18°C
Methyl fructofuranoside (VI)
CH_3OH, HCl (reflux)
Methyl fructopyranoside (VII)

Fig. 23.10 Structure of fructose

23.1.3 Reactions of Glucose

Glucose is produced in plants by photosynthesis using carbon dioxide and water. It is stored as starch and cellulose. Glucose is the biological fuel for the production of energy in living organisms. It is also the metabolic precursor of many biomolecules.

Oxidation to gluconic acid and glucaric acid has already been mentioned (Fig. 23.6). Reduction of the aldehyde function by reagents like sodium borohydride, or by catalytic hydrogenation gives the hexahydric alcohol, glucitol or sorbitol which is also obtained by the reduction of the ketose,

L-sorbose (see Fig. 23.28 for the structures of sorbitol and L-sorbose). The general name of these hexahydric alcohols is *alditol*.

23.1.3.1 Osazone Formation

Glucose reacts as an aldehyde with one equivalent of phenylhydrazine to form a phenylhydrazone. If excess phenylhydrazine is used, it oxidises the secondary alcohol at C2 to the ketone which reacts with another equivalent of phenylhydrazine to give an osazone (Fig. 23.11). Altogether, three molecules of phenylhydrazine are utilised for osazone formation.

    Glucose and mannose are epimeric aldoses and produce the same osazone. The osazone can be converted to the dicarbonyl compound by treating with benzaldehyde and H^+. Benzaldehyde removes the phenyl hydrazine as benzaldehyde phenylhydrazone.

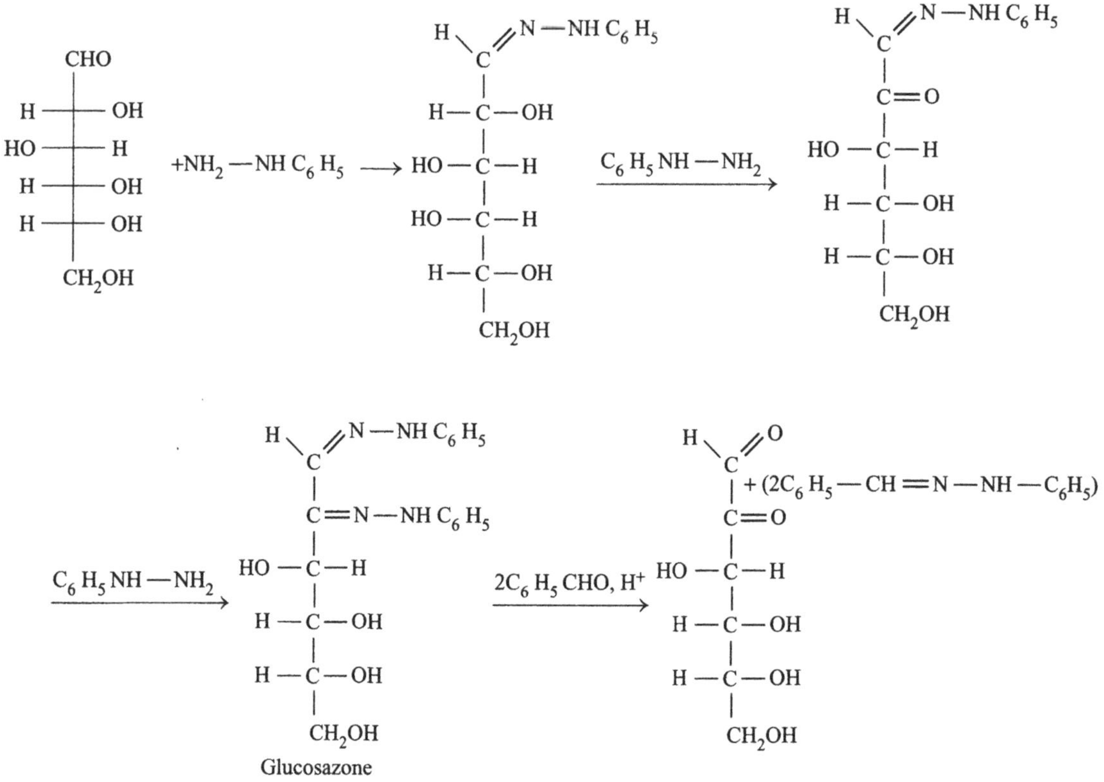

Fig. 23.11   Osazone formation

23.1.3.2 Chain-Lengthening and Chain-Shortening

Chain-lengthening by Kiliani–Fischer synthesis has already been discussed (Fig. 23.7). Chain-shortening can be carried out by Ruff degradation. This is illustrated by the conversion

$$\text{CHO} \xrightarrow[\text{}]{\text{Br}_2, \text{H}_2\text{O}} \text{COOH} \xrightarrow[\text{}]{\text{Ca(OH)}_2} \text{COO}^- \tfrac{1}{2}\text{Ca}^{2+} \xrightarrow[\text{}]{\text{H}_2\text{O}_2, \text{Fe}^{+++}} \text{CHO}$$

$$\begin{array}{c}
\text{CHO} \\
| \\
\text{CHOH} \\
| \\
\text{(CHOH)}_n \\
| \\
\text{CH}_2\text{OH}
\end{array}
\xrightarrow{\text{Br}_2, \text{H}_2\text{O}}
\begin{array}{c}
\text{COOH} \\
| \\
\text{CHOH} \\
| \\
\text{(CHOH)}_n \\
| \\
\text{CH}_2\text{OH}
\end{array}
\xrightarrow{\text{Ca(OH)}_2}
\begin{array}{c}
\text{COO}^- \tfrac{1}{2}\text{Ca}^{2+} \\
| \\
\text{CHOH} \\
| \\
\text{(CHOH)}_n \\
| \\
\text{CH}_2\text{OH}
\end{array}
\xrightarrow{\text{H}_2\text{O}_2, \text{Fe}^{+++}}
\begin{array}{c}
\text{CHO} \\
| \\
\text{(CHOH)}_n + \text{CO}_2 \\
| \\
\text{CH}_2\text{OH}
\end{array}$$

Fig. 23.12 Ruff degradation

of glucose to arabinose (Fig. 23.12). The oxidising agent used in the last step, ferric or ferrous ion and hydrogen peroxide, is known as *Fenton's reagent*.

23.1.3.3 Epimerisation

Epimerisation is the conversion of an aldose to the isomer with the opposite configuration at C2; for example, the conversion of D-glucose to D-mannose. This can be done by treatment with alkali, but under such conditions other centres could be affected. The best method for this is through gluconic acid. Gluconic acid upon treatment with a tertiary amine base like pyridine, isomerises to mannonic acid to form an equilibrium mixture of the two. These are separated and converted to the lactone and then reduced to the aldose using sodium amalgam in the presence of CO_2.

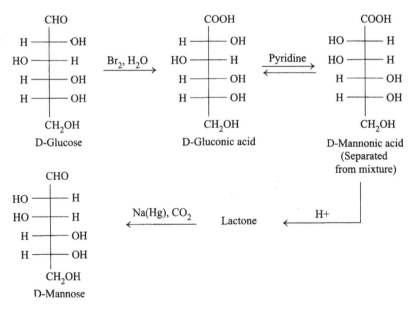

Fig. 23.13 Epimerisation of glucose

23.1.3.4 Acetylation

Glucose upon heating with acetic anhydride gives the pentaacetate. If the reaction is done in the cold, using α or β-glucopyranose, the corresponding α or β acetate is obtained. At higher temperature, a mixture is obtained from either anomer.

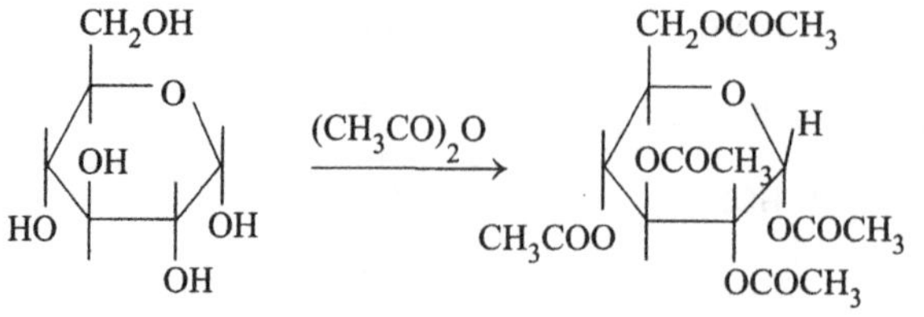

Fig. 23.14 Acetylation of glucose

23.1.3.5 Glycoside Formation

A hemiacetal can be converted to an acetal by reaction with an alcohol in the presence of H^+, (I) → (II) (Fig. 23.15). Glucopyranose reacts with methanol in the presence of HCl to form the acetal, which is called a glucoside (general name, *glycoside*), (III) → (IV) (Fig. 23.15).

Both the α and the β-methyl glucosides are known. Glycosides, like other acetals, are stable in neutral and alkaline media, but are readily hydrolysed by aqueous acids. The products of hydrolysis of the acetals are the parent aldehyde and alcohol. In the case of glycosides, the hydrolysis products are the alcohol – such as methanol – and the parent sugar in the hemiacetal form, see reverse reaction, (IV) to (III) (Fig. 23.15).

Glycosides are not reducing sugars. They are not in equilibrium with the aldehyde form and do not undergo mutarotation. Polysaccharides are glycosides where one sugar molecule is linked by glycoside formation with the OH group of another sugar molecule. Many natural products have sugar units attached to other molecules through glycosidic linkages. These include steroidal glycosides and the plant pigments known as anthocyanins.

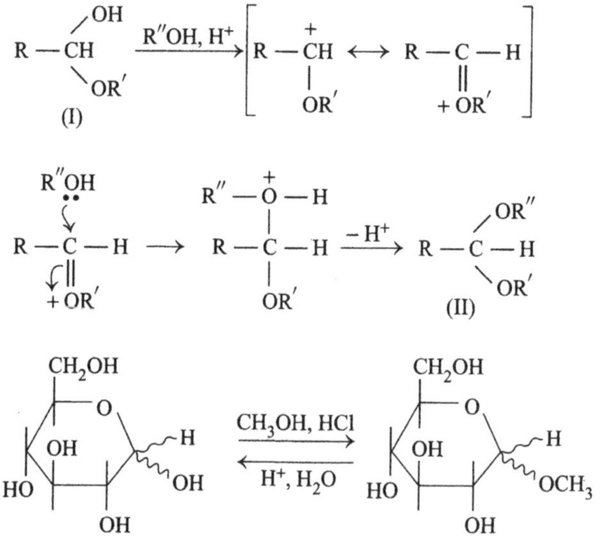

(III) (IV), Methyl D-glucopyranoside

Fig. 23.15 Glycoside formation

23.1.3.6   Ether Formation

Under the conditions of glycoside formation, the alcohol OH groups do not get affected. They can be converted to the methyl ether by treatment with dimethyl sulphate and alkali (this is the Haworth method) in a reaction similar to Williamson's ether synthesis (Fig. 23.16). If the aldehyde function is protected as a glucoside, methylation can be done by the Purdie method, using methyl iodide and silver oxide. If the aldehyde is not protected, it will get oxidised by silver oxide.

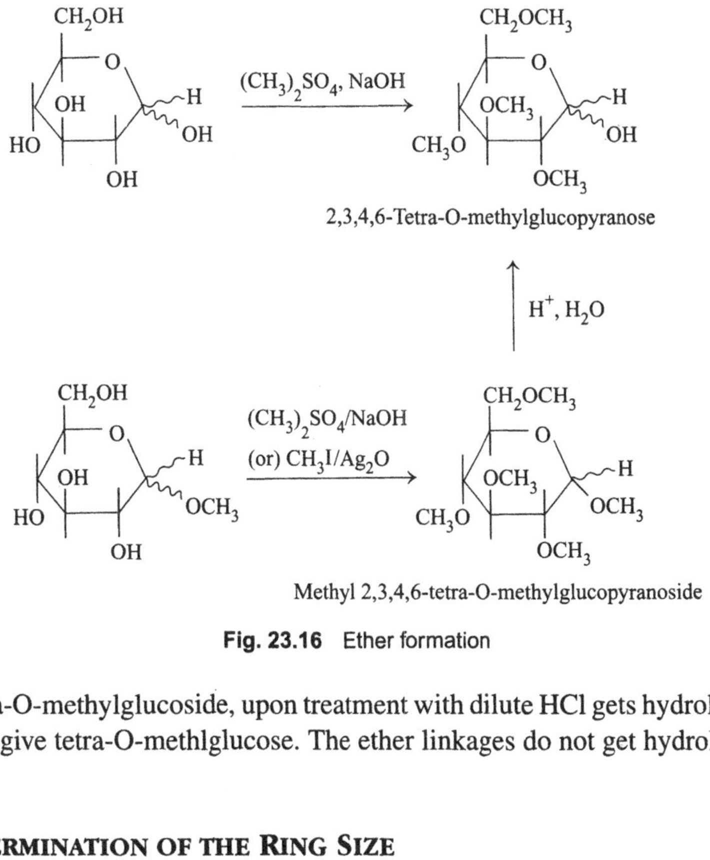

2,3,4,6-Tetra-O-methylglucopyranose

Methyl 2,3,4,6-tetra-O-methylglucopyranoside

Fig. 23.16   Ether formation

Methyl tetra-O-methylglucoside, upon treatment with dilute HCl gets hydrolysed at the acetal linkage alone to give tetra-O-methlglucose. The ether linkages do not get hydrolysed by aqueous acid or alkali.

23.1.4   DETERMINATION OF THE RING SIZE

2,3,4,6-Tetra-O-methylglucose is a hemiacetal and exists in equilibrium with the aldehyde form (Fig. 23.17, structures I and II).

In order to establish the ring size of the sugar molecule, all that needs to be done is to establish the position of the free OH group in the aldehyde (II). This has been done by vigorous oxidation with nitric acid (Fig. 23.17). In the first stage, the aldehyde group gets oxidised to the carboxyl and the free OH to a keto group. This ketoacid (III), undergoes further oxidation resulting in the cleavage of the C—C bond on one or the other side of the keto group. In (III), bond cleavage can take place either at the bond marked (A) or at the bond marked (B). The corresponding products are (IV) + (V) for A-cleavage and (VI) +CO_2 for a B-cleavage. In actual oxidation, a succinic acid derivative (V) and a glutaric acid derivative (VI) were isolated and identified by comparison

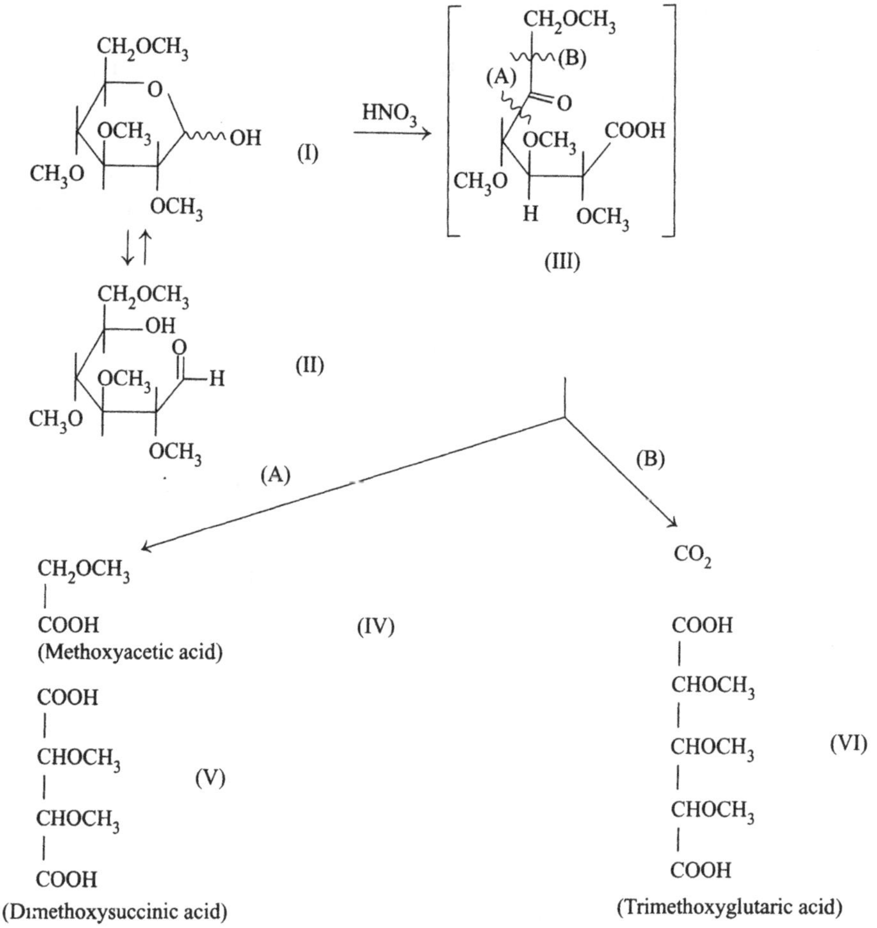

Fig. 23.17 Oxidation of tetra-O-methylglucose: determination of ring size

with known compounds. Hence, the point of attachment of the hemiacetal linkage in glucose was established to be C5, and it is a pyranose or 6-membered ring.

23.2 DISACCHARIDES

Disaccharides are formed by condensation of two monosaccharides with the elimination of water. There are several disaccharides present in nature. Two examples are sucrose formed from a glucose and a fructose molecule, and lactose consisting of a glucose and a galactose molecule (Fig. 23.18 and Fig. 23.19, respectively).

Examination of the structure of sucrose reveals that the OH group of the hemiacetal corresponding to each monosaccharide has condensed with the other so that each one is a glycoside. The linkage is through the aldehyde carbon C1 of glucose and the ketone carbon C2 of fructose. Sucrose is a non-reducing sugar. It can be named as a glucoside or as a fructoside (Fig. 23.18).

In lactose, the linkage is between the aldehyde carbon C1 of galactose and C4 of glucose. It is called a 1,4-linkage. The aldehyde group of the glucose unit in the molecule is free. Lactose is

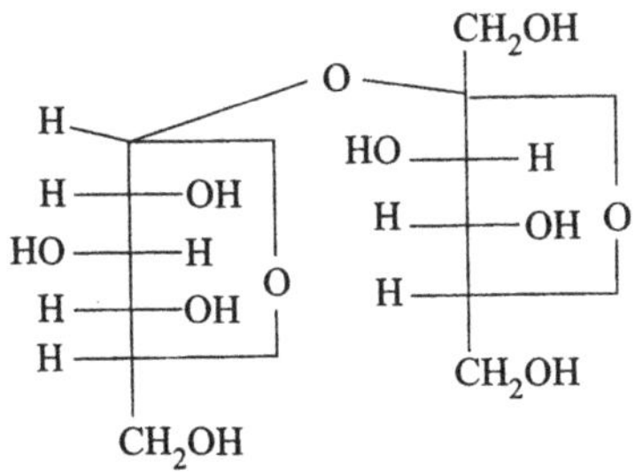

Fischer projection

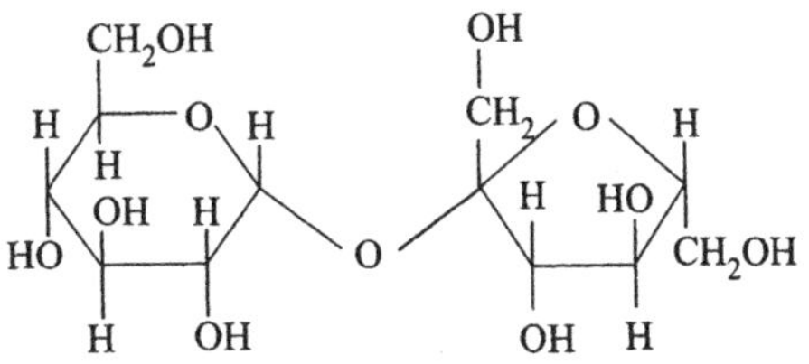

Haworth formula

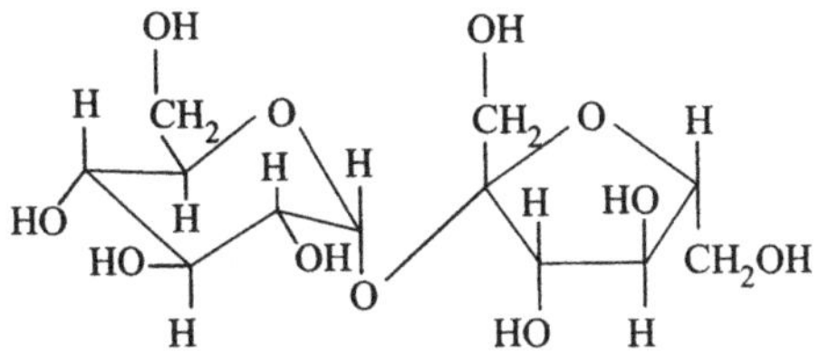

Conformation

α-D-Glucopyranosyl-β-D-fructofuranoside
(or)
β-D-Fructofuraosyl-α-D-glucopyranoside
(sucrose)

Fig. 23.18 Structure of sucrose

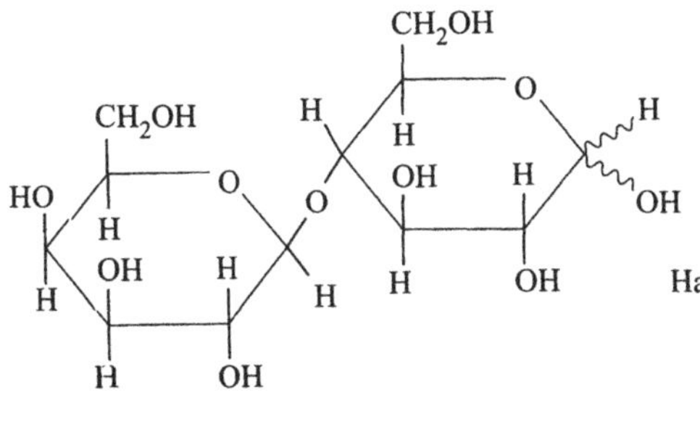

Haworth formula

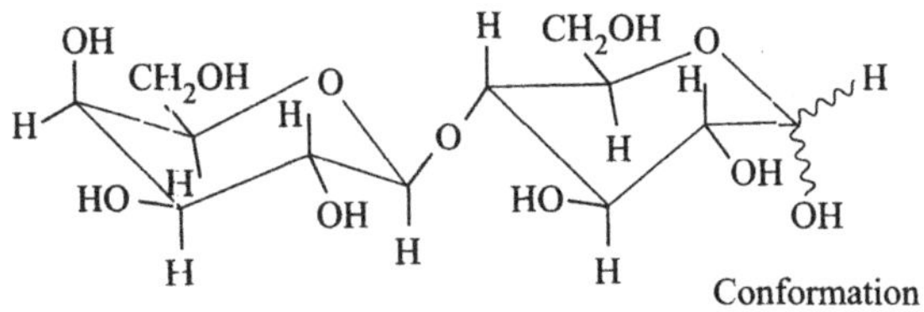

Conformation

4-O-(β-D-Galactopyranosyl)-D-glucopyranose (Lactose)

Fig. 23.19 Structure of lactose

a reducing sugar and can undergo mutarotation. Lactose forms an osazone with phenylhydrazine while sucrose does not.

23.2.1 STRUCTURE OF SUCROSE

Sucrose, being a glycoside of both glucose and fructose, can be hydrolysed by aqueous acids. It can also be hydrolysed by the enzyme invertase. Hydrolysis of sucrose by either means, gives one molecule each of glucose and fructose per molecule of sucrose. The specific rotation of sucrose is +66.5°. The products of hydrolysis, D-(+)-glucose, $[\alpha]_D = +52.7°$ and D-(−)-fructose, $[\alpha]_D = -92.4$, are formed in equal quantities. The resultant mixture has a specific rotation of −19.9°. Since the hydrolysis results in inversion of the sign of rotation, the reaction is called *inversion of sucrose* and the hydrolysed mixture is called *invert sugar*. D-glucose and D-fructose are called *dextrose* and *levulose* respectively, in view of their signs of rotation. The sugar in honey is mostly invert sugar.

The structure of sucrose is based on the following evidence. From the fact that it is a non-reducing sugar, it is inferred that the linkage is between C1 of glucose and C2 of fructose.

Methylation of sucrose with dimethyl sulphate and alkali gives octa-O-methylsucrose which on acid hydrolysis gives 2,3,4,6-tetra-O-methylglucose and 1,3,4,6-tetra-O-methylfructose (Fig. 23.20).

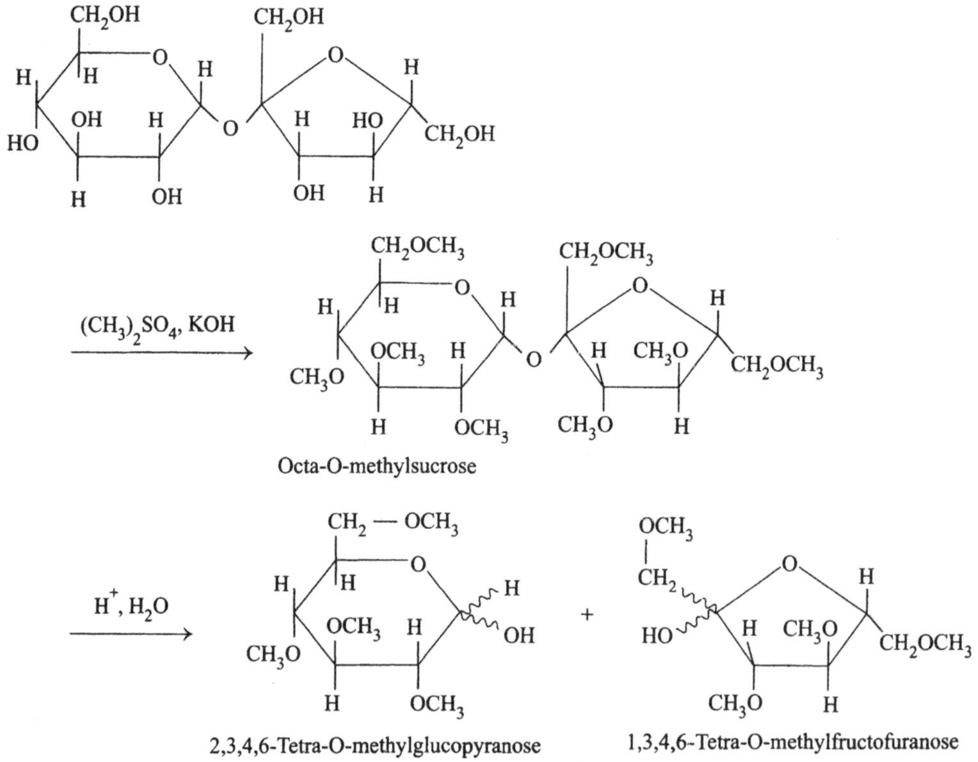

Octa-O-methylsucrose

2,3,4,6-Tetra-O-methylglucopyranose   1,3,4,6-Tetra-O-methylfructofuranose

Fig. 23.20 Methylation of glucose

Periodic acid oxidation of sucrose yields the products shown in Fig. 23.21. Periodic acid is a reagent which selectively oxidises glycols (Chapter 14, Section 14.2.2). In sucrose, there are three $C-C$ bonds which are oxidised and cleaved, consuming three molar equivalents of periodic acid. The isolation of the products as shown, is evidence for the ring size of the monosaccharide units. That the glycosidic linkage is α for glucose and β for fructose has been established taking advantage of the selectivity of enzymes. Maltase, an enzyme specific for α-glucosides hydrolyses sucrose. Hence, sucrose is an α-glucoside. Sucrose is also hydrolysed by another enzyme which is selective for β- fructosides. Hence sucrose is a β-fructoside. The structure has been confirmed by X-ray crystallographic studies and also by the synthesis of sucrose by Lemieux in 1953.

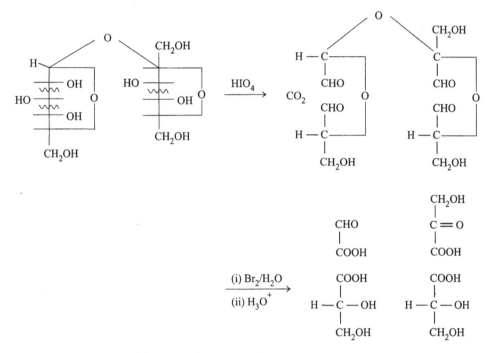

Fig. 23.21 Periodic acid oxidation of sucrose

23.2.1.1 Uses of Sucrose

One of the major uses of sucrose is for edible purposes, as a sweetening agent. Industrially, sucrose is one of the major sources of ethanol through alcoholic fermentation.

23.3 POLYSACCHARIDES: STARCH AND CELLULOSE

23.3.1 STARCH

Starches are polymers of glucose. Starch is the form in which glucose is stored in plants for metabolic purposes. Starch granules, isolated from plant sources, are not completely soluble in water. About 20% dissolves in hot water. This water-soluble fraction is called amylose. About 80% is insoluble in water and is called amylopectin. Upon acid hydrolysis, both fractions give glucose as the only monosaccharide.

23.3.1.1 Amylose

Enzymatic hydrolysis of amylose or soluble starch gives the disaccharide, (+)-maltose. Acid hydrolysis gives D-(+)-glucose. Maltose is a reducing sugar in which one glucose unit is linked by an α-glycosidic linkage to another glucose at the 4-position. It is 4-O-(α-D-glucopyranosyl)-D-glucopyranose. Amylose consists of polymeric chains of glucose units connected by 1,4-glycosidic linkages as in maltose. Methylation of amylose using dimethyl sulphate and alkali followed by acid hydrolysis gives 2,3,5-tri-O-mehylglucose, along with about 0.3% to 0.5% of 2,3,4,6-tetra-O-methylglucose. These facts are consistent with the structure shown in Fig. 23.22.

The chain shown in Fig. 23.22 has a reducing end on the right and a non-reducing end on the left. After complete methylation of amylose followed by hydrolysis, the glucose unit at the non-reducing end gives 2,3,4,6-tetra-O-methylglucose. All the others give 2,3,5-tri-O-methylglucose. The ratio of tetramethylglucose to trimethylglucose enables us to estimate the number of glucose units in the chain, about 200–500, which gives a molecular weight in the range 3000–4000. This is known as *end-group analysis,* for the determination of molecular weights of polymers. Other methods such as osmotic pressure measurement, ultracentrifugation and gel-permeation chromatography are also applicable for the determination of molecular weights. Amylose chains form coils (helices). Soluble starch gives an intense blue colour with iodine. The colour is due to iodine molecules trapped inside the helices.

23.3.1.2 Amylopectin

Complete hydrolysis of amylopectin also gives glucose. The only disaccharide that can be isolated is maltose. However, methylation and hydrolysis gives not only 2,3,6-tri-O-methyglucose (90%) and 2,3,4,6-tetra-O-methylglucose (5%), but also 2,3-di-O-methylglucose (5%). Amylopectin has a highly branched structure with short amylose-like chains joined together through 1,6-linkages [Fig. 23.23, structure (I)]. The highly branched structure is represented in structure (II).

23.3.1.3 Uses of Starch

Starch is one of the major components of our diet. Starch has applications in the manufacture of adhesives, in the textile industry and in pharmaceuticals. Starch from various sources, mainly grains, is used in the manufacture of ethanol by fermentation.

23.3.2 CELLULOSE

Cellulose is the structural carbohydrate of plants. Upon acid hydrolysis, cellulose like starch, gives only glucose. The disaccharide obtained by partial hydrolysis of cellulose is cellobiose which is a

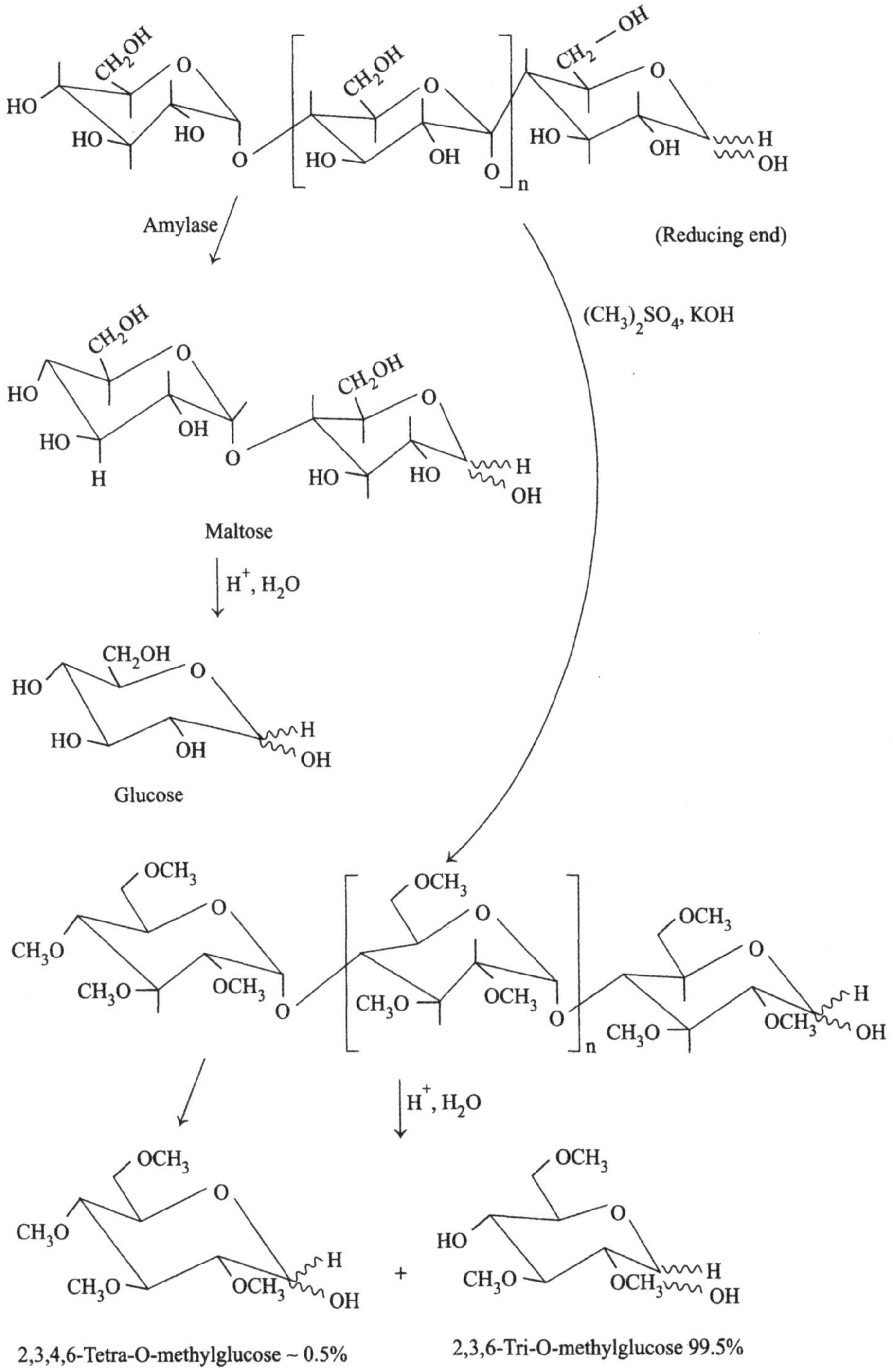

Fig. 23.22 Structure and reactions of amylose

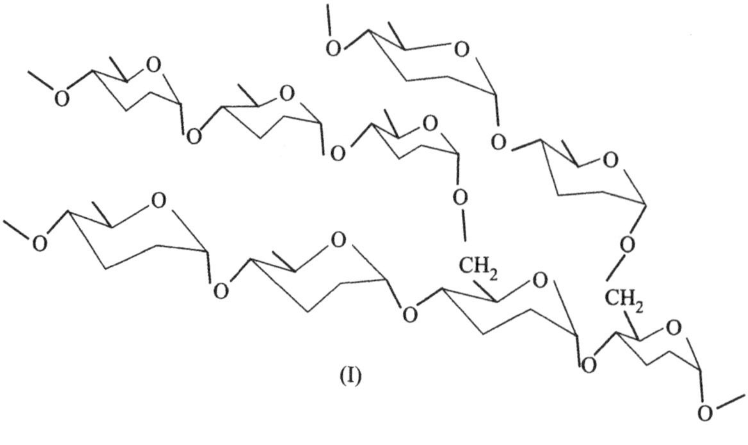

(I)

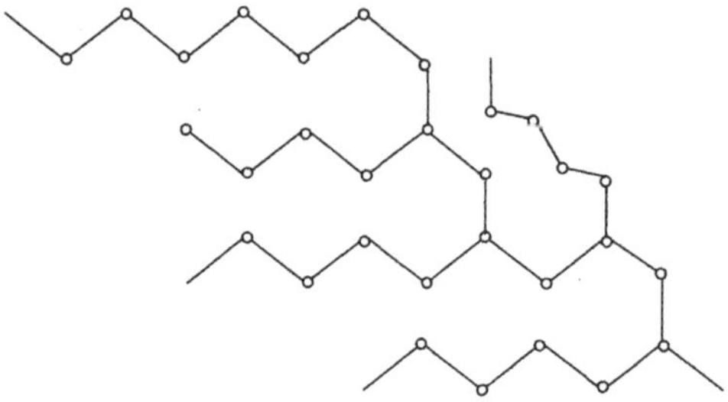

(II)

Fig. 23.23   Structure of amylopectin

reducing sugar made up of two glucose molecules, joined by a 1,4-linkage. The difference between cellobiose and maltose is in the configuration at the glycosidic linkage. Cellobiose is a β-glucoside while maltose is an α-glucoside. Cellulose is made up of glucose units joined by 1,4-β-glycosidic linkages with little or no branching (Fig. 23.24).

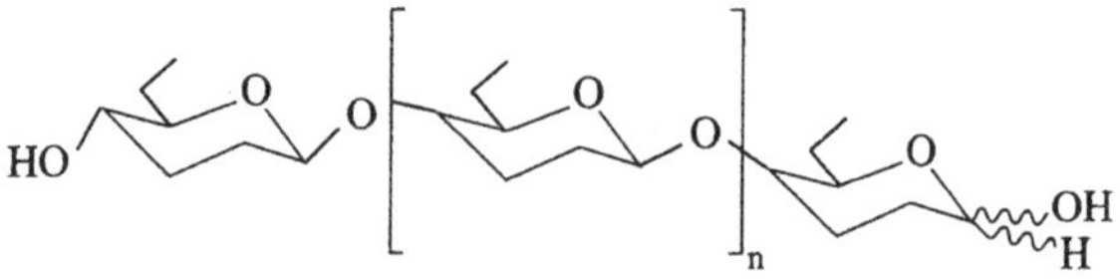

Fig. 23.24   Structure of cellulose

From end-group analysis and physical methods, the molecular weight of cellulose has been estimated to be 250,000 to 1,000,000. In cellulose, the long chains are twisted together into ropes, the chains being held together by hydrogen bonds. This structure is responsible for the fibrous nature of cellulose. Some of the industrially important reactions of cellulose are summarised in Fig. 23.25.

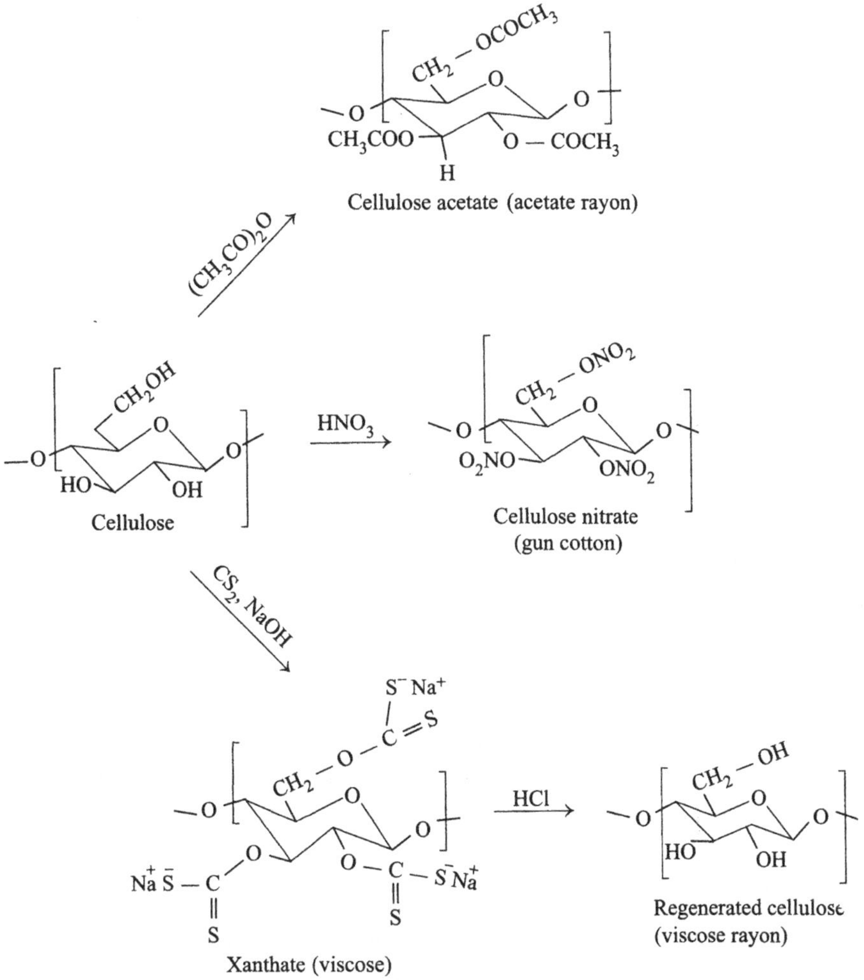

Fig. 23.25 Reactions of cellulose

Rayons are modified cellulose. Cellulose acetate is prepared by acetylation with acetic anhydride. Partially acetylated cellulose is converted to sheets and filaments and is known as *acetate rayon*. Cellulose reacts with carbon disulphide and alkali to form a xanthate; its viscous dispersion in aqueous alkali is known as viscose. When this is forced as filaments into an acid-bath, cellulose is regenerated. This is known as *viscose rayon*. The reaction of cellulose with nitric acid gives *cellulose trinitrate* which under the name *gun cotton*, is used as smokeless powder.

23.4 VITAMINS

Vitamins are chemicals which are essential components of human nutrition, which cannot be synthesised by the body. They have to be present in the diet. They are arbitrarily classified into fat-soluble vitamins and water-soluble vitamins. The former group includes the vitamins A, D, K and E. All the others belonging to the vitamin B family and vitamin C are water soluble. Here, we shall discuss the chemistry of ascorbic acid (vitamin C) – which is related to glucose in structure – and pyridoxine, a vitamin of the B group (which is a pyridine derivative).

23.4.1 Ascorbic Acid (Vitamin C)

Deficiency of vitamin C causes the condition known as scurvy. Vitamin C is the antiscorbutic vitamin. It is present in citrus fruits and various other fruits and vegetables. The structure of ascorbic acid in Fischer projection (I), is given in Fig. 23.26. The 5-membered lactone ring is almost flat as in (II).

Some of the evidence and reactions which lead to this structure are listed in Fig. 23.26.

(i)   It is optically active. The configuration was assigned to be L by studying its transformation to L-threonic acid, (I) $\rightarrow$ (III) $\rightarrow$ (IV)

(ii) It is a monobasic acid. The acidity is due to the enolic OH (cf. acidity of phenol). Dehydroacsorbic acid (III), is not acidic. The presence of the enol is supported by the fact that it gives a colour with neutral ferric chloride, a test characteristic for enols and phenols. It is a strong reducing agent.

(iii) With diazomethane, ascorbic acid forms the dimethyl derivative (V). This is characteristic of acidic OH groups as in phenols and carboxylic acids (Chapter 18, Section 18.3). The remaining OH groups in (V) are methylated by treatment with CH_3I in the presence of silver oxide. This compound (VI), which contains a double bond, upon ozonylysis followed by hydrolysis gives (VII), 3,4-di-O-methyl-L-threonic acid. This shows that the lactone ring in ascorbic acid is 5-membered.

The relationship of ascorbic acid to D-glucose and L-gulose is shown in Fig. 23.27. Ascorbic acid is the enol form of the γ-lactone of 2-keto-L-gulonic acid (IV) [cf. structures (XI) and (XII), Fig. 23.8].

23.4.1.1 Synthesis of Ascorbic Acid

Ascorbic acid has two asymmetric centres, C2 and C3 of D-glucose or C4 and C5 of L-gulose. Synthetic ascorbic acid, in order to be usesful as a vitamin, should have this configuration. Several monosaccharides have this configuration at adjacent carbons. It is advantageous to use one such sugar as the starting material for the synthesis of ascorbic acid to make use of the asymmetry already present. This is an example of making use of the *chiral pool* available in nature for asymmetric synthesis. All the reported syntheses of ascorbic acid use a monosaccharide as the starting material. The most abundant monosaccharide, glucose, has this configuration at C2 and C3 [see structure (I) of glucose written with the aldehyde pointing down and structure (II) of L-gulose in Fig. 23.27].

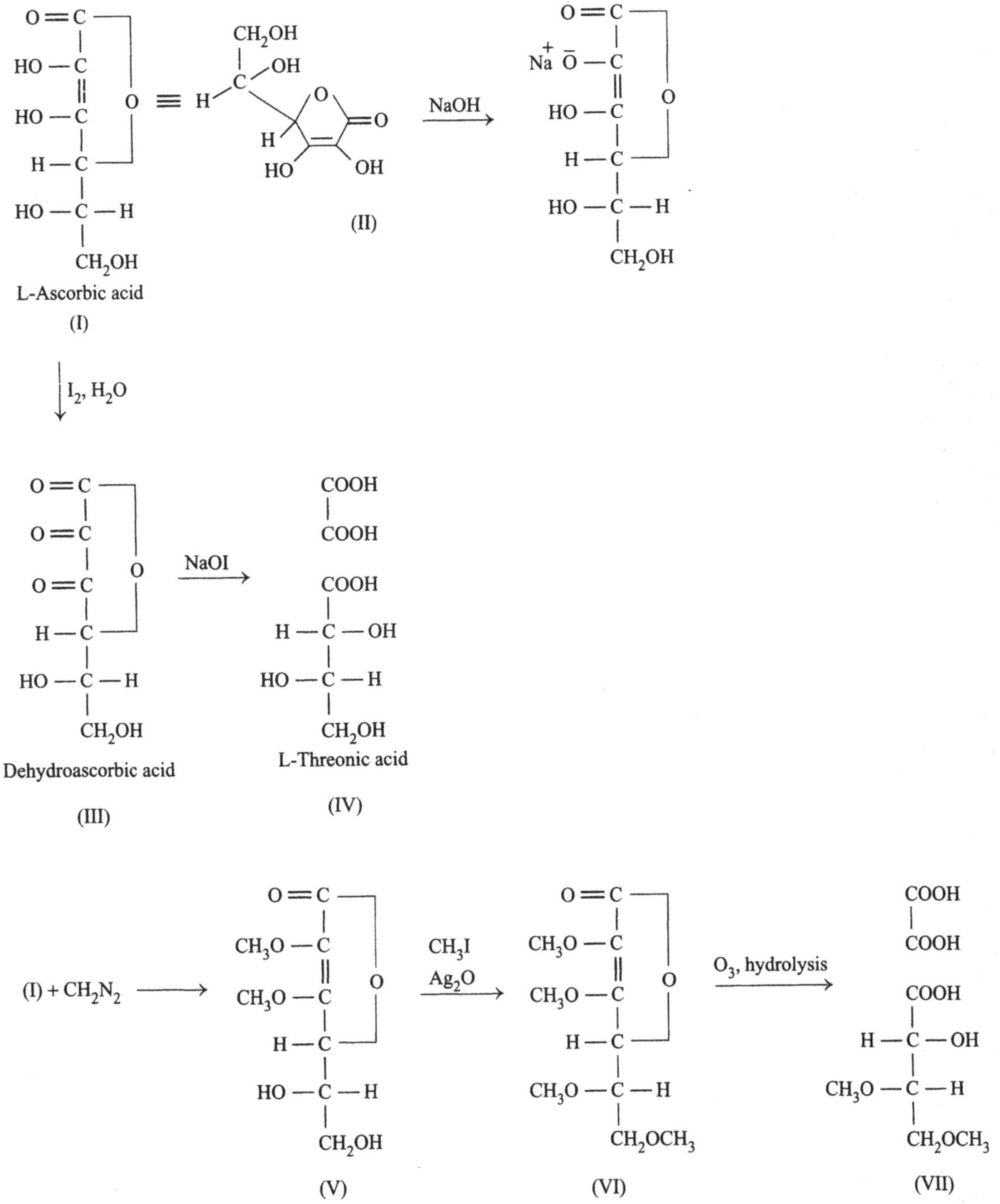

Fig. 23.26 Structure and reactions of L-ascorbic acid

The synthesis of ascorbic acid outlined in Fig. 23.28 is used in the manufacture of ascorbic acid. Glucose (the usual Fischer projection is rotated 180° in structure I) is catalytically hydrogenated to

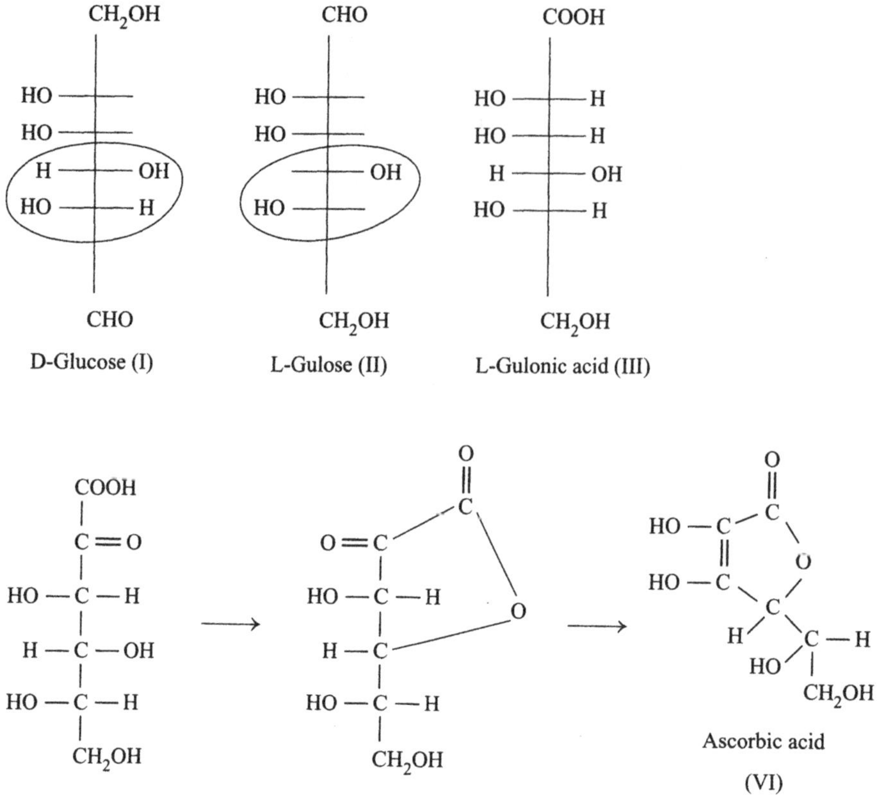

CH₂OH   CHO   COOH

HO —   HO —   HO —— H
HO —   HO —   HO —— H
H —— OH   —— OH   H —— OH
HO —— H   HO —   HO —— H

CHO   CH₂OH   CH₂OH

D-Glucose (I)   L-Gulose (II)   L-Gulonic acid (III)

COOH
|
C=O
|
HO — C — H
|
H — C — OH
|
HO — C — H
|
CH₂OH

2-Keto-L-gulonic acid

(IV)

⟶

2-Keto-L-gulonolactone

(V)

⟶

Ascorbic acid

(VI)

Fig. 23.27 Relationship between ascorbic acid and L-gulonic acid

sorbitol (II). This is oxidised microbially using acetobacter suboxidans selectively at C2 to obtain the ketohexose L-sorbose (III), whose furanose structure is (IV). This is a selective oxidation, which could not have been carried out by the usual chemical oxidising agents. The CH₂OH – marked with an asterisk (∗) at C1 – of sorbose needs to be oxidised to COOH. For this purpose, the other four OH groups should be protected by conversion to cyclic ketals, by condensation with acetone in the presence of mineral acid (Fig. 23.29). Glycols and 1,3-diols form this kind of cyclic ketal, [called isopropylidine derivatives, Fig. 23.29, structures (I) and (II)]. Glucose itself condenses with two molecules of acetone to form the diisopropylidine derivative (III), called diacetone glucose. These are stable in alkaline media. Condensation of sorbose with acetone in the presence of sulphuric acid yields the diisopropylidine derivative (V) (Fig. 23.28). In this reaction, two pairs of OH groups have reacted with a molecule of acetone each to form cyclic ketals. In this synthesis, the next step is the oxidation of the primary alcohol group by alkaline permanganate to COOH. The hydroxyl groups which are protected as ketals do not get affected because the oxidation is carried out in alkaline solution. After the oxidation is complete, the protection is removed by hydrolysis with aqueous

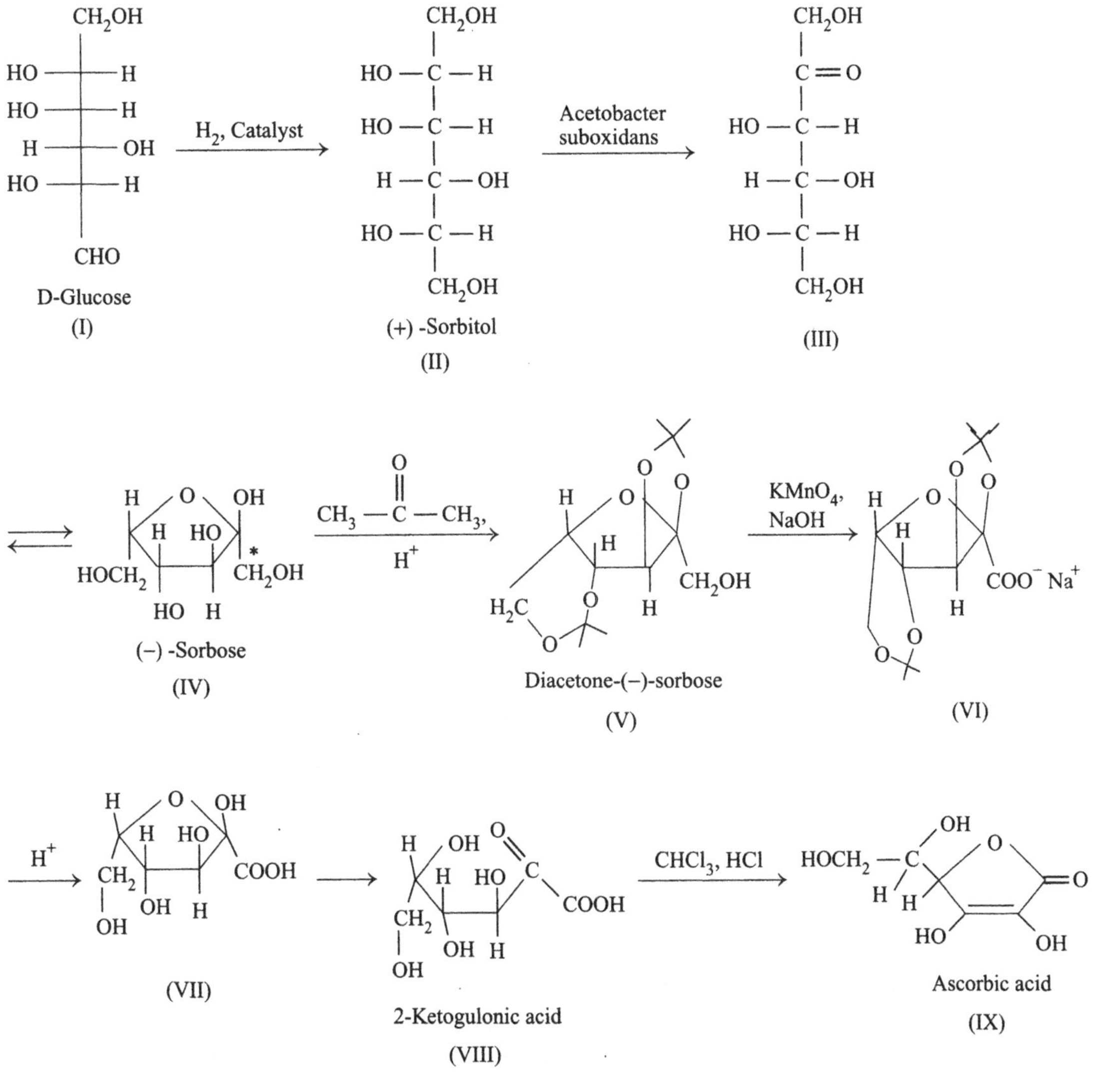

Fig. 23.28 Synthesis of ascorbic acid

acid to release the molecule (VII). The open chain keto form of this cyclic ketal is 2-ketogulonic acid (VIII). Ascorbic acid is the lactone of the enol form of this keto acid. Treatment of (VIII) in chloroform solution with acid results in cyclisation to ascorbic acid (IX) [see the relationship (IV) → (V) → (VI) in Fig. 23.27].

23.4.2 PYRIDOXINE

Pyridoxine, vitamin B6, is present in wheat bran and yeast. The nitrogen in pyridoxine, $C_8H_{11}NO_3$, is a tertiary amine and is part of the pyridine ring. The three oxygens are hydroxylic. It reacts with

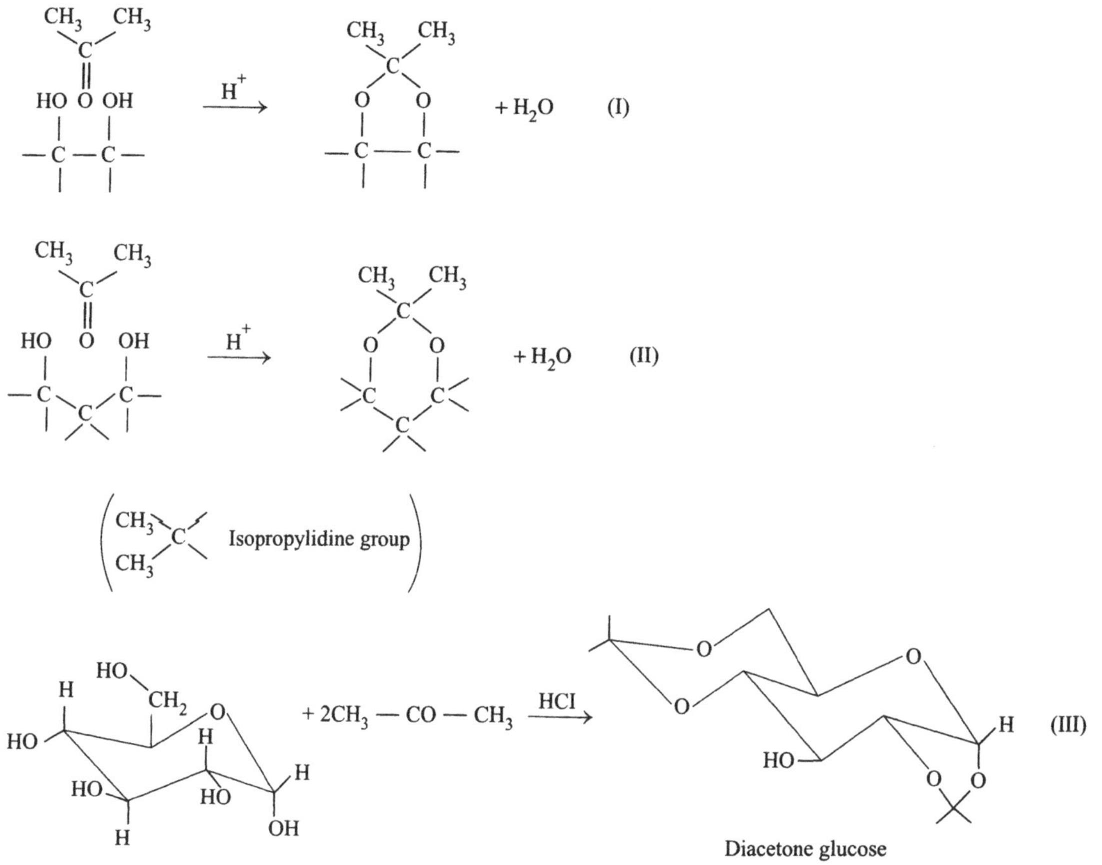

Fig. 23.29 Formation of isopropylidine derivatives

CH_3MgI to release three moles of methane per mole of pyridoxine, confirming the presence of three 'active hydrogens.' This method of estimating the number of active or acidic hydrogens in a molecule is known as the *Zerewitinoff method*. Some of the reactions which helped in assigning the structure (I) to pyridoxine are summarised in Fig. 23.30.

One of the OH groups is phenolic, because it reacts with diazomethane to form a monomethyl ether (II). Presence of the phenolic OH is confirmed also by the ferric chloride colouration test. It is possible to infer from its UV spectrum which is similar to that of 3-hydroxypyridine, that it is a 3-hydroxypyridine derivative. The monomethyl ether reacts with acetic anhydride to form a diacetyl derivative (III). Oxidation of the monomethyl ether (II), with alkaline permanganate under mild conditions gives the tricarboxylic acid (IV). Under vigorous conditions, one of the COOH groups is lost as CO_2 and the anhydride (V), is obtained.

The tricarboxylic acid (IV) produces a blood red colouration with ferric chloride which is characteristic of pyridine-2-carboxylic acid. The dicarboxylic acid obtained from (V) does not give this colouration. Hence the carboxyl group which is lost is in the 2-position. Oxidation of (II) with barium permanganate gives the diacid (VI), which is also obtained by the oxidation

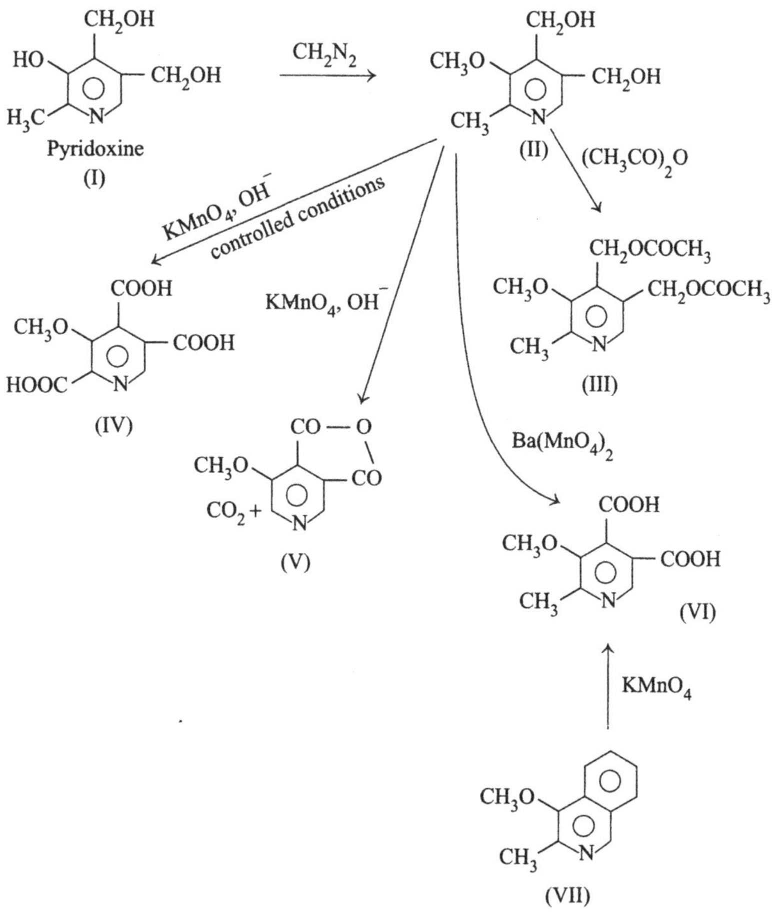

Fig. 23.30 Structure of pyridoxine

of 4-methoxy-3-methylisoquinoline (VII), whose structure is established. This gives the final confirmation of the positions of the groups.

KEY POINTS

- Monosaccharides are polyhydroxy aldehydes or polyhydroxy ketones. An aldohexose with 4 asymmetric centres have 2^4 or 16 optical isomers, which are 8 pairs of enantiomers. According to the convention established by Emil Fischer, when the structures are written in Fischer projection, those compounds in which the OH group on the bottom-most asymmetric carbon is on the right, is designated by the prefix D. There are eight D-aldohexoses and eight L-aldohexoses.

- The aldehyde form is not the preferred form in which the aldose exists. The aldehyde is in equilibrium with the cyclic hemiacetal. There are two cyclic hemiacetals possible, with opposite configurations at the newly created anomeric asymmetric centre, designated α and β. The two forms are at equilibrium with each other in solution along with a very small quantity of the open chain aldehyde. The two anomers, α and β have different specific rotations and when freshly prepared, the solution of either one has its characteristic specific rotation, α-form, $+112.2°$ and the β-form, $+18.7°$. On standing, due to the interconversion and the establishment of equilibrium, the specific rotation of the solution changes till it reaches an intermediate value of $+52.5°$ corresponding to about 38% of the α and about 62% of the β-anomer. The open chain aldehyde is also present, but only in traces. This phenomenon is known as mutarotation.

- Fischer projection is convenient way of representing configurations. The hexagonal representation (Haworth formulae) and representation of the molecule in its chair conformation are also in use. Glucose exists not only in the 6-membered hemiacetal form called the pyranose structure, but also in the 5-membered furanose structure. The latter is present only in trace quantities in solution at equilibrium, but can be prepared in stable form as the glycoside.

- The hemiacetal can be converted to the acetal by the acid catalysed displacement of the anomeric OH by OR, using the alcohol ROH in acid solution, to give the glycoside. Thus, glucose reacts with methanol in the presence of HCl to give the α and the β-methyl glucopyranoside. The glycoside, being an acetal, gets hydrolysed to the hemiacetal by aqueous acid but not by alkali. The alcoholic OH groups of glucose can be converted to the methyl ether by treatment with dimethyl sulphate and alkali. Methyl glucopyranoside reacts with dimethyl sulphate and NaOH to give methyl 2,3,4,6-tetra-O-methylglucopyranoside. These ether linkages cannot be hydrolysed by aqueous acid or alkali.

- The ring size of glucose has been established by periodic acid oxidation of methyl glucoside and identification of the products.

- Glucose and fructose are reducing sugars. They reduce Tollen's reagent and Fehling's solution. In the alkaline medium, the ketose isomerises to the aldose through the enediol. The oxidation product of glucose is 2,3,4,5,6-pentahydroxyhexanoic acid, also called gluconic acid. The general name for such acids is aldonic acid. They readily form a 5-membered and a 6-membered lactone. Oxidation to the aldonic acid is best effected by using buffered bromine water.

- The configuration of glucose was established by Fischer by correlation mainly by using the products of oxidation of the two terminal carbons by HNO_3 to the aldaric acids. Oxidation of glucose with nitric acid gives 2,3,4,5-tetrahydroxyhexanedioic acid, also called glucaric acid.

- Aldose chains can be lengthened by the Kiliani–Fischer synthesis. The additional carbon is introduced as CN by cyanohydrin formation.

- Shortening of the chain can be done by Ruff degradation.

- Conversion of glucose to mannose, which involves inversion of configuration at C2, is called epimerisation. This is best done by the treatment of gluconic acid with pyridine.

- Glucose reacts with two equivalents of phenylhydrazine to give an osazone, glucosazone. In this reaction, the OH at the C2 is also oxidised to the carbonyl group. Epimeric aldoses, like glucose and mannose give the same osazone.

- Disaccharides are formed when two monosaccharides condense with the elimination of water, utilising one OH group of each.

- Sucrose is made up of a glucose and a fructose unit. The two combine by acetal formation involving C1 of glucose and C2 of fructose. It is a glucoside as well as a fructoside and cannot tautomerise to the aldehyde form. Hence sucrose is not a reducing sugar.

- There are other disaccharides which are reducing sugars. An example is lactose. In lactose, C1 of a galactose molecule is condensed with C4 of a glucose molecule. The aldehyde of the glucose unit is still present as the hemiacetal. So lactose can undergo mutarotation. It is a reducing sugar.

- The structure of sucrose has been established by identifying the products of periodic acid oxidation. The linkage at the C1 of glucose is α and the linkage at the C2 of fructose is β. This has been established by making use of the selectivity of enzymes which are capable of hydrolysing sucrose.

- Starch and cellulose are polysaccharides. Both yield only glucose upon complete hydrolysis. Both are formed by 1,4-linkage of glucose units. They differ in the stereochemistry of the glycosidic linkage. In starch, the linkage is α and in cellulose, it is β. Partial hydrolysis of starch gives maltose which is a reducing disaccharide of two glucose units linked by a 1,4-linkage with α configuration for the glucoside linkage of the non-reducing glucose. The disaccharide obtained from cellulose is cellobiose, which is also a reducing sugar. In cellobiose the 1,4-glucose to glucose linkage is β.

- Natural starch is composed of a water-soluble fraction called amylose and a water-insoluble fraction called amylopectin.

- Complete methylation using dimethyl sulphate and alkali converts all the OH groups of starch or cellulose to the methyl ether. Acid hydrolysis of this methylated polysaccharide, gives O-methylglucose in which the OH groups which were originally free remain as methyl ether. Identification of these fragments helps to establish the points of attachment of one glucose unit to another.

- Methylation and hydrolysis of starch gives 2,3,6-tri-O-methylglucose along with about 0.3% to 0.5% of 2,3,4,6-tetra-O-methylglucose. The latter comes from the non-reducing end of the chain. The ratio of the two helps to calculate the number of glucose units in the chain and as a consequence, the molecular weight. This is called end-group analysis. Amylose is a straight chain polysaccharide composed of 200–500 glucose units.

- Methylation and hydrolysis of amylopectin shows that it is highly branched with short amylose chains linked to other amylose chains by 1,6-linkages.

- Cellulose is mostly unbranched with very high molecular weights. The chains are twisted into ropes, giving rise to the fibrous structure.

- Ascorbic acid, vitamin C, is a lactone and is a sugar derivative. Its acidity is due to the presence of the enediol group. The structure of ascorbic acid has been established by degradation reactions. A manufacturing method of ascorbic acid uses glucose as the starting material.

- Pyridoxine is vitamin B6. It is a pyridine derivative. Its structure has been established by conventional chemical methods.

EXERCISES

SECTION I

1. Draw the structures of the following using Fischer projections and if cyclic, using Haworth formula and cyclohexane chair conformation.

    (a) δ-D-mannonolactone
    (b) γ-D-gluconolactone
    (c) Methyl α-D-galactofuranoside
    (d) 2,3,4,6-Tetramethyl α-L-allopyranose

2. Give the structures of the products of the following reactions.

    (a) Lactose + phenylhydrazine →
    (b) Lactose + bromine water →
    (c) Fructose +$(CH_3)_2SO_4$/NaOH →
    (d) Glucose + benzyl alcohol / HCl →
    (e) Sucrose + diethyl sulphate / KOH →
    (f) Product of (d) + aq. HCl →
    (g) 2,3,6-Tri-O-methylglucose + bromine water →
    (h) Galactose +$NaBH_4$ →
    (i) Lactose + $NaBH_4$ →
    (j) Methyl α-D-glucofuranoside +HIO_4 →

3. Which other D-aldoheose will give the same osazone as D-allose?
4. Which are the D-aldohexoses that give optically inactive aldaric acids upon HNO_3 oxidation?

5. Which are the D-aldohexoses that give optically inactive alditols on $NaBH_4$ reduction?
6. Which other aldohexose (of the D or L series) will give the same alditol as D-glucose?
7. Which other aldohexose will give the same aldaric acid as D-glucose?
8. Suggest chemical tests to distinguish between (i) glucose and sucrose, and (ii) sucrose and lactose.

SECTION II

1. Give the structures involved and the mechanism of the mutarotation of maltose.
2. Give the mechanism of epimerisation of glucose in the presence of (i) base and (ii) acid.
3. Give the mechanism of the hydrolysis of methyl glucoside by aq. acid.
4. Give the structure of the product of the reaction of methanol with amylose in the presence of HCl. What will happen when this product is oxidised with periodic acid?
5. Give the structure of the products A, B, and C.
    Pyridoxine + MnO_2 → pyridoxal (A) (the CH_2OH at position 4 is oxidised to aldehyde)

    (A)   + NH_2OH → (B)
    (B)   + H_2 $\xrightarrow{\text{Nickel catalyst}}$ (C), pyridoxamine

6. Give the mechanism of the interconversion in Fig. 23.9, catalysed by (i) acid and (ii) base.
7. What is the product and mechanism of the reaction of meso-2,3-butanediol with acetone in the presence of HCl?
8. What will happen when diacetone glucose [Fig. 23.29, structure (III)] is treated with $(CH_3)_2SO_4$/NaOH and then with aqueous acid?

CHALLENGING QUESTION

1. Today spectroscopy is used more than chemical degradation methods to determine the structure of organic compounds. Discuss what kind of structural information can be obtained from the IR and NMR spectra of pyridoxine.

24 Bioorganic Chemistry – Amino Acids, Proteins and Nucleic Acids

(Natural Products – II)

OBJECTIVES   In this chapter, you will learn about,

- classification, structures and names of amino acids
- essential amino acids
- reactions of amino acids
- acid–base properties, zwitter ion structure, isoelectric point
- general methods of amino acid synthesis
- general methods of peptide synthesis including solid-phase synthesis
- primary structure of proteins
- methods of determination of the structure of peptides
- end-group analysis
- classification of proteins
- secondary and tertiary structures of proteins
- denaturation
- nucleic acids, types, chain components, structure
- biological functions

24.1 AMINO ACIDS

Amino acids are molecules containing the amino group and the carboxyl group. Of particular interest to us are the α-amino acids, $R-CH(NH_2)-COOH$, which are the constituents of proteins. Proteins are condensation polymers of α-amino acids, connected by an amide group formed by elimination of a water molecule. A molecule, $H_2NCH(R)-CO-NHCH(R')-COOH$, is called a dipeptide. The $-CO-NH-$ grouping connecting two amino acids is called a peptide bond. Polypeptides are formed by linking more than two amino acids through peptide bonds. Proteins are polypeptides where the number of amino acids could be very large. Hydrolysis of the protein or the polypeptide releases the amino acids. There are mainly twenty α-amino acids which have been isolated from proteins. A few others, which occur rarely, have also been isolated. They all have the L-configuration and are all primary amines, with the exception of proline—which is a secondary amine. They differ in the structure and functional groups present in the molecule other than the NH_2, attached to the α-carbon. The names, abbreviations, and structures of the twenty amino acids found in proteins are listed in Table 24.1. For each amino acid, two types of abbreviations are given, one with three letters – like Gly for glycine – and another with one letter, like G for glycine.

The simplest amino acid – glycine or aminoacetic acid – is not chiral. All others are L-amino acids (Chapter 21, Fig. 21.5, for the configuration of some amino acids). The neutral amino acids are those containing one basic amino group and one carboxyl group. Those containing two amino groups and one carboxyl are classified as basic amino acids and those containing one amino and two carboxyl groups are classified as acidic amino acids. In our body, amino acids are necessary for protein synthesis. Some amino acids can be synthesised by the body. Those amino acids which cannot be synthesised by the body should be present in the food that we take, usually as constituents of dietary proteins. Such amino acids are called *essential amino acids* [marked with (∗) in Table 24.1].

24.1.1 ACID–BASE PROPERTIES AND ISOELECTRIC POINT

In aqueous solutions of an amino acid, the equilibrium shown in Eq. 24.1 exists.

$$\overset{+}{H_3N} - CHR - COOH \underset{+H^+}{\overset{-H^+}{\rightleftharpoons}} \overset{+}{H_3N} - CHR - COO^- \underset{+H^+}{\overset{-H^+}{\rightleftharpoons}} H_2N - CHR - COO^-$$

Conjugate acid		Zwitter ion		Conjugate base
(acidic solution)	(1)	(isoelectric point)	(2)	(basic solution)

$$(24.1)$$

The concentrations of the three species depend upon the pH. At a particular pH, characteristic of each amino acid, the concentration of the zwitter ion is maximum. This pH is called the isoelectric point. At this pH since the molecule is electrically neutral, it will not migrate in an electric field. The isolectric point is important in the analysis of amino acids by ion-exchange chromatography and by electrophoresis. A neutral amino acid has two K_a values corresponding to the equilibrium steps (1) and (2) of Eq. 24.1.

Amino acids are soluble in water. X-ray evidence shows that in the solid state, amino acids exist in the zwitter-ion form. They have high melting points, in conformity with the dipolar structure.

Amino acids show the typical reactions of primary aliphatic amines and of carboxylic acids (Eq. 24.2). Thus, they can be acylated (Eq. 24.2b), and react with nitrous acid liberating nitrogen and forming hydroxy acids (Eq. 24.2a). They can be esterified upon heating with alcohol and dry HCl (Eq. 24.2c).

Table 24. 1 α-Amino acids present in proteins

Name	Abbreviation	Structural formula	$[\alpha]_D^{25}(H_2O)$	Isoelectric point
Neutral amino acids				
Glycine	Gly (G)	$\begin{array}{c} NH_2 \\ \mid \\ H-CHCO_2H \end{array}$		6.0
Alanine	Ala (A)	$\begin{array}{c} NH_2 \\ \mid \\ H_3C-CHCO_2H \end{array}$	+2.7°	6.1
Valine*	Val (V)	$\begin{array}{c} NH_2 \\ \mid \\ (CH_3)_2CH-CHCO_2H \end{array}$	+6.4°	6.0
Leucine*	Leu (L)	$\begin{array}{c} NH_2 \\ \mid \\ (CH_3)_2CHCH_2-CHCO_2H \end{array}$	−10.8°	6.0
Isoleucine*	Ile (I)	$\begin{array}{c} CH_3 \quad NH_2 \\ \mid \qquad \mid \\ CH_3CH_2CH-CHCO_2H \end{array}$	+11.3°	6.0
Methionine*	Met (M)	$\begin{array}{c} NH_2 \\ \mid \\ CH_3SCH_2CH_2-CHCO_2H \end{array}$	−8.1°	5.7
Proline	Pro (P)	$\begin{array}{c} H_2C \diagdown NH \\ H_2C \diagup \mid \\ H_2C \diagdown CHCO_2H \end{array}$	−85.0°	6.3
Phenylalanine*	Phe (F)	$\begin{array}{c} NH_2 \\ \mid \\ C_6H_5-CH_2-CHCO_2H \end{array}$	−35.1°	5.9
Tryptophan*	Trp (W)	$\begin{array}{c} NH_2 \\ \mid \\ \text{(indole)}-CH_2-CHCO_2H \end{array}$	−31.5°	5.9
Asparagine	Asn (N)	$\begin{array}{c} O \quad\quad NH_2 \\ \parallel \qquad \mid \\ H_2NCCH_2-CHCO_2H \end{array}$	−7.4°	5.4

Table 24. 1 (Continued)

Name	Abbreviation	Structural formula*	$[\alpha]_D^{25}(H_2O)$	Isoelectric point
Glutamine	Gln (Q)	O‖H$_2$NCCH$_2$CH$_2$ — NH$_2$ CHCO$_2$H	+9.1°	5.7
Serine	Ser (S)	NH$_2$ HOCH$_2$ — CHCO$_2$H	−6.8°	5.7
Threonine*	Thr (T)	OH NH$_2$ CH$_3$CH — CHCO$_2$H	−28.3°	5.7
Tyrosine	Tyr (Y)	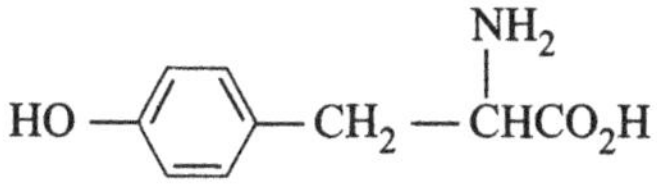 NH$_2$ HO—⟨benzene⟩—CH$_2$ — CHCO$_2$H	−8.6°	5.6
Cysteine	Cys (C)	NH$_2$ HSCH$_2$ — CHCO$_2$H	−9.8°	5.1

Acidic amino acids

Name	Abbreviation	Structural formula*	$[\alpha]_D^{25}(H_2O)$	Isoelectric point
Aspartic acid	Asp (D)	O‖ NH$_2$ HOCCH$_2$ — CHCO$_2$H	+4.7°	3.0
Glutamic acid	Glu (E)	O‖ NH$_2$ HOCCH$_2$CH$_2$ — CHCO$_2$H	+11.5°	3.1

Basic amino acids

Name	Abbreviation	Structural formula*	$[\alpha]_D^{25}(H_2O)$	Isoelectric point
Lysine*	Lys (K)	NH$_2$ H$_2$NCH$_2$CH$_2$CH$_2$CH$_2$ — CHCO$_2$H	+14.6°	9.5
Arginine*	Arg (R)	NH‖ NH$_2$ H$_2$NCNHCH$_2$CH$_2$CH$_2$ — CHCO$_2$H	+12.6°	10.8
Histidine*	His (H)	NH$_2$ imidazole—CH$_2$ — CHCO$_2$H	−39.0°	7.6

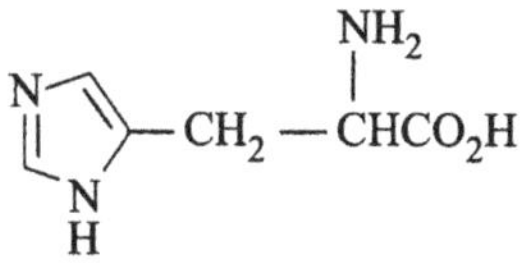

*Essential amino acids

$$
R-CH\begin{smallmatrix}COOH\\NH_2\end{smallmatrix}
\begin{cases}
\xrightarrow{\;HNO_2\;} & R-CH\begin{smallmatrix}COOH\\OH\end{smallmatrix} & \text{(a)}\\[2ex]
\xrightarrow{\;(CH_3CO)_2O\;} & R-CH\begin{smallmatrix}COOH\\NH-COCH_3\end{smallmatrix} & \text{(b)}\\[2ex]
\xrightarrow{\;CH_3OH,\,HCl\;} & R-CH\begin{smallmatrix}COOCH_3\\NH_2\end{smallmatrix} & \text{(c)}
\end{cases}
\qquad (24.2)
$$

24.1.2 SYNTHESIS OF AMINO ACIDS

All the synthetic methods that use achiral reagents and that are listed here, necessarily result in the formation of racemic mixtures. For peptide synthesis or for biological applications where enantiomerically pure amino acids are required, asymmetric synthesis or resolution has to be resorted to. Many amino acids can be isolated from proteins by hydrolysis.

24.1.2.1 Alkylation of Ammonia

α-Haloacids react with ammonia by a nucleophilic displacement of the halide by ammonia (Eq. 24.3a). This is applicable for many simple amino acids.

$$R-CH(Cl)-COOH + NH_3 \longrightarrow R-CH(NH_2)-COOH + HCl \qquad \text{(a)}$$

$$
\text{(phthalimide)}\,N^-K^+ + R-CH(Cl)-COOH \longrightarrow \text{(phthalimide)}N-\overset{\overset{R}{|}}{CH}-COOH \qquad (24.3)
$$

$$
\xrightarrow{\;\text{hydrolysis}\;} H_2N-\overset{\overset{R}{|}}{CH}-COOH \qquad \text{(b)}
$$

Better results are obtained if Gabriel phthalimide synthesis is adopted (Eq. 24.3b) (see Chapter 18 for Gabriel synthesis).

24.1.2.2 Strecker Synthesis

Strecker synthesis involves treating an aldehyde with ammonia and a cyanide. In practice, an equimolar mixture of ammonium chloride and potassium cyanide is used. An amino nitrile is formed which is hydrolysed with aqueous acid to give the amino acid (Eq. 24.4a). The mechanism is outlined in Eq. 24.4b.

$$
R-CHO + NH_4Cl + KCN \longrightarrow R-CH\begin{smallmatrix}CN\\NH_2\end{smallmatrix} \xrightarrow{\;H_3O^+\;} R-CH\begin{smallmatrix}COOH\\NH_2\end{smallmatrix} \qquad \text{(a)}
$$

$$R-CHO + NH_3 \longrightarrow R-\overset{\overset{\displaystyle H}{|}}{C}=NH \xrightarrow{H^+, CN^-} R-\overset{\overset{\displaystyle H}{|}}{C}\overset{NH_2}{\underset{CN}{<}} \qquad (b) \qquad (24.4)$$

24.1.2.3 Malonic Ester Synthesis

There are several ways in which the malonic ester synthesis of carboxylic acids (Chapter 17, Section 17.5) has been adapted for the synthesis of amino acids. One of these involves the use of acetamidomalonic ester (Eq. 24.5).

$$CH_2(COOC_2H_5)_2 + HNO_2 \longrightarrow ON-CH(COOC_2H_5)_2 \rightleftarrows HON=C(COOC_2H_5)_2$$
$$\text{(I)}$$

$$\xrightarrow[\text{(CH}_3\text{CO)}_2\text{O}]{\text{Zn, CH}_3\text{COOH}} \underset{\text{Acetamidomalonic ester}}{CH_3CONH-CH(COOC_2H_5)_2} \xrightarrow{C_2H_5ONa} CH_3CONH-\overset{Na^+}{\underset{}{\bar{C}}}(COOC_2H_5)_2$$
$$\text{(II)} \qquad\qquad\qquad\qquad \text{(III)}$$

$$\xrightarrow{RX} \underset{\underset{\text{(IV)}}{\overset{|}{R}}}{CH_3CONH-C(COOC_2H_5)_2} \xrightarrow{H_3O^+} \left[\underset{\underset{R}{\overset{|}{}}}{H_2N-C(COOH)_2} \right]$$

$$\xrightarrow{-CO_2} \underset{\overset{|}{R}}{H_2N-CH-COOH}$$
$$\text{(V)}$$

$$(24.5)$$

Reaction of diethyl malonate with nitrous acid which results in nitrosation gives (I), in which the nitroso group exists in equilibrium with the oxime tautomer. This is not isolated, but reacted with zinc and acetic acid, and acetic ahydride when (i) the oxime HON=, is reduced to the amine H_2N- and (ii) the amine is acetylated. The product (II), is diethyl acetamidomalonate. The anion of this (III), can be alkylated with the required alkyl halide to obtain (IV). Hydrolysis and decarboxylation of (IV) produces the amino acid (V).

24.2 PROTEINS

24.2.1 PRIMARY STRUCTURE OF PROTEINS

Proteins are formed by the condensation of amino acids through peptide links. The basic structure or *primary structure* of proteins comprises of the sequence of amino acids in the polypeptide chain. The primary structure of the nonapeptide bradykinin – present in blood – is shown in Fig. 24.1.

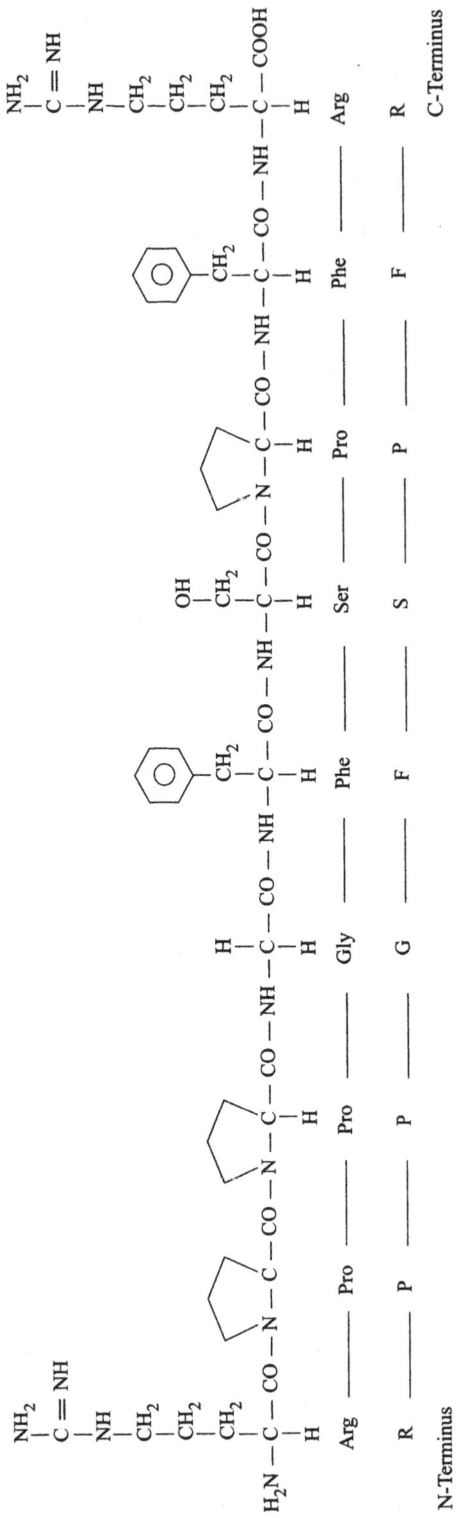

Fig. 24.1 Primary structure of bradykinin

If the polypeptide has no branching or ring structure, one of the component amino acids will have a free α-amino group and another one will have a free carboxyl group. By convention, the sequence is written in such a way that the free amino group is at the left end of the chain and the free COOH is at the right end. The amino end is called the N-terminus and the carboxyl end, the C-terminus. In bradykinin, the N-terminal amino acid and the C-terminal amino acid are both arginine. In many proteins and even in smaller polypeptides containing cysteine, two SH groups – either in the same chain or in different chains – can react by oxidation to form a disulphide link. If the two cysteine molecules are in the same chain, it results in the formation of a cyclic structure as in the case of oxytocin [Fig. 24.2, (a)].

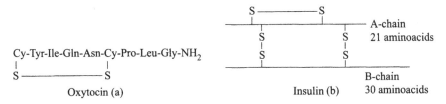

Fig. 24.2   Structure of oxytocin and insulin

The amino acid sequence of insulin – the pancreatic hormone responsible for the control of blood sugar – was worked out by Sanger in 1951–55. This is the first protein whose structure was determined.

Insulin contains two short peptide chains connected by two disulphide bridges [Fig. 24.2, (b)].

The first step in the determination of the structure of a polpepyide is to determine the amino acid composition. This is done by complete hydrolysis and chromatographic analysis. Automated machines are

> Frederic Sanger received the Nobel prize in chemistry for sequencing insulin in 1958. He shared a second Nobel prize in 1980 for his work on nucleic acids.

now available for such analysis. Ninhydrin reaction is used to locate the presence of amino acids in a chromatogram and also to estimate the amino acids quantitatively by colourimetry. All amino acids react with ninhydrin (I) to give the blue coloured product (II) (Fig. 24.3).

The polypeptide is then subjected to partial hydrolysis by chemical and by enzymatic methods. Small peptides such as di-, tri-, or tetrapeptides are isolated from the hydrolysis mixture and their structure determined by end-group analysis methods, mainly by Edman degradation. The issues involved in such an analysis are illustrated by the tripeptide containing one molecule each of alanine, glycine and phenylalanine. Its structure may be any one of the following:

Ala-Gly-Phe	Gly-Ala-Phe
Ala-Phe-Gly	Phe-Ala-Gly
Gly-Phe-Ala	Phe-Gly-Ala

End-group analysis involves finding out which amino acid is at the N-terminus and which is at the C-terminus. If it is a dipeptide, it is enough to determine one end.

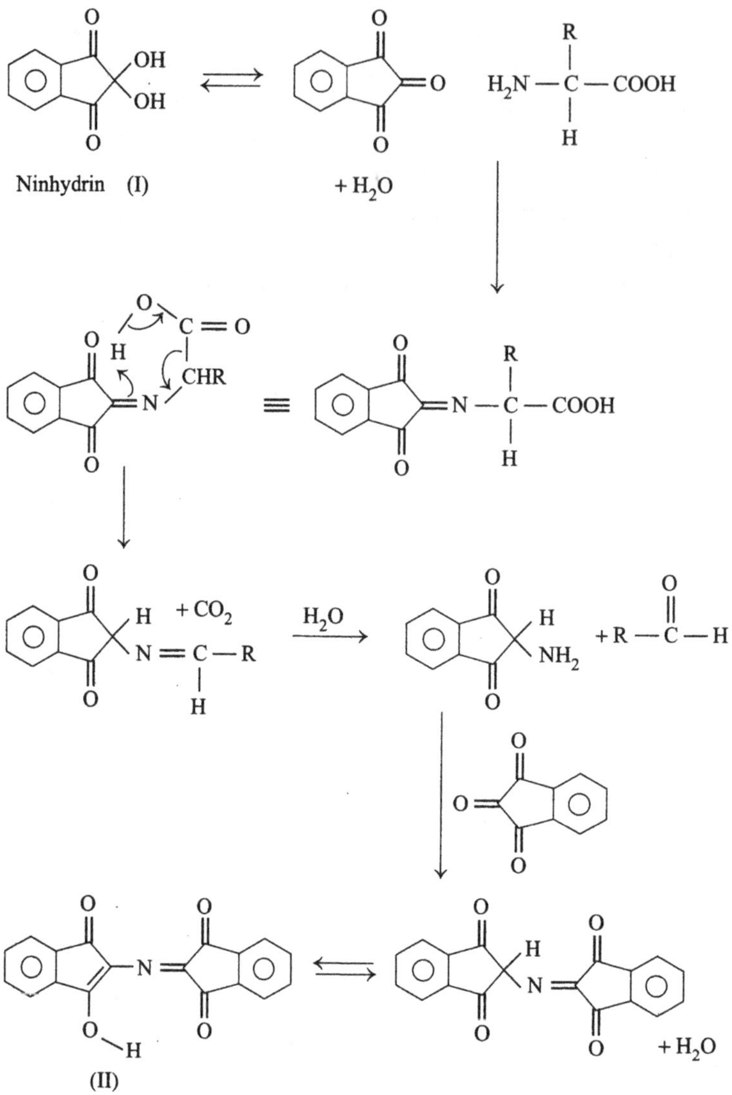

Fig. 24.3 Ninhydrin reaction

24.2.1.1 Sanger's Method for N-terminal Amino Acids (DNP Method)

This is the method developed by Sanger for the determination of the structure of insulin. The N-terminal amino acid of a polypeptide contains a free amino group. When such a polypeptide is treated with 2,4-dinitrofluorobenzene (DNFB), fluorine is displaced by the amine in a nucleophilic displacement reaction (Chapter 9, Section 9.5.3.1). The N-terminal amino acid can be effectively labelled in this manner. Now, acid hydrolysis and identification of the 2,4-dinitrophenylamino acid (DNP amino acid) completes the N-terminal determination [Fig. 24.4 (a)].

Reagents other than DNFB have been developed to label the N-terminal amino acid. One such reagent is 5-dimethylaminonaphthalene-1-sulphonyl chloride, abbreviated as dansyl chloride or DNS-Cl which converts the amino group to a suphonamide [Fig. 24.4, (b)].

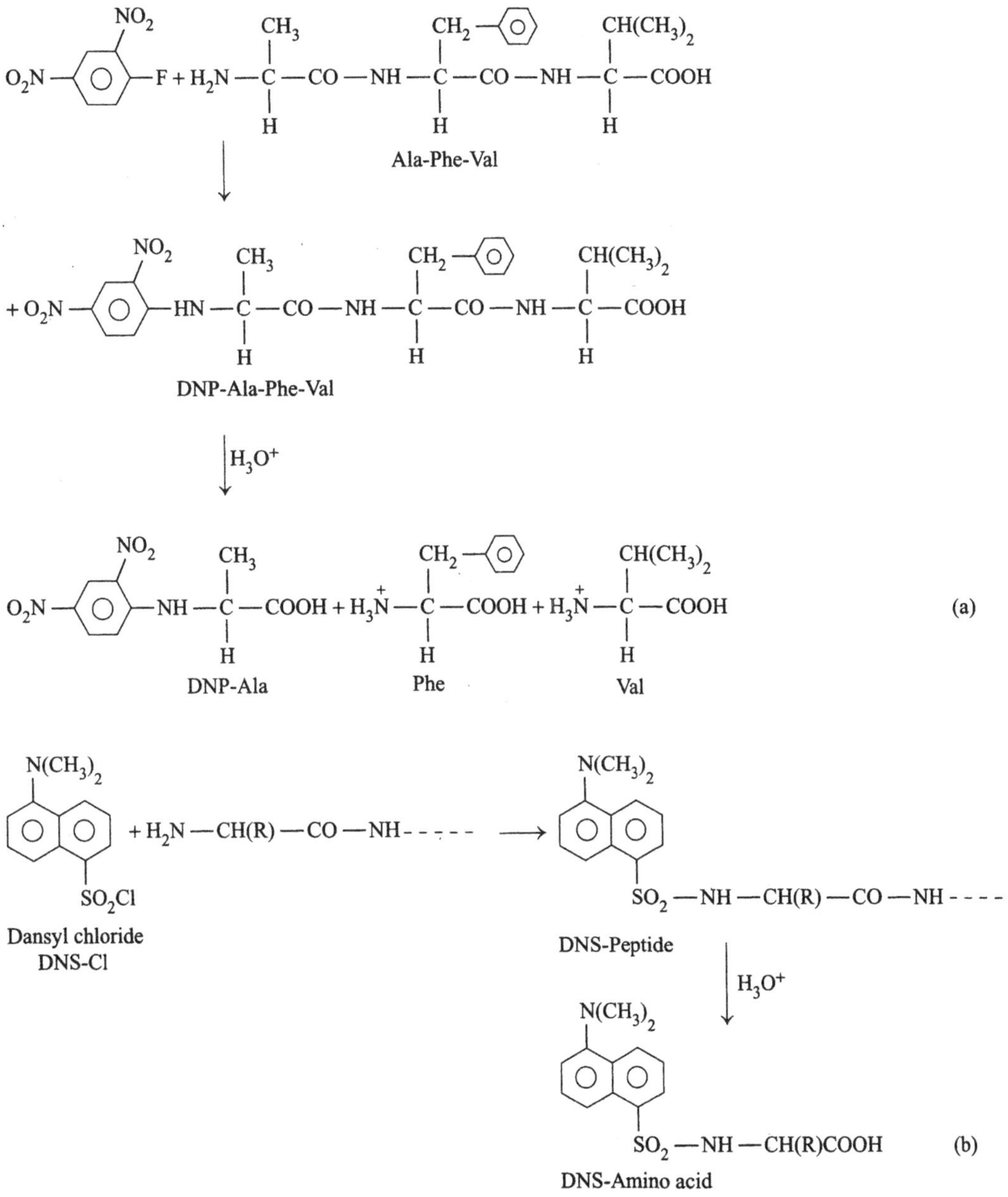

Ala-Phe-Val

DNP-Ala-Phe-Val

H_3O^+

DNP-Ala   Phe   Val   (a)

Dansyl chloride
DNS-Cl

DNS-Peptide

H_3O^+

DNS-Amino acid   (b)

Fig. 24.4 Sanger's method (a) and the use of dansyl chloride (b)

24.2.1.2 Edman Degradation for N-Terminal Amino Acid

One of the drawbacks of Sanger's method is that in the process of removing the labelled N-terminal amino acid, the rest of the chain gets broken up by hydrolysis and cannot be used to proceed further. Today, Sanger's method is only of historical importance. Edman degradation can identify all the amino acids, one at a time, sequentially. This is the most versatile method of determining the amino acid sequence of even large polypeptides (Fig. 24.5).

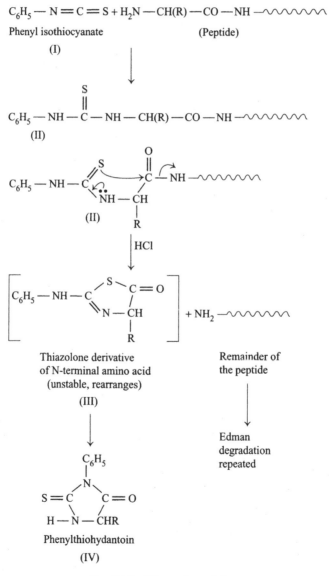

Fig. 24.5 Edman degradation

The reagent for Edman degradation is phenyl isothiocyanate, $C_6H_5-N=C=S$, (I). This reagent reacts with the free amino group of the polypeptide to form a thiourea derivative (II). When this is treated with dry HCl in a non-aqueous solvent, cleavage takes place, whereby the N-terminal amino acid comes out as a thiazolone (III). This thiazolone is unstable and undergoes an acid catalysed rearrangement to give a phenylthiohydantoin, (IV), in which the R group comes from the amino acid, which is identified by comparison with standard samples. The rest of the chain is intact and can be subjected to a second Edman degradation and then to a third and so forth. In principle, the sequencing can be done for the entire polypeptide chain. In practice, upto 20 amino acids can be routinely sequenced. This procedure has been automated. Edman degradation is the most important method for sequencing polypeptides.

24.2.1.3 Akabori Method for C-Terminal Determination

This method involves cleavage of the polypeptide chain by hydrazinolysis (Eq. 24.6).

$$R-\overset{\overset{\displaystyle O}{\|}}{C}-NHR' + H_2N-NH_2 \longrightarrow R-\overset{\overset{\displaystyle O}{\|}}{C}-NHNH_2 + R'NH_2 \qquad \text{(a)}$$

$$H_2N-\overset{\overset{\displaystyle R}{|}}{C}H-CONH-\overset{\overset{\displaystyle R'}{|}}{C}H-CONH-\overset{\overset{\displaystyle R''}{|}}{C}H-COOH + H_2N-NH_2 \text{ (excess)}$$

$$\longrightarrow H_2N-\overset{\overset{\displaystyle R}{|}}{C}H-CONHNH_2 + H_2N-\overset{\overset{\displaystyle R'}{|}}{C}H-CONHNH_2 + H_2N-\overset{\overset{\displaystyle R''}{|}}{C}H-COOH \qquad \text{(b)}$$

$$\quad\quad\quad\text{Hydrazide} \qquad\qquad\qquad \text{Hydrazide} \qquad\qquad \text{Free aminoacid}$$
$$\text{(C-Terminal)}$$

$$(24.6)$$

Hydrazine is a powerful nucleophile which can displace the amine from an amide (Eq. 24.6a). With the tripeptide shown in Eq. 24.6b, the C-terminal amino acid alone is left intact; the other two get converted to the hydrazide. The Akabori method suffers from the same limitation as Sanger's method namely, that only one amino acid, the C-terminal one in this case, in a polypeptide chain can be identified.

Enzymatic methods are also available for sequencing. A group of enzymes called *carboxypeptidases* hydrolyses proteins selectively and sequentially from the carboxy end of the chain.

24.2.1.4 Overlap Method for the Total Structure of the Polypeptide

Once the sequence of amino acids of the small polypeptides – obtained by the hydrolysis of the protein molecule – is known, overlap of structures of segments of two polypeptides is made use of to arrive at the structure of the section comprising the two polypeptides (Fig. 24.6). This is a hypothetical example. Partial hydrolysis of a heptapeptide (I), under investigation gave a number of products. From the hydrolysate, a pentapeptide (II), and a tetrapeptide (III), were isolated and their sequences were determined. The Phe-Ser segment at the C-terminal end of (II) and Phe-Ser at the N-terminal end of (III) are said to 'overlap'. Taking this segment to be common to both (II) and (III), the complete structure was assigned.

Heptapeptide (I) $\longrightarrow$ Polypeptide (II) + Polypeptide (III)

Polypeptide (II) : Ala-Phe-Gly-Phe-Ser

Polypeptide (III) :   Phe-Ser-Glu-Pro

over-
lapping
sequence

∴ Heptapeptide : Ala-Phe-Gly-Phe-Ser-Glu-Pro

(I)

Fig. 24.6 Overlap procedure

24.2.2 PEPTIDE SYNTHESIS

The issues involved in peptide synthesis can be understood by taking a simple example, synthesis of the dipeptide alanylphenylalanine, Ala-Phe. The starting materials are the free amino acids, alanine and phenylalanine. Since stereochemistry is important, L-amino acids are used and mild conditions are to be employed to avoid racemisation of the amino acids. The reaction is a simple amide formation. We want the carboxyl group of alanine and the amino group of phenylalanine to react, not the other way round. Also, self-condensation of alanine or phenylalanine should not take place. Strategies for peptide synthesis are meant to ensure that these conditions are met. To prevent the amino group of alanine from reacting, it has to be protected in such a way that the protecting group can be removed after peptide link formation, without affecting the peptide bond.

Normally, a free carboxyl group does not condense with an amino group. It has to be activated by conversion to a derivative like RCO—X where X is a good leaving group. This is usually carried out by converting the acid to the acyl halide or anhydride. RCO—X can also be an ester, if X is a group like *p*-nitrophenoxy. Another way of condensing the —COOH and —NH$_2$ groups is by using the condensing agent, dicyclohexylcarbodiimide, DCC, which converts the acid in situ to a derivative like RCO—X. If DCC has to be used, the COOH group of the C-terminal amino acid, phenylalanine, should be protected to prevent self-condensation. Several peptide synthesis strategies are available which satisfy these considerations.

24.2.2.1 Protection of the Amino Group

For most usual purposes, acetylation is a satisfactory reaction to protect an amino group. This is unsuitable for peptide synthesis because deprotection or deblocking, which is usually done by hydrolysis, will also lead to the hydrolysis of the peptide bond. Two of the several reagents which have been developed for amino-protection are benzyloxycarbonyl chloride (Z—Cl) and *tert*-butoxycarbonyl chloride (Boc — Cl). The use of benzyloxycarbonylchloride (I), is illustrated in Fig. 24.7, (a).

Reaction of alanine (II), with benzyloxycarbonylchloride (I), gives N-benzyloxycarbonylalanine (III), abbreviated as Z-alanine. The carboxyl group in this molecule is activated by converting it to the acyl chloride (IV), and then condensed with phenylalanine to obtain the N-protected dipeptide (V). A third amino acid can be attached to this at the C-terminal by repeating the activation–condensation steps to prepare a tripeptide. This sequence can be repeated till the required number of amino acid units have been incorporated. The deprotection of the amino end is done by taking advantage of the high reactivity of the benzyloxy grouping. Catalytic hydrogenation inserts two atoms of hydrogen and cleaves the benzyl–oxygen bond. Such a cleavage with simultaneous insertion of hydrogen is called *hydrogenolysis* (cf. hydrolysis) which is characteristic of benzyl–O bonds. Hydrogenolysis can also be done by reduction using sodium and liquid ammonia (Birch reduction). The product of hydrogenolysis is a carbamic acid (VI), which does not exist in the free-state and decomposes liberating carbon dioxide. The dipeptide (VII) is obtained without affecting the peptide bond [Fig. 24.7 (a)].

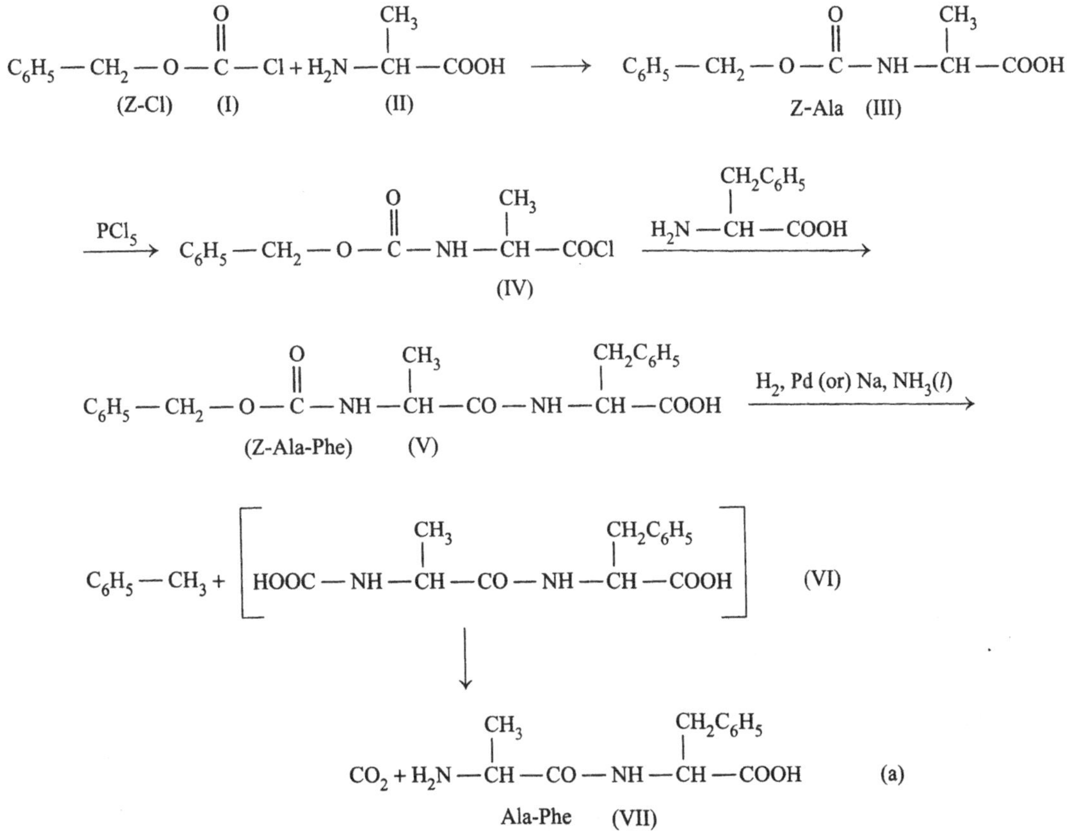

$$C_6H_5-CH_2-O-\overset{\overset{O}{\|}}{C}-Cl+H_2N-\overset{\overset{CH_3}{|}}{CH}-COOH \longrightarrow C_6H_5-CH_2-O-\overset{\overset{O}{\|}}{C}-NH-\overset{\overset{CH_3}{|}}{CH}-COOH$$

(Z-Cl) (I)   (II)   Z-Ala (III)

$$\xrightarrow{PCl_5} C_6H_5-CH_2-O-\overset{\overset{O}{\|}}{C}-NH-\overset{\overset{CH_3}{|}}{CH}-COCl \xrightarrow{\overset{\overset{CH_2C_6H_5}{|}}{H_2N-CH-COOH}}$$

(IV)

$$C_6H_5-CH_2-O-\overset{\overset{O}{\|}}{C}-NH-\overset{\overset{CH_3}{|}}{CH}-CO-NH-\overset{\overset{CH_2C_6H_5}{|}}{CH}-COOH \xrightarrow{H_2, Pd (or) Na, NH_3(l)}$$

(Z-Ala-Phe) (V)

$$C_6H_5-CH_3 + \left[HOOC-NH-\overset{\overset{CH_3}{|}}{CH}-CO-NH-\overset{\overset{CH_2C_6H_5}{|}}{CH}-COOH \right] \quad (VI)$$

$$\downarrow$$

$$CO_2 + H_2N-\overset{\overset{CH_3}{|}}{CH}-CO-NH-\overset{\overset{CH_2C_6H_5}{|}}{CH}-COOH \quad (a)$$

Ala-Phe (VII)

Fig. 24.7 (Continued)

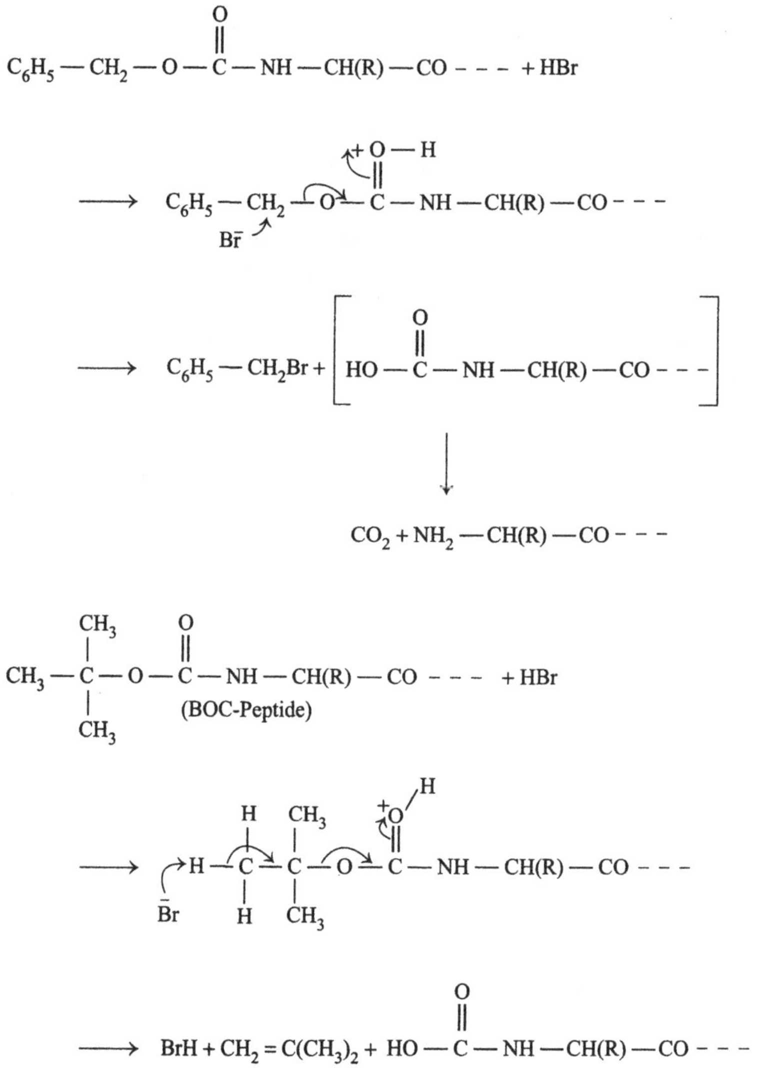

$$C_6H_5 - CH_2 - O - \overset{\overset{\displaystyle O}{\|}}{C} - NH - CH(R) - CO --- + HBr$$

$$\longrightarrow \quad C_6H_5 - CH_2 \overset{+}{-} O \overset{\overset{\displaystyle +O-H}{\|}}{-} C - NH - CH(R) - CO ---$$
$$\overset{}{Br^-}$$

$$\longrightarrow \quad C_6H_5 - CH_2Br + \left[HO - \overset{\overset{\displaystyle O}{\|}}{C} - NH - CH(R) - CO --- \right]$$

$$\downarrow$$

$$CO_2 + NH_2 - CH(R) - CO --- \qquad \text{(b)}$$

$$CH_3 - \overset{\overset{\displaystyle CH_3}{|}}{\underset{\underset{\displaystyle CH_3}{|}}{C}} - O - \overset{\overset{\displaystyle O}{\|}}{C} - NH - CH(R) - CO --- + HBr$$
$$\text{(BOC-Peptide)}$$

$$\longrightarrow \quad H - \overset{\overset{\displaystyle H}{|}}{\underset{\underset{\displaystyle Br}{}}{C}} \doublebond \overset{\overset{\displaystyle CH_3}{|}}{\underset{\underset{\displaystyle CH_3}{|}}{C}} - O \overset{\overset{\displaystyle +O^{\diagup H}}{\|}}{-} C - NH - CH(R) - CO ---$$

$$\longrightarrow \quad BrH + CH_2 = C(CH_3)_2 + HO - \overset{\overset{\displaystyle O}{\|}}{C} - NH - CH(R) - CO --- \quad \text{(c)}$$

Fig. 24.7 Peptide synthesis using N-protected amino acid

Deprotection can also be carried out by treating the Z-peptide with HBr in acetic acid in the cold. Stability of the benzyl cation is the driving force for this selective cleavage by HBr [Fig. 24.7 (b)]. The use of *tert*-butoxycarbonyl chloride is illustrated in Fig. 24.7 (c). After the formation of the Boc-protected polypeptide, deprotection is affected by treatment with HBr. The *tert*-butyl group is eliminated as 2-methylpropene.

24.2.2.2 Activation of the Carboxyl Group

Conversion to the acyl chloride as shown in Fig. 24.7 is not satisfactory in all cases. Another method of activation is the use of *p*-nitrophenyl ester (Fig. 24.8).

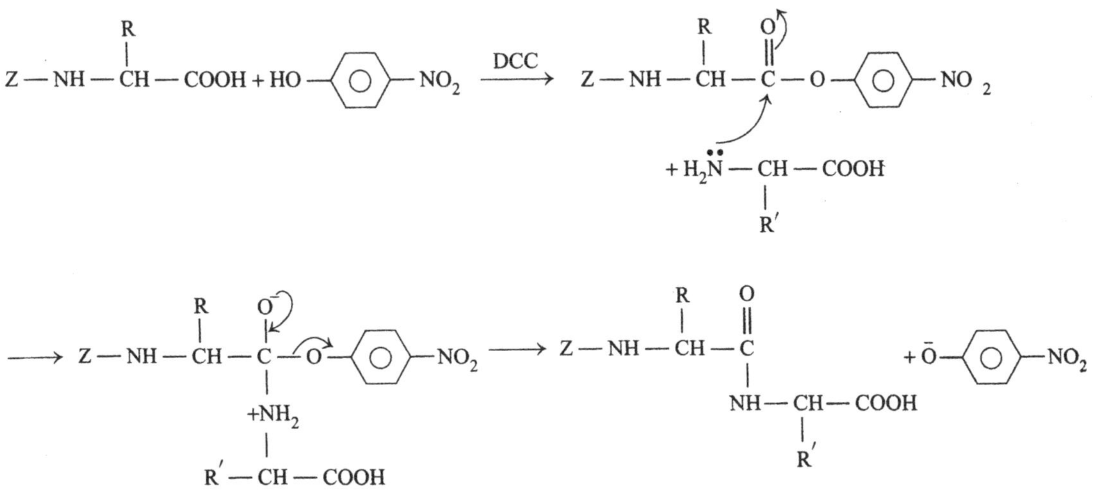

Fig. 24.8 Use of *p*-nitrophenyl ester

p-Nitrophenyl ester is prepared under mild conditions by treating the N-protected amino acid with *p*-nitrophenol in the presence of DCC (vide infra).

24.2.2.3 Condensation Using DCC: Protection of the Carboxyl Group

An amide can be formed directly from the amine and the free carboxylic acid using the condensing agent dicyclohexycarbodiimide, DCC.

This reagent can bring about the condensation of a carboxyl group and an amine to form an amide, and a carboxyl and hydroxyl to form an ester. The activation of the carboxyl takes place in situ by the formation of an adduct with DCC (Fig. 24.9). In order to use this reagent to condense the free COOH of an N-protected amino acid with the NH_2 of another amino acid, the COOH of the latter should be protected. This is done by esterification. The ester should be one that can be hydrolysed or cleaved without affecting the peptide link. *tert*-Butyl ester and benzyl ester are both suitable for this. Both of them can be cleaved by treating with HBr in CF_3COOH. The peptide bond is not affected by this reagent. This reagent will also deprotect the amino group at the N-terminal. If the carboxyl protection is to be retained and the amino group alone has to be deprotected so that one more amino acid can be attached at the N-terminus, *tert*-butyl ester for C-protection and benzyloxy carbonyl for N-protection may be used. The latter can be removed by hydrogenolysis. *tert*-Butyl esters cannot be hydrogenolysed.

24.2.2.4 General Scheme for the Synthesis of a Polypeptide

Application of this methodology to peptide synthesis is outlined in Fig. 24.10. This is one possible scheme. Other N-protection and C-protection strategies are possible. Also, DCC and C-protection can be can be dispensed with by converting the N-protected amino acid to the *p*-nitrophenyl ester or the acyl chloride.

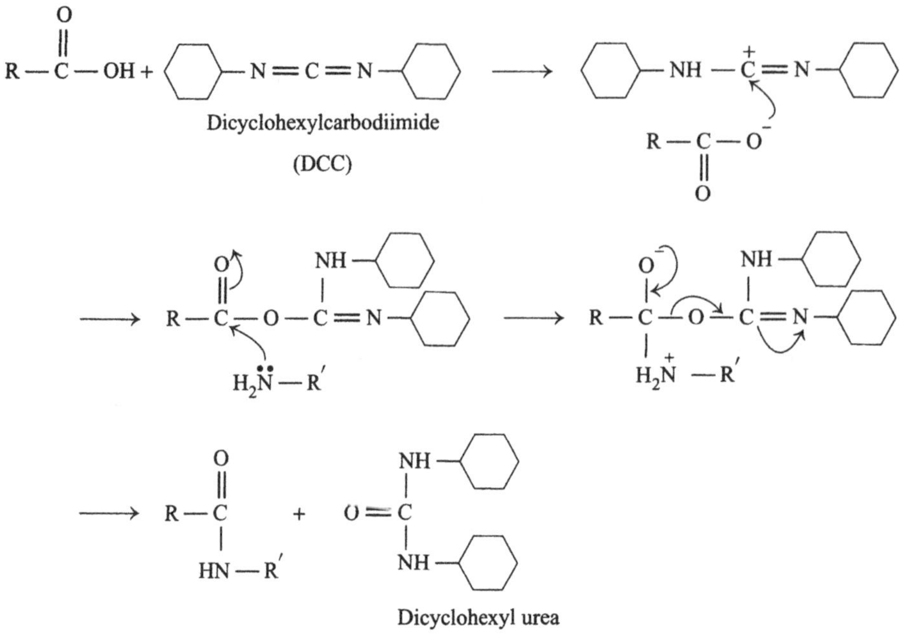

Fig. 24.9   Mechanism of amide formation using DCC

24.2.2.5   Solid-Phase Peptide Synthesis

An important innovation for the synthesis of polypeptides was introduced by Bruce Merrifield in 1962. The technique called solid-phase synthesis, involves the use of benzyl chloride which is part of a polystyrene polymer in the form of insoluble beads. The polymer has $-C_6H_4-CH_2Cl$ groups on the surface, introduced by functionalisation of polystyrene. The C-terminal amino acid is attached to this to form a benzyl ester, which is now in the solid phase. The rest of the steps follow as in Fig. 24.11. At each stage, the resin beads with the attached amino acid are suspended in a suitable solvent, required reagents added and after reaction, other soluble products and excess reagents removed by washing. After the required number of amino acids have been incorporated, the polypeptide is detached from the polymer beads by using the reagent for benzyl ester deprotection, HBr/CF_3COOH. This method has been automated. The importance of solid-phase synthesis of peptides was recognized by the award of the Nobel prize in chemistry to Merrifield in 1984. Solid-phase synthesis techniques have been developed for many other classes of compounds. *Combinatorial chemistry*, which enables preparation of hundreds of related compounds for such applications as screening for pharmaceutical use, is based on solid-phase synthesis.

24.3   CLASSIFICATION

Proteins are classified in different ways. A useful classification is based on properties and functions (Table 24.2).

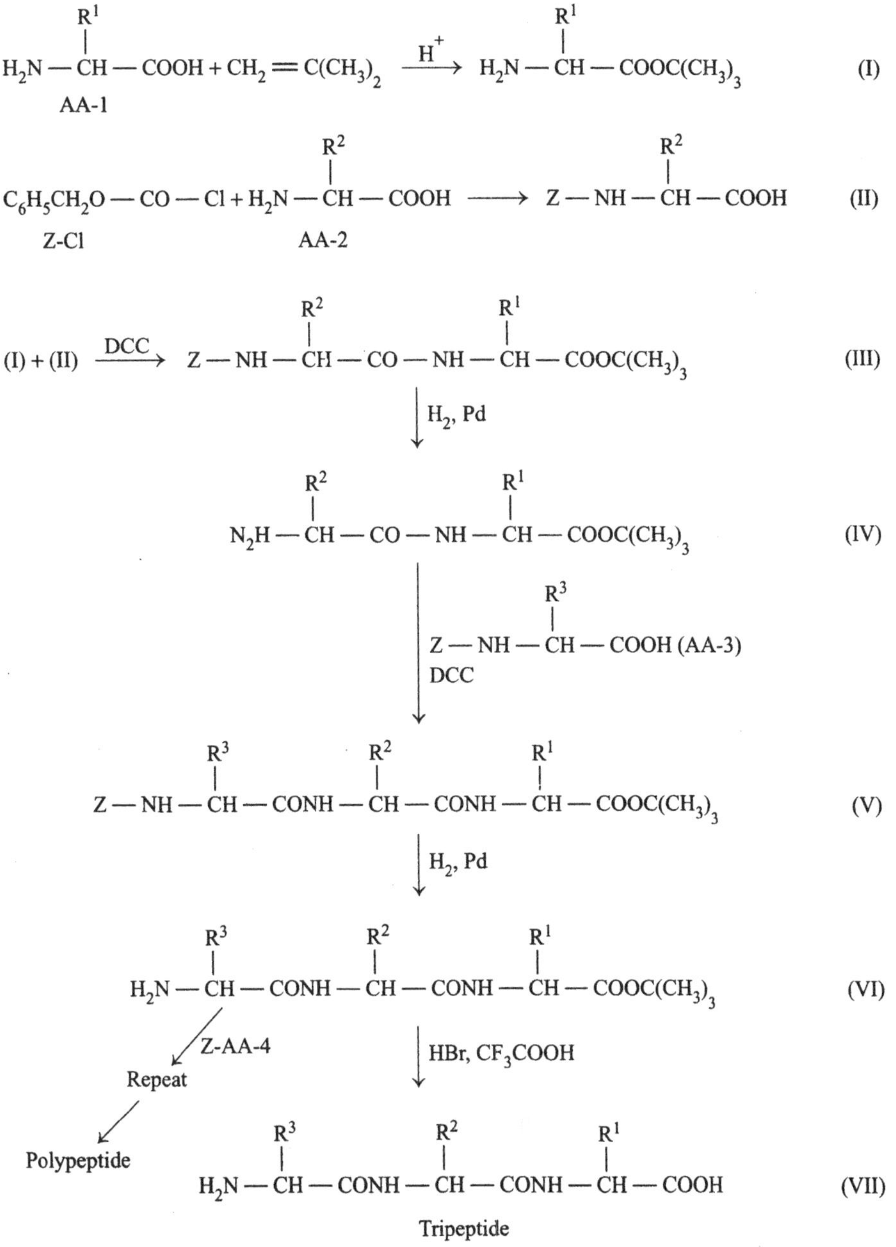

$$H_2N - \overset{\overset{\displaystyle R^1}{|}}{CH} - COOH + CH_2 = C(CH_3)_2 \xrightarrow{\overset{+}{H}} H_2N - \overset{\overset{\displaystyle R^1}{|}}{CH} - COOC(CH_3)_3 \qquad (I)$$

AA-1

$$C_6H_5CH_2O - CO - Cl + H_2N - \overset{\overset{\displaystyle R^2}{|}}{CH} - COOH \longrightarrow Z - NH - \overset{\overset{\displaystyle R^2}{|}}{CH} - COOH \qquad (II)$$

Z-Cl   AA-2

$$(I) + (II) \xrightarrow{DCC} Z - NH - \overset{\overset{\displaystyle R^2}{|}}{CH} - CO - NH - \overset{\overset{\displaystyle R^1}{|}}{CH} - COOC(CH_3)_3 \qquad (III)$$

$$\downarrow H_2, Pd$$

$$N_2H - \overset{\overset{\displaystyle R^2}{|}}{CH} - CO - NH - \overset{\overset{\displaystyle R^1}{|}}{CH} - COOC(CH_3)_3 \qquad (IV)$$

$$\downarrow \begin{array}{c} Z - NH - \overset{\overset{\displaystyle R^3}{|}}{CH} - COOH \ (AA\text{-}3) \\ DCC \end{array}$$

$$Z - NH - \overset{\overset{\displaystyle R^3}{|}}{CH} - CONH - \overset{\overset{\displaystyle R^2}{|}}{CH} - CONH - \overset{\overset{\displaystyle R^1}{|}}{CH} - COOC(CH_3)_3 \qquad (V)$$

$$\downarrow H_2, Pd$$

$$H_2N - \overset{\overset{\displaystyle R^3}{|}}{CH} - CONH - \overset{\overset{\displaystyle R^2}{|}}{CH} - CONH - \overset{\overset{\displaystyle R^1}{|}}{CH} - COOC(CH_3)_3 \qquad (VI)$$

Z-AA-4 ↙   ↓ HBr, CF₃COOH

Repeat

↙

Polypeptide

$$H_2N - \overset{\overset{\displaystyle R^3}{|}}{CH} - CONH - \overset{\overset{\displaystyle R^2}{|}}{CH} - CONH - \overset{\overset{\displaystyle R^1}{|}}{CH} - COOH \qquad (VII)$$

Tripeptide

(AA = amino acid)

Fig. 24.10 A general scheme for the synthesis of polypeptides

24.3.1 SECONDARY STRUCTURE

We are already familiar with the primary structure of proteins, which is the sequence of amino acids that form the protein chain. The chain may assume specific conformations and shapes such as

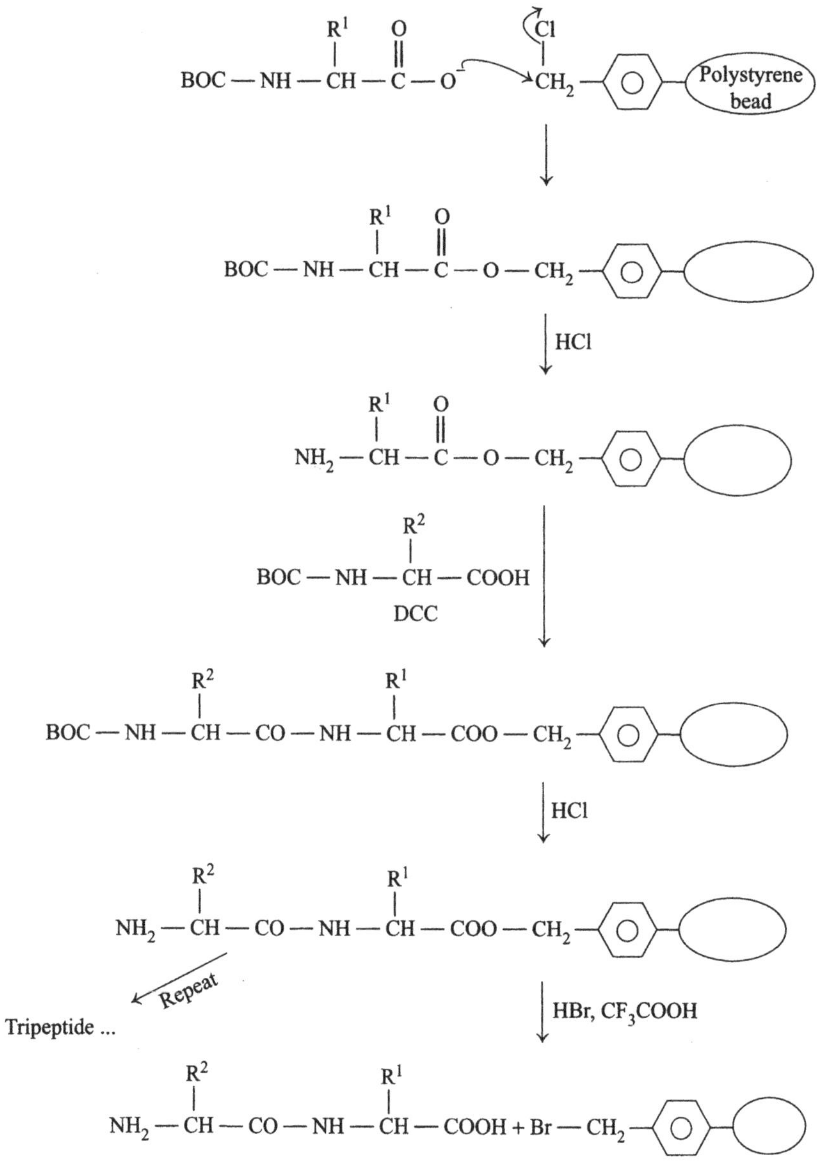

Fig. 24.11 Solid-phase peptide synthesis

coils and sheets. This is called the secondary structure of proteins. Pauling and Corey showed that many proteins assume stable conformations which they classified as the α-helix and the β-pleated sheet (Fig. 24.12). In these, the particular conformations are stabilised by hydrogen bonds. In the sheet structure, the chains may be parallel or anti-parallel.

G N Ramachandran, an Indian Physicist, working at the University of Madras, showed that collagen has a unique triple helical structure. In collagen, the main protein of connective tissues, three left-handed helices are twisted into a right-handed coiled-coil.

Table 24. 2 Classification of proteins

Group	Examples	Remarks
Fibrous proteins (Structural) (insoluble)	Collagens	Present in connective tissues
	Keratins	Present in hair, hoofs, quills, nails
	Elastins	Present in tendons, arteries
Globular proteins (soluble)	Albumins	Present in egg white, serum
	Globulins	Present in serum globulin, tissue globulin
	Histones (basic)	Associated with nucleic acids
	Protamines (basic)	Associated with nucleic acids
Conjugated proteins combined with other non-protein components called prosthetic groups	Nucleoproteins	Combined with nucleic acids
	Lipoproteins	Combined with lipids
	Glyco- and mucoproteins Metalloproteins,	Combined with carbohydrates
	Chromoproteins	As in hemoglobin where the prosthetic groups are heme and iron

24.3.2 TERTIARY STRUCTURE AND DENATURATION

Protein chains, whether helical or sheet-like, assume folded conformations in their native environment. The folded structure is held in place by various intramolecular forces. Hydrogen bonds are the most important of these. In addition, there are dipole–dipole interactions and hydrophobic interactions. A chain containing polar groups like OH and COOH and non-polar hydrocarbon regions and groups, will fold up in such a way that the polar, hydrophilic groups are oriented on the outside in contact with water, and the non-polar hydrocarbon groups which are hydrophobic, are huddled inside the folds, away from water. The folded structure of native proteins is known as tertiary structure. This is very important for the biological functions of the proteins. This is the most stable conformation of the protein in its environment. The tertiary structure can be disrupted by heating or by adding substances or solvents, or by changing the pH which break up the weak forces that are responsible for the folded structure. When that happens, the protein is said to be *denatured*. Denaturation causes the properties to change and results in loss of biological activity. Hardening of egg-white upon heating, and curdling of milk by the addition of acids, are examples of denaturation. Denaturation is generally irreversible, though in some cases when the agent which caused denaturation is removed, the protein may regain its original structure. This is called *renaturation*.

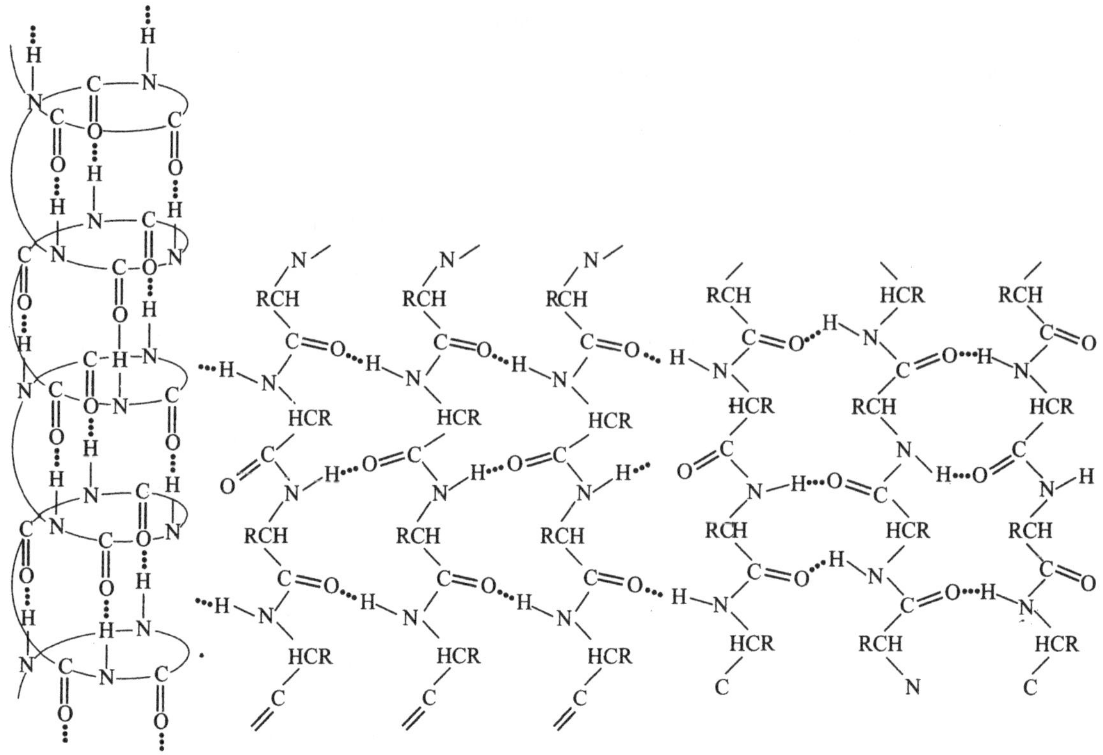

α-Helix
(Right-handed)

β-Pleated structure   (I)
(Parallel)

β-Pleated structure   (II)
(Antiparallel)

Fig. 24.12 α-Helix and β-pleated structures

Some proteins are composed of two or more chains, along with other non-protein components. Hemoglobin is composed of four heme units and four protein chains. Such an assembly is called the *quaternary structure*.

24.4 NUCLEIC ACIDS

24.4.1 STRUCTURE

Nucleic acids are biopolymers of nucleosides. Each nucleoside is made up of a pentose – either D-ribofuranose or D-2-deoxyribofuranose – whose anomeric position is occupied by a heterocyclic base (a purine or a pyrimidine) (Fig. 24.13). The nucleosides are linked by a phosphate bridge connecting the C3 of one sugar with the C5 of the other sugar. A unit consisting of a sugar, a base and a phosphate is called a nucleotide. Nucleotides are phosphoric acid esters of nucleosides.

The polymeric molecules containing ribose units are called ribonucleic acid, RNA. Those made up of deoxyribose units are called deoxyribonucleic acid, DNA. The purine bases in both DNA and RNA are adenine and guanine connected to the C1 of the sugar unit at the nitrogen on

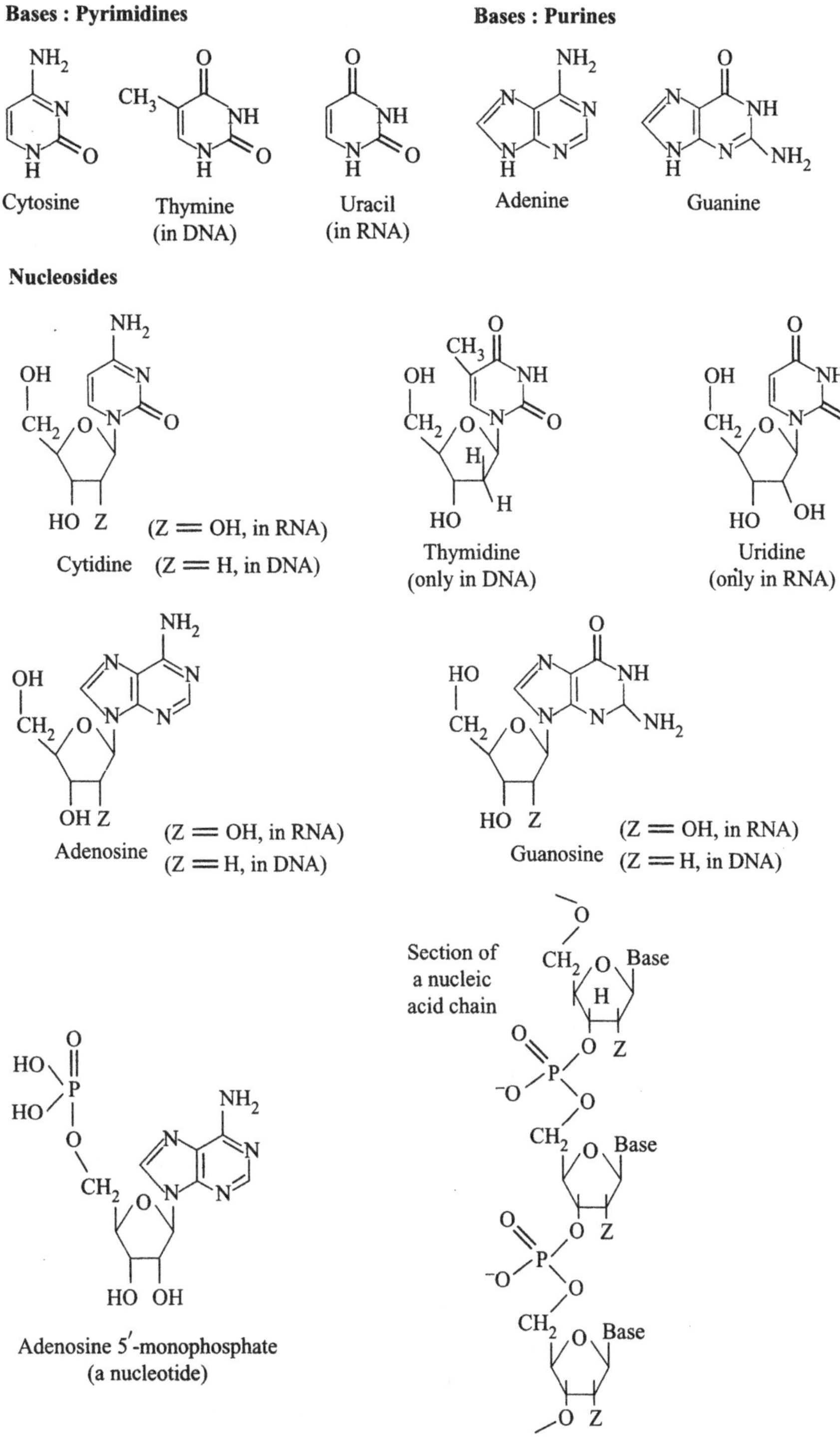

Bases : Pyrimidines

Bases : Purines

Cytosine

Thymine
(in DNA)

Uracil
(in RNA)

Adenine

Guanine

Nucleosides

Cytidine   (Z $=$ H, in DNA)   (Z $=$ OH, in RNA)

Thymidine
(only in DNA)

Uridine
(only in RNA)

Adenosine   (Z $=$ OH, in RNA)
(Z $=$ H, in DNA)

Guanosine   (Z $=$ OH, in RNA)   (Z $=$ H, in DNA)

Section of
a nucleic
acid chain

Adenosine 5′-monophosphate
(a nucleotide)

Fig. 24.13 Bases, nucleosides, nucleotides and nucleic acid

the 9-position. The pyrimidine bases in RNA are cytosine and uracil; in DNA they are cytosine and thymine. The pyrimidines are connected to the C1 of the sugar through the nitrogen at the 1-position. A representative structure of a section of a nucleic acid is shown in Fig. 24.13.

Nucleic acids are the biomolecules which store, transmit and process genetic information. DNA has the unique ability to reproduce itself—*replication*. The double helical structure of DNA is responsible for this. DNA has a helical structure. Two strands of DNA, oriented in an antiparallel manner, are bound to each other by hydrogen bonds to give the double helical structure (Fig. 24.14).

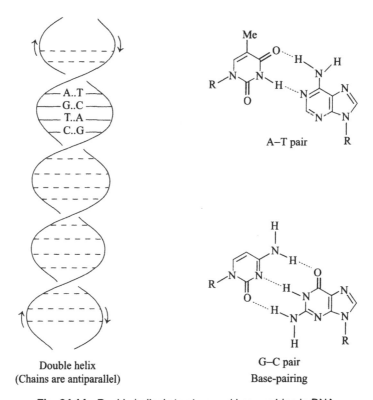

Fig. 24.14 Double helical structure and base-pairing in DNA

The hydrogen bonds between the strands are of a very special nature. A thymine on one chain and an adenine on the other chain are bound to each other by two hydrogen bonds. Similarly cyctosine and guanine are bound to each other by three hydrogen bonds. Thus the bases are paired as, A — T and C — G (Fig. 24.14). In a double-stranded DNA molecule, for every thymine there will be one adenine, and for every cytosine there will be one guanine. The ratio of adenine to thymine and that of guanine to cytosine is close to 1, regardless of the source of the DNA. The two chains are complementary. The structure of one chain is dictated by the structure of the other chain. In a medium where DNA synthesis is taking place, the two chains can dissociate and each chain can grow its own complementary chain. This is the basis of the reproduction of DNA or replication.

24.4.2 BIOLOGICAL FUNCTIONS

Nucleic acids are present in the nuclei of all cells and are the molecules which are responsible for the control of genetic processes. All the characteristics of the organism are encoded in the DNA. The codes for the synthesis of proteins are present in DNA. This is transferred to a class of RNA molecules called *messenger RNA (mRNA)* which carry it outside the nucleus. Following the instruction from the *m*RNA, another class of RNA molecules called *transfer RNA (tRNA)* bring the required amino acids to the ribosome where *ribosomal RNA* molecules *(rRNA)* take over the job of directing protein synthesis.

KEY POINTS

- α-Amino acids occur in nature as the components of proteins. There are twenty amino acids which are most often found in proteins. They are all primary amines except proline, which has a secondary amino group (present as a pyrrolidine ring). Almost all the amino acids from proteins are of L-configuration.

- Amino acids, in the form of proteins, are important components of human nutrition. However, even if some of them are not present in the diet, the body can synthesise them from other metabolic products. This is not true of all the amino acids. The body cannot synthesise a few which have to be necessarily present in the food. These are called essential amino acids.

- Amino acids are amphoteric molecules. In solution, the following equilibrium exists.

$$\overset{+}{H_3N}-CHR-COOH \underset{+H^+}{\overset{-H^+}{\rightleftarrows}} \overset{+}{H_3N}-CHR-COO^- \underset{+H^+}{\overset{-H^+}{\rightleftarrows}} H_2N-CHR-COO^-$$

  The concentration of the species depends upon the pH. The zwitter ion, which is electrically neutral, is maximum at a particular pH characteristic of each amino acid. This pH is called the isoelectric point.

- Amino acids have two K_a values, corresponding to the ammonium ion and the carboxyl group.

- They can be synthesised by conventional methods. These include (i) reaction of ammonia with α-haloacids, (ii) reaction of aldehydes with ammonia and cyanide, followed by hydrolysis (Strecker synthesis), and (iii) malonic ester synthesis using acetamidomalonic ester.

- They can be N-acetylated, esterified and converted to hydroxy acids using nitrous acid.

- Amino acids can be detected and estimated by reaction with ninhydrin which gives a coloured product (the same from all amino acids).

- Condensation of two amino acids gives the amide. In polypeptides, this amide linkage is called a peptide bond. When more than two amino acids are joined by peptide linkages, the products are polypeptides. Proteins are polypeptides containing a very large number of amino acids.

- In a polypeptide chain, one end has a free amino group and the other end has a free carboxyl group. The amino end is called the N-terminus. The carboxyl end is called the C-terminus.

- One of the challenges for the synthesis of polypeptides is to ensure that the correct sequence of amino acids is maintained. The dipeptide, alanylglycine has alanine at the N-terminal and glycine at the C-terminal. That is the *sequence* of amino acids in that dipeptide. If the sequence is reversed, glycylalanine is obtained, which is a different dipeptide. Formation of glycylglycine and alanylalanine by self-condensation should also be prevented.

- The strategies of peptide synthesis involve protection of the amino group and activation of the carboxyl group of one of the amino acids. Such an N-protected, C-activated amino acid will react with the amino group of another amino acid selectively to form a dipeptide.

- After peptide bond formation, the protecting group has to be removed to release the free amino group without affecting the peptide bond. Several reagents have been developed for N-protection which can be deprotected under mild conditions. Two of the important ones are benzyloxycarbonyl chloride and *tert*-butoxycarbonyl chloride, which react with the amino group, to give the corresponding acyl derivatives. After peptide formation, the benzyoxycarbonyl group can be removed by catalytic hydrogenolysis or chemical reduction using sodium and liquid ammonia. Both benzyloxy and *tert*-butoxy protections can be removed by mild acid (HCl or HBr) treatment.

- Carboxyl group can be activated towards amide formation by conversion to acyl halide or better still to the *p*-nitrophenyl ester.

- Carboxy-activation is not required if DCC is used as the condensing agent. However if DCC is used, the carboxyl of the C-terminal acid should be protected by esterification. Benzyl ester is suitable for this. After peptide formation, the ester can be cleaved by catalytic hydrogenolysis or mild acid treatment.

- In the solid-phase peptide synthesis developed by Merrifield, the C-terminal amino acid is anchored to a polystyrene surface and the peptide is built in the solid-phase and finally released from the solid.

- The sequence of amino acids in a polypeptide chain is called its primary structure. In order to determine the primary structure, the polypeptide is hydrolysed and the amino acids analysed to determine its composition. Then the long chain is subjected to partial hydrolysis to obtain shorter polypeptides. The amino acids at the C- and/or N-terminal are identified by end-group analysis. Sanger's reagent and the Akabori reaction are early methods for N-terminal and C-terminal identification respectively. The most widely used method for sequencing is Edman degradation, in which phenyl isothiocyanate is the reagent. Using this reagent, amino acids can be removed one by one from the N-terminal and identified. Fairly long polypeptides can be sequenced by this method.

- Proteins are classified into fibrous proteins and globular proteins. Among these classes, they are further grouped into different types based on properties and functions.

- Secondary structure refers to the presence of helical sections or sheet-like sections, held in place by hydrogen bonds. α-Helix and β-pleated structures have been identified. Collagen has a triple helix structure.

- Protein chains are folded into definite shapes which are their stable conformations in the native state. This is the tertiary structure of proteins. Some proteins are assemblies of more than one chain, with or without non-protein components. Such ensembles are referred to as the quaternary structure.

- The tertiary structure is stabilised by various forces such as hydrogen bonds, dipolar attractions and hydrophobic interactions. Addition of certain chemicals or solvents, heating or change of pH can disrupt these weak bonds and the tertiary structure can collapse. This is called denaturation. Hardening of egg albumin by heat is an example of denaturation.

- Nucleic acids are biopolymers present in cell nuclei which carry genetic information. There are two classes, ribonucleic acid (RNA) which contains ribose and deoxyribonucleic acid, DNA, which contains 2-deoxyribose.

- The individual units are nucleotides and nucleosides.

- A nucleoside consists of a sugar (pentose, either ribose or deoxyribose) to which a base, (a purine or a pyrimidine ring) is attached at the 1-position through a nitrogen of the heterocyclic base. The bases are adenine and guanine (purines), and thymine and cytosine (pyrimidines) in DNA and adenine, guanine, cytosine and uracil in RNA. The nucleosides are strung together by phosphate bridges to form the polymeric nucleic acid. Nucleotides are phosphate esters of nucleosides.

- DNA has a double helical structure. The two helical strands are connected by hydrogen bonds. These hydrogen bonds are very specific. An adenine (A) of one chain is connected to a thymine (T) of the other chain by two hydrogen bonds. Similarly a cytosine (C) on one chain is connected to a guanine (G) of the other chain by three hydrogen bonds. The A—T ratio and the C—G ratio is always 1. This is called base-pairing. The two chains are complementary to each other. This feature is responsible for the unique ability of DNA molecules to replicate.

- DNA present in the cell nucleus, carries the genetic code in its structure. Because of its ability to replicate itself, it transmits the genetic information from parent to offspring. It controls protein synthesis through three types of RNA molecules, messenger (*m*RNA), transfer (*t*RNA) and ribosomal (*r*RNA).

EXERCISES

SECTION I

1. Draw the structures of

    (a) L-Aspartic acid
    (b) Gly-Ala
    (c) Boc-Phe
    (d) N-Benzyloxycarbonylbenzylamine
    (e) Products of the reaction of hydrazine with Phe-Ser-Glu
    (f) Products of the reaction of Phe-Ser-Glu with dansyl chloride

2. Show how the following amino acids can be synthesised.

    (a) Leucine by the Strecker synthesis
    (b) Glutamic acid by the amination of a halo acid by Gabriel synthesis
    (c) Isoleucine by the acetamidomalonic ester route
    (d) Tyrosine by the Strecker synthesis

3. Give the products of the following reactions.

    (a) Tyrosine treated with HCl and sodium nitrite at 0°C
    (b) Phenylalanine treated with benzyloxycarbonyl chloride
    (c) The product of (b) treated with *p*-nitrophenol in the presence of DCC
    (d) The product of (c) treated with alanine
    (e) The product of (d) catalytically reduced using Pd
    (f) Phenylalanine heated with ninhydrin
    (g) Methyl ester of alanine heated with acetic anhydride

SECTION II

1. Explain why *p*-nitrophenyl ester is more suited for peptide synthesis than ethyl ester.
2. When alanyllysine is treated with 2,4-dinitroflourobenzene and the product hydrolysed, two amino acids are obtained, both of which contain a dinitrophenyl group. Explain.
3. The oxime $R_2C=NOH$, or imine $R_2C=NH$, obtained from ketones can be catalytically reduced to obtain the amine R_2CH-NH_2. This is called reductive amination. Show how this reaction can be employed to prepare the following amino acids.

    (a) Phenylalanine
    (b) Aspartic acid
    (c) Serine

4. Suggest a mechanism for the rearrangement of the thiazolone derivative to phenylthiohydantoin in Edman degradation (Fig. 24.5).
5. Outline the mechanism of the condensation of a carboxylic acid and an alcohol in the presence of DCC.
6. Give the mechanism of the reaction of malonic ester with nitrous acid (the first reaction in Eq. 24.5).
7. (a) Give the mechanism of the reaction of the amide with hydrazine (Eq. 24.6a).
    (b) Explain why hydrazine can replace the amine in this reaction.
    (c) Explain why the carboxyl group is not converted to the hydrazide in Eq. 24.6b.

CHALLENGING QUESTION

1. Examine the isoelectric points of various amino acids in Table 24.1. They can be placed in three groups, (i) below 3, (ii) 5–6, (iii) above 7. Explain this classification.

PROJECT

DNA profiling is very much in use for legal and other purposes. Prepare a project report on DNA profiling.

25 Terpenoids and Alkaloids
(Natural Products – III)

OBJECTIVES In this chapter, you will learn about,

- classification of terpenoids
- isoprene rule
- structure and reactions of geraniol, α-terpineol and menthol
- methods of isolation of alkaloids
- general methods of structure determination of alkaloids
- structure and reactions of coniine, piperine and nicotine

25.1 TERPENOIDS

Terpenoids or terpenes are a class of compounds that are found mainly in plants. The simpler ones are liquids, generally with a characteristic fragrance, present in the *essential oils* of plants. Essential oils are volatile oils as distinct from the fatty oils or lipids obtained from oil seeds, which are nonvolatile triglycerides. The adjective, 'essential' is related to *essence* as in essence of flowers, fruits and similar plant parts. They are hydrocarbons or oxygen-containing compounds having hydroxyl, carbonyl and related functional groups. The number of carbons in terpenoids is in multiples of 5, beginning with 10. Those having 10 carbons are called *monoterpenes*, those with 15 carbons are *sesquiterpenes*, those with 20 carbons are *diterpenes*(Fig. 25.1). Many terpenoids on pyrolysis give 2-methyl-1,3-butadiene – also known as isoprene – as one of the products. The carbon skeletons of all terpenoids can be considered to be built up of isoprene units joined head-to-tail. This is known as the *isoprene rule* (Fig. 25.1).

Three monoterepenes are discussed in this chapter. Geraniol is an acyclic monoterpene, α-terpineol and menthol are monocyclic monoterpenes.

25.1.1 GERANIOL

Geraniol is found in rose oil and certain other essential oils. It is an unsaturated alcohol with a molecular formula $C_{10}H_{18}O$. It is isomeric with nerol – another unsaturated alcohol – isolated

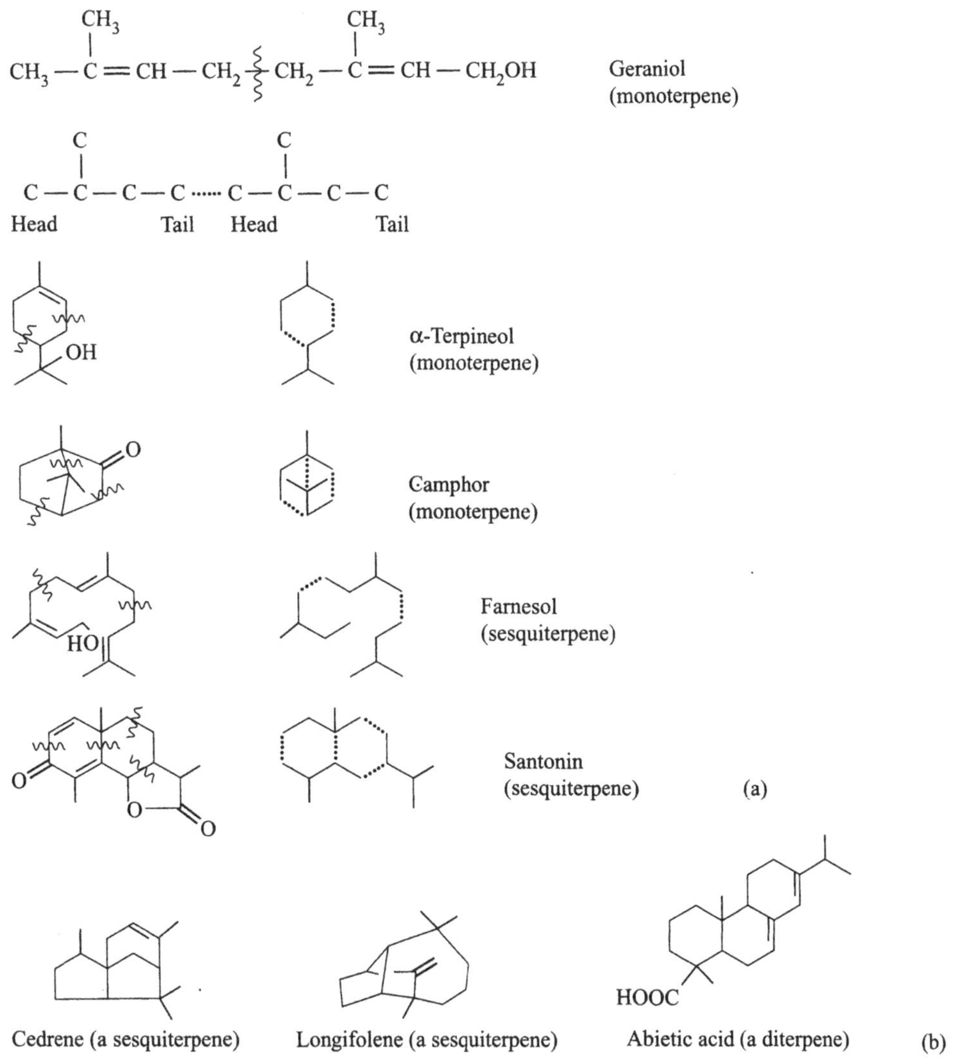

Fig. 25.1 Examples of monoterpenes, sesquiterpenes and diterpenes and illustration of the isoprene rule

from essential oils like neroli oil and bergamot oil. Both of these, upon catalytic hydrogenation take up two molar equivalents of hydrogen to give the same saturated alcohol $C_{10}H_{22}O$, later identified as 3,7-dimethyl-1-octanol. They both add two moles of bromine to give tetrabromides. From these observations, the presence of two double bonds is inferred. A mixture of the two is obtained by the reduction of the aldehyde citral, which is the main constituent of lemon grass oil. There are two isomeric forms of citral, labelled citral-a and citral-b.

The structure of citral-a is first established (Fig. 25.2). By controlled oxidation of citral-a (I), with permanganate followed by chromic acid, acetone (IV), oxalic acid (III), and γ-ketovaleric acid (levulinic acid) (II) are obtained. From these, citral-a is assigned structure (I). Upon reduction, citral-a gives geraniol and citral-b gives nerol. Both alcohols give α-terpineol (VIII), on treatment

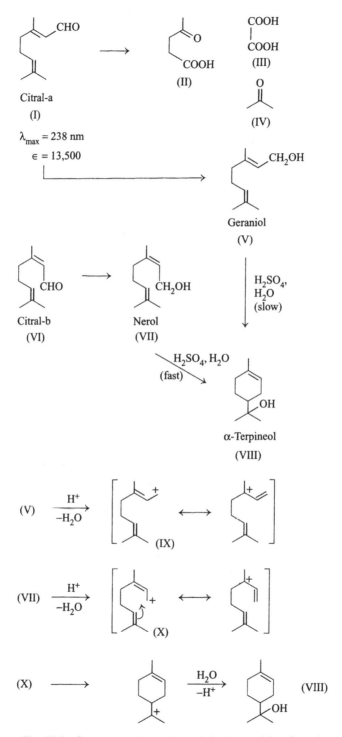

Fig. 25.2 Structure and reactions of citral, geraniol and nerol

with acid. Nerol cyclises about 9 times faster than geraniol. From this observation, the *trans* or E configuration is assigned to geraniol and the *cis* or Z configuration to nerol. Both cyclisations take place through allylic carboactions (IX) and (X). (X) cyclises readily due to its proximity to the 6,7-double bond. The cation (IX) has to isomerise first to (X) before it can cyclise.

25.1.1.1 Spectroscopic Investigation of Geraniol and Citral

The UV spectrum of citral with a maximum at 238 nm is consistent with the α, β-unsaturated aldehyde structure. The IR spectrum of geraniol has a strong band at about $3400 \, cm^{-1}$ for the hyxdoxyl group. The NMR spectrum (Fig. 25.3) is in agreement with the structure.

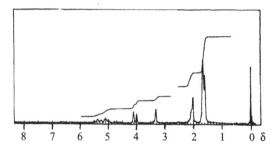

Fig. 25.3 NMR Spectrum of geraniol

Study and interpret the spectrum. The signals fall under the following types: $=C-CH_3$, $=C-CH_2-$, $-OH$, $-CH_2-O-$, $C=CH-$. Some of the signals overlap.

25.1.2 α-Terpineol

The conversion of nerol and geraniol to α-terpineol (I) (Fig. 25.4) in the presence of aqueous sulphuric acid has already been mentioned. α-Terpineol is found in cardamom oil and marjoram oil. It has one double bond as established by the formation of a dibromide by the addition of bromine. The molecular formula, $C_{10}H_{18}O$, suggests one more double bond or a ring. The oxygen should belong to a tertiary alcohol group, because it can be easily dehydrated to a diene, $C_{10}H_{16}$. These evidences suggest a cyclic structure. During the dehydration with sulphuric acid, some *p*-cymene (*p*-isopropyltoluene, IV), is also formed. The diene (II) formed by dehydration rearranges to a cyclohexadiene (III) which then undergoes thermal dehydrogenation to give the aromatic hydrocarbon, *p*-cymene (IV). The fully hydrogenated hydrocarbon from *p*-cymene, 4-isopropyl-1-methylcyclohexane, is called *p*-menthane (V). In terpene chemistry, the numbering of the carbon atoms of the *p*-menthane skeleton is done as in (V) (Fig. 25.4). According to this, α-terpineol is 1-*p*-menthene-8-ol. It is surmised that α-terpineol has a *p*-menthane skeleton with one double bond and one hydroxy group. Some of the reactions which help in assigning the structure of α-terpineol are summarised in Fig. 25.4.

Mild oxidation using 1% alkaline permanganate converts (I) to the trihydroxycompound, (VI). Further oxidation by chromic acid cleaves the cyclohexane ring at the glycol position, to

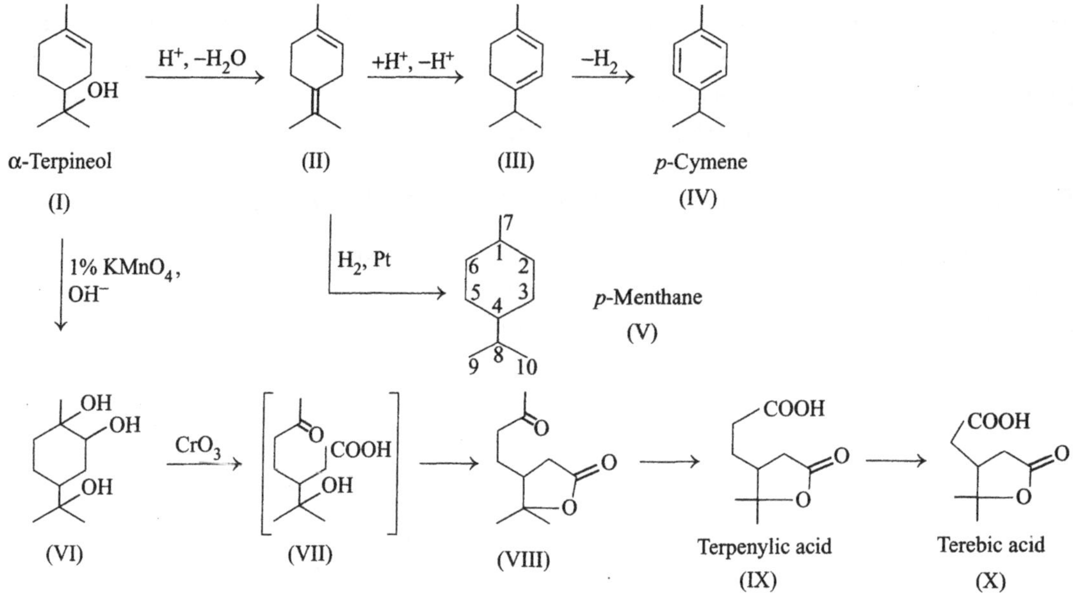

Fig. 25.4 Structure and reactions of α-terpineol

form the ketohydroxy acid (VII), which spontaneously lactonises to (VIII). This can be oxidised in a stepwise manner to terpenylic acid (IX) and terebic acid (X). All these structures are written on the assumption that the structure of α-terpineol is (I). The original workers, (Otto Wallach, German chemist; Nobel prize, 1910) without the benefit of our hind sight, had to work backwards from the known structures of reaction products. The structures of terebic acid and terpenylic acid were established by unambiguous syntheses (British chemist, Sir John Simonsen, 1907). The synthesis of terebic acid is outlined in Fig. 25.5.

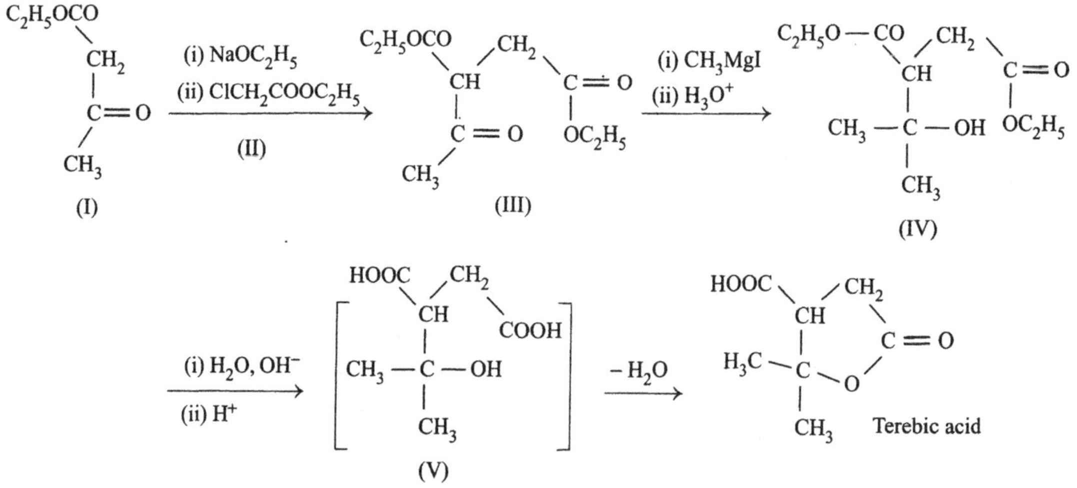

Fig. 25.5 Synthesis of terebic acid

In the first step, ethyl acetoacetate (I) is alkylated using ethyl chloroacetate (II) to prepare the ketodiester (III). This is treated with methylmagnesium iodide to give the tertiary alcohol (IV). After hydrolysis of the ester groups, the resultant diacid (V) spontaneously lactonises to the target molecule—terebic acid. The step involving the Grignard reaction is not as straightforward as it appears. Molecule (III) has, in addition to the keto group, two ester groups which are also capable of reacting with a Grignard reagent. Fortunately, esters are less reactive than ketones and if only the required quantity of Grignard reagent is employed and the reaction conditions are optimised, we can hope to avoid the side reaction involving the ester groups (see question 4, under 'challenging questions'). A more serious problem is that molecule (III) contains an active hydrogen (*can you locate it?*) which will convert methylmagnesium iodide to methane. (The required quantity of Grignard reagent, mentioned above, may not necessarily be one equivalent!)

25.1.3 MENTHOL

Menthol ($C_{10}H_{20}O$) is a crystalline solid with a characteristic smell, present in peppermint oil. Naturally occurring menthol is optically active. It has no double bonds. It contains a secondary alcohol function as inferred from its oxidation to a ketone, $C_{10}H_{18}O$, menthone. We shall assume that the structure is 2-isopropyl-5-methylcyclohexanol or, using terpene nomenclature, 3-hydroxy-*p*-menthane, (I, Fig. 25.6) and provide evidence as we go along. It can be converted

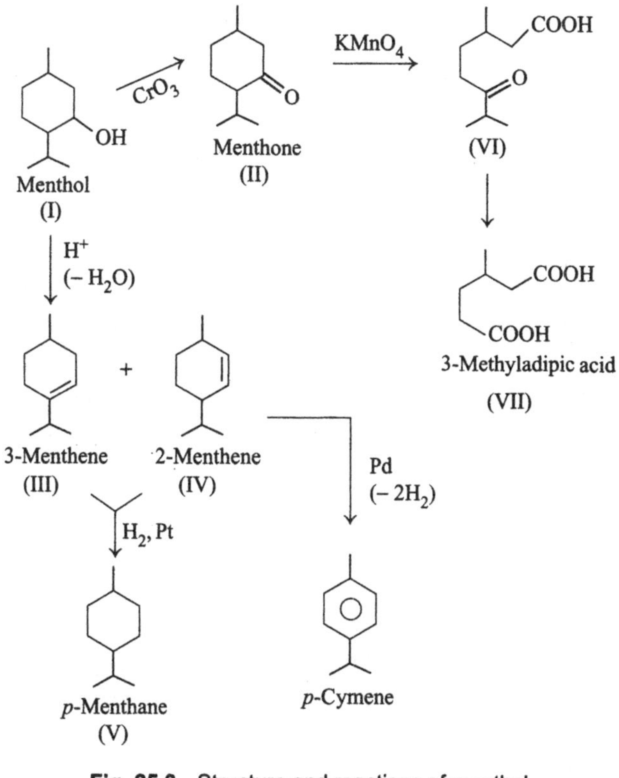

Fig. 25.6 Structure and reactions of menthol

to the ketone, menthone (II) by chromic acid oxidation, and to a mixture of 3-menthene (III) and 2-menthene (IV), by dehydration. The menthenes can be hydrogenated to *p*-menthane (V) and dehydrogenated to *p*-cymene. The position of the hydroxy group has been established by oxidation of menthone to (VI) and then to (VII), which is a known compound—3-methyladipic acid.

25.1.3.1 Stereochemistry of Menthol

Menthol contains three asymmetric carbons and has 8 stereoisomers (4 pairs of enantiomers). The configuration of (–)-menthol in its more stable all-equatorial conformation is (I) (Fig. 25.7).

The configurations of the geometrical isomers, neomenthol (II), isomenthol (IV) and neoisomenthol (V), are shown in their stable conformations, keeping the bulkiest group (isopropyl) in the equatorial position. Two of the isomers (I) and (II), get oxidised to the same ketone—menthone (III). The other two isomers, (IV) and (V), get oxidised to isomenthone (VI). The relative rates of esterification of the four isomers, help in fixing their stereochemistry. Menthol is assigned the all-equatorial configuration and conformation.

Menthol and neomenthol – which differ only in the configuration at the carbinol carbon – can be converted to the corresponding chloro compounds—menthyl chloride (X) and neomenthyl chloride (VII). Their dehydrohalogenation using ethanolic alkali under E2 conditions, throws light on the orientation of the Cl. Neomenthyl chloride, (VII) undergoes fast elimination to give a mixture of 3-menthene (IX) and 2-menthene (VIII), with more of the former as per Saytzeff rule. Menthyl chloride (X) reacts much more slowly and gives only 2-menthene. The preferred orientation of the Cl and H for elimination is when both are axial (See Chapter 12, Section 12.5, Fig. 12.2 for the conformational requirements for elimination). In neomenthyl chloride, in its most stable conformation, VII, the Cl is axial and there are axial hydrogens, one each on C2 and C4, oriented anti to chlorine. Elimination gives both 3-menthene (IX) and 2-menthene (VIII). The former is formed preferentially in about 75% yield, as expected from Saytzeff rule. For menthyl chloride, in its most stable conformation (X), the Cl is equatorial. There are no hydrogens anti to this. True *trans* elimination is not possible. The molecule has to flip over to the unfavourable, all-axial conformation (XI), or some other intermediate conformation, before elimination can take place. This involves extra energy and makes the elimination slower than with neomenthyl chloride. In conformation (XI) there is only one axial hydrogen—on C2. Hence the only alkene obtained is 2-menthene.

25.2 ALKALOIDS

Alkaloid is the name given to a group of basic (*alkali-like*, hence *alkaloid*) compounds containing nitrogen, isolated mostly from plants. All of them are toxic and many have medicinal applications. There are other plant products like caffeine which satisfy these conditions, but are not classified as alkaloids. Alkloids range from simple amines like mescaline (I), to complex molecules like morphine (II), strychnine (III), and reserpine (IV).

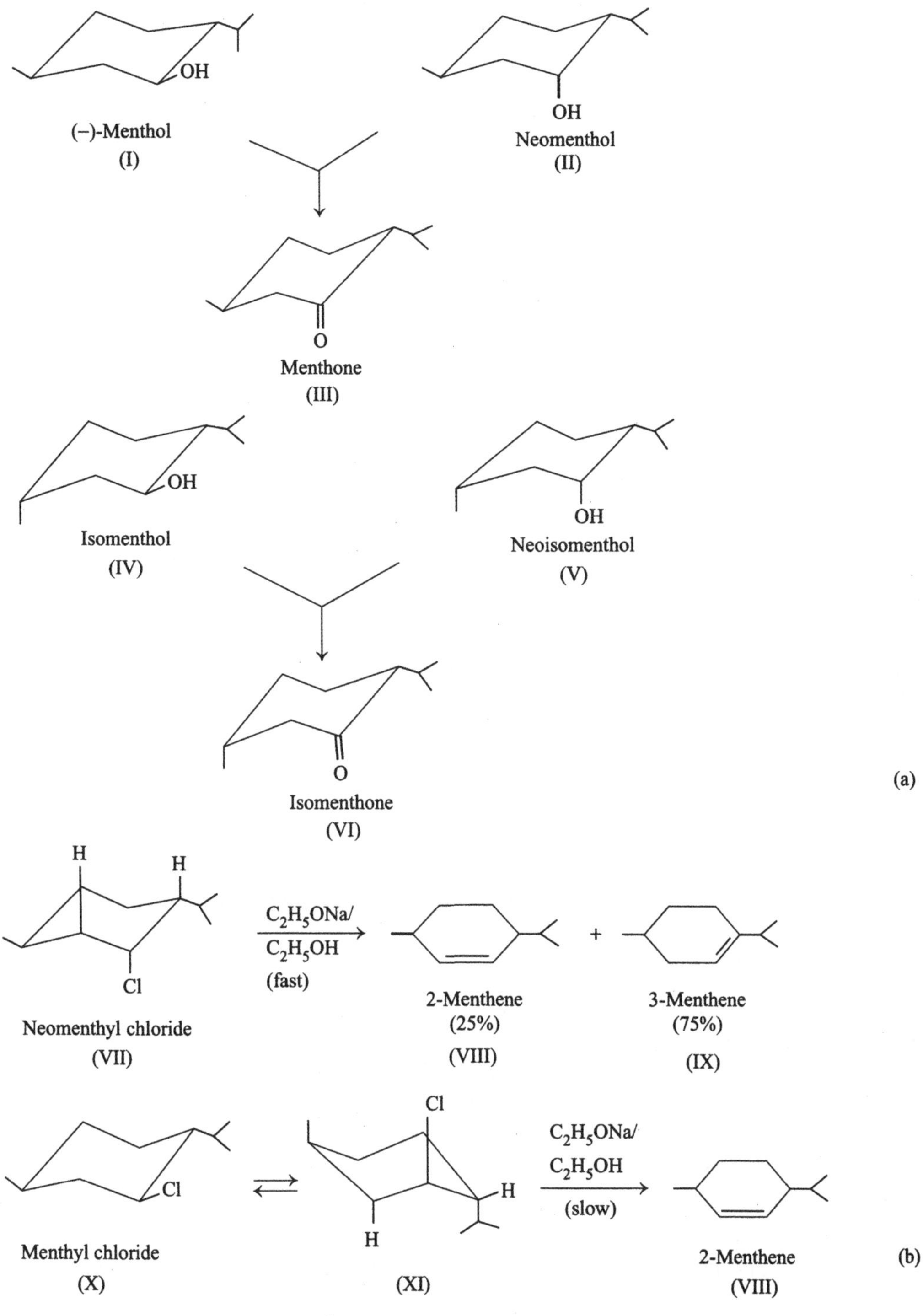

Fig. 25.7 Stereochemistry of menthol

Mescaline
(I)

Reserpine
(IV)

Morphine
(II)

Strychnine
(III)

25.2.1 ISOLATION FROM PLANTS

Alkaloids are present in seeds, roots, bark, leaves and other parts of plants. One species usually contains a mixture of related alkaloids. In the plant they may be present as the salts of organic acids like acetic acid and oxalic acid. The plant part is dried, powdered and extracted with solvents like methanol. After extraction and evaporation of the solvent, the salt is decomposed by adding mineral acid to liberate the free base. Or, the plant material may be directly extracted with dilute mineral acid from which the alkaloid is liberated by neutralisation of the solution. The mixture is separated, mainly by chromatography, to isolate the pure components.

25.2.2 GENERAL METHODS OF STRUCTURE DETERMINATION

25.2.2.1 Chemical Methods

The preliminary steps in the determination of the structure of all unknown compounds – naturally occurring or synthetic – involve qualitative and quantitative analysis to determine the elemental composition, and determination of molecular weight by conventional methods and/or by mass spectrometry. High resolution mass spectrometry not only gives the molecular weight but also the elemental composition. From these data, the molecular formula is arrived at. The next steps involve identification of functional groups, the nature of the rings, and if oxygen atoms are present, whether they are present as alcohols, phenols, ethers, aldehydes, ketones, carboxylic acids, esters, lactones, amides or lactams. All alkaloids contain nitrogen. At least one of the nitrogens will be an amino nitrogen and it has to be determined whether they are primary, secondary, tertiary or quaternary.

Earlier, scientists depended on conventional chemical methods for these identifications. Today spectroscopic investigations are carried out first.

25.2.2.2 OH Groups, Carboxylic Acids, Carbonyl Compounds

OH groups may be phenolic or alcoholic. Acylation reactions such as *acetylation* and *benzoylation* help to identify their presence. Acetylation by acetic anhydride is also used to determine the number of OH groups, by determining the number of moles of acetic anhydride consumed per mole of the unknown compound. The phenolic OH group is identified by ferric chloride colouration reaction. Also, phenols dissolve in alkali and are reprecipitated when CO_2 is bubbled into the solution. The nature of the alcoholic OH is identified by dehydration and oxidation reactions.

The carboxylic acid group is identified by its solubility in alkali and in bicarbonate solution and also by esterification. Esters and amides are identified by hydrolysis. Aldehydes and ketones have characteristic identification tests like oxime or semicarbazone formation.

Active hydrogens – as in OH and NH groups – are estimated by the *Zerewitinoff method* which involves treatment with methylmagnesium iodide followed by volumetric measurement of the methane evolved.

Many alkaloids contain methoxyl groups. They are identified by reaction with HI which liberates methyl iodide. *Zeisel's method* for estimation of the methoxyl goup involves heating a known amount of the compound with HI and passing the CH_3I evolved into silver nitrate solution and gravimetrically estimating the silver iodide precipitated.

25.2.2.3 Nature of Nitrogen

Since nitrogen is present in all alkaloids, several reactions have been adapted specifically to be applied in this field. In the Herzig–Meyer method for $N\!-\!CH_3$ groups, which is similar to Zeisel's method, CH_3I is liberated when the alkaloid is heated with HI. *Hofmann's exhaustive methylation* reaction is very useful in alkaloid research. Amines react with methyl iodide (Eq. 25.1).

$$RNH_2 + CH_3I \rightarrow RNHCH_3 + HI \qquad (a)$$

$$RNHCH_3 + CH_3I \rightarrow RN(CH_3)_2 + HI \qquad (b)$$

$$RN(CH_3)_2 + CH_3I \rightarrow RN^+(CH_3)_3I^- \qquad (c) \qquad (25.1)$$

A primary amine consumes three equivalents of CH_3I, a secondary amine two, and a tertiary amine, one equivalent. The quaternary ammonium iodide is converted to the hydroxide by treatment with moist silver oxide. The hydroxide undergoes Hofmann elimination upon heating (Chapter 12, Section 12.3.2). When the amine is cyclic like piperidine, the course of Hofmann exhaustive methylation occurs in two stages [Fig. 25.8(a)]. After the second stage, a nitrogen-free molecule is obtained. The reaction, where the amine is taken all the way to the quaternary ammonium hydroxide followed by elimination till N-free molecules are obtained is called exhaustive methylation.

If the quaternary ammonium salt does not have β-hydrogens, Hofmann elimination is not possible. Such compounds can be cleaved by the *Emde modification*, in which the compound is hydrogenolysed using sodium amalgam in ethanol or sodium in liquid ammonia or by catalytic hydrogenolysis (H_2/Pd) [Fig. 25.8(b)].

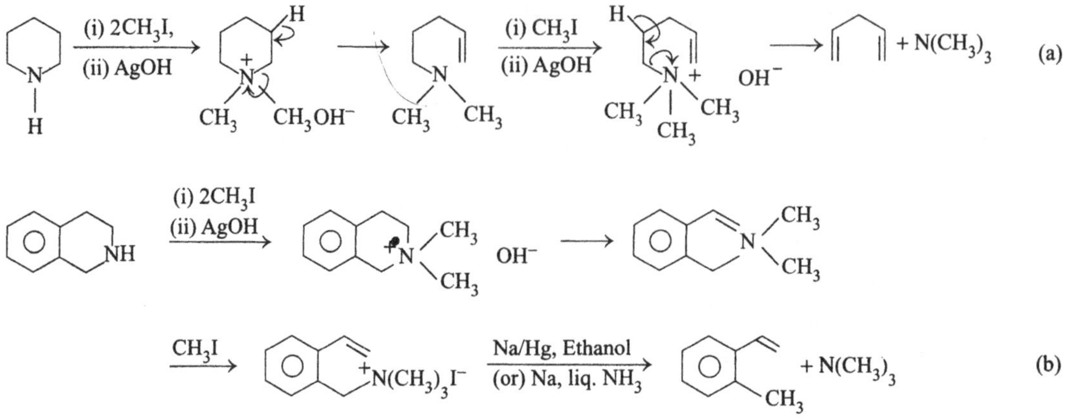

Fig. 25.8 Hofmann exhaustive methylation and Emde modification

Von Braun's method is another reaction for opening cyclic tertiary amines. This is illustrated in Fig. 25.9. Cyanogen bromide is the reagent.

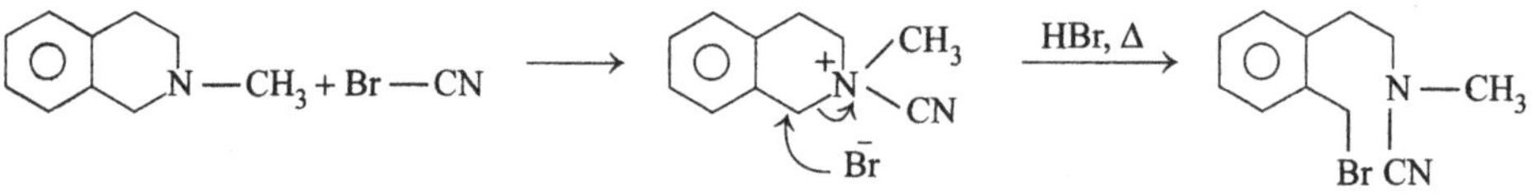

Fig. 25.9 Von Braun's reaction

Methods such as distillation with zinc dust and pyrolysis with zinc chloride, while non-specific, do give valuable information on the nature of the framework and on the presence of aromatic and heterocyclic rings. Addition of bromine, hydrogenation, oxidation using permanganate and chromic acid, and ozonolysis have been extensively used to detect the nature and environment of double bonds.

25.2.2.4 Spectroscopic Methods

Today, spectroscopic methods are the mainstay of structure determination. Chemical methods described in the preceding section were the only tools available to the early investigators. Some of the problems that were faced and solved painstakingly may appear trivial today. Many structures can be and have been solved today by spectroscopic methods, mainly by NMR spectroscopy supported by other techniques like UV, IR and mass spectroscopy, with minimal use of chemical degradation.

25.2.3 CONIINE

Coniine, $C_8H_{17}N$, is one of the alkaloids present in hemlock. It has assured a name for itself in the annals of famous poisons as the lethal component of the hemlock extract which was administered to Socrates. Naturally occurring coniine is optically active. Its structure is quite simple and was determined by chemical degradation [Fig. 25.10(a)].

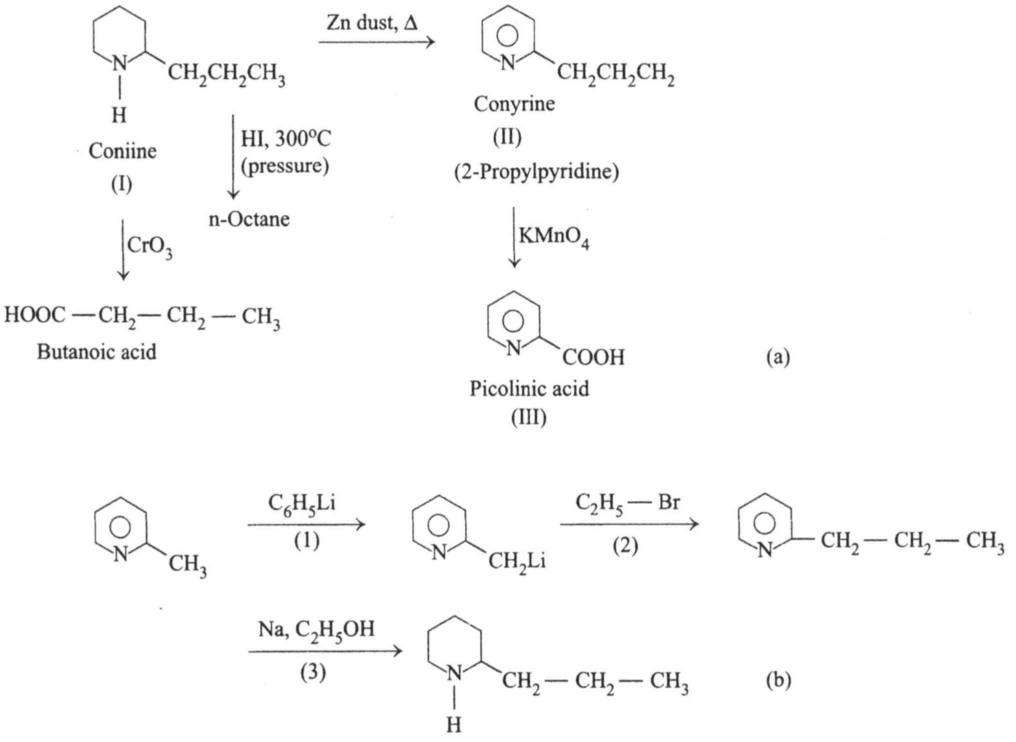

Fig. 25.10 Structure and synthesis of coniine

Coniine (I) upon distillation with zinc dust gives conyrine (II), which is 2-propylpyridine. The three-carbon side chain at the 2-position is further confirmed by the oxidation of conyrine to picolinic acid (III) using permanganate. That the side chain is *n*-propyl and not isopropyl is confirmed by the oxidation of coniine to butanoic acid by chromic acid. (*Will the NMR spectrum of coniine confirm this?*) On heating coniine in a sealed tube with HI upto 300°C, *n*-octane is obtained, which further confirms the absence of branching in the side chain. The synthesis of coniine confirms the structure [Fig. 25.10(b)].

25.2.4 PIPERINE

Piperine, $C_{17}H_{19}NO_3$, is one of the alkaloids present in black pepper. Alkaline hydrolysis of piperine gives a carboxylic acid piperic acid (II), and piperidine. Piperic acid has two double bonds as shown by the fact that it reacts with two moles of bromine to form a tertrabromide. Oxidation by permanganate converts piperic acid first to piperonal (IV) and then to piperonylic acid (V), both known compounds. Piperonylic acid is hydrolysed by HCl to protocatechuic acid (VI). The methylenedioxy group in these compounds – which is the catechol acetal of formaldehyde – is found in many natural products. Hydrolysis of (V) to catechol derivative (VI) is typical of the methylenedioxy group. The structure of piperine is confirmed by the synthesis [Fig. 25.11(b)]. Catechol (VII) is converted to the aldehyde (VIII) by Reimer–Tiemann reaction and then to piperonal (IV), by reaction with diiodomethane and alkali. Claisen–Schmidt reaction of piperonal with acetaldehyde gives the unsaturated aldehyde (IX), which is converted to the target molecule piperic acid (II), by Perkin condensation. (*How can you convert piperic acid to piperine?*)

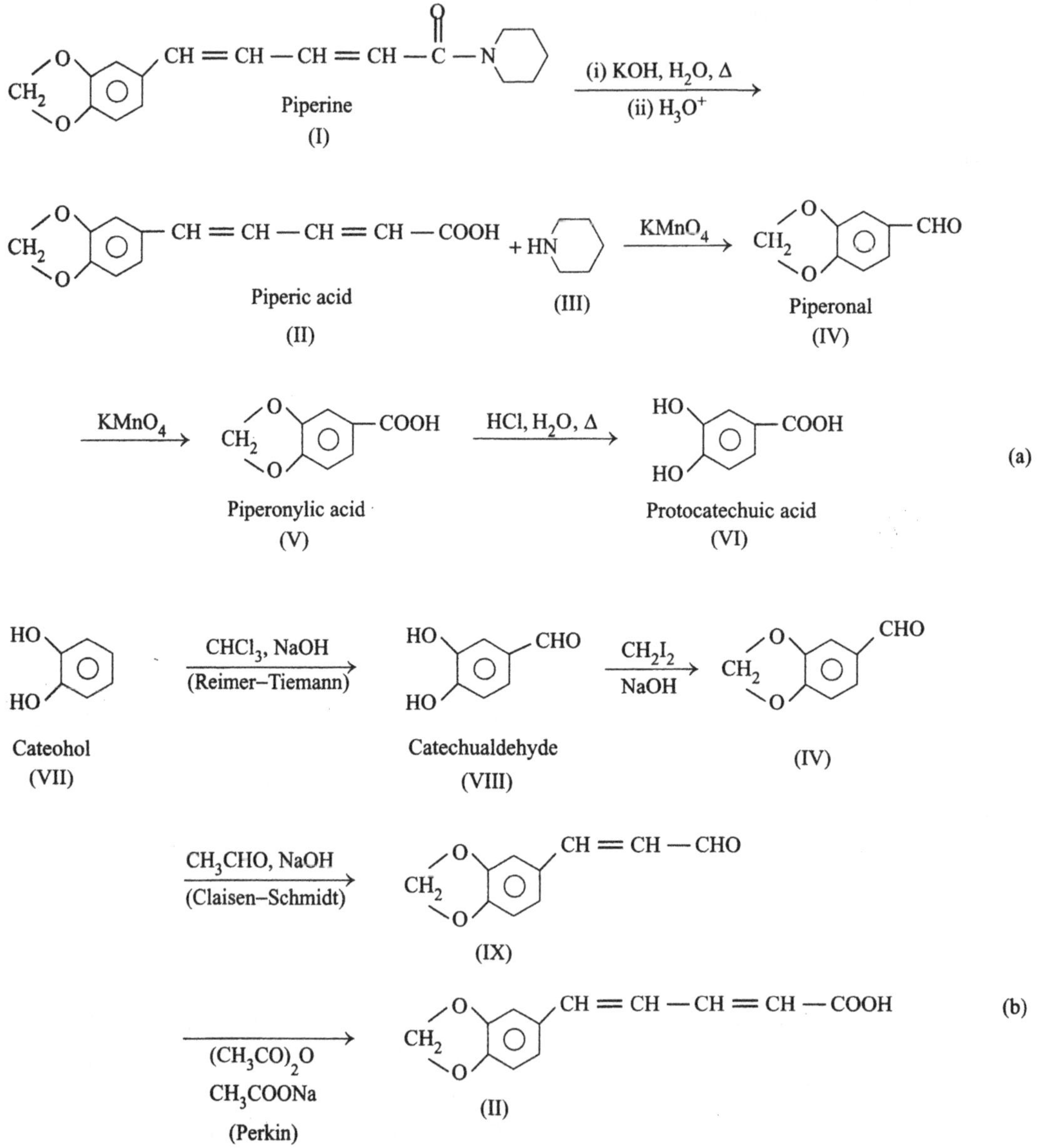

Fig. 25.11   Structures and synthesis of piperine

25.2.5 NICOTINE

Nicotine, $C_{10}H_{14}N_2$, is an optically active liquid, present in the tobacco leaf. (−)-Nicotine, the enantiomer present in tobacco, is much more toxic than (+)-nicotine. The reactions of nicotine (I), which lead to its structure are outlined in Fig. 25.12.

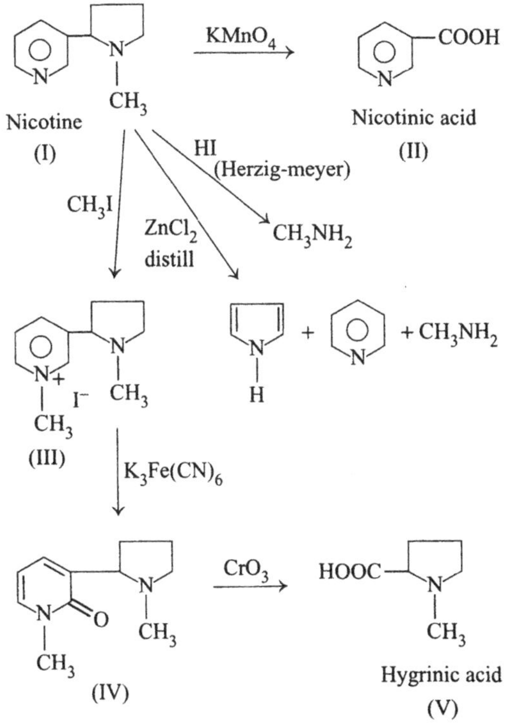

Fig. 25.12 Structure and reactions of nicotine

    Permanganate oxidation gives nicotinic acid (II), which is pyridine-3-carboxylic acid. From this, the structure of nicotine can be written as $(C_5H_4N)–(C_5H_{10}N)$. The first part, $C_5H_4N—$ is 3-pyridyl. That the second part, $C_5H_{10}N$ is N-methyl-2-pyrrolidinyl is established by identifying its degradation products. Herzig–Meyer reaction (heating with HI) gives methylamine, confirming the presence of an N-methyl group. Distillation with zinc chloride gives several products, among which pyridine, pyrrole and methylamine can be identified. This suggests that an N-methylpyrrolidine group is present. The pyrrolidine ring is attached at the 2-position to the 3-position of pyridine. This is confirmed by the products of oxidation of the pyridine ring. In normal oxidations, the pyridine ring is left intact. By oxidising the methiodide (III) with ferricyanide, the N-methylpyridone derivative (IV) is obtained. This is susceptible to oxidation at the 6-membered ring. Oxidation of (IV) by chromic acid gives hygrinic acid (V) which is N-methylpyrrolidine-2-carboxylic acid.

KEY POINTS

- Terpenoids are present in the essential oils of plants. They may be hydrocarbons or may contain oxygen in the form of alcohols, carbonyl compounds, carboxylic acids and their derivatives.

- Their carbon content is in multiples of 5, starting with 10 carbons. C10 terpenes are called monoterpenes, C15 are sesquiterpenes, C20 are diterpenes and so forth.

- The carbon skeletons of terpenes can be divided into 5-carbon units, each corresponding to a 2-methylbutane skeleton, joined head-to-tail. 2-Methylbutane skeleton is present in isoprene, which is 2-methyl-1,3-butadiene. Terpenes structures can be divided into isoprene units, joined head-to-tail. This is known as the isoprene rule.

- Geraniol is a monoterpene alcohol present in many essential oils, notably in rose oil. It is a dienol and is the geometrical isomer of another dienol called nerol. Both on oxidation give citral (an aldehyde), the main component of lemon grass oil. Both cyclise in the presence of dilute sulphuric acid to give α-terpineol. Nerol cyclises faster than geraniol. This helps to establish the stereochemistry of both.

- α-Terpineol is an unsaturated tertiary alcohol having the *p*-menthane skeleton. Its structure has been established by its formation from nerol and by identifying its oxidation products.

- Menthol is a saturated alcohol present in peppermint oil. There are eight stereoisomers, in the form of four pairs of enantiomers.

- Alkaloids are basic, nitrogen containing molecules mainly of plant origin. Most of them are toxic and many have medicinal applications.

- They are extracted from plants using hot methanol or by acid extraction.

- Structures are determined by conventional chemical degradation and by spectroscopic methods.

- Hofmann and Von Braun reactions are useful in the study of cyclic amines.

- Coniine is an alkaloid present in the hemlock plant. Its structure has been established to be 2-propylpiperidine by chemical degradation studies and confirmed by synthesis.

- Piperine is an alkaloid present in black pepper. Its structure has been established by chemical degradation and by synthesis.

- Nicotine is present in tobacco leaves. It is a pyridine derivative with an N-methyl-pyrrolidine side chain. The structure has been established by chemical degradation studies.

EXERCISES

SECTION I

1. Give the products of ozonolysis of

    (a) citral   (b) geraniol

    (c) α-terpineol   (d) 2-menthene

    (e) 3-menthene

2. What is the product of the reaction of piperine with maleic anhydride?
3. Give the structure of the tetrabromide obtained from (a) geraniol and (b) piperine.
4. Identify the 'active hydrogen' present in Structure (III), Fig. 25.5.
5. Refer to Fig. 25.4. Suggest another reagent to convert (a) (I) to (VI) and (b) (VI) to (VII).
6. Locate the isoprene units in the structures of longifolene, cedrene and abietic acid in Fig. 25.1.

SECTION II

1. Give the mechanisms with explanations of steps 1, 2, and 3, in the synthesis of coniine, Fig. 25.10(b).
2. Suggest another synthesis of coniine from picoline, by first condensing the latter with acetaldehyde.
3. Formulate the Hofmann exhaustive methylation of (a) coniine and of (b) nicotine.
4. How can NMR spectroscopy be used to solve the following problems?

    (a) Is the tertiary OH group in α-terpineol on C8 or on C4?

    (b) Is the side chain of coniine, propyl or isopropyl?

    (c) How can you differentiate between 2-menthene and 3-menthene?

CHALLENGING QUESTIONS

1. Interpret the NMR spectrum of geraniol, Fig. 25.3.
2. Suggest a synthesis of terpenylic acid using the same methodology used for terebic acid, Fig. 25.5. (Hint: alkylate with ethyl chloroacetate twice; remove one carboxyl group by decarboxylation.)
3. How can menthol be converted to menthyl chloride and to neomenthyl chloride?
4. Refer to the synthesis of terebic acid, Fig. 25.5. Is it advisable to add the solution of the ketoester, II, dropwise to the solution of methylmagnesium iodide with stirring, or in the reverse manner, that is, add the Grignard reagent dropwise to the ketoester? The choice is based on the need to ensure that the Grignard reagent reacts selectively with the keto group, leaving the ester groups unaffected.

26 Dyes

OBJECTIVES In this chapter, you will learn about,

- classification of dyes
- theory of colour and constitution
- preparation and uses of specific dyes
- azo dyes—Bismark brown and methyl orange
- triphenylmethane dyes—malachite green
- phthalein dyes—phenolphthalein
- xanthen dyes—fluorescein
- anthraquinone dyes—alizarin
- vat dyes—indigo

26.1 INTRODUCTION

Dyes are substances used to impart colour to objects, especially fabrics. In this chapter, we deal mainly with textile dyes. One of the important conditions for a coloured substance to be a dye is that it should be 'fast' which means that the dye should adhere to the fabric and should not be removed during washing even under the mildly alkaline pH of soap. The dye should not be 'fugitive' meaning that the colour should not fade during usage. Dyes are also used in food materials and in preparations like hair dye, cosmetics and pharmaceutical preparations. Almost all dyes are based on aromatic compounds. In the early days, the aromatics for synthetic dye manufacture came from coal tar. Hence they are often referred to as coal tar dyes, though today, many of them may be of petrochemical origin. Suspected environmental and health hazards, especially carcinogenicity, of synthetic dyes are matters of concern today.

26.1.1 COLOUR

Light falling on a substance may be totally absorbed, totally reflected or partially absorbed and the rest reflected. If sunlight is totally absorbed, the surface will appear black. If it is totally reflected, it will appear white. If light is partially absorbed and partially reflected, the substance will have the colour corresponding to the wavelengths of the reflected light.

Absorption of light is due to the electronic excitation of the molecule. The structure of the molecule and the energy levels of the electrons decide which wave lengths are absorbed. We have seen in Chapter 13 that the easiest electronic excitations occur when electrons in the highest occupied molecular orbital (HOMO) are excited to the lowest unoccupied molecular orbitals (LUMO). In simple molecules, this excitation corresponds to wavelengths in the UV region of the electromagnetic spectrum. Absorption in this region does not result in colour. In highly conjugated molecules, the LUMO–HOMO gap is small enough to lead to absorption of wave lengths in the visible region of the spectrum. Since only part of the wave lengths of the visible radiation (sun light) is absorbed and the rest reflected, the reflected light and hence the substance will appear coloured. We do not see the colour of the wave lengths absorbed, but only the colour due to the reflected wave lengths, which is called the complementary colour of the absorbed wave length. Some of the colours corresponding to the visible wave lengths and their complementary colours are listed in Table 26.1.

Table 26. 1 Complementary colours

Wave length (nm)	Colour	Complementary colour
400–420	Violet	Green–yellow
420–490	Blue	Yellow
490–570	Green	Red
570–585	Yellow	Blue
585–647	Orange	Green–blue
647–700	Red	Green

26.1.2 CHROMOPHORES AND AUXOCHROMES

A portion of a molecule which is responsible for electronic excitation in the near UV and visible region is called a *chromophore*. Groups such as conjugated polyenes, conjugated unsaturated carbonyl compounds, aromatics, and azo, nitro and nitroso groups are examples of chromophores. Groups which have unshared pairs of electrons like OH, OR, NH_2, NHR and NR_2, are themselves not chromophores, but when in conjugation with chromophores, contribute to extended conjugation and cause intensification of colour. Such groups are called *auxochromes*. The importance of extended conjugation in causing colour can be seen from Table 26.2.

Table 26. 2 Conjugation and absorption maxima

Linearly fused aromatics	No. of benzene rings	λ max (nm)	Colour
Benzene	1	260	Colourless
Naphthalene	2	280	Colourless
Anthracene	3	375	Colourless
Naphthacene	4	450	Yellow
Pentacene	5	575	Blue

Lycopene, a linear polyene with 10 double bonds in conjugation, with a λ max = 505 nm, is the red pigment responsible for the colour of tomatoes. β-Carotene which is present in carrots and has 11 double bonds in conjugation, is orange in colour.

26.2 DYES

As already mentioned, a dye should be fast. This requires some sort of bonding between the molecules of the dye and the fabric. Cotton, which is cellulose with hydroxyl groups, and silk and wool which are polypeptides, are easy to dye. The polar groups on the fabric bond with the polar dye molecules through hydrogen bonds and other kinds of bonds. Hydrocarbon polymers are difficult to dye.

26.2.1 CLASSIFICATION

Dyes may be classified on the basis of the process involved in dyeing and on the basis of the structural type. Based on the method of application, they are classified as *direct dyes, vat dyes* and *mordant dyes*. Direct dyes are applied directly on the fabric from a hot aqueous solution. Bonding is established through hydrogen bonds or through salt formation. Congo red is an example of a direct azo dye. In vat dyeing, the cloth is soaked in a solution of a precursor of the dye in a vat and then subjected to air oxidation or other treatment. The dye, which is the product of oxidation and is insoluble, is precipitated on the fabric. Indigo is a vat dye. Mordant dyes are applied through chelation with a metal ion. Alizarin is an example of a mordant dye. Dyes come under a few broad structural types.

26.2.1.1 Azo Dyes: Bismark Brown and Methyl Orange

Azo dyes form the largest class of dyes. Aromatic amines undergo diazotisation by nitrous acid in acid medium and the diazonium cation reacts as an electrophile with electron rich aromatic molecules like phenols and amines in a reaction known as coupling. The product is an azo compound (Chapter 18, Section 18.2.3.5). They may be applied as direct dyes or as mordant dyes. One of the earliest synthetic dyes is *Bismark Brown* (Fig. 26.1). This is based on *m*-phenylenediamine

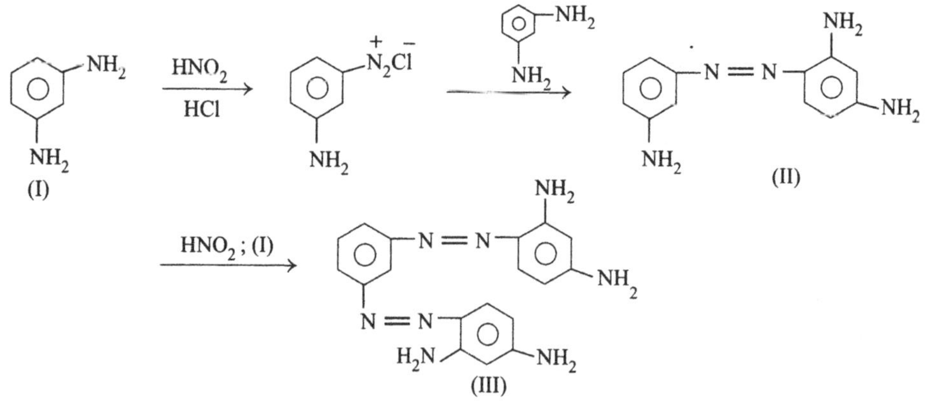

Fig. 26.1 Formation of Bismark brown

(1,3-diaminobenzene) (I) (Chapter 18, Section 18.4.1) which upon diazotisation couples with itself to give a monoazo (II) or a bis-azo (III) coupling product. A mixture of both is known as Bismark brown. Colour formation due to the dye is taken advantage of in the use of *m*-phenylenediamine as the reagent in a sensitive test for detecting and estimating nitrite in water.

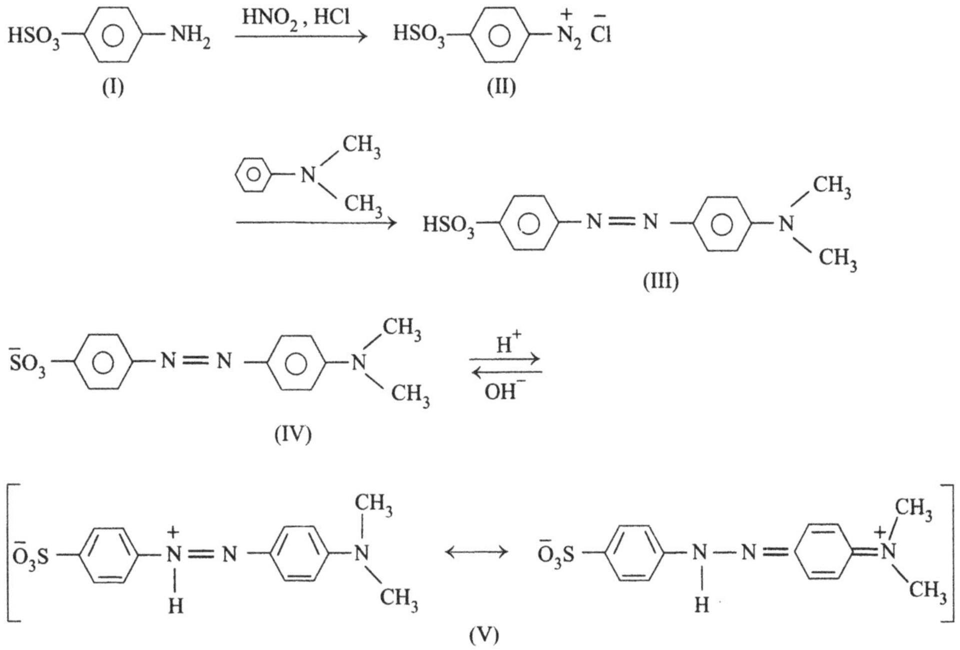

Fig. 26.2 Formation of methyl orange

Another familiar azo dye is *methyl orange* (Fig. 26.2), whose use is more as an acid–base indicator rather than as a dye. Suphanilic acid (I) (Chapter 18, Section 18.4.2) is diazotised and the diazonium salt is coupled with N,N-dimethylaniline to give the dye (III). The anion (IV) formed in alkaline medium is orange in colour. It changes to the zwitter ion (V) which is red in acidic medium.

Methyl orange is a fugitive dye and is not used for fabric dyeing.

26.2.1.2 Triphenylmethane Dyes

Triphenylmethanes with auxochromes like amino or hydroxyl groups, present on one or more of the benzene rings, on oxidation give rise to dyes. The initial product is the colourless (leucobase) which is converted to the dye through an intermediate called the colour base which is the precursor of the dye (Eq. 26.1). This is illustrated with malachite green (Fig. 26.3).

$$\text{Leucobase} \xrightarrow[\text{reduction}]{\text{oxidation}} \text{Colour base} \xrightarrow[\text{alkali}]{\text{acid}} \text{Dye}$$
$$\text{(colourless)} \qquad\qquad \text{(colourless)} \qquad\qquad \text{(coloured)} \qquad (26.1)$$

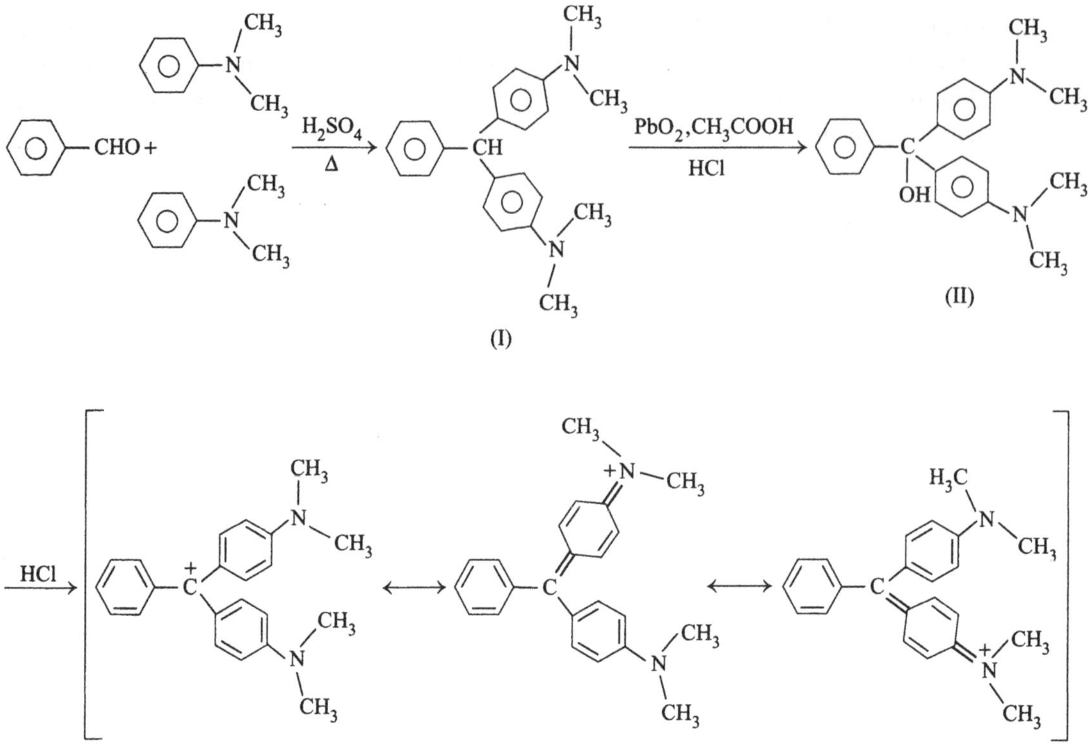

Fig. 26.3 Synthesis of malachite green

Benzaldehyde reacts with two equivalents of dimethylaniline in the presence of suphuric acid to give the triphenylmethane derivative (I). This is colourless and is called the leucobase. Oxidation of (I) with lead dioxide in acid medium gives the colourless carbinol (II), called the colour base. On further treatment with acid, the colour base is converted to the cation (III), which is resonance stabilised. The salt with Cl⁻ as the anion is malachite green, the dye. Malachite green is used as a direct dye for silk and wool and as a mordant dye for cotton. Malachite green is one of many related triphenylmethane dyes. Crystal violet has dimethylamino groups on all three benzene rings.

26.2.1.3 Phthalein Dyes: Phenolphthalein

Phenolphthalein belongs to a group of compounds known as phthaleins, obtained from phthalic acid. It can also be considered to be a triphenylmethane dye. The main use of phenolphthalein is as an acid–base indicator rather than as a dye. Synthesis of phenolphthalein is outlined in Fig. 26.4.

Phenolphthalein (III) is prepared by heating phenol (I) and phthalic anhydride (II) with concentrated sulphuric acid. This reaction – known as phthalein fusion – is a general reaction for phenols. The product (III) is insoluble in water, but soluble in ethanol and in aqueous alkali.

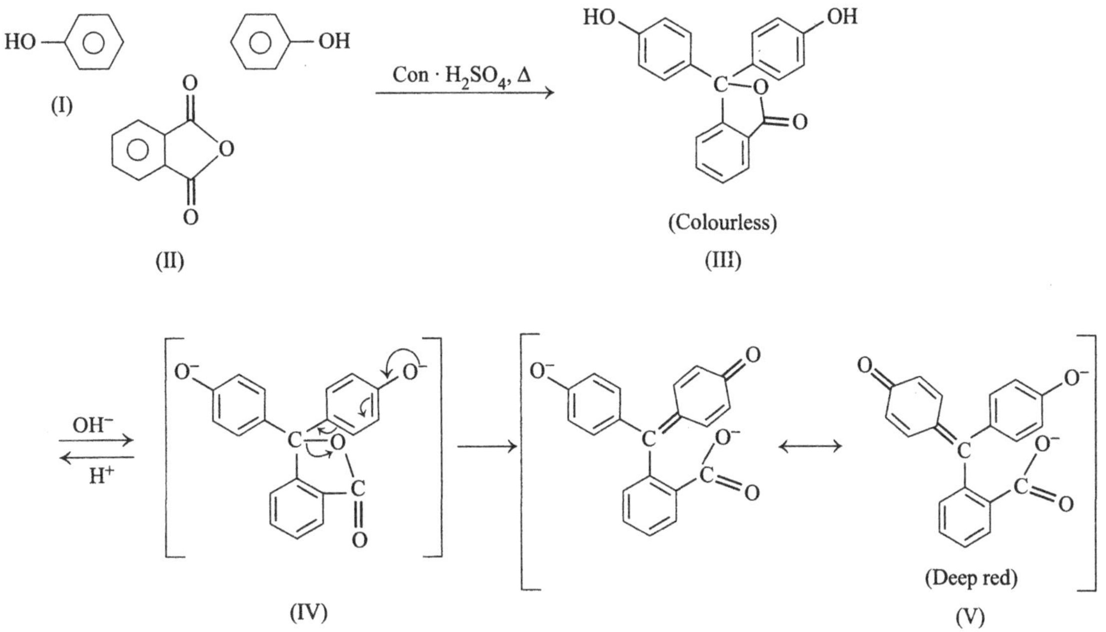

Fig. 26.4 Synthesis of phenolphthalein

In the presence of alkali, the phenolate anion (IV) is converted to the resonance stabilised anion (V) which is deep red in colour. In acid, the protonated form, (III), is colourless.

26.2.1.4 Xanthen Dyes: Fluorescein

When phthalein fusion is done with resorcinol, the product is similar to phenolphthalein, but contains a xanthen ring system. Xanthens are tricyclic compounds, illustrated by (I), (II) and (III) in Fig. 26.5(a). Resorcinol (IV) gives fluorescein (V) [Fig. 26.5(b)] which is soluble in alkali. The anion (VI) rearranges to the resonance stabilised anion (VII) to give a reddish brown solution with a yellowish green fluorescence.

26.2.1.5 Anthraquinoid Dyes: Alizarin

Alizarin is synthesised from anthraquinone (Fig. 26.6).

Sulphonation of anthraquinone (I) gives the sulphonic acid (II). Fusion of the sulphonic acid with the alkali in the presence of an oxidising agent ($KClO_3$) gives alizarin (III). During alkali-fusion, the sulphonic acid group is replaced by OH and the 1-position is oxidised, introducing one more OH group. Alizarin is applied to fabrics as a mordant dye. The fabric is dipped in the solution of the salt of a metal (mordant) like Ba, Cr, Al or Fe. Then the soluble form of the dye is applied. Chelation of the dye with the metal results in a fast dye. The colour depends upon the metal ion used as mordant. Al gives a rose–red colour, ferric iron gives a violet–black colour, chromium gives a brown–violet colour and barium, a blue

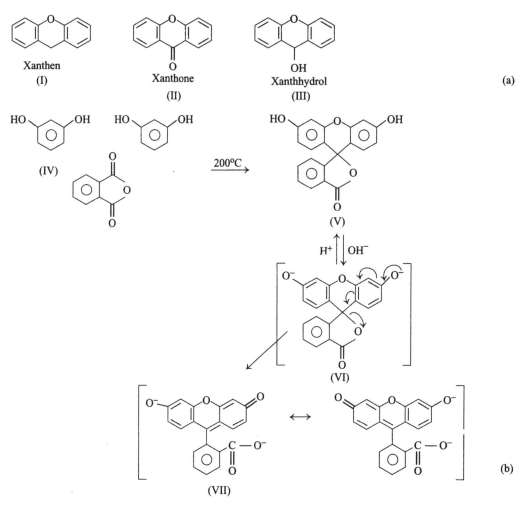

Fig. 26.5 Synthesis of fluorescein

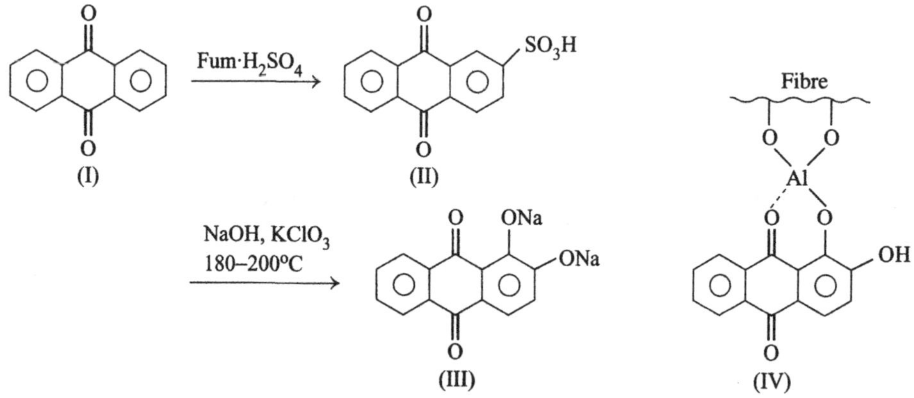

Fig. 26.6 Synthesis of alizarin

colour. The chelated structure of alizarin–aluminium complex on cotton fibre is depicted in (IV) (Fig. 26.6).

26.2.1.6 Vat Dyes: Indigo

Indigo is one of the oldest vegetable dyes known. Its origin is in India. A commercial route for the production of synthetic indigo from coal tar chemicals was introduced towards the end of the nineteenth century. This sounded the death knell of indigo farming in India, a development which had far reaching consequences.

The Portugese called indigo 'anil' after *nila*, meaning blue in sanskrit and *nili*, the sanskrit name of the indigo plant. Aniline was first obtained in 1826 as a product of the dry distillation of Indian indigo.

Indigo is produced from its precursor indoxyl (II), which is present in the plant as the glucoside, indican (I) (Fig. 26.7).

When indigo is produced from the plant, the crushed plant is first allowed to ferment. The enzymes present in the plant bring about the hydrolysis of the glucoside to indoxyl, (II). The hydrolysis can also be carried out using HCl. Natural or synthetic indoxyl is subjected to air-oxidation, when indigo (also called indigotin) is formed. The initially formed *cis*-structure (III), spontaneously isomerises to the *trans* structure (IV). This is insoluble in water. For vat dyeing, the isoluble dye is reduced to the soluble leucobase (V) using sodium hyposulphite. The fabric is soaked in this solution in vats and aerated. The soluble base gets oxidised to the insoluble dye which is deposited on the fabric.

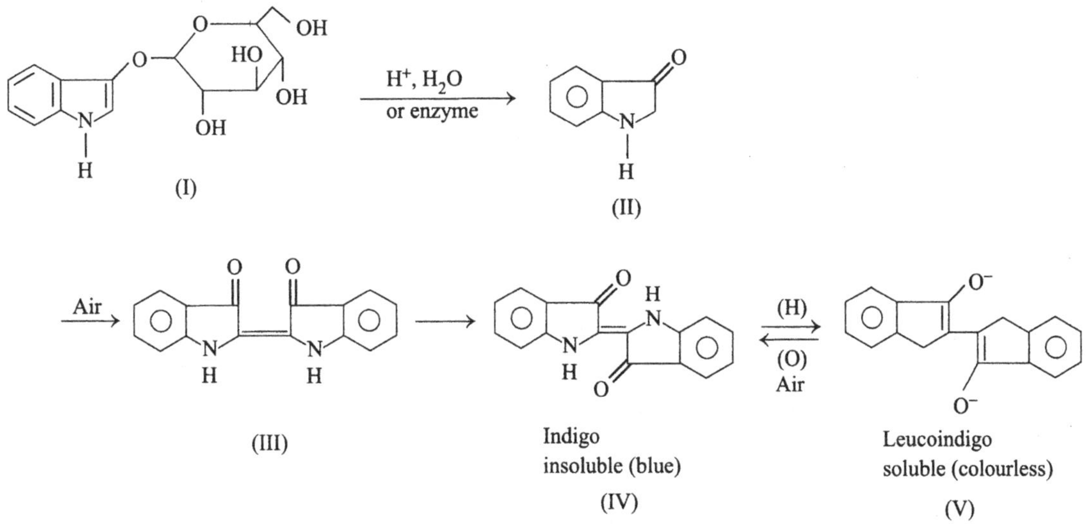

Fig. 26.7 Chemistry of indigo

KEY POINTS

- Dyes are classified as direct dyes, mordant dyes and vat dyes, according the process of dyeing.

- Dyes are classified based on their structure, as azo dyes, triphenylmethane dyes, phthalein dyes, xanthen dyes, anthraquinoid dyes and indigo dyes.

- The colour of a substance is the colour corresponding to the wavelengths of the light reflected by the substance. If all the wavelengths are absorbed, the substance will appear black. If all the light is totally reflected, it will appear white. If some of the wavelengths are absorbed and the rest reflected, the colour of the substance will be the complementary colour of the absorbed wavelengths.

- Absorption of light in the UV–Visible region of the electromagnetic spectrum causes electronic excitation.

- Certain groupings called chromophores, cause absorption in the UV–Visible region. These groups are characterised by conjugation. Extended conjugation tends to reduce the enegy required for electronic excitation and shifts the wave length of absorption to the visible region. Such substances are coloured.

- Azo dyes are the earliest synthetic dyes. Representative examples are Bismark brown obtained by the diazotisation of *m*-phenylenediamine, and methyl orange prepared by the coupling of diazotised sulphanilic acid with N,N-dimethylaniline.

- Triphenylmethanes with groups like amino or hydroxyl on one or more rings, can be oxidised to triphenylcarbinol which, in the presence of an acid give the coloured cation. Malachite green is an example of a triphenylmethane dye.

- Fusion of phenols with phthalic anhydride in the presence of sulphuric acid (phthalein fusion) gives phthaleins. Phenol and phthalic anhydride give phenolphthalein, which is colourless in acid solution, but has an intense red colour in alkaline solution due to the anion.

- When resorcinol is fused with phthalic anhydride in the presence of sulphuric acid, fluorescein is obtained. This can be considered to be a phthalein dye or a xanthen dye. The anion produced in alkaline solution gives intense fluorescence.

- Alizarin is an anthraquinoid dye produced by alkali fusion of anthraquinone-2-sulphonic acid. It is 1,2-dihydroxyanthraquinone. Alizarin is a mordant dye. The colour depends upon the metal ion used as mordant.

- Indigo is one of the earliest known vegetable dyes, and is manufactured synthetically today. It is an indole derivative. Indigo is a vat dye. When the fabric is soaked in the solution of the leucobase and aerated, the insoluble dye gets precipitated on the fibres. The process is done in vats, hence the name vat dye.

EXERCISES

SECTION I

1. Orange II is an azo dye prepared from sulphanilic acid and β-naphthol. Draw its structure.

2. Rhodamine-B is a xanthen dye prepared by phthalein fusion of *m*-diethylaminophenol and phthalic anhydride. Draw its structure. On acid treatment, the lactone ring is opened. Draw the resonance structures of the cation.

3. Indigo isomerises spontaneously from the *cis* configuration to the *trans*. Which is likely to be more stable, why? Suggest a mechanism for the isomerisation.

4. Two syntheses of indoxyl are outlined below. Give the details, including reagents required and reaction conditions.

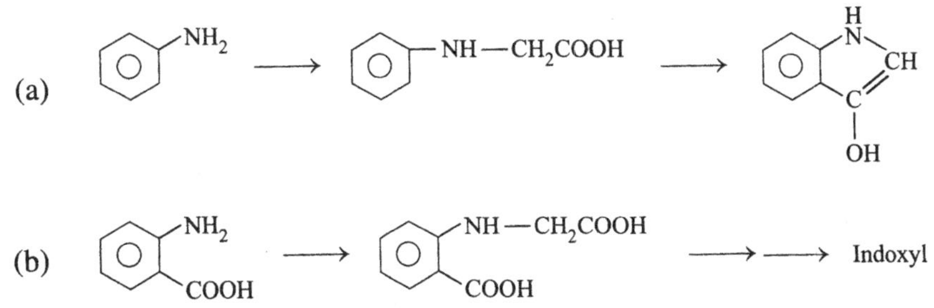

(a)

(b)

PROJECT

Prepare a report on toxicity concerns related to the use of synthetic dyes.

Supplementary Reading

1. Morrison R T and Boyd R N, *Organic Chemistry*, sixth edition, Prentice Hall, 1992.
2. Finar I L, *Organic Chemistry*, Vol. 1, fifth edition, Longman Publishing Group, ELBS edition, 1975.
3. Finar I L, *Stereochemistry and the Chemistry of the Natural Products*, Vol. 2, fifth edition, Longman Publishing Group, ELBS edition, 1975.
4. Carey F A, *Organic Chemistry*, fifth edition, Tata McGraw-Hill, 2005.
5. Solomons G T W, *Organic Chemistry*, eighth edition, Wiley, 2004.
6. McMurrey J F, *Fundamentals of Organic Chemistry*, Vols. 1 and 2, Brooks/Cole Publishing Company, 1996.
7. Pine S H, *Organic Chemistry*, McGraw-Hill, 1987.
8. Hendrikson J B, Pine S H, Cram D J and Hammond G S, *Organic Chemistry*, sixth edition, McGraw-Hill, International Student Edition, 1985.
9. Carey F A and Sundberg R J, *Advanced Organic Chemistry*, Parts A and B, Tata McGraw-Hill, 2003.
10. Eliel E L and Wilen S H, *Stereochemistry of Carbon Compounds*, Wiley, 1994.
11. Eliel E L, *Stereochemistry of Organic Compounds*, Tata McGraw-Hill, 1975.
12. Kemp W, *Organic Spectroscopy*, third edition, W H Freeman & Co, 1991.
13. March J and Smith M B, *March's Advanced Organic Chemistry – Reactions, Mechanisms and Structure*, sixth edition, Wiley Interscience, 2007.
14. Sykes P, *A Guidebook to Mechanisms in Organic Chemistry*, sixth edition, Longman Publishing Group, 1986.
15. Reusch W, *Virtual Textbook of Organic Chemistry* (periodically revised; latest revision, July 2007). Can be viewed at http://www.cem.msu.edu/~reusch/VirtualText/intro1/htm

Index

C

For Product Safety Concerns and Information please contact our EU
representative GPSR@taylorandfrancis.com Taylor & Francis Verlag GmbH,
Kaufingerstraße 24, 80331 München, Germany

Printed and bound by CPI Group (UK) Ltd, Croydon, CR0 4YY
01/05/2025
01858518-0009